This book is the first to give a comprehensive review of the theory, fabrication, characterization, and device applications of abrupt, shallow, and narrow doping profiles in semiconductors. Such doping profiles are a key element in the development of modern semiconductor technology, including silicon very large scale integrated circuits, discrete devices, and optoelectronic devices.

After an introductory chapter setting out the basic theoretical and experimental concepts involved, the fabrication of abrupt and narrow doping profiles by several different techniques, including epitaxial growth, is discussed. The optical, electrical, chemical, and structural techniques for characterizing doping distributions are then presented, followed by several chapters devoted to the inherent physical properties of narrow doping profiles. The latter part of the book deals with particular devices, such as silicon field-effect transistors, and III–V semiconductor devices for electronic and optoelectronic applications.

The book will be of considerable interest to graduate students, researchers, and engineers in the fields of semiconductor physics and microelectronic engineering.

Delta-doping of semiconductors

Edited by

E. F. SCHUBERT

Department of Electrical, Computer and Systems Engineering, Boston University

CAMBRIDGE UNIVERSITY PRESS
Cambridge, New York, Melbourne, Madrid, Cape Town, Singapore, São Paulo

Cambridge University Press
The Edinburgh Building, Cambridge CB2 2RU, UK

Published in the United States of America by Cambridge University Press, New York

www.cambridge.org
Information on this title: www.cambridge.org/9780521482882

First published 1996
This digitally printed first paperback version 2005

A catalogue record for this publication is available from the British Library

Library of Congress Cataloguing in Publication data

Delta-doping of semiconductors / edited by E. F. Schubert.
p. cm.
ISBN 0 521 48288 7 (hardback)
1. Semiconductor – Design and construction. 2. Semiconductor
doping. 3. Molecular beam epitaxy. 4. Field effect transistors.
I. Schubert, E. Fred.
TK7871.85.D457 1995
621.3815′2–dc20 94-48903CIP

ISBN-13 978-0-521-48288-2 hardback
ISBN-10 0-521-48288-7 hardback

ISBN-13 978-0-521-01796-1 paperback
ISBN-10 0-521-01796-3 paperback

Contents

Contributors

M. ASCHE
Paul-Drude-Institut für Festkörperelektronik, Hausvogteiplatz 5–7, D-10117 Berlin, Germany
I. EISELE
Institute of Physics, Faculty of Electrical Engineering, Universität der Bundeswehr München, D-85579 Neubiberg, Germany
L. C. FELDMAN
AT&T Bell Laboratories, 600 Mountain Avenue, Murray Hill, NJ 07974, USA
H.-J. GOSSMANN
AT&T Bell Laboratories, 600 Mountain Avenue, Murray Hill, NJ 07974, USA
R. L. HEADRICK
CHESS, Cornell University, Ithaca, NY 14853, USA
W.-P. HONG
Systems Development Center, Korea Telecom, 17 Woomyeon-Dong, Seocho-Gu, Seoul, Korea
Y. HORIKOSHI
NTT Basic Research Laboratories, Musashino-Shi, Tokyo, 180, Japan
P. M. KOENRAAD
Department of Physics, Eindhoven University of Technology, PO Box 513, 5600 MB Eindhoven, The Netherlands
H. S. LUFTMAN
AT&T Bell Laboratories, 600 Mountain Avenue, Murray Hill, NJ 07974, USA
R. J. MALIK
AT&T Bell Laboratories, 600 Mountain Avenue, Murray Hill, NJ 07974, USA
T. MAKIMOTO
NTT Basic Research Laboratories, 3–1, Morinosato, Wakamiya, Atsugi-shi, Kanagawa, 243–01, Japan
W. T. MASSELINK
Institut für Physik, Humboldt-Universität zu Berlin, Invalidenstraße 110, D-10115 Berlin, Germany
K. NAKAGAWA
Central Research Laboratory, Hitachi Ltd, Kukubunji, Tokyo, 185, Japan
R. C. NEWMAN
Interdisciplinary Research Centre for Semiconductor Materials, Imperial College of Science, Technology and Medicine, Prince Consort Road, London, SW7 2BZ, UK

K. H. PLOOG
Paul-Drude-Institut für Festkörperelektronik, Hausvogteiplatz 5–7, D-10117 Berlin, Germany
C. R. PROETTO
Comision Nacional de Energia Atomica, Centro Atomico Bariloche, 8400 San Carlos de Bariloche (RN), Argentina
D. RICHARDS
Cavendish Laboratory, University of Cambridge, Cambridge, CB3 0HE, UK
D. RITTER
Department of Electrical Engineering, Technion, Haifa 320000, Israel
E. F. SCHUBERT
Department of Electrical, Computer and Systems Engineering, Boston University, 44 Cummington Street, Boston, MA 02215, USA
J. WAGNER
Fraunhofer-Institut für Angewandte Festkörperphysik, Tullastraße 72, D-79108 Freiburg, Germany
B. E. WEIR
AT&T Bell Laboratories, 600 Mountain Avenue, Murray Hill, NJ 07974, USA
K. YAMAGUCHI
Central Research Laboratory, Hitachi Ltd, Kukubunji, Tokyo, 185, Japan
H. YAO
Department of Electrical Engineering, University of Nebraska, Lincoln, NE 68588, USA

Preface

The spatial dimensions of semiconductor device structures have been shrinking since the infancy of semiconductor technology, and will continue to do so in the foreseeable future. The reduction of device dimensions is motivated by higher speeds as well as lower power consumption. Free carriers can traverse smaller dimensions in a shorter time. For example, the base transit time of an electron emitted into the base of a bipolar transistor decreases for thinner base layers. Furthermore, parasitic capacitances are reduced in small device structures. As a consequence, the energy lost in charging and discharging parasitic capacitors is reduced. If, in addition, the operating voltage of such devices is lowered, a further decrease in power consumption results.

Small device structures require that the spatial distribution of dopants is well controlled. Redistribution processes such as diffusion, segregation, drift in an electric field, and other redistribution mechanisms must be understood on a near-atomic length scale. The assessment of such redistribution processes becomes ever more important as the device dimensions shrink.

Delta-doping, spike-doping, and pulse-doping profiles are examples of extremely narrow but well-defined doping profiles. Such profiles are required in semiconductor structures scaled to their practical, economical, and theoretical limits. The fabrication of such profiles, their characterization, and the understanding of their physical properties are necessary steps to taking full advantage of semiconductor doping profiles with atomic level control. In silicon VLSI technology, the depth of the p–n junctions must be controlled within a few tens of angstroms. In optoelectronics, quantum-well lasers require that the location of p–n-junctions be controlled within a few tens of angstroms as well. Any displacement of p–n-junctions over these short distances would have deleterious effects on the device performance.

In the past, native defects, complex defects, and unintentional contaminations in semiconductors have recieved much attention from the research community. *Intentional dopants* and related issues received, surprisingly, much less attention. However, the requirements for the concentration versus depth profiles of intentional impurities are becoming increasingly stringent in modern semiconductor technology. For example, very small semiconductor structures require doping concentrations exceeding 10^{19} cm^{-3}. The limits of the maximum doping concentration in semiconductors have not been well explored, despite a pressing need. I therefore expect that research will be directed towards dopant characteristics, including spatial redistribution effects and high concentration effects.

This book provides the first comprehensive and coherent treatment of extremely narrow doping distributions with very high concentrations. This field is not limited to any particular semiconductor material. I have, therefore, invited contributions on the group-IV semiconductor Si as well as on III–V semiconductors. Both material systems receive equal emphasis.

The book is organized as follows. The introduction (E. F. Schubert) is written in a tutorial style and provides many basic experimental as well as theoretical concepts. In Part Two (C. Proetto), the theoretical framework of the spatial and energetic structure of δ-doped semiconductors is presented. Quantum effects which occur in semiconductors with narrow doping profiles are taken into account. The fabrication of δ-doped structures by epitaxial growth is described in Part Three (Ploog, Makimoto, Horikoshi, Ritter, Eisele, and Gossmann). Doping in III–V, as well as in group-IV, semiconductors during epitaxy is discussed in this part. The epitaxial growth techniques described include molecular-beam epitaxy (MBE), gas source MBE, low-temperature MBE, flux-modulated epitaxy, and solid-phase epitaxy. The characterization of δ-doped semiconductors is the subject of Part Four (Luftman, Schubert, Gossmann, Newman, and Koenraad) and includes secondary-ion mass spectrometry (SIMS), the capacitance-voltage (CV) technique, a discussion of redistribution mechanisms, local vibrational mode spectroscopy (LVM), and a section on the DX center in n-type $Al_xGa_{1-x}As$. Part Five (Yao, Schubert, Wagner, Richards, Masselink, Koenraad, Asche, Headrick, Feldman, and Weir) discusses the physical properties of δ-doped semiconductors and includes carrier transport in the bulk and in quantum wells, optical interband and intraband transitions, and a discussion of ordered doping distributions. Device applications and implications are presented in Part Six (Hong, Schubert, Nakagawa, Yamaguchi, Malik, and Eisele) and include field-effect transistors (FETs), heterostructure FETs, detectors, and doping superlattice devices.

E. F. Schubert
Murray Hill
July 1995

PART ONE

1

Introduction

E. F. SCHUBERT

This book is devoted to semiconductors containing very thin, highly doped layers. Assuming that doping impurities are distributed uniformly within the two-dimensional layer, the doping profile depends on only one spatial dimension. Frequently, such doping distributions are described by the delta (δ-) function. This chapter will introduce the reader to the field of δ-doping. Section 1.1 will address the questions: What are δ-doped semiconductors? How are they fabricated? Why are they made? In Section 1.2, a historical sketch of this subject will be provided. The electronic structure of δ-doped semiconductors will be derived in Section 1.3. Finally, the implications of δ-doping on state-of-the-art microelectronic devices will be discussed in Section 1.4.

1.1 Delta doping: what is it and how is it done?

Delta doping is motivated by the need for well-defined, narrow doping profiles in semiconductors. Doping distributions with high peak concentrations and narrow distribution widths are advantageous for many device applications. The narrowest one-dimensional (1D) doping profile is achieved if doping atoms are confined to a single atomic layer in the host semiconductor. Such a situation is illustrated schematically in Fig. 1.1, which shows a semiconductor substrate, an epitaxial layer, and a dopant sheet embedded in the epitaxial layer. All impurities, shown as dark circles in the inset to Fig. 1.1, occupy substitutional lattice sites in the host semiconductor. In group-IV semiconductors, all lattice sites are identical. In III–V semiconductors, we have the distinct cation and anion lattice sites. Therefore, the doping plane shown in Fig. 1.1 may be either an anion or a cation sublattice plane.

The two parameters characterizing a δ-doping profile are the location of the dopant sheet and the density of doping atoms in the sheet. Assuming that the wafer surface is in the xy-plane of a cartesian coordinate system, that the z-coordinate is measured from the surface to the bulk, and that the dopants are located in the plane at $z = z_d$, then the doping

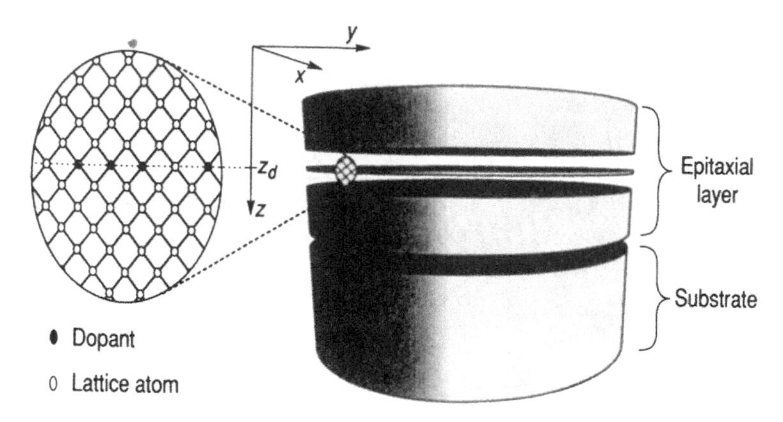

Fig. 1.1. Schematic illustration of a semiconductor substrate and an epitaxial film containing a δ-doping layer. Also shown is a schematic lattice with the impurity atoms being confined to a single atomic plane.

profile is given by

$$N(z) = N^{2D}\delta(z - z_d) \quad (1.1)$$

where the two-dimensional (2D) density, denoted by N^{2D}, is the number of doping atoms in the doping plane per cm^2. Equation (1.1) states that the doping concentration is zero for all locations, except for $z = z_d$. Integration of Eqn. (1.1) yields $\int_{-\infty}^{+\infty} N(z)\,dz = N^{2D}$, that is, the 2D doping density.

The growth of δ-doped semiconductors will be discussed for different epitaxial growth techniques in Chapter 3. Here, we summarize briefly the basic procedure common to all growth techniques. Consider a semiconductor during epitaxial growth. In the *first step* of δ-doping, the epitaxial growth of the semiconductor is suspended, which can be achieved by closing the effusion cells in molecular-beam epitaxy or by valving off the semiconductor precursor flow in vapor-phase epitaxy. We assume that the non-growing semiconductor surface will be atomically flat with no atomic terrace steps on the surface. In the *second step*, the surface is exposed to either a flux of elemental doping atoms or to the flux of a doping precursor. For efficient incorporation, it is desirable that the doping atoms have a sticking probability near unity, or, for the case of a precursor, that the decomposition probability of the precursor is near unity. If the doping atoms do not re-evaporate, then they can form bonds with host semiconductor atoms. Impurity atoms are *chemisorbed* at the semiconductor surface. Ideally, all doping atoms occupy substitutional lattice sites on the semiconductor surface. In the final *third step* of δ-doping, the deposition of dopants is terminated and the epitaxial growth of the host semiconductor is resumed. In the absence of impurity redistribution effects such as diffusion, drift, and segregation, the doping atoms are then confined to one atomic layer of the semiconductor.

However, if the semiconductor surface on which the impurities are deposited is rough and contains steps or terraces, or if thermally stimulated redistribution of impurities does occur during subsequent growth, then the impurities cannot be confined to a single atomic layer. Further details of the epitaxial growth of δ-doping layers using different techniques will be discussed in Part 3.

The density of doping atoms in the doping monolayer can be determined by the dopant flux, which may be a flux of doping atoms or of precursor molecules, and the time that the semiconductor surface is exposed to the dopant flux. Thus, for a given dopant flux, the time t, during which the semiconductor is exposed to the dopant flux, can be varied to achieve the desired 2D density of incorporated dopants. Frequently, the bulk concentration (3D concentration) of a dopant is known for a given set of experimental conditions such as a given growth rate, substrate temperature, effusion cell temperatures, gas flows, etc. If this bulk concentration is denoted by N, then the 2D dopant density obtained during δ-doping is given by

$$N^{2D} = N v_g t \tag{1.2}$$

where N is the bulk concentration (per cm^3) obtained at a given growth rate v_g, and t is the time for which the growth is suspended. Equation (1.2) thus allows one to select the time t required to obtain a doping density N^{2D} provided that the bulk concentration N is known for a given set of growth conditions. Equation (1.2) is valid if (i) dopants deposited on the non-growing surface have the same incorporation probability as dopants incorporated during a continuous growth mode and if (ii) the growth rate of the host semiconductor is not influenced by the doping incorporation. Both conditions are usually met by epitaxial growth systems.

The effective 3D concentration of dopants in the doping sheet of a δ-doped semiconductor can be inferred from the 2D density. The electrical activity of some dopants decreases at high concentrations. Other doping impurities diffuse strongly at high concentrations. It is therefore desirable to estimate the effective 3D concentration of dopants in a 2D doping sheet. In order to calculate the effective 3D concentration, we consider a sheet of dopants with density N^{2D}. The mean distance between the doping atoms is $(N^{2D})^{-1/2}$. Now consider a homogeneously doped semiconductor with *the same mean distance between the dopant atoms.* This semiconductor would have an *equivalent 3D doping concentration* of

$$N^{3D} = (N^{2D})^{3/2}. \tag{1.3}$$

The quantity N^{3D} is the physically relevant 'bulk' concentration of a δ-doped semiconductor. For example, a δ-doping concentration of $N^{2D} = 10^{12}$ cm^{-2} corresponds to an equivalent 3D concentration of $N^{3D} = 10^{18}$ cm^{-3}, as calculated from Eqn. (1.3).

In actual δ-doped semiconductors, dopants may not be confined to a single atomic layer but may be distributed over more than a single layer. Surface roughness and other processes such as diffusion, drift, and segregation may contribute to the doping redistribution. In order to quantify the spread of dopants in δ-doped semiconductors, we first consider, as the simplest approximation, a top-hat distribution, which is illustrated schematically in Fig. 1.2. The width of the doping distribution, Δz_d, is smaller than the lattice constant, a_0. However, for sufficiently strong doping redistribution processes, the distribution width, Δz_d, may become larger than the lattice constant, a_0. For all well-behaved doping

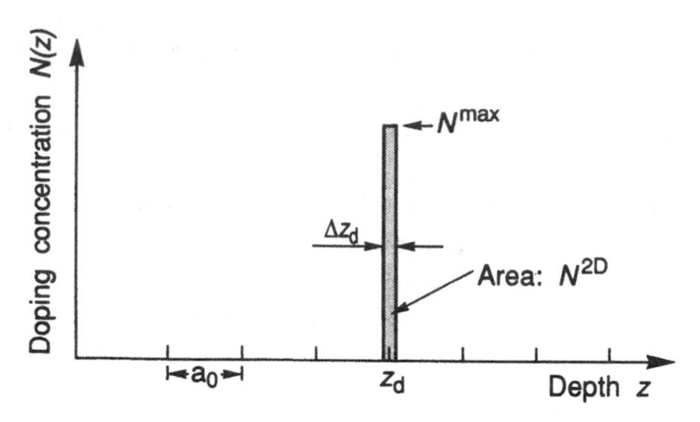

Fig. 1.2. Doping profile of a doping spike located at $z = z_d$ with 2D density N^{2D}, maximum concentration N^{max}, and full-width at half-maximum of Δz_d.

profiles such as a top-hat, a gaussian, or a triangular doping distribution, the maximum concentration and the profile width are related by

$$N^{2D} \approx N^{max} \Delta z_d. \tag{1.4}$$

Considering redistribution processes, the question arises as to whether doping profiles with $\Delta z_d \gtrsim a_0$ can be considered to be δ-function-like. To answer this question, the actual distribution width Δz_d must be compared to other relevant length scales such as the screening length, free-carrier diffusion length, depletion length, free-carrier wavelength, etc. More specifically, the distribution width Δz_d must be smaller than the *shortest* of these length scales. The shortest of the above-mentioned length scales is, in most cases, the free-carrier de Broglie wavelength. The free-carrier de Broglie wavelength decreases with increasing effective mass and the carrier kinetic energy. In practice, the free-carrier de Broglie wavelength is longer than 25 Å, even for carriers with large effective mass and high kinetic energy. We can therefore consider all doping profiles with a distribution width of $\Delta z_d \leqslant 25$ Å to be δ-function-like.

As stated above, δ-doped semiconductors are predominantly fabricated by epitaxial growth. Other doping techniques, such as doping by diffusion and by ion implantation, produce inherently broader doping distributions. However, ion implantation can produce spatially well-confined doping profiles if the implantation depth is very shallow. As an approximate rule, the projected straggle of implanted ions equals the implantation depth. The redistribution of doping ions during postimplantation annealing can be minimized by rapid thermal annealing. Therefore, shallow implantation of impurities with implantation depths of less than 100 Å can produce near-δ-function-like doping profiles.

1.2 Historical review

During the δ-doping process, the non-growing crystal surface is exposed to a flux of doping atoms or doping precursor molecules. These atoms or molecules undergo chemisorption on the non-growing crystal surface. The chemisorption is a desirable effect. However, the chemisorption of impurities on a non-growing semiconductor surface may also be an undesirable process, namely if it occurs before the epitaxial growth is initiated. Before a wafer is introduced into the growth chamber, it is exposed to the ambient, usually clean air. After the introduction of a wafer into the growth system, the wafer is exposed to the ambient of the system, usually a carrier gas or vacuum. The exposure time for the wafer can be several hours, until the actual epitaxial growth is initiated. The incoporation of impurities at the substrate–epilayer interface was first reported by DiLorenzo (1971), who found a high concentration of Si, C, Mg, Fe, Cr, and other impurities at the substrate–epilayer interface. These impurities were incorporated during the substrate cleaning procedure, as well as during the exposure of the wafers to the carrier gas of the vapor-phase growth system. Considering the state-of-the-art of substrate cleaning in 1971, most impurities were probably incorporated during substrate cleaning.

The incorporation of impurities at the substrate–epilayer interface was confirmed by Bass (1979), who showed that adsorption (chemisorption and physisorption) of impurities on the substrate surface can occur in the reaction chamber before epitaxial growth is initiated. During GaAs growth by vapor-phase epitaxy (VPE), hydrogen sulfide (H_2S), silane (SiH_4), and germane (GeH_4) were used as doping precursors (Bass, 1979). It was general practice to initiate the doping precursor flow before the actual GaAs growth was started. This was done to eliminate any transient effects occurring in the H_2S doping line predominantly, and, to a smaller extent, in the silane and germane doping lines. When the purging procedure was done with silane, a pronounced doping spike occurred at the substrate–epitaxial layer interface. This indicated that silicon was strongly chemisorbed on the GaAs substrate surface. Doping profile measurements by the capacitance –voltage (CV) technique indeed revealed the occurrence of a clear doping spike at the substrate–epilayer interface. The measured full-width at half-maximum of the Si doping spikes exceeded 200 Å and the doping profiles can, therefore, not be considered as δ-function-like distributions in the strict sense defined above. Nevertheless, Bass's report (1979) was the first report of doping incorporation on a non-growing semiconductor crystal.

Realizing the possibility of narrow doping distributions, Bass (1979) proposed that the strong surface adsorption of silicon could be used to produce sharp doping spikes which could be used advantageously in IMPATT diodes. A strong adsorption of Si on the GaAs surface was also reported for epitaxial growth by molecular-beam epitaxy (MBE) (Wood *et al.*, 1980). However, the profile widths were >300 Å, that is, broader than those reported by Bass. Wood *et al.* (1980) proposed that the doping technique could be used to synthesize complex free-carrier profiles such as linearly ramped profiles.

The first truly δ-doped semiconductor structure with clear evidence for a narrow doping profile was reported by Schubert *et al.* (1984) and Schubert and Ploog (1985). Capacitance–voltage (CV) measurements on MBE grown GaAs samples revealed a half-width at half-maximum of 20 Å. Assuming that the profiles are symmetric, a full profile width of 40 Å can be deduced from the CV measurements. This profile width is at least five times narrower than the profile widths reported by Bass (1979) and Wood *et al.* (1980) and was thus the first truly δ-doped semiconductor. In the same publication, Schubert and Ploog (1985) reported the first field-effect transistor (FET) using δ-doped GaAs. The authors showed that the δ-doped FET has a narrow free-carrier and dopant distribution and a large gate-breakdown voltage. Such FETs also exhibit reduced short-channel effects due to the narrow distribution of dopants.

The δ-doping technique has been used by Schubert and coworkers in a number of MBE-grown semiconductor devices. Examples are the homojunction FET (Schubert and Ploog, 1985), the high-mobility heterojunction FET (Schubert *et al.*, 1987), light-emitting diodes (LEDs) (Schubert *et al.*, 1985b), and lasers (Schubert *et al.*, 1985a, 1989a). When applied to heterojunction FETs, the δ-doping technique yields the highest free-carrier concentrations. It has been shown that δ-doping distributions are the optimum doping profiles for modulation-doped FETs. For LEDs and lasers, δ-doping has been used in the active region of the structure in which electron–hole recombination occurs. The active region in these optical devices consists of a doping superlattice which allows one to lower the bandgap energy of the active region. Emission wavelengths exceeding 950 nm for a GaAs host have been achieved at room temperature. In addition, lasing has been achieved in δ-doped superlattices (Schubert *et al.*, 1985a and 1989a). Constant lasing wavelengths, as well as tunable wavelengths, have been achieved with δ-doped superlattice lasers.

Delta-function-like doping distributions minimize potential fluctuations originating from random doping atom distributions. Consider a doped slab of concentration N and thickness Δz_d. Assume that the doping atoms are randomly distributed within this slab. Thus, potential fluctuations due to the random dopant distribution will be felt in the vicinity of the slab. One can show that such potential fluctuations, occurring at any point outside the doping slab, are minimized if the thickness of the doped region approaches zero ($\Delta z_d \to 0$), that is, for the δ-doped case (Schubert *et al.*, 1988a). Many characteristics of semiconductor structures are related to potential fluctuations, for example the mobility of free carriers or the luminescence linewidth of doped structures. These characteristics can be optimized by using the δ-doping technique. The mobility of selectivity doped heterostructures is maximized by δ-doping (Schubert *et al.*, 1989b). Free carriers in selectively doped heterostructures are scattered by potential fluctuations caused by the random distribution of remote dopants. Minimizing these fluctuations results in the minimization of scattering by remote ionized dopants. Maximum experimental mobilities have indeed been achieved in δ-doped heterostructures as opposed to homogeneously doped heterostructures (Schubert *et al.*, 1989b).

The minimization of potential fluctuations in δ-doped structures has resulted in a significant improvement of the optical properties of doping superlattices. Quantum-confined interband transitions have been observed in absorption measurements (Schubert *et al.*, 1988b) and luminescence emission experiments (Schubert *et al.*, 1989c). The measured peak energies have been assigned to theoretical transition energies and good agreement has been found between measured and calculated transition energies. Up to the present time, quantum-confined interband transitions have not been reported for homogeneously doped doping superlattices.

The assessment of the spatial distribution of doping atoms in δ-doped structures has been the subject of intense interest, since it is the condition *sine qua non* for δ-doped structures. The early work of Bass (1979) and Wood *et al.* (1980) indicated distribution widths >200 Å. It is likely that the broad profile widths were indeed caused by broad doping distributions rather than a low spatial resolution of the measurement technique. The clear signature of doping redistribution effects of Si in $Al_xGa_{1-x}As$ has also been reported by Lee *et al.* (1985), who concluded that Si δ-function-like profile widths exceed 100 Å.

The spatial localization of doping atoms in δ-doped semiconductors has been investigated using a number of characterization techniques including Auger electron spectroscopy, capacitance–voltage (CV) profiling, secondary ion mass spectrometry (SIMS), and transmission electron microscopy (TEM). The first study on the spatial localization of dopants revealed a 20 Å half-width, half-maximum CV-profile width and thus indicated very good confinement of dopants (Schubert and Ploog, 1985). In order to further study the range of temperatures suitable for the growth of Si δ-doped GaAs, localization was studied as a function of doping density and growth temperature (Schubert *et al.*, 1988c and 1990a). These studies revealed that the CV profile widths are limited by the resolution of the CV technique. Dopants were concluded to be confined to within 15 Å, that is, approximately three lattice constants.

Secondary ion mass spectrometry measurements were employed extensively in order to study the distribution of dopants in δ-doped semiconductors (Zeindl *et al.*, 1987; Beall *et al.*, 1988; Clegg and Beall, 1989; Lanzillotto *et al.*, 1989; Schubert *et al.*, 1990b). The SIMS technique has the highest resolution for shallow doping layers buried 200–500 Å below the semiconductor surface. Low sputtering energies of $\leqslant 3$ keV are advantageously used to achieve high resolution. Studies by different groups have revealed that the widths of the measured doping profiles are limited by the resolution of the SIMS technique. The narrowest profiles display a full-width at half-maximum of only 29 Å (Schubert *et al.*, 1990b). Spatial local-ization of dopants within a layer of thickness $\leqslant 15$ Å was concluded from the SIMS measurements.

The spatial localization and redistribution of Si in GaAs has also been studied by Webb (1989) using Auger electron spectroscopy. The study revealed that segregation of Si towards the GaAs surface during growth at 520 °C was immeasurably small. The Auger technique was shown to have an excellent resolution of approximately 5 Å.

Transmission electron microscopy (TEM) has been employed in the study of spatial localization of dopants in highly δ-doped semiconductors. Zeindl *et al.* (1987) showed that an Sb doping layer of density 5×10^{13} cm^{-2} can be clearly observed as a dark line in the bright-field lattice image of a Si specimen with a $\langle 011 \rangle$ oriented cross-section. The TEM measurements were made under conditions in which strain contrast was weak. The dark line was therefore assumed to be due to scattering of the electrons by heavy Sb atoms. The TEM micrograph indicated that the Sb doping atoms were confined to a layer of approximate thickness 15 Å. Thus, all techniques used to assess the doping distribution in δ-doped semiconductors, namely Auger, CV, SIMS, and TEM, revealed resolution-limited profiles for samples grown under optimized conditions.

The combination of high-resolution assessment techniques and δ-function-like doping profiles offers a sensitive method of measuring the redistribution and diffusion constant of dopants during high-temperature growth and postgrowth annealing. The redistribution of dopants during high-temperature growth can easily be detected by this method (Beall *et al.*, 1989; Schubert *et al.*, 1989). Postgrowth annealing of δ-doped samples for different times and/or temperatures allows one to determine the diffusion constant of dopants as a function of temperature (Schubert *et al.*, 1988d; Beall *et al.*, 1990). This method provides very high sensitivity as well as excellent accuracy for the assessment of diffusion constants.

The growth of δ-doped III–V semiconductors is achieved via the deposition of the doping element or of the doping precursor on the non-growing crystal surface at growth temperature. If this procedure is carried out during Si epitaxial growth, excessive segregation of impurities such as B or Sb results. In order to avoid this problem, Zeindl *et al.* (1987) first evaporated Sb dopants to the desired submonolayer coverage and then covered the doping layer by amorphous Si, all at room temperature. After amorphous growth, the Si wafer was heated to 700 °C to recrystallize the amorphous layer. This procedure is commonly referred to as solid-phase epitaxy. An alternative method of obtaining δ-function-like doping profiles in Si by MBE was demonstrated by Gossmann *et al.* (1990) and Gossman and Schubert (1993), who found that the Si growth temperatures can be reduced to 325 °C where the segregation of Sb and other doping elements is insignificantly small. The authors further showed that the quality of the films as assessed by Hall mobility measurements did not degrade due to the low-temperature growth.

It is commonly assumed that the doping atoms are randomly distributed within the doping plane. Potential fluctuations arise due to the random impurity distribution, which will scatter free carriers and reduce their drift mobility. If it were possible to place the doping atoms in an ordered, non-random arrangement, the carrier mobility would increase. Levi *et al.* (1989) proposed that, under suitable growth conditions, high-density δ-doping may result in partial ordering of the doping atoms. A reduction in the elastic scattering rate by up to a factor of four was predicted. To date, experimental evidence for ordered doping distributions has not been reported. However, Headrick *et al.* (1990 and 1991) reported on δ-doping with one full monolayer coverage of B in Si. The free-carrier mobilities were comparable to highly B-doped bulk Si.

Delta doping can reduce the adverse effects of persistent photoconductivity (PPC) and free-carrier freeze-out which have been found in n-type $Al_xGa_{1-x}As$. These effects have deleterious consequences for many semiconductor devices including microwave transistors. The physical origin of PPC and free-carrier freeze-out is a deep donor called the DX-center. Etienne and Thierry-Mieg (1988) found that the DX-center concentration is strongly reduced in Si δ-doped $Al_{0.3}Ga_{0.7}As$. The reduction is most pronounced at high-Si sheet densities.

Delta-function-like doping profiles were also grown by flow-rate modulation epitaxy (FME) which uses the same precursors as organo-metallic vapor-phase epitaxy (OMVPE). However, FME can grow at typically 100 °C lower temperatures as compared with conventional OMVPE. Sharp Si doping profiles with Si densities of $1 \times 10^{13}\ cm^{-2}$ in GaAs grown at 550 °C have been demonstrated (Kobayashi *et al.*, 1986).

1.3 Basic theory of electronic structure

The free-carrier distribution in semiconductors depends on the distribution of ionized impurities. The free-carrier distribution is instrumental for many properties of semiconductors, including recombination and transport properties. In semiconductors with 'smooth' changes of doping concentration, the free-carrier profile follows the doping profile with good approximation. However, in δ-doped semiconductors which exhibit large doping concentration variations over short distances, the free-carrier distribution is spread out much further than the doping distribution. Here, we first discuss the mathematical representation of doping profiles in δ-doped semiconductors and subsequently discuss the free-carrier distribution in the classical as well as the quantum-mechanical picture.

The Dirac δ-function, $\delta(z)$, is a mathematical distribution whose width is infinitely narrow. The magnitude of $\delta(z)$ is zero for $z \neq 0$ and infinite for $z = 0$. The integral of the function, $\int_{-\infty}^{\infty} \delta(z)\,\mathrm{d}z = 1$, has a value of unity. These are well-known properties of the δ-function. Employment of the δ-function for semiconductor doping profiles implies that the width of the one-dimensional profile is much narrower than *all other relevant length scales.* These relevant length scales include the screening length, the free-carrier diffusion length, and the de Broglie wavelength of a quantized carrier system.

Strickly speaking, the employment of the δ-function for any real physical quantity is problematic. As mentioned above, the δ-function approaches infinity, that is, $\delta(z \to 0) = \infty$, a value which cannot be assumed by any real physical quantity. As will be seen below, however, in the main the *integral* properties of the δ-function enter into calculations of the characteristics of δ-doped semiconductors. For example, the band potential of a semiconductor is obtained by *integration* of the δ-function-like charge distribution using the Poisson equation. Thus, the singularity of the δ-function at $z = 0$ does *not* impose a problem.

Alternatively, the gaussian function can be used to describe very narrow 1D doping distributions. The gaussian function is a continuous function and does not have a singularity. It is well known that the gaussian distribution function is characterized by the *standard deviation.* If the standard deviation approaches zero, then the gaussian function becomes identical to the δ-function. The simplicity and practicality of the δ-function make it the preferred choice for representing narrow doping profiles.

Next we will calculate several elementary quantities of δ-doped semiconductors. As an example, we consider an n-type semiconductor with a 2D donor density $N_{\mathrm{D}}^{\mathrm{2D}}$. We assume that the donor distribution depends on the spatial coordinate z only:

$$N_{\mathrm{D}}(z) = N_{\mathrm{D}}^{\mathrm{2D}}\delta(z - z_{\mathrm{d}}) \tag{1.5}$$

where z_{d} is the location of the plane of doping atoms. The doping profile of Eqn. (1.5) is illustrated schematically in Fig. 1.3(a). Assuming that all donors are ionized, the electrostatic potential created by the donor ions can be calculated using Poisson's equation.

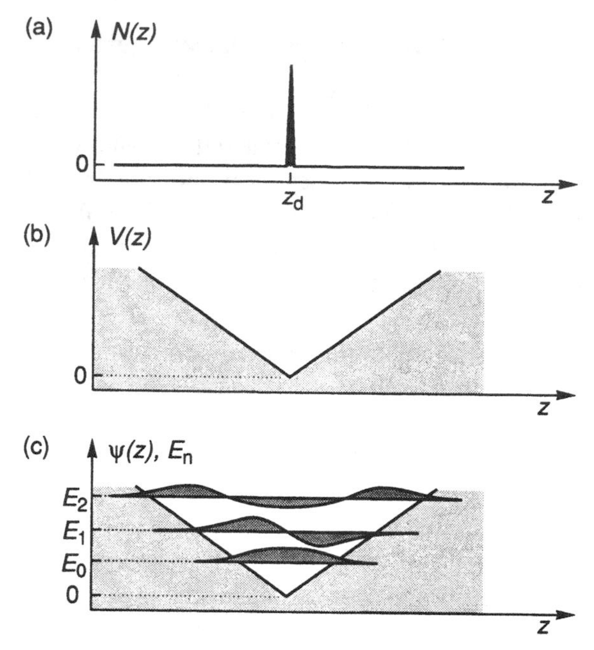

Fig. 1.3. (a) Delta-function-like doping distribution and (b) V-shaped potential well created by a δ-function-like charge distribution. Also shown are (c) the energy levels and wavefunctions of a quantized free-carrier system in a V-shaped well.

One obtains

$$V(z) = \begin{cases} -\dfrac{1}{2}\dfrac{eN_D^{2D}}{\varepsilon}(z - z_d) & \text{for} \quad z \leqslant z_d \\[2ex] +\dfrac{1}{2}\dfrac{eN_D^{2D}}{\varepsilon}(z - z_d) & \text{for} \quad z \geqslant z_d \end{cases} \tag{1.6}$$

where e is the elementary charge. The potential, which is illustrated schematically in Fig. 1.3(b), is (i) symmetric with respect to $z = z_d$, and (ii) linear with respect to z. The potential is V-shaped and we define zero potential at $z = z_d$, that is, $V(z = z_d) = 0$. Note that the donor charge is assumed to be evenly distributed in the donor plane. This assumption is usually referred to as the *jellium model.* The discreteness of impurity charges is neglected in this model.

Next we consider electrons in a V-shaped potential well and we use classical arguments first to describe the electron distribution. Size quantization effects are not taken into account in this classical consideration. Electrons in the well are attracted to the donor plane due to electrostatic interaction with donors. However, random diffusive motion of free electrons will drive electrons away from the donor plane. Thus, free carriers always have a wider spatial distribution as compared with the impurity ion distribution. The variation of majority carrier concentration in semiconductors is known to occur on a length scale of the majority carrier screening length (see, for example, Schubert, 1993). As an example, we consider n-type δ-doped GaAs with a donor density of 10^{12} cm^{-2}, which corresponds to a 3D concentration of 10^{18} cm^{-3}. For this degenerate electron concentration, the Thomas–Fermi screening length can be calculated to be 50 Å (Schubert, 1993). The spatial extent of the electron gas will be twice the screening length, that is, approximately 100 Å.

In a rigorous calculation of the free-carrier distribution in δ-doped semiconductors, size-quantization of the electron gas must be taken into account; that is, the Schrödinger and Poisson equations must be solved simultaneously. Here, we first calculate the electron distribution in a V-shaped quantum well without taking into account the band-bending caused by electrons; that is, the potential well is considered to be strictly V-shaped, as given in Eqn. (1.6). Several methods have been employed to calculate the electron distribution in a V-shaped quantum well, including the mathematically exact solution, the variational method, the WKB method, and a zero-order approximation (Schubert, 1990). In this introductory chapter, we restrict ourselves to the variational method, which provides an excellent means for calculating the spatial and energy structure of δ-doped semiconductors. For rigorous self-consistent calculation, the reader is referred to Chapter 2, by Proetto.

The variational method has been applied to V-shaped quantum wells and yields wave functions and state energies in simple analytic form (Schubert *et al.*, 1989c). The trial function used to approximate the ground-state wavefunction is given by

$$\psi_0(z) = \begin{cases} A_0(1 - \alpha_0 z)\,e^{\alpha_0 z} & z \leqslant 0 \\ A_0(1 + \alpha_0 z)\,e^{-\alpha_0 z} & z \geqslant 0. \end{cases} \tag{1.7}$$

This ground-state wavefunction is of even symmetry and decays exponentially for large absolute values of z. The constant A_0 is determined by the normalization condition $\langle\psi|\psi\rangle = 1$ to be $A_0^2 = 2\alpha_0/5$. The trial parameter α_0 determines the exponential decay of the wave function in the classically forbidden region, that is, beyond the classical turning points. The expectation value of the ground-state energy is given by

$$\langle E_0\rangle = \langle\psi|H|\psi\rangle = \frac{1}{5}\left(\frac{\hbar^2}{2m^*}\alpha_0^2 + \frac{9}{2}\frac{e\mathscr{E}}{\alpha_0}\right). \tag{1.8}$$

Here, H is the hamiltonian operator, $\hbar$ is Planck's constant divided by 2π, m^* is the electron effective mass, and $\mathscr{E}$ is the magnitude of the electric field which can be obtained by the differentiation of Eqn. (1.6), which yields $\mathscr{E} = (1/2)eN_D^{2D}/\varepsilon$. Minimization of $\langle E_0\rangle$ with respect to the trial parameter α_0 yields

$$\alpha_0 = \left(\frac{9}{4}e\mathscr{E}\frac{2m^*}{\hbar^2}\right)^{1/3}. \tag{1.9}$$

Insertion of this result into Eqn. (1.7) yields the normalized ground-state wave function. Insertion into Eqn. (1.8) yields the ground-state energy.

$$E_0 = \frac{3}{10}\left(\frac{9^2}{2}\right)^{1/3}\left(\frac{e^2\hbar^2\mathscr{E}^2}{2m^*}\right)^{1/3}. \tag{1.10}$$

Comparison of this variational result with the mathematically exact result yields that both methods agree to within 1% (Schubert, 1990).

The calculation for the wave function and energy of the first excited state is performed analogously. The trial parameter α_1 is used in the function

$$\psi_1(z) = \begin{cases} A_1 z\, e^{\alpha_1 z} & z \leqslant 0 \\ A_1 z\, e^{-\alpha_1 z} & z \geqslant 0 \end{cases} \tag{1.11}$$

which yields $A_1^2 = 2\alpha_1^3$ and

$$E_1 = \frac{3}{2}\left(\frac{9}{2}\right)^{1/3}\left(\frac{e^2\hbar^2\mathscr{E}^2}{2m^*}\right)^{1/3}. \tag{1.12}$$

The requirements for the second excited-state wave function are: (i) even symmetry; (ii) two nodes; and (iii) exponential decay of the wave function amplitude beyond the classical turning points. The following trial function satisfies these requirements:

$$\psi_2(z) = \begin{cases} A_2(\alpha_2^2 z^2 - 1)(1 - \alpha_2 z)\, e^{\alpha_2 z} & z \leqslant 0 \\ A_2(\alpha_2^2 z^2 - 1)(1 + \alpha_2 z)\, e^{-\alpha_2 z} & z \geqslant 0 \end{cases} \tag{1.13}$$

where the constant $A_2^2 = \frac{4}{63}\alpha_2$ is obtained from the normalization condition. The eigenstate energy of the second excited state is obtained as

$$E_2 = \frac{9}{7}\left(\frac{47}{12}\right)^{2/3}\left(\frac{e^2\hbar^2\mathscr{E}^2}{2m^*}\right)^{1/3}. \tag{1.14}$$

The deviation of this variational result from the exact solution is <2%.

The spatial extent of the free-carrier system is frequently of interest. Since most of the carriers occupy the ground state, its spatial extent is a good approximation of the spatial extent of the entire electron system. We deduce the *spatial extent* of a wave function from the *spatial variance*, σ_z^2, of a wave function, which is given by

$$\sigma_z^2 = \langle z^2 \rangle - \langle z \rangle^2 \tag{1.15}$$

where the $\langle z \rangle$ is the position expectation value. Using the variational function given in Eqn. (1.7), the spatial extent of the ground-state wave function is obtained as

$$z_0 = 2\sigma_z = 2\sqrt{\frac{7}{5}}\left(\frac{4}{9}\frac{\hbar^2}{e\mathscr{E}2m^*}\right)^{1/3} \tag{1.16}$$

where $\mathscr{E} = (1/2)eN_D^{2D}/\varepsilon$. As an example, we use $N_D^{2D} = 1 \times 10^{12}\ \text{cm}^{-2}$ and $m^* = 0.067m_0$ and obtain $z_0 = 80$ Å. Note that this extent (80 Å) is comparable to that deduced from the Thomas–Fermi approach (100 Å) mentioned above.

Ideally, the free-carrier density equals the ionized dopant concentration. The contribution of the free carriers to band-bending cannot be neglected and the potential well will no longer be strictly V-shaped. Whereas still V-shaped in the vicinity of the notch of the potential, the band edges become flat (horizontal) sufficiently far away from the doping sheet. The entire dopant-free-carrier system is neutral and, as a consequence, the band edges are flat at large distances from the dopant sheet.

The self-consistent solution of the spatial and energetic structure of δ-doped semiconductors requires simultaneous solution of the Schrödinger and Poisson equations. The solution is usually done in an iterative way. Initially, a 'reasonable' free-carrier distribution (i.e. ground-state wavefunction) is assumed, from which, using Poisson's equation, a potential is deduced. A new set of wavefunctions is calculated from the potential using the Schrödinger equation. A second iteration of the potential is then calculated using the new free-carrier distribution. This procedure is continued until the calculation converges, that is, the wavefunctions and the potential do not change with further iterations of the calculation.

An example of a self-consistently calculated conduction-band diagram of a δ-doped semiconductor is shown in Fig. 1.4. A single minimum parabolic band with an effective mass of $m^* = 0.067m_0$ and a doping density of 5×10^{12} cm^{-2} were used for the calculation. Figure 1.4 reveals that four subbands are populated. More than 60% of the carriers are

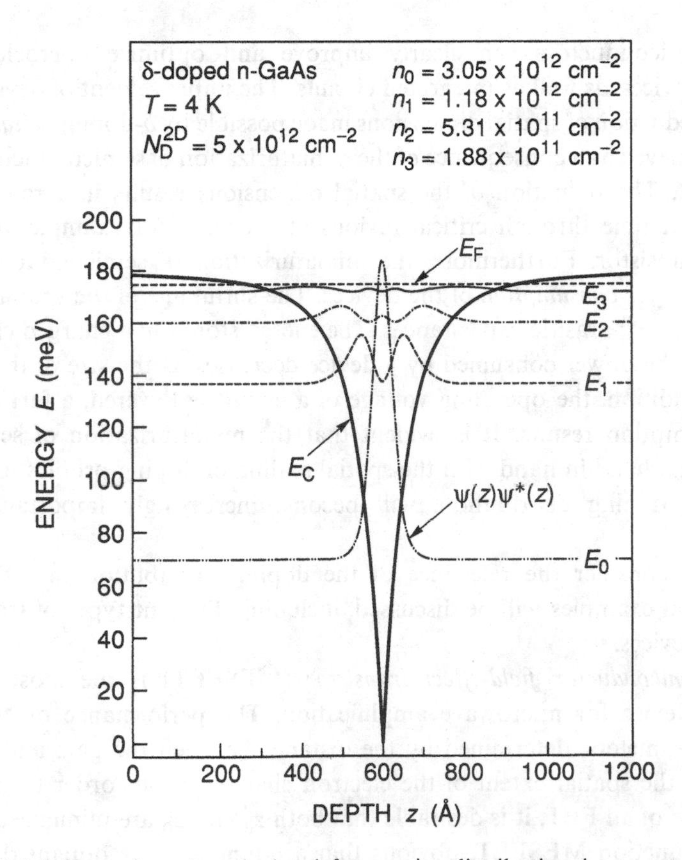

Fig. 1.4. Self-consistently calculated free-carrier distribution in an n-type δ-doped GaAs layer with a 2D doping density of 5×10^{12} cm^{-2}. Four states with energies E_0, E_1, E_2 and E_3 are populated at this doping density.

in the ground state, 24% of the electrons are in the first excited state. The full-width at half-maximum of the ground-state wavefunction is approximately 60 Å.

The electronic structure is qualitatively, as well as quantitatively, different for semiconductors with an anisotropic band structure, such as Si or Ge. These semiconductors have constant energy surfaces shaped like rotational ellipsoids with longitudinal and transverse effective masses. The confinement of a free-carrier gas occurs along one spatial dimension, for example along the z-direction. The effective mass for motion along this direction is commonly referred to as the *confinement mass.* Motion in the doping plane is entirely free and the effective mass associated with motion in these directions is called the *dispersion mass.* The dispersion mass determines the mobility of carriers for motion in the doping plane. The band structure of the host semiconductor must be known in detail in order to differentiate between the confinement mass and the dispersion mass (Gossmann and Schubert, 1993).

1.4 Relevance to microelectronics

Delta-doped semiconductors can clearly improve and optimize microelectronic and optoelectronic devices, as well as integrated circuits. The improvement of device structures is based on the reduction of spatial dimensions made possible by δ-doping. *Higher operating speeds* can be achieved as a consequence of the miniaturization of semiconductor structures: small means fast. The reduction of the spatial dimensions results in a reduction of the free-carrier transit time through critical regions of a device, for example, the base of a heterobipolar transistor. Furthermore, the miniaturization of semiconductor structures results in *lower power consumption* of the devices. The shrinkage of the spatial dimensions reduces, for example, parasitic capacitances. The energy stored in a charged capacitor also decreases. Thus, the power consumed by a device decreases as the size of the structure is reduced. If, in addition, the operating voltage of a circuit is lowered, a further reduction in power consumption results. It is evident that the miniaturization of semiconductor structures must go hand in hand with the spatial scaling of doping profiles. Delta doping, as the ultimate doping distribution, will become increasingly important as scaling progresses.

Next, we will consider the relevance of the doping distribution in different device structures. Several examples will be discussed, including different types of transistors and optoelectronic devices.

The *metal–semiconductor field-effect transistor* (MESFET) is the most widely used semiconductor device for microwave amplification. The performance of MESFETs is, among other parameters, determined by the distance between the gate and the electron channel, z_d, and the spatial extent of the electron channel, z_0. In order to maximize the transconductance of an FET, it is desirable that both z_d and z_0 are minimized. It is, in the case of a homojunction MESFET, obvious that a δ-function-like dopant distribution is the optimum doping profile. As the lateral (along the wafer surface) and the vertical (normal to wafer surface) spatial dimensions of a MESFET are scaled down, the well-known *short-channel effects* occur. These effects manifest themselves as a lack of pinch-off and non-linear current–voltage characteristics in the saturation regime of the transistor. In order to minimize short-channel effects, the gate length, L_g, must be much larger than the gate-to-channel distance and much larger than the extent of the channel, that is, $L_g \gg z_d$ and $L_g \gg z_0$. It is again evident, that, in order to minimize short-channel effects, δ-function-like doping profiles are the optimum doping distribution (see also Chapters 20 by Hong and 21 by Schubert).

The Si *metal-oxide–semiconductor field-effect transistor* (MOSFET) is the device of choice for digital integrated circuits. The majority of circuits manufactured today employ complementary MOS (CMOS) structures based on n- and p-type inversion channel devices. In such inversion channel devices, the conductive channel is induced by the gate potential. In n-type devices, an electron channel is induced at the Si–SiO_2 interface. The Si has p^--type conductivity in n-channel devices. The inversion channel typically extends 100 Å into the Si.

Several reasons make it desirable to have well-controlled doping profiles near the oxide–semiconductor interface. First, the doping in the intimate vicinity of the oxide–semiconductor interface determines the threshold voltage of MOSFETs. The doping distribution along the vertical axis (normal to the wafer surface) determines the shape and location of the induced electron channel. It is therefore desirable to control the doping distribution very accurately. Second, the substrate doping must be enhanced for small gate-length transistors in order to avoid the punch-through effect. The enhancement of the substrate doping leads, however, to an undesired shift in the threshold voltage. This shift can be reduced by a spatially well-controlled p-type doping distribution. Third, short-channel effects are smaller in inversion channel MOSFETs as compared to depletion mode MOSFETs due to the smaller gate-to-channel distance in the former. Nevertheless, if the potential near the oxide–semiconductor interface could be controlled on a 25 Å length scale, then the electron channel would be confined to the region very close to the interface. It is thus highly desirable to control the doping distribution on a near-atomic scale in Si MOSFETs. At the present time, ion implantation is being employed in the commercial manufacture of MOSFET integrated circuits. The ion implanation process, with its inherently broad doping distributions, does not allow for arbitrarily shaped doping profiles. Further improvements of the MOSFET characteristics are therefore expected through the employment of δ-doped structures (see Chapter 22 by Nakagawa and Yamaguchi).

Bipolar transistors and, particularly, *heterojunction bipolar transistors* (HBTs) are high-performance devices with applications in microwave circuits as well as in opto-electronics. HBTs of the n–p–n-type use a heterojunction barrier at the emitter–base interface in order to reduce the hole current into the emitter. In addition, a very thin base is desirable in order to speed up the electron transport through the base layer. The transit time of electrons through the base, which is a fundamental factor determining the speed of HBTs, is proportional to the thickness of the base layer. A thin base layer is therefore advantageous. On the other hand, the base layer must have a high conductivity. The base resistance and the emitter–base capacitance form a parasitic RC-circuit, which imposes another limitation on the speed of an HBT. It is, therefore, desirable to have a highly doped, thin base layer. In the limit of a very thin base thickness, the required doping profiles are δ-function-like.

Doping superlattices have been an active field of research for many years. Even though device structures, including light-emitting diodes (LEDs), lasers, and detectors have been demonstrated, no doping superlattice device is being manufactured commerically at the present time. The modulation of the band-edge potentials in 1D doping superlattices is induced by alternating positive (donor) and negative (acceptor) charge sheets. The highest optical emission intensities in doping superlattices can be achieved for a large potential modulation and for very short superlattice periods. The δ-doping technique is thus ideally suited to meet these requirements of doping superlattices. It has indeed been demonstrated that superior optical properties, including the observation of quantization of interband transitions, can be attained by the employment of the δ-doping technique.

In addition to semiconductor devices, the δ-doping technique can be used beneficially in many other semiconductor structures. Due to the unique ability to shape doping profiles arbitrarily, the technique allows one to *engineer* the doping profiles of semiconductor structures. The careful design and employment of arbitrary doping profiles allows one to optimize semiconductor structures for specific purposes. Examples of such optimized doping profiles can be found in many semiconductor devices, including transistors, lasers, light-emitting diodes, detectors, and solar cells. Heterojunction structures offer unique possibilities to tailor and optimize doping profiles. As an example, we mention the elimination of heterojunction band discontinuities (Schubert *et al.*, 1992). Appropriate doping profiles allow one to eliminate the conduction-band discontinuity or the valence-band discontinuity occurring at the interface between two different semiconductors.

In this introduction, I have shown that well-defined doping profiles have the potential for improving semiconductor technology. As a result we will be able to process, transport, distribute, store, and retrieve information in more efficient ways. The understanding and realization of well-defined doping profiles, which is the purpose of this book, will be beneficial to this process.

References

Bass S. J., *J. Cryst. Growth* **47**, 613 (1979).

Beall R. B., Clegg J. B., Castagné J., Harris J. J., Murray R., and Newman R. C., *Semicond. Sci. Technol.* **4**, 1171 (1989).

Beall R. B., Clegg J. B., and Harris J. J., *Semiconductor Sci. Technol.* **3**, 612 (1988).

Clegg J. B. and Beall R. B., *Surface Interface Anal.* **14**, 307 (1989).

DiLorenzo J. V., *Electrochem. Soc.* **118**, 1645 (1971).

Etienne B. and Thierry-Mieg V., *Appl. Phys. Lett.* **52**, 1237 (1988).

Gossmann H. J. and Schubert E. F., *Crit. Rev. Solid State Mater. Sci.* **18**, 1 (1993).

Gossmann H. J., Schubert E. F., Eaglesham D. J., and Cerullo M., *Appl. Phys. Lett.* **57**, 2440 (1990).

Headrick R. L., Weir B. E., Levi A. F. J., Eaglesham D. J., and Feldman L. C., *Appl. Phys. Lett.* **57**, 2779 (1990).

Headrick R. L., Weir B. E., Levi A. F. J., Freer B., Bevk J., and Feldman L. C., *J. Vac. Sci. Technol.* **A9**, 2269 (1991).

Kobayashi N., Makimoto T., and Horikoshi Y., *Jpn J. Appl. Phys.* **25**, L746 (1986).
Lanzillotto A.-M., Santos M., and Shayegan M., *Appl. Phys. Lett.* **55**, 1445 (1989).
Lee H., Schaff W. J., Wicks G. W., Eastman L. F., and Calawa A. R., *Inst. Phys. Conf. Ser.* **74**, 321 (1985).
Levi A. F. J., McCall S. L., and Platzman P. M., *Appl. Phys. Lett.* **54**, 940 (1989).
Schubert E. F. (1990) *Optical and Quantum Electronics* **22**, S141 (1990).
Schubert E. F., *Doping in III–V Semiconductors* (Cambridge University Press, 1993).
Schubert E. F. and Ploog K., *Jpn. J. Appl. Phys.* **24**, L608 (1985).
Schubert E. F., Cunningham J. E., Tsang W. T., and Timp G. L., *Appl. Phys. Lett.* **51**, 1170 (1987).
Schubert E. F., Fischer A., Horikoshi Y., and Ploog K., *Appl. Phys. Lett.* **47**, 219 (1985a).
Schubert E. F., Fischer A., and Ploog K., *Electronics Letters* **21**, 411 (1985b).
Schubert E. F., Harris T. D., Cunningham J. E., and Jan W., *Phys. Rev.* **B39**, 11011 (1989c).
Schubert E. F., Harris T. D., and Cunningham J. E., *Appl. Phys. Lett.* **53**, 2208 (1988a).
Schubert E. F., Kopf R. F., Kuo J. M., Luftman H. S., and Garbinski P. A., *Appl. Phys. Lett.* **57**, 497 (1990a).
Schubert E. F., Luftman H. S., Kopf R. F., Headrick R. L., and Kuo J. M., *Appl. Phys. Lett.* **57**, 1799 (1990b).
Schubert E. F., Pfeiffer L., West K. W., and Izabelle A., *Appl. Phys. Lett.* **54**, 1350 (1989b).
Schubert E. F., Ploog K., Fischer A., and Horikoshi Y. (1984) *Semiconductor Devices with at Least One Monoatomic Layer of Doping Atoms.* US Patent 4 882 609, foreign application priority date Nov. 19 (1984).
Schubert E. F., Stark J., Chiu T. H., and Tell B., *Appl. Phys. Lett.* **53**, 293 (1988d).
Schubert E. F., Stark J., Ullrich B., and Cunningham J. E., *Appl. Phys. Lett.* **52**, 1508 (1988c).
Schubert E. F., Tu C. W., Kopf R. F., Kuo J. M., and Lunardi L. M., *Appl. Phys. Lett.* **54**, 2592 (1989).
Schubert E. F., Tu L.-W., Zydzik G. J., Kopf R. F., Benvenuti A., and Pinto M. R., *Appl. Phys. Lett.* **60**, 466 (1992).
Schubert E. F., Ullrich B., Harris T. D., and Cunningham J. E., *Phys. Rev.* **B38**, 8305 (1988b).
Schubert E. F., van der Ziel J. P., Cunningham J. E., and Harris T. D., *Appl. Phys. Lett.* **55**, 757 (1989a).
Webb C., *Appl. Phys. Lett.* **54**, 2091 (1989).
Wood C. E. C., Metze G., Berry J., and Eastman L. F., *J. Appl. Phys.* **51**, 383 (1980).
Zeindl H. P., Wegehaupt T., Eisele I., Oppolzer H., Reisinger H., Tempel G., and Koch F., *Appl. Phys. Lett.* **50**, 1164 (1987).

PART TWO

Theory

2

Electronic structure of delta-doped semiconductors

C. R. PROETTO

2.1 Introduction

Doping semiconductors with donor or acceptor impurities has traditionally been an essential step in most of the basic research and industrial applications of these materials. Over the last few years, and in pace with the continuous reduction of the spatial dimensions of semiconductor heterostructures and devices, the size of dopant distributions has been decreased dramatically. One of the more promising doping techniques at the atomic level is clearly delta (δ)-doping, where the dopants are confined to one or a few atomic monolayers. This technique represents the ultimate control of a dopant profile and will certainly play an important role in future quantum-electronic and photonic-device research.

The purpose of this chapter is to discuss the electronic structure of δ-doped semiconductors. All the calculations will be done within the envelope function framework of the effective-mass approximation theory, which could be expected to work even better than in the case of compositional superlattices, where one has to deal with interfaces between two different semiconductors (Bastard, 1988).

In Section 2.2 we describe the general model that will be applied in subsequent sections to the analysis of several particular cases. Once the usual approximation of replacing the random distribution of impurity point charges within the two-dimensional doping layers by a uniform two-dimensional charge density has been assumed, the problem reduces to the solution of a well-defined set of one-dimensional second-order differential equations.

Section 2.3, mostly analytical, is devoted to the study of depleted (no free carriers in the system) quantum wells and superlattices. We start with the case of a large period asymmetric sawtooth superlattice, which could be realized by alternating planes with p and n dopants (the so-called n–i–p–i configuration) such that the distance from n to p planes in the growth direction is not the same as the distance from p to n planes. The familiar eigenvalues and eigenfunctions of the symmetric triangular quantum well are obtained as a particular case of this general situation. By decreasing the superlattice period, we move from the isolated quantum well or 'atomic' regime to the superlattice regime, where the coupling between adjacent quantum wells is considerable and a Kronig–Penney type of model should be used.

Section 2.4, mostly numerical, deals with the more complicated problem of the subband structure of undepleted quantum wells and superlattices. Due to the presence of free carriers, a self-consistent solution of the Schrödinger and Poisson equations is necessary, a task that has been accomplished for different situations of increasing complexity: n-type δ-layers in GaAs, n-type δ-layers in Si, and p-type δ-layers in GaAs. These self-consistent subband structures could be used as the starting point to obtain other physical characteristics of δ-doped layers, such as transport and optical properties, which will be discussed in Part 5.

Finally, Section 2.5 is devoted to the conclusions and, at least from the theoretical perspective, possible lines of future work.

2.2 General model

Our calculations are based on two major approximations. We assume (i) that the effective-mass approximation is valid, so that we can neglect the periodic lattice potential and use the effective mass and the dielectric constant of the host semiconductor, and (ii) that the positive (negative) charge of the ionized donor (acceptor) impurities is homogeneously smeared out in the plane perpendicular to the growth direction (jellium approximation), neglecting the point-charge character of dopants and the spatial potential fluctuations which result from the random distribution of impurities in the doped planes.

The envelope function approximation within the effective-mass theory, as applied to the case of compositional heterostructures, quantum wells, and superlattices, needs the extra assumption that the periodic parts of the Bloch functions are the same in each kind of layer which constitutes the heterostructure; obviously, this condition is automatically and exactly satisfied in the case of δ-doped layers. Quite recently (Mäder and Baldereschi, 1992a, 1992b), the validity of the effective-mass approximation has been tested, by comparing with the results of a more microscopic theory (the empirical tight-binding Koster–Slater approach) applied to the specific case of ultrathin isoelectronic intralayers in semiconductors, and excellent agreement has been found between both approaches.

The second assumption, while almost invariably made in the calculations of the electronic structure of δ-doped layers, is much more difficult to justify and constitutes one of the open problems in the theoretical understanding of these structures. The problem is particularly acute for the compensated systems, while for non-compensated ones, at sufficiently high doping concentrations, one can assume that the impurity-carrier system is on the metallic side of the metal–insulator transition. Also, it has been shown that statistical potential fluctuations are minimized by employing a doping profile consisting of a train of δ-functions (Schubert *et al.*, 1988b). It should be noted, however, that recent experiments have studied the low density case where the subband description does not apply (Koch, 1993), and one is forced to consider each impurity layer as a two-dimensional array of impurities at random positions. The theoretical understanding of this configuration is just beginning to emerge (Efros *et al.*, 1990; Ferreyra and Proetto, 1991; Levin *et al.*, 1991; Metzner *et al.*, 1992; Sobkowicz *et al.*, 1992; Fritzsche and Schirmacher, 1993).

For the general case of arbitrary concentrations of donor and acceptor impurities, we have to solve the Poisson and Schrödinger equations in a self-consistent way: the presence of free carriers (electrons or holes) changes the quantum well or superlattice potential (in the Hartree approximation), which in turn disturbs the charge distribution. The Poisson equation is conveniently written as

$$\frac{d^2}{dz^2} V(z) = \frac{4\pi e}{\varepsilon} \rho(z) \tag{2.1}$$

where $V(z)$ is the superlattice potential, e is the electron charge ($e > 0$), ε is the static dielectric constant corresponding to the host semiconductor, and $\rho(z)$ is the charge density given by

$$\rho(z) = eN_D^{2D} \sum_{z_D} \delta(z - z_D) - eN_A^{2D} \sum_{z_A} \delta(z - z_A) - en(z) + ep(z) \tag{2.2}$$

where N_D^{2D} (N_A^{2D}) is the two-dimensional concentration of ionized donor (acceptor) impurities, z_D and z_A are the coordinates along z of the donor and acceptor ionized impurity sheets, and $n(z)$ ($p(z)$) is the three-dimensional electron (hole) density at point z.

Note that two simplifications are already implicit in writing Eqns. (2.1) and (2.2): in the first place, and as a consequence of assumption (ii), the averaging of the ionized impurity charges in the x–y plane renders an in principle three-dimensional Poisson equation a one-dimensional problem as posed by Eqn. (2.1). Second, we have assumed a δ-function profile for the distribution of impurities in the z-direction; this approximation, however, is not essential and will be relaxed in the self-consistent calculations of Section 2.4. For the model used here, one has

$$n(z) = \frac{2}{\Omega} \sum_{n\mathbf{k}} f[\mu - E_n^e(\mathbf{k})] \, |\psi_{nk_z}^e(z)|^2 \tag{2.3}$$

and

$$p(z) = \frac{1}{\Omega} \sum_{n\mathbf{k}} \sum_{i=1}^{4} f[\mu - E_n^h(\mathbf{k})] \, |\psi_{in\mathbf{k}}^h(z)|^2 \tag{2.4}$$

where the factor 2 in Eqn. (2.3) stands for the electron spin degeneracy, $f(x)$ is the Fermi–Dirac function with chemical potential μ, and Ω is the sample area. Due to the translational symmetry in the x–y plane of our system, the electron and hole wave functions can be written as a plane-wave (with wave vector $\mathbf{k}_\parallel$) along this plane times a wave function along z, given by the solution of the respective electron:

$$\left[-\frac{\hbar^2}{2m_z}\frac{d^2}{dz^2}+V(z)\right]\psi^{e}_{nk_z}(z)=E^{e}_{n}(\mathbf{k})\psi^{e}_{nk_z}(z) \tag{2.5}$$

and hole

$$\sum_{j=1}^{4}[H^{0}_{ij}(k_x,k_y,\hat{k}_z)+V(z)\delta_{ij}]\psi^{h}_{jn\mathbf{k}}(z)=E^{h}_{n}(\mathbf{k})\psi^{h}_{in\mathbf{k}}(z)\quad(i=1,2,3,4) \tag{2.6}$$

Schrödinger equations.

In Eqn. (2.5), m_z is the effective mass associated with motion along the growth direction, and $\psi^{e}_{nk_z}(z)$ is the electron envelope wave function corresponding to subband index n and superlattice wave vector k_z with eigenvalue $E^{e}_{n}(\mathbf{k})$. For the particular cases of GaAs and Si the conduction-band bottom is parabolic and

$$E^{e}_{n}(\mathbf{k})=E^{e}_{n}(k_z)+\frac{\hbar^2}{2}\left(\frac{k_x^2}{m_x}+\frac{k_y^2}{m_y}\right) \tag{2.7}$$

where $E^{e}_{n}(k_z)$ is the superlattice dispersion relation that will be the subject of Sections 2.3 and 2.4, and the second term on the right-hand side is the kinetic energy associated with the free motion in the x–y plane. Bulk GaAs has a single isotropic conduction-band minimum at the Γ point of the Brillouin zone, and, consequently, equal effective masses along the three directions: $m_x = m_y = m_z = 0.067\ m_0$ (m_0 being the free-electron mass). The case of Si is more complicated, as its conduction band has six equivalent minima in the $\langle 100\rangle$ directions of momentum space, each of them placed at 85% of the way to the Brillouin-zone boundaries. The Fermi surfaces of the minima are rotational ellipsoids, the long axis corresponding to the longitudinal effective mass $m_l = 0.916\ m_0$, while the short axis defines the transverse effective mass $m_t = 0.190\ m_0$ (Ando *et al.*, 1982). For our choice of quantum confinement along z, $m_z = m_l$ for the two conduction-band minima along the z-axis in reciprocal space, while $m_z = m_t$ for the four remaining conduction-band minima. As $m_t \ll m_l$, these four valleys are expected to have higher values of $E^{e}_{n}(k_z)$ in Eqn. (2.5) than the two valleys whose long axis, corresponding to the heavy effective mass m_l, is perpendicular to the doped plane.

Equation (2.6) is the 4×4 effective-mass matrix eigenvalue problem for holes, with $\psi^{h}_{in\mathbf{k}}(z)$ being the four-component envelope wave function corresponding to subband index n and wave vector $\mathbf{k}$ with eigenvalue $E^{h}_{n}(\mathbf{k})$, and $\hat{k}_z$ is the operator $-i\hbar d/dz$.

The 4×4 Luttinger matrix (the kinetic energy) is given by (Luttinger, 1956)

$$H^0=\begin{Bmatrix} P+Q & R & -S & 0\\ R^* & P-Q & 0 & S\\ -S^* & 0 & P-Q & R\\ 0 & S^* & R^* & P+Q\end{Bmatrix} \tag{2.8}$$

where

$$P \pm Q = \frac{\hbar^2}{2m_0}[(\gamma_1 \mp 2\gamma_2)\hat{k}_z^2 + (\gamma_1 \pm \gamma_2)k_\parallel^2] \tag{2.9a}$$

$$R = \sqrt{3}\frac{\hbar^2}{2m_0}[\gamma_2(k_x^2 - k_y^2) - 2i\gamma_3 k_x k_y] \tag{2.9b}$$

$$S = \sqrt{3}\frac{\hbar^2}{m_0}\gamma_3(k_x - ik_y)\hat{k}_z \tag{2.9c}$$

with $k_\parallel^2 = k_x^2 + k_y^2$, and $\gamma_1, \gamma_2, \gamma_3$ the Luttinger parameters appropriate for the bulk material. In writing H^0 we have neglected very small linear k terms caused by the lack of inversion symmetry of the GaAs zincblende structure (Broido and Sham, 1985).

For the general treatment of the valence bands, one should use the 6×6 Luttinger–Kohn $\mathbf{k}\cdot\mathbf{p}$ Hamiltonian (Luttinger and Kohn, 1955). This retains the three highest (doubly degenerate) valence bands and describes the dispersion of, and interactions between, these bands up to order k^2. In the presence of spin–orbit coupling, the six-fold degeneracy at Γ_8 is split into a four-fold and a two-fold degeneracy. The four-fold degenerate state corresponds to total angular momentum $J = \frac{3}{2}$, while the two-fold degenerate state (the spin-split-off state) corresponds to the $J = \frac{1}{2}$ multiplet. The spin–orbit splitting Δ is about 340 meV in GaAs so the spin-split-off band can be ignored for energies close to the band edge, and consequently the original 6×6 Hamiltonian reduces to the 4×4 Luttinger Hamiltonian, Eqn. (2.8). For the case of Si, however, Δ is about 44 meV, and one should retain the full 6×6 Hamiltonian. When $k_\parallel = 0$, the Luttinger Hamiltonian becomes diagonal ($R = S = 0$), and allows us to define $m_z = m_0/(\gamma_1 \mp 2\gamma_2)$, where the $-$ $(+)$ sign corresponds to the heavy (light) holes. For the case of Si, the spin-split-off band should also be included, with an effective mass $m = m_0/\gamma_1$ (O'Reilly, 1989).

As our attention is confined to an electrically neutral superlattice, the integral over one period h of $\rho(z)$ should be zero:

$$\int_{-h/2}^{h/2} \rho(z)\,\mathrm{d}z = 0 \tag{2.10}$$

which can be rewritten as

$$N_D^{2D} - N_A^{2D} - \frac{2}{\Omega}\sum_{n,\mathbf{k}} f[\mu - E_n^e(\mathbf{k})] + \frac{1}{\Omega}\sum_{n,\mathbf{k}} f[\mu - E_n^h(\mathbf{k})] = 0 \tag{2.11}$$

using Eqn. (2.2). For given values of N_D^{2D} and N_A^{2D} this equation determines the chemical potential μ. Equation (2.11), together with Eqns. (2.1), (2.5), and (2.6) are the basic expressions that should be used in a self-consistent way. The set of second-order differential equations, (2.1), (2.5), and (2.6), needs to be complemented with the corresponding boundary conditions; they will be discussed for different particular cases in the following sections.

2.3 Depleted quantum wells and superlattices

Superlattices with a periodic potential can be generated by alternating n-type and p-type doping in an otherwise homogeneous semiconductor. If $N_D^{2D} = N_A^{2D} \equiv N^{2D}$, such a doping superlattice is depleted of free carriers ($n(z) = p(z) \equiv 0$ in Eqn. (2.2)), as all the electrons will recombine (in principle) with the holes. If a donor sheet is separated from the two adjacent acceptor sheets by two unequal (equal) distances, an asymmetric (symmetric) sawtooth structure is obtained, as shown in Fig. 2.1(a) and 2.1(b). One of the main features of these systems is the indirect gap in real space, which leads to long excited carrier recombination times (Ploog and Döhler, 1983; Döhler, 1986). It has been shown that asymmetric δ-doped superlattices such as that of Fig. 2.1(a) exhibit novel photovoltaic effects and low-intensity non-linear optical properties (Glass *et al.*, 1989). Other interesting optical effects which have been observed in δ-doped superlattices include absorption

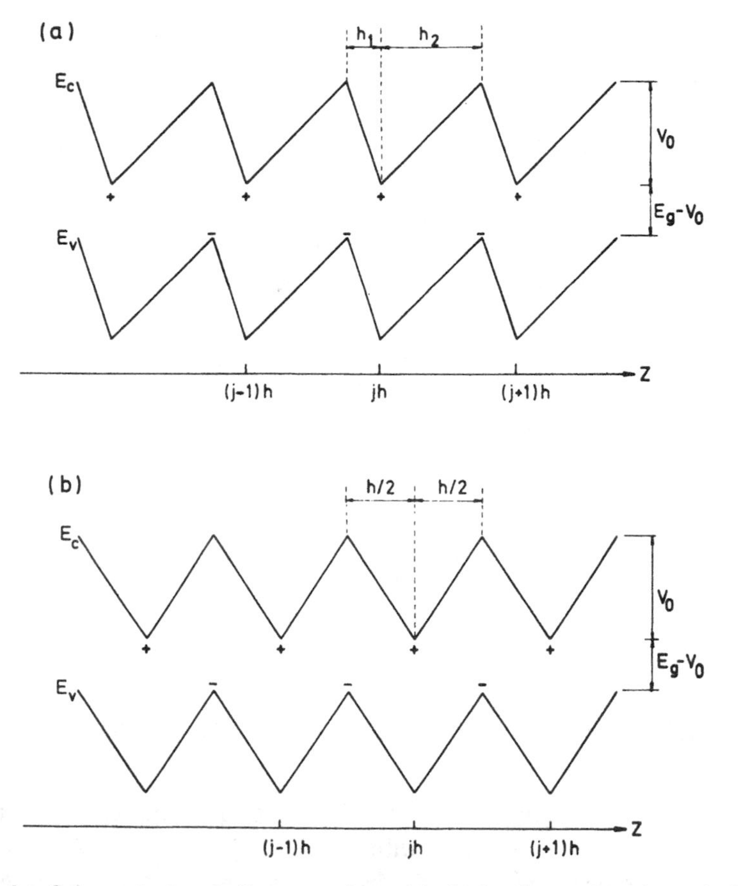

Fig. 2.1. Schematic band diagram of an (a) depleted asymmetric and (b) depleted symmetric δ-doped semiconductor. E_g denotes the bulk band-gap energy. E_C and E_V are the conduction- and valence-band edges, respectively, and V_0 is the potential modulation.

Table 2.1. *Numerical values of effective mass* (m_z), *effective Bohr radius* ($a^* = \varepsilon\hbar^2/m_z e^2$), *and effective Rydberg* ($Ry^* = e^4 m_z/2\varepsilon^2\hbar^2$) *associated with size quantization along z; ε is the static dielectric constant, and its value is 12.5 (11.5) for GaAs (Si). The Luttinger parameters which determine the heavy-, light-hole, and spin-split-off hole effective masses are* $\gamma_1 = 6.85$ (4.285), $\gamma_2 = 2.1$ (0.339), *and* $\gamma_3 = 2.9$ (1.446) *for GaAs (Si)*

	GaAs			Si				
	Electrons	Heavy-holes	Light-holes	Electrons (m_l)	Electrons (m_t)	Heavy-holes	Light-holes	Spin-split-off holes
$m_z(m_0)$	0.067	0.377	0.090	0.916	0.190	0.277	0.201	0.233
a^* (Å)	98.73	17.55	73.50	6.64	32.03	21.97	30.28	26.12
Ry^* (meV)	5.83	32.83	7.84	94.24	19.55	28.50	20.68	23.97

and luminescence at wavelengths longer than the GaAs band gap (Romanov and Orlov, 1973), and quantization of electron and hole states confined in two-dimensional regions (Schubert *et al.*, 1988a).

By replacing the charge density, Eqn. (2.2), (with $n(z) = p(z) \equiv 0$) in the Poisson equation, Eqn. (2.1), one obtains the superlattice potential

$$V(z) = \frac{4\pi e^2 N^{2D}}{\varepsilon}\begin{cases} h_1(z - jh)/h & 0 \leqslant z - jh \leqslant h_2 \\ h_2(-z + jh)/h & -h_1 \leqslant z - jh \leqslant 0 \end{cases} \tag{2.12}$$

with jh ($j = 0, \pm 1, \pm 2, \ldots$) corresponding to the ionized donor impurity layers' positions along z and $h_1 + h_2 = h$ (see Fig. 2.1(a)); note that the symmetric configuration is recovered when $h_1 = h_2 = h/2$.

We first consider a depleted asymmetric superlattice with negligible quantum mechanical coupling between adjacent wells. In this case, the individual V-shaped wells can be considered to be isolated from each other, the k_z dependence of the subband structure is negligible, and the superlattice solutions are identical to those of the isolated well. We will refer to this limit as the multiple-quantum well (MQW) regime or 'atomic' limit.

(*a*) *Asymmetric quantum well*

The Schrödinger equation for an electron or hole (at $k_\parallel = 0$) confined to move in a triangular asymmetric potential is, according to Eqns. (2.5) and (2.6), given by

$$\left[-\frac{\hbar^2}{2m_z}\frac{d^2}{dz^2} + V(z)\right]\psi_n(z) = E_n\psi_n(z) \tag{2.13}$$

where the indices e, h, and k_z have been dropped for the sake of simplicity, m_z is one of the effective masses of the first row of Table 2.1 (depending on which semiconductor and particle one is considering), and

$$V(z) = \frac{4\pi e^2 N^{2D}}{\varepsilon} \begin{cases} h_1 z/h & z \geqslant 0 \\ -h_2 z/h & z \leqslant 0 \end{cases} \tag{2.14}$$

which can be considered as the limit of Eqn. (2.12) when h, h_1, and h_2 are very large (as compared with the wave-function extension along z), but the ratios h_1/h and h_2/h are kept finite.

Measuring energies in units of the effective $Ry^* = m_z e^4/2\varepsilon^2\hbar^2$, lengths in units of the effective Bohr radius $a^* = \varepsilon\hbar^2/m_z e^2$, and defining the dimensionless parameters

$$\nu_+ = \frac{2h_1}{h}\nu, \quad \nu_- = \frac{2h_2}{h}\nu, \quad \nu = 4\pi a^{*2} N^{2D}, \tag{2.15}$$

the exact solution of Eqns. (2.13) and (2.14) is

$$\psi_n(\tilde{z}) = \begin{cases} aAi[\nu_+^{1/3}(\tilde{z} - \varepsilon_n/\nu_+)] + bBi[\nu_+^{1/3}(\tilde{z} - \varepsilon_n/\nu_+)] & \tilde{z} \geqslant 0 \\ cAi[\nu_-^{1/3}(-\tilde{z} - \varepsilon_n/\nu_-)] + dBi[\nu_-^{1/3}(-\tilde{z} - \varepsilon_n/\nu_-)] & \tilde{z} \leqslant 0 \end{cases} \tag{2.16}$$

where Ai and Bi are the Airy functions (Antosiewicz, 1970) and a, b, c, d are coefficients that will be determined from the boundary and normalization conditions; note that $\tilde{z} = z/a^*$ and $\varepsilon_n = E_n/Ry^*$ are dimensionless magnitudes. The condition that the wave function should be finite when $|\tilde{z}| \to \infty$ and the behavior of Bi at large arguments implies $b = d = 0$.

The remaining boundary conditions are the continuity of the wave function

$$\psi_n(0^+) = \psi_n(0^-) \Rightarrow aAi(-\varepsilon_n/\nu_+^{2/3}) = cAi(-\varepsilon_n/\nu^{2/3}) \tag{2.17}$$

and its derivative

$$\psi_n'(0^+) = \psi_n'(0^-) \Rightarrow a\nu_+^{1/3} Ai'(-\varepsilon_n/\nu_+^{2/3}) = -c\nu^{1/3} Ai'(-\varepsilon_n/\nu^{2/3}) \tag{2.18}$$

at $\tilde{z} = 0$. Due to the fact that the effective mass is the same on both sides of $\tilde{z} = 0$, it does not affect the boundary conditions; this should be contrasted with the case of compositional superlattices, where the effective mass has a jump at each interface and appears explicitly in the boundary conditions related to the derivative of the envelope wave function (BenDaniel and Duke, 1966).

Elimination of a and c from Eqns. (2.17) and (2.18) yields the expression

$$Ai(-\varepsilon_n/\nu_+^{2/3})Ai'(-\varepsilon_n/\nu_-^{2/3}) + (\nu_+/\nu_-)^{1/3} Ai'(-\varepsilon_n/\nu_+^{2/3})Ai(-\varepsilon_n/\nu_-^{2/3}) = 0 \tag{2.19}$$

which is an implicit equation for the eigenvalues ε_n, whose solutions should be sought numerically. Replacing ν_+ and ν_- by their expressions given in Eqn. (2.15), it is easy to see that the scaled eigenvalues $\varepsilon_n/\nu^{2/3}$ are only functions of the ratio h_1/h_2. As Eqn. (2.19) is invariant under the change $h_1 \to h_2$ and vice versa, it is sufficient to consider the case where $h_1/h_2 \leqslant 1$.

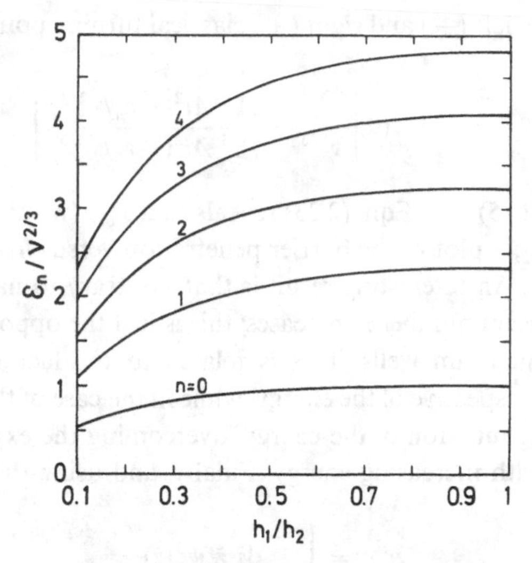

Fig. 2.2. Lowest eigenvalues ε_n for a triangular asymmetric quantum well as a function of the 'asymmetry' parameter h_1/h_2. The results are invariant under the exchange of h_1 and h_2.

In Fig. 2.2 we show the five lowest eigenvalues of Eqn. (2.19) as a function of h_1/h_2. As the limit $h_1/h_2 \ll 1$ is incompatible with the hypothesis of isolated quantum wells, the lower limit of the x-axis has been chosen as 0.1. It is interesting to note the monotonic increase of the eigenvalues $\varepsilon_n/v^{2/3}$ when one moves from the asymmetric to the symmetric situation, where the confinement reaches its maximum strength. The lower the level, the faster it approaches the symmetric limit; this is due to the fact that when $h_1/h_2 \to 0$, the eigenvalues $\varepsilon_n \to 0$, and then the ground state is closest to its symmetric value, followed by the first excited state, etc.

Having obtained the eigenvalues, we now turn to the eigenfunctions. Imposing the normalization condition

$$\int_{-\infty}^{\infty} \mathrm{d}\tilde{z}\, \psi_n^2(\tilde{z}) = 1 \tag{2.20}$$

on the wave function, Eqn. (2.16), we obtain for the normalization constant

$$a^2 = \left\{ \left(\frac{1}{v_+} + \frac{1}{v_-} \right) [\varepsilon_n Ai^2(-\varepsilon_n/v_+^{2/3}) + v_+^{2/3} Ai'^2(-\varepsilon_n/v_+^{2/3})] \right\}^{-1} \tag{2.21}$$

where Eqns. (2.17) and (2.18) have been used in intermediate steps. Some useful integrals of the Airy functions necessary to obtain the explicit result Eqn. (2.21) are given by Stern (1972).

Once the eigenvalues and eigenfunctions have been determined, it is instructive to consider some physical questions connected with the problem at hand. One of these concerns the integrated probability of finding the particle in the classical forbidden region:

$$P_{\mathrm{b},n} = \int_{-\infty}^{-\tilde{z}_{\mathrm{t},n}^-} \mathrm{d}\tilde{z}\, \psi_n^2(\tilde{z}) + \int_{\tilde{z}_{\mathrm{t},n}^+}^{\infty} \mathrm{d}\tilde{z}\, \psi_n^2(\tilde{z}) \tag{2.22}$$

where $\tilde{z}^{\pm}_{t,n} = \varepsilon_n/v_{\pm}$ are the left (−) and right (+) classical turning points. Using Eqn. (2.16), we obtain the result

$$P_{b,n} = a^2 Ai'^2(0)\left[\frac{1}{v_+^{1/3}} + \frac{1}{v_-^{1/3}}\frac{Ai^2(-\varepsilon_n/v_+^{2/3})}{Ai^2(-\varepsilon_n/v_-^{2/3})}\right]. \tag{2.23}$$

Substitution of Eqn. (2.15) into Eqn. (2.23) reveals that $P_{b,n}$ (without any scaling with v) is just a function of h_1/h_2; a plot of the barrier penetration versus h_1/h_2 for the five lowest states is given in Fig. 2.3. An interesting result is that, contrary to naïve expectations, $P_{b,n}$ decreases when the quantum number n increases; this is just the opposite to what one finds for the compositional quantum wells. This is related to the fact that in the last case, $z_{t,n} = L$ (L = well size) irrespective of the energy, while in the case of the V-shaped quantum wells, $z_{t,n}$ is an increasing function of the energy, overcoming the expected increase in the penetration associated with increasing energy. Finally, and using the definition

$$\langle \tilde{z}^p \rangle_n = \int_{-\infty}^{\infty} d\tilde{z}\, \tilde{z}^p \psi_n^2(\tilde{z}) \tag{2.24}$$

we have determined $\langle \tilde{z} \rangle_n$ and $\langle \tilde{z}^2 \rangle_n$, which are relevant quantities as they provide some insight into the spatial extension of the wave function. From Eqns. (2.16) and (2.24), one obtains

$$\langle \tilde{z} \rangle_n = \frac{2}{3}a^2\left(\frac{1}{v_+} + \frac{1}{v_-}\right)\left[\varepsilon_n^2\left(\frac{1}{v_+} - \frac{1}{v_-}\right)Ai^2\left(-\frac{\varepsilon_n}{v_+^{2/3}}\right) + v_+^{2/3}\varepsilon_n\left(\frac{1}{v_+} - \frac{1}{v_-}\right)Ai'^2\left(-\frac{\varepsilon_n}{v_+^{2/3}}\right)\right.$$
$$\left. - \frac{v_+^{1/3}}{2} Ai\left(-\frac{\varepsilon_n}{v_+^{2/3}}\right)Ai'\left(-\frac{\varepsilon_n}{v_+^{2/3}}\right)\right] \tag{2.25}$$

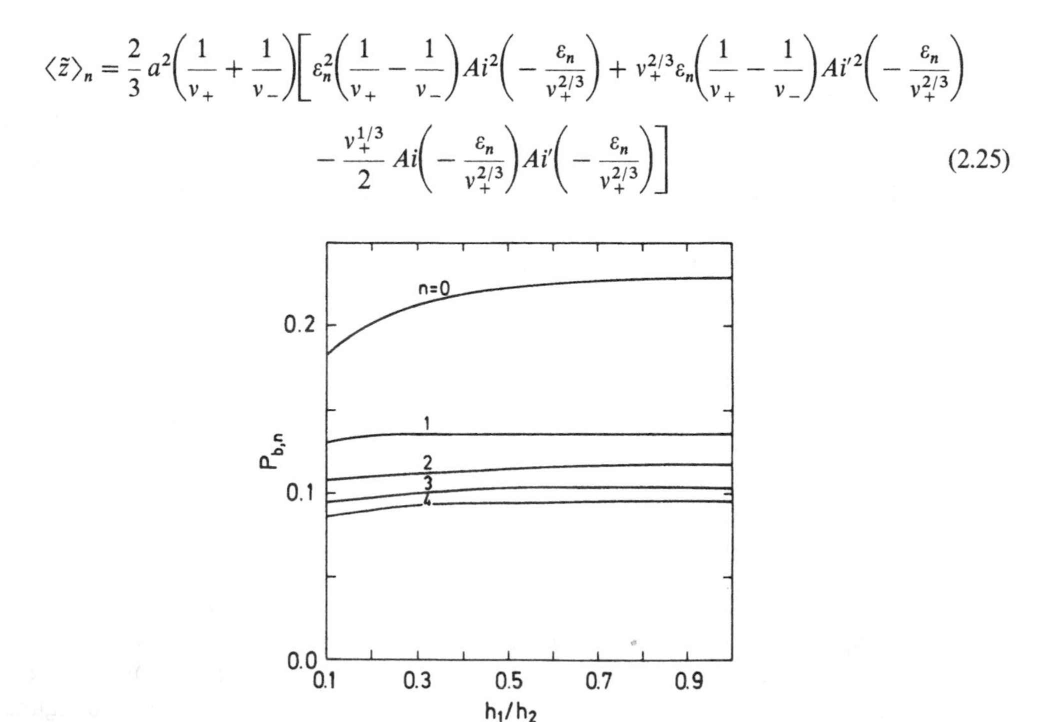

Fig. 2.3. Probability of finding the particle in the classical forbidden region $P_{b,n}$ for the lowest states of a triangular asymmetric quantum well.

and

$$\langle \tilde{z}^2 \rangle_n = \frac{8}{15} a^2 \left\{ \left[\varepsilon_n^3 \left(\frac{1}{\nu_+^3} + \frac{1}{\nu_-^3} \right) + \frac{3}{8} \left(\frac{1}{\nu_+} + \frac{1}{\nu_-} \right) \right] Ai^2 \left(-\frac{\varepsilon_n}{\nu_+^{2/3}} \right) + \nu_+^{2/3} \varepsilon_n^2 \left(\frac{1}{\nu_+^3} + \frac{1}{\nu_-^3} \right) \right.$$
$$\left. \times Ai'^2 \left(-\frac{\varepsilon_n}{\nu_+^{2/3}} \right) - \frac{\nu_+^{1/3}}{2} \varepsilon_n \left(\frac{1}{\nu_+^2} - \frac{1}{\nu_-^2} \right) Ai \left(-\frac{\varepsilon_n}{\nu_+^{2/3}} \right) Ai' \left(-\frac{\varepsilon_n}{\nu_+^{2/3}} \right) \right\}. \tag{2.26}$$

Using Eqns. (2.17) and (2.18), one can check that $\langle \tilde{z} \rangle_n$ changes sign under the exchange $h_1 \to h_2$, and that it is identically zero when $h_1 = h_2$, as it should be by the symmetry of the potential. In a quite analogous way, it is easy to see that $\langle \tilde{z}^2 \rangle_n$ is invariant under the change of h_1 with h_2, and that it remains finite in the symmetric case $h_1 = h_2$. The numerical solution of Eqns. (2.25) and (2.26) produces the results shown in Figs. 2.4 and 2.5 respectively.

It should be emphasized that the scaled results presented in Figs. 2.2–2.5 are valid for any δ-doped compensated semiconductor in the isolated quantum well regime. Given a particular system, microscopic parameters such as the effective mass, dielectric constant, and two-dimensional impurity concentrations determine ν, while the asymmetry parameter h_1/h_2 is given by the position of the impurity layers.

In the numerical results presented in this and the following sections, we sometimes need to generate Airy functions of arbitrary arguments with precision. There is a very accurate algorithm for generating the Airy functions based on Chebyshev's series approximations (Prince, 1975); we use this method in our calculations. Contrary to widespread belief, the calculation of the exact eigenvalues and eigenfunctions in terms of the Airy functions is not a difficult task, and their use should be encouraged in order to avoid unnecessary

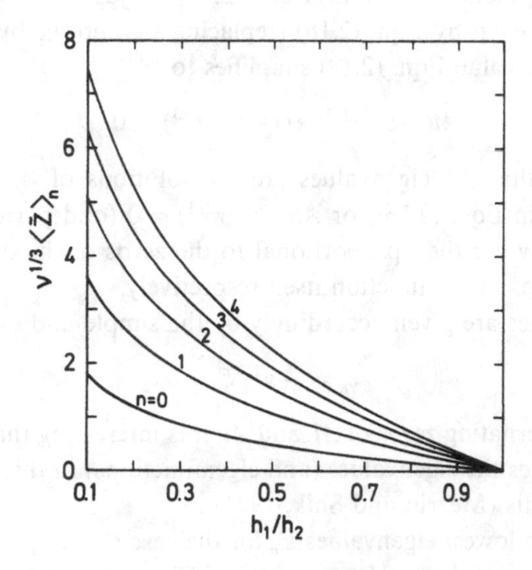

Fig. 2.4. Coordinate mean value $\langle \tilde{z} \rangle_n$ versus h_1/h_2 for the first five states of a triangular asymmetric quantum well.

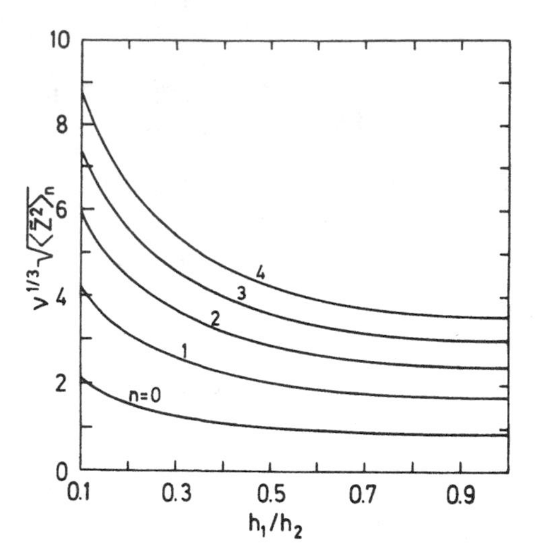

Fig. 2.5. Mean square deviation of the coordinate as a function of h_1/h_2 for the lowest states of a triangular asymmetric quantum well.

approximations. They also provide a useful basis for the study of excitonic effects (Proetto, 1990) in sawtooth superlattices.

(*b*) *Symmetric quantum well*

When the donor and acceptor ionized impurity layers are equidistant, $h_1 = h_2 = h/2$, $v_+ = v_- = v$ and the potential $V(z)$ in Eqn. (2.14) is just proportional to $|z|$. The eigenfunctions are still given by Eqn. (2.16) (replacing v_+ and v_- by v and again taking $b = d = 0$), but the eigenvalue Eqn. (2.19) simplifies to

$$Ai(-\varepsilon_n/v^{2/3})Ai'(-\varepsilon_n/v^{2/3}) = 0 \tag{2.27}$$

which in turn implies that the eigenvalues are the solutions of $Ai'(-\varepsilon_n/v^{2/3}) = 0$ (even parity solutions, $a = c$ in Eqn. (2.16)) or $Ai(-\varepsilon_n/v^{2/3}) = 0$ (odd parity solutions, $a = -c$ in Eqn. (2.16)), and they are then proportional to the zeros of the derivative of the Airy function and to the zeros of the function itself respectively.

The energy eigenvalues are given accordingly by the simple and exact relation

$$\varepsilon_n = -v^{2/3}\xi_n \tag{2.28}$$

where the ξ_n are the alternating zeros of Ai' and Ai; it is interesting that a single parameter $v = 4\pi a^{*2}N^{2D}$ determines the eigenvalues (and eigenfunctions) in this limit of isolated and symmetric quantum wells (Mezrin and Shik, 1991).

Using Eqn. (2.28), the lowest eigenvalues E_n for the case of δ-doped layers in GaAs and Si have been calculated, and the results are given in Figs. 2.6 and 2.7 respectively; in both cases, the top panel corresponds to electrons and the bottom one to holes. The origin of

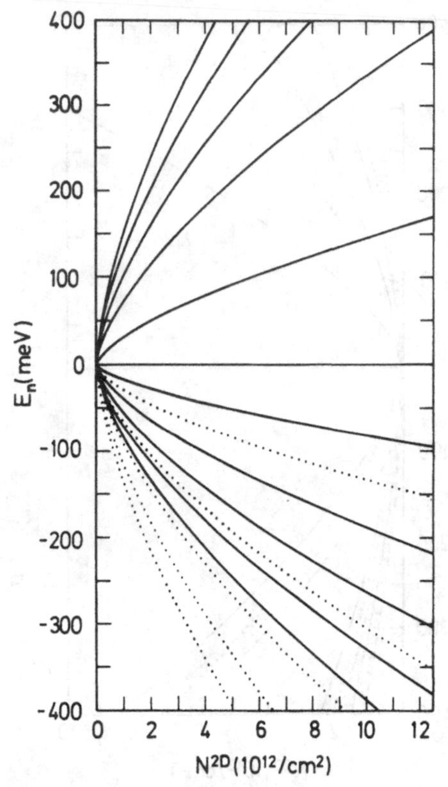

Fig. 2.6. $k_{\parallel} = 0$ level structure for a compensated GaAs n–i–p–i δ-doped superlattice in the multiple-quantum-well regime. Upper panel: electrons. Bottom panel: full lines, heavy holes; dotted lines, light holes. Only the five lowest levels of each type are included.

the energy for electrons (holes) has been taken as the value of $V(z)$ at the donor (acceptor) planes. While this particular choice of energy origin is convenient for a graphical display of the results, one should not conclude that the electron and hole subbands move apart when N^{2D} increases, as precisely the opposite is true; this will be discussed in detail below. The different choice of the horizontal scale N^{2D} for GaAs and Si is dictated by the experimental fact that much higher impurity concentrations are needed in the latter in order to observe subband formation (Eisele, 1989; Gossmann and Schubert, 1993). The lower panel of Fig. 2.7 clearly shows that for a proper treatment of the Si valence subbands, the full 6 × 6 Luttinger Hamiltonian should be employed.

Substitution of Eqn. (2.28) into Eqn. (2.21) yields the normalization constant

$$a^2 = \frac{\nu^{1/3}}{2}[-\xi_n Ai^2(\xi_n) + Ai'^2(\xi_n)]^{-1} \tag{2.29}$$

which reduces to

$$a^2 = -\frac{\nu^{1/3}}{2\xi_{2p}}\frac{1}{Ai^2(\xi_{2p})} \tag{2.30}$$

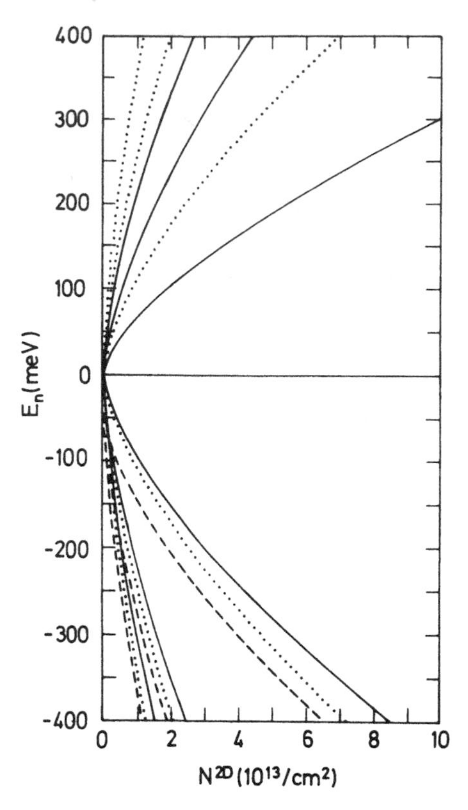

Fig. 2.7. $k_{\parallel} = 0$ level structure for a compensated Si n–i–p–i δ-doped superlattice in the multiple-quantum-well regime. Upper panel: full lines, heavy electrons ($m_z = m_l$); dotted lines, light electrons ($m_z = m_t$). Bottom panel: full lines, heavy holes; dotted lines, light holes; and dashed lines, spin-split-off holes. Only the three lowest levels of each type are included.

for even parity solutions and

$$a^2 = \frac{\nu^{1/3}}{2} \frac{1}{Ai'^2(\xi_{2p+1})} \tag{2.31}$$

for odd parity solutions.

From Eqn. (2.23), the probability of finding the particle in the barriers is given by

$$P_{b,2p} = -\frac{1}{\xi_{2p}} \frac{Ai'^2(0)}{Ai^2(\xi_{2p})} \tag{2.32}$$

and

$$P_{b,2p+1} = \frac{Ai'^2(0)}{Ai'^2(\xi_{2p+1})} \tag{2.33}$$

for the even and odd parity states respectively. Note that they are universal numbers, independent even of the single parameter ν, and consequently also of the impurity concentration, effective mass, dielectric constant, etc. This should be contrasted with the case of compositional quantum wells (typically $Al_xGa_{1-x}As/GaAs/Al_xGa_{1-x}As$), where the probability of finding the particle in the barriers depends on well size, barrier height, effective mass, etc.

On inspection of Eqn. (2.25) it can easily be seen that $\langle\tilde{z}\rangle_n \equiv 0$ (as it should be, by the symmetry of $V(z)$), while Eqn. (2.26) yields

$$\langle\tilde{z}^2\rangle_{2p} = -\frac{1}{5\xi_{2p}\nu^{2/3}}\left(1 - \frac{8}{3}\xi_{2p}^3\right) \tag{2.34}$$

and

$$\langle\tilde{z}^2\rangle_{2p+1} = \frac{8}{15\nu^{2/3}}\xi_{2p+1}^2 \tag{2.35}$$

for the mean square deviation of the coordinate. Note that for highly excited states $-\xi_{2p} \gg 1$ and both equations give the same result. For easy reference, we summarize all the analytical results obtained for the asymmetric and symmetric quantum wells in Table 2.2, while in Table 2.3, we have included the numerical constants that, for a given ν, enable trivial calculation of the eigenvalues, barrier penetration, and mean square deviation of the coordinate. Most of the analytical results given for the symmetric triangular quantum well were previously obtained by López (1972).

At the beginning of the present section, and by taking $N_D^{2D} = N_A^{2D}$, we restricted ourselves to the depleted or compensated situation; next, and by increasing the superlattice period, we considered the superlattice in the multiple quantum well regime. We can now test the consistency of these assumptions in the light of the preceding results.

In the first place, because the band-edge modulation for the symmetric well is given by

$$V_0 = \frac{eFh}{2} = \frac{\pi e^2 N^{2D}}{\varepsilon} h \tag{2.36}$$

where F is the built-in electric field, and since the condition of weak coupling among the wells is fulfilled when $E_n \ll V_0$, with E_n independent of h (see Eqn. (2.28)), it is clear that if h is too small the condition of weak coupling will be violated; from Eqns. (2.28) and (2.36) we obtain

$$h \gg h_n = -2\xi_n\left(\frac{a^*}{4\pi N^{2D}}\right)^{1/3} \tag{2.37}$$

as the condition for the isolated quantum well approximation. However, h cannot be arbitrarily large, as can be seen by defining the energy gap of a doping superlattice in the MQW regime as the energy difference between the lowest conduction subband states and the highest valence subband states, as follows:

$$E_g^{MQW} = E_g + E_0^e + E_0^{hh} - \frac{\pi e^2 N^{2D}}{\varepsilon} h \tag{2.38}$$

Table 2.2. *Summary of exact results for the asymmetric and symmetric triangular quantum wells*

	Asymmetric quantum well	Symmetric quantum well
$V(\tilde{z})/Ry^*$	$v_+\tilde{z} \quad \tilde{z} \geqslant 0$ $-v_-\tilde{z} \quad \tilde{z} \leqslant 0$	$v\tilde{z} \quad \tilde{z} \geqslant 0$ $-v\tilde{z} \quad \tilde{z} \leqslant 0$
$\psi_n(\tilde{z})$	$N_n Ai\left[v_+^{1/3}\left(\tilde{z} - \frac{\varepsilon_n}{v_+}\right)\right] \quad \tilde{z} \geqslant 0$ $Ai\left(-\frac{\varepsilon_n}{v_+^{2/3}}\right)Ai^{-1}\left(-\frac{\varepsilon_n}{v_-^{2/3}}\right)N_n Ai\left[v_-^{1/3}\left(\tilde{z} - \frac{\varepsilon_n}{v_-}\right)\right] \quad \tilde{z} \leqslant 0$ with $N_n = \left\{\left(\frac{1}{v_+} + \frac{1}{v_-}\right)\left[\varepsilon_n Ai^2\left(-\frac{\varepsilon_n}{v_+^{2/3}}\right) + v_+^{2/3} Ai'^2\left(-\frac{\varepsilon_n}{v_+^{2/3}}\right)\right]\right\}^{-1/2}$	$N_n Ai\left[v^{1/3}\left(\tilde{z} - \frac{\varepsilon_n}{v}\right)\right] \quad \tilde{z} \geqslant 0$ $(-1)^n N_n Ai\left[v^{1/3}\left(\tilde{z} - \frac{\varepsilon_n}{v}\right)\right] \quad \tilde{z} \leqslant 0$ with $N_{2p} = \frac{v^{1/6}}{\sqrt{(-2\xi_{2p})}} \lvert Ai^{-1}(\xi_{2p})\rvert$ $N_{2p+1} = \frac{v^{1/6}}{\sqrt{2}} \lvert Ai'^{-1}(\xi_{2p+1})\rvert$
ε_n	The solutions of $Ai\left(-\frac{\varepsilon_n}{v_+^{2/3}}\right)Ai'\left(-\frac{\varepsilon_n}{v_-^{2/3}}\right) + \left(\frac{v_+}{v_-}\right)^{1/3} Ai'\left(-\frac{\varepsilon_n}{v_+^{2/3}}\right)Ai\left(-\frac{\varepsilon_n}{v_-^{2/3}}\right) = 0$	$\varepsilon_n = -v^{2/3}\xi_n$
$P_{b,n}$	$N_n^2 Ai'^2(0)\left[\frac{1}{v_+^{1/3}} + \frac{1}{v_-^{1/3}} \frac{Ai^2(-\varepsilon_n/v_+^{2/3})}{Ai^2(-\varepsilon_n/v_-^{2/3})}\right]$	Even symmetry states: $-\frac{1}{\xi_{2p}} \frac{Ai'^2(0)}{Ai^2(\xi_{2p})}$ Odd symmetry states: $\frac{Ai'^2(0)}{Ai'^2(\xi_{2p+1})}$
$\langle\tilde{z}\rangle_n$	$\frac{2}{3} N_n^2\left(\frac{1}{v_+} + \frac{1}{v_-}\right)\left[\varepsilon_n^2\left(\frac{1}{v_+} - \frac{1}{v_-}\right)Ai^2\left(-\frac{\varepsilon_n}{v_+^{2/3}}\right) + v_+^{2/3}\varepsilon_n\left(\frac{1}{v_+} - \frac{1}{v_-}\right)Ai'^2\left(-\frac{\varepsilon_n}{v_+^{2/3}}\right) - \frac{v_+^{1/3}}{2} Ai\left(-\frac{\varepsilon_n}{v_+^{2/3}}\right)Ai'\left(-\frac{\varepsilon_n}{v_+^{2/3}}\right)\right]$	0
$\langle\tilde{z}^2\rangle_n$	$\frac{8}{15} N_n^2\left\{\left[\varepsilon_n^3\left(\frac{1}{v_+^3} + \frac{1}{v_-^3}\right) + \frac{3}{8}\left(\frac{1}{v_+} + \frac{1}{v_-}\right)\right]Ai^2\left(-\frac{\varepsilon_n}{v_+^{2/3}}\right) + v_+^{2/3}\varepsilon_n^2\left(\frac{1}{v_+^3} + \frac{1}{v_-^3}\right)Ai'^2\left(-\frac{\varepsilon_n}{v_+^{2/3}}\right) - \frac{v_+^{1/3}}{2}\varepsilon_n\left(\frac{1}{v_+^2} - \frac{1}{v_-^2}\right)Ai\left(-\frac{\varepsilon_n}{v_+^{2/3}}\right)Ai'\left(-\frac{\varepsilon_n}{v_+^{2/3}}\right)\right\}$	Even symmetry states: $-\frac{1}{5v^{2/3}\xi_{2p}}\left(1 - \frac{8}{3}\xi_{2p}^3\right)$ Odd symmetry states: $\frac{8}{15v^{2/3}}\xi_{2p+1}^2$

Table 2.3. *Symmetric quantum well numerical constants from which, for a given n* ($0 \leqslant n \leqslant 7$) *and v, the corresponding exact eigenvalues, barrier penetration, and mean square deviation of the coordinate can be obtained. The numbers in parentheses are the corresponding variational results obtained from the trial wave functions of Table 2.4*

n	$-\xi_n = \varepsilon_n/v^{2/3}$	$Ai(\xi_n)$	$Ai'(\xi_n)$	$P_{b,n}$	$v^{1/3}\sqrt{\langle \tilde{z}^2 \rangle_n}$
0	1.0188	0.5357	0	0.23	0.87
	(1.0302)			(0.22)	(0.90)
1	2.3381	0	0.7012	0.14	1.71
	(2.4764)			(0.17)	(1.91)
2	3.2482	−0.4190	0	0.12	2.26
	(3.1946)			(0.12)	(2.32)
3	4.0879	0	−0.8031	0.104	2.99
4	4.8202	0.3804	0	0.096	3.51
5	5.5206	0	0.8652	0.089	4.03
6	6.1633	−0.3579	0	0.085	4.50
7	6.7867	0	−0.9108	0.080	4.96

where E_0^e and E_0^{hh} are the respective electron and heavy-hole ground-state energies, and the last term on the right-hand side is the potential amplitude V_0. The important feature of Eqn. (2.38) is that E_g^{MQW} is a decreasing function of N^{2D} and h, which allows us to define a maximum value of h (h_{max}) from the condition $E_g^{MQW} = 0$, using Eqn. (2.38) we obtain

$$h_{max} = \frac{E_g + E_0^e + E_0^{hh}}{\pi e^2 N^{2D}/\varepsilon}. \tag{2.39}$$

If $h > h_{max}$, the system will have a transition from a semiconductor to a semimetallic configuration, as a consequence of the transfer of electrons from the valence to the conduction band. In this case, the MQW will not be depleted of free carriers, and the analysis is more complicated, as will be discussed in Section 2.4. Note that for small N^{2D}, h_{max} diverges like $1/N^{2D}$, while for large N^{2D}, h_{max} decreases like $(1/N^{2D})^{1/3}$.

Summarizing, a doping superlattice can be considered as a collection of depleted and isolated wells when its superlattice period fulfills the following inequality:

$$-2\xi_n\left(\frac{a^*}{4\pi N^{2D}}\right)^{1/3} \ll h < \frac{E_g + E_0^e + E_0^{hh}}{\pi e^2 N^{2D}/\varepsilon} \tag{2.40}$$

which for a given semiconductor and N^{2D} delimits a range of allowed values of h. A graphical representation of this inequality is given in Fig. 2.8 for GaAs and in Fig. 2.9 for Si, where the left- and right-hand-side terms of Eqn. (2.40) are plotted as a function of N^{2D}; for small N^{2D}, the $1/N^{2D}$ divergence of Eqn. (2.39) overcomes the $(1/N^{2D})^{1/3}$ divergence

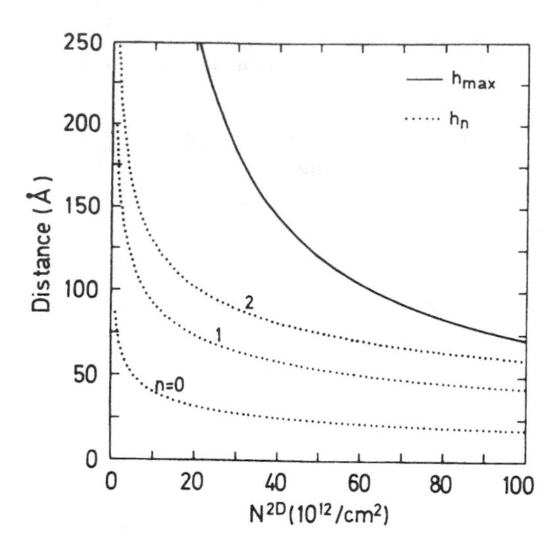

Fig. 2.8. Graphical representation of the GaAs constraints for the simultaneous validity of the multiple-quantum-well approximation and depleted configuration. The minimum distance h_n is defined by Eqn. (2.37) in the text, and corresponds to the electrons; if the condition $h \gg h_n$ is fulfilled for electrons, it is also satisfied for the heavier heavy and light holes (with the same n). The maximum distance h_{max} is given by Eqn. (2.39).

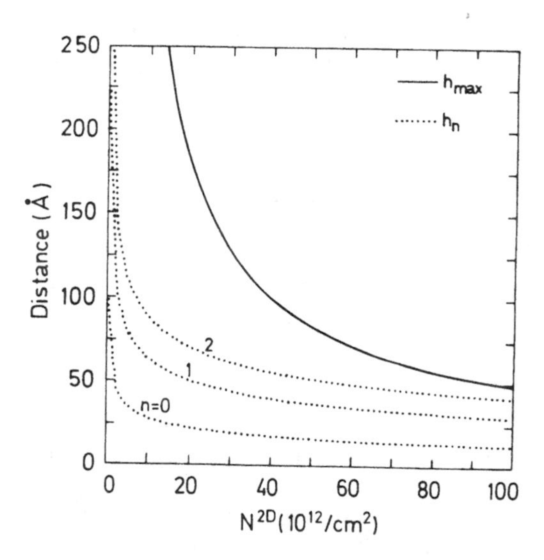

Fig. 2.9. Same as Fig. 2.8, but for Si. The minimum distance h_n corresponds to the electrons with $m_z = m_t$. If the condition $h \gg h_n$ is fulfilled for these electrons, it would be satisfied even better by the (same n) electrons with $m_z = m_l$, the heavy, light, and spin-split-off hole, all of them with heavier effective masses.

Table 2.4. *Summary of results for the three lowest states of a symmetric triangular well using variational wave functions,* $\tilde{z}_{t,n} = \varepsilon_n/v$

$\psi_n(\tilde{z})$	$\psi_0(\tilde{z}) = N_0(1 + \alpha_0\lvert\tilde{z}\rvert)\,\mathrm{e}^{-\alpha_0\lvert\tilde{z}\rvert}$, $N_0 = (2\alpha_0/5)^{1/2}$, $\alpha_0 = (9/4)^{1/3}v^{1/3}$
	$\psi_1(\tilde{z}) = N_1\lvert\tilde{z}\rvert\,\mathrm{e}^{-\alpha_1\lvert\tilde{z}\rvert}$, $N_1 = (2\alpha_1^3)^{1/2}$, $\alpha_1 = (3/4)^{1/3}v^{1/3}$
	$\psi_2(\tilde{z}) = N_2(1 + \alpha_2\lvert\tilde{z}\rvert)(\alpha_2^2\tilde{z}^2 - 1)\,\mathrm{e}^{-\alpha_2\lvert\tilde{z}\rvert}$, $N_2 = (2\alpha_2/63)^{1/2}$, $\alpha_2 = (47/12)^{1/3}v^{1/3}$
ε_n	$\varepsilon_0 = \frac{3}{10}(\frac{81}{2})^{1/3}v^{2/3}$
	$\varepsilon_1 = \frac{3}{2}(\frac{9}{2})^{1/3}v^{2/3}$
	$\varepsilon_n = \frac{9}{7}(\frac{47}{12})^{2/3}v^{2/3}$
$P_{b,n}$	$P_{b,0} = \frac{2}{5}\mathrm{e}^{-2\alpha_0\tilde{z}_{t,0}}(\frac{5}{2} + 3\alpha_0\tilde{z}_{t,0} + \alpha_0^2\tilde{z}_{t,0}^2)$
	$P_{b,1} = 2\,\mathrm{e}^{-2\alpha_1\tilde{z}_{t,1}}(\frac{1}{2} + \alpha_1\tilde{z}_{t,1} + \alpha_1^2\tilde{z}_{t,1}^2)$
	$P_{b,2} = \frac{4}{63}\,\mathrm{e}^{-2\alpha_2\tilde{z}_{t,2}}(\frac{33}{4} + \frac{59}{2}\alpha_2\tilde{z}_{t,2} + \frac{55}{2}\alpha_2^2\tilde{z}_{t,2}^2 + 19\alpha_2^3\tilde{z}_{t,2}^3 + \frac{23}{2}\alpha_2^4\tilde{z}_{t,2}^4 + 5\alpha_2^5\tilde{z}_2^5\tilde{z}_{t,2}^5 + \alpha_2^6\tilde{z}_{t,2}^6)$
$\langle\tilde{z}^2\rangle_n$	$\langle\tilde{z}^2\rangle_0 = \frac{7}{5}(\frac{4}{9})^{2/3}/v^{2/3}$
	$\langle\tilde{z}^2\rangle_1 = (48)^{1/3}/v^{2/3}$
	$\langle\tilde{z}^2\rangle_2 = \frac{842}{63}(\frac{12}{47})^{2/3}/v^{2/3}$

of Eqn. (2.37), and the inequality is easily satisfied. However, when N^{2D} increases, both Eqn. (2.37) and Eqn. (2.39) decrease as $(1/N^{2D})^{1/3}$, and the range of allowed values of h is quite small and eventually disappears. As can be seen from these diagrams, there is ample room for the design of sawtooth superlattices in the depleted MQW regime for the physically relevant impurity concentration range of GaAs ($10^{12}/\mathrm{cm}^2 - 10^{13}/\mathrm{cm}^2$) and Si ($10^{13}/\mathrm{cm}^2 - 10^{14}/\mathrm{cm}^2$).

While most of the physical properties of symmetric quantum wells can be calculated exactly using the eigenfunctions of Eqn. (2.16), variational wave functions are sometimes preferred because of their simplicity. In particular, Schubert *et al.* (1989) have proposed a set of trial wave functions for the three lowest states. The analytical expressions for the normalized variational wave functions, energies, barrier penetration, and mean square deviation of the coordinate are summarized in Table 2.4.

These trial functions use the fact that ψ_0 and ψ_2 should have even spatial symmetry, while ψ_1 is an odd function; they also fulfill the constraint that $\psi_n(z)$ should have n nodes. Note, however, that ψ_0 and ψ_2 are not orthogonal. The wave function $\psi_1(z)$ is identical to the Fang and Howard (1966) wave function for positive z. To test the accuracy of these variational functions, we have made a comparison with the exact solutions obtained previously, and the results are given in parentheses in Table 2.3. The good overall agreement between exact and variational results proves that the latter are an effective compromise between accuracy and simplicity. The lack of orthogonality between ψ_0 and ψ_2 explains the fact that ε_2 (exact) $>\varepsilon_2$ (variational). A major drawback of these variational calculations is that it becomes increasingly cumbersome to write trial wave functions for the excited states. The Wentzel–Kramers–Brillouin (WKB) approximation has been also employed in order to obtain the eigenvalues of a V-shaped quantum well (Schubert, 1990).

(c) *Symmetric superlattice*

When the assumption of negligible tunneling between adjacent quantum wells is violated (by decreasing the superlattice period, for example), the preceding results should be generalized as one moves from the isolated or multiple-quantum-well regime to the superlattice regime, where the matrix element of wave functions centered at two first-neighbor quantum wells ('hopping') becomes appreciable.

The Schrödinger equation for electrons and holes (at $k_{\parallel} = 0$) is, according to Eqns. (2.5) and (2.6),

$$\left[-\frac{\hbar^2}{2m_z}\frac{d^2}{dz^2} + V(z)\right]\psi_{nk_z}(z) = E_n(k_z)\psi_{nk_z} \tag{2.41}$$

where the periodic superlattice potential $V(z)$ is given by Eqn. (2.12) (with $h_1 = h_2 = h/2$). The exact solution in the interval $-h/2 \leqslant z \leqslant h/2$ is still given by Eqn. (2.16), but now the boundary conditions are different, as the superlattice wave functions should satisfy the Bloch theorem

$$\psi_{nk_z}(z + h) = e^{ik_z h}\psi_{nk_z}(z) \tag{2.42}$$

Accordingly, one should impose the following set of boundary conditions on the solutions of Eqn. (2.16):

$$\psi_{nk_z}(0^+) = \psi_{nk_z}(0^-) \tag{2.43a}$$

$$\psi'_{nk_z}(0^+) = \psi'_{nk_z}(0^-) \tag{2.43b}$$

$$\psi_{nk_z}(h/2^-) = e^{ik_z h}\psi_{nk_z}(-h/2^+) \tag{2.43c}$$

$$\psi'_{nk_z}(h/2^-) = e^{ik_z h}\psi'_{nk_z}(-h/2^+) \tag{2.43d}$$

where the last two equations follow from Eqn. (2.42). Substituting Eqn. (2.16) into Eqn. (2.43), we obtain a set of four homogeneous linear equations for the unknown coefficients a, b, c, d; from the condition that for a non-trivial solution these coefficients should not be zero simultaneously, one arrives at a 4×4 determinant, whose expansion yields the eigenvalue equation

$$\begin{aligned}\cos(k_z h) = -\pi^2[&(A_1B_2 - A_2B_1)(A'_1B'_2 - A'_2B'_1)\\ &+ (A_1B'_2 - A'_2B_1)(A'_1B_2 - A_2B'_1)]\end{aligned} \tag{2.44}$$

where we have used the definitions $A_1 = Ai[y(0)]$, $B_1 = Bi[y(0)]$, $A_2 = Ai[y(h/2)]$, $B_2 = Bi[y(h/2)]$, and $y(z) = \nu^{1/3}[z/a^* - E_n(k_z)/(\nu Ry^*)]$.

The dispersion relation $E_n(k_z)$ versus k_z can be readily obtained from Eqn. (2.44). An analysis of Eqn. (2.44) in the limit $h \to \infty$ is easy and instructive; the asymptotic expansions of the Airy functions are (Antosiewicz, 1970)

$$Ai(x) \approx \frac{1}{2\sqrt{\pi}} x^{-1/4} e^{-2x^{3/2}/3} \tag{2.45a}$$

$$Ai'(x) \approx \frac{1}{2\sqrt{\pi}} x^{1/4} e^{-2x^{3/2}/3} \tag{2.45b}$$

$$Bi(x) \approx \frac{1}{2\sqrt{\pi}} x^{-1/4} e^{2x^{3/2}/3} \tag{2.45c}$$

$$Bi'(x) \approx \frac{1}{2\sqrt{\pi}} x^{1/4} e^{2x^{3/2}/3} \tag{2.45d}$$

Inspection of Eqn. (2.44) in the limit where Eqn. (2.45) applies reveals that the right-hand side of the eigenvalue equation will diverge as the product $B_2 B_2'$, unless the multiplicative coefficient in front of it vanishes. This occurs if

$$A_1 A_1' = 0 \tag{2.46}$$

which coincides with Eqn. (2.27) for a particle confined to move in an isolated symmetric quantum well, as expected. Using (2.16), the system Eqn. (2.43), and the normalization condition, the coefficients a, b, c, d, and consequently the eigenfunctions, can be obtained (Ferreyra and Proetto, 1990).

Even though we have derived analytical expressions for the eigenvalue equation and wave functions, the resulting equations are rather complicated. Simpler results can be obtained, however, if the analysis is limited to the edge energies of each subband. The subband-edge and corresponding envelope wave functions can be useful for analyzing the superlattice properties. A similar analysis has been carried out for the case of compositional $GaAs/Al_xGa_{1-x}As$ superlattices by Cho and Prucnal (1987).

To proceed, we first rewrite Eqn. (2.44) as follows:

$$\cos(k_z h)[(A_1 B_1' - A_1' B_1)(A_2 B_2' - A_2' B_2)] + (A_1 B_2 - A_2 B_1)(A_1' B_2' - A_2' B_1') + (A_1 B_2' - A_2' B_1)(A_1' B_2 - A_2 B_1') = 0 \tag{2.47}$$

where the Wronskian relation satisfied by the Airy functions:

$$Ai(z) Bi'(z) - Ai'(z) Bi(z) = 1/\pi \tag{2.48}$$

has been used twice to eliminate a factor of π^2.

When k_z takes the subband-edge superlattice values 0 or $\pm\pi/h$, Eqn. (2.47) simplifies considerably; for $k_z = 0$ it reduces to

$$(A_1 B_2 - A_2 B_1)(A_1' B_2' - A_2' B_1') = 0 \tag{2.49}$$

Table 2.5. $\mathbf{k}_{\parallel} = 0$ *subband-edge eigenvalue equations and corresponding eigenfunctions for a depleted δ-doped symmetric superlattice of period h. The parameters α and β are defined by* $\alpha = a^*E_n(k_z)/Ry^*v$ *and* $\beta = v^{1/3}/a^*$*; the remaining symbols are defined in the text*

	$k_z = 0$	$k_z = \pm\pi/h$
Even-index subbands $n = 0, 2, 4, \ldots$	$a = c$, $b = d$, $b/a = -A_1'/B_1'$ with $a = \left[\frac{\pi^2\beta}{2}\frac{B_1'^2B_2'^2}{\beta(h/2-\alpha)B_1'^2+\alpha\beta B_2'^2}\right]^{1/2}$ Eigenvalue equation $A_1'B_2' - A_2'B_1' = 0$	$a = c$, $b = d$, $b/a = -A_1'/B_1'$ with $a = \left(\frac{\pi^2\beta}{2}\frac{B_1'^2B_2^2}{\alpha\beta B_2^2 - B_1'^2}\right)^{1/2}$ Eigenvalue equation $A_1'B_2 - A_2B_1' = 0$
Odd-index subbands $n = 1, 3, 5, \ldots$	$a = -c$, $b = -d$, $b/a = -A_1/B_1$ with $a = \left(\frac{\pi^2\beta}{2}\frac{B_1^2B_2^2}{-B_1^2+B_2^2}\right)^{1/2}$ Eigenvalue equation $A_1B_2 - A_2B_1 = 0$	$a = -c$, $b = -d$, $b/a = -A_1/B_1$ with $a = \left[\frac{\pi^2\beta}{2}\frac{B_1^2B_2'^2}{B_2'^2 - \beta(\alpha - h/2)B_1^2}\right]^{1/2}$ Eigenvalue equation $A_1B_2' - A_2'B_1 = 0$

which in turn implies

$$A_1'B_2' - A_2'B_1' = 0 \tag{2.50}$$

or

$$A_1B_2 - A_2B_1 = 0. \tag{2.51}$$

Using Eqn. (2.47) again, but particularizing this time for the second subband-edge superlattice wave vector $k_z = \pm\pi/h$, one obtains

$$(A_1B_2' - A_2'B_1)(A_1'B_2 - A_2B_1') = 0 \tag{2.52}$$

which is equivalent to the eigenvalue equations

$$A_1B_2' - A_2'B_1 = 0 \tag{2.53}$$

or

$$A_1'B_2 - A_2B_1' = 0. \tag{2.54}$$

In a quite analogous way one can prove that the wave functions corresponding to the edge energies have definite parity (Voisin *et al.*, 1984; Ferreyra and Proetto, 1990). For easy reference, Table 2.5 summarizes the eigenvalue equations and normalized eigenfunctions for all the subband-edge energies.

It should be noted that the four subband-edge eigenvalue equations are different functions (of energy) from the general result of Eqn. (2.44). They only give the same answer (same eigenvalue) for the particular case of subband-edge energies. The same relation applies between the subband-edge wave-function coefficients and the general coefficients.

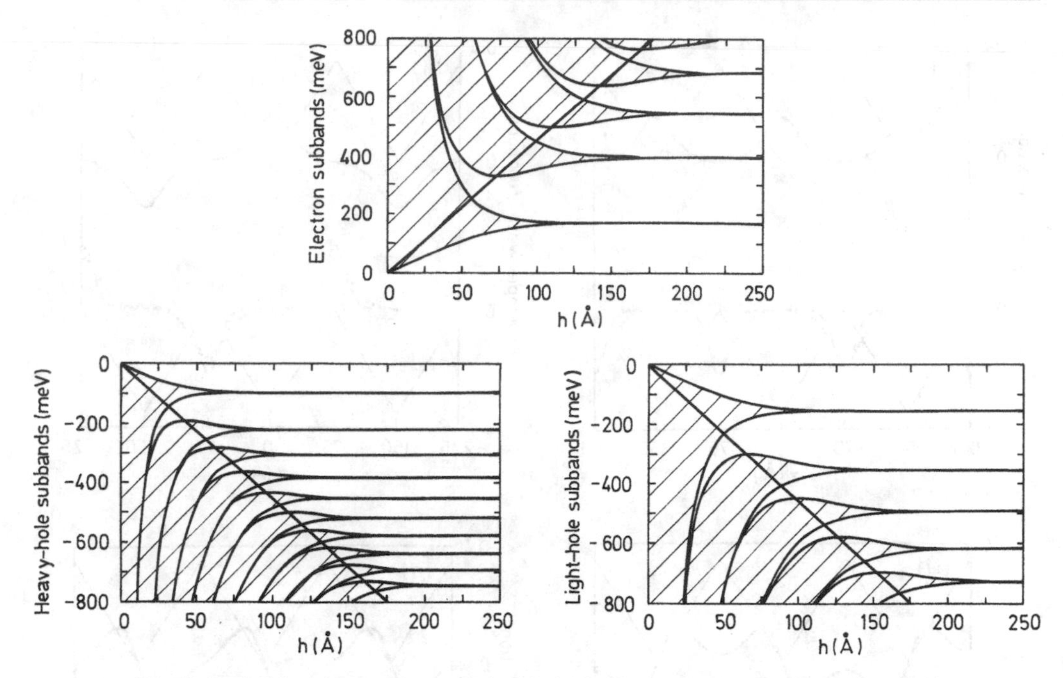

Fig. 2.10. Subband energies and bandwidths for a GaAs depleted δ-doped superlattice as a function of period h. The straight line corresponds to the potential modulation V_0 and $N^{2D} = 1.25 \times 10^{13}/\text{cm}^2$.

As an illustration of the application of the expressions obtained in this section, we have calculated, for the particular case of GaAs, the $k_\parallel = 0$ electron, heavy-hole, and light-hole energy subbands as a function of the period h for a given value of δ-doping concentration $N^{2D} = 1.25 \times 10^{13}/\text{cm}^2$.

The corresponding results are given in Fig. 2.10: shaded areas define the allowed or subband energies, and the straight line corresponds to the potential modulation V_0. Note that for electrons we have chosen the potential at the donor planes as the origin of energies, while for the holes the origin of energies corresponds to the value of $V(z)$ at the acceptor planes. The essential feature of these results is the progressive narrowing of the subband width as the period increases, as a result of the growing difficulty that the particles have in 'jumping' to an adjacent well. For a given subband, and increasing h, the atomic limit is first reached by the heavy holes, next by the light holes, and finally by the electrons; this is a consequence of their different effective masses: m_z (electrons) $<$ m_z (light holes) $<$ m_z (heavy holes), see Table 2.1.

It is instructive to check the consistency of the approximate criteria given in the preceding section on the validity of the isolated well approximation and the rigorous result of the present section. From Fig. 2.8, for $N^{2D} \approx 10^{13}/\text{cm}^2$ and $h \approx 150$ Å, the condition $h \gg h_n$ is fulfilled by the $n = 0$ and $n = 1$ quantum well states, but not for the state with $n = 2$. This is in complete agreement with the results of Fig. 2.10, where we observe that, for $h = 150$ Å, the first two electron subbands are in the atomic limit, while the third one has an appreciable bandwidth.

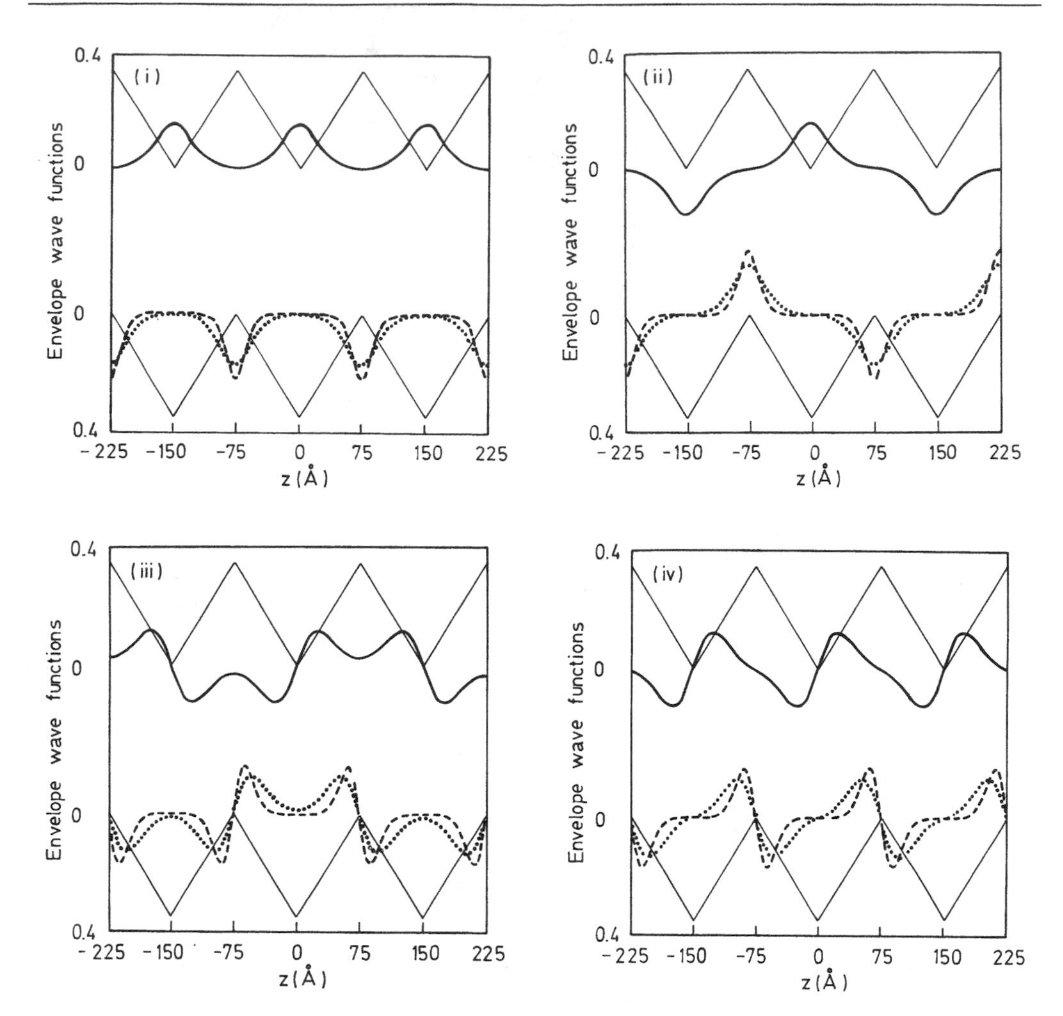

Fig. 2.11. Subband-edge normalized envelope wave functions for a GaAs depleted δ-doped superlattice with $N^{2D} = 1.25 \times 10^{13}/\text{cm}^2$ and $h = 150$ Å. Case (i), $n = 0$ (minimum); case (ii), $n = 0$ (maximum); case (iii), $n = 1$ (minimum); case (iv), $n = 1$ (maximum). Full lines (electrons), dashed lines (heavy holes), dotted lines (light holes).

Fig. 2.11 displays the subband-edge normalized envelope wave functions for the following situations:

(i) Minimum energy of the first ($n = 0$) electron (solid line), heavy-hole (dashed line), and light-hole (dotted line) subbands. For this particular case, all wave functions are even in the unit cell (as are all the even-index band-edge wave functions), but they are also even from cell to cell (see Table 2.5). Note that, as a consequence of the indirect gap in the real space characteristic of doping superlattices, the electronic charge distribution is displaced a distance $h/2$ from the heavy-hole and light-hole distributions. The different degree of localization of the electrons, heavy holes and light holes is again a direct consequence of their different effective masses.

(ii) Maximum energy of the first ($n = 0$) electron, heavy-hole, and light-hole subbands. The wave functions are even in the unit cell, but they are now odd from cell to cell, unlike the preceding case. This means that the wave function (and, consequently, the charge distribution) must be very small in the barrier regions; this condition induces no large changes in the charge distribution for the present value of N^{2D}, because the wave function is already very small around such regions.

(iii) Minimum energy of the first excited ($n = 1$) electron, heavy-, and light-hole subbands.

(iv) Maximum energy of the first excited ($n = 1$) electron, heavy-, and light-hole subbands.

In the last two cases the wave functions are odd in the unit cell (as are all odd-index band-edge wave functions), but while in (iii) they are also odd from cell to cell, in (iv) the wave functions are even from cell to cell. As we move towards higher-energy subbands, we approach the 'nearly-free-particle' limit (Ashcroft and Mermin, 1976), where the periodic potential of the superlattice is a small perturbation compared with the kinetic energy of the particle.

2.4 Undepleted quantum wells and superlattices

As soon as $N_D^{2D} \neq N_A^{2D}$ (and this is experimentally unavoidable, at least to a certain degree) the system is non-compensated and the mobile particles (electrons if $N_D^{2D} > N_A^{2D}$, holes if $N_D^{2D} < N_A^{2D}$) screen and modify the perfect V-shaped potential considered until now. Under these conditions, the subband structure is given by the self-consistent solution of the Poisson and Schrödinger equations for the system at hand, which usually has to be obtained by numerical methods.

The simplest approach to this problem is to apply the quasiclassical or Thomas–Fermi approximation (Ioriatti, 1990; Mezrin and Shik, 1991; Mezrin *et al.*, 1992), which is valid when the number of bound states in the potential well of the δ-layer is large enough ($\gg 1$). Within this scheme, the density is taken to be that of a free-electron gas in thermal equilibrium with the local value of the potential. The functional dependence of the potential is then determined from Poisson's equation, which becomes a non-linear differential equation for the self-consistent potential, to be solved subject to the appropriate boundary conditions. However, by treating the carriers as a free, three-dimensional Fermi gas, one thereby ignores the quantization of electron states affected by the confining potential, which clearly raises the question of the applicability of this method. The approximation seems to work well for the case of n-type δ-doped GaAs, where several subbands are occupied for the physically relevant range of impurity concentration $N_D^{2D} \approx 10^{12}/\text{cm}^2$–$10^{13}/\text{cm}^2$, and $N_A^{2D} \approx 0$.

In what follows, we describe and apply, to three different physical systems, a more accurate and general method (Ando and Mori, 1979) which allows the self-consistent computation of electronic states of heterostructures, quantum wells, and superlattices for any compositional and doping profile, with the inclusion of many-body effects within the local density-functional theory (Lundqvist and March, 1983). The treatment accounts for composition dependence of confinement barrier heights, dielectric constant and effective mass, and, if necessary, the effects of non-parabolicity; a tutorial description of the method has been given by Fiorentini (1990). We start with the simplest of the uncompensated systems: the case of n-type δ-doping in GaAs (Zrenner *et al.*, 1988; Degani, 1991; Reboredo and Proetto, 1992).

(a) *n-type δ-doping in GaAs*

In this section, we restrict ourselves to the case of n-type non-compensated superlattices. This means that for a given two-dimensional concentration of ionized donor impurities $N_{\mathrm{D}}^{2\mathrm{D}}$, the concentration of acceptor impurities $N_{\mathrm{A}}^{2\mathrm{D}}$ is allowed to change in the interval $0 \leqslant N_{\mathrm{A}}^{2\mathrm{D}} \leqslant N_{\mathrm{D}}^{2\mathrm{D}}$, and, consequently, no holes are present in the system. The case where $N_{\mathrm{D}}^{2\mathrm{D}} = 0$, $N_{\mathrm{A}}^{2\mathrm{D}} \neq 0$ (p-type δ-doping) will be analyzed below.

The relevant equations for this section are the Poisson and electron Schrödinger equations (Eqns. (2.1) and (2.5) respectively). The charge density $\rho(z)$, which enters into the Poisson equation is given by Eqn. (2.2), where $n(z)$ is defined by Eqn. (2.3) and $p(z) \equiv 0$.

Performing the integration over $\mathbf{k}_{\parallel}$, the electron density of Eqn. (2.3) can be rewritten as

$$n(z) = \frac{m^* k_{\mathrm{B}} T}{\pi \hbar^2} \sum_{n, k_z} \ln\{1 + \exp - [E_n(k_z) - \mu]/k_{\mathrm{B}}T\} \, |\psi_{nk_z}(z)|^2 \tag{2.55}$$

which, in the zero temperature limit, reduces to

$$n(z) = \frac{m^*}{\pi \hbar^2} \sum_{n, k_z} [\mu - E_n(k_z)] \theta[\mu - E_n(k_z)] \, |\psi_{nk_z}(z)|^2 \tag{2.56}$$

where m^* is the isotropic effective mass for conduction electrons in GaAs, and $\theta(x)$ is the Heaviside step function.

Under these conditions, the neutrality constraint given by Eqn. (2.11) simplifies to

$$N_{\mathrm{D}}^{2\mathrm{D}} - N_{\mathrm{A}}^{2\mathrm{D}} = \frac{m^*}{\pi \hbar^2} \sum_{n, k_z} [\mu - E_n(k_z)] \theta[\mu - E_n(k_z)] \tag{2.57}$$

which, for given values of $N_{\mathrm{D}}^{2\mathrm{D}}$ and $N_{\mathrm{A}}^{2\mathrm{D}}$, determines the chemical potential μ. Equation (2.57), together with Eqns. (2.1) and (2.5) are the basic expressions that should be used in a self-consistent way.

As the Hartree approximation usually overestimates the Coulomb repulsion, we have included the exchange-correlation contribution $V_{xc}(z)$ in the definition of the superlattice potential

$$V_{eff}(z) = V(z) + V_{xc}[n(z)] \tag{2.58}$$

within the formalism of the density-functional approximation (Hohenberg and Kohn, 1964; Kohn and Sham, 1965; Sham and Kohn, 1966). In this formulation, many-body effects are taken into account by introducing an exchange-correlation potential $V_{xc}[n(z)]$ which is given by a functional derivative of the exchange-correlation part of the ground-state energy with respect to the particle density. In the simplest approximation (the local density approximation or LDA) the exchange-correlation potential energy is approximated by the exchange-correlation contribution to the chemical potential of a homogeneous gas having a uniform density which is equal to the local density $n(z)$ of the inhomogeneous system. It is understood that $V_{eff}(z)$ should be inserted into the Schrödinger equation as the superlattice potential when many-body effects are included, while only $V(z)$ enters into the Poisson equation.

Among the many possibilities for $V_{xc}(z)$ that are available in the literature (Wigner 1938; Gunnarsson and Lundqvist, 1976; Moruzzi *et al.*, 1978), we have chosen the parametrized expression of Perdew and Zunger (1981) which seems to give accurate results over a wide range of densities. The explicit expression is

$$V_{xc}[n(z)] = \left[-\frac{1}{\alpha\pi r_s} + \gamma \frac{(1 + \frac{7}{6}\beta_1\sqrt{r_s} + \frac{4}{3}\beta_2 r_s)}{(1 + \beta_1\sqrt{r_s} + \beta_2 r_s)^2} \right] \frac{m^* e^4}{\hbar^2 \varepsilon^2} \tag{2.59}$$

where $r_s(z) = [3/(4\pi a^{*3} n(z))]^{1/3} \geqslant 1$. If $r_s(z) \leqslant 1$ the following expression should be used:

$$V_{xc}[n(z)] = \left[-\frac{1}{\alpha\pi r_s} + A \ln r_s + \left(B - \frac{A}{3} \right) + \frac{2}{3} C r_s \ln r_s + \frac{1}{3}(2D - C) r_s \right] \frac{m^* e^4}{\hbar^2 \varepsilon^2} \tag{2.60}$$

with the constants in Eqns. (2.59) and (2.60) taking the following values: $\alpha = 0.521$, $\gamma = -0.1423$, $\beta_1 = 1.0529$, $\beta_2 = 0.3334$, $A = 0.0311$, $B = -0.0480$, $C = 0.0020$, and $D = -0.0116$.

As for the boundary conditions, the Schrödinger equation should satisfy the usual boundary conditions for Bloch waves. We also adopt the following boundary conditions for the Poisson equation:

$$V(jh/2) = 0 \tag{2.61a}$$

$$V(z) = V(z + h) \tag{2.61b}$$

where Eqn. (2.61a) is equivalent to choosing the origin of energies at the position of the acceptor planes.

Following Ando and Mori (1979), we perform a plane-wave expansion of the periodic part of $\psi_{nk_z}(z)$:

$$\psi_{nk_z}(z) = e^{ik_z z} \sum_{l=-\infty}^{\infty} c_l^{(n)}(k_z)\, e^{i2\pi l z/h}. \tag{2.62}$$

The coefficients $c_l^{(n)}(k_z)$ satisfy the orthogonality relation

$$\sum_{l=-\infty}^{\infty} c_l^{(n)}(k_z) c_l^{(m)}(k_z) = \delta_{n,m} \tag{2.63}$$

which follows from the orthonormality of the $\psi_{nk_z}(z)$.

Substituting Eqn. (2.62) in Eqn. (2.5) we obtain the reciprocal space version of the electron Schrödinger equation:

$$\frac{\hbar^2}{2m^*}\left(k_z + \frac{2\pi l}{h}\right)^2 c_l^{(n)}(k_z) + \sum_{l'=-\infty}^{\infty} c_{l'}^{(n)}(k_z) V_{l-l'} = E_n(k_z) c_l^{(n)}(k_z) \tag{2.64}$$

where V_l is the Fourier transform of $V_{\text{eff}}(z)$:

$$V_{\text{eff}}(z) = \sum_{l=-\infty}^{\infty} V_l \, e^{i2\pi lz/h} \tag{2.65}$$

and is given by

$$V_l = \frac{he^2}{\pi l^2 \varepsilon}[N_D^{2D} - (-1)^l N_A^{2D}] - \frac{4\pi e^2}{\varepsilon}\left(\frac{h}{2\pi l}\right)^2 n_l + V_{xc}^l \tag{2.66}$$

where n_l and V_{xc}^l are the corresponding Fourier transforms of the electronic density and exchange-correlation potential contributions respectively. Using Eqns. (2.55) and (2.62) it is possible to obtain an analytical expression for n_l in terms of the coefficients $c_l^{(n)}(k_z)$. The Fourier transform of $V_{xc}[n(z)]$, as a consequence of the complex dependence on $n(z)$, must be obtained numerically. This Fourier transform method reduces the solution of the coupled set of Schrödinger and Poisson equations to the matrix diagonalization implied in Eqn. (2.64), in order to obtain the eigenvalues $E_n(k_z)$ and the eigenvectors $c_l^{(n)}(k_z)$; besides, it also avoids the explicit solution of the Poisson equation. All the numerical results that follow correspond to the zero temperature limit $T = 0$, and unless otherwise stated, the (in principle) infinite matrix Hamiltonian, Eqn. (2.64), has been truncated to a finite one with $l_{\max} = 25$, which gives a matrix size of 51×51 (for each k_z).

In Fig. 2.12 we show the self-consistent potential $V_{\text{eff}}(z)$ for the case of a period $h = 150$ Å, $N_D^{2D} = 5 \times 10^{12}/\text{cm}^2$, and N_A^{2D}, taking values from N_D^{2D} (compensated case, bottom curve) to zero (only donor impurities present in the superlattice, upper curve) in steps of $0.5 \times 10^{12}/\text{cm}^2$. In the case of homogeneously doped superlattices (Zeller *et al.*, 1982; Ruden and Döhler, 1983; Choquette *et al.*, 1991), the presence of free electrons neutralizes the uniform jellium background, giving rise to flat regions in the self-consistent potential. However, in the present case of δ-doping, the screening length of the electrons is much larger than the width of the dopant planes, and consequently no flat regions are obtained even at the completely uncompensated situation $N_A^{2D} = 0$. Note that this remains valid even in the less ideal case of δ-doping planes of finite width, as long as the characteristic wave length of the electrons (the spatial extent of the ground-state function, for example) remains larger than this width (Koch and Zrenner, 1989). As N_A^{2D} decreases and charge accumulates in the wells, the potential depth reduces considerably ($\approx 50\%$).

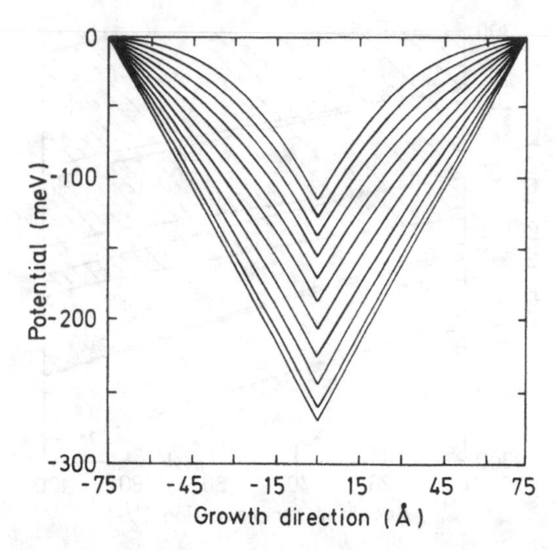

Fig. 2.12. Self-consistent potentials for a GaAs superlattice with $h = 150$ Å, $N_D^{2D} = 5 \times 10^{12}/cm^2$ and N_A^{2D} taking values from N_D^{2D} (bottom curve) to zero (upper curve) in steps of $0.5 \times 10^{12}/cm^2$.

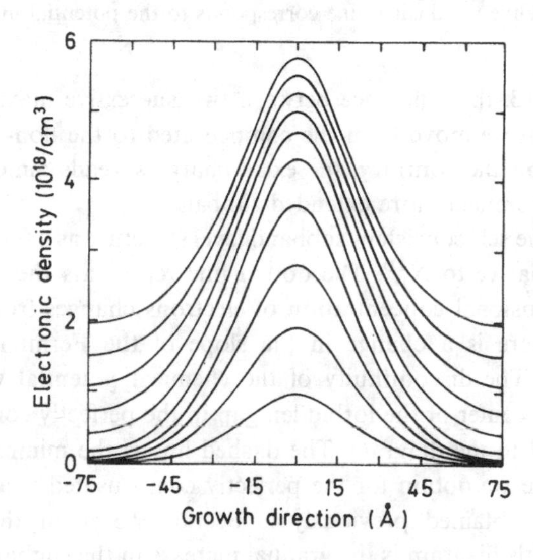

Fig. 2.13. Electronic densities of the same superlattice of Fig. 2.12. The bottom curve corresponds to $N_A^{2D} = 4.5 \times 10^{12}/cm^2$, while the upper curve is the result when $N_A^{2D} = 0$, with the two-dimensional acceptor concentration decreasing in steps of $0.5 \times 10^{12}/cm^2$.

The corresponding electronic densities are shown in Fig. 2.13. Notice the sizeable accumulation of charge in the classically forbidden region in the barriers as the acceptor concentration decreases from N_D^{2D} to zero. Close to the compensated situation, most of the non-compensated electrons concentrate around the central region, as they populate the ground-state subband, which is strongly localized near the well center. This effect is

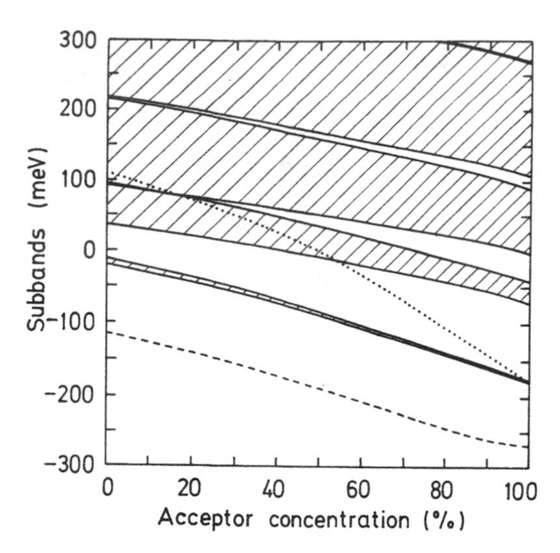

Fig. 2.14. Subband electronic structure for an n-type GaAs superlattice as a function of N_A^{2D}, for $h = 150$ Å, and $N_D^{2D} = 5 \times 10^{12}/\text{cm}^2$. Shaded areas represent allowed energies (subbands), the dotted line gives the Fermi level position while the dashed line corresponds to the potential minimum.

clearly seen in Fig. 2.13: the difference between two successive maxima in the electronic density decreases when we move from the compensated to the non-compensated regime. In the latter regime, on the contrary, the extra charge spreads almost uniformly, as the electrons now populate much more extended subbands.

Figure 2.14 shows the self-consistent subband level structure as a function of the acceptor concentration N_A^{2D} relative to N_D^{2D}. The dotted line represents the change in the Fermi level as the two-dimensional concentration of electrons changes from N_D^{2D} ($N_A^{2D} = 0$) to zero ($N_A^{2D} = N_D^{2D}$); there is a change in the slope of the Fermi level when it passes through a miniband. The discontinuity of the chemical potential when $N_A^{2D} = N_D^{2D}$ (it should be close to the center of the forbidden gap in the perfectly compensated situation) has not been included in the diagram. The dashed line is the minimum of the potential. The subband structure we obtain for the perfectly compensated situation coincides with the analytical results obtained previously in Section 2.3(c) for this special case. The important feature of this diagram is the gradual increase in the subband width as we move away from the perfectly compensated situation; this is a consequence of the progressive softening of the barriers between wells as charge accumulates in the superlattice (see Fig. 2.12).

To determine the relevance (or lack) of the many-body corrections we have repeated the calculations with $V_{xc} \equiv 0$, for $N_A^{2D} = 2.5 \times 10^{12}/\text{cm}^2$. For this acceptor concentration, the values of the first three gaps are 73.2 meV, 15.0 meV, and 9.0 meV, respectively, while without exchange and correlation the corresponding results are 68.2 meV, 12.6 meV, and 8.8 meV. From this we conclude that the many-body corrections should be included in order to obtain reliable quantitative results.

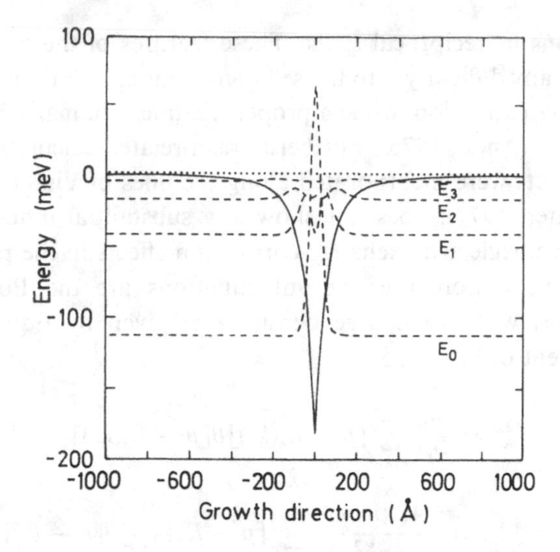

Fig. 2.15. Self-consistent potential (full line), level structure and associated densities (dashed lines), and Fermi level (dotted line) for a GaAs superlattice with $h = 4000$ Å, $N_D^{2D} = 5 \times 10^{12}/\text{cm}^2$, $N_A^{2D} = 0$. The densities are normalized to the population of each state.

One of the more attractive features of the present approach is that by the straightforward procedure of increasing the period h one can move from the superlattice regime to the multiple-quantum-well or atomic regime, where each quantum well is effectively disconnected from its neighboring wells; an example of this situation is shown in Fig. 2.15. The self-consistent potential and wave functions correspond to a superlattice with $h = 4000$ Å, $N_D^{2D} = 5 \times 10^{12}/\text{cm}^2$, and $N_A^{2D} = 0$. A comparison with the corresponding (same concentration) results of Figs. 2.12, 2.13 and 2.14 makes all the differences between the superlattice and the atomic regime evident: shallow potentials, extended charge distribution, and dispersive subbands in the former; deeper potentials, localized electronic charge, and non-dispersive subbands in the latter. A minor drawback of the calculations in this regime is that the potential sharpness increases inversely with period size, so the number of Fourier components retained in Eqn. (2.64) should be increased as compared with the superlattice regime ($l_{max} = 60$ has been used in the calculation). The results are in qualitative agreement with previous calculations for isolated n-type δ-doped planes (Zrenner *et al.*, 1988), with small differences which we attribute to our inclusion of many-body effects, their inclusion of non-parabolicity, and different values of parameters such as the dielectric constant, effective mass, etc.

(b) n-type δ-doping in Si

The case of n-type δ-doped layers in Si is rather more complicated than in GaAs, as the bulk conduction band bottom of Si has six symmetry-related anisotropic minima at points in the ⟨100⟩ directions in reciprocal space. These features of the band structure can be incorporated without any difficulty into the self-consistent calculations as long as one stays within the Hartree approximation, while a proper treatment of many-body effects requires a more refined analysis (Ando, 1976). Furthermore, a related calculation by Eisele (1989), where many-body effects were incorporated along the lines of Vinter's work on n-type Si inversion layers (Vinter, 1977), does not show any substantial influence on the results. Consequently, we have neglected exchange-correlation effects in the present case.

As in the preceding section, the relevant equations are the Poisson and electron Schrödinger equations, with the charge density $\rho(z)$ given by Eqn. (2.2), with $N_A^{2D} = p(z) \equiv 0$. The equivalent of Eqn. (2.57) is

$$N_D^{2D} - N_A^{2D} = \frac{2m_t}{\pi\hbar^2} \sum_{n,k_z} [\mu - E_n(k_z)]\,\theta[\mu - E_n(k_z)] + \frac{4\sqrt{(m_t m_l)}}{\pi\hbar^2} \sum_{n',k_z} [\mu - E_{n'}(k_z)]\,\theta[\mu - E_{n'}(k_z)] \tag{2.67}$$

where the first term on the right-hand side corresponds to the population of the subbands associated with the largest effective mass $m_z = m_l$, while the second gives the population of the subbands corresponding to $m_z = m_t$. The extra factors 2 and 4 are the respective valley degeneracies, and we denote the energies of the first (second) ladder by E_n ($E_{n'}$).

As an example of the subband structure of n-type δ-doped Si, in Fig. 2.16 we illustrate results for several typical N_D^{2D} concentrations, for the fully non-compensated regime

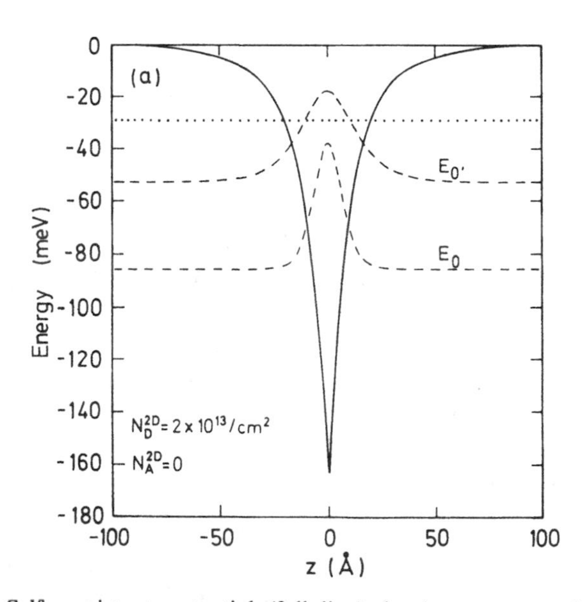

Fig. 2.16. Self-consistent potential (full line), level structure and associated densities (dashed lines), and Fermi level (dotted line) for an n-type δ-doped Si superlattice, with $h = 200$ Å. The densities are normalized to the population of each state.

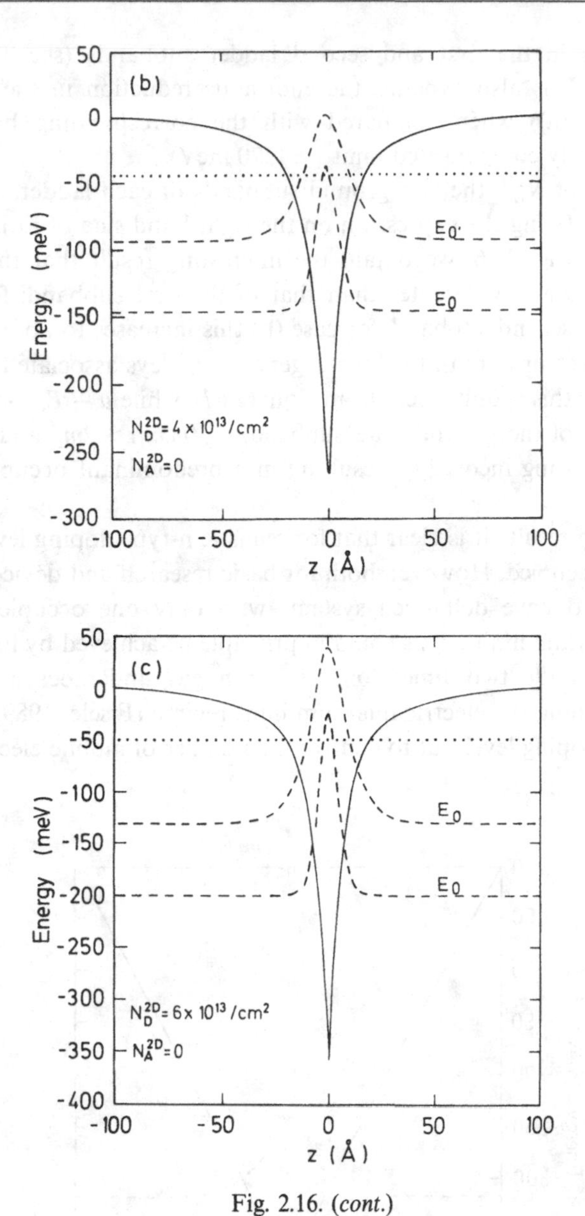

Fig. 2.16. (*cont.*)

$N_A^{2D} = 0$. Case (a) corresponds to $N_D^{2D} = 2 \times 10^{13}/\mathrm{cm}^2$, case (b) to $N_D^{2D} = 4 \times 10^{13}/\mathrm{cm}^2$, and case (c) to $N_D^{2D} = 6 \times 10^{13}/\mathrm{cm}^2$. In all three cases $h = 200$ Å, and $T = 0$.

While in the preceding case of n-type doping in GaAs, we were forced to choose a period of several thousand angstroms (see Fig. 2.15) in order to reach the isolated quantum-well limit, in the case of Si such configuration is obtained for much smaller values of h ($\approx 100 - 200$ Å). This is clearly related to the fact that the screening length of electrons in Si is much smaller than in GaAs due to their larger effective masses. Taking the effective Bohr radius as a rough measure of this screening length, we have $a^*(\mathrm{Si}) \approx 6.64$ Å and

32.03 Å for electrons in the first and second ladder subbands (see Table 2.1), while $a^*(\text{GaAs}) \approx 98.73$ Å. This also explains the enormous reduction in the potential depth (≈ 160 meV for case (a)) when compared with the corresponding bare depth $V_0 = \pi e^2 h N^{2D}/\varepsilon$ of the exactly compensated limit (≈ 1570 meV).

For the chosen set of N_D^{2D}, the two ground subbands of each ladder, with energies E_0 and $E_{0'}$ are occupied. Using the expression on the right-hand side of Eqn. (2.67), and the subband structure of Fig. 2.16, we obtain the interesting result that the population of the second subband is always greater than that of the first subband: for case (a) 60% of the charge is in the second subband, for case (b) this increases to about 64%, while for case (c) 66% of the electrons are in the four-degenerate valleys associated with the second ladder. The reason for this is quite clear from Eqn. (2.67): while $\mu - E_0 > \mu - E_{0'}$ and this favors the occupancy of the ground-state subband, $4\sqrt{(m_t m_l)} > 2m_t$, and this more than compensates the preceding inequality, resulting in a predominant occupancy of the first excited subband.

From the preceding results it is clear that for realistic n-type doping levels in Si, two or more subbands are occupied. However, both for basic research and device applications, it would be desirable to have delta-well systems with only one occupied subband (the so-called electric quantum limit). This could in principle be achieved by low doping levels, but in the case of Si the two-dimensional electron gas undergoes a metal–insulator transition before reaching the electric quantum limit regime (Eisele, 1989). An alternative is to retain a strong doping level but to control the number of mobile electrons in the well by other methods.

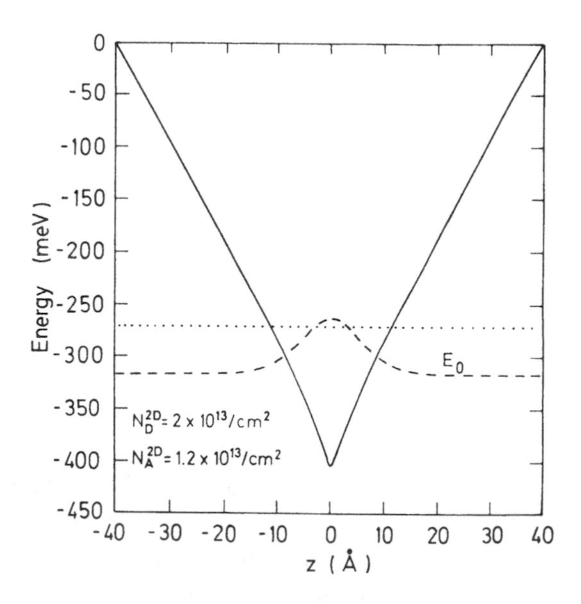

Fig. 2.17. Self-consistent potential (full line), level structure and associated density (dashed line), and Fermi level (dotted line) for an n-type δ-doped Si superlattice partially charge depleted. The period $h = 80$ Å.

One way of achieving charge depletion is by homogeneously p-doping the host material in which a donor delta is grown (Johnson and MacKinnon, 1991); while this method works well for narrow-gap semiconductors such as InSb, its applicability to the present case of Si is doubtful. Accordingly, we have charge depleted the quantum well by introducing a finite concentration N_A^{2D} at the acceptor planes. The results given in Fig. 2.17 correspond to $N_D^{2D} = 2 \times 10^{13}/\mathrm{cm}^2$, $N_A^{2D} = 1.2 \times 10^{13}/\mathrm{cm}^2$, and $h = 80$ Å. The 60% charge depletion gives the desired result of only one occupied subband: the second subband (empty, not included) lies about 3 meV above the Fermi level. Note, first, even for this small value of h the system behaves as a set of isolated quantum wells, and, second, the weak screening provided by the remaining electrons, as the self-consistent potential depth is not wholly different from its bare fully compensated value $V_0 \approx 630$ meV.

(c) *p-type δ-doping in GaAs*

While the main body of the experimental work on δ-doping has been devoted to the n-type case, only very recently have single acceptor (Be) doping spikes in GaAs also been investigated in detail. Schubert *et al.* (1990) essentially concentrate on the characterization of Be δ-doping layers in GaAs: a spatial localization of Be in δ-doped GaAs within a few lattice constants (<20 Å) for concentrations $N_A^{2D} < 10^{14}/\mathrm{cm}^2$ is reported from capacitance–voltage profiles and secondary ion mass spectroscopy. The work by Wagner *et al.* (1991) and Richards *et al.* (1993), aimed at giving a microscopic characterization of the two-dimensional hole gas (2DHG) formed around the acceptor impurity layer, reported photoluminescence studies which yielded information about the subband hole structure.

On the theoretical side, while the self-consistent treatment of the two-dimensional electron gas (2DEG) at the n-type doping spikes, as discussed in the two preceding subsections, is by now an almost standard calculation, much less is known about the properties of the equivalent 2DHG associated with p-type doping spikes. We are aware of only two very recent self-consistent calculations on the hole subband structure of acceptor δ-doped GaAs (Reboredo and Proetto, 1993; Richards *et al.*, 1993). One reason for the scarcity of theoretical studies on the 2DHG is that in order to give a rigorous description of the valence subbands one must deal with the full 4 × 4 matrix Luttinger Hamiltonian, which is quite demanding from the computational viewpoint, as compared with the scalar problem posed by the 2DEG. A second problem complicating the study of 2DHG is that it is not yet clear to what extent it is necessary to include many-body corrections in order to give a quantitative description of the subband structure. We show below that the inclusion of many-body effects is important and leads to good agreement with experimental results.

We consider a periodic sequence of p-type δ-doped planes, immersed in an otherwise homogeneous GaAs host matrix, grown in an [001] direction, which we take along the quantization axis z. The relevant equations are now the Poisson and hole Schrödinger equations (Eqns. (2.1) and (2.6) respectively). The charge density $\rho(z)$ which enters into the Poisson equation is given by Eqn. (2.2), with $p(z)$ defined by Eqn. (2.4) and $N_D^{2D} = n(z) \equiv 0$.

The 4 × 4 matrix, Eqn. (2.8), can be block diagonalized by a change of basis into two independent matrices H^{U} and H^{L}:

$$\tilde{H}^0 = \begin{Bmatrix} H^{\mathrm{U}} & 0 \\ 0 & H^{\mathrm{L}} \end{Bmatrix} \tag{2.68}$$

with the explicit expressions

$$H^{\sigma} = \begin{bmatrix} P \pm Q & \tilde{R} \\ \tilde{R}^* & P \mp Q \end{bmatrix}$$

$$\tilde{R} = |R| - \mathrm{i}\hat{k}_z|S| \tag{2.69}$$

where $\sigma = \mathrm{U(L)}$ refers to the upper (lower) $\pm$ signs. The unitary matrix that block diagonalizes H^0 depends on k_x and k_y but is independent of $\hat{k}_z$; the explicit expression for the matrix can be found elsewhere (Broido and Sham, 1985). The upper- and lower-block envelope functions satisfy

$$\sum_{j=1}^{2} [H^{\sigma}_{i,j}(k_x, k_y, \hat{k}_z) + V(z)\delta_{i,j}]\psi^{\sigma}_{jn\mathbf{k}}(z) = E^{\sigma}_n(\mathbf{k})\psi^{\sigma}_{in\mathbf{k}}(z) \quad (i = 1, 2) \tag{2.70}$$

where $\psi^{\sigma}_{in\mathbf{k}}(z)$ are now two component envelope wave functions.

As in preceding sections, the superlattice potential $V(z)$ is given by the sum of the potential of the ionized impurities $V_{\mathrm{I}}(z)$ and the Hartree potential $V_{\mathrm{H}}(z)$

$$V(z) = V_{\mathrm{I}}(z) + V_{\mathrm{H}}(z) \tag{2.71}$$

and is determined self-consistently from the Schrödinger, Eqn. (2.70), and Poisson equations. We choose the following boundary conditions for the latter:

$$V(jh) = 0 \tag{2.72}$$

and

$$V(z) = V(z + h) \tag{2.73}$$

where Eqn. (2.72) is equivalent to adopting the value of the self-consistent potential at the mid-point between two consecutive acceptor planes as the origin of energies. While an analytical solution of Eqn. (2.70) is not possible, the structure of H^{U} and H^{L} suggests the following numerical strategy. We first observe that, upon neglecting the off-diagonal terms, Eqn. (2.70) is split into two (each doubly degenerate) single component equations:

$$\left[-\frac{\hbar^2}{2m_0}(\gamma_1 \mp 2\gamma_2)\frac{\mathrm{d}^2}{\mathrm{d}z^2} + V(z)\right]\Phi^{\substack{\mathrm{hh}\\ \mathrm{lh}}}_{n,k_z}(z) = E^{\substack{\mathrm{hh}\\ \mathrm{lh}}}_n(k_z)\Phi^{\substack{\mathrm{hh}\\ \mathrm{lh}}}_{n,k_z}(z) \tag{2.74}$$

where

$$E^{\substack{\mathrm{hh}\\ \mathrm{lh}}}_n(k_z) = E^{\substack{\mathrm{hh}\\ \mathrm{lh}}}_n(\mathbf{k}) - \frac{\hbar^2}{2m_0}(\gamma_1 \pm \gamma_2)k_{\parallel}^2 \tag{2.75}$$

where the superscripts hh and lh denote heavy and light holes respectively.

We will refer to Eqn. (2.74) as the parabolic approximation, as all the non-parabolicities embodied in the complete Eqn. (2.70) are lost in this approximation. An attractive feature of Eqn. (2.74) is that, as a consequence of the neglect of the non-diagonal terms, the mixing between heavy and light holes is zero, and the 2DHG splits into two perfectly defined components, one heavy, with perpendicular (parallel) mass $m_z^{\mathrm{hh}} = m_0/(\gamma_1 - 2\gamma_2)$ ($m_\parallel^{\mathrm{hh}} = m_0/(\gamma_1 + \gamma_2)$), and a light component, with perpendicular (parallel) mass $m_z^{\mathrm{lh}} = m_0/(\gamma_1 + 2\gamma_2)$ ($m_\parallel^{\mathrm{lh}} = m_0/(\gamma_1 - \gamma_2)$). For later use, we give the explicit parabolic version of the neutrality constraint, Eqn. (2.11), at zero temperature, which by performing the sum analytically on $\mathbf{k}_\parallel$ becomes

$$N_{\mathrm{A}}^{2\mathrm{D}} = \frac{m_\parallel^{\mathrm{hh}}}{\pi\hbar^2} \sum_{n,k_z} [\mu - E_n^{\mathrm{hh}}(k_z)]\,\theta[\mu - E_n^{\mathrm{hh}}(k_z)] + \frac{m_\parallel^{\mathrm{lh}}}{\pi\hbar^2} \sum_{n,k_z} [\mu - E_n^{\mathrm{lh}}(k_z)]\,\theta[\mu - E_n^{\mathrm{lh}}(k_z)] \tag{2.76}$$

where $\theta(x)$ is the Heaviside step function.

The self-consistent solutions of Eqn. (2.74) are obtained from a plane-wave expansion of the periodic part of $\Phi_{n,k_z}^{\mathrm{hh}/\mathrm{lh}}(z)$, which essentially transforms Eqn. (2.74) into a typical matrix eigenvalue problem. This step of the calculation was explained in detail in subsection 2.4(a), the only difference being that in the present case $V_{\mathrm{H}}(z)$ is determined by the sum of the heavy- and light-hole densities.

The self-consistent solution of Eqn. (2.74) yields an infinite set of doubly degenerate heavy- and light-hole subband energies and wave functions, and we use this basis to expand the solutions $\psi_{in\mathbf{k}}^{\sigma}(z)$ of the exact eigenvalue, Eqn. (2.70). This method is quite efficient in the numerical sense: as the parabolic basis is close to the solution of the full Eqn. (2.70), this allows us to obtain excellent convergence with a relatively small intermediate basis. Further technical details on the method can be found in the work by Reboredo and Proetto (1993).

All the numerical results that follow correspond to the zero temperature limit and $N_{\mathrm{A}}^{2\mathrm{D}} = 8 \times 10^{12}/\mathrm{cm}^2$; the values of γ_1, γ_2, γ_3, and ε for bulk GaAs are given in the caption to Table 2.1. While our approach is valid for any value of the period h, we restrict the present calculation to the atomic limit by taking $h = 1000$ Å and including only the states at $k_z = 0$, where k_z is the Bloch index in the z-direction.

As a first rough approximation to the present problem, one can solve Eqn. (2.74) by neglecting any screening effect ($V_{\mathrm{H}}(z) \equiv 0$ in Eqn. (2.71)), in which case the potential $V(z)$ is just the bare V-shaped potential associated with a uniformly charged plane, whose exact analytical solution, Eqn. (2.28), gives two occupied subbands, one heavy and one light, with an energy splitting of about 43 meV. With these subband energies, from Eqn. (2.76) it is easy to obtain a Fermi level that lies 90 meV above the ground heavy-hole subband.

When screening (in the Hartree approximation) is included in the calculation, the self-consistent solution of Eqn. (2.74) gives the potential and level structure displayed in Fig. 2.18: the softening of the non-screened V-shaped potential now yields six occupied subbands (four heavy, two light), an energy splitting of about 21.5 meV between the first heavy- and light-hole subbands, and the Fermi level lies 66 meV above the ground-state

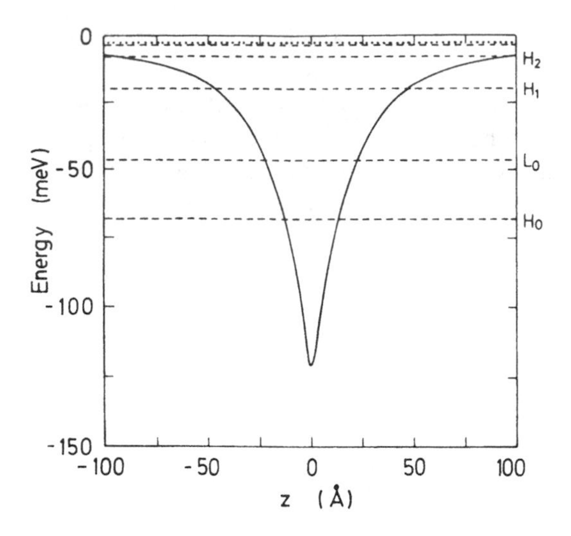

Fig. 2.18. Parabolic self-consistent potential (full line), subband structure at $k_{\parallel} = 0$ (dashed lines), and Fermi level (dotted line) in the Hartree approximation for a p-type δ-doped layer in GaAs. While there are in principle six subbands below the Fermi level, essentially all the charge is concentrated in the lowest heavy-hole (H_0) and light-hole (L_0) subbands.

heavy-hole subband. Using Eqn. (2.76) again, but this time substituting the level structure of Fig. 2.18, it is simple to show that most of the holes are distributed almost equally between the two lowest subbands. This is because what really matters for the calculation of the population density in Eqn. (2.76) is the energy difference between the Fermi level or chemical potential and the subband energy times the effective parallel mass; as $\mu - E_0^{\mathrm{hh}} > \mu - E_0^{\mathrm{lh}}$, while $m_{\parallel}^{\mathrm{hh}} < m_{\parallel}^{\mathrm{l}}$, there is a compensation between these two factors and each of the two lowest subbands accommodates approximately half of the total density.

Finally, the self-consistent solution of Eqn. (2.70), which includes screening and non-parabolicities, gives the results presented in Fig. 2.19: two occupied subbands, an energy splitting of about 20 meV between the two occupied subbands, and about 34 meV from the ground-state subband to the Fermi level.

From the comparison between Figs. 2.18 and 2.19 it is clear that while the self-consistent potential and subband energy levels are not very sensitive to the non-parabolicities, the Fermi level decreases dramatically (32 meV) when the full Luttinger Hamiltonian is included in the calculation. This is related to the increase in the ground-state in-plane effective mass over its value $m_0/(\gamma_1 + \gamma_2)$ in the parabolic approximation: being the ground state of the system, the only possible contribution of the off-diagonal elements is to lower its energy, thereby increasing its in-plane mass. This in turn implies that the ground-state subband is able to accommodate more carriers, thereby lowering the Fermi level. For the set of parameters corresponding to Fig. 2.19, about 82% (18%) of the holes are in the ground-state (first excited-state) subband.

A clear example of these strong non-parabolicities is given in Fig. 2.20, where the in-plane

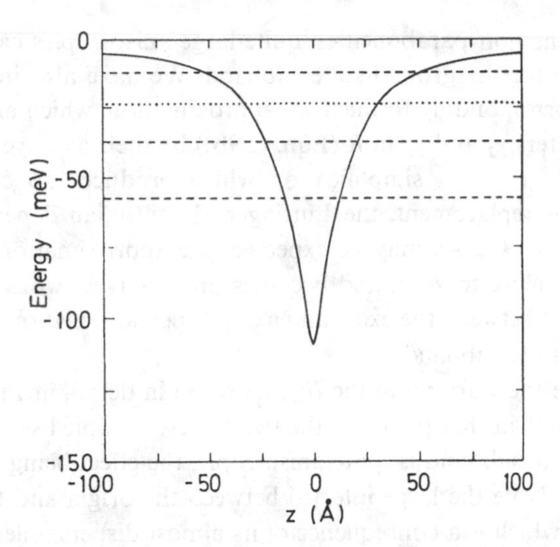

Fig. 2.19. Luttinger self-consistent potential (full line), subband level structure at $k_{\parallel} = 0$ (dashed lines), and Fermi level (dotted line) in the Hartree approximation.

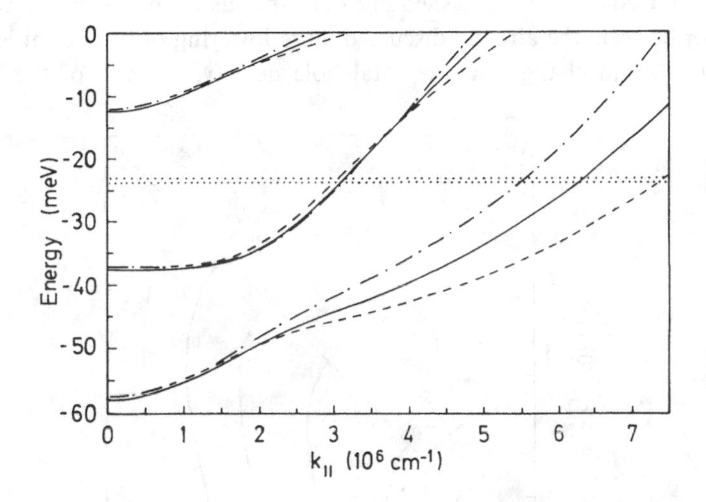

Fig. 2.20. In-plane dispersion relations for the three lowest subbands of p-type δ-doped GaAs with $N_A^{2D} = 8 \times 10^{12}/\text{cm}^2$. The solid, dashed, and dot–dashed lines represent subband energies in the axial approximation [11] and [01] directions respectively. The straight dotted lines give the Fermi level, the upper (lower) corresponding to the axial (Luttinger) approximation.

dispersion relations corresponding to the three lowest eigenvalues of the preceding diagram are displayed. While self-consistent calculations for holes in heterojunctions (Broido and Sham, 1985) and compositional quantum wells (Ando, 1985) have reversed sign for the effective mass associated with the first excited subband, in the present case this effect is quite weak. Beyond the non-parabolicities, quite large anisotropies can clearly be seen in Fig. 2.20, particularly for the ground-state subband. We have also included the in-plane dispersion relation corresponding to the axial approximation, which amounts to replacing the Luttinger parameters γ_2 and γ_3 in R (Eqn. (2.9b)) by their average (Broido and Sham, 1985; Richards *et al.*, 1993), a simplification which produces no change in H^0 when $k_x = k_y = 0$. With this replacement, the Luttinger Hamiltonian depends only on $k_\parallel$, the magnitude of vector (k_x, k_y). As may be expected, the approximation is quite good (but not exact) when $k_\parallel$ is close to zero, and becomes progressively worse for large values of $k_\parallel$, where the difference between the exact and axial dispersion relations amounts to several meV for the ground-state subband.

In order to analyze the warping in the (k_x, k_y) plane in detail, in Fig. 2.21 we illustrate constant-energy surfaces in that plane for the two lowest occupied subbands. As discussed above, the ground-state subband is quite anisotropic, the effect being rather small for the first excited subband. Note the large interval between the origin and the first contour for the second subband, which is a consequence of its almost dispersionless energy versus the $k_\parallel$ relation (very large effective mass) close to $k_\parallel = 0$ of Fig. 2.20.

The hole density distributions associated with the potentials and energy levels of the preceding diagrams are displayed in Fig. 2.22: the dotted line corresponds to the parabolic self-consistent solutions, while the dashed line corresponds to the Luttinger self-consistent solution. As can be seen, the already-discussed large lowering of the Fermi level has been achieved by very small changes in the total hole density, in spite of the fact that the

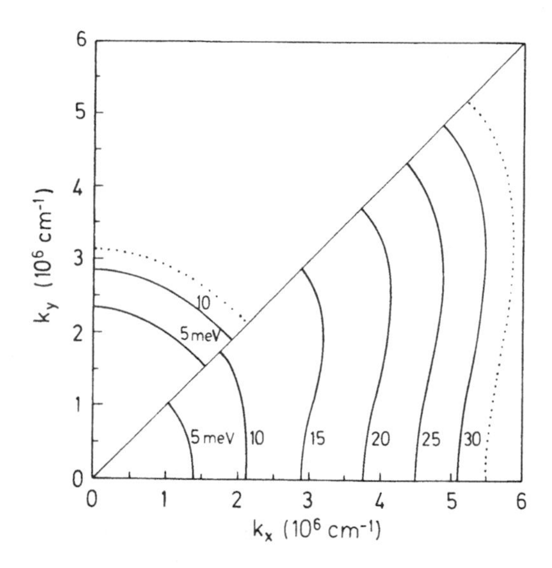

Fig. 2.21. Constant energy lines in the (k_x, k_y) space of the two lowest subbands. The lower (upper) panel corresponds to the ground (first-excited) subband. The solid lines represent equi-energy contours, with the dotted lines corresponding to the Fermi energy. The energy interval for adjacent contours is chosen to be 5 meV, and the origin of energy lies at the $k_\parallel = 0$ energy of each subband.

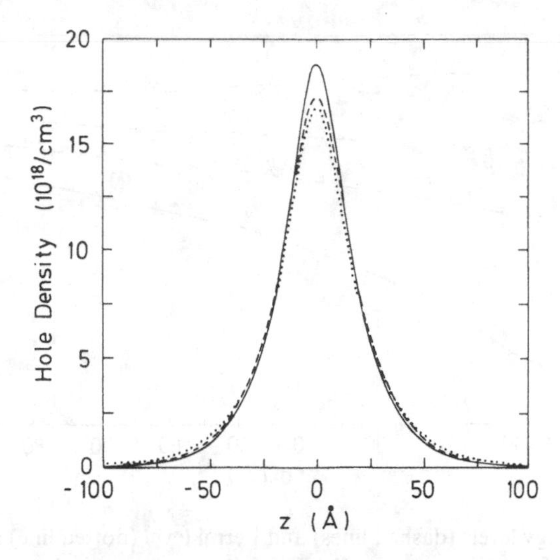

Fig. 2.22. Total hole densities: dotted line, parabolic self-consistent (Hartree) approximation; dashed line, Luttinger self-consistent (Hartree) approximation; full line, parabolic self-consistent (Hartree plus exchange-correlation) approximation.

ground-state subband has a much greater occupancy in the Luttinger self-consistent approximation than in the parabolic one.

With the purpose of investigating the dependence of the results obtained until now on the assumption of a perfectly sharply defined doped plane, we have given a finite width to the doping plane. Accordingly, we generalize Eqn. (2.2) to

$$\rho(z) = -e \frac{N_A^{2D}}{\sigma\sqrt{\pi}} \sum_{z_A} e^{-(z-z_A)^2/\sigma^2} + ep(z) \tag{2.77}$$

replacing the δ-function profile for the distribution of impurities with a normalized Gaussian-function profile of width $\sigma/\sqrt{2}$. Figure 2.23 shows the corresponding results obtained with the full Luttinger Hamiltonian for the $k_{\parallel} = 0$ subband level structure, Fermi level, and bottom of the self-consistent potential as functions of the spread of the dopant profile. Due to our choice of origin of energy as the value of the self-consistent potential at $z = jh$, which is different (in principle) for each value of σ, only differences of energy should be compared in Fig. 2.23, and not absolute values. As expected, the increasing width of the doping profile leads to a decrease in the subband spacing, as the self-consistent potential becomes rather shallow. As a consequence of this flattening of the potential, the number of occupied subbands increases from two to three, the transition taking place for $\sigma \approx 50$ Å. The subband structure is almost insensitive to spreads smaller than 20 Å; this is quite consistent with the hole density distributions of Fig. 2.22. It also implies that for the samples with $N_A^{2D} < 10^{14}/\text{cm}^2$ studied by Schubert *et al.* (1990), with estimated dopant spreads smaller than 20 Å, a δ-function impurity distribution profile is an essentially correct assumption, as far as properties related to the hole subband structure are concerned.

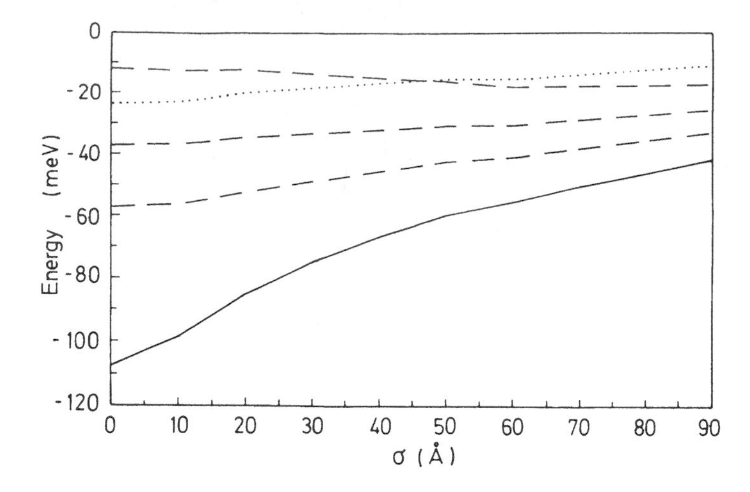

Fig. 2.23. Energy levels (dashed lines) and Fermi level (dotted line) as function of the width of the Gaussian profile. The full line corresponds to the bottom of the potential.

Until now, we have neglected many-body effects beyond the Hartree approximation for the 2DHG. As discussed above, an attractive way of studying the exchange-correlation effect on the subband structure of heterostructures and quantum wells is to use the density-functional method.

However, there are some problems in the choice of V_{xc} in our system. While for the parabolic and isotropic gas of spin $\frac{1}{2}$ particles there are several simple and quite accurate parametrizations of the V_{xc} within the LDA in the literature, we are not aware of any equivalent parametrization for the four-component, anisotropic, and non-parabolic hole gas. Faced with this difficulty, and encouraged by the almost identical results for the charge density and subband energy levels of the parabolic self-consistent and Luttinger self-consistent calculations, we have included the many-body corrections in the parabolic approximation.

While further details can be found in the work by Reboredo and Proetto (1993), the main results are that the inclusion of many-body effects decreases the number of occupied subbands from six (Fig. 2.18) to three (two heavy, one light), but, what is more important, it increases the energy splitting between the ground-state heavy- and light-hole subbands to about 28 meV. The full line of Fig. 2.22 shows the corresponding hole density distribution: the inclusion of many-body effects decreases the Coulomb repulsion between holes with the same spin, and the holes concentrate closer to the impurity charged plane.

A comparison of these theoretical results with available experimental data is not conclusive for several reasons. Gilinsky *et al.* (1991) studied the photoluminescence (PL) spectra of δ-doped p-type GaAs and ascribes one of the observed PL bands to radiative recombination of spatially separated photoelectrons with holes occupying size-quantized levels of the δ-layer potential well. The broad PL bands (presumably due to the fact that the potential confining the holes is repulsive for the photogenerated electrons) precludes a quantitative comparison with the calculated subband structure. Wagner *et al.* (1991), and Richards *et al.* (1993), obtain a significant enhancement of the 2DHG emission intensity by placing the acceptor doping spike in the center of an $Al_xGa_{1-x}As/GaAs/Al_xGa_{1-x}As$ quantum well 600 Å wide. Note that the $Al_xGa_{1-x}As$ barriers, while quite important from the experimental viewpoint, are irrelevant for the calculation of the hole subband structure, as the hole density profile shown in Fig. 2.22 is much narrower than the well width. However, it seems that there is some discrepancy between both sets of data. At variance with Wagner *et al.* (1991), in the experiment by Richards *et al.* (1993) the PL spectra were excited with circularly polarized light, so that by analyzing the degree of circular polarization of the emitted light it is possible to discriminate between emission from heavy- and light-hole subbands. While we have a sizeable discrepancy with the first set of data (even including the many-body corrections), quite good agreement is obtained between the above-calculated hole subband structure and the latter set of experimental results. More experiments are clearly needed in order to clarify this as yet unsettled issue.

2.5 Conclusions

In this chapter we have focused on the subject of the electronic structure of δ-doped semiconductors. The results presented can be regarded as a useful example of the versatility of the envelope function approximation within the framework of the effective-mass theory, as applied to the case of quantum wells and semiconductor superlattices.

It was found convenient to divide the 'universe' of sawtooth doping superlattices in depleted and undepleted configurations, each of them corresponding to qualitatively different situations. In the first (second) case, the donor and acceptor impurity concentrations are equal (unequal), and, consequently, mobile particles are absent (left-out) after recombination of electrons with holes.

The key feature of the depleted or compensated configuration is the piecewise linear dependence of the confining potential $V(z)$ which opens the way towards the derivation of the exact subband structure for several relevant situations: symmetric and asymmetric triangular quantum wells, and symmetric sawtooth superlattices. Beyond the eigenvalues, the analysis was extended to a variety of wave-function properties, such as barrier penetration and wave-function extension. A comparison between exact and variational results for the case of symmetric triangular quantum wells proves the latter to be a good compromise between accuracy and simplicity.

Once the system is uncompensated, the presence of mobile particles (electrons or/and holes) screens and modifies the perfectly V-shaped confining potential and one is forced to obtain a self-consistent numerical solution of the coupled Poisson and Schrödinger equations. This task has been accomplished for different situations of increasing complexity: n-type δ-layers in GaAs, n-type δ-layers in Si, and p-type δ-layers in GaAs. Many-body effects beyond the Hartree approximation (exchange and correlation) were included in the first and last system (within the framework of the density-functional formalism), and in both cases were shown to be quantitatively important. We have found that for realistic n-type doping levels in Si, typically two subbands are occupied, the lowest one associated with electrons with the largest effective mass m_l, while the upper corresponds to the smaller effective mass m_t. Introducing a finite concentration N_A^{2D} at the acceptor planes we analyzed the charge depletion of this system, and found that it is possible to move towards a regime where only one subband is occupied; this is important both for basic research and the technological applications of these systems. The properties of the two-dimensional hole gas were analyzed for the particular case of p-type δ-doping in GaAs, and we found good agreement between very recent studies using photoluminescence spectroscopy on the hole subband structure and the self-consistent results.

Much experimental and theoretical work remains to be done on the properties of the low-density insulating phase, where the impurity-related disorder should be incorporated from the very beginning. Full understanding of this phase is essential in order to justify the validity of the 'jellium' approximation on which all the material presented in this chapter relies.

Acknowledgments

The author would like to express his gratitude to J. M. Ferreyra and F. A. Reboredo, who produced much of the material presented here. He also wishes to thank J. Wagner for making available to him new experimental results prior to publication, and V. Grunfeld for a careful reading of the manuscript.

References

Ando T., *Phys. Rev.* **B13**, 3468 (1976).
Ando, T., *J. Phys. Soc. Japan* **54**, 1528 (1985).
Ando T. and Mori S., *J. Phys. Soc. Japan* **47**, 1518 (1979).
Ando T., Fowler A. B., and Stern F., *Reviews of Modern Physics* **54**, 437 (1982).
Antosiewicz H. A., in *Handbook of Mathematical Functions*, edited by M. Abramowitz and I. A. Stegun (Dover Publications, Inc., New York, page 446, 1970).
Ashcroft N. W. and Mermin N. D., *Solid State Physics* (Holt, Rinehart, and Winston, New York, 1976).
Bastard G., in *Wave Mechanics Applied to Semiconductor Heterostructures* (Les Editions de Physique, Les Ulis, 1988).
BenDaniel D. J. and Duke C. B., *Phys. Rev.* **152**, 683 (1966).
Broido D. A. and Sham L. J., *Phys. Rev.* **B31**, 888 (1985).
Cho H. and Prucnal P. R., *Phys. Rev.* **B36**, 3237 (1987).
Choquette K. D., Misemer D. K., and McCaughan L., *Phys. Rev.* **B43**, 7040 (1991).
Degani M. H., *Phys. Rev.* **B44**, 5580 (1991).
Döhler G. H., *IEEE J. Quantum Electron.* **QE-22**, 1682 (1986).
Efros A. L., Pikus F. G., and Samsonidze G. G., *Phys. Rev.* **B41**, 8295 (1990).
Eisele I., *Superlattices and Microstructures* **1**, 123 (1989).
Fang F. F. and Howard W. E., *Phys. Rev. Lett.* **16**, 797 (1966).
Ferreyra J. M. and Proetto C. R., *Phys. Rev.* **B42**, 5657 (1990).

Ferreyra J. M. and Proetto C. R., *Phys. Rev.* **B44**, 11231 (1991).
Fiorentini V., *Semicond. Sci. Technol.* **5**, 211 (1990).
Fritzsche H. T. and Schirmacher W., *Europhys. Lett.* **21**, 67 (1993).
Gilinsky A. M., Zhuravlev K. S., Lubyshev D. I., Migal V. P., Preobrazhenskii V. V., and Semiagin B. R., *Superlattices and Microstructures* **10**, 399 (1991).
Glass A. M., Schubert E. F., Wilson B. A., Bonner C. E., Cunningham J. E., Olsøn D. H., and Jan W., *Appl. Phys. Lett.* **54**, 2247 (1989).
Gossmann H. J. and Schubert E. F., *Critical Reviews in Solid State and Materials Science* **8**, 1 (1993).
Gunnarsson O. and Lundqvist B. I., *Phys. Rev.* **B13** 4274 (1976).
Hohenberg P. and Kohn W., *Phys. Rev.* **136**, B864 (1964).
Ioriatti L., *Phys. Rev.* **B41**, 8340 (1990).
Johnson E. A. and MacKinnon A., *Superlattices and Microstructures* **9**, 441 (1991).
Koch F. and Zrenner A., *Mater. Sci. Eng.* **B1**, 221 (1989).
Koch F., *Physica* **B184**, 298 (1993).
Kohn W. and Sham L. J., *Phys. Rev.* **140**, A1133 (1965).
Levin E. I., Raikh M. E., and Shklovskii B. I., *Phys. Rev.* **B44**, 11281 (1991).
López A., On a Complete Set of Functions Associated with a Peculiar Eigenvalue Problem, internal report CAB/1972/1.
Lundqvist S. and March N. H. (eds.), *Theory of the Inhomogeneous Electron Gas* (Plenum, New York, 1983).
Luttinger J. M. and Kohn W., *Phys. Rev.* **97**, 869 (1955).
Luttinger J. M., *Phys. Rev.* **102**, 1030 (1956).
Mäder K. A. and Baldereschi A., in: *Proceedings of the Int. Meeting on the Optics of Excitons in Confined Systems*, edited by A. D'Andrea, R. del Sole, R. Girlanda, and A. Quattropani (Institute of Physics, Bristol, p. 341, 1992a).
Mäder K. A. and Baldereschi A., *Mat. Res. Soc. Symp.* **240**, 597 (1992b).
Metzner C., Beyer H. J., and G. H. Döhler, *Phys. Rev.* **B46**, 4128 (1992).
Mezrin O. and Shik A., *Superlattices and Microstructures* **10**, 107 (1991). Note that their definition of the parameter ν differs by a factor of 4π from ours.
Mezrin O. A., Shik A. Y., and Mezrin V. O., *Semicond. Sci. Technol.* **7**, 664 (1992).
Moruzzi V. L., Janak J. F., and Williams A. R., *Calculated Electronic Properties of Metals* (Pergamon, New York, 1978).
O'Reilly E. P., *Semicond. Sci. Technol.* **4**, 121 (1989).
Perdew J. and Zunger A., *Phys. Rev.* **B23**, 5048 (1981).
Ploog K. and Döhler G. H., *Adv. Phys.* **32**, 285 (1983).
Prince P. J., *ACM Trans. Math. Software* **1**, 372 (1975).
Proetto C. R., *Phys. Rev.* **B41**, 6036 (1990).
Reboredo F. A. and Proetto C. R., *Solid State Comm.* **81**, 163 (1992).
Reboredo F. A. and Proetto C. R., *Phys. Rev.* **B47**, 4655 (1993).
Richards D., Wagner J., Schneider H., Hendorfer G., Maier M., Fischer A., and Ploog K., *Phys. Rev.* **B47**, 9629 (1993).
Romanov Y. A. and Orlov L. K., *Sov. Phys. Semicond.* **7**, 182 (1973).
Ruden P. and Döhler G. H., *Phys. Rev.* **B27**, 3538 (1983).
Schubert E. F., Ullrich B., Harris T. D., and Cunningham J. E., *Phys. Rev.* **B38**, 8305 (1988a).
Schubert E. F., Harris T. D., and Cunningham J. E., *Appl. Phys. Lett.* **53**, 2208 (1988b).
Schubert E. F., Harris T. D., Cunningham J. E., and Jan W., *Phys. Rev.* **B39**, 11011 (1989).
Schubert E. F., *Opt. Quantum Electron.* **22**, S141 (1990).
Schubert E. F., Kuo J. M., Kopf R. F., Luftman H. S., Hopkins L. C., and Sauer N. J., *J. Appl. Phys.* **67**, 1969 (1990).
Sham L. J. and Kohn W., *Phys. Rev.* **145**, 561 (1966).
Sobkowicz P., Wilamowski Z., and Kossut J., *Semicond. Sci. Technol.* **7**, 1155 (1992).

Stern F., *Phys. Rev.* **B5**, 4891 (1972).
Vinter B., *Phys. Rev.* **B15**, 3947 (1977).
Voisin P., Bastard G., and Voos M., *Phys. Rev.* **B29**, 935 (1984).
Wagner J., Ruiz A., and Ploog K., *Phys. Rev.* **B43**, 12134 (1991).
Wigner E. P., *Phys. Rev.* **46**, 1002 (1934); *Trans. Faraday Soc.* **34**, 678 (1938).
Zeller Ch., Vinter B., and Abstreiter G., *Phys. Rev.* **B26**, 2124 (1982).
Zrenner A., Koch F., and Ploog K., *Surf. Sci.* **196**, 671 (1988).

PART THREE

Epitaxial growth

3

Recent progress in delta-like confinement of impurities in GaAs: Ultrahigh-density incorporation and generation of quantum wires and dots by self-organization

KLAUS H. PLOOG

3.1 Introduction

The concept of delta-doping (also called monolayer, planar, or sheet doping) of semiconductors implies that the distribution of impurity atoms is actually confined to just one lattice plane. This confinement of the impurity distribution to a few (or ideally one) lattice planes leads to a variety of novel electronic phenomena, depending on the chemical nature of the impurity (i.e. isoelectronic or dopant impurity) and on the density (i.e. a small fraction of, or even a complete, lattice plane). Both the majority and minority carrier properties are affected in a distinct manner. Although the concept of delta-doping of semiconductors was demonstrated more than a decade ago [1], the ongoing challenge for the crystal grower is to control the confinement of impurity distribution to exactly one lattice plane, for example the (001) plane in cubic semiconductors grown along [001]. In addition, the most recent challenges are (i) to control the lateral distribution of the impurity atoms within the lattice plane in order to create semiconductor quantum wire and quantum dot structures and (ii) to increase the density of impurity atoms to as much as a completely filled lattice plane.

In this chapter we review some recent achievements in the latter two fields using GaAs as the prototype host lattice for both isoelectronic and dopant impurities and employing elemental-source molecular beam epitaxy (MBE) for layer growth. Silicon is the most thoroughly studied impurity for delta-doping of GaAs, and besides Al the elements In and Sb have been investigated as isoelectronic impurities in GaAs nominally confined to one lattice plane. Here we restrict ourselves to Si as the donor impurity and In as the isoelectronic impurity in GaAs. Some of the intriguing results obtained in this field can serve as guidelines for other combinations of materials. Both these impurity species have in common that their atomic radii differ considerably from those of the constituents of the GaAs host lattice, hence at high impurity densities elastic strain in the vicinity of their location will be generated.

This chapter is organized as follows. We begin in Section 3.2 with a brief account of the status of Si-delta-doping in GaAs and of In and Al monolayers in GaAs fabricated by the standard method of suspension of crystal growth by MBE and controlled deposition of the impurities on the non-growing (001) surface. We continue in Section 3.3 with a discussion of the laterally ordered incorporation of the impurity atoms along the step edges of vicinal (001)GaAs surfaces and the feasibility of fabricating quantum wires and quantum dots by self-organized epitaxial growth. In Section 3.4 we present novel approaches to increasing the fraction of the lattice sites occupied by the electrically active impurity atoms toward the limit of a completely filled monolayer. The conclusion is then given in Section 3.5.

3.2 Status of monolayer impurity incorporation in GaAs

The scaling of the concept of artificially layered semiconductor structures down to its ultimate physical limit of less than the lattice constant of the host material is driven by the need for further miniaturization of semiconductor devices and by the search for novel quantum-size effects in fundamental research. The confinement of the impurity atoms to one lattice plane of the host material requires an atomically flat starting surface having no steps and, in addition, negligible impurity diffusion (segregation) normal to this surface. For MBE growth of the III–V semiconductor matrix along [001] three growth stages are in general required to form the structure [1, 2, 3]. In the first stage, a buffer layer is grown to generate the starting surface as smooth as possible, which in the case of GaAs is monitored via the (2×4) reconstruction in the reflection high-energy electron diffraction (RHEED) pattern. In the second stage, the group-III-flux impinging on the substrate is suspended while the group-V-flux is maintained, and the surface is then exposed to the impurity flux for a controlled amount of time. During the impurity deposition there are distinct changes in the surface reconstruction, for example from (2×4) to (3×1) for Si-delta-doping of GaAs at $T_s \approx 550\ ^\circ\mathrm{C}$ with intermediate structures depending on the Si coverage [4]. Finally, the third stage of growth involves terminating the impurity flux and restoring the group-III-flux so that the previously deposited impurity atoms are epitaxially overgrown with the host material.

Ideally, the original arrangement of the surface layer of the second stage with the impurity atoms on lattice sites should not change during this overgrowth. However, although this ideal surface sheet of impurity atoms on lattice sites may have been stable, it very often does not remain so once it is rebonded to form part of a three-dimensional (3D) solid. As a consequence, during overgrowth, diffusion and surface segregation of the impurity atoms occur, the extent of which depends on the detailed growth conditions (see [3] and [5] and references therein).

In the case of delta-*doping*, the mechanism of saturation of the free-carrier concentration is another important issue in addition to the origin of the broadening of the impurity profile. It has been well established for several years that the monotonic relationship between carrier concentration, as measured by the Hall effect, and dopant density, as deduced from the impurity flux and measured by secondary ion mass spectroscopy (SIMS),

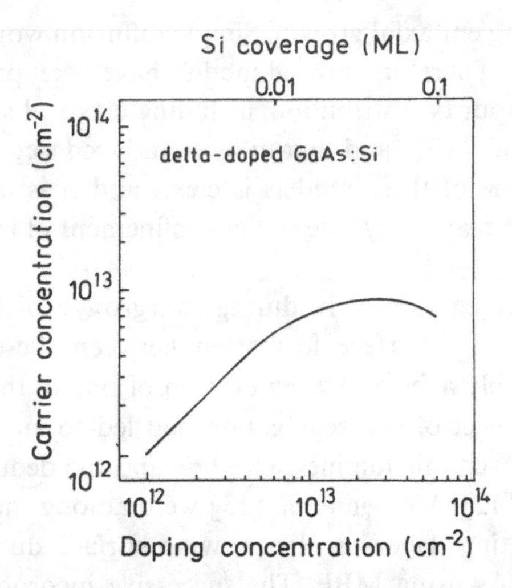

Fig. 3.1. Sheet carrier concentration as a function of Si-delta-doping concentration measured at room temperature. Continuous impurity flux used for delta-doping.

saturates below 10^{13} cm^{-2} for GaAs:Si [6] (Fig. 3.1), and that this saturation-induced limit can be extended to slightly higher values through intentional broadening of the dopant profile [7]. (It should be noted that, due to the multiple subband conduction in delta-doped semiconductors, the Hall effect is weighted towards the contribution of the higher mobility subbands, and hence does not give an accurate measure of the total carrier density per dopant sheet in the sample.) In the prototype example of Si-delta-doping of GaAs, measurements of the localized vibrational modes (LVM) by infrared (IR) absorption [8] and Raman scattering [9], combined with SIMS and electrical characterization, have revealed that several factors contribute to the observed saturation behavior with increasing Si content: (i) decrease in Si_{Ga} donors, (ii) compensation by Si_{As} acceptors or Si–X complexes with acceptor behavior, and (iii) Si_{Ga}–Si_{As} nearest neighbor pair formation [6]. In addition, the influence of DX centers on the electrical properties has to be taken into account. Although GaAs:Si is the most thoroughly studied delta-doping system, the dependence of all these factors contributing to the saturation upon the growth conditions is at present still not well understood.

The actual spatial width of the impurity distribution in delta-doping is determined by segregation effects [3, 5]. Using MBE, low growth temperatures (about 400 °C for GaAs:Si) and low impurity concentrations (about 0.001 monolayer (ML)) are required to approach the ideal situation of confinement to one lattice plane. However, the restriction to rather low growth temperatures leads to the introduction of significant concentrations of native defects in the host lattice, which affect the electronic properties adversely, and for many applications in devices, as well as in fundamental research, dopant concentrations in the range of 0.01 ML or more are desirable. Investigations of the broadening of (001) delta-doping profiles [9] have revealed a preferential migration of impurities toward the semiconductor surface during epitaxial growth. Simple diffusion would result in symmetric (Gaussian-type) broadening. Therefore, several models have been proposed to account for the observed asymmetric impurity distribution, including classical surface segregation [9], solubility-limited segregation [10], and Fermi-level induced segregation [11]. As yet, however, the validity of none of these models is clear, and it is important to note that segregation phenomena also play a key role in the confinement of isoelectronic impurities, as discussed next.

Detailed studies of the segregation of In during overgrowth of 1 ML InAs in a GaAs matrix revealed [12] that the interface formation between these two different III–V semiconductors is considerably affected by segregation of one of the constituent elements on the growing surface. Neglect of this segregation had led to an erroneous relationship between the redshift of the excitonic luminescence line and the deduced width of the InAs region in the GaAs matrix [12]. Moison *et al.* [13] were among the first to show directly the build-up of an In 'floating' layer on the growing surface during overgrowth of an ultrathin InAs layer by GaAs using MBE. The successive incorporation of In from the floating layer into the GaAs overlayer leads to a compositional grading at this 'upper' interface, the extension of which was determined to be 20 Å for continous overgrowth [14, 15]. In the case of pulsed overgrowth, a spreadout of In as far as 300 Å into the GaAs cap layer was found [16].

For a long time this strong segregation has prevented the formation of perfectly confined isoelectronic impurity sheets, not only in the highly strained system of InAs in a GaAs matrix (7.26% lattice mismatch) but also in the very familiar AlAs/GaAs system [17]. Periodic structures composed of regular arrays of nominally 1 ML (001)AlAs interspersed in a GaAs matrix do show distinct two-dimensional (2D) localized vibrational modes [18], pronounced Franz–Keldysh oscillations at low electric fields, and Wannier–Stark localization at high fields [19], and hence reveal the electric-field induced transition from 3D to 2D behavior. However, a more detailed investigation of the actual spatial distribution of the 1 ML AlAs 'barriers' in the GaAs matrix by Raman scattering [19] has indicated a broadening of the AlAs profile over several monolayers due to segregation. Very recently, we found [20] that this segregation of Ga over many lattice planes during growth of AlAs on GaAs (the well-established 'normal' interface) can be probed directly in real time by an additional phase shift in the RHEED intensity oscillations. In the following sections we will show that this segregation of both dopant and isoelectronic impurities can be partially overcome by a special MBE growth mode.

3.3 Laterally ordered incorporation of impurity sheets along step edges

Periodic arrays of ordered steps with a height of one lattice plane or 'monolayer' ($\triangleq a/2$) and flat terraces having a width of $w = a/\sqrt{2} \tan \phi$, where a is the lattice constant, are easily generated on the low-index (001)GaAs surface by a slight misorientation of angle ϕ in the [110] or the [100] direction (Fig. 3.2). Such an ordered step arrangement on vicinal (001)GaAs substrates was first proposed by Petroff *et al.* [21] in order to fabricate

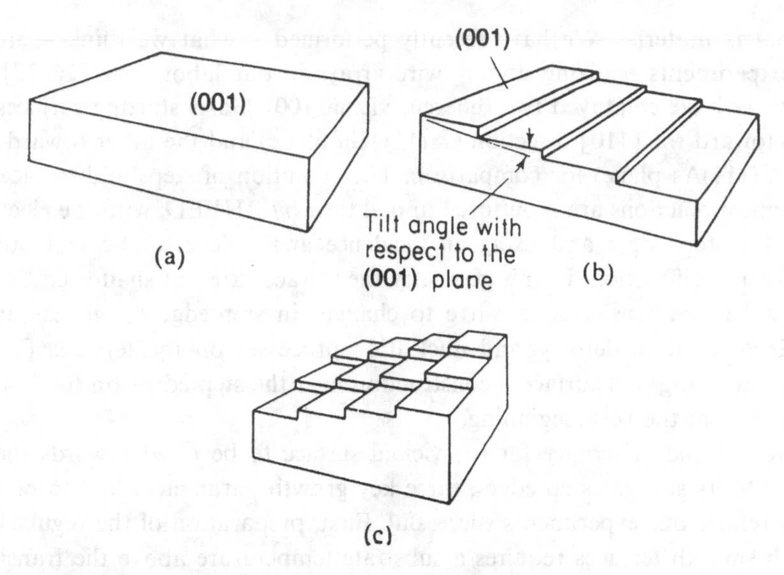

Fig. 3.2. Schematic illustration of generation of ordered steps and flat terraces on (001)GaAs tilted towards the [110] or [100] direction.

quantum wire structures in 'lateral' or 'tilted' GaAs/Al_xGa_{1-x}As superlattices directly by epitaxial growth. However, in a more detailed investigation of the step-flow growth on these vicinal (001) planes, which is essential for the formation of well-ordered lateral superlattices, Horikoshi *et al.* [22] have shown that there is a distinct anisotropy of the growth front for the steps created by misorientation towards [110] or [$\bar{1}$10] due to the given [$\bar{1}$10] orientation of the arsenic dimers on the (2 × 4) reconstructed As-stable surfaces (terraces). (For clarity it should be noted that the steps running along [110], i.e. misorientation in the [$\bar{1}$10] direction, are also called 'As-terminated' steps, and the steps running along [$\bar{1}$10], i.e. misorientation in the [110] direction, are called 'Ga-terminated' steps, according to the atoms forming the edges of the terraces.) Recent scanning tunneling microscopy (STM) data [23] have confirmed that the As-terminated steps on the vicinal (001)GaAs surface are very ragged (although they display better step ordering), whereas the Ga-terminated steps form rather straight ledge profiles. This anisotropy of the growth fronts on the vicinal (001) substrates has made the successful fabrication of well-ordered lateral GaAs/Al_xGa_{1-x}As superlattices extremely difficult [24], and it has to be taken into account in all discussions of ordered incorporation of impurity atoms in the following section.

3.3.1 Fabrication of doping wires in GaAs

Engineering by epitaxy of doping wires in semiconductors [25] is based on the preferential incorporation of the dopant impurities along step edges which requires (i) well-ordered vicinal surfaces with straight ledge profiles, (ii) sufficient migration lengths of the impurity atoms on the terraces, and (iii) negligible distortion of the growth front during overgrowth with the matrix material. We have recently performed – what we think – are the first successful experiments realizing doping wire arrays in our laboratory [26, 27]. In these experiments we have employed two different vicinal (001)GaAs starting surfaces tilted by 2°, the one toward the [110] direction (≙(111)Ga plane) and the other toward the [$\bar{1}$10] direction (≙(1$\bar{1}$1)As plane) for comparison. The evolution of steps and terraces and the surface chemical reactions are monitored in real time by RHEED, with the electron beam parallel to the step edges and using an incidence angle close to the first out-of-phase condition for the diffraction. In this geometry the terraces are not shadowed by steps and the specular beam intensity is sensitive to changes in step-edge roughness and strain-induced effects, adatom density, and nucleation processes on the terraces [28]. Hence, any Si-induced changes in surface reconstruction near the step edges on the lower terrace can be traced from the very beginning.

In addition to the preference for the vicinal surface to be tilted towards the (111)Ga plane owing to its straight step edges, three key growth parameters had to be optimized in order to render our experiments successful. First, preparation of the regularly stepped surface with smooth terraces requires a substrate temperature above the transition from a 2D nucleation and step propagation mechanism to the desired step flow mechanism at T_{crit}, which can be monitored by the change from an oscillating to a constant RHEED intensity [29]. The value of T_{crit} ranges from 500 to 630 °C and depends on the angle and orientation of the misorientation [26]. Second, the terraces on the vicinal surface have to be carefully smoothed by successive deposition of several ML GaAs above T_{crit} with growth interruptions of several minutes and long-time annealing so that the number of undesired holes and islands is minimized. The existence of well-ordered step arrays is evidenced by the clear splitting of the RHEED streaks into inclined slashes [28]. Third, the Si is deposited above T_{crit} and is supplied *not* continuously but in pulses of a certain duration interrupted by two times longer annealing periods. This interruption allows the Si atoms to migrate onto the terraces and arrange themselves along the step edges. Finally, overgrowth of the ordered Si distribution is performed at a temperature below T_{crit} in order to minimize adverse effects on the morphology of the growth front.

The results of our experiments are summarized in Figs. 3.3–3.5 [26, 27]. In Fig. 3.3 we show the behavior of the RHEED intensity during Si deposition in pulses of 90 s duration and 180 s interruption using a flux of $1 \times 10^{11}\ \mathrm{cm^{-2}\,s^{-1}}$ ($\approx 1.7 \times 10^{-4}\ \mathrm{ML\,s^{-1}}$) and a schematic illustration of the Si incorporation along a step edge. The RHEED intensity (Fig. 3.3(a)) decreases non-linearly with the Si coverage and exhibits a certain recovery when the Si flux is interrupted. When reaching a coverage of 0.041 ML the intensity increases drastically and at the same time a (3 × 2) reconstruction develops, as evidenced by the evolution of half-order spots in the [$\bar{1}$10] azimuth and of third-order spots in the [110] azimuth. These findings are consistent with the model of the Si covered vicinal surface shown in Fig. 3.3(b) and Fig. 3.4. The twofold periodicity in the [110] direction is due to the formation of areas with dimerized Si atoms on Ga sites (at the given growth conditions Si is incorporated only as a donor, as confirmed by independent electrical measurements on reference samples, even for concentrations as high as 2×10^{13} atoms $\mathrm{cm^{-2}}$). The threefold periodicity in the [$\bar{1}$10] direction is produced by an ordered array of Si dimers

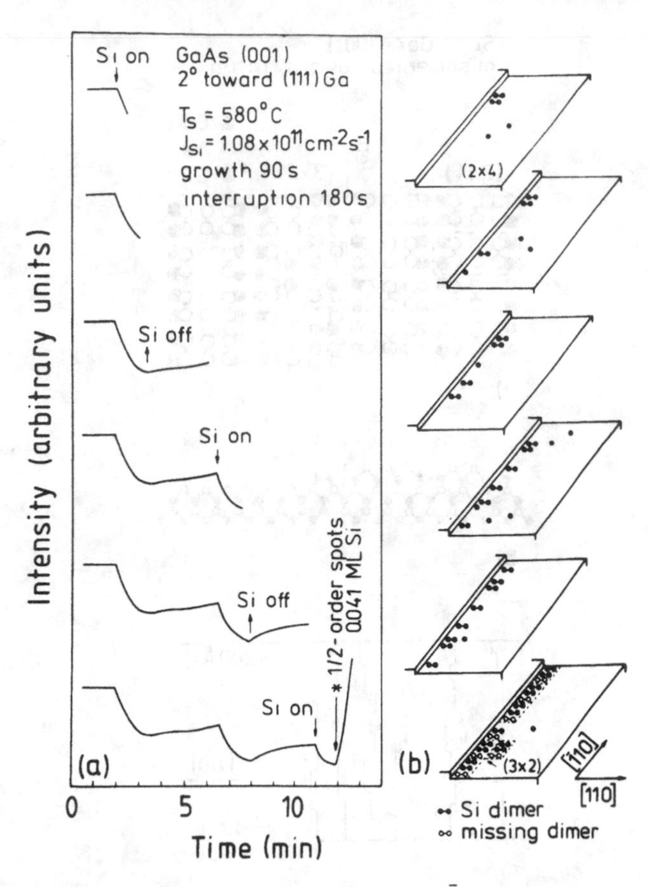

Fig. 3.3. (a) RHEED intensity recorded in the [$\bar{1}$10] azimuth during interrupted Si deposition at 580 °C on vicinal (001)GaAs tilted by 2° towards (111)Ga and (b) model for incorporation of Si as (3 × 2) units along the step edge.

in (3 × 2) units which consist of two Si dimers and one missing dimer per unit mesh. The fractional-order spots in the RHEED pattern thus appear after a certain completion of (3 × 2) units along the step edge and the attachment of (3 × 2) units in the second row. In the case of ideal wire-like incorporation of Si dimers the first stripe of (3 × 2) units should be completed at a Si coverage of $\frac{4}{3}\theta_{crit}(\hat{=}0.066$ ML). In the experiments [27], a lower value of 0.041 ML was found. This discrepancy can be explained by the neglect of any step-edge roughness in the ideal model. In addition, in the actual experiments a second stripe of (3 × 2) units started to develop before the first one was completed. It should be mentioned here that a single Si layer forming an ideal (3 × 2) structure is completed after deposition of 0.67 ML Si (with 1 ML equivalent to 6.25×10^{14} atoms cm^{-2}). In reference experiments [27] we have indeed observed the transition from the (3 × 2) to a distorted (1 × 3) structure at coverages of about 0.6 ML Si. This can be explained by the 90° rotation of the dangling bond direction when going from the Ga to the As plane, and it thus indicates the incorporation of Si atoms above a critical concentration in a second layer.

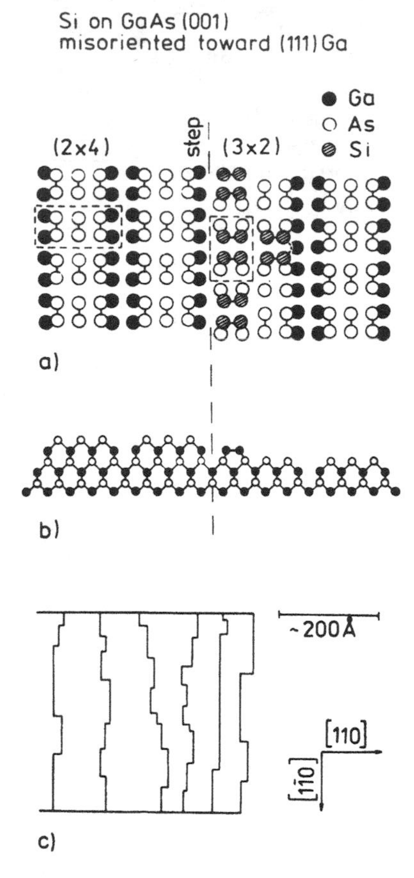

Fig. 3.4. Schematic model of (2 × 4) reconstructed vicinal (001)GaAs misoriented toward (111)Ga with Si atoms attached in (3 × 2) symmetry along the step edge in top (a) and side view (b); (c) is a schematic representation of the step-edge roughness derived from STM studies [23].

Our conclusion of a wire-like incorporation of the Si atoms at step edges is thus based on the following observations (Fig. 3.3(a)). First, the RHEED intensity during Si deposition decreases non-linearly. Second, the Si-induced reconstruction and the final intensity increase appear at a coverage which is correlated with the number of sites at the step edges. The attachment of migrating Si atoms as dimers at the step edges generates kinks. The kink density increases sublinearly with the coverage because with increasing density a rising number of dimers become direct neighbors. This accounts for the decrease in observed sublinear intensity. In addition, strain effects are important. The completion of a stripe of (3 × 2) units along the step edge to a high degree and the attachment of (3 × 2) units in a second stripe leads to the threefold and twofold periodicities observed in the [$\bar{1}$10] and the [110] directions respectively. The abrupt change in reconstruction and strain accompanying this process account for the final intensity increase.

The RHEED intensity behavior during Si deposition becomes very different when the terraces on the vicinal (001)GaAs surface are not smooth or when the terrace width w becomes larger than the Si migration length (i.e. at small misorientation angle ϕ or for Si deposition without flux interruption). In these cases the RHEED intensity decreases linearly, and the intensity minimum and appearance of fractional-order spots are correlated neither with the amount and direction of misorientation nor with the critical coverage. The linear intensity decrease indicates that the Si adatoms form a lattice gas of increasing density. Clustering of the Si atoms on the terraces can then lead to the formation of islands with a (3 × 2) structure, evidenced by the occurrence of fractional-order spots in the respective azimuths.

In principle, a wire-like incorporation of (3 × 2) Si units along step edges would also be possible on vicinal (001) surfaces tilted toward the [$\bar{1}$10] direction (≙(1$\bar{1}$1)As plane), as indicated in Fig. 3.5. However, the As-terminated steps on these surfaces are very ragged and thus render a long-range ordering of the wire-like incorporation of impurity atoms

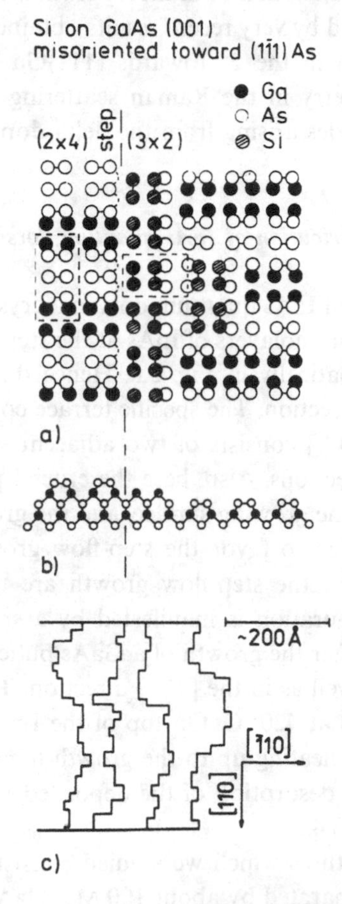

Fig. 3.5. Schematic model of (2 × 4) reconstructed vicinal (001)GaAs misoriented toward (111)As with Si atoms attached in (3 × 2) symmetry along the step edge in top (a) and side view (b); (c) is a schematic representation of the step-edge roughness derived from STM studies [23].

much more difficult. On the other hand, the critical terrace width for wire-like Si incorporation is much larger for misorientation toward $(1\bar{1}1)$As than toward (111)Ga, because the preferential path for Ga, as well as that for Si adatom diffusion, is along the [110] direction, that is, normal to the As-terminated steps for the tilt toward the $(1\bar{1}1)$As plane but parallel to the Ga-terminated steps for the misorientation toward the (111)Ga plane [30].

After the wire-like incorporation of the impurity atoms along step edges, this ordered arrangement has to be stabilized during overgrowth with the GaAs host material when the impurity atoms are rebonded to become part of a 3D solid. Our studies have shown [26, 27] that several tens of lattice planes of GaAs have to be deposited at temperatures below T_{crit} in order that the growth front morphology of the stepped surface is not adversely affected. Under these conditions the Si segregation is also reduced (see also Section 3.4.1).

The conclusion on a self-organization of Si incorporation along step edges drawn from the RHEED studies is confirmed by very recent results obtained by Raman scattering [31]. The wire-like Si incorporation in the 2° towards (111)Ga misoriented sample induces a distinct polarization asymmetry in the Raman scattering intensity of collective inter-subband plasmon–phonon modes arising from the delta-doping layer.

3.3.2 Fabrication of InAs quantum dots in GaAs

The concept of the fabrication of InAs quantum dots in a crystalline GaAs matrix is based on the deposition of fractional monolayers of InAs at the step edges of terraced (001)GaAs surfaces [32], as shown schematically in Fig. 3.6. Hence this approach is similar to the one described in the preceding section. The specific terrace configuration of a vicinal (001) GaAs surface tilted toward [100] consists of two adjacent staircases with steps running along the [110] and $[1\bar{1}0]$ directions. Also, here the crucial point is the preparation of a well-defined step structure on the growing surface, and the growth conditions of the GaAs matrix have thus to be adjusted to favor the step-flow growth mode. Both the terrace distribution on the surface and the step-flow growth are monitored by RHEED. The evolution of the terrace configuration is manifested by a splitting of the reconstruction streaks [28], which develops after the growth of a GaAs buffer at 580 °C with the electron beam aligned in the [110] as well as in the $[\bar{1}10]$ direction. This splitting exists even after the deposition of 0.8 ML InAs at 420 °C. On top of the InAs insertion, three ML GaAs are deposited at 420 °C before heating up to the growth temperature of the GaAs matrix (580 °C) in order to avoid the desorption of the deposited In. For further details on the growth procedure see Section 3.4.2.

The InAs/GaAs heterostructures which we studied consist of ten periods of fractional InAs monolayers which are separated by about 100 ML GaAs (see Fig. 3.6). Our detailed X-ray diffraction experiments [32] have shown that the lattice mismatch of 7.26% between InAs and GaAs is totally accomodated by elastic biaxial compression in the (001) plane. These measurements have also revealed that in samples misoriented by 3.2° and 6.4° toward

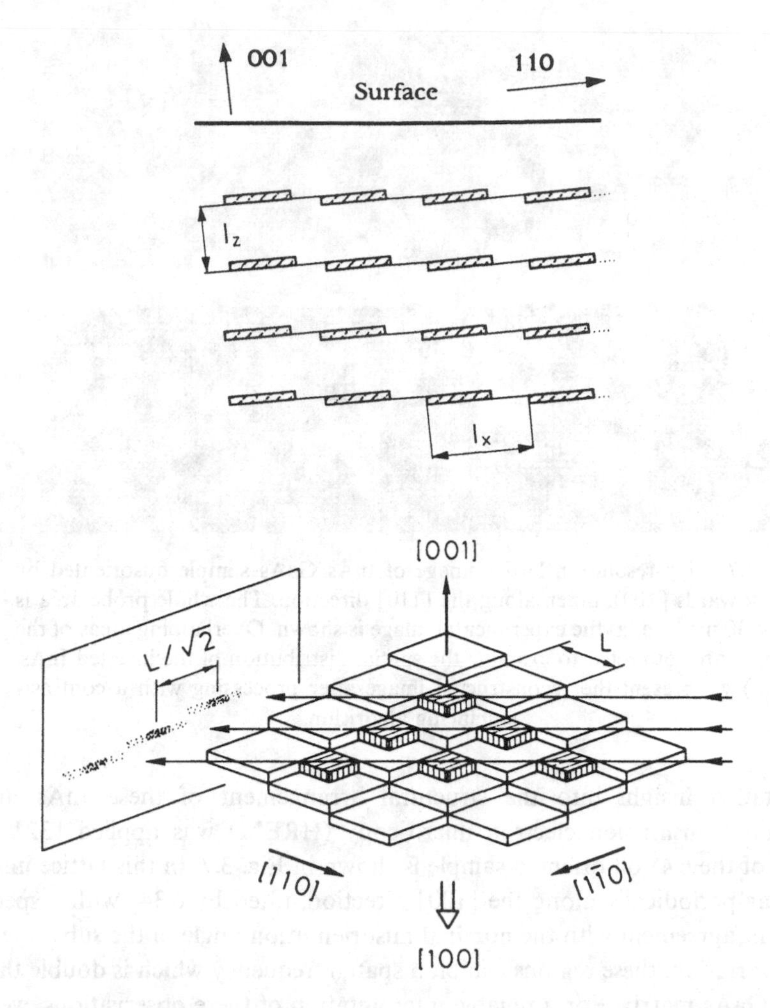

Fig. 3.6. *Lower part*: Sketch of InAs clusters (shaded boxes) grown at the step edges of the terraces formed on the vicinal (001)GaAs surface tilted towards [100]. The long arrows intersecting the clusters represent the electron beam of the transmission electron microscope. The resulting projection of the clusters along the [110] direction is depicted on the left. *Upper part*: Schematic cross-section of the 3D array of InAs quantum dots in GaAs matrix.

[100] the average In coverage of the terraces is 0.3 ML per period. In addition to the vertical periodicity we have also determined the lateral periodicity of these InAs quantum dot arrays by triple crystal X-ray diffractometry [33]. The lateral periodicity of the array along [100] was found to be 110 Å compared to a value of 100 Å deduced from the misorientation angle of the sample. This ideal lateral periodicity extends over about 100 cells of the array corresponding to 1000 Å, which demonstrates the excellent structural quality of these quantum dot arrays fabricated by self-organized epitaxial growth.

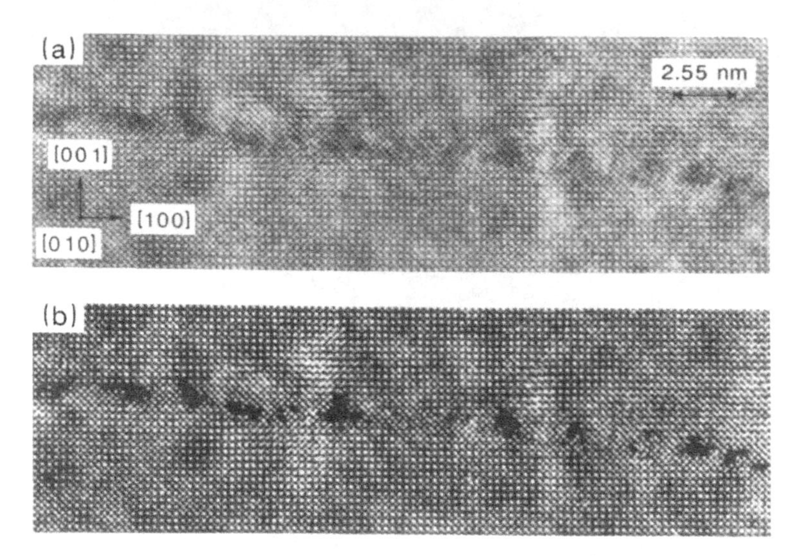

Fig. 3.7. High-resolution lattice image of InAs/GaAs sample misoriented by 6.4° towards [100], taken along the [110] direction. The whole probe area is $10 \times 30\ nm^2$. In (a) the experimental image is shown. Overlapping areas of the sample are connected to examine the spatial distribution of the inserted InAs. In (b) we present the reconstructed image after processing with a contrast-enhancing algorithm.

For a detailed insight into the structural arrangement of these InAs insertions, high-resolution transmission electron microscopy (HREM) was applied [32], and the lattice image of the 6.4° off-oriented sample is shown in Fig. 3.7. In this lattice image dark regions appear periodically along the [100] direction, tilted by 6.34° with respect to the (001) planes, in agreement with the nominal misorientation angle of the substrate. Besides the contrast variation, these regions exhibit a spatial frequency which is double that of the surrounding GaAs matrix. For a reliable interpretation of these observations, we have (i) processed the micrograph using an algorithm which enhances the contrast between the dark regions and the surrounding matrix, and (ii) performed extensive image simulations for a range of defocus values and specimen thicknesses [32]. Note that the special extent of the black regions is systematically larger than the actual physical size of the InAs location. The precise quantitative determination of the actual cluster size is prevented by the inherent fluctuation of the terrace configuration on real surfaces. However, image simulations have shown that the observed cluster size in both cases represents an upper limit for the actual one. The actual dimension in both the vertical and lateral directions is indeed in the subnanometer range. Two important conclusions can be deduced from these lattice images. First, isolated InAs clusters are formed within the crystalline GaAs matrix. The high periodicity of their distribution is clear evidence for the step-edge nucleation of In adatoms. The period extracted from the lattice image is typically 4–5 lattice constants ($\approx$22.7–28.3 Å), in agreement with the mean terrace width of $w = 35$ Å given by the measured tilt angle of 6.34°. Biatomic steps are thus distributed homogeneously along the tilt direction. Second, the interfaces between InAs and GaAs are free from any defects. This finding is important considering the large lattice mismatch between the two materials.

In terms of electronic properties the existence of isolated InAs quantum dots in a crystalline GaAs matrix manifests itself in distinct optical properties. We have found [32] that the InAs insertions realized in these samples represent a genuine 0D system in the sense that the center-of-mass motion of the exciton is localized by the InAs dots. Hence, the spontaneous emission is governed by a decay rate characteristic of atomic dipole transitions, in excellent agreement with our results.

3.4 Ultrahigh-density incorporation of impurity sheets

When the concentration of impurity atoms confined to one lattice plane is increased to the limit of one completed monolayer, not only the nucleation process but also relaxation phenomena and redistribution effects play important roles because of the difference in chemical nature and atomic radius of the impurity species compared to the host lattice. Nucleation, relaxation and redistribution do all depend on the specific growth conditions and they have a strong influence on the structural perfection of the host lattice and on the resulting electronic properties. This illustrates that we are just at the very beginning of understanding and controlling the ultrahigh-density incorporation of impurity sheets in GaAs. On the other hand, in this density range we can more easily employ X-ray diffraction and high-resolution transmission electron microscopy to study the structural properties of the impurity sheets and the surrounding crystalline matrix in detail.

3.4.1 Ultrahigh-density Si doping sheets in GaAs

The incorporation of coherently strained Si monolayers in a crystalline GaAs matrix is not only important for the ultimate limit of Si-delta-doping with carrier densities beyond 5×10^{13} cm^{-2}, it also yields insights into the fundamental problems of heteroepitaxy of Si on GaAs and GaAs on Si where the effects of both lattice mismatch and difference in polarity dominate any possible defect formation in this system. In addition, ideas, such as controlling the Fermi level at GaAs surfaces [34] and tuning the band line-up at GaAs/$Al_xGa_{1-x}As$ interfaces [35], need the realization of commensurate Si/GaAs interfaces.

Based on our recent finding that the Si nucleation on (001)GaAs takes place via the formation of Si nanoclusters in a regular arrangement [36], we have further developed our concept of fabricating doping wires by Si incorporation along ordered step edges of vicinal (001)GaAs surfaces, in order to control the site occupancy of the Si atoms at coverages approaching one monolayer also [37]. In the approach we use (001)GaAs substrate misoriented by 2° towards the (111)Ga plane and the same procedure to smoothen the terrace planes as was described in Section 3.3.1. The most important improvement compared to previous attempts is that the Si is deposited with a periodically interrupted flux of (1–2) × 10^{11} cm^{-2} s^{-1} at a substrate temperature of 590 °C. This allows the Si atoms to migrate to the step edges and to form ordered stripes composed of (3 × 2) units, as monitored by RHEED (see Fig. 3.3). The ordered impurity sheet is then capped by 10–100 nm GaAs using a lower growth temperature of 540 °C in order that the ordered impurity distribution on the terraces does not deteriorate.

The results of conductivity and Hall effect measurements on single delta-doped GaAs:Si layers with 2D impurity concentrations ranging from 1 × 10^{13} to 4 × 10^{14} cm^{-2} are summarized in Fig. 3.8. This concentration range corresponds to a Si coverage of 0.017–0.61 ML when we assume all Si atoms to be confined to one (001)Ga plane having a density of 6.26 × 10^{14} atoms cm^{-2}. The plot of the measured sheet carrier concentration versus incorporated impurities clearly shows that the pulsed delta-doping technique produces much higher free-carrier concentrations than does the continuous delta-doping method. In contrast to the results obtained by other authors [6, 7] with continuous delta-doping at substrate temperatures $400 < T_s < 590$ °C, there is no rapid decrease in free-carrier density beyond 10^{13} Si atoms cm^{-2}. A slight reduction in the fraction of Si atoms incorporated as donors is observed beyond 5 × 10^{13} atoms cm^{-2}. The maximum free-carrier density of 8 × 10^{13} cm^{-2} achieved with 3.8 × 10^{14} Si atoms cm^{-2} is about one order of magnitude higher than that with conventional delta-doping.

Deeper insight into the incorporation mechanism leading to these excellent results is obtained from the concomitant RHEED studies (see Fig. 3.3, 3.4 and 3.9). The specular beam intensity recorded during deposition of 0.61 ML Si (≙4 × 10^{14} atoms cm^{-2}) is depicted in Fig. 3.9 as a function of the Si coverage. Also indicated are the changes of surface reconstruction. For a Si coverage up to 0.12 ML the interpretation of the RHEED data leads to the same conclusion as is given in Section 3.3.1. Hence, the change in the

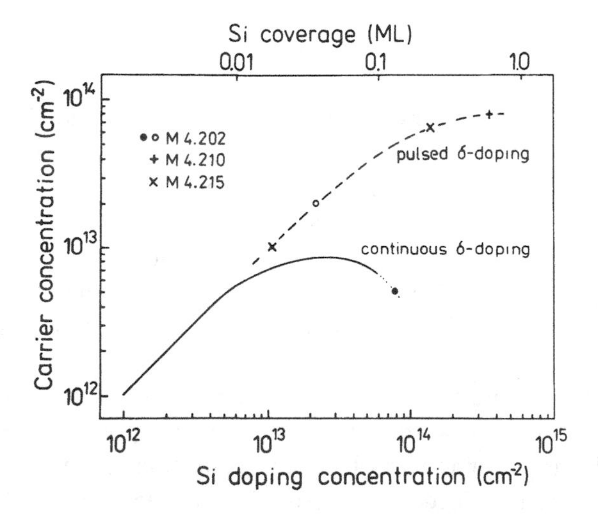

Fig. 3.8. Sheet carrier concentration as a function of the Si-delta-doping for continuous (full line, ref. 7) and for pulsed delta-doping (dashed line, substrate misorientation 2° toward (111)Ga) at 590 °C on (001)GaAs.

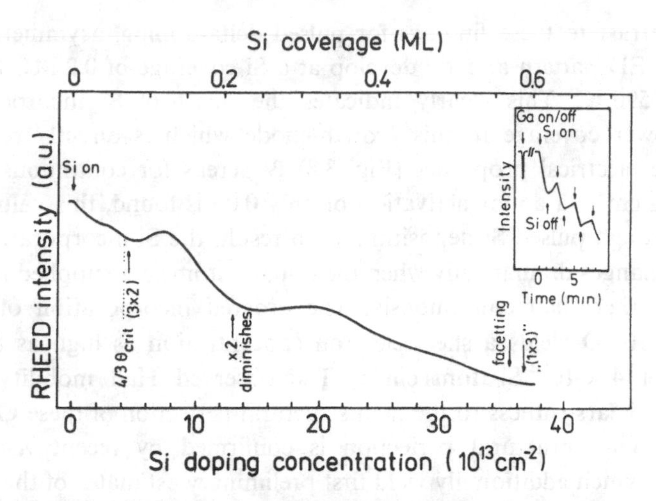

Fig. 3.9. Specular RHEED beam intensity as a function of the Si coverage for interrupted Si deposition at 590 °C on vicinal (001)GaAs misoriented by 2° toward (111)Ga. The growth and interruption times were 60 s and 90 s respectively. The changes in reconstruction are indicated at the corresponding coverages. $\frac{4}{3}\theta_{crit}$ is the coverage at which the first stripe of (3 × 2) units according to the model shown in Fig. 3.4(b) is completed. In the inset a real-time intensity measurement for the first three doping pulses is shown.

dominant reconstruction, as well as the appearance of a (local) intensity minimum after reaching a certain Si coverage and the intensity recovery behavior after growth interruption (inset of Fig. 3.9), indicate the ordering process of the Si atoms on the terraces and their arrangement as dimers in (3 × 2) units along the step edges (Figs. 3.3 and 3.4). The intensity increase after the minimum is due to an abrupt change in strain and in the structure factor after completion of the first stripe of (3 × 2) units. According to the model described in Section 3.3.1 the minimum is expected at a coverage of $\frac{4}{3}\theta_{crit}$, where the critical coverage θ_{crit} is the number of Ga sites at the step edges given by the misorientation. This yields an expected coverage of 0.067 ML ($\triangleq$4.2 × 10^{13} atoms cm^{-2}). In excellent agreement with this model, the intensity minimum and the occurrence of half-order spots arising from the (3 × 2) reconstruction are observed at a coverage of about 0.075 ML Si. Our assumption of confinement of the attached Si atoms in the (001)Ga plane is consistent with the surprisingly high donor activity of 0.8 found in this concentration range, although any possible redistribution of the Si atoms during overgrowth is neglected. When we increase the Si coverage above 0.12 ML ($\triangleq$7.5 × 10^{13} atoms cm^{-2}), the specular beam intensity decreases continuously due to a deterioration of the (3 × 2) structure. A clear decrease in the half-order spot intensity occurs at Si coverages above 0.2 ML. Finally, facetting and third-order spots are observed above 0.6 ML Si, which indicate the incorporation of larger quantities of Si atoms in a second layer, that is, in the As plane. The 90° rotation of the dangling bond direction when going from the Ga to the As plane accounts for the transition from the (3 × 2) to a distorted (1 × 3) structure.

In marked contrast to these findings for pulsed delta-doping, asymmetric third-order spots in the RHEED pattern already develop at a Si coverage of 0.2 ML for continuous delta-doping at 590 °C. This clearly indicates the onset of Si incorporation on As sites at much lower coverage in this growth mode which is directly reflected by the differences in the electrical properties (Fig. 3.8). Whereas for continuous deposition of 8×10^{13} Si atoms cm^{-2} a donor activation of only 0.06 is found, this value increases by a factor of 10 through pulsed Si deposition. As a result, the Si incorporation mechanism in delta-doping changes dramatically when the dopant atoms are supplied in pulses to the growth surface rather than continuously. The ordered incorporation of Si at 590 °C monitored by RHEED yields a sheet electron concentration as high as 8×10^{13} cm^{-2} with a supply of 4×10^{14} Si atoms cm^{-2}. The observed Hall mobility in excess of 1000 cm^2 V^{-1} s^{-1} bears witness to the high structural perfection of these GaAs–Si–GaAs heterostructures. This structural perfection is confirmed by recent X-ray diffraction experiments [38], which additionally yield first preliminary estimates of the broadening of the impurity profile in the range 120–150 Å during overgrowth due to segregation. More detailed work on the actual width of the impurity profile when approaching 1 ML occupancy is underway in our laboratory.

Before we developed the described technique of pulsed delta-doping we had studied nucleation, relaxation, and redistribution of (001) Si layers of different thickness (1–13 Å) in a GaAs matrix by X-ray diffraction and HREM [36]. These studies also clearly revealed that with continuous Si supply the nucleation of Si on (001)GaAs takes place via the formation of Si nanoclusters in a highly regular arrangement, as shown in Figs. 3.10 and 3.11. In order to analyze the high-resolution lattice images taken from [110]-oriented cross-sections of these GaAs/Si/GaAs heterostructures, we had to develop a new technique [39]. In brief, the experimental lattice images are Fourier-filtered and processed by an algorithm which maximizes the local contrast variations in the image. An ideal GaAs lattice without the Si layer, generated from a portion of the experimental lattice image, is then extrapolated over the entire experimental lattice image, and the absolute difference in atom positions is determined and represented by a 2D vector field. The strain pattern generated in this way directly visualizes the internal lattice distortion induced by the strained Si insertion [39].

In the lattice images of Figs. 3.10 and 3.11 taken from GaAs samples containing a 1-Å and a 13-Å-thick (001) Si layer, the experimental imaging conditions are chosen such that a clear contrast between Si and GaAs is observed. It should be noted that the average value of the actual thickness of the Si impurity sheet is determined by X-ray diffraction [40], with an experimental uncertainty of about 5% (see below). The characteristic features associated with the presence of epitaxial Si are the single-spaced (bright spots) and the half-spaced lattice fringes shown in Figs. 3.10 and 3.11, respectively. The regions exhibiting bright spots in Fig. 3.10(a) appear periodically along the layer plane with a lateral spacing of ten $[1\bar{1}0]$ lattice constants ($\triangleq 40$ Å). In between, the lattice image is that of GaAs. In Fig. 3.11(a) a clear half-spacing is observed only on the left side, while on the right side a GaAs-like lattice image appears. The atomic displacement vectors in the corresponding strain patterns depicted in Figs. 3.10(b) and 3.11(b) directly represent the lattice distortion

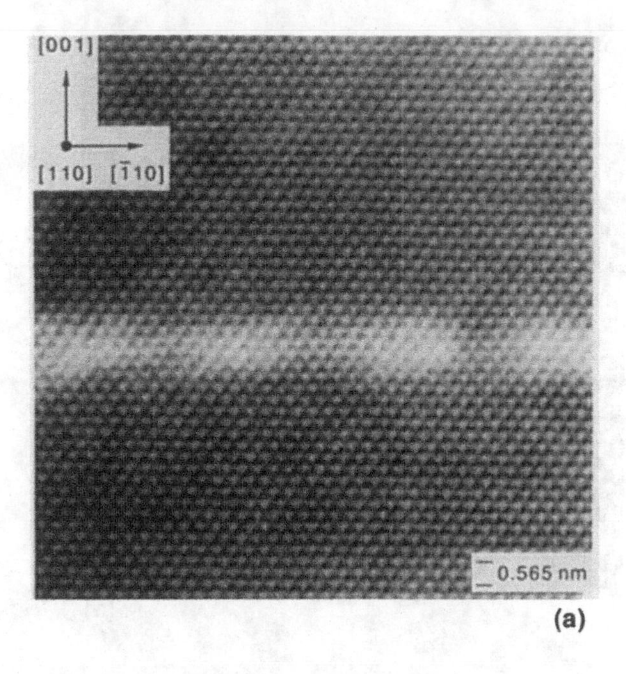

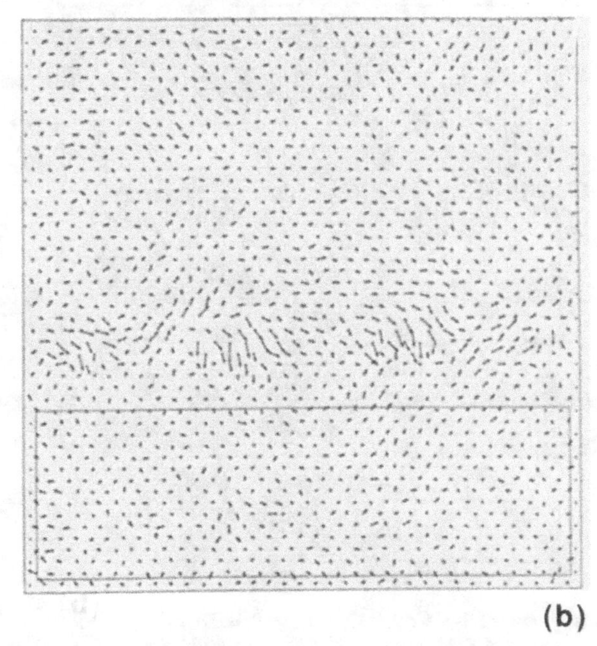

Fig. 3.10. Experimental lattice image (a) and strain pattern (b) of a sample containing a 1-Å-thick Si film in GaAs. Note the coincidence between the Si-related contrast (bright spots) apparent in (a) with the local lattice distortion in (b). (The rectangle in (b) surrounds the area used for the generation of the ideal GaAs lattice.)

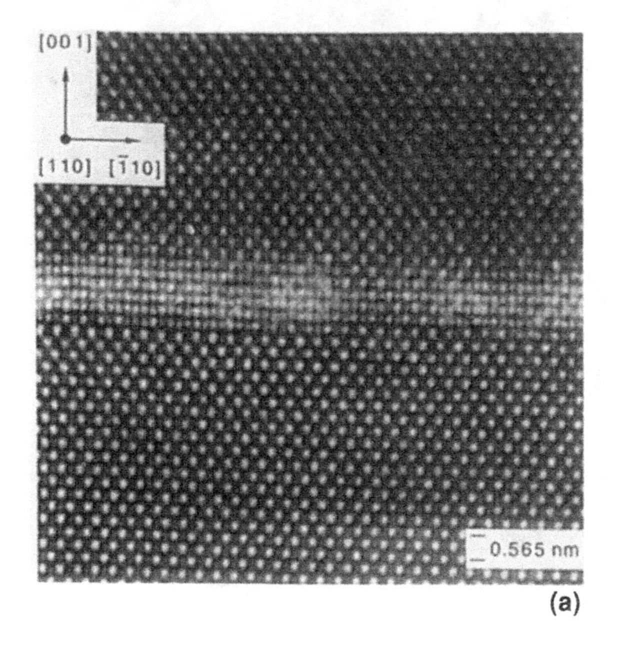

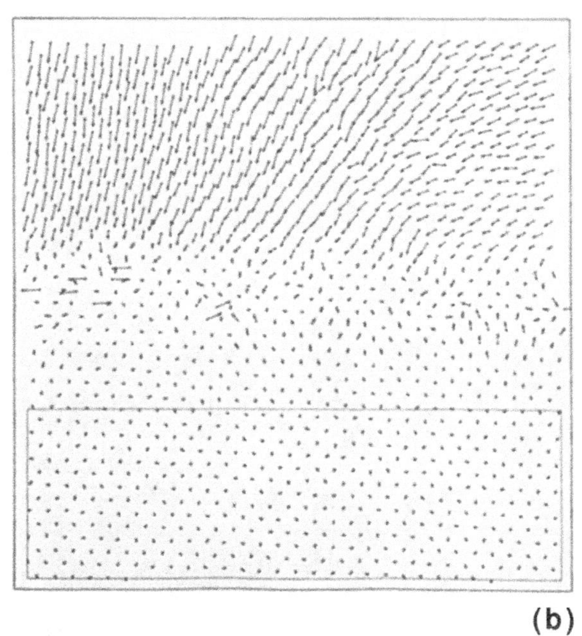

Fig. 3.11. Experimental lattice image (a) and strain pattern (b) of a sample containing a 13-Å-thick Si film in GaAs. Note the coincidence between the Si-related contrast (half-spaced lattice fringes) apparent in (a) with the resulting shift of the atom positions in the GaAs cap layer in (b). (The rectangle in (b) surrounds the area used for the generation of the ideal GaAs lattice.)

induced by the strained Si insertion [39]. In the case of the 1-Å-thick Si film a local lattice distortion coincident with the Si-related contrast is observed. Thus, Fig. 3.10(b) shows that in the initial stage of growth Si nucleates via the formation of nanoclusters, whose interfaces remain in alignment with the surrounding GaAs matrix. We note that this observation confirms a recent speculation by Tanino *et al.* [41], who attributed the anomalous shape and spectral position of Raman resonances from ultrathin Si layers to 3D phonon confinement. This nucleation process may be related to the incorporation behavior of Si along step edges, as discussed in Section 3.3.1. Although the (001)GaAs substrates used here are not tilted, there are always steps and these steps provide nucleation sites for Si atoms. With successive growth the clusters are expected to coalesce to finally form a continuous film. In fact, the strain pattern of the 13-Å-thick Si film (Fig. 3.11(b)) shows that the film is continuous, as the Si film here leads to a global net shift of the GaAs lattice planes above the film. However, the magnitude of this shift varies laterally, being highest where the Si-related contrast (half-spacing) is observed. This finding shows that the 13-Å-thick Si film is partially intermixed with GaAs, thus forming a $(Si_2)_x(GaAs)_{1-x}$ alloy [42] with laterally fluctuating Si content.

In Fig. 3.12 we show the lattice image of the sample containing the 13-Å-thick Si film on a larger scale. The laterally fluctuating contrast is also apparent in this picture. The fluctuation is far less regular than the one associated with the cluster nucleation (Fig. 3.10(a)), and the characteristic length scale is about 100 Å instead of 40 Å. Both these observations indicate that the intermixing phenomenon is independent of the nucleation process. Indeed, as will be shown below, this intermixing rather stems from a redistribution of the epitaxial Si upon overgrowth. Another interesting observation in Fig. 3.12 is the

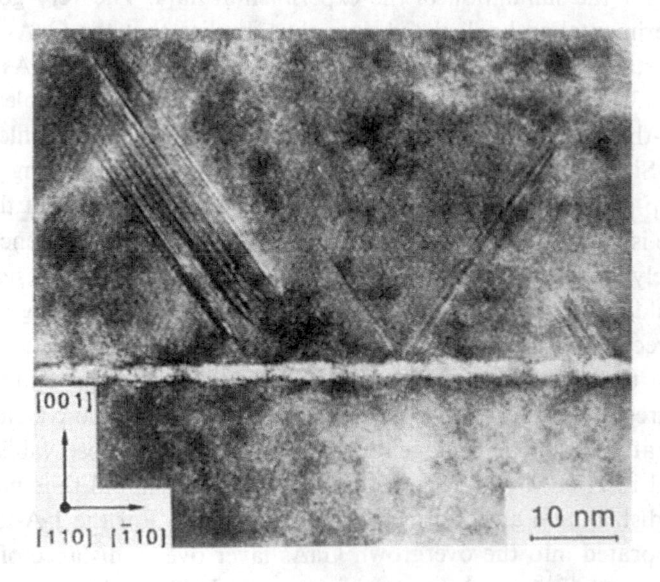

Fig. 3.12. Experimental lattice image of the sample of Fig. 3.11 taken from a larger area. Note the contrast fluctuation within the layer and the appearance of stacking faults extending from the Si layer.

appearence of stacking faults in the GaAs cap layer. A Burger vector analysis of the lattice image in the direct vicinity of the stacking faults arising from the epitaxial film shows that they partially relieve the strain of the Si layer. Complete dislocations, on the other hand, are not observed. The formation of partial dislocations associated with the presence of twins and stacking faults is an energetically favorable strain relief path for tensile strained films in general.

At this point we should briefly introduce the method for determining the actual thickness of the Si layer inserted in the GaAs matrix with high accuracy by X-ray diffraction [40]. The method is based on the interference of X-ray wave fields within the layered crystals producing distinct interference patterns around the Bragg reflections. In Fig. 3.13(a) we show the calculated diffraction patterns using different (001)Si layer thicknesses t_i in the GaAs matrix ranging from 1 to 6 ML (ML = 1.315 Å) and keeping the GaAs cap layer thickness constant at $t_c = 2000$ Å. The important features in the calculated spectra are: (i) the amplitude of the interference peaks depends strongly on the thickness t_i; (ii) the interference peak positions shift with respect to the GaAs substrate peak as a function of t_i; (iii) the diffraction patterns are asymmetric with respect to the GaAs peak. These features permit high accuracy (better than 0.2 Å) determination by X-ray diffraction of the interface, the layer thickness and the crystalline quality of the capping layer (which is related to the structural properties of the interface layer). In Fig. 3.13(b) some experimental and simulated diffraction patterns of different GaAs/Si/GaAs heterostructures are shown. The values obtained for the Si layer thickness t_i are average values, that is, the Si layer thickness is not necessarily perfectly uniform, but may fluctuate slightly around an average value. It should be noted, however, that no smoothing or disorder parameter (static Debye–Waller factor) was used for the simulation of the experimental data. The very good agreement between the experimental and calculated data thus implies that the GaAs capping layer is of excellent crystalline quality, suggesting a sharp and coherent Si/GaAs interface.

Finally, in Fig. 3.14 we present the SIMS depth profiles of two samples containing a 1-Å- and a 7-Å-thick Si film. For comparison, the SIMS depth profile of a heavily (7×10^{13} cm^{-2}) Si delta-doped sample is added which was grown using continuous Si supply. The sharp spike at z_0 corresponds to the Si sheet. The shape of these spikes on the substrate side is identical to that obtained from the delta-doped reference sample, thus being limited solely by the instrumental depth resolution. On the surface side, in contrast, pronounced shoulders are resolved in the Si depth distribution. Silicon migration thus only occurs in the direction of growth, which demonstrates that the migration is caused by segregation rather than by diffusion. We recall that the dilute incorporation of Si in GaAs caused by the segregation process cannot be detected by HREM, where an actual Si/GaAs alloy (Si concentration in the percentage range) is required for an observable contrast. The contrast observed in HREM (see Figs. 3.10(a) and 3.11(a)) should thus not be confused with the dopant distribution as performed by Liu *et al.* [43]. For the 1-Å-thick Si film, Si atoms are incorporated into the overgrown GaAs layer over a distance of 400 Å with a concentration of about 10^{19} cm^{-3}. In a previous study [44] we showed that this top-hat profile results from a Fermi-level controlled segregation process, where the incorporated fraction of the segregating Si atoms satisfies the solubility limit of Si in GaAs. In contrast,

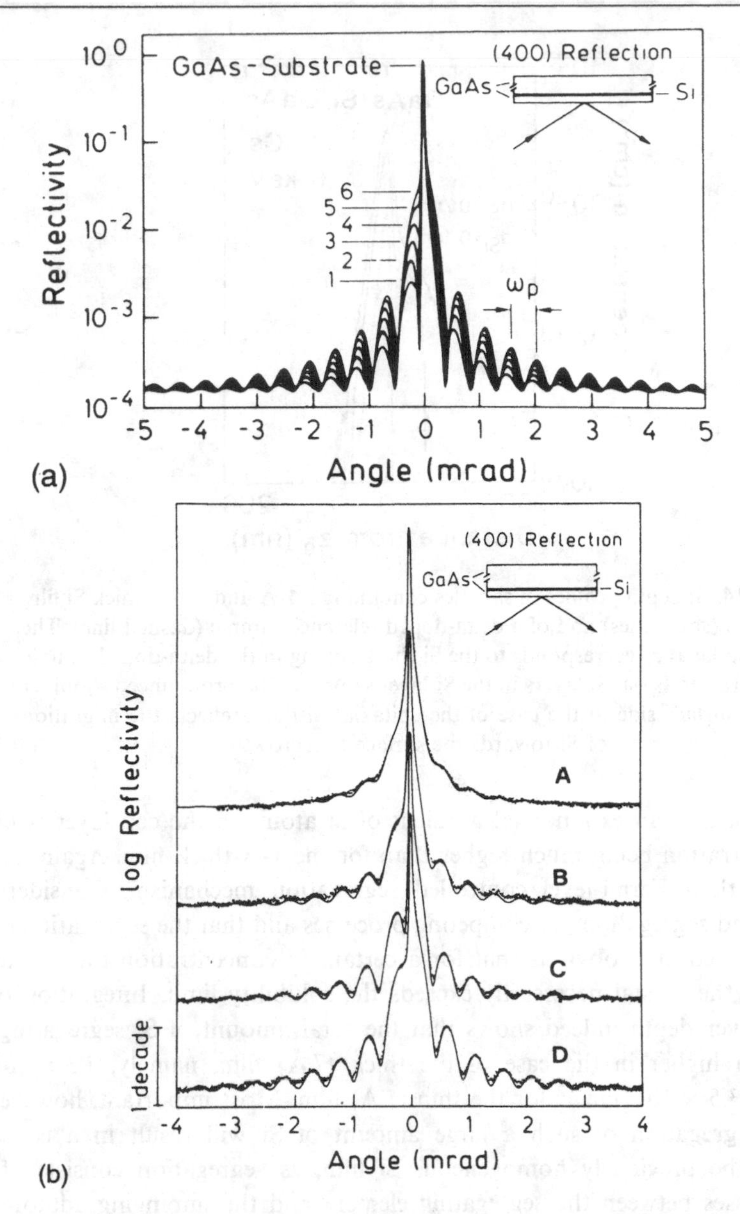

Fig. 3.13. (a) Calculated X-ray diffraction patterns in the vicinity of the (004)GaAs reflection ($\lambda = 1.54$ Å). The GaAs cap layer is for all patterns $t_c = 2000$ Å. The thickness of the Si interface layer ranges from 1 to 6 monolayers. In these calculations coherent GaAs/Si and Si/GaAs interfaces are assumed. (b) Experimental (dotted lines) and simulated (solid lines) double-crystal X-ray diffraction patterns in the vicinity of the symmetric (004)GaAs reflection. The GaAs cap layer thickness is between 1950 and 2000 Å for all samples. The Si interface layer thickness t_i is 0.05, 0.10, 0.20, and 0.65 nm for the curves A, B, C, and D respectively. The well-pronounced interference fringes indicate the excellent structural quality of the whole GaAs/Si/GaAs heterostructure.

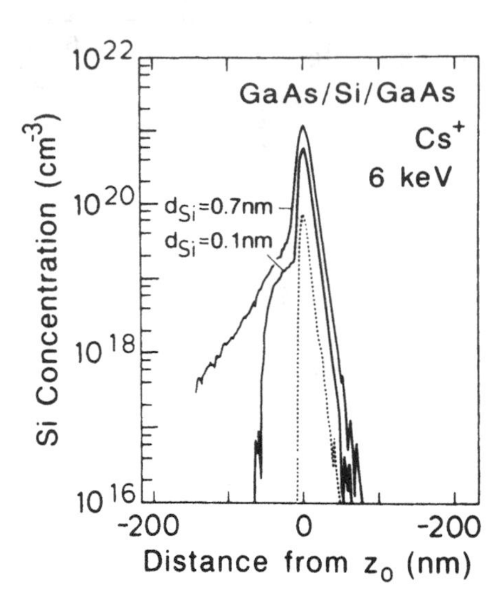

Fig. 3.14. Si depth profiles of samples containing a 1-Å and a 7-Å-thick Si film in GaAs (solid lines) and of a delta-doped reference sample (dashed line). The sharp spike at z_0 corresponds to the Si sheet doping in the delta-doped sample and to the epitaxial Si layers in the Si-layer samples. The pronounced shoulder on the surface side in the case of the epitaxial Si films reflects the migration of Si towards the surface overgrowth.

for the 7-Å-thick film an exponential gradient of Si atoms in the cap layer is observed, with the concentration being much higher than for the 1-Å-thick film. Again, this result is consistent with a Fermi-level controlled segregation mechanism. Considering that incorporation and segregation are competing processes and that the segregation efficiency is kinetically limited, it is obvious that for a certain Si concentration the amount of Si incorporated in the crystal necessarily exceeds the solubility limit. Integration of the Si concentration over depth indeed shows that the total amount of Si segregating on the surface is much higher in the case of the thick (7 Å) film, namely, $1.3 \times 10^{14}\,\mathrm{cm}^{-2}$ compared with $3.5 \times 10^{13}\,\mathrm{cm}^{-2}$ for the thin (1 Å) film. Most important, however, is the fact that the segregation of such a large amount of Si will result in a considerable intermixing of the previously homogeneous Si film, as segregation consists of atomic exchange processes between the segregating element and the impinging adatoms of the overlayer.

The experiments described in the last paragraphs have revealed that Si nucleation on (001)GaAs occurs via the formation of regularly arranged nanoclusters even on surfaces without regular step arrays. However, continuous Si deposition leads to nucleation of Si in the second (001) layer (i.e. the As layer) before a homogeneous 1 ML Si film has been formed. Then, with increasing thickness the Si clusters coalesce forming a homogeneous Si film. Upon overgrowth of the Si film by GaAs a fraction of the deposited Si segregates on the growth surface. For thin Si films (about one monolayer) the segregating Si atoms are incorporated such as to satisfy the solubility limit of Si in GaAs. For thicker films (several monolayers), however, a much higher fraction of Si segregates, and the amount of incorporated Si exceeds the solubility limit due to the kinetic limitations of the segregation process. The atomic exchange reactions associated with segregation lead to a strong intermixing of these thicker Si films, as observed in the lattice image. Silicon films thicker than 10 Å are found to relieve strain by the generation of extensive stacking faults rather than by complete dislocations. Since the redistribution as well as the segregation of Si in the GaAs matrix depend strongly on the distinct growth conditions, additional studies of these phenomena are necessary.

3.4.2 InAs monolayers in crystalline GaAs matrix

The InAs/GaAs materials system is a particularly challenging candidate for the study of heteroepitaxial growth owing to the large (7.26%) lattice mismatch between the constituents and the absence of any polarity problem as for the Si/GaAs system. However, during overgrowth of InAs with GaAs a considerable fraction of the In segregates forming a floating In layer on the growth surface which is gradually dissolved in the overgrowing GaAs, thus preventing the formation of a well-defined abrupt InAs/GaAs interface. The formation of GaAs/InAs and InAs/GaAs interfaces is therefore governed by a unique balance of surface and interface energy as well as by kinetic processes. We have recently developed specific strategies to control the formation of these interfaces individually using conventional elemental-source MBE [12]. The key parameters are a sequence of growth interruptions and modulations of the substrate temperature during interface build-up. In particular, the temperature modulation results in a complete flash-off of the In floating layer prior to continuous GaAs overgrowth. The interfaces of the strained InAs insertions are then in alignment with the surrounding GaAs matrix and hence are free of defects, as determined by high-resolution X-ray diffraction and transmission electron microscopy.

In Fig. 3.15 we illustrate a scheme for the growth procedure of 1 ML InAs in a GaAs matrix. The substrate temperatures are referred to the evaporation of the native oxide from the (001)GaAs substrates at 580 °C and to the transition from the As- to the Ga-stabilized reconstruction at 630 °C, both monitored by RHEED. Prior to deposition of the InAs/GaAs heterostructure, a 1–2 μm GaAs buffer layer is grown at 580 °C (not shown in the diagram) in order to improve the surface smoothness. The GaAs matrix embedding the InAs insertions is grown at 540 °C. The temperature is then lowered to 420 °C for InAs deposition. The InAs layer is deposited in increments of 0.6 ML interrupted by annealing the growth surface at 420°C under As_4 for 120 s. Then, before heating up again to 540 °C for GaAs overgrowth, 3 ML GaAs are deposited at 420 °C on top of the InAs layer in order to prevent desorption of the deposited In. The thickness of this covering GaAs layer is an essential parameter for the perfection of the structure. Growth is then interrupted in order to raise the substrate temperature to 540 °C before GaAs growth is continued. The effective deposition rates of InAs are accurately determined after growth by high-resolution X-ray diffraction (XRD) and are compared with the independently calibrated growth rates of bulk InAs layers.

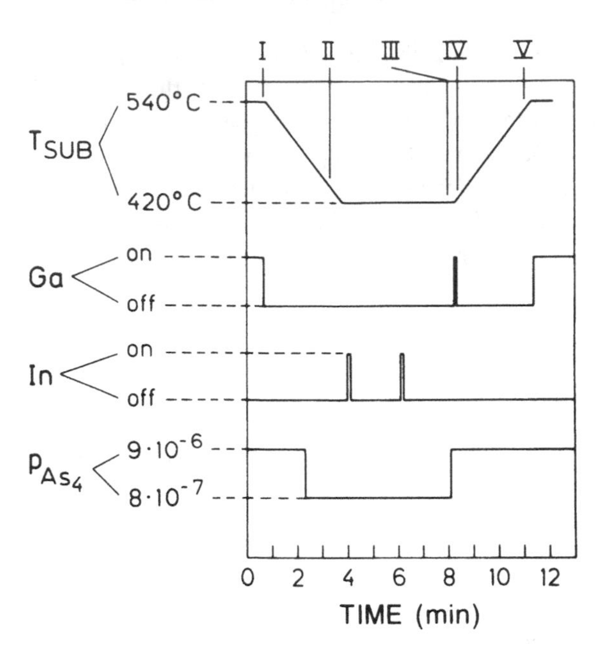

Fig. 3.15. Schematic diagram of the temperature and flux switching during deposition of 1.2 ML InAs on GaAs and subsequent overgrowth with GaAs. Roman numbers refer to RHEED patterns taken during growth which are displayed in Fig. 3.16.

The RHEED patterns taken during growth of 1.2 ML InAs in Fig. 3.15 are shown in Fig. 3.16. When the temperature is lowered for InAs deposition, the original (2 × 4) reconstruction of GaAs (I) changes to the $c(4 \times 4)$ reconstruction (II) arising from excess As_4 adsorbed on the surface. A (2 × 3) reconstruction with diffuse half-order streaks appears after opening the In shutter for the first 0.6 ML. After the second 0.6 ML InAs the reconstruction streaks first disappear, but evolve again during annealing (III). This temporal evolution of the reconstruction reflects the dynamic recovery of a smooth 2D growth front, that is, a well-ordered surface, as long as the thickness of the strained InAs layer is below the critical value of 3.6 ML [12]. When the Ga shutter is opened, the reconstruction disappears and we observe a (1 × 1) pattern (IV). Finally, during subsequent growth interruption and heating, a well-defined (2 × 4) γ-reconstruction appears again at 520 °C (V). The occurrence of a (1 × 1) pattern in the first stage of GaAs overgrowth (IV) is in contrast to the $c(4 \times 4)$ reconstruction of the static (non-growing) GaAs surface at this temperature and As_4 flux. However, the observed long (1 × 1) diffraction streaks indicate that no islanding of GaAs on InAs occurs. The low growth temperature of 420 °C and the high deposition rate of 1 ML GaAs per second obviously impose kinetic limitations on the film morphology, thus preventing relaxation to the equilibrium configuration (agglomerated morphology). Instead, the absence of any reconstruction indicates the existence of surface-active species on the growth surface, that is, a segregated In floating layer. The recovery of the (2 × 4) reconstruction

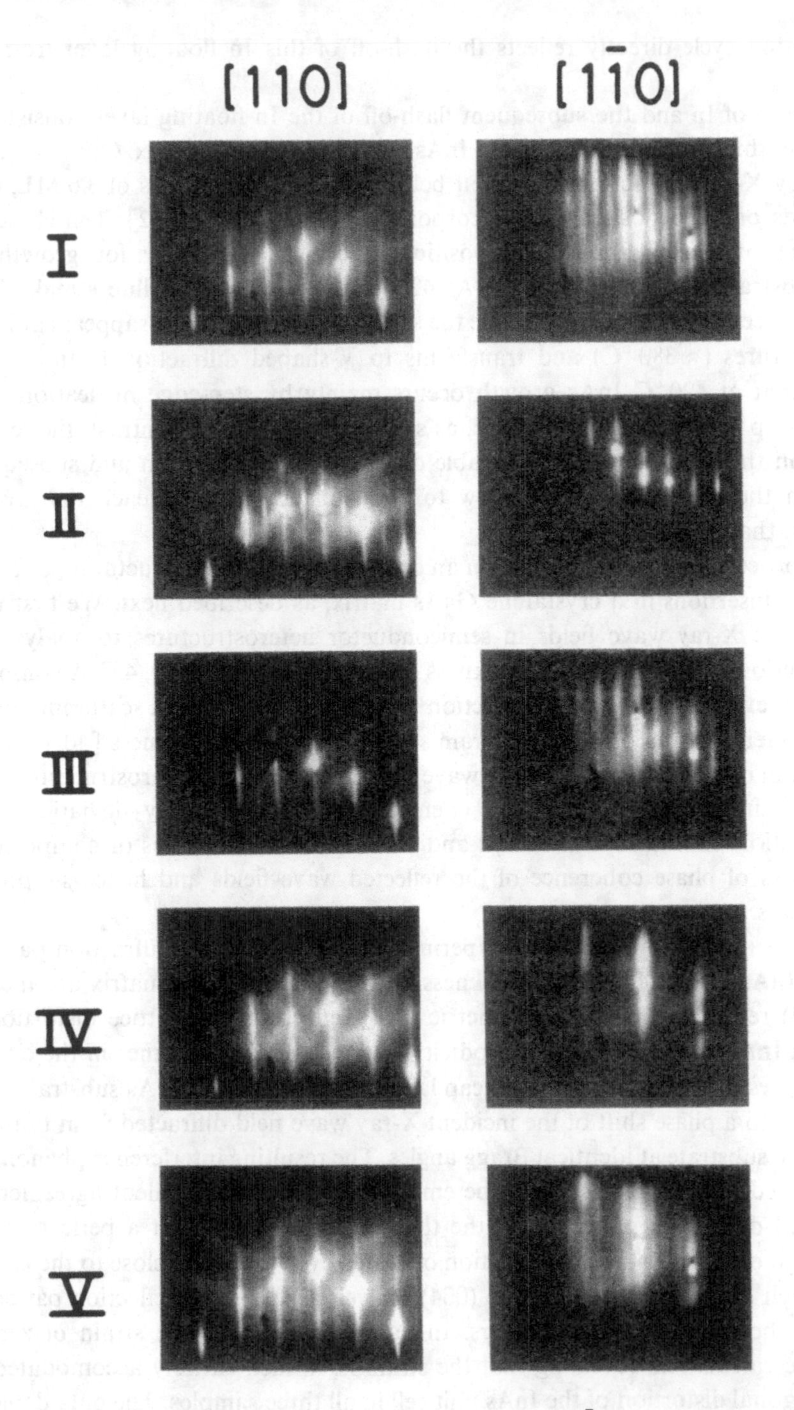

Fig. 3.16. RHEED patterns taken along [110] and [1$\bar{1}$0] during the deposition of 1.2 ML InAs on GaAs and subsequent overgrowth with GaAs. Roman numbers refer to Fig. 3.15. For details see text.

during the heating cycle directly reflects the flash-off of this In floating layer from the surface.

The segregation of In and the subsequent flash-off of the In floating layer consistently reduce the actual thickness of the embedded InAs layer in the GaAs matrix. Our systematic investigations by X-ray diffraction show that below the critical thickness of 3.6 ML, only about two-thirds of the deposited In is incorporated on the surface [12]. The choice of the temperature of 420 °C for InAs deposition is critical, especially for growth on misoriented substrates (see Section 3.3.2). At 420 °C the RHEED satellite streaks from vicinal (001) surfaces are well resolved, while the splitting of the streaks disappears at lower growth temperatures ($\approx$380 °C) and transforms to V-shaped diffraction features. This clearly shows that at 420 °C InAs growth occurs mainly by step-edge nucleation, thus preserving the step structure on the surface, as shown in Fig. 3.6. In contrast, the terrace configuration on the surface becomes unstable due to terrace nucleation and subsequent islanding, when the temperature is too low to allow In adatoms to reach the favored nucleation sites, that is, the step edges.

The growth procedure described results in an exceptional degree of structural perfection of strained InAs insertions in a crystalline GaAs matrix, as described next. We first used the interference of X-ray wave fields in semiconductor heterostructures to analyze the structural perfection of layers having a thickness of only 1 ML [12, 32, 40, 45]. A computer simulation of the experimental Bragg diffraction pattern using dynamical scattering theory allows precise determination of both the strain state and the layer thickness [40, 45]. The multiple reflection of phase-coherent X-ray wave fields inside perfect heterostructures gives rise to complex interference phenomena (Pendellösung fringes). Any deviation from structural perfection, that is, crystal defects and fluctuations of interfaces or composition, results in the loss of phase coherence of the reflected wave fields and hence suppresses interference effects.

In Fig. 3.17 we show a comparison of experimental and theoretical diffraction patterns of three single InAs layers of different thickness in a crystalline GaAs matrix around the symmetric (004) reflection and the asymmetric (224) reflection. The lattice distortion of the sandwiched InAs layer breaks the periodicity of the (001) lattice planes in the crystal, that is, it decouples the lattice of the GaAs cap layer from that of the GaAs substrate. This decoupling leads to a phase shift of the incident X-ray wave field diffracted from both the cap layer and the substrate at identical Bragg angles. The resulting interference phenomena are directly related to the strain state of the embedded layer. The excellent agreement of the experimental diffraction pattern with the theoretical simulation for a perfect crystal confirms the high degree of interface perfection obtained even for layers close to the critical thickness. In each of the three samples the (004), as well as the (224), reflection pattern is reproduced by the same set of parameters. In particular, an in-plane strain of zero is assumed for the calculation, revealing that the strain is, in fact, entirely accomodated by the elastic tetragonal distortion of the InAs unit cell in all three samples. The data depicted in Fig. 3.17 allow determination of the actual thickness of the embedded InAs layers below the critical thickness (i.e. 2.6 ML). From these experiments it becomes evident that there is a distinct difference of about 30% between the deposited and the actually incorporated

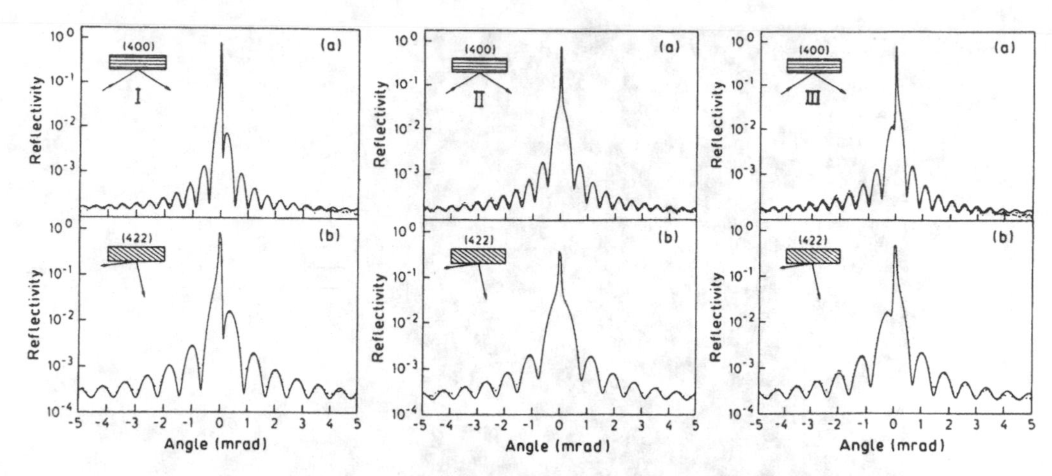

Fig. 3.17. Comparison of experimental (dotted line) and theoretical diffraction patterns obtained from three GaAs samples containing single InAs layers of 0.8 ML (I), 1.6 ML (II), and 2.4 ML (III) thickness in the vicinity of the symmetric (004) and asymmetric (224) reflection.

amount of In which corresponds directly to the fraction of In which is desorbed during the heating cycle, as discussed earlier. A further analysis of the diffraction patterns depicted in Fig. 3.17 has revealed that the GaAs matrix surrounding the InAs insertions is indeed free of In incorporated from the segregation effect [12].

The integral structural information obtained from XRD is complemented by the local probe provided by the focused electron beam used in high-resolution electron microscopy. In order to obtain lattice images free from artefacts, the experimental micrographs are reconstructed by Fourier filtering with a maximum transferred spatial frequency of 34 $Å^{-1}$. Furthermore, extensive image simulations are performed which verify that, for sufficient ranges of defocus value (−200 to −500 Å) and specimen thickness (80–150 Å), the positions of the large dark spots in the reconstructed lattice images correspond directly to the projection of columns of InAs dumbbells. The electron micrographs thus provide accurate quantitative information on the atomic-scale morphology of the interfaces [39]. In Figs. 3.18 and 3.19 we show high-resolution Fourier-filtered lattice images obtained from samples containing 3 ML InAs and 1 ML InAs, respectively, together with the corresponding vector representations of the relative shift in atom position [46]. For 3 ML InAs (Fig. 3.18) the overall shift of the lattice in the [001] direction is determined to be 1.24 ± 0.01 Å. This distortion is shared among four lattice planes (Fig. 3.18(b)), corresponding to an InAs film of 3 ML. The overall strain is distributed in the lattice image according to 20:40:30:10. Assuming a homogeneous distribution of the strain over each of the 3 ML (ideally the strain distribution would be 22:33:33:12) [46], the strain per ML is determined to be $(7.06 \pm 0.1)\%$, close to the value expected from elasticity theory. In the sample with an InAs layer thickness of 1 ML the overall shift of the lattice in the [001] direction is determined to be 0.58 ± 0.01 Å. This distortion originates from 1 ML, as demonstrated by the fact that two displacement vectors with a length ratio of 60:40

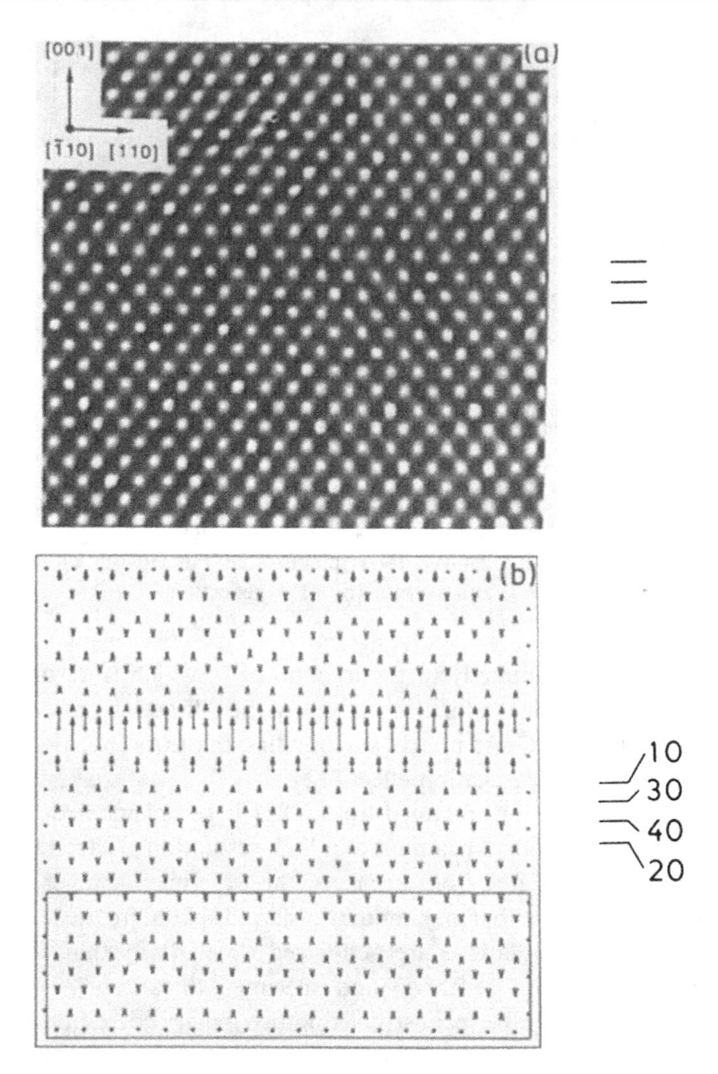

Fig. 3.18. (a) Fourier-filtered experimental lattice image of 3 ML InAs in GaAs. Bars indicate planes containing In. (b) Vector representation of the relative shift in position of each ML with respect to its immediate neighbor. The magnitude of the vectors is amplified by a factor of 5. Numbers give the relative magnitude of the displacement vectors in percentages.

are detected (Fig. 3.19(b)). This corresponds to a strain of $(12.6 \pm 0.3)\%$, significantly larger than that expected from elasticity theory.

The first important result of our detailed HREM studies is that below the critical layer thickness of about 3 ML the interfaces of the strained InAs layers are in alignment with the surrounding GaAs matrix, despite the large lattice mismatch of 7.26%, and hence are free of structural defects. The second important result is [46] that a 3-ML-thick InAs film can already be considered as an elastic continuum, that is, the bulk case establishes quite

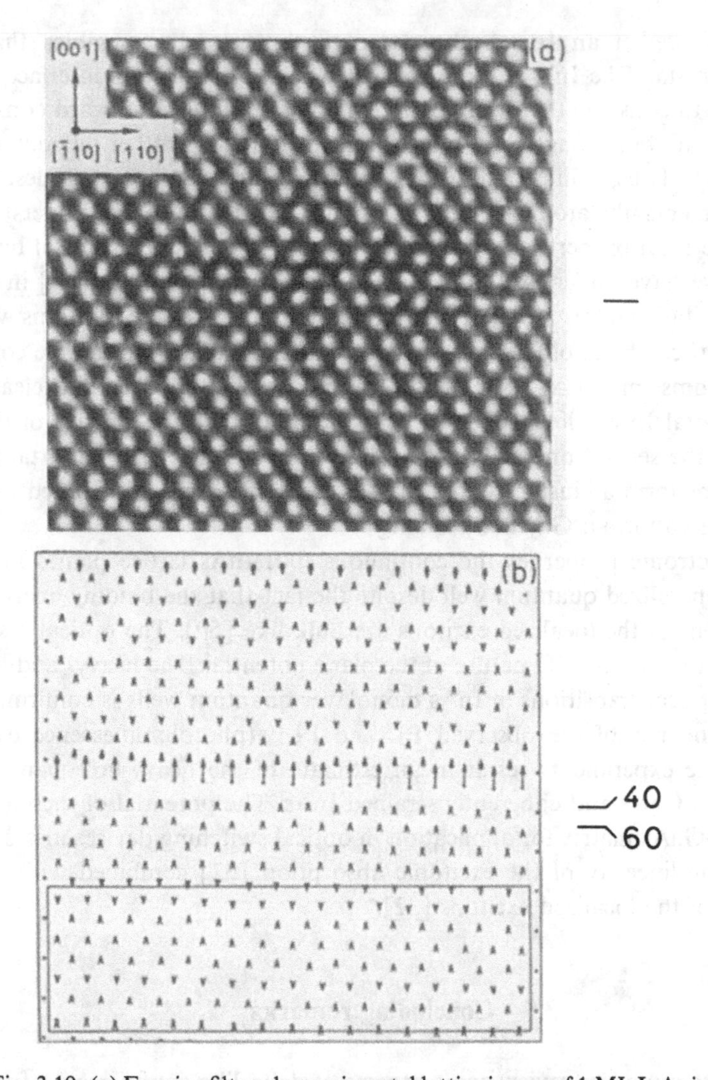

Fig. 3.19. (a) Fourier-filtered experimental lattice image of 1 ML InAs in GaAs. The bar indicates the plane containing In. (b) Vector representation of the relative shift in position of each ML with respect to its immediate neighbor. The magnitude of the vectors is amplified by a factor of 5. Numbers give the relative magnitude of the displacement vectors in percentages. The cross-section is oriented along [1̄10] instead of along [110], resulting in the inversion of the length ratio of the two displacement vectors at the interface with respect to the image simulation.

rapidly. However, a significant deviation from elasticity theory occurs in the ultimate limit of a monoatomic layer. In this case, the measured tetragonal distortion of the unit cell corresponds to an In—As bond length of 2.62 Å, exactly equal to that of unstrained bulk InAs. This experimental result confirms recent *ab initio* total-energy calculations [47]. By explicitly considering the present case of a single In lattice plane buried in a GaAs matrix, these calculations predict an In—As bond length that closely resembles that of an unstrained bulk crystal. The In—As bonds directly at the InAs/GaAs interface are thus stretched in order to conserve their bulk bond length. The tendency toward conservation of the unstrained bulk bond length has in fact been suggested for rather different chemical environments [48]. Thus, while films of a few ML thickness are well described by macroscopic arguments, the atomic configuration at the interface can be understood only by considering the local properties of the crystal lattice, namely, the chemical bond.

Very recently we have undertaken X-ray standing wave experiments [49] in order to obtain more quantitative information on the actual distribution of the In atoms within the respective (001) lattice planes of the GaAs matrix. Measurements on a sample comprising 6.864×10^{14} In atoms cm^{-2} ($\triangleq 1.1$ ML as determined by X-ray diffraction) clearly show that 70% of the total In are located in the first (001) lattice plane on top of the GaAs substrate, 20% in the second one and 10% in the third one. It is also important to note that these In atoms form a cluster, that is, they are not randomly distributed over these three lattice planes to form a $Ga_xIn_{1-x}As$ alloy.

In terms of electronic properties the continuous (001)InAs lattice planes in a GaAs matrix act as a generalized quantum well despite the fact that the binding energy as well as the spatial extent of the localized excitons are bulk-like [50]. The optical response of the system is governed by the 2D nature of the plane potential. The heavy- and light-hole character of the optical transitions in InAs monolayer quantum wells is confirmed by the polarization dependence of the observed PL and PLE (photoluminescence excitation) spectra [51]. These experiments result in an estimate of the heavy-hole band offset of 0.3–0.5 eV between GaAs and coherently strained InAs. The potential of these (001)InAs lattice planes in a GaAs matrix for application in optical switching devices arises from the observed large non-linearity of the excitonic absorption [52] combined with the short radiative lifetime of the localized excitons [32].

3.5 Concluding remarks

This article has attempted to review some aspects of delta-like confinement of impurities in GaAs which are – in the opinion of the author – the most challenging ones for the crystal grower. For both the generation of quantum wires and dots by self-organized growth and the ultrahigh-density incorporation up to one complete monolayer, the nucleation, the redistribution, and the relaxation are the important processes for the structural integrity and hence for the resulting electronic properties.

Most of the recent progress has been made as a result of increased understanding of the nucleation process. This progress has led to the successful alignment of the impurity atoms along regular step-edge arrays generated on vicinal (001)GaAs. The ordering of the impurity atoms on the growth surface yields a distinct surface reconstruction which can be monitored in real time by RHEED. However, as the surface impurity atoms have to reach the appropriate lattice sites, the impurity atoms have to be supplied not continuously but in pulses with long interruptions to allow for sufficient surface migration (or smoothing of the terraces). This high degree of growth control for impurity incorporation has made feasible the direct fabrication of quantum wire and dot structures by epitaxial growth.

When the impurity coverage of the growth surface is increased to reach one monolayer, the redistribution of the impurity atoms during overgrowth with the matrix and the strain relaxation become more and more important. The redistribution includes both the change of lattice sites, for example Si from the Ga to the As site, leading to free-carrier compensation in monolayer doping, and the exchange of atoms leading to segregation normal to the growth face. Far less is currently known about the microscopic mechanisms of these processes. To control these phenomena at the atomic level we must develop methods for monitoring the phenomena in real time, that is, during crystal growth, and for determining the actual site distribution after the growth. For the latter task a combination of high-resolution X-ray diffraction, X-ray standing wave techniques, and HREM might be useful.

Acknowledgements

Stimulating discussions with L. Däweritz are gratefully acknowledged. Part of this work was sponsored by the Bundesministerium für Forschung und Technologie of the Federal Republic of Germany.

References

[1] S. Bass, *J. Cryst. Growth* **47**, 613 (1979); C. E. C. Wood, G. M. Metze, J. D. Berry, and L. F. Eastman, *J. Appl. Phys.* **51**, 383 (1980).
[2] K. Ploog, M. Hauser, and A. Fischer, *Appl. Phys.* **A45**, 233 (1988).
[3] E. F. Schubert, *J. Vac. Sci. Technol.* **A8**, 2980 (1990).
[4] G. E. Crook, O. Brandt, L. Tapfer, and K. Ploog, *J. Vac. Sci. Technol.* **B10**, 841 (1992); M. R. Fahy, M. J. Ashwin, J. J. Harris, R. C. Newman, and B. A. Joyce, *Appl. Phys. Lett.* **61**, 1805 (1992).
[5] J. J. Harris, J. B. Clegg, R. B. Beall, J. Castagné, K. Woodbridge, and C. Roberts, *J. Cryst. Growth* **111**, 239 (1991).
[6] M. J. Ashwin, M. Fahy, J. J. Harris, R. C. Newman, D. A. Sansom, R. Addinall, D. S. McPhail, and V. K. M. Sharma, *J. Appl. Phys.* **73**, 633 (1993).
[7] K. Köhler, P. Ganser, and M. Maier, *J. Cryst. Growth* **127**, 720 (1993).
[8] R. C. Newman, in *Festkörperprobleme* (*Advances in Solid State Physics*) Vol. XXV, ed. P. Grosse (Vieweg, Braunschweig, 1985), p. 605.
[9] J. Wagner, M. Ramsteiner, W. Stolz, M. Hauser, and K. Ploog, *Appl. Phys. Lett.* **55**, 978 (1989).
[10] M. Santos, T. Sajoto, A. Zrenner, and M. Shayegan, *Appl. Phys. Lett.* **53**, 2504 (1988).
[11] E. F. Schubert, J. M. Kuo, R. F. Kopf, A. S. Jordan, H. S. Luftman, and L. C. Hopkins, *Phys. Rev.* **B42**, 1363 (1990).
[12] O. Brandt, K. Ploog, L. Tapfer, M. Hohenstein, R. Bierwolf, and F. Phillipp, *Phys. Rev.* **B45**, 8443 (1992).
[13] J. M. Moison, C. Guille, F. Houzay, F. Barthe, and M. Van Rompany, *Phys. Rev.* **B40**, 6149 (1989).
[14] C. Guille, F. Houzay, J. M. Moison, and F. Barthe, *Surface Sci.* **189/190**, 1041 (1987).

[15] J.-M. Gérard and J.-Y. Marzin, *Phys. Rev.* **B45**, 6313 (1992).
[16] H. Yamaguchi and Y. Horikoshi, *J. Appl. Phys.* **68**, 1610 (1990).
[17] B. Jusserand and F. Mollot, *Appl. Phys. Lett.* **61**, 423 (1992).
[18] H. Ono and T. Baba, *Mater. Sci. Forum* **83–87**, 1409 (1992).
[19] H. Schneider, J. Wagner, K. Ploog, A. Fischer, and K. Fujiwara, *J. Cryst. Growth* **127**, 836 (1993).
[20] W. Braun and K. Ploog, *J. Appl. Phys.* **75**, 1993 (1994).
[21] P. M. Petroff, A. C. Gossard, and W. Wiegmann, *Appl. Phys. Lett.* **45**, 620 (1984).
[22] Y. Horikoshi, H. Yamaguchi, F. Briones, and M. Kawashima, *J. Cryst. Growth* **105**, 326 (1990).
[23] M. D. Pashley, K. Haberern, and J. M. Gaines, *Appl. Phys. Lett.* **58**, 406 (1991).
[24] M. Sundaram, S. A. Chalmers, P. F. Hopkins, and A. C. Gossard, *Science* **254**, 1326 (1991).
[25] G. E. W. Bauer and A. A. van Gorkum, in *Science and Engineering of One- and Zero-Dimensional Semiconductors*, eds. S. E. Beaumont and C. M. Sotomayor Torres (Plenum, New York, 1990), p. 133.
[26] L. Däweritz, K. Hagenstein, and P. Schützendübe, *J. Vac. Sci. Technol.* **A11**, 1802 (1993).
[27] L. Däweritz and H. Kostial, *Appl. Phys.* **A58**, 81 (1994).
[28] M. G. Lagally, D. E. Savage, and M. C. Tringides, in *Reflection High Energy Electron Diffraction and Reflection Electron Imaging of Surfaces*, eds. P. K. Larsen and P. J. Dobson (Plenum, New York, 1988), p. 427.
[29] J. H. Neave, P. J. Dobson, B. A. Joyce, and J. Zhang, *Appl. Phys. Lett.* **47**, 100 (1985).
[30] K. Shiraishi, *Appl. Phys. Lett.* **60**, 1363 (1992).
[31] M. Ramsteiner, J. Wagner, G. Jungk, D. Behr, L. Däweritz, and R. Hey, *Appl. Phys. Lett.* **64**, 490 (1994).
[32] O. Brandt, L. Tapfer, K. Ploog, R. Bierwolf, M. Hohenstein, F. Phillipp, H. Lage, and A. Heberle, *Phys. Rev.* **B44**, 8043 (1991).
[33] B. Jenichen, K. Ploog, and O. Brandt, *Appl. Phys. Lett.* **63**, 156 (1993).
[34] G. G. Fountain, S. V. Hattangady, D. J. Vitkavage, R. A. Rudder, and R. J. Markunas, *Electron. Lett.* **24**, 1135 (1988); S. Tiware, S. L. Wright, and J. Batex, *IEEE Electron Device Lett.* **EDL-9**, 488 (1988); H. Hasegawa, M. Akazawa, K. Matsuzaki, H. Ishii, and H. Ohno, *Jpn J. Appl. Phys.* **27**, L 2265 (1988).
[35] L. Sorba, G. Bratina, G. Ceccone, A. Antonini, J. F. Walker, M. Micovic, and A. Franciosi, *Phys. Rev.* **B43**, 2450 (1991); M. Peressi, S. Baroni, and R. Resta, *Phys. Rev.* **B43**, 7347 (1991); M. Akazawa, H. Hasegawa, H. Tomozawa, and H. Fujikura, *Jpn J. Appl. Phys.* **31**, L1012 (1992).
[36] O. Brandt, G. Crook, K. Ploog, R. Bierworf, M. Hohenstein, M. Maier, and J. Wagner, *Jpn J. Appl. Phys.* **32**, L24 (1993).
[37] L. Däweritz, H. Kostial, and R. Hey, *J. Cryst. Growth* **150**, 214 (1995).
[38] B. Jenichen, private communication.
[39] R. Bierworf, M. Hohenstein, F. Phillipp, O. Brandt, G. E. Crook, and K. Ploog, *Ultramicroscopy* **49**, 273 (1993).
[40] L. Tapfer, G. E. Crook, O. Brandt, and K. Ploog, *Appl. Surf. Sci.* **56–58**, 650 (1992).
[41] H. Tanino, H. Kawanami, and H. Matsuhata, *Appl. Phys. Lett.* **60**, 1978 (1992).
[42] This particular composition is required for stoichiometry reasons, see T. Sudersena Rao, K. Nozawa, and Y. Horikoshi, *Jpn J. Appl. Phys.* **30**, 547 (1991).
[43] D. G. Liu, J. C. Fan, C. P. Lee, C. M. Tsai, K. H. Chang, T. L. Lee, and L. J. Chen, *Appl. Phys. Lett.* **60**, 2628 (1991).
[44] O. Brandt, G. E. Crook, K. Ploog, J. Wagner, and M. Maier, *Appl. Phys. Lett.*, **59**, 2730 (1991).
[45] L. Tapfer, M. Ospelt, and H. von Känel, *J. Appl. Phys.* **67**, 1298 (1990).
[46] O. Brandt, K. Ploog, R. Bierwolf, and M. Hohenstein, *Phys. Rev. Lett.* **68**, 1339 (1992).
[47] K. Shiraishi and E. Yamaguchi, *Phys. Rev.* **B42**, 3064 (1990).

[48] D. J. Chadi, *Phys. Rev. Lett.* **43**, 43 (1979); S. A. Chambers and V. A. Loebs, *Phys. Rev. Lett.* **63**, 640 (1989); *Phys. Rev.* **B42**, 5109 (1990); J. C. Mikkelsen, Jr and J. B. Boyce, *Phys. Rev Lett.* **49**, 1412 (1982); *Phys. Rev.* **B28**, 7130 (1983); R. G. Dandrea, J. E. Bernard, S.-H. Wei, and A. Zunger, *Phys. Rev. Lett.* **64**, 36 (1990).
[49] C. Giannini, L. Tapfer, S. Logomarsino, J. C. Boulliard, A. Taccoen, B. Capelle, M. Ilg, O. Brandt, and K. Ploog, *Phys. Rev.* **B48**, 11496 (1993).
[50] R. Cingolani, O. Brandt, L. Tapfer, G. Scamarcio, G. C. La Rocca, and K. Ploog, *Phys. Rev.* **B42**, 3209 (1990); O. Brandt, H. Lage, and K. Ploog, *Phys. Rev.* **B43**, 14285 (1991).
[51] O. Brandt. H. Lage, and K. Ploog, *Phys. Rev.* **B45**, 4217 (1992); O. Brandt, H. Lage, and K. Ploog, *Appl. Phys. Lett.* **59**, 576 (1991).
[52] O. Brandt, G. C. La Rocca, A. Heberle, A. Ruiz, and K. Ploog, *Phys. Rev.* **B45**, 3803 (1992).

4

Flow-rate modulation epitaxy (FME) of III–V semiconductors

T. MAKIMOTO AND Y. HORIKOSHI

4.1 Introduction

Optoelectronic devices, such as quantum-well lasers and two-dimensional electron gas (2DEG) field-effect transistors, have been grown in the AlGaAs/GaAs heterojunction system. Sharp heterointerfaces are necessary to improve the characteristics of these devices and to develop new quantum-effect devices. Furthermore, a heavily doped p-type layer with a well-defined doping profile is essential to the achievement of the ultrahigh-speed performance of heterojunction bipolar transistors (HBTs), which will not give the expected device performance unless the diffusion of p-type impurities from the GaAs base layer to the AlGaAs emitter layer is well suppressed.

Metalorganic chemical vapor deposition (MOCVD) has been shown to be a useful growth method that can produce thin semiconductor epitaxial layers of reasonable quality for device applications. The growth temperature, especially for AlGaAs (above 700 °C), is usually at least 100 °C higher than in molecular beam epitaxy (MBE), because a lower MOCVD growth temperature causes the electrical and optical properties of AlGaAs epitaxial layers and the flatness of the AlGaAs/GaAs heterointerfaces to deteriorate. The MOCVD growth temperatures for high-quality epitaxial layers are very high, so Zn, Be, and Si impurities in GaAs or AlGaAs diffuse during growth. For these impurities, lower temperatures are desirable in order to grow extremely narrow doping profiles such as δ-doping.

Recently, a new epitaxial growth method, flow-rate modulation epitaxy (FME) (Kobayashi *et al.*, 1989), has been developed by modifying the conventional MOCVD method. Flow-rate modulation epitaxy reduces the growth temperature while maintaining the crystal quality and the interface flatness. In this chapter, we describe the growth and δ-doping methods in FME, and emphasize that δ-doping is suitable for the investigation of the doping mechanism as well as for the study of low-dimensional electron physics and for the improvement of device performance.

4.2 Growth mechanism in FME

Flow-rate modulation epitaxy is a modified form of the conventional MOCVD method and is based on the alternate supply of group III organometals and arsine (AsH_3). In this method, group III organometals are completely decomposed to supply group III metallic atoms on the growing surface. Horikoshi *et al.* (1986) reported that the surface migration of Ga and Al atoms is enhanced under As-free or low-pressure As conditions during MBE growth of GaAs and AlAs, which reduces the growth temperature and produces a flatter AlGaAs/GaAs heterointerface. The method of alternately supplying Ga and As atoms is called migration-enhanced epitaxy (MEE). The FME method is an application of MEE to the conventional MOCVD method. Figure 4.1 illustrates the characteristics of FME for GaAs growth. Triethylgallium (TEG) and AsH_3 are used for undoped GaAs layers. Flow-rate modulation epitaxy is characterized by three points: (1) In conventional MOCVD growth, additional compounds are formed between TEG and AsH_3 (Putz *et al.*, 1983 and Voronin *et al.*, 1983). These additional compounds are thermally stable. Therefore, when the growth temperature is decreased, they remain undecomposed and may be incorporated into the growing layers and deteriorate crystal quality. The formation of such compounds is suppressed by FME, because of the alternate supply of source gases. This results in high-quality GaAs layers at lower growth temperatures. (2) In FME, the Ga source gas is completely decomposed to supply Ga metallic atoms on the growing surface. For this purpose, TEG is used instead of trimethylgallium (TMG), because TEG decomposes at lower temperatures. The migration of Ga atoms is enhanced under AsH_3-free or low-AsH_3 conditions. This reduces the growth temperature without deteriorating crystal quality and produces a flatter interface. The alternate supply of source gases

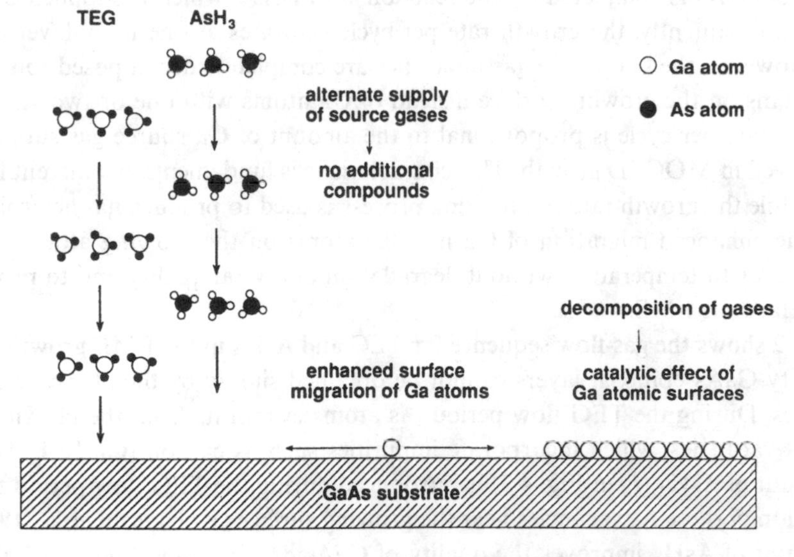

Fig. 4.1. Characteristics of FME for GaAs growth. TEG and AsH_3 are used for undoped GaAs layers. The FME method is characterized by three points, as shown in this diagram.

is expected to enhance the migration of surface adatoms and also to solve the problem of additional compound formation. (3) During FME growth, Ga and As atomic surfaces appear alternately at the growing surface. These atomic surfaces have very different characteristics for capturing impurity atoms. The Ga atomic surface has a catalytic effect that does not appear during conventional MOCVD growth; this allows AsH_3 to be decomposed easily even at lower temperatures (Lays and Veenvliet, 1981; Nishizawa and Kurabayashi, 1983). The catalytic effect of the Ga atomic surface also enhances the silane (SiH_4) cracking efficiency, as described later, resulting in heavy δ-doping at low growth temperatures.

Atomic layer epitaxy (ALE) (Suntola and Antson, 1974; Nishizawa *et al.*, 1985; Badair *et al.*, 1985; Usui and Sunakawa, 1986; and Doi *et al.*, 1986) is also based on an alternate supply of source gases, but the growth mechanism of ALE is fundamentally different from that of FME. In ALE, the growth rate per cycle is not proportional to the amount of group III organometals supplied, and it saturates at one monolayer per cycle when the amount supplied is large enough (the growth rate self-limiting process). The growth rate self-limiting process needs group III organometals which do not completely decompose on the growing surface, so methyl-compounds such as TMG have often been used instead of ethyl-compounds such as TEG. The most probable mechanism of the growth rate self-limiting process is as follows. In the ALE growth of GaAs, the Ga source gas molecules are not completely decomposed so that Ga atoms with one or two substituents, such as methyl or halogen group molecules, are adsorbed on the growing surface. Excess source gas molecules supplied over a one-molecular layer coverage of the growing surface tend not to be adsorbed on the surface covered by methyl or halogen group molecules because the adsorption between source molecules and this modified surface is very weak. The formation of GaAs is completed by the reaction with AsH_3, which is supplied during the next step. Consequently, the growth rate per cycle saturates at one monolayer per cycle. In FME, however, the Ga source gas molecules are completely decomposed to supply Ga metallic atoms on the growing surface instead of Ga atoms with one or two substituents. The growth rate per cycle is proportional to the amount of Ga source gas supplied, as is often observed in MOCVD growth. This characteristic is fundamentally different from that of ALE. While the growth rate self-limiting process is used to produce flat heterointerfaces in ALE, the enhanced migration of Ga metallic atoms on the growing atoms is used to reduce the growth temperature without degradation of crystal quality and to produce flat heterointerfaces in FME.

Figure 4.2 shows the gas-flow sequence for TEG and AsH_3 in the FME growth of GaAs. High-quality GaAs epitaxial layers cannot be obtained simply by the alternate supply of source gases. During the TEG flow period, As atoms evaporate from the growing surface to form As vacancies, which incorporate impurities such as carbon (C). In FME, a very small amount of AsH_3 (r_0 in Fig. 4.2) is introduced during the TEG flow period to reduce the formation of these As vacancies near the growing surface (Kobayashi *et al.*, 1985). This small amount of AsH_3 improves the quality of GaAs dramatically. Figure 4.3 shows the effect of AsH_3 during the TEG flow period on the electrical properties at 77 K. The growth temperature was 550 °C. One cycle consisted of 1 s of TEG flow with or without

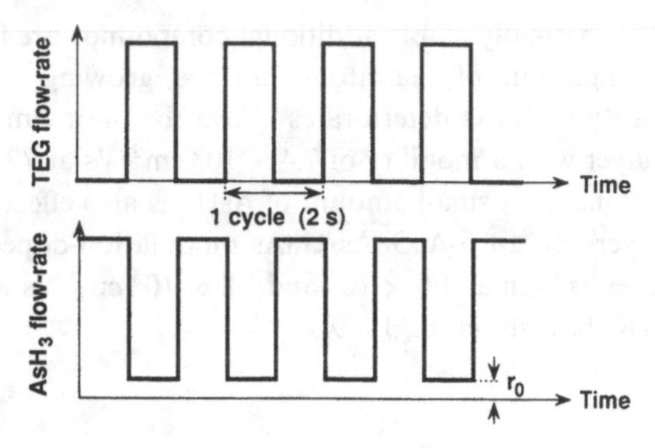

Fig. 4.2. Gas-flow sequences of TEG and AsH_3 in FME growth of GaAs.

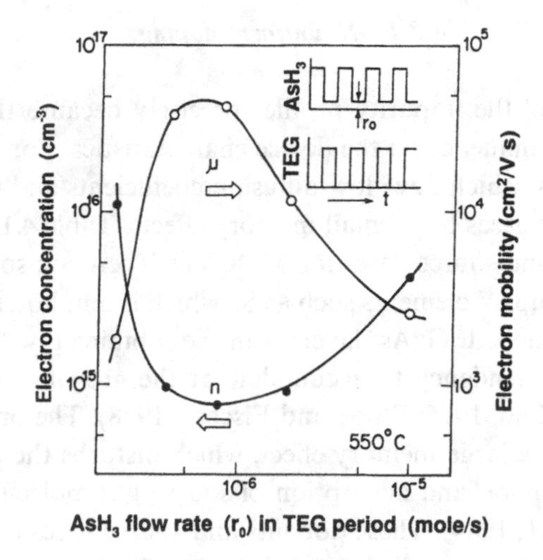

Fig. 4.3. Effect of AsH_3 during the TEG flow period on the electrical properties at 77 K for GaAs layers grown at 550 °C.

a small amount of AsH_3, followed by 1 s of AsH_3 flow, and the growth rate was adjusted to one GaAs monolayer per cycle. Without AsH_3, an undoped GaAs layer shows a low electron mobility of 2×10^3 cm^2/Vs with a residual impurity concentration as high as 1×10^{16} cm^{-3}. This is ascribed to the incorporation of impurities into As vacancies formed during the TEG flow period. With a small amount of AsH_3, the electron mobility increases drastically while the impurity concentration decreases. With the optimum r_0 value, the mobility is 5×10^4 cm^2/Vs and the impurity concentration is 8×10^{14} cm^{-3}. Supplying AsH_3 during the TEG flow period suppresses the formation of the As vacancies and the incorporation of impurities. The optimum r_0 value is only a few percent of the amount needed to grow an As atomic surface, so this small amount of AsH_3 has little influence on reducing the migration enhancement of Ga atoms on the growing surface or on forming additional compounds. However, when the amount of AsH_3 is above this optimum, crystal quality begins to deteriorate, indicating that the growth becomes similar to the conventional

MOCVD mode, that is, thermally stable additional compounds are formed between TEG and AsH_3, and the migration of Ga atoms on the growing surface is suppressed. Consequently, the quality of GaAs deteriorates above the optimum r_0 value. Recently, a higher-quality GaAs layer with a mobility of 7.4×10^4 cm^2/Vs at 77 K has been obtained under optimum conditions. This small amount of AsH_3 is also effective for improving the quality of the AlAs layers. In an n-AlGaAs/GaAs modulation-doped structure grown at 550 °C, 2DEG mobilities as high as 1.0×10^5 and 1.8×10^5 cm^2/Vs were obtained at 77 K and 6 K respectively (Kobayashi *et al.*, 1989).

4.3 N-type impurity doping and δ-doping

4.3.1 Si uniform doping

It is important to control the impurity profiles precisely because the distribution of the impurity has a profound influence on the device characteristics. For this purpose, we have to use suitable impurities, which have low-diffusion coefficients and negligible segregation effects, and whose source gases have small memory effects. Table 4.1 shows the characteristics of n-type dopants and source gases for MOCVD. There are some problems with Sn and Se. Unlike other group IV elements such as Si, which is amphoteric, Sn acts as a donor in GaAs, and heavily doped GaAs layers can be obtained with Sn. However, the disadvantage of Sn is its tendency to accumulate at the growing surface during growth (the segregation effect) (Cho, 1975; Ploog and Fischer, 1978). The problem with Se is that its source gas, H_2Se, has a large memory effect, which disturbs the control of the doping profiles due to the desorption and adsorption of source gas molecules on the inside wall of the reactor (Lewis *et al.*, 1984). These dopants and source gases cannot produce abrupt doping profiles, so they are not applicable for δ-doping. Therefore, Si and SiH_4 are widely used as an n-type dopant and a source gas, respectively, for growing δ-doped structures. This section describes the Si doping mechanism in FME.

Table 4.1. *Characteristics of n-type dopants and source gases for MOCVD*

Dopants	Source gases	Characteristics
Sn	$Sn(CH_3)_3$	Segregation
S	H_2S	Low doping efficiency Large diffusion coefficient Memory effect
Se	H_2Se	Memory effect
Si	SiH_4	Small diffusion coefficient Negligible memory effect

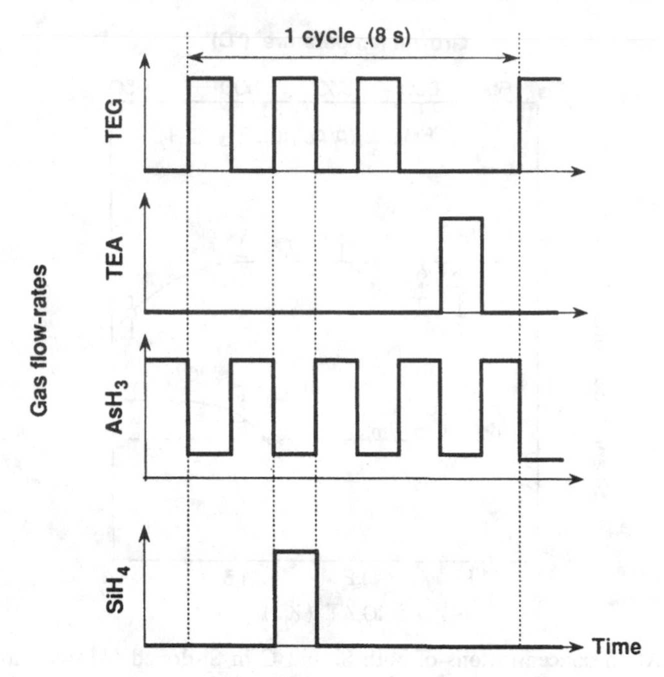

Fig. 4.4. Gas-flow sequences of TEG, TEA, and AsH_3 for an $(AlAs)_1(GaAs)_3$ ordered alloy grown by repeating the growth of three GaAs monolayers followed by one AlAs monolayer. The source gas, SiH_4, was introduced with TEG, and Si was doped into the middle layer of the three GaAs monolayers.

Ga atomic surfaces never appear in the conventional MOCVD method, because the growth proceeds under As-stabilized conditions. During FME growth, Ga- and As-stable surfaces appear alternately on the growing surface, so it is possible to introduce SiH_4 into either the Ga- or As-stable surface and to investigate the characteristics of each surface. Figure 4.4 shows the gas-flow sequences of TEG, triethylaluminum (TEA), and AsH_3 for an $(AlAs)_1(GaAs)_3$ ordered alloy, which is grown by repeating the growth of three GaAs monolayers followed by one AlAs monolayer. The source gas, SiH_4, was introduced with TEG, and Si was introduced into the middle layer of the three GaAs monolayers. Figure 4.5 shows the atom concentrations of both Si and C in Si-doped $(AlAs)_1(GaAs)_3$ layers as a function of the growth temperature (Makimoto *et al.*, 1986). They were measured by secondary ion mass spectroscopy (SIMS). The Si doping concentration was constant above 500 °C. Figure 4.6 shows the growth temperature dependence of the carrier concentration in Si-doped $(AlAs)_1(GaAs)_3$, obtained by Hall measurement at room temperature. Above 525 °C, the carrier concentration was almost constant and coincided very well with the Si atom concentration shown in Fig. 4.5. In conventional MOCVD, Veuhoff *et al.* (1985) reported the Si doping efficiency of SiH_4 decomposition (see Fig. 4.6). The temperature-independent incorporation of Si in FME is due to the catalytic decomposition of SiH_4 on the Ga-stable surface. The results given in Fig. 4.6 also indicate that Si doping with a reasonable concentration is possible even when the growth temperature is as low as 550 °C.

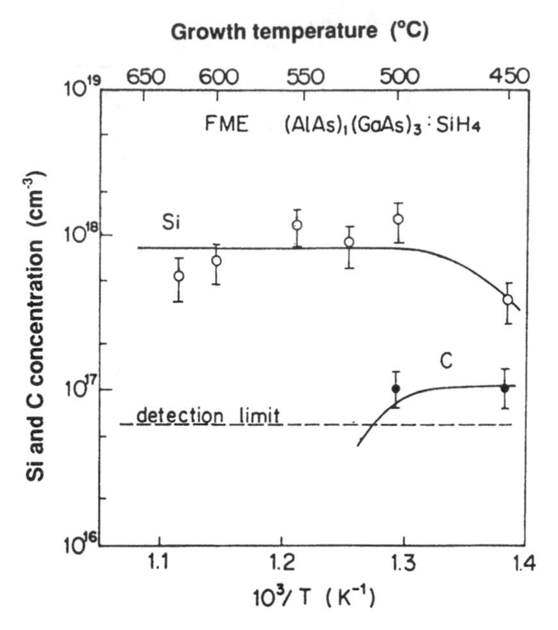

Fig. 4.5. Atom concentrations of both Si and C in Si-doped $(AlAs)_1(GaAs)_3$ layers as a function of growth temperature. They were measured by SIMS.

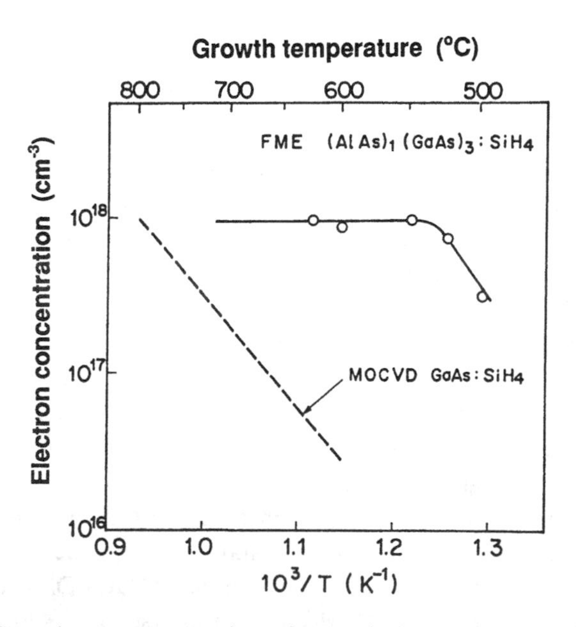

Fig. 4.6. Growth temperature dependence of the carrier concentration in Si-doped $(AlAs)_1(GaAs)_3$, obtained by Hall measurement at room temperature. The broken line shows Si doping efficiency in conventional MOCVD (Veuhoff *et al.*, 1985).

4.3.2 *Si δ-doping in the Ga or As atomic surface*

The characteristics of the Ga atomic surface were investigated directly by comparing the capacitance–voltage (*C–V*) profile of Si δ-doped structures grown by supplying the same amount of SiH_4 selectively, either to the Ga or As atomic surface. The Si δ-doping was studied in three different modes: (i) doping on an As atomic surface, (ii) doping on a Ga atomic surface, and (iii) doping during the Ga atomic plane formation. The gas-flow sequences used for this experiment are shown in Fig. 4.7. The Si doping on the As atomic surface was performed by supplying a pulse of SiH_4 flow (with a small amount of AsH_3) immediately following an AsH_3 flow pulse, as shown in Fig. 4.7(a). For the Si doping on the Ga atomic surface, a SiH_4 flow pulse was added following a TEG flow pulse, as shown in Fig. 4.7(b). For the Si doping during the Ga atomic plane formation, a SiH_4 flow pulse was added to a TEG flow pulse with a small amount of AsH_3, as shown in Fig. 4.7(c).

Figure 4.8 shows a typical room temperature *C–V* profile of a structure which contains three Si δ-doped layers every 330 nm, as shown in the inset in Fig. 4.8 (Kobayashi *et al.*, 1986). The first and third peaks from the surface are carrier profiles originating from Si δ-doping on the As atomic surfaces. The observed carrier profiles are almost symmetrical, indicating that there is a negligible memory effect. In addition, doping reproducibility is fairly good because the first and third peaks have almost the same intensity. The sheet electron concentration (n^{2D}) was estimated by integrating the observed *C–V* profiles. Although the same amount of SiH_4 was supplied on the As and Ga atomic surfaces, the

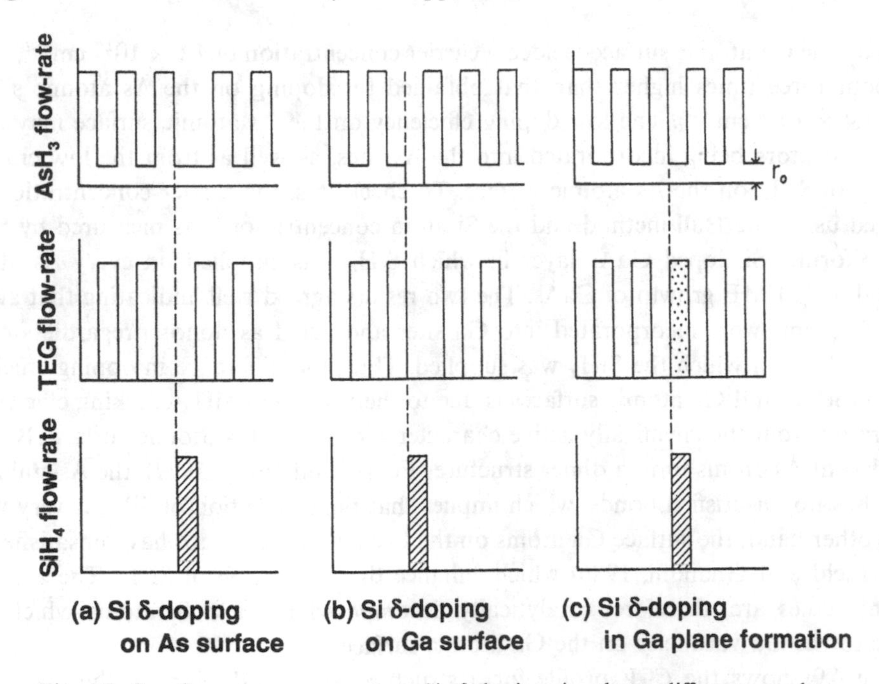

Fig. 4.7. Gas-flow sequences using Si δ-doping in three different modes: (a) doping on an As atomic surface; (b) doping on a Ga atomic surface; and (c) doping during Ga atomic plane formation.

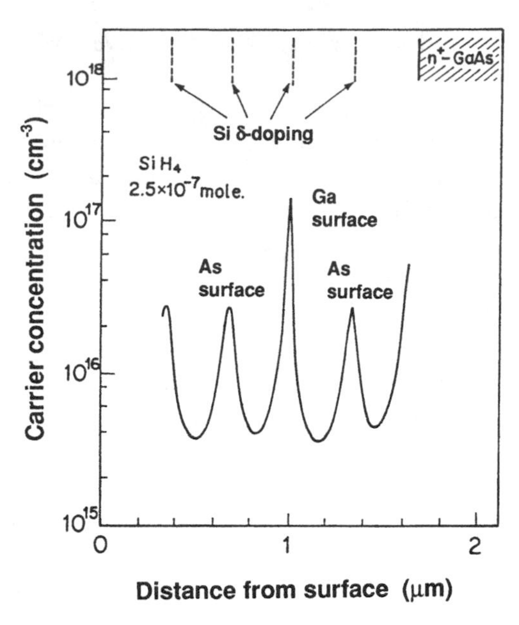

Fig. 4.8. Typical room temperature *C–V* profile of a structure which contains three Si δ-doped layers every 330 nm. The structure is shown in the inset. The first and third peaks from the surface are carrier profiles originating from Si δ-doping on the As atomic surfaces.

doping on the Ga atomic surface yielded a carrier concentration of 4.1×10^{11} cm^{-2}, which was about three times higher than that obtained for doping on the As atomic surface ($n^{2D} = 1.4 \times 10^{11}$ cm^{-2}). The low doping efficiency on the As atomic surface may result from Si acceptors being incorporated into the As sites, as well as from the low cracking efficiency of SiH_4 on the As atomic surface. To check this, the carrier concentration was measured using the Hall method and the Si atom concentration was measured by SIMS for a uniformly Si doped GaAs layer in which SiH_4 was supplied on every As atomic surface during FME growth of GaAs. The two results agreed well, indicating that almost all the Si atoms were incorporated into Ga sites and acted as donors, regardless of the atomic surface on which the SiH_4 was supplied. Thus, the difference in doping efficiency between the As and Ga atomic surfaces is due to their different SiH_4 cracking efficiencies, which result from the chemically active characteristics of the Ga atomic surface. Because two adjacent As atoms form a dimer structure (Foxon and Joyce, 1977), the As-stabilized surface has no unsatisfied bonds, which implies that the adsorption of SiH_4 is very weak. On the other hand, the surface Ga atoms on the Ga-stabilized surface have unsatisfied sp^2 bonds (Field and Ghandhi, 1986) which enhance the adsorption of SiH_4. The adsorbed SiH_4 molecules are, therefore, catalytically decomposed to yield Si atoms which then migrate to the Ga vacancies on the Ga atomic surface.

Figure 4.9 shows the *C–V* profile for a structure for Si δ-doping on the Ga atomic surface and during the Ga plane formation. The doping during the Ga plane formation was four times as efficient as the doping on the Ga atomic surface. The enhanced

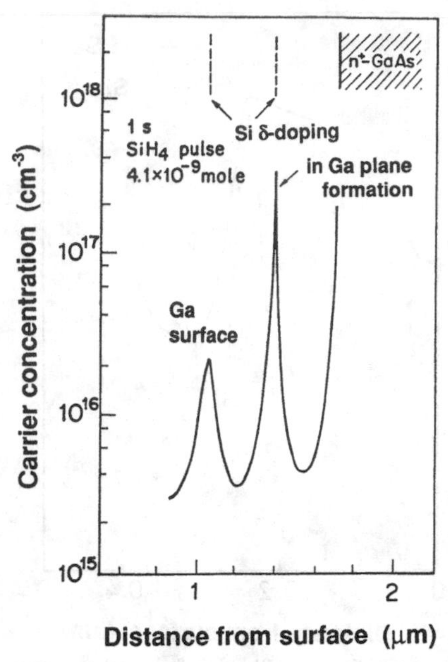

Fig. 4.9. *C–V* profile for a structure for Si δ-doping on the Ga atomic surface and during Ga plane formation. Doping during Ga plane formation was four times as efficient as doping on the Ga atomic surface.

doping efficiency in this mode is explained as follows. During the growth of the Ga atomic plane, there exist many Ga steps, which are chemically very active (Neave *et al.*, 1985). In this mode, SiH_4 is strongly adsorbed on Ga atoms and is catalytically decomposed immediately, and then Si atoms are effectively incorporated into the high density of the Ga vacancy sites.

4.3.3 δ-doping with Si and Se

In order to investigate the controllability of Si and Se doping profiles, we grew δ-doped structures with these dopants. Figure 4.10 shows room temperature *C–V* profiles for the δ-doped structures with Si and Se. The source gases for Si and Se were SiH_4 and H_2Se respectively. For the Se δ-doping, the increase in the carrier concentration around 0.4 μm is due to the effect of the n^+-GaAs substrate. For the Si δ-doping, the observed profile is symmetrical, indicating that the memory effect of SiH_4 and the segregation effect of Si are negligible. In contrast, for the Se δ-doping, the observed profile is asymmetrical and broadens towards the surface. Considering the few reports on the segregation effects of Se in GaAs during MBE growth, it is probably due to the memory effect of H_2Se. Thus, using H_2Se, it is difficult to obtain a sharp doping profile, so H_2Se is not a suitable source gas for δ-doping. Compared with H_2Se, SiH_4 has a negligible memory effect, so SiH_4 is an adequate source gas.

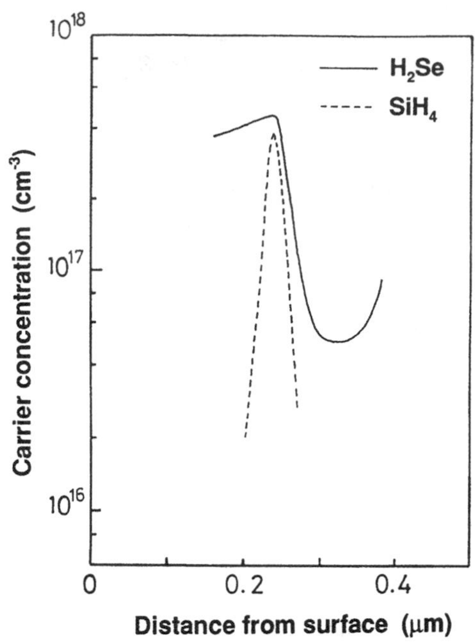

Fig. 4.10. Room temperature $C-V$ profiles for δ-doped structures with Si and Se. Source gases for Si and Se are SiH_4 and H_2Se respectively.

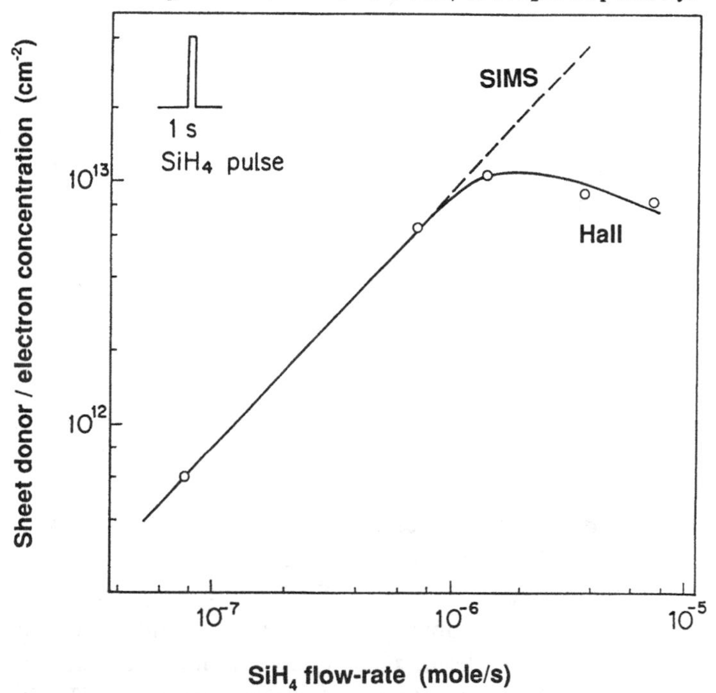

Fig. 4.11. Sheet electron concentration (n^{2D}) measured as a function of the supplied SiH_4 flow rate. The n^{2D} value was estimated using Hall measurement at room temperature for a 0.5-μm GaAs layer with one Si δ-doping in the middle using the gas-flow sequence as shown in Fig. 4.7(c).

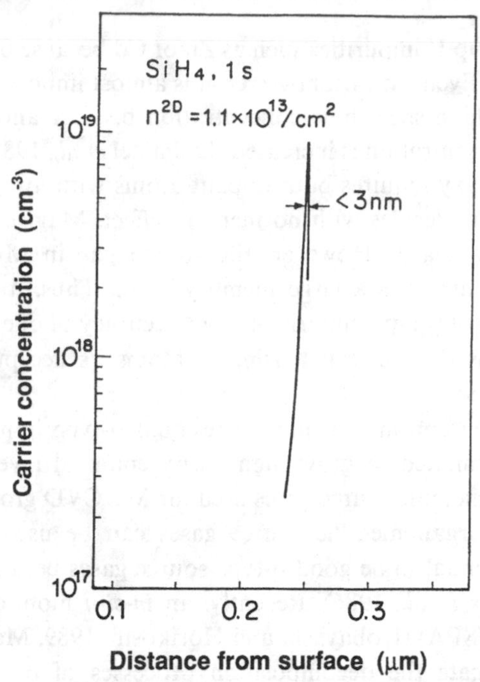

Fig. 4.12. *C–V* profile for a Si δ-doped structure with extremely high n^{2D}. The FWHM value is as narrow as 3 nm, which is similar to the calculated value reported by Schubert and Ploog (1986).

In order to investigate the upper limit of the sheet electron concentration, n^{2D} was measured as a function of the supplied SiH_4 flow rate (Fig. 4.11). The n^{2D} value was estimated from Hall measurements at room temperature for a 0.5 μm GaAs layer with one Si δ-doped layer in the middle using the gas-flow sequence shown in Fig. 4.7(c). In conventional MBE, δ-doping was performed by stopping the growth and evaporating Si for a few minutes. Note that this was performed by only a 1 s SiH_4 flow pulse without stopping the growth process in FME. The n^{2D} value increased up to 8×10^{12} cm^{-2} in proportion to the SiH_4 flow rate. It reached a maximum value of 1.1×10^{13} cm^{-2} and saturated. Figure 4.12 shows the *C–V* profile for a Si δ-doped structure with extremely high n^{2D}. The full-width at half-maximum (FWHM) is as narrow as 3 nm, which is similar to the calculated value reported by Schubert and Ploog (1986). Considering that the FWHM depends on the carrier concentration (i.e. the screening lengths), Si atoms diffuse very little during FME growth. Thus, Si is an adequate dopant in GaAs, and an almost ideal δ-doped structure is grown by FME.

4.4 P-type δ-doping

As described above, abrupt impurity profiles have been demonstrated for n-type impurities using Si. For p-type doping of GaAs, Be provides much more abrupt carrier distribution profiles than any other group II impurities such as Zn or Cd because of its smaller diffusion coefficient (Ilegems, 1977). Even with Be, however, it is almost impossible to attain an ideal δ-doped structure with a high sheet hole concentration because anomalous Be diffusion occurs when its doping concentration is increased (Duhamel *et al.*, 1981). An abrupt doping profile in vapor phase epitaxy requires both dopant atoms with low-diffusion coefficients and the use of source gas molecules with no memory effect. Magnesium exhibits a fairly low-diffusion coefficient in GaAs. However, the source gas in MOCVD is bis-cyclopentadienyl magnesium, which has a large memory effect. Thus, it is rather difficult to control the doping profiles of p-type impurities to an accuracy of one atomic layer. In this section, p-type δ-doping is demonstrated using C atoms as acceptors, and its doping mechanism will be discussed.

Carbon atoms have been considered only as residual p-type impurities in GaAs and AlGaAs that must be eliminated to grow high-purity epitaxial layers (Stringfellow and Linnebach, 1980). Organometallic source gases used for MOCVD growth include C atoms themselves. Therefore, if organometallic source gases can be used intentionally for C doping, they have the potential to be good p-type source gases because of their negligible memory effect (M. Weyers *et al.*, 1986). Recently, an *in-situ* monitoring method, called surface photo-absorption (SPA) (Kobayashi and Horikoshi, 1989; Makimoto *et al.*, 1990), has been used to investigate the decomposition processes of organometal molecules. Makimoto *et al.* (1990) reported that TEG decomposes almost completely on the growing surface above 500 °C, while TMG does not completely decompose even at 630 °C, and that methyl-Ga species are adsorbed on the growing surface under AsH_3-free conditions. Thus, C δ-doping is possible between 500 and 630 °C, using trimethyl and triethyl organometals for the C-doped and the undoped layers respectively.

4.4.1 C-doping in GaAs

In FME, C-doped p^+-GaAs layers are grown by an alternate supply of TMG and AsH_3, which enhances high C-doping efficiency. Methyl-Ga species are formed efficiently on the growing surface when TMG is supplied under AsH_3-free conditions, and C atoms from the methyl groups replace As atoms on the growing surface to be incorporated into the epitaxial layer as acceptors, as described later. Uniformly C-doped GaAs layers were grown by alternately supplying TMG and AsH_3. The hole concentration of these layers evaluated by Hall measurement agreed well with the C-atom concentration evaluated by SIMS of up to 2.4×10^{19} cm^{-3}, indicating that the activation ratio of the C acceptor is unity up to this concentration range.

Next, C δ-doping was performed in order to observe the memory effect of trimethyl organometals and to evaluate the diffusion coefficient of C in GaAs (Kobayashi *et al.*, 1987). The gas-flow sequences used for the C δ-doping are shown in Fig. 4.13. For undoped GaAs growth, one cycle consisted of 1 s of TEG flow with a small amount of AsH_3 followed by 1 s of AsH_3 flow. During growth of the C δ-doped layer, the supply of TEG and a small amount of AsH_3 were stopped, and a 1 s flow pulse of TMG or trimethylaluminum

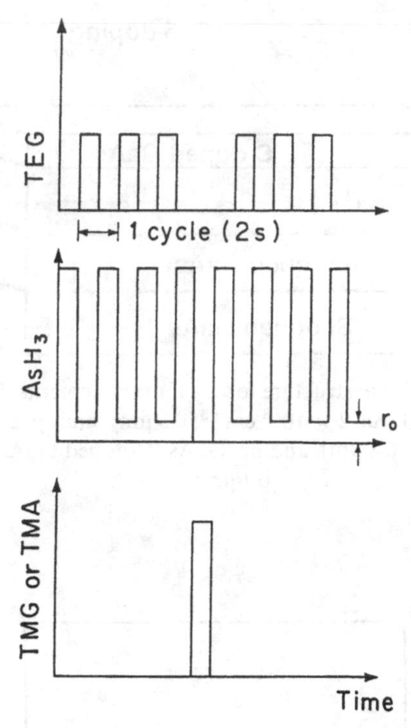

Fig. 4.13. Gas-flow sequences used for C δ-doping. For undoped GaAs growth, one cycle consisted of 1 s of TEG flow with a small amount of AsH_3 followed by 1 s of AsH_3 flow. During growth of the C δ-doped layer, the supply of TEG and a small amount of AsH_3 were stopped, and a 1 s pulse of TMG or TMA was supplied.

(TMA) was supplied. Figure 4.14 illustrates a schematic diode structure for *C–V* measurements. The diode was composed of n^+-GaAs (Si-doped GaAs, 2×10^{18} cm^{-3}, 0.5 μm), undoped GaAs with a few C δ-doped layers (0.5–0.8 μm), and p^+-GaAs (C-doped GaAs, 1×10^{19} cm^{-3}, 0.1 μm).

Figure 4.15 shows a typical room-temperature *C–V* profile of GaAs containing three C δ-doped layers spaced at 0.2 μm intervals. The first peak near the n^+-GaAs layer was not observed due to depletion. The observed peaks are symmetrical and have the same shape as each other, indicating that the memory effect seen in Se δ-doping is negligible, and that the doping is reproducible. Trimethylaluminum is more difficult to decompose than TMG, so a higher C-doping concentration is expected. Figure 4.16 shows the *C–V* profile of GaAs with a single C δ-doped layer formed by TMA. The C atoms are more efficiently incorporated into the epitaxial layer when TMA is used instead of TMG for C-doping. The observed *C–V* profile is very sharp, with an FWHM of 5.8 nm, which is similar to that for a Si δ-doped structure with an equivalent sheet carrier concentration. The FWHM value depends on the sheet carrier concentration, that is the screening length. Thus, C atoms diffuse little during growth at 550 °C and are confined within one monolayer.

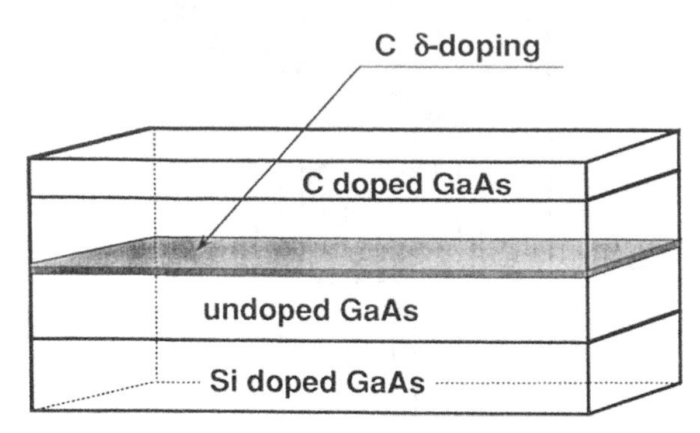

Fig. 4.14. Schematic diode structure for $C-V$ measurements. It is composed of n^+-GaAs (Si-doped GaAs, 2×10^{18} cm^{-3}, 0.5 μm), undoped GaAs with a few C δ-doped layers (0.5–0.8 μm), and p^+-GaAs (C-doped GaAs, 1×10^{19} cm^{-3}, 0.1 μm).

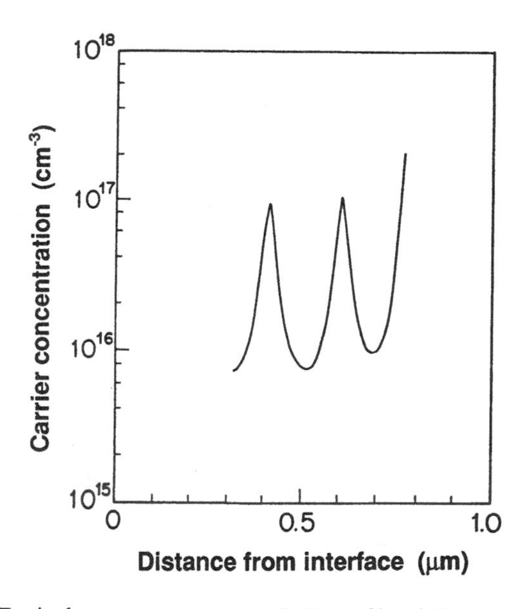

Fig. 4.15. Typical room-temperature $C-V$ profile of GaAs containing three C δ-doped layers spaced at 0.2 μm intervals.

The sample was annealed for 1 hr at 800 °C to estimate the diffusion coefficient for C atoms in GaAs. The broken line in Fig. 4.16 shows the $C-V$ profile for the annealed sample. After annealing, the peak height decreased to 3.2×10^{17} cm^{-3} and the FWHM value broadened to 43 nm. The diffusion coefficient estimated from this broadened $C-V$ profile was 2×10^{-16} cm^2/s, which is much lower than those of other p-type impurities (Ilegems,

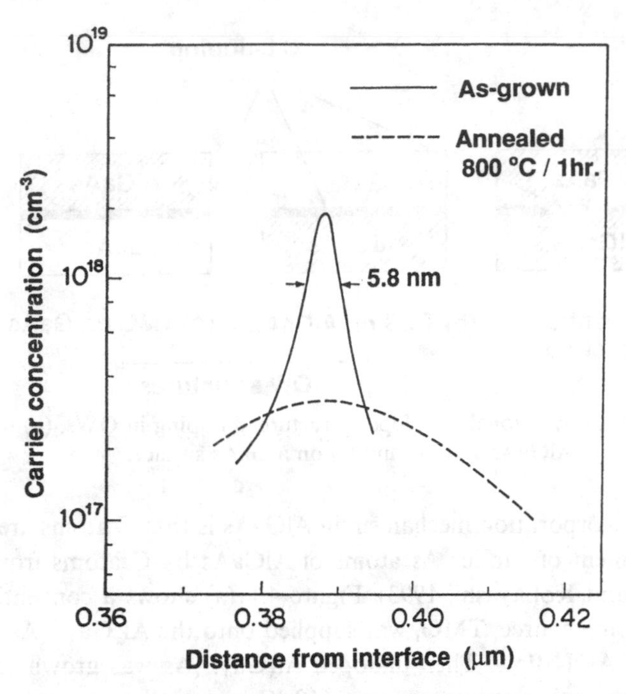

Fig. 4.16. *C–V* profile of GaAs with a single C δ-doped layer formed by TMA. The broken line shows the *C–V* profile for the annealed sample.

1977). The sheet hole concentration estimated by integrating the observed *C–V* profile, however, did not change from before to after annealing, indicating that almost all C atoms still act as acceptors even after annealing. These results indicate that C is a promising p-type dopant in GaAs.

4.4.2 C δ-doping in AlGaAs using TMG and its doping mechanism

C-doping in AlGaAs is used for p-AlGaAs graded base layers in HBTs (Ito and Makimoto, 1991) and p-AlGaAs/GaAs modulation doped structures (Makimoto and Chang, 1992). It has been studied using metalorganic molecular beam epitaxy (Yamada *et al.*, 1989) and MOCVD (Cunningham *et al.*, 1990). To grow heavily C-doped layers, these growth methods usually use rather low growth temperatures and low V/III ratios. Therefore, so far it has been considered that the C incorporation mechanism is that C atoms are incorporated into the epitaxial layer through the formation of metal carbides due to incomplete reaction between trimethyl sources (TMG and/or TMA) and AsH_3 on the growing surface. In this section, C δ-doping is performed by supplying TMG on AlGaAs surfaces under AsH_3-free conditions. The doping mechanism is studied by comparing the doping efficiency between the TMG supply onto the GaAs surface and the TMG supply onto the AlGaAs surface. Based on the experimental result that C incorporation increases with increasing Al composition of the surface layer onto which TMG is supplied, we

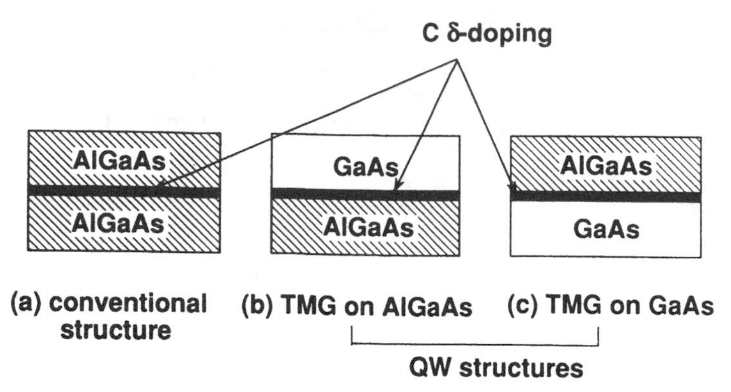

Fig. 4.17. (a) Conventional C δ-doped structure. δ-doping in QWs, (b) on an AlGaAs surface, and (c) on a GaAs surface.

propose that the C incorporation mechanism in AlGaAs is that C atoms are incorporated through the replacement of surface As atoms of AlGaAs by C atoms from the methyl-groups (Makimoto and Kobayashi, 1993). Figure 4.17(a) shows a conventional δ-doped structure. The C-doping source, TMG, was supplied onto the $Al_xGa_{1-x}As$ ($x = 0$ or 0.5) surface, followed by AsH_3 flow. Then undoped $Al_xGa_{1-x}As$ was grown to cover the C δ-doped layer. The growth temperature was 610 °C and Hall measurements were used to evaluate the sheet hole concentration (p^{2D}). For conventional δ-doping, p^{2D} was 1.0×10^{13} cm^{-2} for $Al_{0.5}Ga_{0.5}As$ and 6.0×10^{11} cm^{-2} for GaAs. To clarify how the doping concentration depends on the surface layer and on the covering material, the structures shown in Figs. 4.17(b) and (c) were grown. These are quantum-well (QW) structures. The C δ-doping was performed on the AlGaAs surface in Fig. 4.17(b) and on the GaAs surface in Fig. 4.17(c). In Fig. 4.17(b), the covering material was undoped GaAs. For this structure, p^{2D} was 1.1×10^{13} cm^{-2}, which is almost the same as the value obtained for the conventional δ-doping into $Al_{0.5}Ga_{0.5}As$, as shown in Fig. 4.17(a). In Fig. 4.17 (c), the C δ-doping was performed on the GaAs surface and the covering material was undoped $Al_{0.5}Ga_{0.5}As$. In this experiment, p^{2D} was 7.0×10^{11} cm^{-2}, which is almost the same as the value obtained for the δ-doping into GaAs. These results indicate that the doping concentration does not depend on the covering material but only on the surface layer, onto which TMG is supplied, and that an AlGaAs surface enhances the C-incorporation.

Figure 4.18 shows the dependence of p^{2D} on the Al composition of the surface layer. Open circles show p^{2D} obtained for the conventional structures in which C δ-doping was performed between $Al_xGa_{1-x}As$ layers with the same composition. Closed circles are for C δ-doping onto $Al_xGa_{1-x}As$ surfaces covered with GaAs, which corresponds to the δ-doping into GaAs-on-$Al_xGa_{1-x}As$ heterointerfaces of QW structures. As mentioned before, if the underlying layer is the same material, p^{2D} obtained for the conventional structure is almost the same value as that obtained for the QW structure. As the Al composition increases, the number of incorporated C atoms increases. The line in Fig. 4.18 has a slope of 1, so it shows that p^{2D} is proportional to the Al composition of the surface

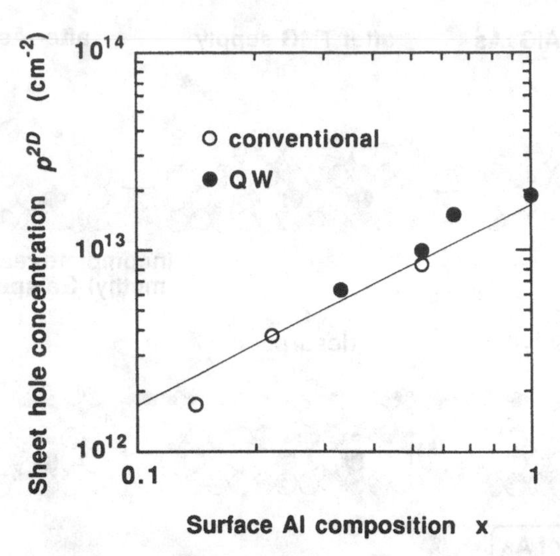

Fig. 4.18. Dependence of p^{2D} on the Al composition of the surface layer. Open and closed circles show p^{2D} obtained for conventional structures and QW structures respectively. The line has a slope of 1, showing that p^{2D} is proportional to the Al composition of the surface onto which TMG is supplied.

onto which TMG is supplied. The C incorporation rate into the AlAs surface is 30 times higher than that into GaAs. These results indicate that C atoms were incorporated into the AlGaAs surface, and that the As atoms of the AlGaAs surface were partially replaced by C atoms from methyl-groups. A high p^{2D} value of 2×10^{13} cm^{-2} was obtained for the AlAs surface and 3.1% of As atoms on the AlAs surface were replaced by C atoms, indicating that TMG does not completely decompose at 610 °C and that there co-exist Ga atoms and methyl-Ga species on the growing surface. These results coincide with the results obtained by SPA (Makimoto *et al.*, 1990).

Let us consider the C incorporation mechanism in AlGaAs. As shown in Fig. 4.19, there are two possible surface reactions for the C-doping mechanisms when TMG is supplied on the AlGaAs surface. One doping mechanism is the incomplete reaction between these methyl-Ga species and AsH_3 molecules on the surface (Fig. 4.19(a)). This may be the dominant mechanism at low growth temperatures because of the higher density of methyl-Ga species and lower reactivity of AsH_3 as the temperature decreases. In this case, the C impurities are incorporated into upper As sites of the Ga layer formed by TMG. Figure 4.19(b) shows another mechanism, in which the surface As atoms of the AlGaAs layer are replaced by C atoms from the methyl-groups of TMG supplied. Therefore, C atoms are incorporated into the lower As sites of the Ga layer formed by TMG. Since the bond-strength of Al—C is expected to be stronger than that of Ga—C, the preferable incorporation sites may be the As sites bonded with the lower Al atoms. The AsH_3 flow-rate and the decomposition rate after TMG supply in this experiment may be sufficient to complete the reaction between AsH_3 and methyl-Ga species. Considering the dependence of p^{2D} on the surface Al composition, the replacement reaction of the surface As atoms by

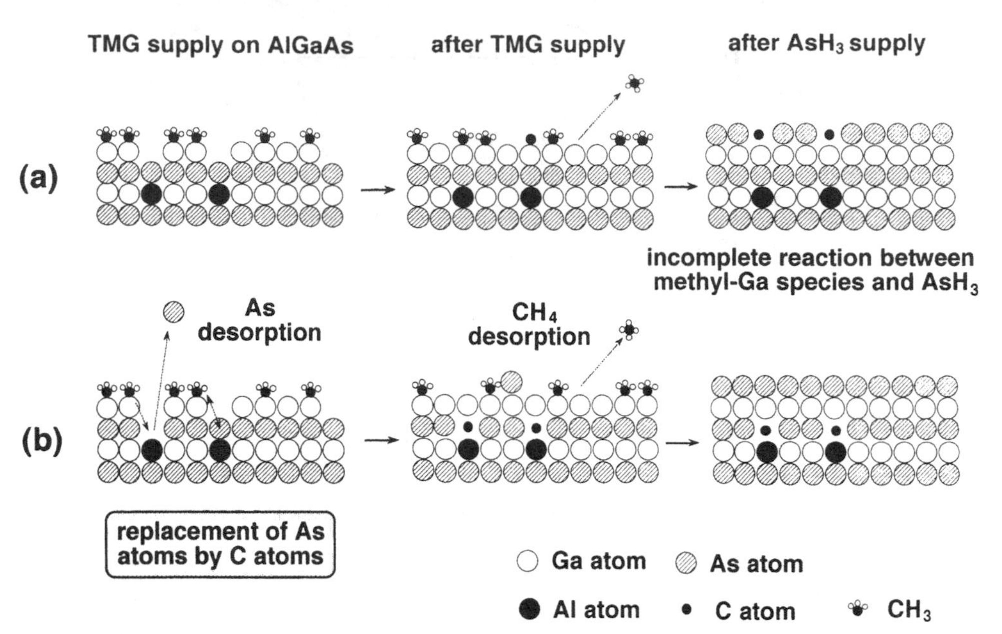

Fig. 4.19. Surface reactions for the C doping mechanisms when TMG is supplied on the AlGaAs surface. (a) Incomplete reaction between methyl-Ga species and AsH_3 molecules. (b) Replacement of the surface As atoms by C atoms from the methyl groups during a TMG supply period. The C incorporation sites are the As sites bonded with the lower Al atoms.

C atoms from methyl-groups of TMG is the dominant doping mechanism in this growth condition. To grow high-quality epitaxial layers, this replacement reaction between As and C atoms should be suppressed. In FME, a very small amount of AsH_3 (r_0 in Fig. 4.2) is introduced during the TEG flow period. This small amount of AsH_3 is considered to be effective in suppressing the above replacement reaction.

From the C–V measurement of the δ-doped structure in $Al_{0.32}Ga_{0.68}As$, the observed peaks are symmetrical and have the same shape, indicating that the memory effect was negligible and that the δ-doping was reproducible. The observed FWHM value is similar to that shown in Fig. 4.16. Considering that the FWHM is determined by the screening length, the diffusion of the C atoms in AlGaAs is not serious, even at a growth temperature of 610 °C. These results indicate that C is also a promising p-type dopant in AlGaAs.

4.5 Summary

N- and p-type δ-doped structures in GaAs or AlGaAs were grown by FME, which reduces the growth temperature while maintaining crystal quality and interface flatness. Silicon and Carbon were used as n- and p-type impurities respectively. During FME growth, Ga- and As-stabilized surfaces appeared alternately on the growing surface, so it is possible to introduce SiH_4 into either a Ga- or As-stabilized surface and to investigate the characteristics of each surface. The SiH_4 cracking efficiency was greatly enhanced when it was supplied during the Ga plane formation, due to the catalytic effect of the Ga atomic surface. Using this favorable characteristic, it is possible to grow heavily Si δ-doped structures without growth interruption, regardless of the relatively low growth temperatures. The p-type dopant in FME is C. For p-type δ-doped structures, undoped and C-doped layers are grown using triethyl and trimethyl organometals respectively. The $C-V$ profiles of the C δ-doped structures before and after annealing shows that the diffusion coefficient of C in GaAs is smaller than those of other p-type impurities, and that doping gases, that is, trimethyl organometals, have a negligible memory effect. Thus, it is possible to grow almost ideal p-type δ-doped structures with C impurities using FME. Furthermore, C δ-doping in AlGaAs reveals the doping mechanism.

We have introduced the FME method, which is a modified form of MOCVD, and demonstrated that δ-doping is suitable for the investigation of doping mechanisms, as well as for the study of low-dimensional electron physics and for the improvement of device performance.

References

Bedair S. M., Tishler M. A., Katsuyama T. and El-Masry N. A., 1985 *Appl. Phys. Lett.* **47**, 51.
Cho A. Y., 1975 *J. Appl. Phys.* **46**, 1733.
Cunningham B. T., Baker J. E., and Stillman G. E., 1990 *Appl. Phys. Lett.* **56**, 836.
Doi A., Aoyagi Y. and Namba S., 1986 *Appl. Phys. Lett.* **48**, 1787.
Duhamel N., Henoc P., Alexandre F. and Rao E. V. K., 1981 *Appl. Phys. Lett.* **39**, 49.
Field R. J. and Ghandhi S. K., 1986 *J. Cryst. Growth* **74**, 551.
Foxon C. T. and Joyce B. A., 1977 *Surf. Sci.* **64**, 293.
Horikoshi Y., Kawashima M. and Yamaguchi H., 1986 *Jpn J. Appl. Phys.* **25**, L868.
Ilegems M., 1977 *J. Appl. Phys.* **48**, 1278.
Ito H. and Makimoto T., 1991 *Appl. Phys. Lett.* **58**, 2770.
Kobayashi N., Makimoto T. and Horikoshi Y., 1986 *Jpn J. Appl. Phys.* **25**, L746.
Kobayashi N., Makimoto T. and Horikoshi Y., 1987 *Appl. Phys. Lett.* **50**, 1435.
Kobayashi N. and Horikoshi Y., 1989 *Jpn J. Appl. Phys.* **28**, L1880.
Kobayashi N., Makimoto T., Yamauchi Y., and Horikoshi Y., 1989 *J. Appl. Phys.* **66**, 640.
Lays M. R. and Veenvliet H., 1981 *J. Cryst. Growth* **55**, 145.
Lewis C. R., Ludowise M. J. and Dietze W. T., 1984 *J. Electron. Matter.* **13**, 447.
Makimoto T., Kobayashi N., and Horikoshi Y., 1986 *Jpn J. Appl. Phys.* **25**, L513.
Makimoto T., Yamauchi Y., Kobayashi N. and Horikoshi Y., 1990a *Jpn J. Appl. Phys.* **29**, L207.
Makimoto T., Yamauchi Y., Kobayashi N. and Horikoshi Y., 1990b *Jpn J. Appl. Phys.* **29**, L645.
Makimoto T. and Chang S. S., 1992 *Jpn J. Appl. Phys.* **31**, L797.
Makimoto T. and Kobayashi N., 1993 *Jpn J. Appl. Phys.* **32**, L1300.
Neave J. H., Dobson P. J., Joyce B. A. and Zhang J., 1985 *Appl. Phys. Lett.* **47**, 100.
Nishizawa J. and Kurabayashi T., 1983 *J. Electrochem. Soc.* **130**, 413.
Nishizawa J., Abe H. and Kurabayashi T., 1985 *J. Electrochem. Soc.* **132**, 1197.
Putz N., Korec J., Heyen M. and Balk P., 1983 *Proc. 4th European Conf. Chemical Vapor Deposition*, J. Bloem *et al.* (eds.), Eindhoven, Netherlands, 31 May–2 June, pp. 103–9.
Ploog K. and Fischer A., 1978 *J. Vac. Sci. Technol.* **15**, 255.
Schubert E. F. and Ploog K., 1986 *Jpn J. Appl. Phys.* **25**, 966.
Stringfellow G. B. and Linnebach R., 1980 *J. Appl. Phys.* **51**, 2212.

Suntola T. and Antson J., 1974 Finnish Patent No. 52359.
Usui A. and Sunakawa H., 1986 *Jpn J. Appl. Phys.* **25**, L212.
Veuhoff E., Kuech T. F. and Meyerson B. S., 1985 *J. Electrochem. Soc.* **132**, 1958.
Voronin V. A., Prochorov V. A., Plahotnaja L. S. and Chuchmarev S. K., 1983 *Proc. 4th European Conf. Chemical Vapor Deposition*, J. Bloem *et al.* (eds.), Eindhoven, Netherlands, 31 May–2 June, pp. 235–42.
Weyers M., Putz N., Heinecke H., Heyen M., Luth H. and Balk P., 1986 *J. Electron. Matter* **15**, 57.
Yamada T., Tokumitsu E., Saito K., Akatsuka T., Miyauchi M., Konagai M. and Takahashi K., 1989 *J. Cryst. Growth* **95**, 145.

5

Gas source molecular beam epitaxy (MBE) of delta-doped III–V semiconductors

DAN RITTER

Introduction

Delta (δ) doping in semiconductors can be achieved using several crystal growth techniques; it is interesting because of its potential applications in semiconductor devices, and as a tool for studying doping behavior during epitaxy. This chapter reviews experiments on δ doping of III–V semiconductors which were carried out using gas source molecular beam epitaxy (GSMBE). The experiments described here provide information on the behavior of δ doped layers grown by GSMBE, as well as giving more general information on doping during GSMBE.

The GSMBE methods are useful for growing high quality III–V compound semiconductor heterostructures and devices. A comprehensive review of the methods and the properties of the grown layers and devices was recently given by Panish and Temkin (1993). Gas source molecular beam epitaxy was originally developed to overcome some of the difficulties encountered in conventional molecular beam epitaxy (MBE), but was later found to have several other important advantages. In conventional MBE systems all sources of group III and group V atoms are solid, and the molecular beams are generated by evaporating the solids from effusion cells. Such systems will be referred to in this chapter as elemental source MBE (ESMBE). In the variants of GSMBE some or all of the molecular beams are generated by injecting simple gaseous compounds of the required elements into the growth chamber.

The imputus to replace the elemental sources in MBE with gaseous chemicals was the need to achieve well controlled beams of phosphorus and phosphorus plus arsenic for MBE growth of InP based compounds. The ternary and quaternary InGaAsP compounds, lattice matched to InP, are among the most important III–V semiconductors because of their applications infiber optic communications and potential applications in high speed electronics. With an elemental phosphorus source it is very difficult to obtain a well-controlled beam of phosphorus atoms because of the high vapor pressure of solid P, because solid P forms allotropes having different vapor pressures and solid P vaporizes to form P_4 molecules, which are relatively inert to interactions with the growing surface. The latter leads to unacceptably high P_4/In ratios (about 100) during epitaxy (Asahi *et al.*, 1981). Recently, however, it has been reported that P-containing compounds have been grown successfully by ESMBE using a beam of P_2 derived from solid P in effusion cells equipped with an additional high temperature cracking stage (Baillargeon *et al.*, 1994).

To overcome the solid P problem in ESMBE a molecular beam of P_2 can be obtained by cracking phosphine gas using a thermal cracker cell (Panish, 1980). It was found that at the temperature needed for efficient cracking of PH_3 the cracking products are mainly P_2 and H_2 molecules and that the P_2/In ratio needed for the growth of high quality InP can be as low as unity (Panish and Sumski, 1984). It was also realized that in an MBE system that was modified to allow the use of PH_3 there are advantages in using arsine as a source for As instead of solid As. The advantages are that both hydrides can be cracked in the same cracker cell so that the As/P ratio across the wafer is uniform, and because the gas sources are external, source replenishment is accomplished while the MBE system is under vacuum. These merits have led several groups to use this version of GSMBE for the growth of III–V semiconductors that do not contain phosphorus, such as GaAs/GaAlAs, and GaInAs. In this chapter GSMBE systems that use hydrides as group V sources, and solid group III sources, will be referred to as hydride source MBEs (HSMBEs).

The main modification required to convert an ESMBE reactor into an HSMBE reactor is the installation of additional pumps with a large pumping speed in order to maintain low pressure during growth. In such modified MBE systems it was natural to attempt to use gaseous metalorganic group III sources as well (Veuhoff *et al.*, 1991, Vogjdani *et al.*, 1982) to replace the elemental group III sources. Such efforts were successful (Tsang, 1990). This variant of GSMBE is referred to here as metal organic MBE (MOMBE) and is sometimes referred to as chemical beam epitaxy (CBE). The advantages of the MOMBE method are that all sources are external to the vacuum reactor, and all group III molecular beams originate from the same injector, assuring better unformity across the wafer. By contrast with the group V gases the metalorganic group III compounds are thermally cracked on the crystal surface, and precracking at the source is not needed.

The MOMBE method was developed historically from the ESMBE technique, using the same large vacuum chambers. However, for systems designed exclusively for MOMBE, dramatic simplification and size reduction can be achieved, mostly because the combined group III source axis can be normal to the surface of the growing crystal, and because there is no longer a need to rotate the substrate. A compact MOMBE reactor is simpler to operate and to maintain, and, naturally, is also much less expensive. Recently, a prototype of such a system was tested at AT&T Bell Laboratories by Hamm *et al.* (1994a), and it was demonstrated that the material quality is as good as in the standard MOMBE systems. A schematic drawing of the compact MOMBE reactor is shown in Fig. 5.1.

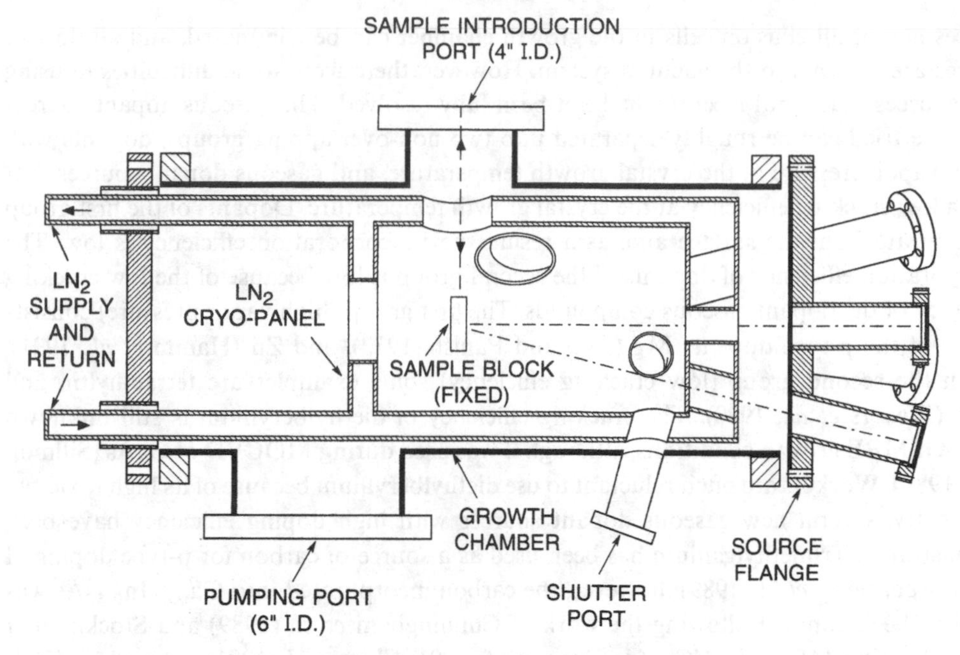

Fig. 5.1. Schematic representation of a compact MOMBE growth chamber. The chamber was designed with the aim of decreasing the size of the vacuum chamber to the minimum needed (Hamm *et al.*, 1994a).

Dopant sources in GSMBE

One of the main advantages of the GSMBE techniques is the ability to achieve the high dopant concentration and abrupt dopant profiles required for high speed electronic devices. For example, electron concentrations up to 10^{20} cm^{-3} were obtained by doping $Ga_{0.47}In_{0.53}As$ with Sn (Panish *et al.*, 1990), and similar hole concentrations were obtained by using Be (Hamm *et al.*, 1989). The high doping levels and abrupt profiles can be obtained because the growth temperature during GSMBE is lower than is usually possible with other epitaxy methods. For example, the growth temperature of InP based compounds during MOMBE is about 500 °C, whereas the same materials are grown at about 600 °C by the metalorganic chemical vapor deposition (MOCVD) technique. The lower growth temperature during MOMBE is possible because the hydrides are precracked and the reactant is the dimer of the group V element, whereas in MOCVD the hydrides are cracked on the substrate. In fact, cracking of the hydrides on the substrate is inefficient in GSMBE, apparently because of the very short residence time of the hydride molecule on the crystal surface. However, it was demonstrated recently that GaAs can be grown using arsine without precracking (Park *et al.*, 1993).

In GSMBE it is possible to use either solid dopant sources evaporated from effusion cells, as in ESMBE, or gaseous dopant sources that decompose on the substrate, as in MOCVD. Both methods are currently in use. The advantage of using gaseous dopant sources is that all effusion cells in the growth chamber can be eliminated, and all dopant sources are external to the vacuum system. However, there were some difficulties in using such sources that, until recently, had not been fully resolved. The gaseous dopant sources that were tried can be roughly separated into two non-overlapping groups: dopants with a high vapor pressure at the crystal growth temperature, and gaseous dopant sources that have a low cracking efficiency at the crystal growth temperature. Dopants of the first group re-evaporate from the surface and, as a result, their incorporation efficiency is low. The incorporation efficiency of dopants of the second group is low because of the low cracking efficiency of the dopant gaseous compounds. The first group (high vapor pressure) consists mainly of the p-type dopants Mg (Cho and Panish, 1972), and Zn (Hamm *et al.*, 1991), and in the second group (low cracking efficiency) some examples are tetraethyltin and silane (Weyers *et al.*, 1990). The cracking efficiency of diethylberyllium is still unknown under GSMBE growth conditions, although it was used during MOCVD of GaAs (Sillmon *et al.*, 1989). Workers are often reluctant to use diethylberyllium because of its high toxicity.

Recently, several new gaseous dopant sources with high doping efficiency have been demonstrated. Trimethylgallium has been used as a source of carbon for p-type doping of GaAs (Abernathy *et al.*, 1989); however, the carbon incorporated into $Ga_{0.47}In_{0.53}As$ was found to be a donor. Following the work of Cunningham *et al.* (1989) and Stockman *et al.* (1992) using MOCVD, CCl_4 (de Lyon *et al.*, 1991; Chin *et al.*, 1991; Song *et al.*, 1994) and CBr_4 (Hamm, 1994b) were shown to be highly efficient sources for p-type C doping of GaAs, GaInP, and $Ga_{0.47}In_{0.53}As$. Jackson *et al.* (1994a) have found that $SiBr_4$ is an efficient source of Si, and ultra high Si doping levels with abrupt profiles were achieved in InP, as shown in Fig. 5.2. An interesting possibility (so far not reported for GSMBE) is the use of such compounds as Fe $(CO)_5$ and Fe $(C_5H_5)_2$ for Fe doping of InP and GaAs to achieve semi-insulating material.

δ doping during GSMBE of GaAs

The properties of δ doped GaAs layers grown by HSMBE are similar to those grown by ESMBE because the growth precursors in both methods are similar. Thus the reader interested in δ doping of GaAs during HSMBE is also referred to Chapter 3, by Ploog, which describes doping during ESMBE. There are, however, some differences between the methods, as outlined below. These differences, along with experimental data comparing δ doping during HSMBE and ESMBE have led one group to the conclusion that the properties of δ doped GaAs layers gown by HSMBE are superior to those grown by ESMBE (Cunningham *et al.*, 1990; Ourmazd *et al.*, 1990). In my opinion more experiments are needed to clarify this point.

In both HSMBE and ESMBE growth of GaAs, the group III precursors are Ga atoms derived from solid sources. The difference between the methods is that the arsenic atoms are obtained from a solid source in ESMBE, and from arsine in HSMBE. In ESMBE of GaAs the arsenic growth precursors are either As_4 or As_2 molecules. The tetramer is obtained from a simple effusion cell operated at relatively low temperatures, and the

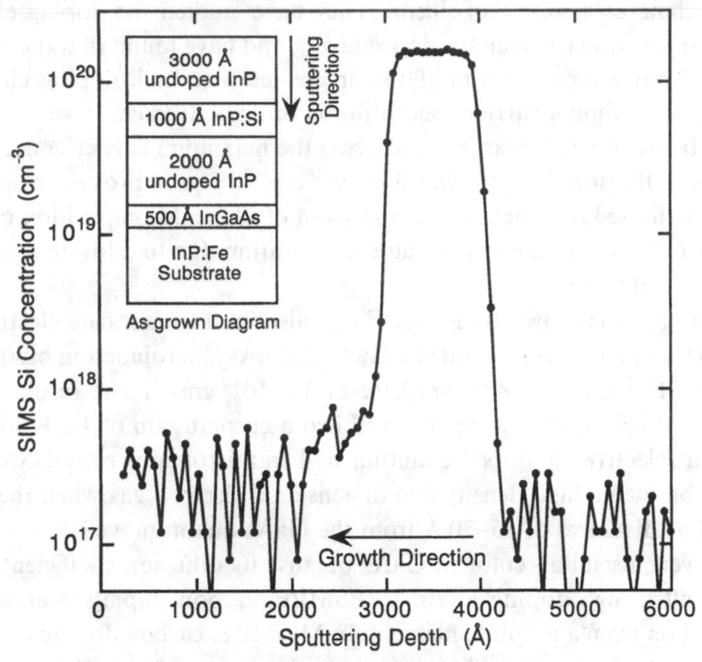

Fig. 5.2. SIMS depth profile showing a 1000 Å Si doping pulse in InP grown by HSMBE using $SiBr_4$ (Jackson *et al.*, 1994b).

dimer may be obtained if an additional high temperature cracker is added to the effusion cell. In HSMBE, the group V growth precursors are always dimer molecules because at the high temperature needed to crack arsine efficiently (about 950 °C) the ratio of As_4/As_2 molecules is very small (Panish and Sumski, 1984). An additional difference between the methods is that the cracking of arsine produces hydrogen molecules and a small amount of atomic hydrogen. It is not clear at present if the hydrogen has any role in the growth process during HSMBE of GaAs, although it has been suggested that the hydrogen saturates dangling bonds during growth, leading to the pinning of the Fermi level (Cunningham *et al.*, 1990). However, to the best of my knowledge no detailed studies on the role of hydrogen in HSMBE have yet been reported.

The most widely used dopants in HSMBE of GaAs are Be and Si, as is the case in ESMBE. Delta doping experiments during GSMBE using both dopants were reported by Cunningham *et al.* (1990). They found that the silicon diffusion versus inverse temperature relationship shows a two component Arrhenius plot in which activation energies for diffusion differ by an energy equal to the bandgap of GaAs. This effect was later modeled by Cunningham *et al.* (1991) via a non-equilibrated concentration of Ga vacancies generated at the δ position during anneal. Chiu *et al.* (1988) investigated Si δ doping of GaAs during MOMBE. They obtained narrow dopant distributions at 500 °C. However, at the same growth temperature, narrow dopant distributions are also obtained during ESMBE of GaAs (Schubert *et al.*, 1990).

Sandhu *et al.* (1992) have used the δ doping technique to improve the doping efficiency of

GaAs using disilane as a source of silicon. They have studied the doping efficiency as a function of the separation between the doped layers, and have found that the incorporation efficiency is 100% at a separation of 100 Å or greater. The bulk doping efficiency using disilane is 60%. A maximum carrier concentration of 7.5×10^{18} cm^{-3} was obtained when the separation between layers was 50 Å, whereas the maximum carrier concentration that was obtained in bulk doped layers was 3×10^{18} cm^{-3}. The improvement in the doping efficiency was attributed to either the enhancement of silane decomposition on the arsenic stabilized surface, or to a reduction in autocompensation due to a lower concentration of vacancies during δ doping.

Delta doping of GaAs grown by HSMBE was also used to fabricate electronic devices. Goossen *et al.* (1991) have demonstrated a GaAs/AlGaAs heterojunction bipolar transistor grown by HSMBE, with a Be δ doped base at 3×10^{14} cm^{-2}. The resulting base sheet resistance was 270 Ω/square. The device exhibited a current gain of 14. Kuo *et al.* (1988) have reported a selectively δ doped quantum well transistor grown by HSMBE using Si doping. They obtained a high density two dimensional electron gas when the doped layer was positioned at distances of 15–70 Å from the GaAs quantum well.

Carbon is a very useful acceptor in GaAs because its diffusion coefficient is negligibly small, even at ultra high doping levels. In ESMBE, carbon doping is accomplished by evaporating carbon from a graphite filament. In MOMBE, carbon doping can be achieved using gaseous carbon sources. The most common source for carbon doping during MOMBE is trimethylgallium, which serves both as a source of Ga and of carbon. For the growth of GaAs with minimal C incorporation, triethylgallium is used. With trimethylgallium, doping levels close to 10^{20} cm^{-3} were obtained, and this technique is widely used for doping of the base layer in GaAs/GaAlAs heterojunction bipolar transistors (Abernathy *et al.*, 1989).

Because the most frequently used carbon source in MOMBE is trimethylgallium, which is also a source of Ga, carbon δ doping cannot be carried out during a growth interruption using this source. However, δ doping-like profiles were achieved by heavily carbon doping one monolayer of GaAs. Using this technique a carbon doping profile with a FWHM of 50 Å was obtained having a peak concentration of 7×10^{19} cm^{-3}, as can be seen in Fig. 5.3. As mentioned above, CCl_4 and CBr_4 have recently been found to be efficient C sources, and these new sources can be used for depositing C on the surface during a gorwth interruption.

Beryllium δ doping of $Ga_{0.47}In_{0.53}As$ and InP during MOMBE

Beryllium is the most widely used p-type dopant in MBE of III–V compounds. During ESMBE growth it has a sticking coefficient of unity due to its very low vapor pressure at the typical growth temperatures of III–V semiconductors (Ilegems, 1977). Other p-type dopants such as Mg and Zn have a much higher vapor pressure, and are therefore not often used during MBE growth. By contrast to the case of Be doping during ESMBE, in MOMBE Be can be removed from the surface by the organic compounds which are

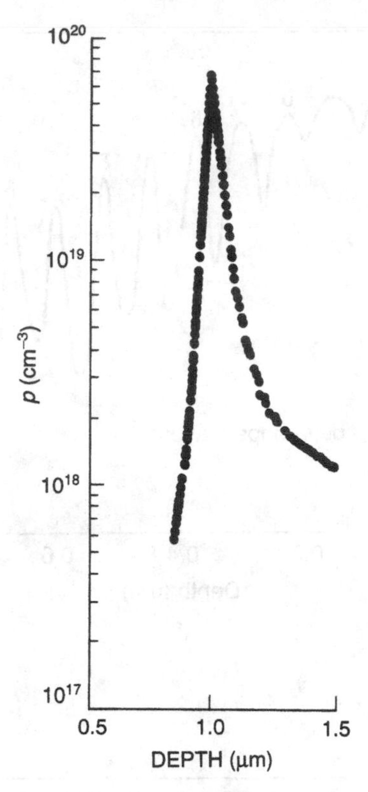

Fig. 5.3. Polaron profile of a carbon δ-doped GaAs layer grown by MOMBE using trimethylgallium as a carbon source (Abernathy *et al.*, 1989).

generated by the decomposition of the metalorganic compounds (Ritter *et al.*, 1991, 1993). As a result, the incorporation efficiency of Be during MOMBE growth is lower than unity. Beryllium δ doping experiments have contributed more information on this effect.

Figure 5.4(a) and (b) depicts the Be SIMS profile of InP and GaInAs samples, grown by MOMBE at a substrate temperature of 510 °C and a growth rate of 1.2 μm/hr. The Be was derived from a conventional effusion cell. Both layers consisted of several pairs of a 200-Å-thick Be doped layer followed by a δ doped layer. The δ doped layers were grown by depositing Be on the surface during a growth interruption, with no additional pause in growth before starting the Be deposition, or after the Be deposition. During the growth of each pair of layers the Be oven temperature was kept constant, and identical doses of Be were deposited on the surface (or growing surface) for the δ doped and bulk doped layers. The duration of the growth interruption for δ doping was thus the time needed to grow a 200-Å-thick layer. A 500-Å-thick undoped spacer layer was grown between the bulk doped and the δ doped layer. The Be oven temperature was increased for each consecutive pair.

The GaInAs sample consisted of six pairs of bulk δ doped layers grown at Be oven temperatures of 830 °C–1130 °C with 50 °C intervals. For the highest Be concentration the

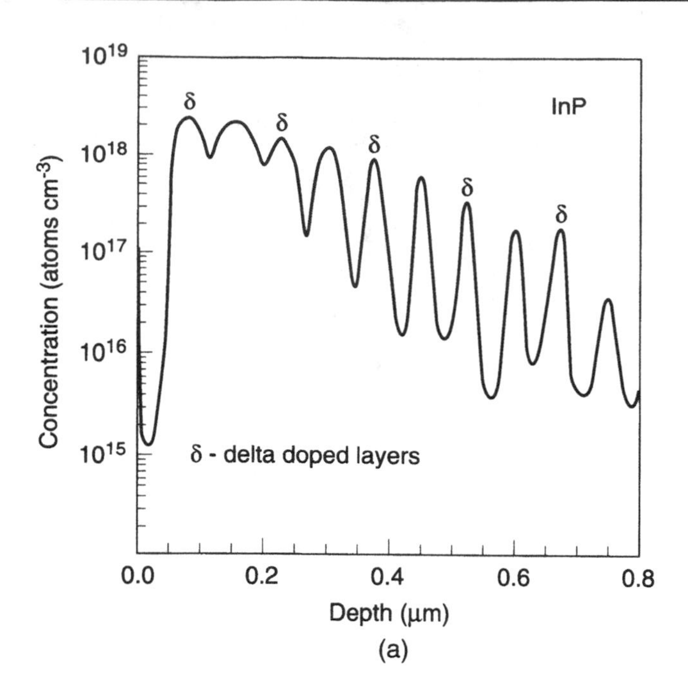

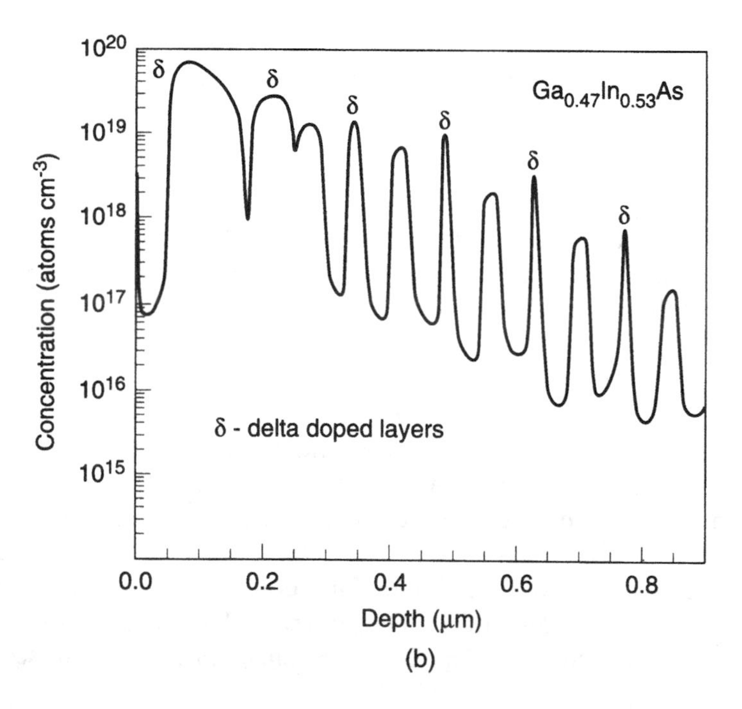

Fig. 5.4. Beryllium SIMS profile of a $Ga_{0.47}In_{0.53}As$ (a) and InP (b) layer with δ-doped and bulk doped layers.

500-Å-thick spacer layer was not sufficient to prevent merging of the δ doped and bulk doped regions, due to the onset of fast diffusion of Be at concentrations close to 10^{20} cm^{-3} (Panish *et al.*, 1991). The InP sample consisted of five pairs of bulk δ doped layers grown at Be oven temperatures of 830 °C–1080 °C with 50 °C intervals. It can clearly be seen that the diffusion of Be in InP is much faster than in GaInAs, as was found in bulk doping experiments, which demonstrated that the onset of fast diffusion in InP occurs at a Be concentration of about 2×10^{18} cm^{-3} (Panish *et al.*, 1991; Veuhoff *et al.*, 1991). Considerable widening of the δ doped regions can also be observed.

The dose of Be atoms incorporated in each doped layer, as measured by SIMS, is given in Fig. 5.5 as a function of the Be oven temperature. It can be seen that in the GaInAs layers, the Be dose per pulse in the δ doped layers is larger by a factor of about two than that in the bulk doped layers. Since the bulk doped layers and the δ doped layers were exposed to identical Be doses, this clearly demonstrates that Be atoms were removed during the growth of the bulk doped InP and GaInAs layers.

From Fig. 5.5 it can also be seen that although the InP and GaInAs samples were exposed to identical Be doses, the Be concentration in InP is lower than in GaInAs for both the bulk doped and δ doped layers. This result agrees with previous data, and was also verified by Hall measurements (Ritter *et al.*, 1991). Assuming that some Be is lost from the surface as a result of the reaction of Be with trimethylgallium or methyl radicals to form diethylberyllium, which evaporates, the lower incorporation efficiency of Be in

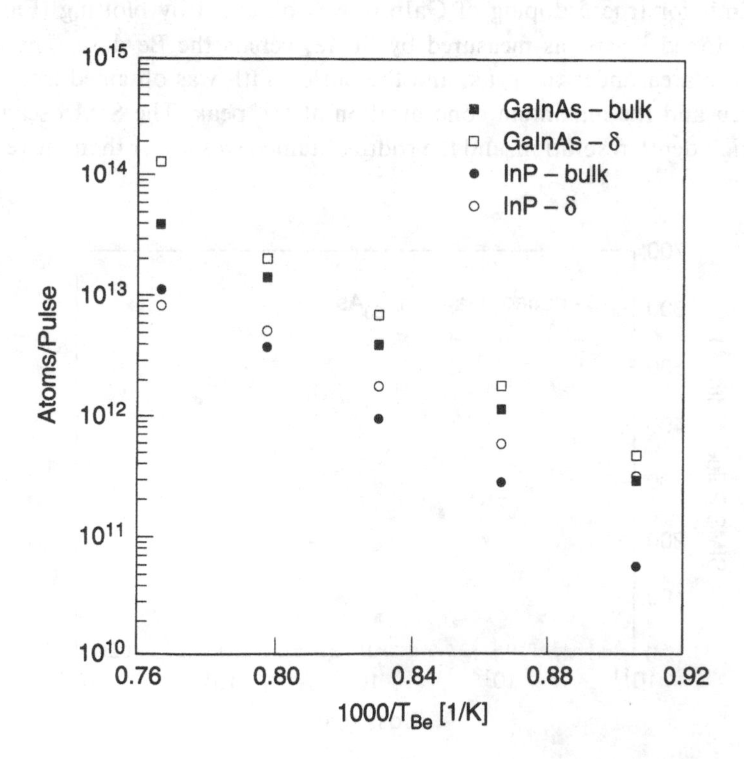

Fig. 5.5. Beryllium dose per pulse versus inverse Be cell temperature.

InP can be explained as follows. The metalorganic source of Ga is triethylgallium. During the growth of InP, the concentration of trimethylindium on the growing surface is higher by a factor of about two than it is during the growth of GaInAs at the same growth rate. Assuming that the diethylgallium or ethyl radicals are relatively slow to remove Be, and noting that for the removal of one Be atom two organic radicals are needed, the removal rate of Be atoms during the growth of InP may be expected to be higher by as much as a factor of about four than during the growth of GaInAs.

The removal of Be from the surface makes it more difficult to achieve well controlled Be δ doping during MOMBE because the partial pressure of the organic compounds during growth interruptions may vary. This can be seen in Fig. 5.5, where the data for the δ doped layers is seen to have a larger scatter than that for the bulk doped layers. The partial pressure of the organic compounds is higher during growth than during a growth interruption, but it is also more reproducible. Thus, the incorporation efficiency of Be during δ doping is higher than during bulk doping, but bulk doping is more reproducible. The removal of Be during δ doping takes place during the growth interruption (by the residual gases in the growth chamber) and also when growth is resumed (before the Be on the surface is incorporated into the bulk). The average time for which a Be atom on the surface is exposed to organic gases in the growth chamber is about the time it takes to grow one monolayer both during bulk doping, and when growth is resumed after δ doping.

The upper limit for true δ doping of GaInAs was obtained by plotting (Fig. 5.6) the width of the δ doped layers, as measured by SIMS, versus the Be dose. The dose was obtained from the area under the peak, and the pulse width was obtained from the ratio between the area and the maximum concentration at the peak. The SIMS scan was not optimized for high depth resolution, and the width obtained was larger than the real dopant

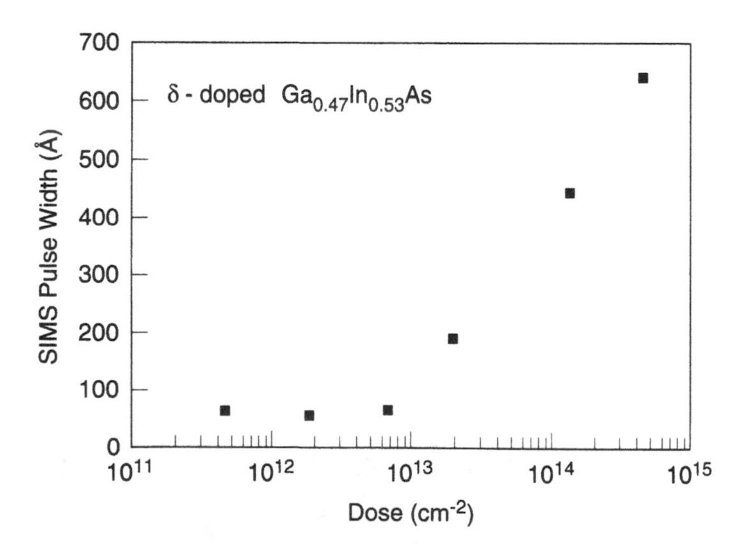

Fig. 5.6. Pulse width versus Be dose.

distribution, as explained below. Nevertheless, a clear onset for broadening can be observed in Fig. 5.6 at a dose of about $7 \times 10^{12}\ cm^{-2}$.

The true width of the δ doped layers can be estimated using data obtained from bulk doping experiments. Panish *et al.* (1991) found that there is an onset of fast diffusion of Be in GaInAs at a concentration that depends on the growth temperature. Below this concentration no diffusion of Be was observed in GaInAs. At a growth temperature of 510 °C, the onset of fast diffusion occurs at a concentration of about $5–7 \times 10^{19}\ cm^{-3}$, as was also found by Ritter *et al.* (1991). Assuming that the concentration of Be atoms in the doped layer at which the onset of broadening takes place (which corresponds to a dose of $7 \times 10^{12}\ cm^{-2}$) is $7 \times 10^{19}\ cm^{-3}$, one obtains a thickness of 10 Å for the Be distribution in the δ doped layer. It thus seems that the Be atoms are confined to about one monolayer.

For InP true δ doping is not possible even at low doses, as seen in Fig. 5.4. It is well documented that acceptors in InP diffuse extremely fast at concentrations above $2 \times 10^{18}\ cm^{-3}$ (Ritter *et al.*, 1991, 1993; Veuhoff *et al.*, 1991). For a monolayer-thick δ doped layer this corresponds to a dose of about $10^{11}\ cm^{-2}$, which is lower than the range of doses attempted here.

Summary

The understanding of doping during GSMBE and the search for new and better dopant sources are important tasks for investigators who seek to establish GSMBE as a leading growth technique of III–V semiconductors. It is the excellent doping capability during GSMBE that is the main advantage of GSMBE over other growth methods. In this chapter the possibilities offered by the δ doping method were reviewed as well as other approaches for doping during GSMBE. It is my belief that gas source dopants will soon become the most widely used dopant sources during GSMBE.

Acknowledgment

The author wishes to thank M. B. Panish for helping to prepare the manuscript.

References

Abernathy C. R., Pearton S. J., Caruso R., Ren F., Kovalchik J. (1989) *Appl. Phys. Lett.* **55**, 1750.
Asahi H., Kawamura Y., Ikeda M., Okamoto H. (1981) *J. Appl. Phys.* **52**, 2852.
Baillargeon J. N., Cho A. Y., and Fischer R. J. (1994) *Proceedings of the Sixth International Conference on InP and Related Materials, Santa Barbara*, p. 148.
Chin T. P., Kirchner P. D., Woodall J. M., and Tu C. T. (1991) *Appl. Phys. Lett.* **59**, 2865.
Chiu T. H., Cunningham J. E., and Tell B. (1988) *Appl. Phys. Lett.* **64**, 1578.
Cho A. Y. and Panish M. B. (1972) *Appl. Phys. Lett.* **43**, 5118.
Cunningham B. T., Haase M. J., McCollum M. J., Baker J. E., and Stillman G. E. (1989) *Appl. Phys. Lett.* **54**, 1905.

Cunningham J. E., Chiu T. H., Tell B., and Jan W. (1990) *J. Vac. Sci. Technol.* **B8**, 157.
Cunningham J. E., Chiu T. H., Jan W., and Kuo Y. (1991) *Appl. Phys. Lett.* **59**, 1452.
Goossen K. W., Cunningham J. E., Kuo T. Y., Jan W. Y., and Fonstad C. G. (1991) *Appl. Phys. Lett.* **59**, 682.
Hamm R. A., Panish M. B., Nottenburg R. N., Chen Y. K., and Humphrey D. A. (1989) *Appl. Phys. Lett.* **54**, 2586.
Hamm R. A., Ritter D., Temkin H., and Panish M. B. (1991) *Appl. Phys. Lett.* **58**, 2378.
Hamm R. A., Ritter D., and Temkin, H. (1994a), to be published in *J. Vac. Sci. Technol.*
Hamm R. A. (1994b), unpublished.
Ilegems M. (1977) *J. Appl. Phys.* **48**, 1278.
Jackson S. L., Thomas S., Fresina M. T., Ahmari D. A., Baker J. E., and Stillman G. E. (1994a) *Proceedings of the Sixth International Conference on InP and Related Materials, Santa Barbara*, p. 57.
Jackson S. L., Fresina M. T., Baker J. E., Stillman G. E. (1994b) *Appl. Phys. Lett.* **64**, 2867.
Kuo T. Y., Cunningham J. E., Schubert E. F., Tsang W. T., Chiu T. H., Ren F., and Fonstad C. G. (1988) *J. Appl. Phys.* **64**, 3324.
de Lyon T. J., Buchan N. I., Kirchner P. D., Woodall J. M., McInturff D. T., Scilla G. J., and Cardone F. (1991) *Journal of Crystal Growth* **111**, 564.
Ourmazd A., Cunningham J., Jan W., and Rentschler A. (1990) *Appl. Phys. Lett.* **56**, 854.
Panish M. B. (1980) *J. Electrochem. Soc.* **127**, 2729.
Panish M. B. and Sumski S. (1984) *J. Appl. Phys.* **55**, 3571.
Panish M. B., Hamm R. A., Hopkins L. C., Chu S. N. G. (1990) *Appl. Phys. Lett.* **56**, 1137.
Panish M. B., Hamm R. A., Ritter D., Luftman H. S., and Cotell C. M. (1991) *J. Cryst. Growth* **112**, 343.
Panish M. B. and Temkin H. (1993) *Gas Source Molecular Beam Epitaxy: Growth and Properties of Phosphorus Containing III–V Heterostructures*, Springer-Verlag, Berlin.
Park S. J., Ro J. R., Sim J. K., and Lee E. H. *Proceedings of ICCBE-4* (1993).
Ritter D., Hamm R. A., Panish M. B., and Geva M. (1991) *Proceedings of the Third International Conference on InP and Related materials*, Cardiff.
Ritter D., Hamm R. A., Panish M. B., and Geva M. (1993) *Apl. Phys. Lett.* **63**, 1543.
Sandhu A., Yamaura S., Okamoto N., Ando H., and Fujii T. (1992) *Electronics Letters* **28**, 1188.
Schubert E. F., Kuo J. M., Kopf R. F., Luftman H. S., Hopkins L. C., and Sauer N. J. (1990) *J. Appl. Phys.* **67**, 1969.
Sillmon R. S., Hues S. M., Gaskil D. K., Bottka N., and Thomson P. E. (1989) *J. Electronic Materials* **18**, 501.
Song J. I., Hong B. W. P., Palstrom C. J., and Chough K. B. (1994) *Proceedings of the Sixth International Conference on InP and Related Materials*, Santa Barbara, p. 523.
Stockman S. A., Hanson A. W., and Stillman G. E. (1992) *Appl. Phys. Lett.* **60**, 2903.
Tsang W. T., Sudho A. S., Yang L., Camarda R., Leibenguth R. (1989) *Appl. Phys. Lett.* **54**, 2336.
Tsang W. T. (1990) *J. Cryst. Growth* **105**, 1.
Veuhoff E., Baumeister H., Rieger J., Gorgel M., and Treichler R., (1991), *Proceedings of the Third International Conference on InP and Related materials*, Cardiff.
Vogjdani N., Lamarchand A., and Paradan M. (1982) *J. Phys. Colloq.* C5 **43**, 339.
Weyers M., Musolf J., Marx D., Kohl A., and Balk P. (1990) *J. Cryst. Growth* **105**, 383.

6

Solid phase epitaxy for delta-doping in silicon

I. EISELE

Introduction

For modern silicon devices abrupt heterostructures and extremely sharp doping profiles are essential in order to achieve small dimensions. For new device concepts the control of thickness as well as stoichiometric composition, even on an atomic scale, is necessary because it allows the creation of artificial structures that exhibit novel physical properties [1]. The growth of abrupt doping profiles or delta-doping profiles (δ-doping) is one important part of this task. The most established technique for depositing layers with monolayer thickness control is molecular beam epitaxy.

The well-known standard doping techniques in silicon are epitaxy, diffusion, and ion implantation. From these methods, only epitaxy under ultra-high vacuum (UHV) condition fulfils the requirements for two-dimensional monolayer growth [2]. However, for silicon the well-controlled incorporation of doping atoms during epitaxial growth was not possible for a long time. Several problems arise due to low sticking coefficients of the doping materials [3–8] and segregation effects [4–12] at the growing surface (see Fig. 6.1). As a result, the incorporation probabilities are quite low and the doping profiles are significantly broadened, even at growth temperatures below 700 °C [13, 14]. For δ-doping profiles segregation has to be reduced significantly, which can be achieved by lowering the growth temperature. In addition, a decreasing flux of the impinging silicon atoms lowers the transition temperature between thermodynamic equilibrium and the kinetically limited segregation regime [9].

Besides segregation of doping atoms, the parameters temperature and silicon flux also determine the growth mode, which changes from single crystalline to amorphous with decreasing temperature and increasing flux [15, 16]. For the fabrication of abrupt doping transitions and δ-doping, therefore, either low temperature epitaxial growth (substrate temperatures slightly below 400 °C) or amorphous growth (with typical growth at room temperature) followed by an annealing step for recrystallization can be used. The latter

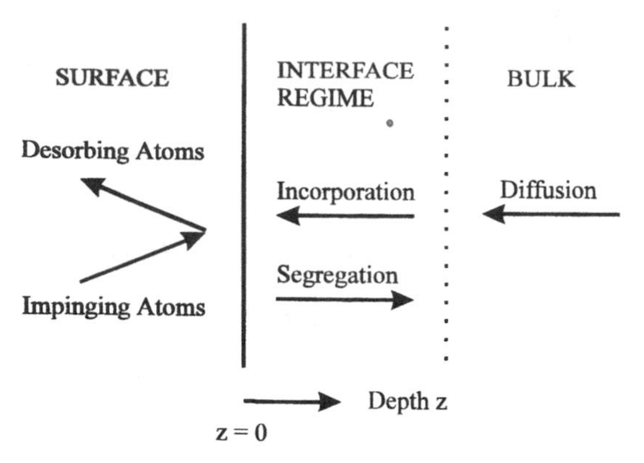

Fig. 6.1. Schematic diagram of the growing surface during silicon MBE.

process is called solid phase epitaxy (SPE) and its application for the growth of δ-doping layers in silicon will be reviewed in this chapter.

Solid phase epitaxy of silicon

Solid phase epitaxy (SPE) describes the process by which amorphous films crystallize epitaxially on a single crystalline substrate at temperatures below the melting point. In the case of silicon, SPE growth becomes noticeable above a temperature of about 500 °C [17]. The SPE growth rate fits an Arrhenius-type expression with an activation energy of about 3 eV [18]. The activation energy is the same for all substrate orientations, while the SPE growth rates differ significantly (see Fig. 6.2).

In SPE the above-mentioned problems connected with doping crystalline MBE layers can be largely avoided by a two-step process. At first, silicon and the dopant material are deposited in an ultra-high vacuum onto the substrate at a temperature where segregation can be neglected. For practical purposes, room temperature is chosen, resulting in amorphous films. In subsequent annealing steps, epitaxial crystallization takes place [17, 19, 20].

The vacuum requirements in the case of SPE become very stringent. The low deposition temperatures enhance undesired impurity incorporation because the sticking coefficient approaches one. In addition, the vacuum-deposited amorphous silicon films usually contain microscopic voids throughout the material which significantly increase the probability of impurity absorption [21, 22]. The absorbed contaminations reduce the SPE growth rate and lead to heavily disturbed films.

Cleanliness of the amorphous crystalline interface is another basic requirement for successful SPE growth. It has been shown that by decreasing the oxygen concentration at the substrate surface, the crystallization mode changes from polycrystalline to layer-by-layer SPE growth [23]. Because *ex-situ* cleaning procedures do not usually eliminate all

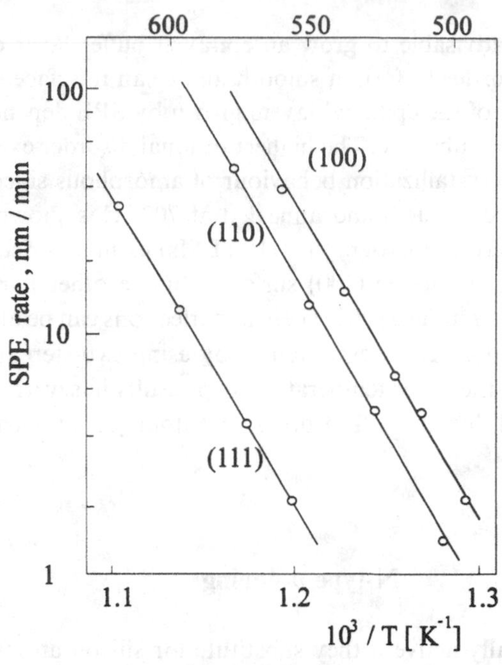

Fig. 6.2. Temperature dependence of the SPE rates for (100), (110) and (111) silicon substrate orientations [25].

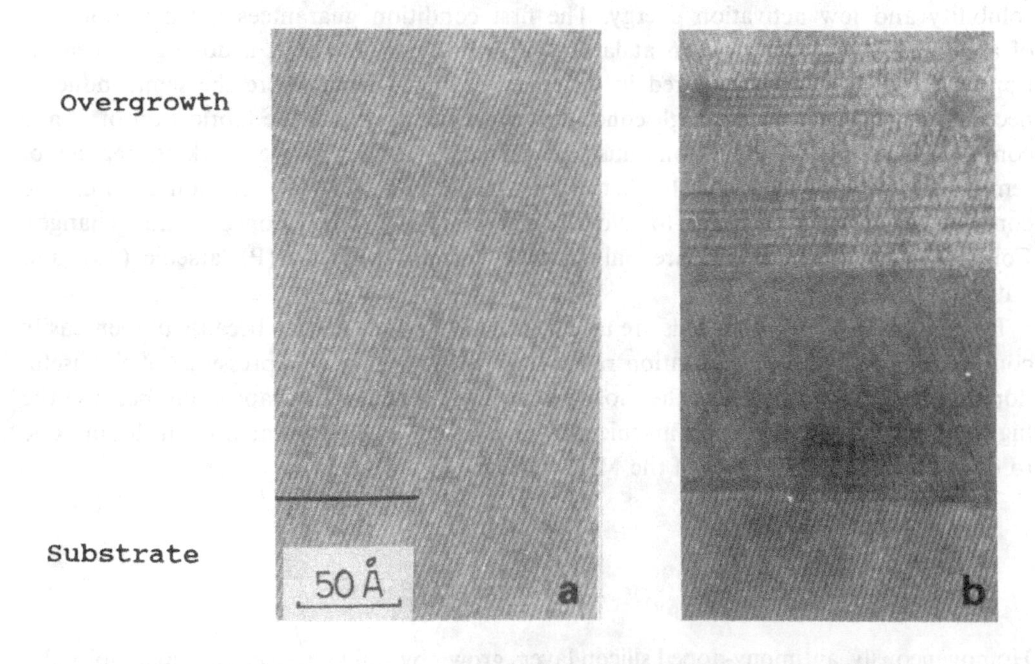

Fig. 6.3. High resolution TEM pictures of recrystallized silicon layers grown on (a) (100) and (b) (111) substrates at room temperature. Recrystallization temperature: $T = 700\,°C$. The (111) surface orientation shows twin formation [26].

contaminations [24], it is advisable to grow an epitaxial buffer layer onto the substrate under UHV conditions in order to form a smooth and clean interface.

The structural perfection of the epitaxial layers grown by SPE depends strongly on the orientation of the underlying substrate. The highest residual disorder exists for films grown on (111) substrates. The recrystallization behaviour of amorphous silicon films deposited on (100)- and (111)-oriented surfaces and annealed at 700 °C is shown in Fig. 6.3. High resolution transmission electron micrographs (HRTEMs) exhibit perfect crystalline structure without extended defects for the (100) surface. On the other hand, a distinct twin formation occurs for the (111) surface [26]. These imperfections can be annealed at elevated temperatures. The best results have been obtained by using two-step annealing at 600 °C and 900 °C [27]. However, the latter temperature step results in severe diffusion problems for the doping material and thus the SPE process for δ-doping is restricted to (100) surface orientations.

N-type δ-doping

Doping atoms are electrically active if they substitute for silicon atoms at regular lattice sites. Only in this case can the excess electron of group-V doping elements move freely in the conduction band. The common requirements for n-type doping materials are high solubility and low activation energy. The first condition guarantees the incorporation of a high doping concentration at lattice sites. For n-type silicon a doping concentration $>2 \times 10^{19}\ \mathrm{cm}^{-3}$ is required in order to reach the state where the semiconductor becomes degenerate. Such a high concentration is necessary for the fabrication of ohmic contacts. The second condition must be fulfilled because in the working regime of semiconductor devices (typically between -40 °C and $+70$ °C) all donors must be completely activated in order to avoid temperature dependent concentration changes. For silicon, both conditions are only fulfilled for phosphorous (P), arsenic (As), and antimony (Sb).

For silicon MBE, effusion cells are usually used as doping sources because of their easily controllable and stable evaporation rates. Comparing the vapour pressures of the useful doping materials, antimony is the most promising candidate for evaporation because the high vapour pressures of solid arsenic and phosphorous cause unintentional doping due to desorption from hot parts of the MBE system.

Antimony

Homogeneously, antimony-doped silicon layers grown by solid phase epitaxy in a molecular beam epitaxial growth chamber are well characterized [18, 28, 29]. The advantage of SPE growth is the enhanced maximum carrier concentration of $2\text{–}3 \times 20^{20}\ \mathrm{cm}^{-3}$, which exceeds the equilibrium solubility limit at 700 °C ($2 \times 10^{19}\ \mathrm{cm}^{-3}$) by a factor of about 10.

The first demonstration of any δ-doping in silicon was achieved with Sb and was reported by Zeindl *et al.* [30] in 1987; the demonstration employed SPE according to the following procedure. On top of a (100)-oriented substrate a 400-nm-thick epitaxial buffer is grown at 700 °C in order to obtain a clean and atomically smooth surface. Afterwards, the substrate is cooled down to room temperature and, by opening the shutter of the Sb effusion cell, the desired coverage of the doping material is deposited. The adsorbed dopants are covered by a 3 nm amorphous silicon cap which is then recrystallized by heating to 700 °C, where the epitaxial growth is continued. It turns out that by using this sequence segregation can be avoided and an almost complete incorporation of the doping material at lattice sites is guaranteed over a wide doping range. A different approach has been reported by van Gorkum *et al.*, who deposited the doping sheet as well as the complete silicon overgrowth up to 80 nm at room temperature [31]. Afterwards, they carried out a recrystallization step followed by high temperature annealing at 750 °C for 10 min. They also studied the adsorption and desorption kinetics of Sb on (100) silicon substrates extensively in order to produce uniform sheet doping concentrations [33]. Keeping the substrate temperature between 550 and 700 °C, it was found that a saturated and stable amount of Sb is adsorbed on the surface. This can be understood by considering the growth mode of Sb on (100) silicon. Up to a critical coverage of approximately one monolayer an ad-atomlayer of Sb forms at the silicon surface whereas excess Sb atoms form islands [33]. Apparently, the Si—Sb bond is stronger than the Sb—Sb bond because desorption from islands already occurs at temperatures >150 °C, whereas the ad-layer desorbs only above 600 °C [34]. Annealing between 150 and 600 °C will therefore lead to a nearly temperature independent spatially uniform monolayer coverage. The uniformity of the surface coverage is better than 2% over a three-inch wafer [35].

The existence of antimony clusters has been demonstrated when the desorption step above 150 °C is not carried out [36]. A high resolution TEM picture of a Sb δ-layer is shown in Fig. 6.4, where the dopant and the amorphous silicon cap layer were both deposited at room temperature without intermediate annealing above 150 °C. Those precipitates with a diameter of about 5 nm are crystalline, and from the lattice constant one can conclude that they are pure Sb.

Recently, it has been found that a maximum Sb surface coverage of 0.82 ML can be achieved at 300 °C. [37]. Assuming an areal density of 6.81×10^{14} cm^{-2} atoms for the (100) silicon surface, this corresponds to a dopant sheet density of 5.6×10^{14} cm^{-2}. Almost arbitrary values between 0 and 1 ML can be adjusted by subsequent annealing above 600 °C for a given time [32]. The partial desorption, which allows the reduction of the saturated Sb layer, is depicted in Fig. 6.5 for a temperature of 750 °C. After the desired Sb coverage, an amorphous silicon cap of about 3–8 nm is deposited at close to room temperature and is recrystallized at about 550 °C. The δ-doped layer is usually covered by crystalline silicon, which is deposited to the desired thickness at conventional growth temperatures.

The existence of surface phases has been reported for complete monolayer coverage. Low energy electron diffraction (LEED) revealed a regular $(\sqrt{3} \times \sqrt{3})$–R 30° Sb structure

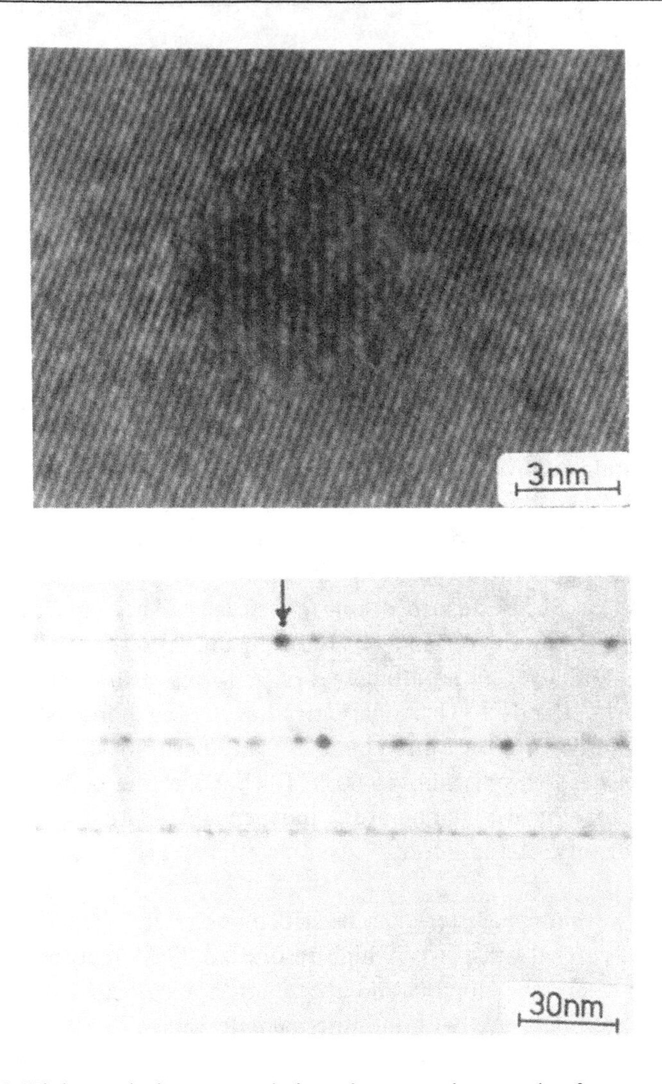

Fig. 6.4. High resolution transmission electron micrograph of an antimony precipitate after recrystallization [36].

on (111) surfaces, whereas for (100) orientations the surface phase is disordered, with only faint spots of a 2 × 1 reconstruction [38]. It has been demonstrated that these surface phases are preserved if they are buried into silicon by an SPE process. Unfortunately, the overgrowth of the (111) surface phase leads to a relatively low crystal quality due to twin formation [16]. Therefore, for neither of the two surface orientations, could the periodic two-dimensional arrangement of the doping atoms be investigated with respect to electrical sheet properties.

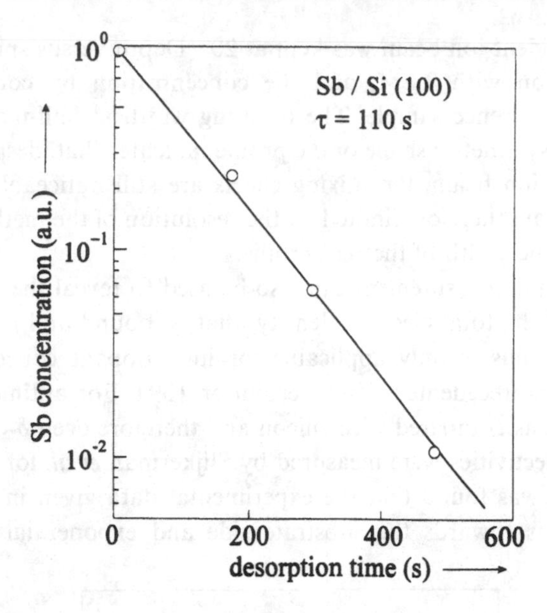

Fig. 6.5. Relative Sb concentration as measured by XPS as a function of desorption time at a substrate temperature of 750 °C [32].

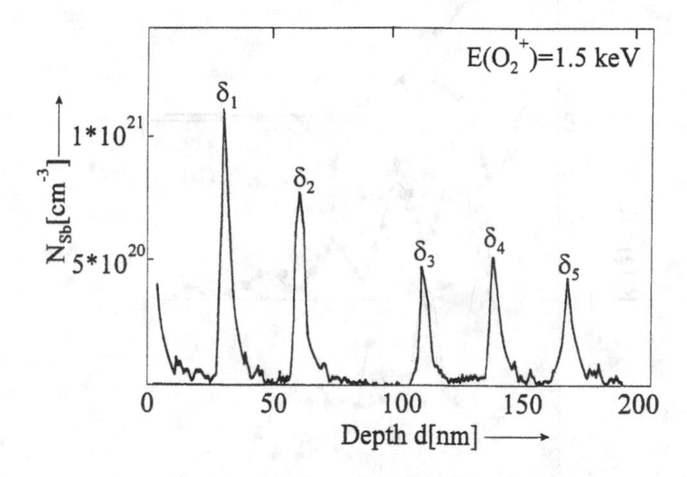

Fig. 6.6. Secondary ion mass spectrometry (SIMS) depth profiles for five Sb δ-layers with different sheet concentrations.

Concentration profiles

Several methods have been successfully applied for characterization of the concentration profiles. An example of a secondary ion mass spectrometry (SIMS) profile of five Sb-doped δ-layers is depicted in Fig. 6.6. For the analysis, O_2^+ primary ions at 1.5 keV were used. For a better depth resolution, that is, reduced knock-on and ion mixing effects, the angle between sample and incident ion beam was kept at 20°. Depth versus sputtering time was calibrated by comparison with TEM, and the concentration by comparison with a homogeneously doped reference sample. The resulting width at half-maximum amounts to 3 nm. However, the asymmetric shape of the profile indicates that, despite the low angle of the incident primary ion beam, ion mixing effects are still noticeable. Secondary ion mass spectrometry data are therefore limited by the resolution of the method and can only give an upper limit for the width of the real profile.

Specular X-ray reflection measurement can also be used to reveal the Sb δ-profile. This probes the derivative of the total electron density, that is, bound and free electrons, as a function of depth, and thus is only applicable for high dopant concentrations but, if applicable, it provides unprecedented depth resolution [39]. For antimony, the electron density is quite different as compared with silicon and therefore even δ-doping layers can be investigated. The reflectivities were measured by Slijkerman *et al.* for a Sb coverage of 0.7 monolayers [40]. It was found that the experimental data given in Fig. 6.7 are best fitted if abrupt functions towards the substrate side and exponential or half-gaussian

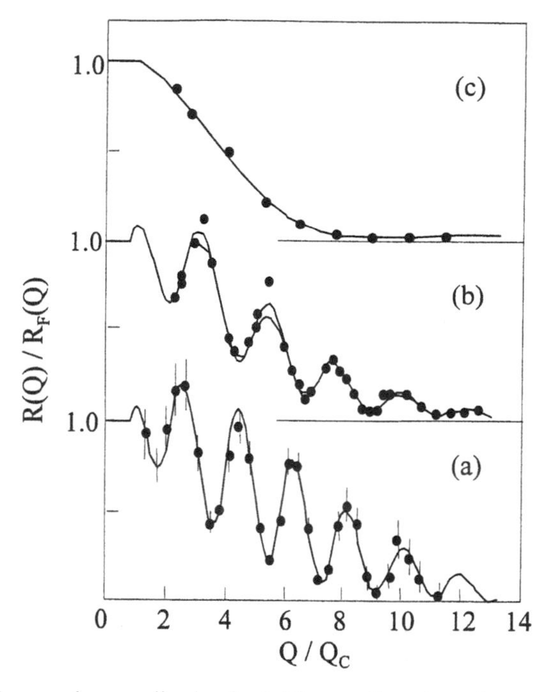

Fig. 6.7. Measured normalized reflectivities as a function of normalized momentum transfer (Q_c corresponds to the momentum transfer for total reflection). (a) (100) Si substrates without δ-layer. The solid line is the best fit for only a native oxide on top of the surface. (b) Sample with an Sb δ-layer and the silicon cap recrystallized by SPE. The solid line is the best fit for an exponential Sb profile at a depth of 7.35 nm with a decay length of 1.01 nm. The dashed line is a fit for a decay length of 1.38 nm. (c) Sample with a Sb δ-layer underneath amorphous silicon. The solid curve is the best fit for an exponential profile at a depth of 9.4 nm with a decay length of 0.57 nm [40].

profiles towards the α–Si cap are assumed. In addition, the density profile arising from the native SiO_2 at the surface has to be considered. Its contribution is depicted in Fig. 6.7(a). After deposition of the amorphous silicon cap the standard deviation was determined to be 0.57 nm and broadened to 1.01 nm after crystallization at 530 °C for 3 min.

High resolution ion scattering spectrometry is another technique for monitoring the morphology and crystal quality of δ-doped Sb layers. Experiments with primary 98.7 keV protons have been carried out *in-situ* by Slijkerman *et al.* [41] during all stages of preparation. Figure 6.8 shows the Sb and Si Rutherford backscattering (RBS) peaks taken along ⟨111⟩ directions. The saturation coverage was obtained by Sb deposition at a temperature of 670 °C. According to the area of the Sb peak a 0.65 ML coverage was achieved. After depositing a 3.1-nm-thick amorphous silicon cap at room temperature the Sb peak shifts to lower energies because the doping layer is now buried. The broadening is caused by thickness variations in the α–Si overlayer and by energy straggling of the ion beam. From an analysis of the peak shapes it can be concluded that the Sb atoms have not diffused into the α–Si. After crystallization of the α–Si cap at 480 °C and 3 min annealing at 530 °C the peak shape does not change significantly except for a small fraction at the surface energy. This means that a small portion of Sb has segregated to the surface.

A depth resolution down to atomic dimensions can also be obtained by high-resolution transmission electron microscopy. Because the mass ratio between antimony ($m = 121$) and silicon ($m = 28$) is quite large the method gives a sufficient contrast if more than 0.1%

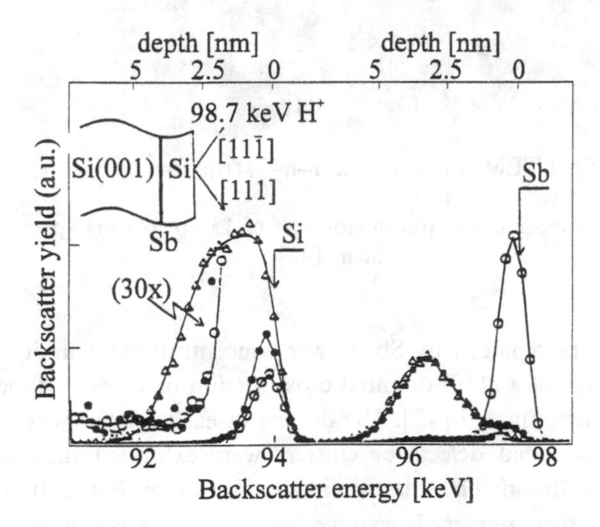

Fig. 6.8. High resolution RBS spectra taken for alignment along [111] direction according to the inset. Successive stages of preparation are shown: (○) after deposition of 0.65 ML Sb, (Δ) after deposition of a 3.1 nm amorphous silicon cap at room temperature, (●) after recrystallization of the α–Si layer at 530 °C, (■) corresponds to the Sb signal for random beam incidence and detection directions. The arrows indicate the surface peak positions [41].

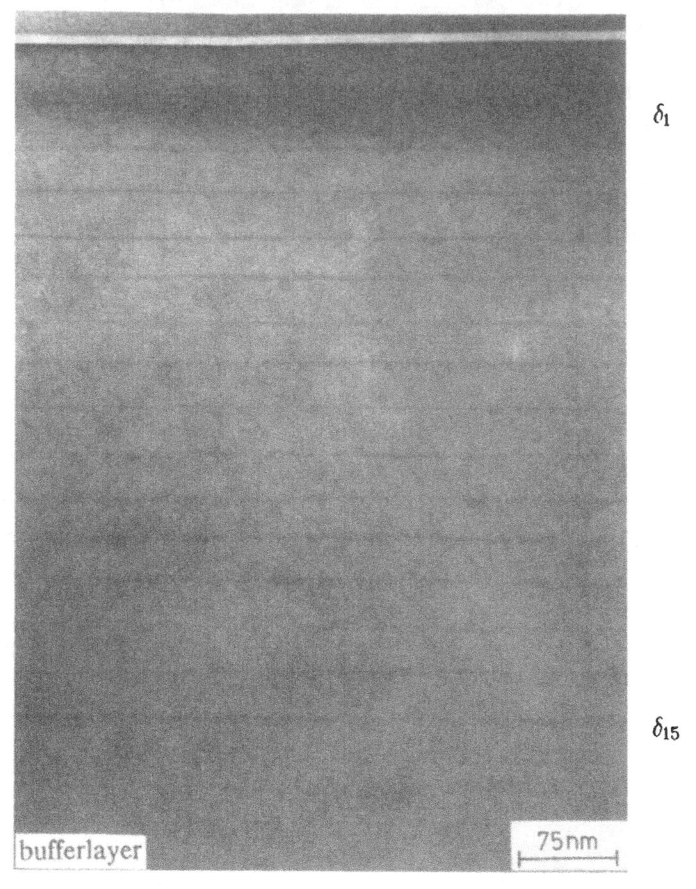

Bright field TEM image of a n–i–n–i structure

Fig. 6.9. Transmission electron micrograph of 15 Sb δ-layers spaced 50 nm apart [42].

of the silicon atoms are replaced by Sb. Lower concentrations cannot be detected. The bright field TEM image of a (110)-oriented cross section of 15 Sb δ-doped layers spaced 50 nm apart is shown in Fig. 6.9 [42]. The doping in each δ-layer was determined to be $2 \times 10^{14}\,\mathrm{cm}^{-2}$. No extended defects or clusters were exhibited throughout the whole sample. The high resolution TEM lattice image shown in Fig. 6.10 confirms perfect crystallinity [26] for one δ-layer. It can be seen that the width of the doped layer amounts to only about 2 nm. This width is partly caused by surface roughness, which can be estimated from the interface between the crystalline silicon and the native oxide. Therefore, these results are in good agreement with results from other analytical techniques.

In conclusion, all analysis techniques prove that by using SPE the width of n-type δ-layers can be confined on an atomic scale.

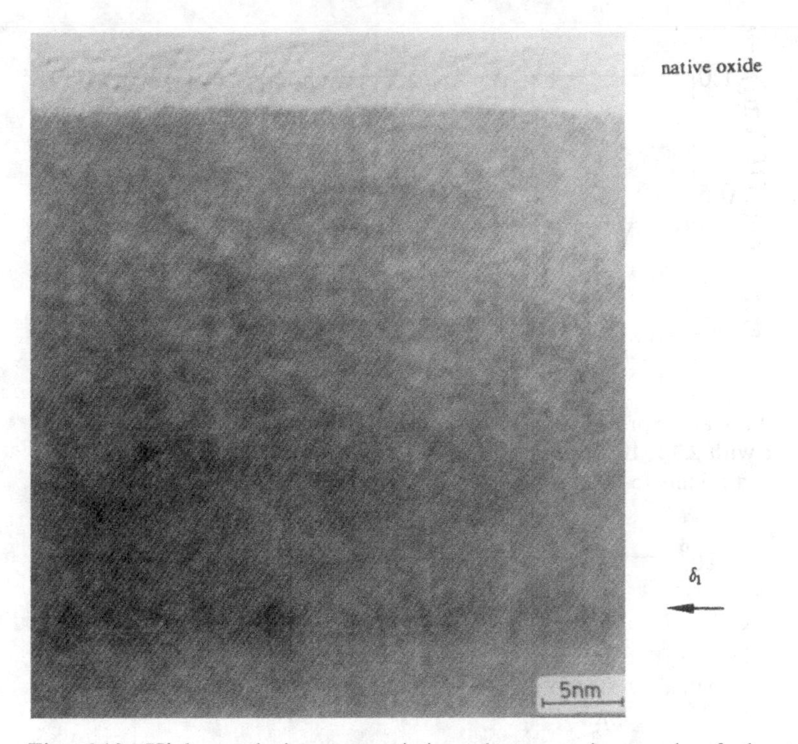

Fig. 6.10. High resolution transmission electron micrograph of the top antimony-doped δ-layer from Fig. 6.9.

Electrical activation and conductivity

The above-mentioned analysis techniques basically detect the number and spatial distribution of antimony atoms but not the electrical activation of the dopant material. This can be achieved by using capacitance–voltage (CV) profiles which resolve the impurity distribution from the voltage dependent capacity of a Schottky diode. Figure 6.11 shows a carrier concentration profile for a Sb layer with a coverage of 2.3×10^{-3} ML [32]. The δ-layer was covered by a 60 nm top layer of silicon. The increase at 260 nm denotes the interface between the (100) silicon substrate and the epitaxial buffer layer. The full width at half-maximum of the profile amounts to about 18 nm. For δ-layer concentrations above 3.5×10^{12} cm^{-2}, avalanche breakdown occurs which obscures the interpretation of the experimental results.

Because it is difficult to determine the profile of the electrically active dopant concentration, the integral value, that is, the sheet concentration in the δ-layer, is usually investigated. The resistivity ρ_{XX} of the δ-layer, the concentration of the free electrons n^{2D}, and the mobility μ can be obtained simultaneously from Hall measurements using van der Pauw geometry.

The resulting sheet resistivity ρ_{XX} as a function of the measured free electron concentration n^{2D} is compiled in Fig. 6.12 for various temperatures [26, 32]. Each point corresponds

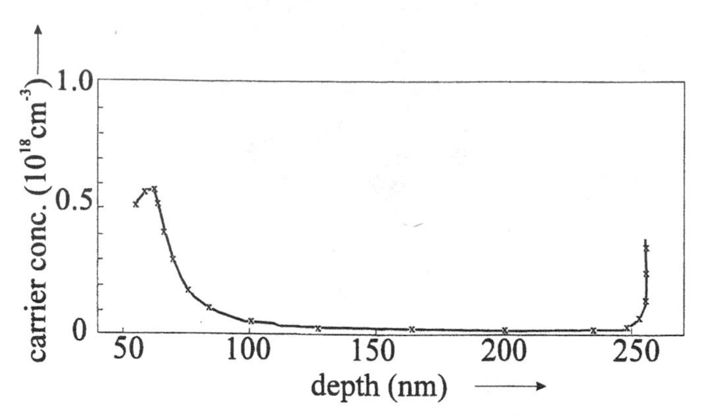

Fig. 6.11. Carrier profile determined from *CV* profiles for an antimony-doped δ-layer with 2.3×10^{-3} monolayer coverage. The thickness of the silicon cap amounts to 60 nm. The epitaxial layer starts at 260 nm [32].

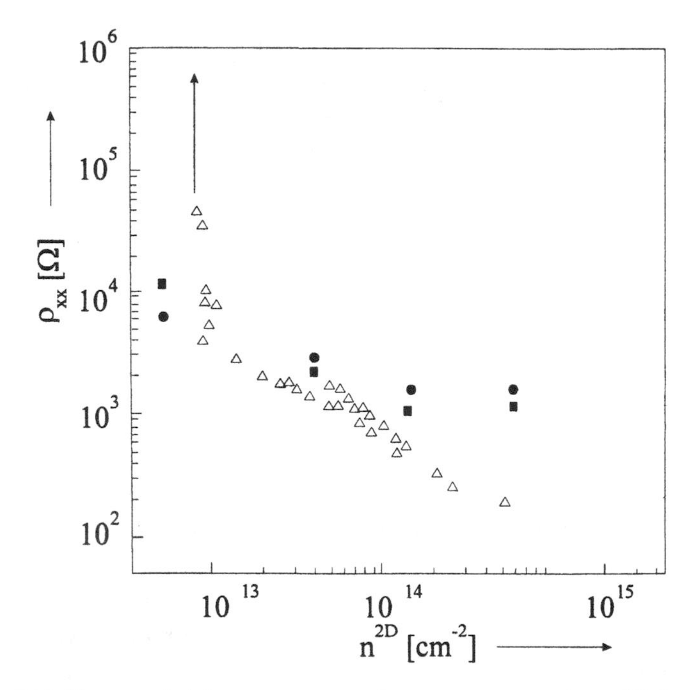

Fig. 6.12. Sheet resistivity ρ_{XX} versus electron concentration n^{2D} for single antimony δ-layers at 4.2 K (Δ), 77 K (■), and 300 K (●) [26, 32].

to a single δ-layer embedded in a nominally undoped background material. Substrate materials with 100–1000 Ωcm have been used. It should be noted that the Hall densities n^{2D} do not necessarily correspond exactly to the electron concentration in the δ-layer because confinement effects are ignored and the Hall factor is assumed to be unity. In reality the electronic levels form a set of two-dimensional subbands in a V-shaped potential well. From self consistent calculations of the Schrödinger and Poisson equations it is known that at higher concentration values two subbands with different degeneracy and density of states effective mass are occupied.

For high doping concentrations the system of free charge carriers in the δ-layer becomes degenerate, that is, due to the overlap of the Bohr radii no freeze out effects at lower temperatures take place. In this case almost no temperature dependence can be observed. At a concentration slightly below 10^{13} cm^{-2} the sharp rise in resistivity at 4.2 K indicates the transition between the non-degenerate and the degenerate states (metal–insulator transition). The average separation between the doping atoms is about 3 nm for the critical density. The upper concentration values exceed 10^{14} cm^{-2}, which are by far the highest values reported to date for any two-dimensional electron gas.

The resulting Hall mobilities are plotted in Fig. 6.13 versus electron concentration. A pronounced temperature dependence exists in the non-degenerate regime, and at 4.2 K no measurements are possible for concentrations below the metal–insulator transition. For the degenerate case the mobility is independent of temperature and almost independent of concentration. This is the same behaviour as for the three-dimensional case. Unfortunately, the maximum mobility value of 100 cm^2/Vs at 4.2 K is too low to carry out further investigations under high magnetic fields. A first approach to increasing the mobility is indicated by the square at $n^{2D} = 8 \times 10^{13}$ cm^{-2}. Here two δ-layers separated by 10 nm have been measured [26]. Due to the overlap of the V-shaped potentials the electrons in higher subbands are not localized in one Coulomb well and the average distance from the scattering centre increases. As a result of this modulation doping, the mobility is increased to 170 cm^2/Vs.

In order to obtain information about the active fraction of antimony atoms the density of free electrons n^{2D} as determined by Hall measurements must be compared with data

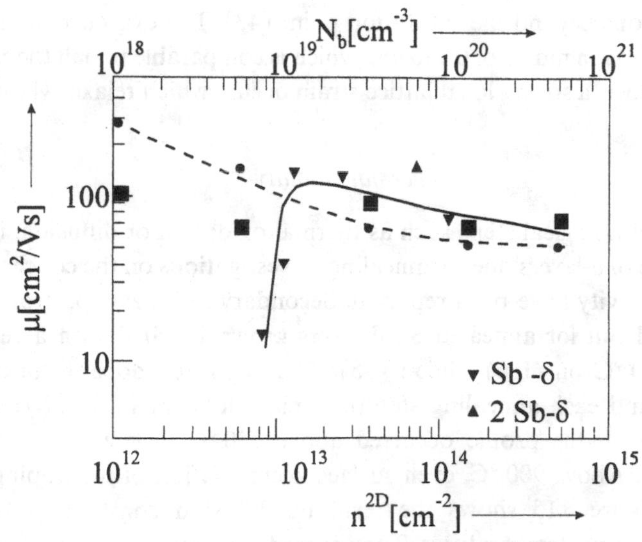

Fig. 6.13. Mobility μ versus electron concentration n^{2D} for single antimony δ-layers at 4.2 K (▼), 77 K (■), and 300 K (●) [26, 32]. The single point (▲) denotes the mobility of a double delta.

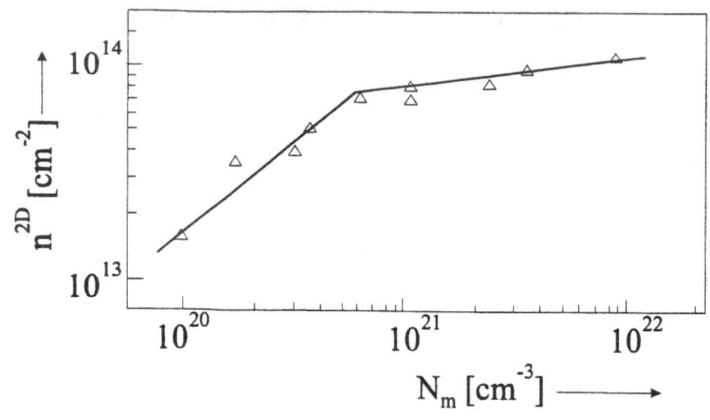

Fig. 6.14. Two-dimensional Hall density n^{2D} versus nominal concentration N_m in the maximum of the SIMS profile for antimony δ-layers [42].

on concentration profiles. In Fig. 6.14 n^{2D} is plotted against the nominal concentration in the maximum of SIMS profiles [42]. The Hall measurements were again taken at 4.2 K. The linear slope between 10^{13} cm^{-2} and 9×10^{13} cm^{-2} shows almost a 1:1 correlation between the two measurement techniques, corresponding to 100% activation. This is well above the solubility limit of Sb in Si which was determined to be 2×10^{19} cm^{-3} at 700 °C, corresponding to 7.4×10^{12} cm^{-2} for the two-dimensional case. Above 10^{14} cm^{-2} the electrical measurements deviate significantly from the SIMS data. However, the incorporation model, as well as the TEM analysis, suggests that no clustering occurs for an areal electron density between 10^{14} and 5×10^{14} cm^{-2}. Thus the reduced activity probably arises from point defects. These could be interstitial antimony atoms or the existence of Sb—V^{2-} vacancy complexes, which are known to exist for the three-dimensional case at doping concentrations beyond the saturation point [43]. The existence of vacancies could be due to the large Sb ion radius of 0.136 nm, which is comparable to half the Si—Si distance of 0.118 nm. Therefore, a strong local lattice strain occurs which relaxes via vacancies [44].

Thermal stability

For subsequent technological steps such as thermal oxidation or diffusion, it is important to study the change of δ-layers due to annealing. Investigations on the concentration profile and the sheet resisitivity have been reported. Secondary ion mass spectrometry measurements were carried out for annealed Sb δ-layers grown by SPE with a recrystallization temperature of 550 °C on (100) silicon [45]. The deposited dopant concentration was 4.1×10^{13} cm^{-2}, and each annealing step took place for 1 hr in an N_2/H_2 atmosphere. Significant changes in the profile occurred above 750 °C, where the authors observed enhanced diffusion. Above 900 °C, even surface accumulation of the doping material has been observed. Figure 6.15 shows the resulting diffusion constants as a function of concentration at various temperatures. They exceed the bulk values by at least one order of magnitude. For concentrations above 10^{18} cm^{-3} the diffusion constant becomes strongly concentration dependent, which the authors attribute to oxide generation during annealing.

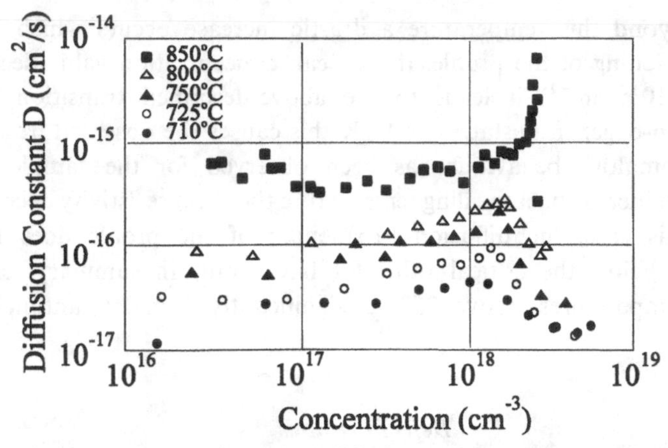

Fig. 6.15. Concentration-dependent diffusion constants D for one Sb δ-layer with a nominal deposition of 4.1×10^{13} cm^{-2} [45]. The bulk value at 800 °C amounts to 1.5×10^{-18} cm^2/s.

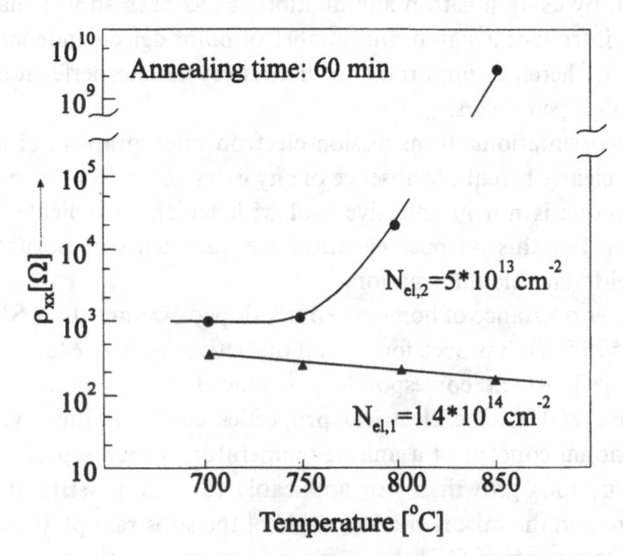

Fig. 6.16. Sheet resistivity ρ_{XX} as a function of annealing temperature for two Sb δ-layers with nominal depositions of 5×10^{13} cm^{-2} and 1.4×10^{14} cm^{-2} respectively [46]. The annealing time was 1 hr. The resistivity measurements were carried out at 4.2 K.

Another reasonable explanation could be enhanced electric field drift for doping concentration signficantly exceeding the intrinsic concentration.

The variation of the sheet resistivity ρ_{XX} with annealing for SPE grown δ-layers has also been investigated [46]. Delta-layers with a nominal antimony deposition of 5×10^{13} cm^{-2} and 1.4×10^{14} cm^{-2} have been grown by SPE and annealed in nitrogen for 1 hr at different temperatures. The resulting sheet resistivity measured at 4.2 K is plotted in Fig. 6.16. For the lower concentration value, almost no resistivity change can be seen up to 750 °C. Beyond this temperature a drastic increase occurs which is related to the diffusion broadening of the profile. If the peak concentration value decreases below a critical value (10^{13} cm^{-2}) it leads to the above-described transition between the degenerate and non-degenerate states. At 4.2 K this causes the observed freeze out of free electrons. An anomalous behaviour has been observed for the sample with higher concentration: with increasing annealing temperature the sheet resistivity decreases slightly. Apparently, in this case, the diffusion broadening of the profile does not decrease the concentration below the critical value for freeze out. In summary, one can state that annealing temperatures above 750 °C significantly broaden antimony δ-doping profiles.

Defect densities

It is known that the SPE process can give rise to various defects. For (111) surface orientations, extended defects occur during the crystallization process due to twin formation [16], and, by using positron annihilation, it has been shown that the recrystallization process can introduce a significant number of point defects independent of surface orientation [47]. It is therefore important to discuss crystalline perfection in connection with the growth of δ-doped layers.

For (100) surface orientations, transmission electron micrographs (TEM), as depicted in Fig. 6.9 and 6.10, clearly reveal the absence of any extended defects such as dislocations. However, this technique is not an effective tool with which to indicate the presence of vacancy-type defects. For this purpose electrical measurements or positron annihilation experiments can yield valuable information.

Evaluating the sheet resistance of homogeneously doped 500-nm-thick SPE layers which were crystallized at 575 °C, it has been found that the activation value for Sb varies between only 0.2 and 0.7 [29], which corresponds to reduced crystal quality. By additional post-annealing above 800 °C, the electrical properties could be improved significantly. However, the traditional concept of a unique temperature which separates the regime of crystalline and amorphous growth is not applicable to silicon MBE. It turns out that epitaxial growth rate and the misorientation angle of the substrate play an important role in critical epitaxial temperature [15]. Even more important is the fact that an epitaxial thickness t_{epi} exists which grows prior to the formation of the amorphous phase [48]. For a growth rate of about 0.1 nm/s and a growth temperature of 100 °C, t_{epi} already amounts to several nanometers. Because this is a reasonable temperature for the growth of the silicon cap after deposition of the δ-doping concentration, an additional post-growth anneal at 600–700 °C should be enough to create a perfect crystalline structure without point defects. This has been clearly demonstrated [49] for a growth temperature of 220 °C and an annealing temperature of 600 °C. This is in agreement with the electrical measurements depicted in Fig. 6.14, where a one-to-one correlation between nominal and electrically active dopant concentration in the δ-layer is depicted up to a sheet concentration of about 10^{14} cm^{-2}. These results clearly differ from the data of 500-nm-thick SPE layers.

P-type δ-doping

High solubility in silicon and energy levels close to the valence band are the necessary requirements for p-type dopants for the same reasons as have been discussed for n-type doping. The most preferable candidate is boron (B) but if it is evaporated in elemental form it requires extremely high evaporation temperatures. This means that electron beam evaporators or special effusion cells are necessary [50–53]. Gallium (Ga), the second candidate, is much easier to evaporate but poses severe segregation and incorporation problems [50].

Gallium

Solid phase epitaxy for the growth of homogeneously doped gallium layers in silicon is a process that is well suited to overcoming the equilibrium solubility limits [54, 55]. Whereas the equilibrium value at 700 °C amounts to 9×10^{18} cm^{-3} the maximum carrier concentration in SPE grown films is about 2×10^{20} cm^{-3}. Gallium-doped δ-layers can be grown by the same procedure as described for Sb [36]. The total incorporated Ga concentration can again be determined by SIMS measurements. A depth profile of three Ga δ-layers with nominal deposition of 4×10^{14} cm^{-2} (δ_1), 6×10^{13} cm^{-2} (δ_2), and 1×10^{13} cm^{-2} (δ_3) is depicted in Fig. 6.17 [36]. The data were taken for an energy of 7.5 keV for the primary O_2^+ ions and an incidence angle of 90° between ion beam and sample surface. The width of 10 nm at half-maximum is therefore comparable to the antimony results. It is astonishing that the δ-layer with the highest nominal deposition (δ_1) shows the lowest peak height. To investigate this phenomenon further, the peak concentrations of single δ-layers have been plotted as a function of monolayer coverage (Fig. 6.18). Despite the increasing deposition, the peak concentration did not exceed 10^{20} cm^{-3}. A comparison of these results with transmission electron micrographs helps in gaining a better understanding

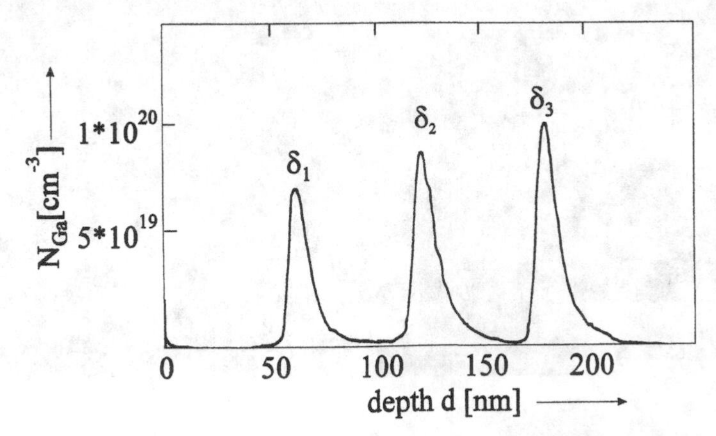

Fig. 6.17. SIMS depth profile of three Ga δ-layers with nominal deposition of 4×10^{14} cm^{-2} (δ_1), 6×10^{13} cm^{-2} (δ_2), and 1×10^{13} cm^{-2} (δ_3) [36].

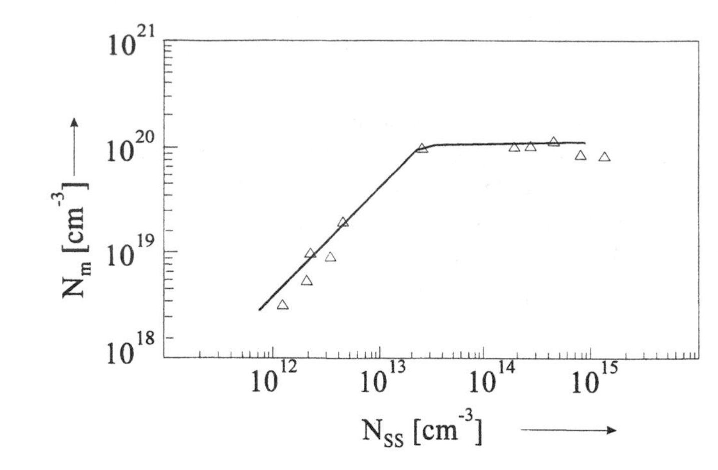

Fig. 6.18. Peak concentration N_m of gallium atom determined by SIMS as a function of surface coverage N_{SS} [46].

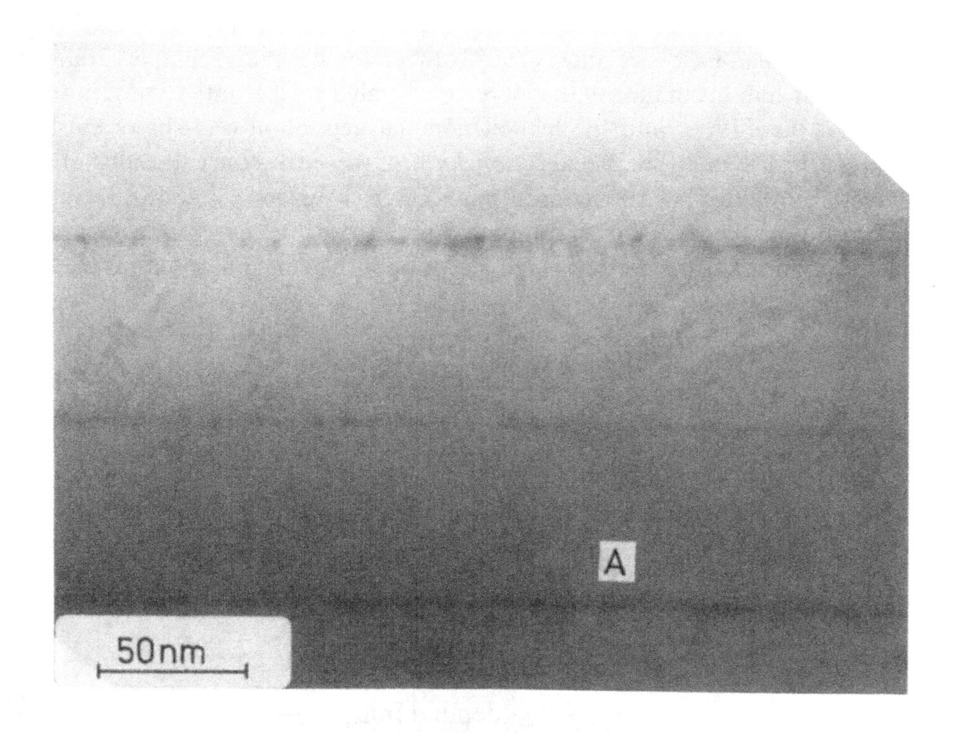

Fig. 6.19. Bright field TEM image of three Ga δ-layers with concentrations according to Fig. 6.17 [36].

of the incorporation behaviour of gallium. In Fig. 6.19 the bright field TEM image of the three Ga δ-layers with concentrations according to Fig. 6.17 is depicted. The strong tendency of gallium to form clusters or islands is evident. Even in layer δ_3, where only 10^{13} cm^{-2} gallium atoms have been predeposited, a few precipitates can be identified. The width of the profile amounts to 3 nm and, despite the clusters, no dislocations or extended defects can be detected. The increasing cluster density with increasing surface coverage in δ_1 and δ_2 prove that no direct correlation between the SIMS concentrations and the coverage is possible because the ionization coefficient of pure gallium clusters is lower than that of gallium incorporated into a silicon host crystal. Apparently, above 2×10^{13} cm^{-2} most of the additional gallium is accumulated in clusters or is desorbed again.

This has been confirmed by electrical measurements. The hole concentration was measured by electrochemical capacitance–voltage depth profiling at room temperature [46]. Figure 6.20 depicts the concentration profile of a p^+p modulation doping with 15 Ga δ-layers of equal concentration (nominal 10^{14} cm^{-2}) spaced 100 nm apart. To enhance the depth resolution, which is mainly determined by the width of the depletion zone, a relatively high dopant concentration of 8×10^{17} cm^{-3} was chosen for the region between each individual δ-layer. The maximum concentration of electrically active dopants in each particular δ-layer amounts to an areal density of 1.4×10^{13} cm^{-2}. The decreasing peak concentration towards the substrate is caused by the limits of the measuring set up, that is, inhomogeneous etching during depth profiling. The full width at half-maximum amounts to about 20 nm. In conclusion, one can state that δ-doping profiles with Ga as p-type doping material can be grown by SPE methods but the incorporation of atoms on lattice sites is limited to about $1–2 \times 10^{13}$ cm^{-2}.

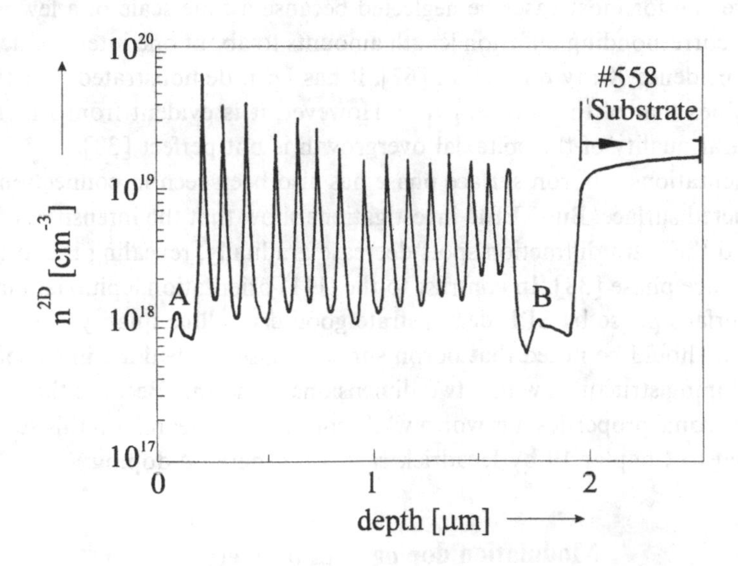

Fig. 6.20. Capacitance voltage (CV) depth profile of a p^+p modulated structure with Ga δ-layers spaced 100 nm apart [46]. The electrical active concentration n^{2D} is plotted versus depth.

Boron

The solid phase epitaxial growth of boron (B) has been studied and for an equilibrium solubility limit of 1×10^{19} cm^{-3} at 700 °C an enhanced maximum carrier concentration of 2×10^{20} cm^{-3} has been achieved [56]. However, the segregation coefficient of boron in the crystalline growth regime is nowhere near as high as for antimony and gallium [12]. It is therefore possible to grow sharp doping profiles without SPE. The first p-type δ-doping structures evaporating boron from an elemental source have been reported by Mattey *et al.* [57, 58] using epitaxial temperatures. Below 600 °C, accumulation and phase formation are kinetically limited and no significant broading of the δ-doping profiles occurs for (100) surface orientations [59]. In order to obtain sharp δ-profiles it is sufficient to interrupt the silicon flux at growth temperatures below 600 °C, and to deposit the desired boron concentration. Afterwards, the silicon growth can be obtained without any temperature changes. Lowering of the growth temperature to 400 °C even allows δ-concentrations up to 10^{14} cm^{-2}. An SPE process with an amorphous silicon cap is not required in this case. One interesting application of SPE is the formation of ordered surface phases where boron atoms substitute silicon atoms and form a two-dimensional lattice. Boron is known to induce a ($\sqrt{3} \times \sqrt{3}$) R 30° reconstruction on a Si (111) surface [18, 60, 61]. It can be achieved on a 7×7 reconstructed surface by depositing B at 750 °C. The desired $\sqrt{3}$ structure is observed at a coverage of one-third of a monolayer and it is assumed that boron atoms are substituted for silicon atoms. Recently, it has been shown that this surface phase can be partly preserved if it is covered by 5–25 nm of amorphous silicon which is then recrystallized at temperatures around 600 °C [61, 62]. The diffusion broadening at this temperature can for most cases be neglected because a time scale of a few minutes is enough and the corresponding diffusion length amounts to about one interatomic distance. Using grazing incidence X-ray diffraction [62], it has been demonstrated that the buried structure resembles that of the surface phase. However, it is evident from LEED studies that the structural quality of the epitaxial overgrowth is not perfect [38].

For (100) orientations a boron surface phase has also been seen in connection with the 2×1 reconstructed surface. But LEED investigations show that the intensities of both the fundamental and the extra diffraction spots decrease gradually, revealing the formation of a disordered surface phase [38]. In contrast to the (111) orientation, epitaxial films grown on top of the surface phase by SPE demonstrate good crystalline quality.

In summary, it should be noted that boron surface phases embedded in silicon by SPE might allow δ-doping structures with a two dimensional ordering. Because this would lead to improved electronic properties it is worth while continuing research on this subject. (The reader is referred to Chapter 19 by Headrick *et al.* on ordered δ-doping.)

Modulation doping with δ-layers

Superlattices with a periodic potential can be created by alternating n- and p-type δ-layers. For compensated non-degenerate doping, the structure is completely depleted of free charge carriers because electrons and holes recombine with each other. The resulting potential is a symmetric or asymmetric sawtooth depending on the relative positions of the δ-layers. With these structures, interesting band gap engineering effects, such as variation of the effective band gap, can be obtained. For degenerate δ-doping superlattices the potential is quite different because the dopant concentration is equal to the free charge carrier concentration. The Fermi level lies above the band edge and quantization effects due to the δ-doping have to be considered [63].

Delta-doped n–i–p–i structures have been grown with Sb and Ga, as well as Sb and B. The SIMS depth profile of a n(Sb–δ)–i–p(Ga–δ)–i structure is presented in Fig. 6.21 [42]. The spacing between the δ-layers is about 100 nm. The measurements have been carried out with 7.5 keV O_2^+ primary ions. The incidence angle between ion beam and surface plane was kept at 90°. Each individual δ-layer can be clearly resolved, and the incorporated doping concentrations agree well with the design parameters. The decay of the peak maxima and the increase of the full width at half-maximum towards the substrate can be explained by ion mixing and knock-on effects.

Great care has to be taken to avoid unintentional doping. A weak concentration of the opposite dopant can be detected in each doped region. Despite closed effusion cells there is always a certain partial pressure of both doping elements inside the MBE chamber. Whenever an SPE process is carried out near room temperature this will cause an incorporation of the opposite doping material. For the data given in Fig. 6.21 this effect is below a few percent and can be neglected in the calculation of the charge carrier concentration.

The optimum dopant combination for δ-type modulation doping seems to be Sb and B. Sb δ-profiles are grown by SPE whereas the B δ-profiles are fabricated by low temperature MBE. In this way the advantage of sharp n-type doping with almost arbitrary doping concentrations is combined with p-type δ-doping at elevated temperatures.

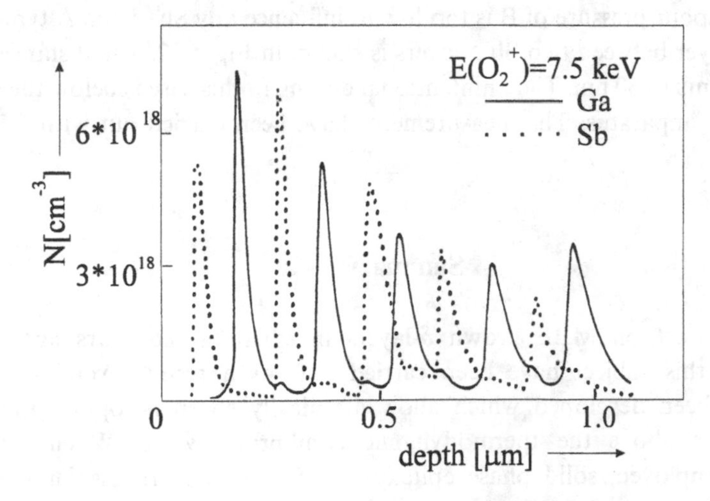

Fig. 6.21. SIMS depth profile of an n–i–p–i structure with Sb and Ga δ-layers [42].

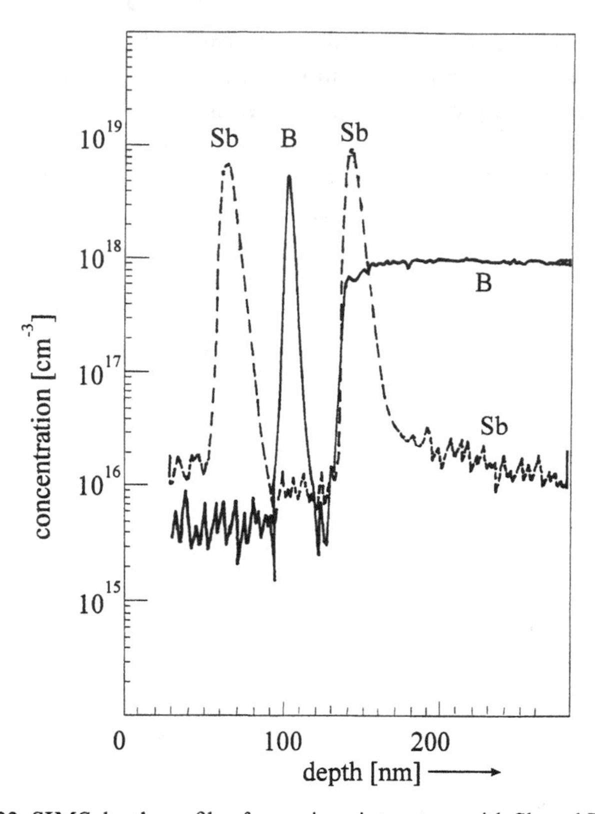

Fig. 6.22. SIMS depth profile of an n–i–p–i structure with Sb and B δ-layers.

Unintentional doping of the boron doped δ-layers is therefore almost eliminated. On the other hand the vapour pressure of B is too low to influence any Sb–delta. A typical SIMS profile of a B δ-layer between two Sb δ-layers is shown in Fig. 6.22. The distance between the δ-layers amounts to 60 nm. The unintentional doping in this case is below the detection limit of the SIMS apparatus. The measurements have been carried out with 7.5 keV O_2^+ primary ions.

Summary

Since the first report on MBE grown δ-layers in silicon some years ago, intensive investigations of this subject have been carried out. As a result several new growth techniques have been developed which allow atomically sharp δ-doping profiles with dopant solubilities above the thermodynamic equilibrium value. When the correct parameters are employed, solid phase epitaxy is a favourable process for overcoming segregation and clustering effects. Perfect δ-layers can be grown which form a basic element for advanced silicon technology as well as being novel device concepts.

References

[1] For a review see P. Dhez and C. Weisbuch, *Physics, Fabrication, and Applications of Multilayered Structures*, NATO ASI Series B; *Physics* Vol. 182, Plenum Press (1988).
[2] W. K. Burton, N. Cabrera, and F. C. Frank, *Phil. Trans. Roy. Soc.* A **143**, 299 (1951).
[3] Y. Ota, *J. Electrochem. Soc.* **126**, 1761 (1976).
[4] U. König, E. Kasper, and H. J. Herzog, *J. Crystal Growth* **52**, 151 (1981).
[5] J. C. Bean, *Appl. Phys. Letters* **33**, 654 (1978).
[6] G. E. Becker and J. C. Bean, *J. Appl. Phys.* **48**, 3395 (1977).
[7] S. S. Iyer, R. A. Metzger, and F. G. Allen, *J. Appl. Phys.* **52**, 5608 (1981).
[8] R. A. Metzger and F. G. Allen, *J. Appl. Phys.* **55**, 931 (1984).
[9] H. Jorke, *Surf. Sci.* **193**, 569 (1988).
[10] R. A. A. Kubiak, W. Y. Leong, and E. H. C. Parker, *Appl. Phys. Letters* **44**, 878 (1984).
[11] T. Tatsumi, H. Hirayama, and N. Aizaki, Japan. *J. Appl. Phys.* **27**, L954 (1988).
[12] E. de Frésart, K. L. Wang, and S. S. Rhee, *Appl. Phys. Lett.* **53**, 48 (1988).
[13] R. A. Metzger and F. G. Allen, *J. Appl. Phys.* **55**, 931 (1984).
[14] S. A. Barnett and J. E. Greene, *Surf. Science* **151**, 67 (1985).
[15] V. Fuenzalida and I. Eisele, *J. Cryst. Growth* **74**, 597 (1986).
[16] H. P. Zeindl, V. Fuenzalida, J. Messarosch, I. Eisele, H. Oppolzer, and V. Huber, *J. Cryst. Growth* **81**, 231 (1987).
[17] S. T. Picreaux. *Defects in Semiconductors*, Vol. 2, North Holland, p. 135 (1981).
[18] A. V. Zotov and V. V. Korobtsov, *J. of Crystal Growth* **98**, 519 (1989).
[19] J. A. Roth and C. L. Anderson, *Appl. Phys. Lett.* **31**, 689 (1977).
[20] D. Streit, R. A. Metzger, and F. G. Allen, *Appl. Phys. Lett.* **44**, 234 (1984).
[21] J. C. Bean and J. M. Poate, *Appl. Phys. Letters* **36**, 59 (1980).
[22] C. W. Magee, J. C. Bean, G. Foti, and J. M. Poate, *Thin Solid Films* **81**, 1 (1981).
[23] M. von Allmen, S. L. Lau, J. W. Mayer, and W. F. Tseng, *Appl. Phys. Lett.* **35**, 280 (1979).
[24] P. J. Grunthaner, F. J. Grunthaner, R. W. Fathauer, T. L. Lin, M. M. Hecht, L. D. Bell, W. J. Kaiser, F. D. Schowengerdt, and J. H. Mazur, *Thin Solid Films* **183**, 197 (1989).
[25] I. G. Kaverina, V. V. Korobtsov, V. G. Zavodinskij, and A. V. Zotov, *Phys. Status Solidi* (a) 82, 345 (1984).
[26] I. Eisele, *Appl. Surf. Science* **36**, 39 (1989).
[27] S. S. Lau, J. W. Mayer, and W. Tseng in: *Handbook of Semiconductors*, Vol. 3, ed. S. P. Keller, North Holland (1980).
[28] D. Streit, E. D. Ahlers, and F. G. Allen, *J. Vac. Sci. Technol.* B **5**, 752 (1987).
[29] A. Casel, H. Kibbel, and F. Schäffler, *Thin Solid Films* **183**, 351 (1990).
[30] H. P. Zeindl, T. Wegehaupt, I. Eisele, H. Oppolzer, H. Reisinger, G. Tempel, and F. Koch, *Appl. Phys. Lett.* **50**, 1164 (1987).
[31] A. A. van Gorkum, K. Nakagawa, and Y. Shiraki, *Jpn J. Appl. Phys.* **26**, L1933 (1987).
[32] A. A. van Gorkum, K. Nakagawa, and Y. Shiraki, *J. Appl. Phys.* **65**, 2485 (1989).
[33] R. A. Metzger and F. G. Allen, *Surf. Sci.* **137**, 397 (1984).
[34] S. A. Barnett and J. E. Greene, *Surf. Sci.* **151**, 67 (1985).
[35] H. Jorke, H. Kibbel, F. Schäffler, A. Casel, H. J. Herzog, and E. Kasper, *Appl. Phys. Lett.* **54**, 819 (1989).
[36] H. P. Zeindl, T. Wegehaupt, and I. Eisele, *Thin Solid Films* **184**, 21 (1990).
[37] W. F. J. Slijkerman, P. M. Zagwijn, J. F. van der Veen, D. J. Gravesteijn, and G. F. A. van de Walle, *Surf. Sci.* **262**, 25 (1992).
[38] A. V. Zotov and V. G. Lifshits, *J. of Crystal Growth* **121**, 88 (1992).
[39] A. Braslan, P. S. Pershan, G. Swislow, B. M. Ocko, and J. Als-Nielsen, *Phys. Rev.* A **38**, 2457 (1988).

[40] W. F. J. Slijkerman, J. M. Gay, P. M. Zagwijn, J. F. van der Veen, J. E. Macdonald, A. A. Williams, D. J. Gravesteijn, and G. F. A. van de Walle, *J. Appl. Phys.* **68**, 5105 (1990).
[41] W. F. J. Slijkerman, P. M. Zagwijn, J. F. van der Veen, A. A. van Gorkum, and G. F. A. van der Walle, *Appl. Phys. Lett.* **55**, 963 (1989).
[42] H. P. Zeindl, E. Hammerl, W. Kiunke, and I. Eisele, *J. of Electronic Materials* **19**, 1119 (1990).
[43] A. N. Larsen, F. T. Pedersen, G. Weyer, R. Galloni, R. Rizzoli, and A. Armigliato, *J. Appl. Phys.* **59**, 1908 (1986).
[44] T. J. van Netten, K. Stapel, and L. Niesen, *J. Phys. Colloq.* C **47**, 1049 (1986).
[45] S. J. Fukatsu, S. Kubo, Y. Shiraki, and R. Ito, *Appl. Phys. Lett.* **58**, 1152 (1991).
[46] H. P. Zeindl, Ph.D. Thesis, Universität der Bundeswehr München (1989).
[47] H. Schut, A. van Veen, G. F. A. van de Walle, and A. A. van Gorkum, *J. Appl. Phys.* **70**, 3003 (1991).
[48] H. J. Gossmann and D. J. Eaglesham in *Semiconductor Interfaces, Microstructures and Devices*, Zhe Chuan Feng, ed., Institute of Physics Pub., Bristol (1993).
[49] H. J. Gossmann, P. Asoka-Kumar, T. C. Leung, B. Nielsen, K. G. Lynn, F. C. Unterwald, and L. C. Feldmann, *Appl. Phys. Lett.* **61**, 540 (1992).
[50] C. P. Parry, S. M. Newstead, R. D. Barlow, P. Augustus, R. A. A. Kubiak, M. G. Dowsett, T. E. Whall, and E. H. C. Parker, *Appl. Phys. Lett.* **58**, 481 (1991).
[51] S. Andrien, J. A. Chroboczek, Y. Campidelli, E. André, and F. A. d'Avitaya, *J. Vac. Sci. Technol.* B **6**, 835 (1988).
[52] M. W. Denhoff, *J. Vac. Sci. Technol.* B **8**, 1035 (1990).
[53] K. Nakagawa, M. Miyao, and Y. Shiraki, *Jpn J. Appl. Phys.* **27**, L2013 (1988).
[54] L. Vescan, E. Kasper, O. Meyer, and M. Maier, *J. Cryst. Growth* **73**, 482 (1985).
[55] A. Casel, H. Jorke, E. Kasper, and H. Kibbel, *Appl. Phys. Lett.* **48**, 922 (1986).
[56] J. Caber, H. Ishiwara, and S. Furukawa, *Jpn J. of Appl. Phys.* **21**, L712 (1982).
[57] N. L. Mattey, M. G. Dowsett, E. H. C. Parker, T. E. Whall, S. Taylor, and J. F. Zhang, *Appl. Phys. Lett.* **57**, 1648 (1990).
[58] N. L. Mattey, M. Hopkinson, R. F. Houghton, M. G. Dowsett, D. S. McPhail, T. E. Whall, E. H. C. Parker, G. R. Booker, and J. Whitehurst, *Thin Solid Films* **184**, 15 (1990).
[59] C. P. Parry, R. A. A. Kubiak, S. M. Newstead, E. H. C. Parker, and T. E. Whall, *Mat. Res. Symp. Proc.* **220**, 103 (1991).
[60] H. Horayama, T. Tatsumi, and N. Aizaki, *Surf. Sci.* **193**, L47 (1988).
[61] V. V. Korobtsov, V. G. Lifshits, and A. V. Zotov, *Surf. Sci.* **195**, 466 (1988).
[62] K. Akimoto, H. Hirosawa, T. Tatsumi, H. Hirayama, J. Mizuki, and J. Matsui, *Appl. Phys. Lett.* **56**, 1225 (1990).
[63] I. Eisele, *Superlattices and Microstructures* **6**, 123 (1989).

7

Low temperature molecular beam epitaxy (MBE) of silicon

H.-J. GOSSMANN

Delta (δ)-doping in Si has long been hindered by the Si-specific so-called 'doping problem': all common dopants in Si exhibit surface segregation at temperatures above the epitaxial temperature. Temperatures low enough to suppress surface segregation kinetically are also below the epitaxial temperature, that is, growth under these conditions has been believed to result in amorphous films, with δ-like dopant distributions but with unsatisfactory electrical quality. In this chapter we will show that a solution exists to the Si-doping problem, which we call 'low temperature molecular beam epitaxy' (LT-MBE). It is based on the recent realization that for growth of Si on Si(100) the concept of an epitaxial temperature is not appropriate. Instead, Si always grows epitaxially, although only up to a certain thickness, h_{epi}. However, films with thicknesses above h_{epi} can be grown by executing rapid thermal anneal cycles at appropriate times. Low temperature molecular beam epitaxy thus allows epitaxial growth at essentially arbitrary temperatures and in particular at temperatures where segregation is completely suppressed. The films are of excellent quality, as demonstrated using a variety of experimental results, obtained from electrical measurements and positron annihilation spectroscopy. Finally, we will present some results obtained from δ-doped films grown by LT-MBE, demonstrating some advantages of δ-doping such as enhanced carrier mobility.

7.1 Growth of δ-doped Si films

7.1.1 Choice of a growth method

A necessary, although certainly not sufficient, condition for the practical realization of δ-doped structures is control of diffusion and segregation during growth. For Si, this requirement has represented a formidable obstacle for quite some time and is illustrated in the form of an overview in Fig. 7.1. Consider a representative δ-doped structure,

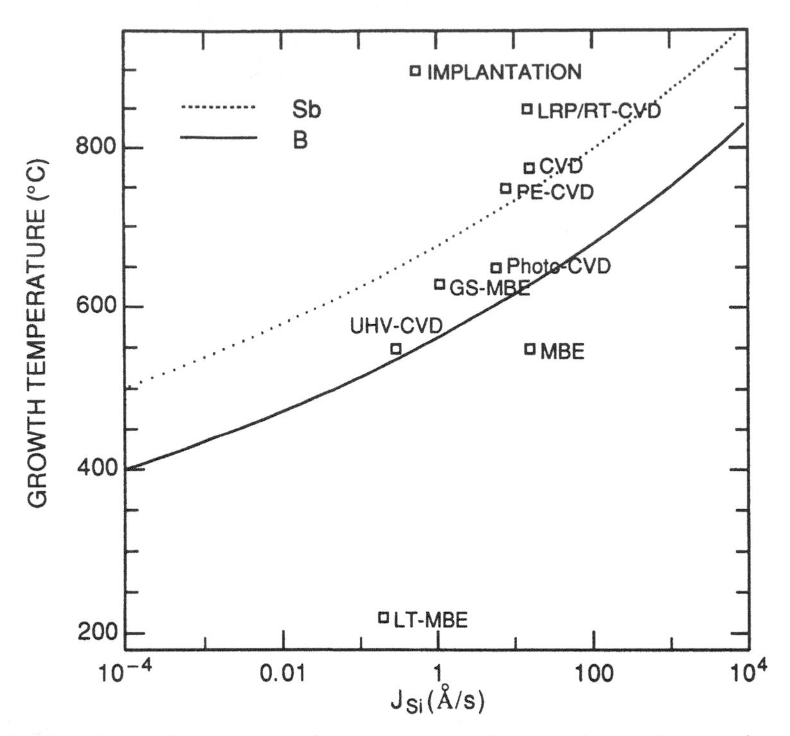

Fig. 7.1. Upper limit on growth temperatures, given by the requirement for a spread of less than 10 Å FWHM of a δ-layer buried under 1000 Å of Si, as a function of Si growth-rate, J_{Si}. The solid line is applicable to B, the dotted one to Sb. Extrinsic diffusion corresponding to a concentration of 10^{20} cm^{-3} is assumed. Also indicated are growth-rate/growth-temperature couples for various growth techniques.

consisting of a single δ-layer covered with 1000 Å Si. We demand that during growth of the cap layer the spread of the δ-layer due to diffusion be limited to 10 Å. For the dopants B (solid line) and Sb (dotted line), Fig. 7.1 shows the maximum temperature permissible during growth of such a structure as a function of growth-rate, J_{Si}. The appropriate bulk diffusion coefficients have been obtained from an extrapolation of literature values (see Chapter 11 on dopant diffusion and segregation in δ-doped Si films). In order to account for the enhancement due to concentration-dependent effects, a concentration of 10^{20} cm^{-3}, constant throughout the diffusion, was assumed. This is certainly exceeded in the initial stages of diffusion for δ-layers with sheet concentrations of the order 10^{14} cm^{-2}, and Fig. 7.1 thus places upper limits on the maximum growth-temperature.

Table 7.1 lists various growth techniques together with the lowest temperature at which high-quality epitaxial growth with a significant growth-rate (at least 0.2 Å/s) has been reported. The term ‘high-quality’ is ideally synonymous with ‘device-quality’, as specified in terms of pn-junction ideality factors or minority carrier lifetimes. Unfortunately, complete electrical device data are not available for all techniques. In this case the particular

Table 7.1. *Parameters of common Si epitaxial growth techniques and selected characteristics of resulting devices. T_s denotes the substrate temperature, J_{Si} the growth rate, τ the minority carrier lifetime, and n the pn-junction ideality factor achieved under these circumstances. (CVD: chemical vapor deposition; LRP/RT: limited reaction processing/rapid thermal; GS: gas source; MBE: molecular beam epitaxy; LT: low temperature)*

Method	T_s(°C)	J_{Si}(Å/s)	$\tau(\mu s)$	n
CVD[a]	775	16	480	1.01
LRP/RT–CVD[b]	850	15	>1	1.05
UHV–CVD[c]	550	≈0.3		1.00–1.05
GS–MBE[d]	≥630	1.1		1.03
PE–CVD[e]	750	8		1.10
Photo–CVD[f]	650	6		1.09
MBE[g]	550	16	6	
LT-MBE[h]	220	0.2		1.05
Implantation[i]	900	0.5[k]	0.01	
Implantation[j]	900	0.5[k]	10	

[a] Burger and Reif (1987).
[b] Sturm *et al.* (1986), Green *et al.* (1989, 1990).
[c] Meyerson *et al.* (1987).
[d] Hirayama *et al.* (1988, 1990).
[e] Burger and Reif (1985).
[f] Yamazaki *et al.* (1990).
[g] Thomas and Francombe (1968).
[h] Gossmann *et al.* (1993).
[i] High dose implant, $>5 \times 10^{13}$ cm^{-2}, Mac Rae (1971).
[j] Low dose implant, $<5 \times 10^{13}$ cm^{-2}, Pickar (1975).
[k] Equivalent deposition rate required to grow a 1000 Å film during a typical 30 min anneal of the ion-implanted film.

entry is omitted. Techniques for which neither ideality factors nor lifetimes have been reported, for example solid phase epitaxy, are not listed. Although ion implantation is not strictly a growth technique, it is included for its widespread use in the production of doped films.

Each growth technique listed in Table 7.1 is represented in Fig. 7.1 as a square at the appropriate growth-rate/growth-temperature. It is quite clear that ion implantation is not compatible with the demanded broadening independent of the concentration and the type of dopant. For Sb, ultra-high-vacuum chemical vapor deposition (UHV-CVD), gas source molecular beam epitaxy (GS-MBE), photo-CVD, molecular beam epitaxy (MBE), and low temperature (LT) MBE conform to the requirements; for B only MBE and LT-MBE offer the necessary combination of growth-rate and growth-temperature, while UHV-CVD is marginal. Although a certain element of arbitrariness has entered through the choice of smearing, and layer thickness, the trend is nevertheless clear. The realization of the full potential of δ-doping, without imposing limits with respect to the type of dopant or concentration, requires an ultra-high-vacuum (UHV) environment, as present in UHV-CVD and MBE. Only then can the growth-temperature be lowered sufficiently, while material quality is maintained. Molecular beam epitaxy is the method of choice, since it is a physical deposition method and does not suffer from the complications of chemical vapor deposition for which deposition rates are dependent on the substrate temperature, species, or co-deposited species, such as dopants.

7.1.2 The doping problem in Si molecular beam epitaxy

7.1.2.1 Dopant segregation

Conceptually, and in practice, the easiest way to dope films during physical vapor deposition is co-evaporative doping: the dopant atoms are deposited together with the matrix atoms. Appropriate evaporation rates and the opening and closing of shutters should then effect almost arbitrary dopant concentration profiles. This is, however, not true for Si. At the typical growth-temperatures of $\geqslant 400$ °C, dopant incorporation during MBE of Si is governed by severe surface segregation and low incorporation probabilities. This is a well-known phenomenon and has become known as the 'Si-doping problem' (Thomas and Francombe 1969; Kuznetsov and Postnikov 1969; Tolomasov *et al.* 1971; Kuznetsov and Postnikov 1974; Postnikov and Kuznetsov 1975; Bennet and Parish 1975; Postnikov and Perov 1977; Perov and Postnikov 1977; Becker and Bean 1977; Ota 1977; König *et al.* 1979; Kuznetsov *et al.* 1979).

The tendency of a dopant to segregate is characterized by the segregation ratio, $r_{\mathrm{D,A}}$, defined as

$$r_{\mathrm{D,A}} = \frac{N^{\mathrm{2D}}_{\mathrm{surf}}}{N^{\mathrm{2D}}_{(klm)}} \frac{N_{\mathrm{Si}}}{N_{\mathrm{D,A}}}, \tag{7.1}$$

with $N^{\mathrm{2D}}_{\mathrm{surf}}$ denoting the dopant surface coverage, $N^{\mathrm{2D}}_{(klm)}$ the areal density of one monolayer at the crystal surface (for Si $N^{\mathrm{2D}}_{(100)} = 6.78 \times 10^{14}\ \mathrm{cm}^{-2}$), $N_{\mathrm{D,A}}$ the dopant concentration, and $N_{\mathrm{Si}} = 4.99 \times 10^{22}\ \mathrm{cm}^{-3}$ the atomic density of Si. If all dopant atoms are incorporated and none segregate, their relative concentration on the surface is the same as in the bulk, that is, $N^{\mathrm{2D}}_{\mathrm{surf}}/N^{\mathrm{2D}}_{(klm)} = N_{\mathrm{D,A}}/N_{\mathrm{Si}}$, and $r_{\mathrm{D,A}} = 1$. This is the desired behavior during thin film growth. If, on the other hand, the dopant atoms do segregate, their relative concentration is smaller in the bulk than on the surface, $N^{\mathrm{2D}}_{\mathrm{surf}}/N^{\mathrm{2D}}_{(klm)} > N_{\mathrm{D,A}}/N_{\mathrm{Si}}$, leading to $r_{\mathrm{D,A}} > 1$. Then the dopant concentration at the beginning of growth will be significantly below the value given by $N_{\mathrm{Si}} J_{\mathrm{D,A}}/J_{\mathrm{Si}}$, where J_{Si} denotes the Si and $J_{\mathrm{D,A}}$ the dopant flux, until $N^{\mathrm{2D}}_{\mathrm{surf}}$ has reached a sufficient value. Likewise, after the dopant flux has been shut off, the dopant surface coverage slowly depletes, leaving a trail of dopant in the nominally undoped part of the film.

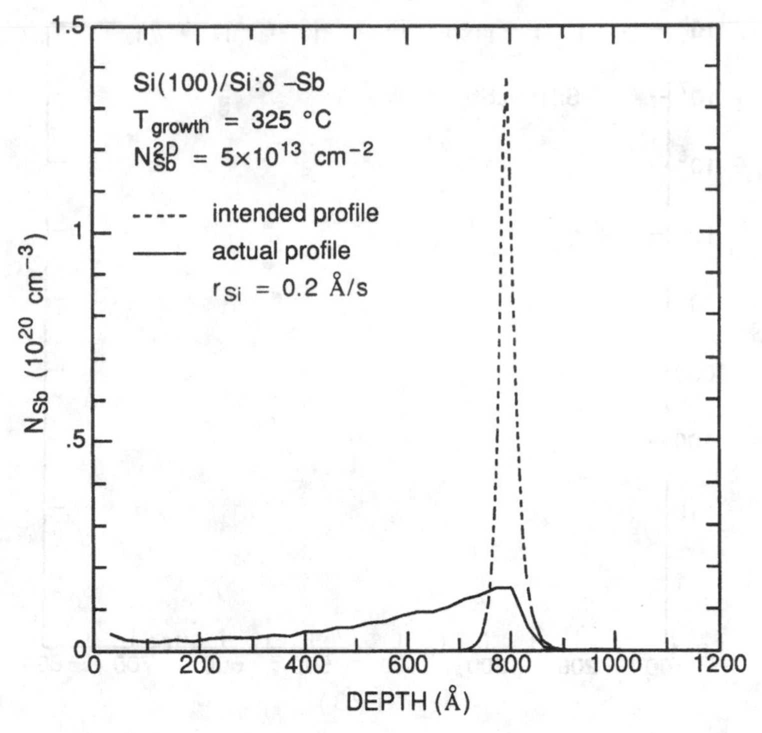

Fig. 7.2. Intended (dashed line) and actual (solid line) concentration profile of Sb in a Si film grown by MBE at a growth-temperature of $T_s = 325$ °C and a growth-rate of $r = 0.2$ Å/s on Si(100).

The dopant atoms for a δ-doped structure are deposited during a growth interruption, that is, the Si flux is shut off until the desired dopant surface coverage has built up. In a concentration vs depth profile the edge of the δ closest to the substrate is thus not broadened by segregation, irrespective of the value of $r_{D,A}$. This is in contrast with the edge closest to the surface: if $r_{D,A} > 1$ then the dopant atoms do not completely incorporate after completion of the δ when the Si flux is turned back on again, resulting in a 'tail' towards the surface. The process of segregation thus produces an asymmetric dopant concentration profile instead of the desired pulse-like shape. This asymmetry and the associated broadening can be quite large, as Fig. 7.2 illustrates for Sb.

Figure 7.3 shows experimental values for the segregation ratio of Sb on Si(100). For a growth-rate of 3 Å/s (squares, Jorke 1988) and above a growth-temperature of $\approx$450 °C, $r_{Sb} \gg 1$. Segregation is a thermally activated process and can be kinetically suppressed at sufficiently low temperatures. From Fig. 7.3 we conclude that at a rate of $J_{Si} = 3$ Å/s 'sufficiently low' implies 450 °C. Indeed, as the triangles show, the segregation ratio becomes unity in the vicinity of 200 °C, even for a much smaller growth-rate of 0.2 Å/s (Gossmann *et al.* 1990, 1991, 1993).

Doping is central to any kind of electronic device, and the doping problem has been studied extensively: for Ga by Becker and Bean (1977), Nakagawa *et al.* (1988), and Zagwijn

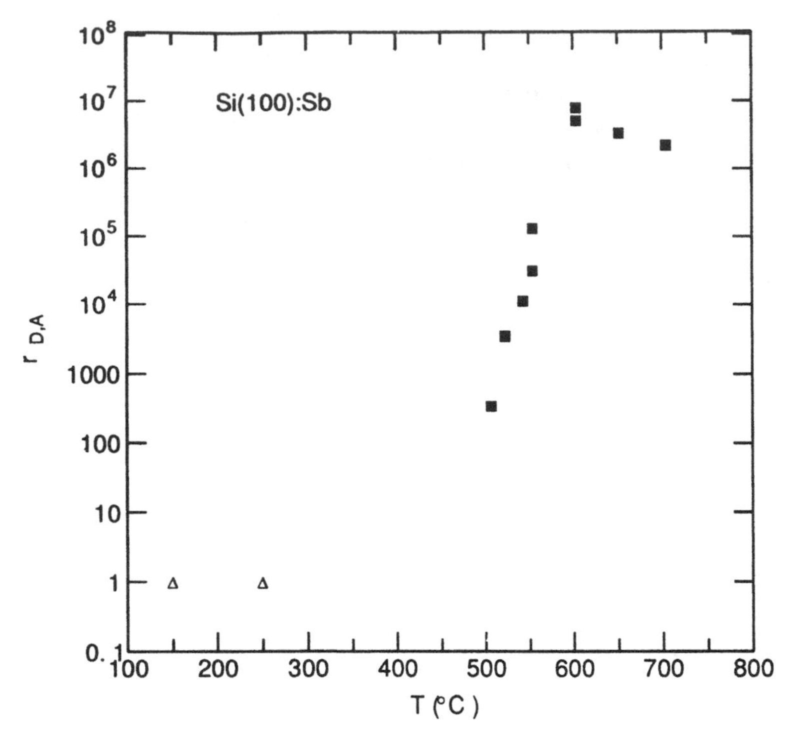

Fig. 7.3. Segregation ratio, $r_{D,A}$, as a function of growth-temperature for Sb doping of Si(100). Squares denote experimental results taken from Jorke (1988) for a growth-rate of 3 Å/s, the triangles are from Gossmann *et al.* (1993) for 0.2 Å/s.

et al. (1991); for In by Knall *et al.* (1984, 1989); for P by Friess *et al.* (1992); and for Sb by Ota (1977), Bean (1978), König *et al.* (1979, 1981), Tabe and Kajiyama (1983), Metzger and Allen (1984a, 1984b), Barnett and Greene (1985), Nakagawa *et al.* (1988), and Jorke (1988). Segregation also takes place during doping with B, although it is less severe (Kubiak *et al.* 1984; Kubiak *et al.* 1985; Ostrom and Allen 1986; Tatsumi *et al.* 1987; Tatsumi *et al.* 1988; Hirayama *et al.* 1988; Andrieu *et al.* 1988; de Frésart *et al.* 1988; Lin *et al.* 1989; Jackman *et al.* 1988; Jackman *et al.* 1989; Headrick *et al.* 1989; Parry *et al.* 1991). Several groups have developed models which allow the calculation of dopant incorporation in Si (Jorke 1988; Tabe and Kajiyama 1983; Metzger and Allen 1984a, 1984b; Barnett and Greene 1985; Barnett *et al.* 1986; Ni *et al.* 1989). The models are based on a one-dimensional rate theory leading to a system of coupled rate equations that can be solved analytically. Their free parameters are determined by fits to experimental dopant profiles. For Sb in particular, calculated and experimental dopant concentration profiles agree quite well, provided the dopant concentration is sufficiently small. At sheet concentrations $\geqslant 10^{14}$ cm^{-2} the Coulomb repulsion of the dopants away from each other can no longer be neglected and leads to qualitatively different behavior of the segregation (Karunasiri *et al.* 1994).

This will be discussed in more detail in Chapter 11 on dopant diffusion and segregation in δ-doped Si films.

7.1.2.2 *The epitaxial temperature*

The traditional view of epitaxial growth by physical vapor deposition methods, such as MBE, is illustrated in Fig. 7.4 (see e.g. Venables and Price 1975). It assumes that there is an 'epitaxial temperature' that, for growth of a film with a finite thickness in finite time, separates the regime of epitaxial, single-crystalline growth from the regime of amorphous growth. Consider a substrate with N_{ad} adsorption sites per unit area. The distance between neighboring sites is then of the order of $d \approx 1/\sqrt{N_{ad}}$. For a deposition rate, J, the time it takes to build up one monolayer will be $\Delta t = N_{ad}/J$. Arriving atoms will stay where they adsorb if the mean path, $\langle x \rangle \approx \sqrt{(D_{surf}\Delta t)}$, traveled due to surface diffusion during time Δt is smaller than d, where D_{surf} is the surface diffusion coefficient. This then implies that $J \gg N_{ad}^2 D_{surf}$ and for atoms with directional bonds, such as Si, the film will be amorphous. Conversely, if diffusion is fast enough, $J \ll N_{ad}^2 D_{surf}$, atoms have sufficient time to find correct lattice sites before being locked in, and the film will be epitaxial. Since D_{surf} is temperature-dependent, we can define the epitaxial temperature, T_{epi}, implicitly by setting $J = N_{ad}^2 D_{surf}(T_{epi})$. Surface diffusion is a thermally activated process and thus the dependence of T_{epi} on J will only be weak. Thus there should be a temperature, unique to

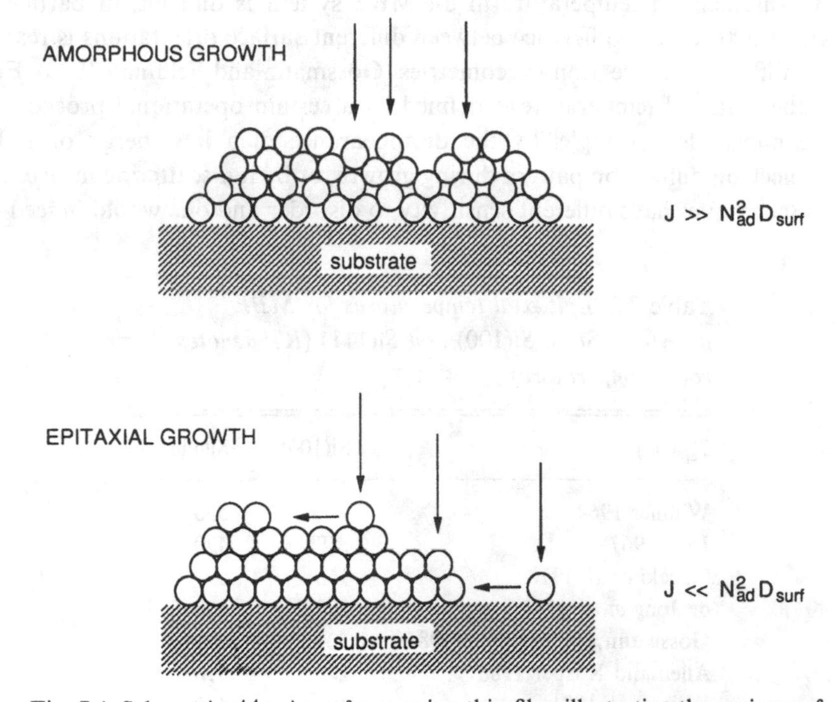

Fig. 7.4. Schematic side view of a growing thin film, illustrating the regimes of amorphous and epitaxial growth (after Venables and Price, 1975).

a particular thin film system, that parameterizes growth: if the growth-temperature $T_{growth} < T_{epi}$, the resulting film is amorphous, and if $T_{growth} > T_{epi}$ the film is single-crystalline. It is worth while emphasizing here that in the former case the film will be amorphous *throughout* its whole thickness and in particular at the substrate/film interface, while in the latter case the film will be *completely* single-crystalline, in particular at its free surface.

As we have seen in the preceding section, the behavior of dopants in Si-MBE at growth-temperatures above a certain critical value T_{cs} is dominated by surface segregation and low incorporation probabilities, making it impossible to achieve abrupt doping profiles. Surface segregation is a thermally activated phenomenon and as such is susceptible to suppression by lowering the growth-temperature. However, within the framework of the epitaxial temperature, this can only be done successfully if $T_{growth} > T_{epi}$ is satisfied simultaneously. For this reason, and to minimize interdiffusion in buried heterostructures, a significant amount of effort has been spent by many groups to determine T_{epi} for Si. The results for Si on Si(100) and Si(111) are summarized in Table 7.2.

It was recognized early that poor vacuum conditions and starting surfaces that are not cleaned in an optimum way yield high epitaxial temperatures (Widmer 1964; Tannenbaum and Povilonis 1964; Thomas and Francombe 1969; Jona 1967; Weisberg 1968; Abbink *et al.* 1968; Thomas and Francombe 1971). Table 7.2 thus lists only data obtained from growth in UHV. Nevertheless, large variations in T_{epi} are observed. A significant fraction of the variations can certainly be attributed to errors in the quoted temperature; the accurate measurement of temperature in an MBE system is difficult, in particular for temperatures $\lesssim 500$ °C. The difference between different surface orientations is real and is rooted in the different surface atomic geometries (Gossmann and Feldman 1985). Further, in all cases the epitaxial temperature is defined by a certain operational procedure, that is, it is determined, for example, by the disappearance of a low-energy or reflection high-energy electron diffraction pattern during growth, or by ion-scattering measurements. The various techniques have different sensitivity to disorder and one would indeed expect

Table 7.2. *Epitaxial temperatures for MBE growth of Si on Si(100) and Si(111) (RT denotes room temperature)*

T_{epi}(°C)	Si(100)	Si(111)
Widmer 1964	—	550
Jona 1967	RT	420
Shiraki *et al.* 1978	165	—
de Jong *et al.* 1983	200	600
Gossmann and Feldman 1985	300	520
Allen and Kasper 1988	240	490
Weir *et al.* 1991	100	400

some variation in epitaxial temperatures for this reason alone. (A more fundamental reason for the variations lies in the fact that the epitaxial temperature concept is not appropriate; we will discuss this further below.) Contrary to the trend in Table 7.2 and again from an operational point of view, it is traditionally believed that epitaxial Si MBE films on Si(100) that are suitable for devices are obtained only for growth-temperatures $T_{epi} \geqslant 400$ °C (Shiraki 1986). In order to grow epitaxial films with abrupt dopant profiles we have to require that $T_{epi} < T_{growth} < T_{cs}$, an inequality that clearly cannot be met, for example, for Sb, where $T_{cs} \geqslant 300$ °C (see Fig. 7.2).

The only solution to this dilemma appears to be the relaxation of the requirement of *epitaxial* growth. The growth of doped films is separated into two steps. First, deposition takes place at very low temperatures where segregation is suppressed kinetically. For practical reasons, room temperature is usually chosen, and the films are thus amorphous. In the second step, the amorphous material is regrown in an annealing cycle. The resulting growth technique is known as solid phase epitaxy (SPE) and has been described in Chapter 6 by I. Eisele. Since the dopants are buried when the film is crystallized, surface segregation does not come into play and bulk diffusion is generally small at regrowth-temperatures of the order of 600 °C, although diffusion transients exist and limit the achievable sharpness of the dopant profiles (see Chapter 11 on dopant diffusion and segregation in δ-doped Si films). Further, Si films doped by SPE exhibit reduced carrier mobility and a dopant activation of less than 100% (Vescan *et al.* 1985; Casel *et al.* 1986; Casel *et al.* 1990), related to the open-volume defects that are present in SPE films with concentrations of the order of 10^{18}–10^{19} cm^{-3} (Schut *et al.* 1991; Asoka-Kumar *et al.* 1993). Solid phase epitaxy is thus not an ideal solution to the doping problem.

7.2 Low temperature molecular beam epitaxy

7.2.1 *The epitaxial thickness*

In the preceding section we reviewed the traditional view of epitaxial growth and the concept of the epitaxial temperature that separates the regimes of crystalline and amorphous growth. In this context one expects that for $T_{growth} < T_{epi}$ the film will be amorphous *throughout* its whole thickness, which includes in particular the zone immediately at the substrate/film interface, while for $T_{growth} > T_{epi}$ the film will be *completely* crystalline, in particular at its free surface, irrespective of its thickness. The traditional standard question in a thin film growth experiment is 'will the film be crystalline or amorphous at the chosen growth-temperature?' Figure 7.5 shows that the answer to this question is 'neither'. Silicon was deposited at ≈200 °C onto a buffer layer, grown at high temperature on a Si(100) substrate. To mark the position at which low temperature growth commenced, 6 Å of Ge, visible as a line in the diagram, were deposited on top of the buffer layer (Gossmann and Eaglesham 1991a). The cross-sectional electron micrograph shown in Fig. 7.5 was taken under diffraction conditions that make crystalline material appear bright and disordered regions black. It

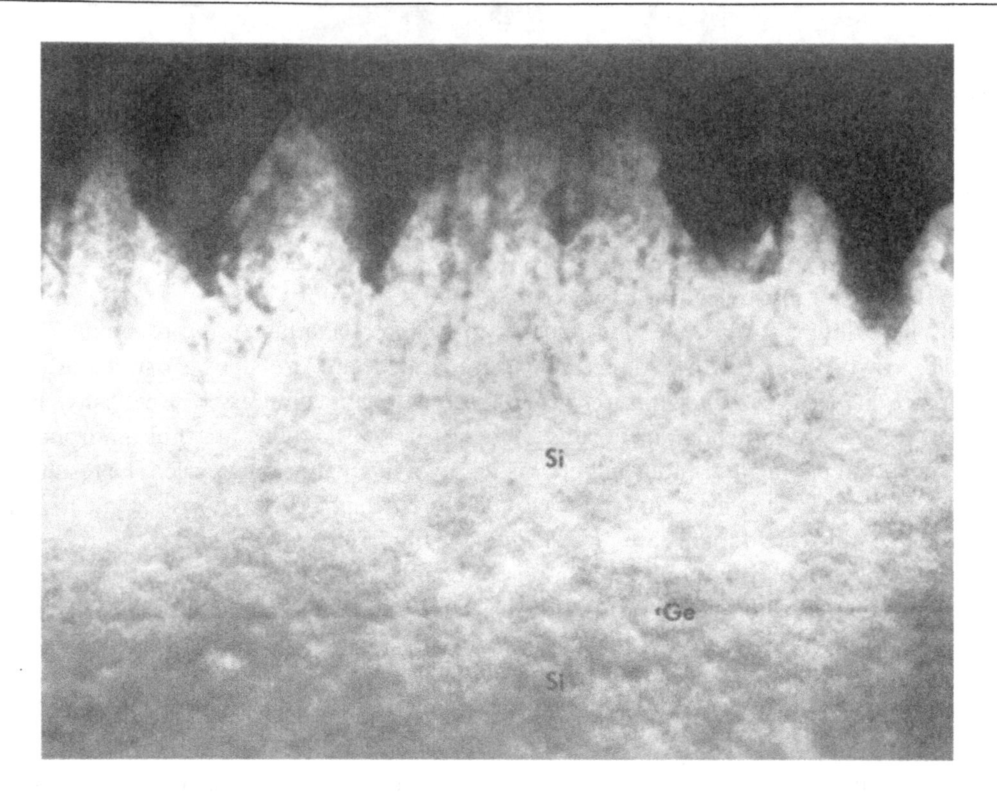

Fig. 7.5. Cross-sectional transmission electron micrograph of a thin film of Si deposited at ≈200 °C on a Si(100) substrate by MBE. To mark the position at which low-temperature growth commenced, 6 Å of Ge, visible as a line in the diagram, was deposited on top of the buffer layer (Gossmann and Eaglesham 1991a).

is quite clear from the micrograph that growth initially proceeds epitaxially, but that after a certain thickness h_{epi} epitaxy breaks down so that, after a transition region where crystalline and amorphous phases co-exist, the film turns completely amorphous. It is also quite clear that the epitaxial temperature concept is not appropriate for describing this situation (Eaglesham *et al.* 1990; Gossmann and Eaglesham 1991a, 1993).

The ragged boundary between the bright and dark regions in Fig. 7.5 is *not* due to surface roughness. Instead, the amorphous phase nucleates at slightly different thicknesses at different places along the growth front. Figure 7.6 shows an example of growth at ≈50 °C, where growth was stopped before h_{epi} was reached (Gossmann and Eaglesham 1991b). Again, the line is due to a Ge marker layer that indicates the beginning of low temperature growth.

The epitaxial thickness is related to temperature by an Arrhenius relation and is plotted in Fig. 7.7 for Si(100)/Si at a growth-rate of 0.7 Å/s (squares) and 50 Å/s (triangles). Straight lines are the results of fits to each data set, giving activation energies of 0.35 eV and 0.43 eV respectively (Eaglesham *et al.* 1990; Gossmann and

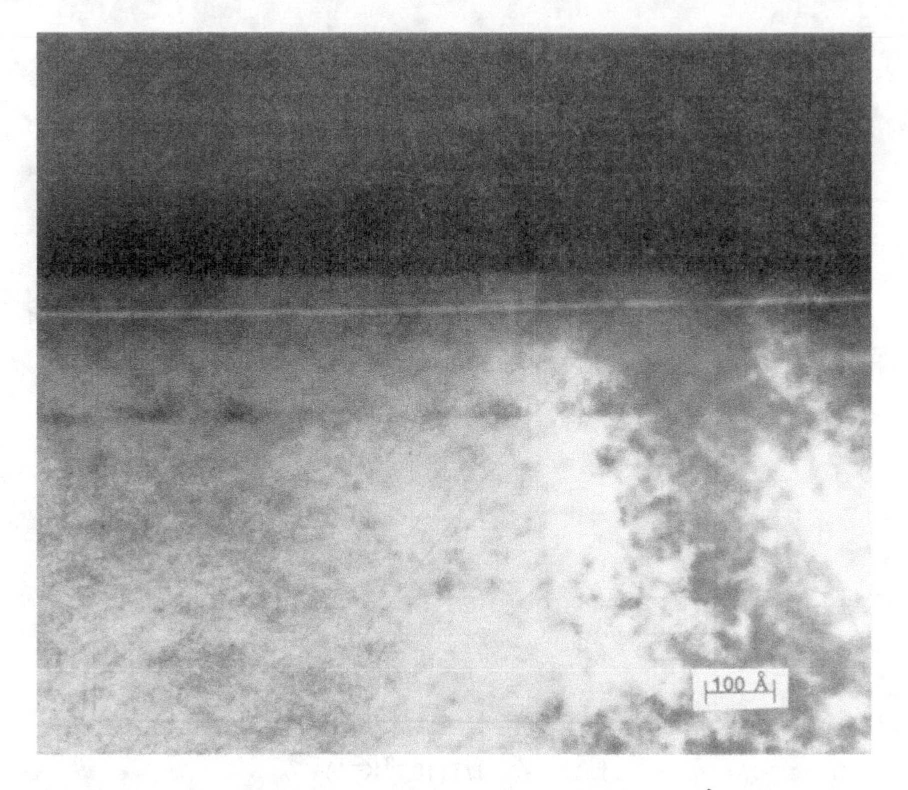

Fig. 7.6. Cross-sectional transmission electron micrograph of 50 Å Si deposited at 50 °C on a Si(100) substrate by MBE. To mark the position at which low-temperature growth commenced, 6 Å of Ge, visible as a line in the diagram, was deposited on top of the substrate (Gossmann and Eaglesham 1991b).

Eaglesham 1993). Growth of epitaxial layers of thickness $m \times h_{epi}$, $m = 1, 2, \ldots$, is accomplished by interrupting growth and annealing the sample at a temperature $T_{RTA} > 500$ °C for 100 s every h_{epi} (Eaglesham *et al.* 1990). The exact value of T_{RTA} is fairly critical to the defect concentration in the film. This issue will be discussed further in Section 7.2.3.2.

The reason for the breakdown of epitaxy at h_{epi} has not been established conclusively. Any influence of the Ge marker layer can readily be excluded (Eaglesham *et al.* 1990). Detailed analysis of electron micrographs further excludes the possibility that the breakdown of epitaxy is due to a continuous increase in defect density up to some critical value (Gossmann and Eaglesham 1993). This is also shown explicitly by electrical data and positron annihilation spectroscopy, discussed in Section 7.2.3. Instead, the available evidence points to microscopic surface roughness as the limiting factor (Eaglesham *et al.* 1990; Gossmann and Eaglesham 1991a). Reflection high-energy electron diffraction (RHEED) intensity oscillations, for example, show a decrease in amplitude during growth at low temperatures (Aarts and Larsen 1988; Gossmann 1991). This damping is related to

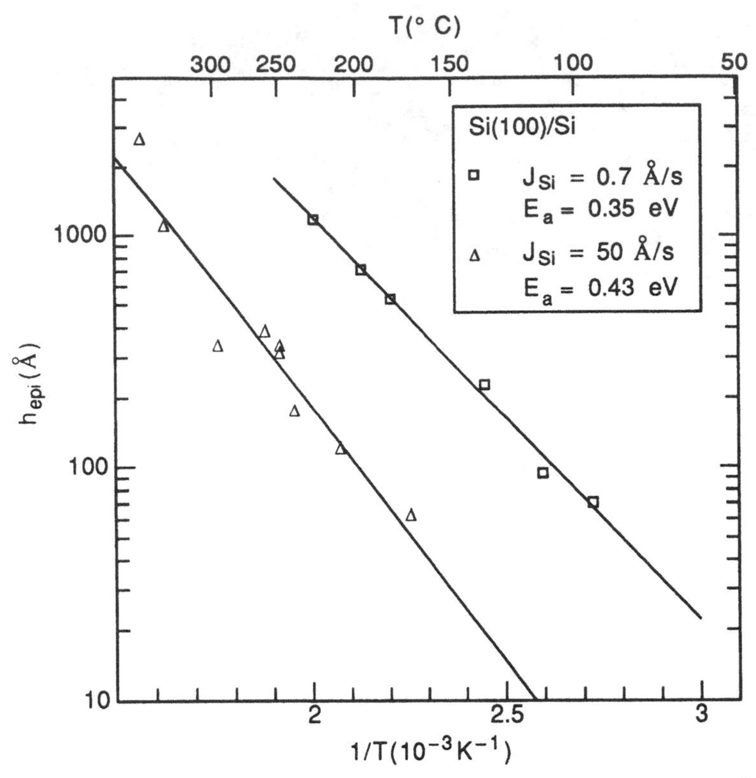

Fig. 7.7. The epitaxial thickness h_{epi} for Si(100)/Si at a growth-rate of 0.7 Å/s (squares) and 50 Å/s (triangles) as a function of growth-temperature. Straight lines are the results of fits to each data set, giving the activation energies indicated in the diagram (Eaglesham *et al.* 1990, Eaglesham and Gossmann 1992).

incomplete filling of a monolayer before the next one is started, leading to an increase in surface roughness (Cohen *et al.* 1989). Figure 7.8(a) shows oscillations in the intensity of the (00) RHEED reflex during growth of Si on Si(100) at ≈50 °C. The oscillations damp very quickly and are no longer observable after about 25 monolayers. In contrast, at the growth-temperature of 220 °C shown in Fig. 7.8(b), the dampening is much smaller and oscillations can be observed for hundreds of periods. Note that for these temperatures the epitaxial thicknesses are of the order of 50 Å and 1000 Å respectively. We conclude that the number of oscillations that are visible for a given growth-temperature scales with temperature in the same way as does h_{epi}. The thickness at which the RHEED intensity oscillations become unobservable is related to the microscopic surface roughness at that point during growth; the RHEED results are thus consistent with the view that the breakdown of epitaxy at h_{epi} is due to the surface exceeding some critical roughness on an atomic scale. Eaglesham *et al.* (1991) have confirmed this roughening by electron microscopy.

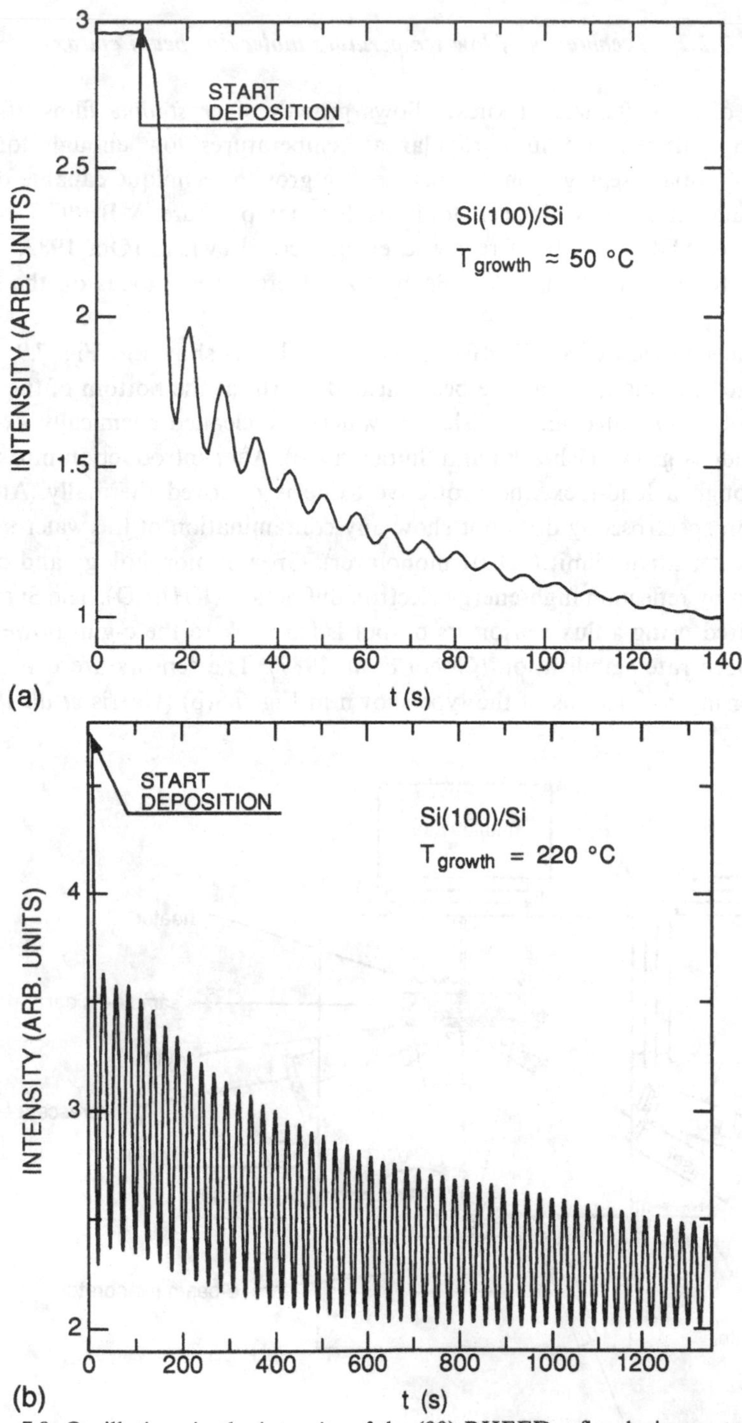

Fig. 7.8. Oscillations in the intensity of the (00) RHEED reflex during growth of Si on Si(100) at (a) ≈50 °C and (b) at 220 °C. Electrons of 19 keV were incident at an angle of ≈0.5°, about 10° off a ⟨100⟩ aximuthal direction.

7.2.2 *Techniques of low temperature molecular beam epitaxy*

The concept of an epitaxial thickness allows growth of *crystalline* films at essentially arbitrary temperatures, and in particular at temperatures low enough for complete suppression of dopant segregation. It thus yields a growth technique capable of resolving the doping dilemma; we will refer to it as low temperature MBE (LT-MBE). The fundamentals of MBE have been reviewed extensively elsewhere (Ota 1983; Allen 1985; Shiraki 1986; Kasper and Bean 1988; Bean 1993). Here we will focus on the specifics of LT-MBE.

A schematic side view of a LT-MBE growth chamber is shown in Fig. 7.9. Float-zone Si is evaporated thermally from an e-beam heated hearth at the bottom of the ultra-high-vacuum chamber. Oriented and polished Si wafers are cleaned chemically *ex-situ* and a protective oxide is grown (Ishizaka and Shiraki 1986). After introduction into the growth chamber through a load-lock, the protective oxide is removed thermally. At this point Auger electron spectroscopy does not show any contamination of the wafer surface with O or C at the sensitivity limit ($\approx$0.01 monolayer). Growth morphology and crystallinity are monitored by reflection high-energy electron diffraction (RHEED). The Si evaporation rate is measured using a flux sensor; its output is fed back to the e-gun power supply in order to achieve rate stabilization (Gogol *et al.* 1987). The sensors are calibrated using RHEED intensity oscillations of the type shown in Fig. 7.8(b) (Harris *et al.* 1981; Neave

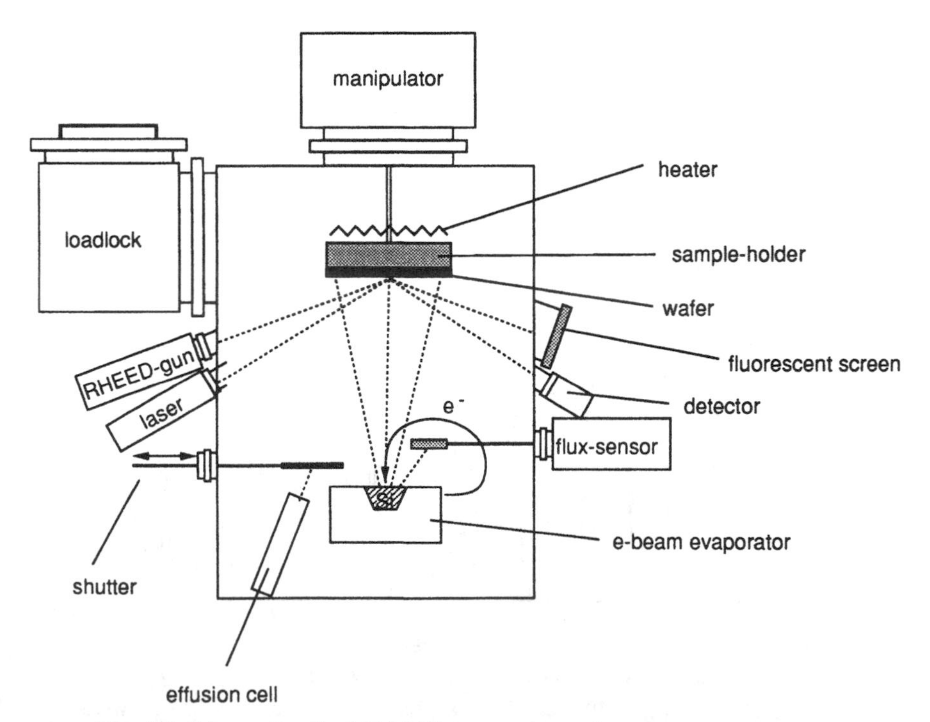

Fig. 7.9. Schematic of a LT-MBE system (pumping systems not shown).

et al. 1983). The drift of this calibration as a function of evaporation time, essentially due to the varying source geometry during source depletion, can also be measured in this way, and can thus be predicted and accounted for.

Dopant atoms are evaporated thermally from appropriate effusion sources. Sb is employed for n-type doping, elemental B for p-type doping. The latter requires a special high-temperature cell, capable of operating at 1800 °C. Other p-type doping schemes, utilizing the B-compounds B_2O_3 or HBO_2, have been proposed by Tatsumi *et al.* (1987), and Ostrom and Allen (1986). They are not compatible with LT-MBE since these compounds have to react on the wafer surface to liberate elemental B (Lin *et al.* 1989, 1990):

$$2HBO_2 + 2Si \rightarrow 2B + 2SiO_2 + H_2 \tag{7.2a}$$

and

$$2B_2O_3 + 3Si \rightarrow 4B + 3SiO_2, \tag{7.2b}$$

for HBO_2 and B_2O_3, respectively, followed by

$$SiO_2 + Si \rightarrow 2SiO. \tag{7.3}$$

The removal of the SiO_2 byproduct via Eqn. (7.3) is required to avoid oxygen incorporation exceeding 10^{19} cm^{-3}. The reaction described in Eqn. (7.3) takes place at substrate temperatures of the order of 650–750 °C, leading to significant profile broadening (Tatsumi *et al.* 1987, 1988; de Frésart *et al.* 1988; Lin *et al.* 1989, 1990; Jackman *et al.* 1988, 1989; Tuppen *et al.* 1988).

The conventional approaches of temperature measurement in ultra-high-vacuum, infrared pyrometry and thermocouple measurements cannot be used in the temperature regime of LT-MBE under the restrictions of an ultra-clean environment. On the other hand, as we will discuss in Section 7.2.3.2, accurate knowledge of temperature is crucial to the growth of high-quality films by LT-MBE. A very accurate determination of temperature, even below room temperature, is however possible by laser-interferometry, described in detail by Donnelly and McCaulley (1990). Basically, the interference fringes produced by laser light reflected from the front- and back-sides of a double-sided polished wafer are counted as a function of wafer temperature. They are due to the temperature dependence of the index of refraction and, to a lesser extent, the thermal expansion of the wafer. Employing 1.52 μm laser light incident at 50°, one 2π phase shift corresponds to about 5 K; a precision of 1 K is thus easily achieved. The accuracy is essentially limited by the error in the measurement of the wafer thickness, about ±2 μm, which translates into ±2 K at 550 °C for a 0.5-mm-thick wafer. Absorption eventually causes the interference between light reflected from the front- and back-sides of the wafer to die out; at the chosen wavelength fringes are observable up to temperatures of about 750 °C. An optical pyrometer operating at 0.8–1.1 μm is used for temperatures above 700 °C. It is calibrated in the temperature range 600–750 °C by using the interferometrically determined temperature as a reference.

7.2.3 *Growth defects*

7.2.3.1 *Dopant activity and carrier mobility*

The electrical activity of dopants and the carrier mobility in homogeneously doped films grown by LT-MBE are shown in Fig. 7.10(a, b) (n-type, Sb) and Fig. 7.11(a, b) (p-type, B). In both diagrams, panels (a) show the majority carrier concentration as a function of atomic concentration; the solid lines represent 100% activation. Panels (b) show the mobility as a function of atomic concentration; the solid lines represent literature values, for As in Fig. 7.10(b) and B in Fig. 7.11(b) (Masetti *et al.* 1983). All films were grown at a temperature of $T_{\mathrm{growth}} = 270\ °\mathrm{C}$.

Carrier concentration and mobility were determined by the Hall technique in the van-der-Pauw (1958) geometry. The relationship between the Hall mobility, $\mu_{n,p}^{\mathrm{H}}$, and the

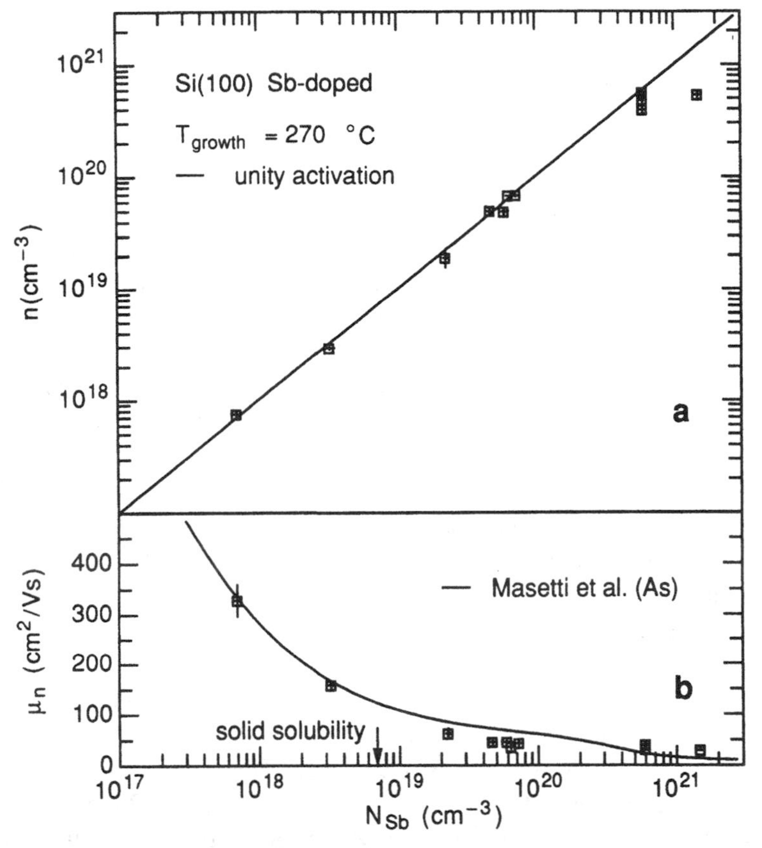

Fig. 7.10. Concentration (a) and mobility (b) of n-type carriers in Sb-doped films (squares) at room temperature, as a function of Sb atomic concentration. The solid line in (a) corresponds to unity activation. The arrow indicates the solid solubility limit of Sb in Si. The solid line in (b) represents the bulk mobility of n-type Si doped with As (Masetti *et al.* 1983). A Hall factor of 1.00 was assumed.

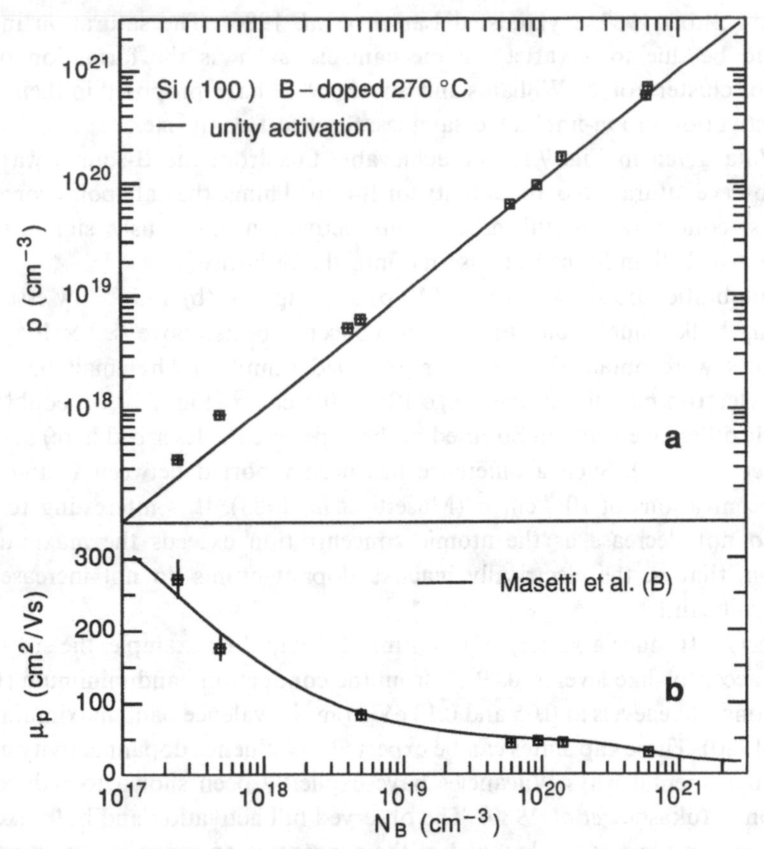

Fig. 7.11. Concentration (a) and mobility (b) of p-type carriers in B-doped films (squares) at room temperature, as a function of B atomic concentration. The solid line in (a) corresponds to unity activation. The solid line in (b) represents the bulk mobility of p-type Si doped with B (Masetti *et al.* 1983). A Hall factor of 0.75 was assumed (Lin *et al.* 1981).

drift mobility, $\mu_{n,p}$, as well as between the Hall concentration, n^{H}, p^{H}, and the carrier concentration, n, p, is established via the Hall factor, $R^{\mathrm{H}}_{n,p}$. For n-type carriers

$$\mu_n = \frac{\mu_n^{\mathrm{H}}}{R_n^{\mathrm{H}}}, \quad n = n^{\mathrm{H}} R_n^{\mathrm{H}}, \tag{7.4}$$

and correspondingly for p-type carriers (Sze, 1981). The Hall factor has been determined for uniformly doped films as $R_n^{\mathrm{H}} = 1.0$ for n-type doping, and as $R_p^{\mathrm{H}} = 0.75$ for p-type doping (Lin *et al.* 1981).

As can be seen from Fig. 7.10(a) and Fig. 7.11(a), complete activation is achieved over basically the whole investigated concentration range. For Sb, saturation of the carrier concentration takes place at $N_{\mathrm{Sb}} \geqslant 6 \times 10^{20}$, two orders of magnitude above the estimated solid solubility limit at the growth-temperature (Trumbore 1960). This is also at least 50% higher than the maximum fully electrically active concentration achieved by ion implantation

(Williams and Short 1982; Nylandsted Larsen *et al.* 1986). The saturation in electrical activity could be due to a variety of mechanisms, such as the formation of metallic precipitates or clusters, or, as Williams and Short (1982) have proposed in their discussion of dopant activation in ion-implanted samples, Sb-defect complexes.

For the data given in Fig. 7.11 the achievable flux from the B-source was not high enough to observe saturation of the activity for B-doped films; the data point corresponding to the highest concentration still indicates full activation. Obviously, significantly more electrically active B than Sb can be inserted into the Si lattice.

The drift mobilities are shown in Fig. 7.10(b) and Fig. 7.11(b) and follow essentially the corresponding bulk values (solid lines). (For concentrations above $\approx 2 \times 10^{20}\,\mathrm{cm}^{-3}$ the As bulk values were obtained from laser annealed samples.) The small but significant deviation in electron mobility in the range 10^{19}–$10^{20}\,\mathrm{cm}^{-3}$ (Fig. 7.10(b)) could be due to some intrinsic difference between Sb (used in the experiments described here) and As (used by Masetti *et al.* 1983). Such a difference has been reported between P and As above electron concentrations of $10^{19}\,\mathrm{cm}^{-3}$ (Masetti *et al.* 1983). It is interesting to note that mobilities do not decrease as the atomic concentration exceeds the maximum carrier concentration, that is, the electrically inactive dopant atoms do not increase impurity scattering significantly.

Point defects introduce a variety of levels into the gap. For example, the single vacancy in Si has an acceptor-like level at 0.09 eV from the conduction band minimum (Kimerling 1976), and donor-like levels at 0.05 and 0.13 eV from the valence band maximum (Watkins and Troxell 1980). These gap states can be expected to influence dopant activity and carrier mobility in detrimental ways. Vacancies have explicitly been shown to reduce the hole concentration (Mukashev *et al.* 1980). The observed full activation and bulk-like mobility of LT-MBE grown films at the low end of the concentration range investigated is thus a strong indicator that the concentration of electrically active defects in the films is well below $10^{17}\,\mathrm{cm}^{-3}$.

7.2.3.2 Vacancy-like defects and the influence of T_{RTA}

The line shape of the radiation emitted from annihilating positrons in the surface region of a solid (positron annihilation spectroscopy, PAS) gives quantitative, depth-resolved information on defect concentrations (Schultz and Lynn 1988). In particular, the method is very sensitive to vacancy-like defects. A sensitivity limit of $\approx 5 \times 10^{15}\,\mathrm{cm}^{-3}$ for di-vacancies in Si is easily achievable (Asoka-Kumar *et al.* 1993). It thus represents a unique way of surveying film quality as a function of growth-temperature, impurity concentration, or post-growth-processing parameters in a defect concentration region inaccessible to TEM or ion scattering.

In a typical positron annihilation experiment a beam of 0–20 keV positrons is directed towards the target. The positrons thermalize at a depth related to their primary energy and annihilate. The main quantity detected is the line shape of the emitted annihilation line at 511 keV. Usually, it is quantified by the so-called *S*-parameter, the ratio of the counts in a narrow window around the center energy of 511 keV and the total counts in the annihilation line. If a positron thermalizes in a perfect crystal region and annihilates there, the overlap with core electrons is larger than if the positron becomes trapped and annihilates in an open-volume defect, for example a di-vacancy. Consequently, the line will be broader (the S-parameter smaller) in the former than in the latter case. Depth-sensitive information is obtained by recording the S-parameter, normalized to its bulk-value, S_b, as a function of positron energy. Defects are present in those regions where S/S_b exceeds a value of unity.

Figure 7.12 shows the normalized S-parameters for 2000-Å-thick Si films grown on Si(100) as a function of positron energy in panels (b)–(d). A depth scale is indicated at the top. The growth-temperature is indicated as T_{growth}. Also shown for comparison in panel (a) is the result for a plain Si wafer (Gossmann *et al.* 1992; Asoka-Kumar *et al.* 1993). The value of $S/S_b = 1$ over the whole energy- (or depth-) range indicates the high quality of standard Si-wafers, as expected, and serves as a baseline for the MBE samples. For the data in panel (b) growth took place at 475 °C. The positron data is indistinguishable from the bulk Si sample, placing an upper limit of $\approx 5 \times 10^{15}$ cm^{-3}, the sensitivity of the experiment, on the concentration of open-volume defects. For the films in panels (c) and (d) growth took place at a temperature of 220 °C. To prevent these films from becoming amorphous, 2 min rapid thermal anneals to a peak temperature of T_{RTA} were executed every $\approx$300 Å. A typical temperature–time profile for $T_{RTA} = 600$ °C is shown in Fig. 7.13, together with the heater power. The thermal cycles are executed without growth-interruption, that is, the shutter stays open (O) all the time.

Figure 7.12(c) and (d) illustrates that the correct value of T_{RTA} is of considerable importance. A RTA temperature of 600 °C gives results that are indistinguishable from the data of the virgin substrate shown in Fig. 7.12(a), whereas $T_{RTA} = 450$ °C gives rise to a S-parameter significantly above the bulk-value in a region of the film close to the surface. A quantitative analysis of the S-parameter values is possible by considering the diffusion and trapping of positrons in a sample with a defect concentration profile $c(z)$, where z measures the depth into the sample (for details see Schultz and Lynn 1988; Gossmann *et al.* 1992; Asoka-Kumar *et al.* 1993). In order to keep the number of free parameters finite, a certain shape has to be assumed for the defect profile. The simplest assumption is that of a defect concentration that is assumed constant throughout the whole film:

$$c(z) = c_2 \qquad 0 \leqslant z \leqslant t, \tag{7.5a}$$

with t being the thickness of the film, or a two-level model, given by

$$c(z) = \begin{cases} c_1 & 0 \leqslant z < \zeta \\ c_2 & \zeta \leqslant z \leqslant t \end{cases}. \tag{7.5b}$$

The concentrations c_1, c_2, and the depth ζ are fit to the S-parameter data. The results of fits to the data in Fig. 7.13 and to other sets are summarized in Table 7.3. Also shown are defect concentrations reported by Schut *et al.* (1991) in films grown by MBE. It is obvious that MBE and LT-MBE growth leads to material with vacancy-like defects below the sensitivity level of the positron annihilation experiment, 5×10^{15} cm^{-3} at room

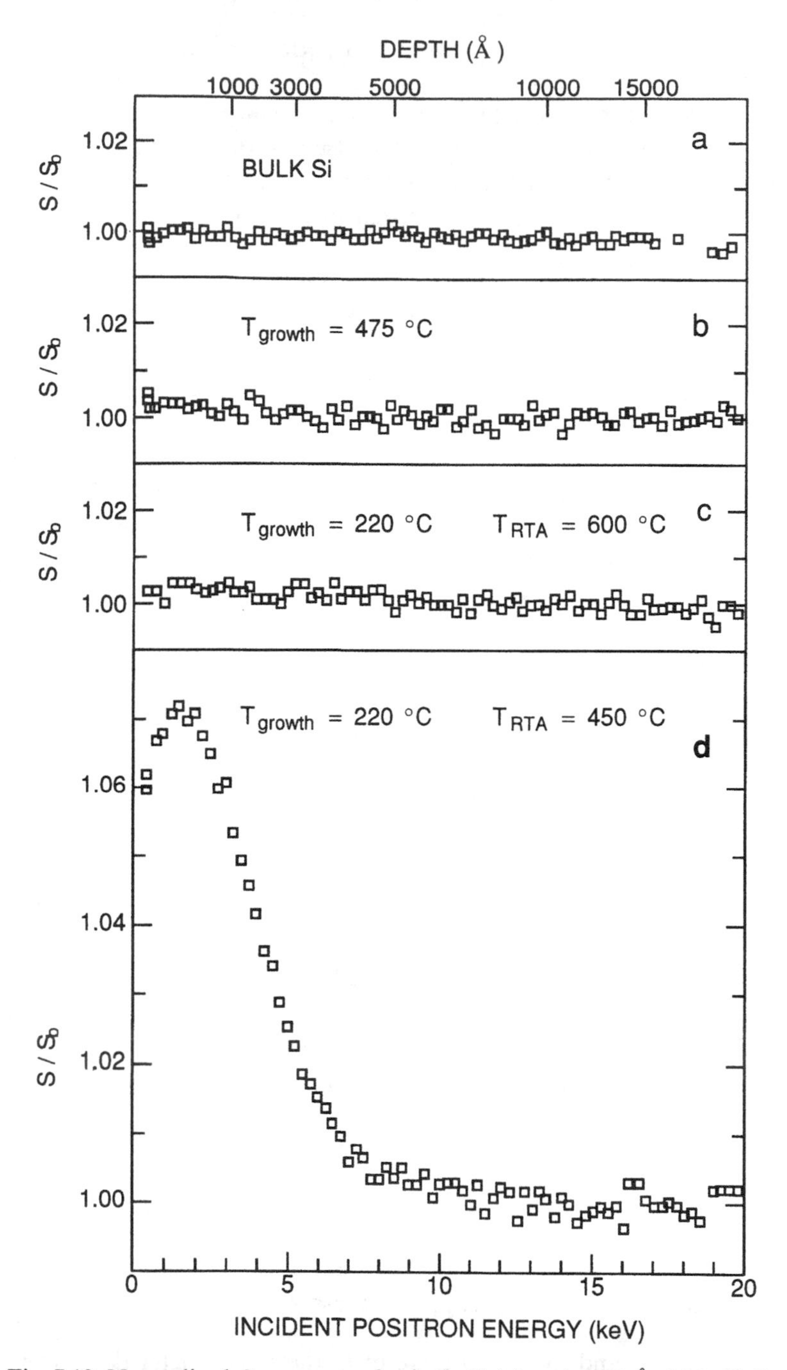

Fig. 7.12. Normalized S-parameter for bulk Si (a) and 2000 Å thick Si films (b)–(d) grown by MBE at growth-temperatures T_{growth}. For the films shown in (c) and (d), rapid thermal anneals were executed every ≈ 300 Å to a peak temperature T_{RTA}. The temperature vs time profile for the film in (c) is shown in Fig. 7.13 (Gossmann *et al.* 1992, Asoka-Kumar *et al.* 1993).

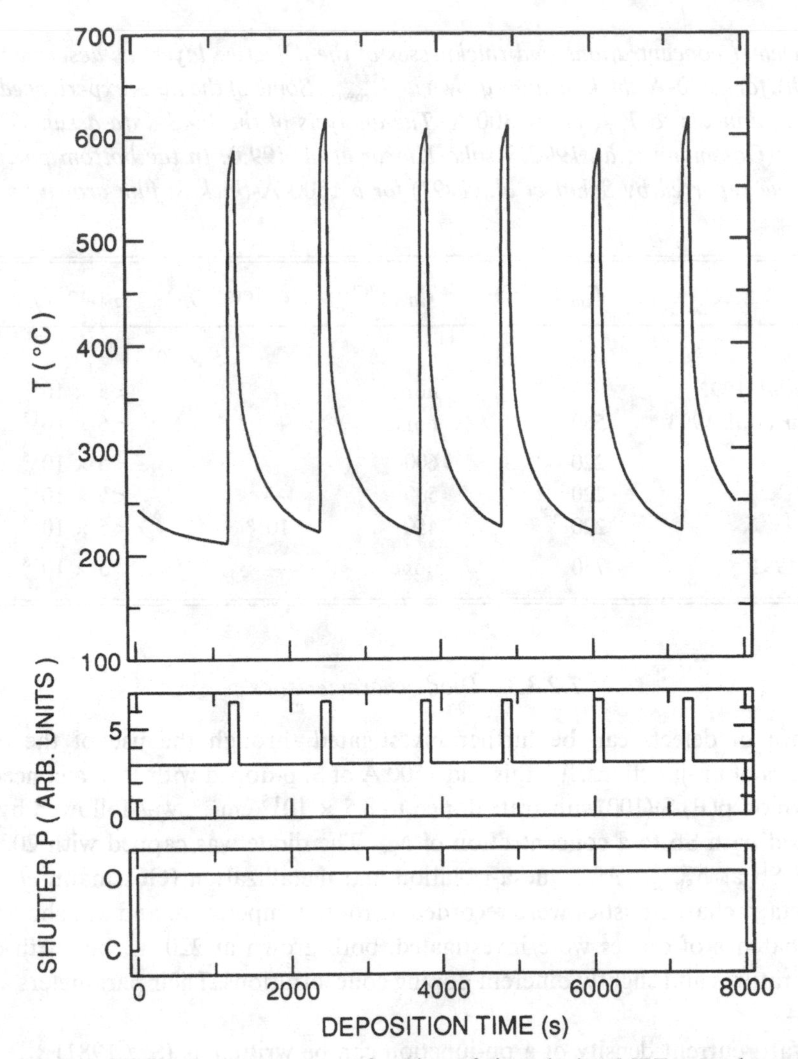

Fig. 7.13. Wafer-temperature as a function of time during MBE growth at a growth-temperature T_{growth} = 220 °C with 2 min rapid thermal anneals to 600 °C. The heater power, *P*, is indicated. No growth interruption took place during the flashes, that is, the shutter stayed open (O) all the time.

temperature, provided the correct growth parameters are used. (1) The influence of T_{RTA} on the crystalline quality of the film is significant. (2) The substrate preparation must also play a significant role in achieving high-quality material: Schut *et al.* (1991) have reported positron annihilation data on films grown by MBE at 730 °C. Despite that rather high temperature, they observe open-volume defect levels of 3×10^{18} cm^{-3}. The authors point out that their substrate cleaning technique, consisting of a HF-dip prior to introduction of the wafer into the growth chamber, followed by an anneal at 880 °C, leads to the formation of carbide precipitates.

Table 7.3. *Defect concentrations and thicknesses of the defective layer, as described by Eqn.* (7.5a, b), *for* 2000-Å-*thick Si-films grown at* T_{growth}. *Some of the films experienced* 2 *min rapid thermal anneals to* T_{RTA} *every* 300 Å. *The analysis of the PAS-data assumed di-vacancies.* (*Gossmann et al. 1992, Asoka-Kumar et al. 1993*). *In the bottom part of the table the value reported by Schut et al.* (*1991*) *for a* 1500-Å-*thick Si film grown by MBE is also listed*

Reference	T_{growth} (°C)	T_{RTA} (°C)	c_1 (cm^{-3})	c_2 (cm^{-3})	ζ (Å)
	bulk		$<5 \times 10^{15}$		N/A
Gossmann *et al.* 1992,	475	none	—	$<5 \times 10^{15}$	—
Asoka-Kumar et al. 1993	560	none	—	$<5 \times 10^{15}$	—
	220	600	—	$<5 \times 10^{15}$	—
	220	500	—	$<5 \times 10^{15}$	—
	220	450	10^{18}	$<5 \times 10^{15}$	1100
Schut *et al.* 1991	730	none	—	3×10^{18}	—

7.2.3.3 *Diode characteristics*

The presence of defects can be further investigated through the use of the electrical characteristics of pn-junctions. To this end 1500 Å of Si p-doped with B to a concentration of N_{B}, grown on p(B) Si(100) substrate doped to 8.5×10^{18} cm^{-3}, was followed by 1500 Å of Si n-doped with Sb to a concentration of N_{Sb}. The diode was capped with 200 Å of Si doped with Sb at $N_{\mathrm{Sb,cap}}$. After mesa-isolation and metallization (Gossmann *et al.* 1993) current–voltage characteristics were recorded at room temperature and are shown in Fig. 7.14. Two batches of diodes were investigated, both grown at 220 °C, but with different RTA temperatures and slightly different doping concentrations. Their parameters are listed in Table 7.4.

The forward current density of a pn-junction can be written as (Sze 1981)

$$j_{\mathrm{F}} = j_0 \, \mathrm{e}^{qV_{\mathrm{F}}/nkT}, \tag{7.6}$$

where V_{F} is the forward voltage, T the temperature, and j_0, q, and k are a constant, the elementary charge, and Boltzmann's constant respectively. The factor n is called the 'ideality factor': if carrier transport in the junction is solely due to diffusion, that is, if there are no electrically active defects of any kind, $n = 1$. If transport is dominated by recombination, that is, in the presence of large numbers of defects, $n = 2$. As can be seen from Fig. 7.14, the diode of batch 1 with $T_{\mathrm{RTA}} = 450$ °C has a significantly larger ideality factor than does the diode of batch 2 ($T_{\mathrm{RTA}} = 600$ °C). In fact, the numerical value of $n = 1.94$ for $T_{\mathrm{RTA}} = 450$ °C indicates that in this case the forward current is dominated by recombination effects, that is, the material exhibits a large number of defects. In contrast, for $T_{\mathrm{RTA}} = 600$ °C, we obtain the practically perfect value of $n = 1.05$.

Table 7.4. *Growth parameters of diodes used for the electrical characterization of LT-MBE films.* T_{RTA} *denotes the RTA temperature,* N_B *and* N_{Sb} *the dopant concentration of the p- and n-side of the junction, respectively, and* $N_{Sb, cap}$ *the Sb concentration in the cap.*

Batch	T_{RTA} (°C)	N_{Sb} (cm^{-3})	N_B (cm^{-3})	$N_{Sb, cap}$ (cm^{-3})
1	450	2.3×10^{17}	4.8×10^{17}	3×10^{19}
2	600	9.4×10^{17}	2.4×10^{17}	15×10^{19}

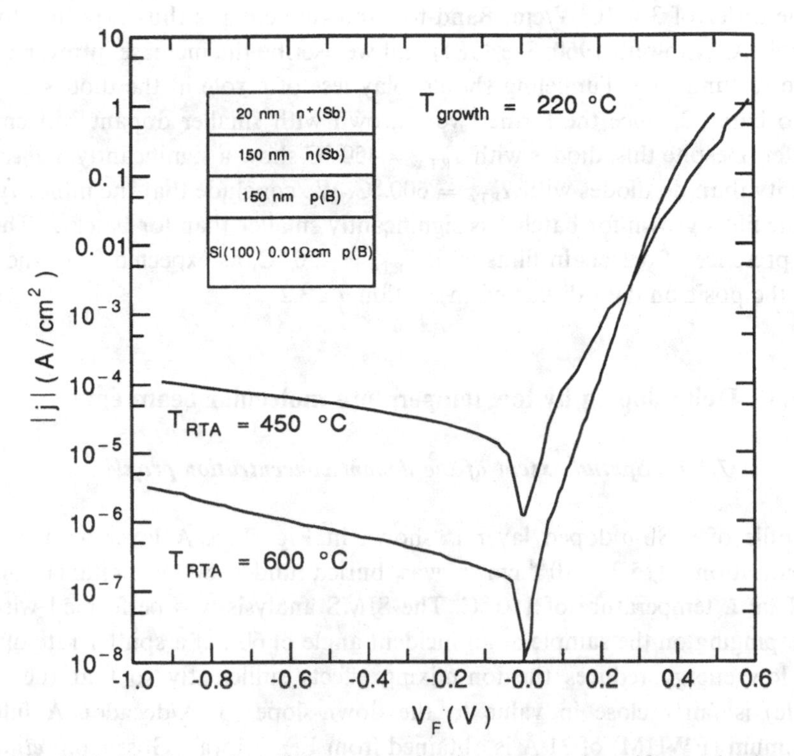

Fig. 7.14. Absolute value of the current density as a function of forward voltage for a shallow pn-junction grown by LT-MBE. The diode structure is shown schematically as inset. Both diodes were grown at the same temperature of 220 °C, but differed in the value of T_{RTA}, as indicated in the diagram.

In the absence of quantum-mechanical tunneling and at voltages below breakdown of the junction due to avalanche or thermal effects, the reverse current density can be written as (Sze 1981)

$$j_R = \frac{qWn_i}{\tau}, \tag{7.7}$$

where W is the width of the depletion region, n_i is the intrinsic carrier concentration, and

τ is the minority carrier lifetime. Note that Eqn. (7.7) assumes that the doping concentration is sufficiently high for the generation current in the depletion region to dominate. Equation (7.7) also implies that the reverse current is independent of reverse voltage. As can be seen from Fig. 7.14, this is not the case for the diode grown at $T_{RTA} = 600$ °C. Instead, the reverse current increases with reverse voltage in an approximately exponential fashion. Such behavior could in principle be due to electrically active defects; however, the excellent ideality factor argues against this explanation. Alternatively, the dependence of the reverse current on reverse voltage could be due to tunneling. As we will see below, LT-MBE allows the growth of very abrupt dopant transitions and, as can be seen from Table 7.4, both sides of the junction are fairly highly doped. Estimates of the field at the junction yield values of the order of 3×10^5 V/cm. Band-to-band tunneling is thus expected to play a significant role (Chynoweth 1960; Sze 1981) and we ascribe the increase in reverse current with voltage to tunneling. Tunneling should play less of a role in the diodes of batch 1 compared to batch 2, since the former were grown with smaller dopant concentrations than the latter. Despite this, diodes with $T_{RTA} = 450$ °C show a significantly *higher* reverse current density than do diodes with $T_{RTA} = 600$ °C. We conclude that the minority carrier lifetime in the films grown for batch 1 is significantly smaller than for batch 2. This again implies the presence of defects in films with $T_{RTA} = 450$ °C, as expected from the ideality factors and the positron data discussed in Section 7.2.3.2.

7.3 Delta-doping by low temperature molecular beam epitaxy

7.3.1 Spatial extent of the dopant concentration profile

A SIMS-profile of a Sb-δ-doped layer is shown in Fig. 7.15. A layer with an atomic sheet concentration of 5.2×10^{14} cm^{-2} was buried under 800 Å epitaxial Si grown by LT-MBE at a temperature of 150 °C. The SIMS analysis was performed with 3 keV Cs^+ ions, impinging on the sample at an incident angle of 60°, at a sputter rate of 1.5 Å/s. This fairly low energy reduces the ion-mixing effects sufficiently so that the up-slope (25 Å/decade) is fairly close in value to the down-slope (37 Å/decade). A full-width-at-half-maximum (FWHM) of 31 Å is obtained from Fig. 7.15(a) (Gossmann *et al.* 1993). However, this is only an upper limit on the width of the actual dopant distribution: at an energy of 1.5 keV the resolution of the SIMS setup was determined as 15 Å FWHM on a sample consisting of one monolayer of Al in GaAs (Luftman *et al.* 1991). The resolution is worse at the higher energy used for the data given in Fig. 7.15(a). Subtracting the resolution quadratically we thus obtain an upper limit on the width of the dopant distribution of 27 Å.

Results for 1.2×10^{14} cm^{-2} B atoms covered with 600 Å Si are shown in Fig. 7.15(b). The structure was grown at 220 °C. Secondary ion mass spectrometry analysis was performed with a 2.0 keV O_2^+ beam at 60° incidence and a sputter rate of 0.4 Å/s. Despite this low energy of the sputter beam, ion-mixing effects still led to a significant asymmetry in the peak, that is, tailing towards the substrate side. The observed FWHM of 40 Å is

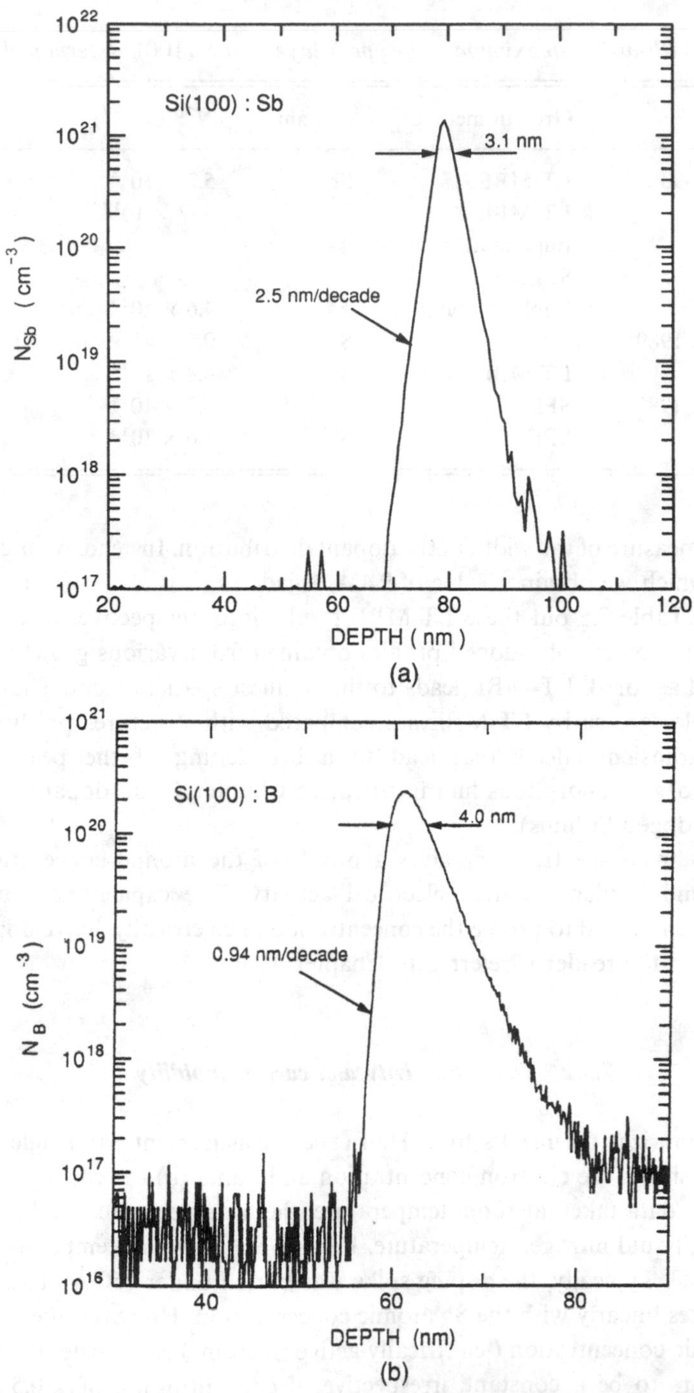

Fig. 7.15. Secondary ion mass spectrometry (SIMS) depth profile of (a) an Sb-δ-doping layer with an atomic sheet concentration of 5.2×10^{14} cm^{-2}, and (b) a B-δ-doping layer with an atomic sheet concentration of 1.2×10^{14} cm^{-2}, grown by LT-MBE.

Table 7.5. *Full-width-at-half-maximum of n-type δ-layers in Si(100), determined by SIMS*

Reference	Growth method	Dopant	N^{2D} (cm^{-2})	FWHM (Å)
Gossmann *et al.* 1993	LT-MBE	Sb	5.2×10^{14}	27
Friess *et al.* 1992	LT-MBE	P	$5\text{–}7 \times 10^{13}$	380
Djebbar *et al.* 1990	Implantation	As		87
Eisele 1989	SPE	Sb		80
Denhoff *et al.* 1989	Implantation	As	4.6×10^{13}	190
van Gorkum *et al.* 1989	SPE	Sb	$0.13\text{–}4.1 \times 10^{14}$	120
Jorke *et al.* 1989	LT-MBE	Sb	6.8×10^{14}	82
van Gorkum *et al.* 1987	SPE	Sb	6.8×10^{14}	120
Zeindl *et al.* 1987	SPE	Sb	1.6×10^{13}	140

thus not a good measure of the width of the dopant distribution. Instead, we use the leading edge slope, for which we obtain a value of 9.4 Å/decade.

Table 7.5 and Table 7.6 put these LT-MBE results into perspective. In each table we compare the spatial extent of δ-doped profiles obtained from various growth methods.

For Sb as well as for B, LT-MBE leads to the smallest spatial extent. That this is true even when samples grown by LT-MBE are compared with structures produced by SPE is due to the transient effects that lead to a broadening of the peak during the recrystallization of the amorphous film in SPE (see Chapter 11 on dopant diffusion and segregation in δ-doped Si films).

Secondary ion mass spectrometry gives a profile of the atomic concentration of the dopant atoms, independent of their electrical activity. The capacitance–voltage (CV) technique can be employed to profile the concentration of electrically active dopants. For a detailed discussion the reader is referred to Chapter 9.

7.3.2 Dopant activity and carrier mobility

Figure 7.16 summarizes the results from Hall effect measurements on single Sb-δ-doped layers. Panel (a) shows the electron concentration and panel (b) the carrier mobility. The squares represent data taken at room temperature; the triangles were obtained from Hall measurements at liquid nitrogen temperature. Films were grown at temperatures between 150 °C and 270 °C. Typically, the doping spike was buried under 750 Å of Si. The carrier concentration rises linearly with the Sb atomic concentration. However, the ratio between carrier and atomic concentration ('electrically active fraction') is consistently below unity. In fact, it appears to be a constant, irrespective of concentration, of $\approx$0.5 and is also independent of temperature in the range between room temperature and liquid nitrogen temperature.

This apparent reduction in electrical activity is not due to defects but, rather, it is a

Table 7.6. *Full-width-at-half-maximum and up-slope of p-type δ-layers in Si(100), determined by SIMS*

Reference	Growth method	Dopant	N^{2D} (cm^{-2})	Up-slope (Å/dec)
Gossmann *et al.* 1993	LT-MBE	B	1.2×10^{14}	9.4
Vink *et al.* 1990	CVD	B	6.2×10^{14}	>13[b]
Mattey *et al.* 1990	MBE	B	1×10^{13}	27[a]
Jorke and Kibbel 1990	MBE	B	5×10^{13}	95

[a] Extrapolated to a sputter beam energy of 0 eV.
[b] Ignores tailing towards the surface at a concentration about two orders of magnitude below the peak concentration.

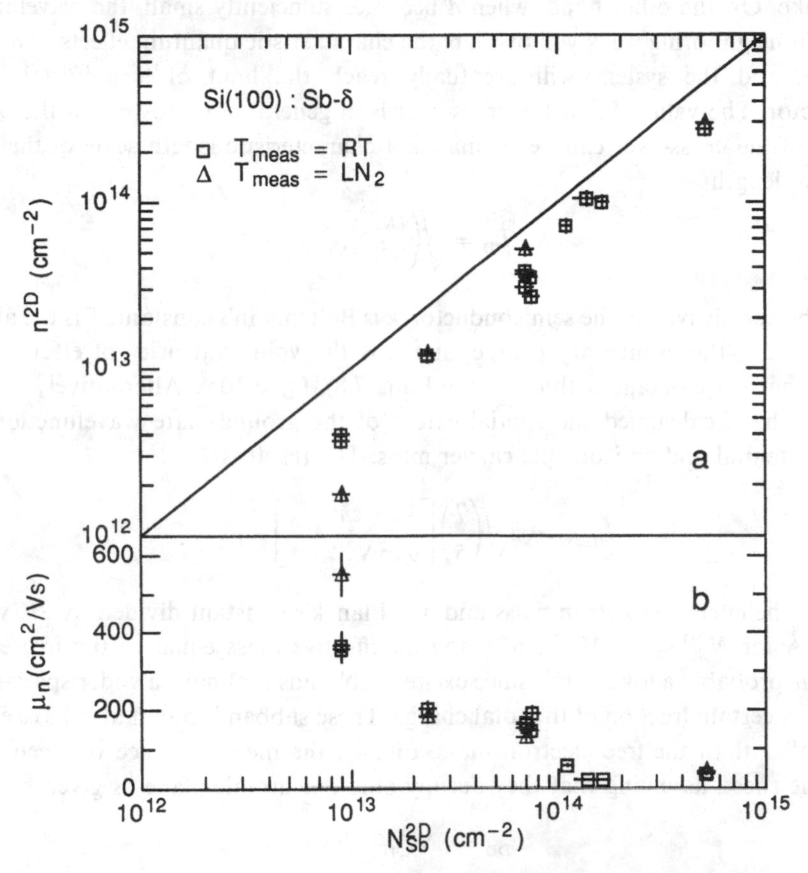

Fig. 7.16. Sheet concentration (a) and mobility (b) of n-type carriers in Sb-δ-doped Si films at room temperature (squares), and liquid nitrogen temperature (triangles) as a function of atomic sheet concentration. The solid line in (a) corresponds to unity activation. A Hall factor of 1.00 is assumed.

consequence of the confinement of carriers in the doping spike (Gossmann and Unterwald 1993). For a δ-doped structure, the wavefunction of a carrier is always wider than the dopant distribution in the direction along the sample normal. Consequently, there is a finite probability that the carriers experience the undoped semiconductor material adjacent to the δ-layer. The carrier mobility in a δ-doped film of sheet concentration $N_{A,D}^{2D}$ should therefore be larger than the mobility in the equivalent, uniformly doped film with a dopant concentration $N_{A,D}$. If we demand that the mean distance between dopant atoms in the equivalent uniformly doped film be the same as in the plane of the dopant sheet in the δ-doped film, then (Schubert *et al.* 1987)

$$N_{A,D} = (N_{A,D}^{2D})^{3/2}. \tag{7.8}$$

Consider, next, a δ-doped superlattice, that is, a train of doping spikes, each spike being a distance d from the next one. As long as d is much larger than the extent of the carrier spilling, the carrier mobility of the doping-superlattice (DSL) will be the same as that of a single spike. On the other hand, when d becomes sufficiently small, the wavefunctions of carriers in neighboring wells will overlap, the characteristic quantum effects of δ-doping will vanish, and the system will eventually reach the limit of a uniformly doped semiconductor. The value of d at the cross-over is in general not known, but the order of magnitude of the cross-over can be estimated: a characteristic length scale of the system is the Debye length

$$L_D = \sqrt{\left(\frac{\varepsilon kT}{e^2 n}\right)}, \tag{7.9}$$

where ε is the permittivity of the semiconductor, k is Boltzmann's constant, T is the absolute temperature, e is the elementary charge, and n is the volume density of electrons. For $N_{Sb}^{2D} = 10^{13}$ cm^{-2} we obtain, with the aid of Eqn. (7.8), $L_D \approx 10$ Å. Alternatively, Schubert *et al.* (1990) have calculated the spatial extent of the ground state wavefunction for a V-shaped potential and an isotropic carrier mass. The result is

$$L_{GSW} = 2\sqrt{\left(\frac{7}{5}\right)}\left[\frac{4}{9}\frac{\varepsilon\hbar^2}{e^2 N_{A,D}^{2D} m^*}\right]^{1/3}, \tag{7.10}$$

where m^* is the effective electron mass and $\hbar$ is Planck's constant divided by 2π, yielding $L_{GSW} \approx 20$ Å for $N_{Sb}^{2D} = 1 \times 10^{13}$ cm^{-2} and an effective mass equal to the free electron mass. This is probably a lower limit since excited subbands that have a wider spatial extent might carry a certain fraction of the total charge. These subbands might also have effective masses smaller than the free electron mass. Finally, the mean distance between dopant atoms in the sheet, assuming that they occupy only one atomic plane, is given by

$$L_{DD} = [N_{A,D}^{2D}]^{-1/2} \tag{7.11}$$

which yields $L_{DD} = \approx 30$ Å for the above example.

We thus expect that for periods of the order of a few nm the δ-DSL has properties that resemble those of a uniformly doped film much more than those of a film with a single

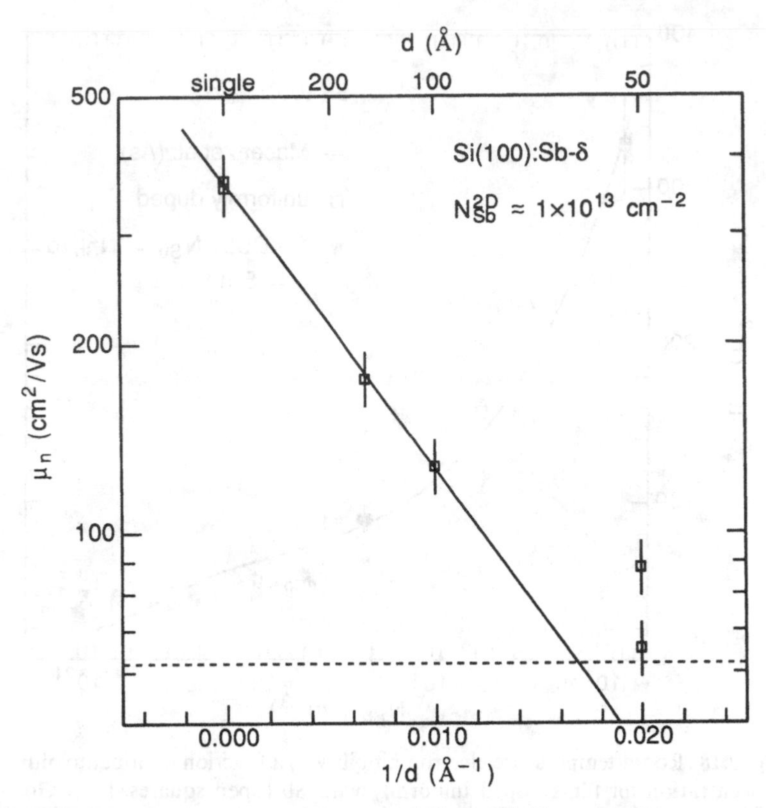

Fig. 7.17. Room-temperature electron mobility of Si films containing one and 5–10 Sb doping spikes (δ-doping superlattice) as a function of the inverse distance between spikes. (A single spike is assigned $1/d = 0$.) Each spike has a sheet concentration of $N_{Sb}^{2D} \approx 1 \times 10^{13}$ cm^{-2}. The dashed line represents the mobility of an equivalent, uniformly doped film (Eqn. (7.8), $N_{Sb} = 3 \times 10^{19}$ cm^{-3}). The solid line is an exponential fit to the data for $1/d > 50$ Å. A Hall factor of 1.00 is assumed.

doping spike. The experimental results confirm this expectation. Figure 7.17 shows the room temperature electron mobility for films containing one and 5–10 doping spikes as a function of the inverse period $1/d$. (A single spike is assigned $1/d = 0$.) Each spike has a sheet concentration of $N_{Sb}^{2D} \approx 1 \times 10^{13}$ cm^{-2}. The dashed line represents the mobility of an equivalent uniformly doped film (Eqn. (7.8), $N_{Sb} = 3 \times 10^{19}$ cm^{-3}, from Fig. 7.10). The mobility in the film with one spike is much larger than in the uniformly doped sample, as expected (Schubert *et al.* 1987). This is still true, although to a lesser extent, for the DSLs with periods of 200 and 100 Å. The mobility appears to decrease exponentially with $1/d$ (solid line) until the superlattice with $d = 50$ Å breaks away from the exponential decay, close to the mobility of the equivalent, uniformly doped film. Apparently for $N_{Sb}^{2D} = 10^{13}$ cm^{-2} the overlap of the electron wavefunctions is sufficient at $d = 50$ Å for the δ-DSL to resemble a uniformly doped film fairly closely, at least as far as the mobility

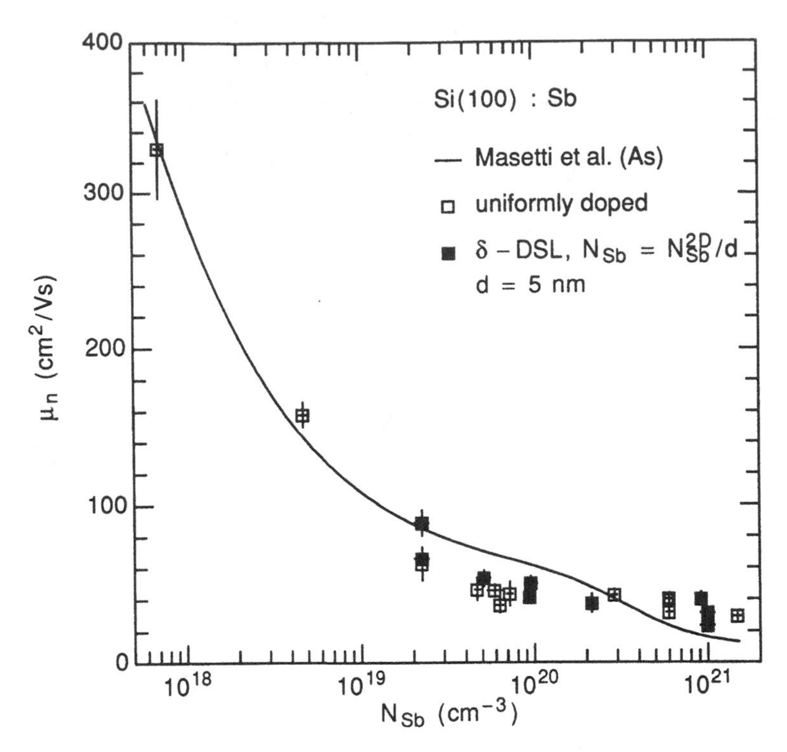

Fig. 7.18. Room-temperature electron mobility as a function of dopant volume concentration for films doped uniformly with Sb (open squares, from Gossmann *et al.*, 1993), and of Sb-δ-doping superlattices with a distance of $d = 50$ Å between spikes. Two-dimensional concentration N_{Sb}^{2D} is converted to volume concentration using Eqn. (7.12). A Hall factor of 1.00 is assumed. The solid line represents the result of Masetti *et al.* (1983).

is concerned. In this limit the equivalent volume concentration of a δ-DSL is given uniquely by

$$N_{Sb} = \frac{N_{Sb}^{2D}}{d}. \tag{7.12}$$

Applying Eqn. (7.12) to a variety of δ-DSLs with $d = 50$ Å results in the filled squares shown in Fig. 7.18. Excellent agreement of the mobilities is reached between δ-doped and uniformly doped samples (open squares) grown under identical circumstances. Note that at the high-concentration end, the mobilities are significantly higher than the values obtained by Masetti *et al.* (1983) on laser-annealed specimens, which confirms earlier results by Jorke and Kibbel (1989) on δ-DSLs with $N_{Sb}^{2D} = 6.78 \times 10^{14}$ cm^{-2}.

The above results on the carrier mobility in δ-DSLs indicate that the material is indistinguishable from uniformly doped samples, for which excellent material quality was established in preceding sections. This implies that the observed incomplete electrical activation in films with single doping spikes is not due to growth defects but must be ascribed to an artefact introduced into the Hall measurement by the confinement of the

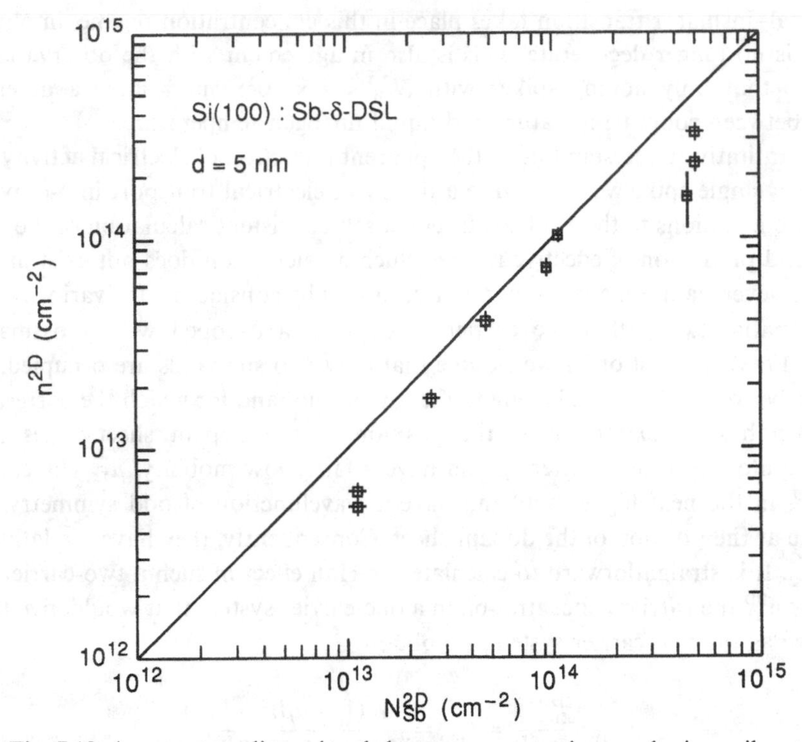

Fig. 7.19. Average two-dimensional electron concentration per doping spike at room temperature as a function of average dopant sheet concentration per spike for Sb-δ-doping superlattices with a distance of $d = 50$ Å between spikes. The solid line represents full electrical activation. A Hall factor of 1.00 is assumed.

carriers. Since Fig. 7.17 shows that these effects have been significantly reduced in δ-DSLs with a period of 50 Å, the electrical activity of the dopants in those δ-DSLs should be much closer to 100%. This is indeed observed, as can be seen from Fig. 7.19, which shows the average two-dimensional electron concentration per doping spike as a function of the average sheet atomic concentration per spike. We note that depletion due to surface pinning of the Fermi level is small, even for the smallest sheet concentrations. Assuming pinning in mid-gap, a single δ-layer at a depth of 750 Å would be depleted by a sheet charge of $\approx 5 \times 10^{11}$ cm^{-2}. The effect is of the same order of magnitude for the doping superlattices. Although the doping spike closest to the surface might be depleted by $\approx 8 \times 10^{12}$ cm^{-2} for a DSL with period 50 Å, there are typically ten such spikes in the DSL. Nevertheless, appropriate corrections have been applied to the data shown in Fig. 7.18; they were always smaller than 10%. The activity decrease visible in Fig. 7.18 at a concentration of $N^{2D}_{Sb} \approx 5 \times 10^{14}$ cm^{-2} in each spike is due to the deterioration of the crystalline quality observed for those concentrations. The maximum fully active sheet concentration thus appears to be of the order of 10^{14} cm^{-2}. The deviation from 100% activity below 10^{13} cm^{-2} may be due to freeze-out of carriers. Eisele (1989) has reported that at a temperature of

4.2 K a metal–insulator transition takes place in this concentration regime, implying that the system is no longer degenerate. This is also in agreement with the observation given in Fig. 7.16 that only doping spikes with $N_{Sb}^{2D} < 1 \times 10^{13}$ cm^{-2} show a difference in activation between room temperature and liquid nitrogen temperature.

A full, quantitative understanding of the apparent reduction of electrical activity in films doped with a single spike would require a theory of electrical transport in δ-doped films, including modifications to the Hall coefficient, a self-consistent calculation of the subband structure, and prediction of effective masses. Such a calculation does not exist at present. One can, however, gain some semi-quantitative insight by considering the various subbands and their spatial extent that the carriers occupy in a δ-doped well (Gossmann and Unterwald 1993). To first order we assume that only two subbands are occupied. Then a certain number of carriers, n_1^{2D}, belong to the lowest subband for which the corresponding wavefunction has an extremum at the position of the dopant sheet. These carriers experience quite significant scattering and have a fairly low mobility, μ_1. The carriers, of density n_2^{2D}, in the next higher subband have a wavefunction of odd symmetry, that is, with a node at the position of the dopant sheet. Consequently, they have a relatively high mobility, μ_2. It is straightforward to calculate the Hall effect in such a two-carrier system. Defining by n_{eff}^{2D} the carrier concentration in a one-carrier system that would give the same Hall voltage as the two-carrier system, we obtain

$$\frac{n_{eff}^{2D}}{n_1^{2D} + n_2^{2D}} = f_{app} = \frac{(1 + \alpha\beta)^2}{(1 + \alpha)(1 + \alpha\beta^2)}. \tag{7.13}$$

Here, $\alpha = n_1^{2D}/n_2^{2D}$ and $\beta = \mu_1/\mu_2$. In the context of Eqn. (7.13), the abscissa in Fig. 7.16 corresponds to n_{eff}^{2D} and $n_1^{2D} + n_2^{2D}$ is the total atomic concentration, N_{Sb}^{2D}. Thus the right-hand side of Eqn. (7.13) is also equal to the apparent electrical activity, f_{app}. Assuming $\alpha = 10$ and $\beta = 0.1$ gives $f_{app} = 0.33$ from Eqn. (7.13), quite close to the experimental value from Fig. 7.16, in particular considering the crudeness of the approximations. The apparent electrical activity approaches 1.0 for $\beta \rightarrow 1$, that is, the Hall measurements would indicate full activiation for films where the enhancement of mobility away from the doping plane is small or absent, for example due to growth defects. Such a situation exists for Sb-δ-doped films grown by SPE and, indeed, the Hall activity of these films is unity (van Gorkum *et al.* 1989).

The electrical characterization of single B-δ-doped layers yields results very similar to Sb-doping (Gossmann and Unterwald 1993). As Fig. 7.20(a) shows, the concentration of holes appears significantly below unity with the exception of the data clustered around $N_B^{2D} = 10^{14}$ cm^{-2}. In analogy with Sb, this is due to the confinement of the carriers in the δ-doped well, although the details must certainly differ from Sb-doping. Figure 7.20(a) indicates that the B-dopant atoms are completely ionized up to $N_B^{2D} \approx 2 \times 10^{14}$ cm^{-2}, about a factor of two more than in the case of Sb. The activation then declines to about 50% at $N_B^{2D} \approx 5 \times 10^{14}$ cm^{-2}. The corresponding hole mobilities are shown in Fig. 7.20(b).

Figure 7.21(a, b) compares the mobilities achievable by δ-doping with those of equivalent uniformly doped films (Gossmann and Unterwald 1993). The mobility as a function of

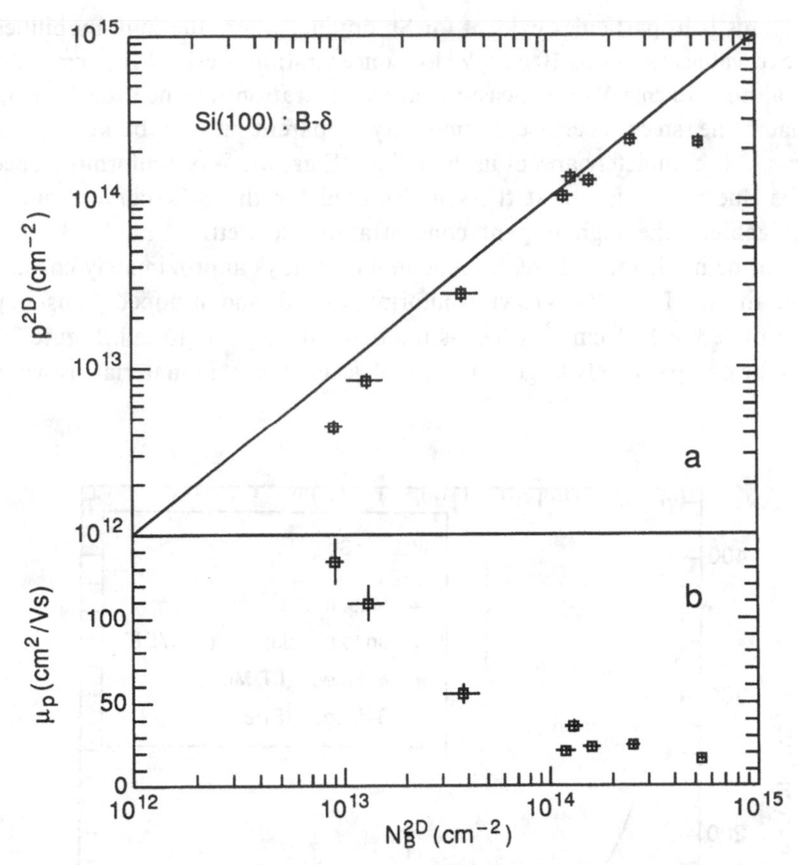

Fig. 7.20. Sheet concentration (a) and mobility (b) of p-type carriers in B-δ-doped films at room temperature, as a function of atomic sheet concentration (squares). The solid line in (a) corresponds to unity activation. A Hall factor of 0.75 is assumed.

volume concentration is shown in each panel, for doping with Sb in (a), with B in (b). Two-dimensional concentrations were converted to volume concentrations using Eqn. (7.8), where applicable. Shown are data from single doping spikes grown by LT-MBE (filled squares) and SPE (filled circles, from van Gorkum *et al.* (1989) for Sb-doping in Fig. 7.21(a), from Headrick *et al.* (1990) for B-doping in Fig. 7.21(b)). Also shown as solid lines are bulk mobility data up to the maximum reported concentrations (Masetti 1983), 5×10^{21} cm^{-3} for n-type-doping (As) and 1.2×10^{21} cm^{-3} for p-type-doping (B), as well as results from uniformly doped LT-MBE grown films (open squares).

Figure 7.21(a) and (b) clearly shows that δ-doping obviously has two advantages (Schubert *et al.* 1987). (1) Mobilities are significantly higher than in equivalent, uniformly doped films. This conclusion is irrespective of whether Eqn. (7.8) or Eqn. (7.12) (with d as the width of the spike from Section 7.3.1) is used to convert sheet to volume concentrations, and agrees with results by Carns *et al.* (1993). (2) Up to one order of magnitude higher volume concentrations are achievable than in uniformly doped films, without deterioration in mobilities. This is in particular evident for Sb-doping, where the bulk mobilities (solid line) decrease to values as low as 10 cm^2/Vs for concentrations beyond 10^{21} cm^{-3}, whereas for δ-doping almost 40 cm^2/Vs are measured at concentrations of one order of magnitude higher. In fact, the steep decrease in mobility apparent in the bulk curve beyond $\approx 3 \times 10^{20}$ cm^{-3} is completely absent in the LT-MBE grown δ- or uniformly doped films. This could be due to the fact that the samples used for the bulk curve required laser annealing to achieve the high dopant concentration (Masetti *et al.* 1983). A similar observation can be made for B: here, too, the mobility stays approximately constant with concentration in the LT-MBE grown, uniformly-doped and δ-doped films beyond a concentration of $\approx 3 \times 10^{20}$ cm^{-3}, whereas the bulk curve starts to fall. Figure 7.21 also demonstrates the comparatively large number of defects present in material grown by SPE:

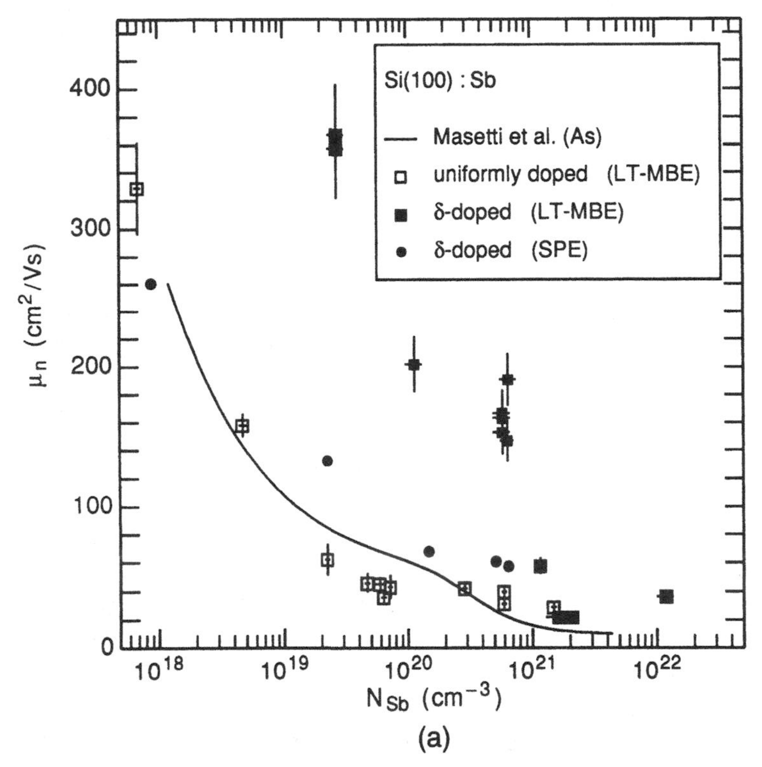

Fig. 7.21. Room temperature mobility of electrons (a) and holes (b) in films δ-doped with a single spike with Sb (a) and **B** (b) (filled squares). The two-dimensional concentration of a spike has been converted to a volume concentration using Eqn. (7.8). For comparison, included are the mobilities of uniformly doped films grown by LT-MBE (open squares), of δ-doped films grown by SPE (filled circles, from van Gorkum *et al.*, 1989 in (a) and from Headrick *et al.*, 1990 in (b)), as well as the results obtained by Masetti *et al.* (1983) on uniformly doped bulk samples (solid lines). A Hall factor of 1.00 (Sb) and 0.75 (B) is assumed.

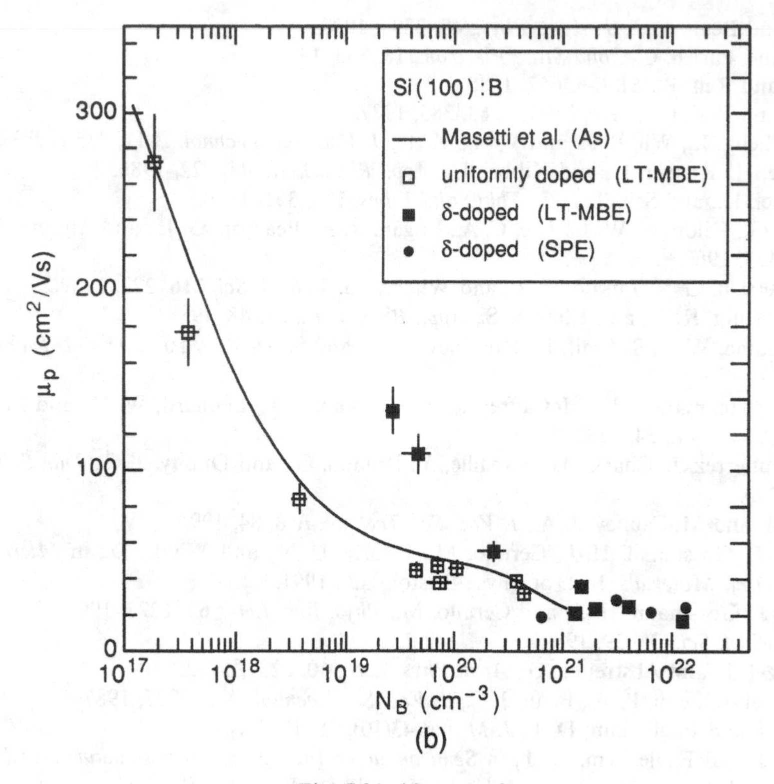

Fig. 7.21. (*Continued*).

the mobilities in Sb-δ-doped films grown by SPE are much smaller than in the films grown by LT-MBE. This effect becomes masked at concentrations beyond $\approx 5 \times 10^{20}$ cm^{-3} (corresponding to $N_{Sb}^{2D} = 5 \times 10^{13}$ cm^{-2}), as the B-data illustrate.

References

Aarts, J. and Larsen, P. K., in *RHEED and Reflection Electron Imaging of Surfaces*, Larsen, P. K. and Dobson, P. J., eds., Plenum, New York, 1988, p. 449.

Abbink, H. C., Broudy, R. M., and McCarthy, G. P., *J. Appl. Phys.*, **39**, 4673, 1968.

Allen, F. G., *Proc. Electrochem. Soc.*, **85–7**, 3, 1985.

Allen, F. and Kasper, E., in *Silicon Molecular Beam Epitaxy*, Vol. I, Kasper, E. and Bean, J. C., eds., CRC, Boca Raton, 1988, p. 65.

Andrieu, S., Chroboczek, J. A., Campidelli, Y., André, E., and d'Avitaya, F. A., *J. Vac. Sci. Technol.* **B 6**, 835, 1988.

Asoka-Kumar, P., Gossmann, T. C., Unterwald, F. C., Feldman, L. C., Leung, T. C., Au, H. L., Talyanski, V., Nielson, B., and Lynn, K. G., *Phys. Rev.* B **48**, 5345, 1993.

Barnett, S. A. and Greene, J. E., *Surf. Sci.*, **151**, 67, 1985.

Barnett, S. A., Winters, H. F., and Greene, J. E., *Surf. Sci.*, **165**, 303, 1986.

Bean, J. C., *Appl. Phys. Lett.*, **33**, 654, 1978.

Bean, J. C., in *Proc. NATO Adv. Study Inst. Multicomponent and Multilayered Thin Films for Advanced Microtechnologies, NATO ASI Series, Series E, No. 234*, Auciello, O. (ed.) (Kluwer 1993).

Becker, G. E. and Bean, J. C., *J. Appl. Phys.*, **48**, 3395, 1977.
Bennett, R. J. and Parish, C., *Solid State Electron.*, **18**, 833, 1975.
Burger, W. R. and Reif, R., *EDL*-**6**, 652, 1985.
Burger, W. R. and Reif, R., *J. Appl. Phys.*, **63**, 383, 1987.
Carns, T. K., Zheng, X., Wu, B. J., and Wang, K. L., *J. Vac. Sci. Technol.* B **11**, 885 (1993).
Casel, A., Jorke, H., Kasper, E., and Kibbel, H., *Appl. Phys. Lett.*, **48**, 922, 1986.
Casel, A., Kibbel, H., and Schäffler, F., *Thin Solid Films*, **183**, 351, 1990.
Chynoweth, A. G., Feldman, W. L., Lee, C. A., Logan, R. A., Pearson, G. L., and Aigrain, P., *Phys. Rev.* **118**, 425, 1960.
Cohen, P. I., Petrich, G. S., Pukite, P. R., and Whaley, G. J., *Surf. Sci.* **216**, 222, 1989.
de Frésart, E., Wang, K. L., and Rhee, S. S., *Appl. Phys. Lett.*, **53**, 48, 1988.
de Jong, T., Douma, W. A. S., Smit, L., Korablev, V. V., and Saris, F. W., *J. Vac. Sci. Technol.* B, **1**, 888, 1983.
Denhoff, M. W., Jackman, T. E., McCaffrey, J. P., Jackman, J. A., Lennard, W. N., and Massoumi, G., *Appl. Phys. Lett.*, **54**, 1332, 1989.
Djebbar, N., Gutierrez, J., Charki, H., Vapaille, A., Prudon, G., and Dupuy, J. C., *Thin Solid Films*, **184**, 37, 1990.
Donnelly, V. M. and McCaulley, J. A., *J. Vac. Sci. Technol.* A **8**, 84, 1990.
Eaglesham, D. J., Gossmann, H.-J., Cerullo, M., Pfeiffer, L. N., and Windt, D., in *Microscopy of Semiconducting Materials*, Inst. of Phys., Bristol, UK, 1991.
Eaglesham, D. J., Gossmann, H.-J., and Cerullo, M., *Phys. Rev. Lett.*, **65**, 1227, 1990.
Eisele, I., *Appl. Surf. Sci.*, **36**, 39, 1989.
Friess, E., Nützel, J., and Abstreiter, G., *Appl. Phys. Lett.*, **60**, 2237, 1992.
Gogol, C. A., Deutschman, R. A., Bean, J. C., *J. Vac. Sci. Technol.* A **5**, 2077, 1987.
Gossmann, H.-J. and Eaglesham, D. J., *J. Mater.* **43**(10), 28, 1991a.
Gossmann, H.-J. and Eaglesham, D. J., in *Semiconductor Interfaces, Microstructures, and Devices: Properties and Applications*, Feng, Z. C., ed., Inst. Phys., Bristol, UK, 1993.
Gossmann, H.-J., unpublished 1991.
Gossmann, H.-J., Asoka-Kumar, P., Leung, T. C., Nielsen, B., Lynn, K. G., Unterwald, F. C., and Feldman, L. C., *Appl. Phys. Lett.*, **61**, 540, 1992.
Gossmann, H.-J., Schubert, E. F., Eaglesham, D. J., and Cerullo, M., *Appl. Phys. Lett.*, **57**, 2440, 1990.
Gossmann, H.-J., Schubert, E. F., Eaglesham, D. J., and Cerullo, M., in *Proceedings of the 2nd International Conference on Electronic Materials*, Materials Research Society, Pittsburg, 1991, p. 451.
Gossmann, H.-J., Unterwald, F. C., and Luftman, H. S., *J. Appl. Phys.* **73**, 8237 (1993).
Gossmann, H.-J., and Eaglesham, D. J., unpublished, 1991b.
Gossmann, H.-J. and Feldman, L. C., *Phys. Rev.* B, **32**, 6, 1985.
Gossmann, H.-J. and Unterwald, F. C., *Phys. Rev.* B **47**, 12618 (1993).
Green, M. L., Brasen, D., Geva, M., Reents, W., Stevie, F., and Temkin, H., *J. Electron. Mat.*, **19**, 1015, 1990.
Green, M. L., Brasen, D., Luftman, H., and Kannan, V. C., *J. Appl. Phys.*, **65**, 2588, 1989.
Harris, J. J., Joyce, B. A., and Dobson, P. J., *Surf. Sci.*, **103**, L90, 1981.
Headrick, R. L., Feldman, L. C., and Robinson, I. K., *Appl. Phys. Lett.*, **55**, 442, 1989.
Headrick, R. L., Weir, B. E., Levi, A. F. J., Eaglesham, D. J., and Feldman, L. C., *Appl. Phys. Lett.*, **57**, 2779, 1990.
Hirayama, H., Hiroi, M., Koyama, K., and Tatsumi, T., *J. Cryst. Growth*, **105**, 46, 1990.
Hirayama, H., Tatsumi, T., and Aizaki, N., *Appl. Phys. Lett.*, **52**, 1335, 1988.
Hirayama, H., Tatsumi, T., and Aizaki, N., *Appl. Phys., Lett.*, **52**, 1484, 1988.
Ishizaka, B. and Shiraki, Y., *J. Electrochem. Soc.*, **133**, 666, 1986.
Jackman, T. E., Houghton, D. C., Jackman, J. A., and Rockett, A., *J. Appl. Phys.*, **66**, 1984, 1989.

Jackman, T. E., Houghton, D. C., Denhoff, M. W., Kechang, S., McCaffrey, J., Jackman, J. A., and Tuppen, C. G., *Appl. Phys. Lett.*, **53**, 877, 1988.
Jona, F., in *Proc. of the 13th Sagamore Army Materials Research Conf.*, Burke, J. J., Reed, N. L., and Weiss, V., eds., Syracuse Univ., Syracuse, 1967, p. 399.
Jorke, H., *Surf. Sci.*, **193**, 569, 1988.
Jorke, H., Kibbel, H., Schäffler, F., Casel, A., Herzog, H.-J., and Kasper, E., *Appl. Phys. Lett.*, **54**, 81, 1989.
Jorke, H. and Kibbel, H., *Thin Solid Films*, **183**, 323, 1989.
Jorke, H. and Kibbel, H., *Appl. Phys. Lett.*, **57**, 1763, 1990.
Karunasiri, R. P. U., Gilmer, G. H., and Gossmann H.-J., *Surf. Sci.*, **317**, 361, 1994.
Kasper, E. and Bean, J. C., *Silicon Molecular Beam Epitaxy*, Vols. I and II, CRC, Boca Raton, 1988.
Kimerling, L. C., *IEEE Trans. Nucl. Sci.* NS-**23**, 1497, 1976.
Knall, J., Barnett, S. A., Sundgren, J.-E., and Greene, J. E., *Surf. Sci.*, **209**, 314, 1989.
Knall, J., Sundgren, J.-E., Greene, J. E., Rockett, A., and Barnett, S. A., *Appl. Phys. Lett.*, **45**, 689, 1984.
König, U., Kasper, E., and Herzog, H.-J., *J. Cryst. Growth*, **52**, 151, 1981.
König, U., Kibbel, H., and Kasper, E., *J. Vac. Sci. Technol.*, **16**, 985, 1979.
Kubiak, R. A. A., Leong, W. Y., and Parker, E. H. C., *Appl. Phys. Lett.*, **44**, 878, 1984.
Kubiak, R. A. A., Leong, W. Y., and Parker, E. H. C., *Proc. Electrochem. Soc.*, **85–7**, 169, 1985.
Kuznetsov, V. P. and Postnikov, V. V., *Sov. Phys. Crystallogr.*, **14**, 441, 1969.
Kuznetsov, V. P. and Postnikov, V. V., *Sov. Phys. Crystallogr.*, **19**, 211, 1974.
Kuznetsov, V. P., Tolomasov, V. A., and Tumanova, A. N., *Sov. Phys. Crystallogr.*, **24**, 588, 1979.
Lin, J. F., Li, S. S., Linares, L. C., and Teng, K. W., *Solid St. Electron.*, **24**, 827, 1981.
Lin, T. L., Fathauer, R. W., and Grunthaner, P. J., *Appl. Phys. Lett.*, **55**, 795, 1989.
Lin, T. L., Fathauer, R. W., and Grunthaner, P. J., *Thin Solid Films*, **184**, 31, 1990.
Luftman, H. S., Schubert, E. F., and Kopf, R. F., unpublished 1991.
Mac Rae, A. U., in *Ion Implantation*, Eisen, F. H., and Chadderton, L. T., eds., Gordon and Breach, New York, 1971, p. 363.
Masetti, G., Severi, M., and Solmi, S., *IEEE Trans. Electron Dev.*, ED-**30**, 764, 1983.
Mattey, N. L., Dowsett, M. G., Parker, E. H. C., Whall, T. E., Taylor, S., and Zhang, J. F., *Appl. Phys. Lett.*, **57**, 1648, 1990.
Metzger, R. A. and Allen, F. G., *J. Appl. Phys.*, **55**, 931, 1984a.
Metzger, R. A. and Allen, F. G., *Surf. Sci.*, **137**, 397, 1984b.
Meyerson, B. S., LeGoues, F. K., Nguyen, T. N., and Harame, D. L., *Appl. Phys. Lett.*, **50**, 113, 1987.
Mukashev, B. N., Kolodin, L. G., Nussupov, K. H., Spitsyn, A. V., and Vavilov, V. S., *Rad. Effects* **46**, 79, 1980.
Nakagawa, K., Miyao, M., and Shiraki, Y., *Jpn J. Appl. Phys.*, **27**, L2013, 1988.
Neave, J. H., Joyce, B. A., Dobson, P. J., and Norton, N., *Appl. Phys.* A **31**, 1, 1983.
Ni, W.-X., Knall, J., Hasan, M. A., Hansson, G. V., Sundgren, J.-E., Barnett, S. A., Markert, L. C., and Greene, J. E., *Phys. Rev.* B, **40**, 10449, 1989.
Nylandsted Larsen, A., Pedersen, F. T., and Weyer, G., *J. Appl. Phys.*, **59**, 1908, 1986.
Ostrom, R. M. and Allen, F. G., *Appl. Phys. Lett.*, **48**, 221, 1986.
Ota, Y., *J. Electrochem. Soc.*, **124**, 1795, 1977.
Ota, Y., *Thin Solid Films*, **106**, 3, 1983.
Parry, C. P., Newstead, S. M., Barlow, R. D., Augustus, P., Kubiak, R. A. A., Dowsett, M. G., Whall, T. E., and Parker, E. H. C., *Appl. Phys. Lett.*, **58**, 481, 1991.
Perov, A. S. and Postnikov, V. V., *Sov. Phys. Crystallogr.*, **22**, 475, 1977.
Pickar, K. A., *Appl. Solid State Sci.*, **5**, 151, 1975.
Postnikov, V. V. and Kuznetsov, V. P., *Sov. Phys. Crystallogr.*, **20**, 71, 1975.
Postnikov, V. V. and Perov, A. S., *Sov. Phys. Crystallogr.*, **22**, 350, 1977.
Schubert, E. F., Cunningham, J. E., and Tsang, W. T., *Solid State Communications* **63**, 591, 1987.

Schubert, E. F., Kopf, R. F., Kuo, J. M., Luftman, H. S., and Garbinski, P. A., *Appl. Phys. Lett.*, **57**, 497, 1990.
Schultz, P. J. and Lynn, K. G., *Rev. Mod. Phys.*, **60**, 701, 1988.
Schut, H., van Veen, A., van de Walle, G. F. A., and van Gorkum, A. A., *J. Appl. Phys.*, **70**, 3003, 1991.
Shiraki, Y., Katayama, Y., Kobayashi, K. L. I., and Komatsubara, K. F., *J. Cryst. Growth*, **45**, 287, 1978.
Shiraki, Y., *Prog. Crystal Growth Charact.*, **12**, 45, 1986.
Sturm, J. C., Gronet, C. M., and Gibbons, J. F., *J. Appl. Phys.*, **59**, 4180, 1986.
Sze, S. M., *Physics of Semiconductor Devices*, Wiley, New York, 1981.
Tabe, M. and Kajiyama, K., *Jpn J. Appl. Phys.*, **22**, 423, 1983.
Tannenbaum Handelman, E. and Povilonis, E. I., *J. Electrochem. Soc.*, **111**, 201, 1964.
Tatsumi, T., Hirayama, H., and Aizaki, N., *Appl. Phys. Lett.*, **50**, 1234, 1987.
Tatsumi, T., Hirayama, H., and Aizaki, N., *Jpn J. Appl. Phys.*, **27**, L954, 1988.
Thomas, R. N. and Francombe, M. H., *Appl. Phys. Lett.*, **13**, 270, 1968.
Thomas, R. N. and Francombe, M. H., *Solid State Electron.*, **12**, 799, 1969.
Thomas, R. N. and Francombe, M. H., *Surf. Sci.*, **25**, 357, 1971.
Tolomasov, V. A., Abrasimova, L. N., and Gorshenin, G. N., *Sov. Phys. Crystallogr.*, **15**, 1076, 1971.
Trumbore, F. A., *Bell System Tech. J.* **39**, 210, 1960.
Tuppen, C. G., Prior, K. A., Gibbings, C. J., Houghton, D. C., and Jackman, T. E., *J. Appl. Phys.*, **64**, 2751, 1988.
van der Pauw, L. J., *Philips Res. Rep.* **13**, 1, 1958.
van Gorkum, A. A., Nakagawa, K., and Shiraki, Y., *Jpn J. Appl. Phys.*, **26**, L1933, 1987.
van Gorkum, A. A., Nakagawa, K., and Shiraki, Y., *J. Appl. Phys.*, **65**, 2485, 1989.
Venables, J. A. and Price, G. L., in *Epitaxial Growth*, Matthews, J. W., ed., Academic, New York, 1975, Part B, p. 381.
Vescan, L., Kasper, E., Meyer, O., and Maier, M., *J. Cryst. Growth*, **73**, 482, 1985.
Vink, A. T., Roksnoer, P. J., Maes, J. W. F. M., Vriezema, C. J., van Ijzendoorn, L. J., and Zalm, P. C., *Jpn J. Appl. Phys.*, **29**, L2307, 1990.
Watkins, G. D. and Troxell, J. R., *Phys. Rev. Lett.* **44**, 593, 1980.
Weir, B. E., Freer, B. S., Headrick, R. L., Eaglesham, D. J., Gilmer, G. H., Bevk, J., and Feldman, L. C., *Appl. Phys. Lett.*, **59**, 204, 1991.
Weisberg, L. R., *J. Appl. Phys.*, **38**, 4535, 1968.
Widmer, H., *Appl. Phys. Lett.*, **5**, 108, 1964.
Williams, J. S. and Short, K. T., *J. Appl. Phys.*, **53**, 8663, 1982.
Yamazaki, T., Watanabe, S., and Ito, T., *J. Electrochem. Soc.*, **137**, 313, 1990.
Zagwijn, P. M., Erokhin, Y. N., Slikjerman, W. F. J., van der Veen, J. F., van de Walle, G. F. A., Gravesteijn, D. J., and van Gorkum, A. A., *Appl. Phys. Lett.*, **59**, 1461, 1991.
Zeindl, H. P., Wegehaupt, T., Eisele, I., Oppolzer, H., Reisinger, H., Tempel, G., and Koch, F., *Appl. Phys. Lett.*, **50**, 1164, 1987.

PART FOUR

Characterization

8

Secondary ion mass spectrometry of delta-doped semiconductors

HENRY S. LUFTMAN

Experimental studies of delta-doped structures rely at some point on verifying that the intention of having a quantity of dopant atoms confined to a very narrow region within a matrix of semiconductor atoms has in fact been achieved. Two challenges are quickly apparent – the need to identify impurity atoms at concentrations of the order of parts per million, and to do so in a manner that also identifies the location of these atoms within the material to a high accuracy. The 'chemical' knowledge of atom detection complements the 'electrical' knowledge from charge carrier detection discussed in other chapters. A less than anticipated value of charge carrier density may reflect insufficient dopant atom inclusion, incomplete activation of the dopant atoms, or the presence of unintended impurities that act as either traps or counter-dopants. Electric characteristics that depend upon the confinement of the dopants will be affected by broadening the intended dopant profile by either the method of deposition or subsequent diffusion during later processing steps. Secondary ion mass spectrometry, SIMS, is a 'chemical' analytical technique that is particularly well suited to obtaining this information. It is capable of sub-ppm detection and identification of most elements and, in some applications, gives depth information accurate to less than 3 nm. In SIMS, ions with a kinetic energy that is generally between 1 and 15 keV bombard the surface of the material to be analyzed. Some of this energy results in the ejection of atoms and atom clusters from the surface of the specimen (Fig. 8.1). A small fraction of these sputtered-off particles leave this area ionized, either positively or negatively. These 'secondary ions' are then identified using a mass spectrometer and counted as a function of their mass. Over time, the specimen is gradually eroded. Data is collected as secondary ion counts of a given mass versus time which can in principle be converted to concentration versus depth by the use of standards and calibration of the erosion rate. (General sources on SIMS include Benninghoven *et al.* (1987); Wilson *et al.* (1989); and Wittmaack (1992).)

The purpose of this chapter is primarily to acquaint those interested in delta-doping with the potential contributions that SIMS analysis can make, and its limitations. An overview of factors leading to the sensitivity of SIMS is given in the following section,

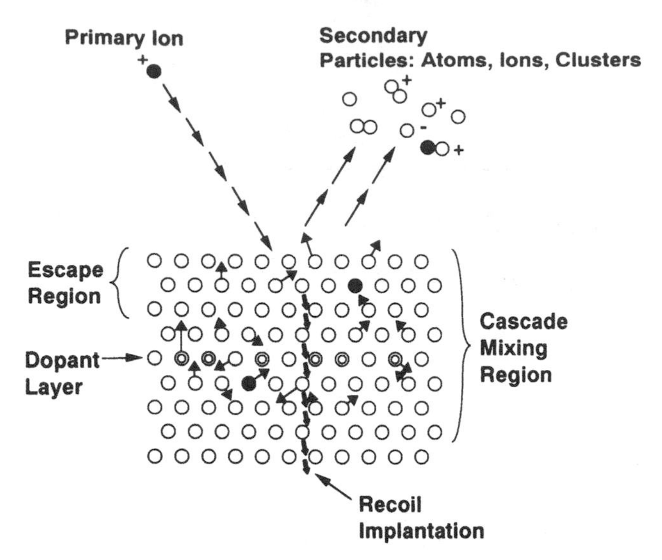

Fig. 8.1. The SIMS process. Incoming primary ion generates a variety of secondary particles from a narrow escape region, while causing some redistribution of sample atoms through a cascade mixing region.

followed by a brief discussion of instrumentation. The sources of depth resolution limitations will then be discussed in conjunction with those operating conditions that are most advantageous. Finally, several SIMS analyses of delta-doped systems will be shown.

8.1 The process

Much of the physics of SIMS can be understood by its 'basic equation'. For a given analysis a specimen consists of some matrix material **M** (e.g. GaAs) containing an impurity of interest **A** (e.g. Be), whose fractional concentration $c(\mathbf{A}, z)$ varies with depth from the surface plane. When bombarded by a primary ion beam of current I_p, the signal intensity generated from a particular charge state **q** of **A** is given by

$$I^{\mathbf{q}}(\mathbf{A}) = I_p Y c(\mathbf{A}, z)\alpha^{\mathbf{q}}(\mathbf{A}) f^{\mathbf{q}}(\mathbf{A}), \tag{8.1}$$

where Y is the specimen sputter yield, $\alpha^{\mathbf{q}}(\mathbf{A})$ is the probability of sputtered **A** appearing in charge state **q**, and $f^{\mathbf{q}}(\mathbf{A})$ is the instrumental transmission and detection efficiency.

The sputter yield, Y, is dependent upon the mass and kinetic energy of the bombarding species, the angle of impact of the primary ion beam to the normal of the specimen surface, and the nature of the matrix species **M**, including the masses and bond energies of its constituent atoms. Yields are typically in the range 0.1–10 atoms per primary ion. Primary ion currents typically used in SIMS range from 0.1 to 100 A/m^2 (e.g. a 100 nA ion beam rastered over a 500 × 500 μm area corresponds to 0.4 A/m^2 or 2.5 × 10^{14} ions/cm^2/s). Normal erosion rates in SIMS vary from 0.1 to 10 nm/s. Other considerations concerning the sputtering process will be discussed later in the context of depth resolution. The instrumental detection efficiency depends in part upon the identity of the secondary ion species – its mass and energy distribution. The collection efficiency of the equipment's extraction fields, the band width of its energy filter, and the particle detector quality are under some control by the experimenter. Efficiency is often traded off to improve mass, or lateral or depth resolution. Values from 0.01 to 10% are common.

The factor $\alpha^q(\mathbf{A})$ is perhaps the most subtle in Eqn. (8.1). When species **A** is sputtered from the specimen it will leave in a variety of forms. Most of it, usually at least 99%, will leave as neutral atoms, $\mathbf{A}^0$. Some will be sputtered as neutral atomic clusters such as $\mathbf{A}_n^0$, $\mathbf{A}_n\mathbf{M}_m^0$ (where **M** are atoms from the matrix material) or $\mathbf{A}_n\mathbf{X}_m^0$ (where **X** are atoms from the primary ion beam that implanted into the matrix or that recombined with **A** after it left the specimen surface). A small fraction of **A** will sputter as either monatomic singly charged ions ($\mathbf{A}^+$ or $\mathbf{A}^-$), as multiply charged ions, or as parts of cluster ions (e.g. $\mathbf{A}_n^{\pm 2}$, $\mathbf{MA}^\pm$, $\mathbf{XA}^\pm$). For example, one can detect Si in GaAs using a Cs^+ primary ion beam by monitoring species such as Si^-, Si^+, $SiAs^-$, $SiGa^+$, $SiCs^+$, and Si_2^-.

The factor $\alpha^q(\mathbf{A})$ is sensitive to many conditions. For the simplest case of a monatomic ion $\mathbf{A}^+(\mathbf{A}^-)$, its probability of formation will depend upon the ionization potential (electron affinity) of **A** and the propensity of the matrix **M** to keep (release) electrons, related to its work function. To enhance the creation of positive secondary ions the use of an O_2^+ primary ion source is common, as the implanted electronegative **O** increases the matrix work function, thus increasing the probability of stripping electrons from sputtered atoms. Similarly, the use of a Cs^+ beam enhances the yield of negative secondary ions as electropositive Cs decreases the work function of the matrix. The yield of monatomic ions is also dependent upon the efficiency of the competing processes that generate polyatomic ions, which occasionally have greater yields than for the monatomic species. For example, one often obtains a greater yield and improved SIMS sensitivity if one monitors Si in GaAs as $SiAs^-$ as opposed to Si^-. Boron in Si can be followed by monitoring the secondary ion B^+ when using an O_2^+ primary beam, but good sensitivity is also achievable following BSi^- with a Cs^+ primary beam. The best sensitivity for Zn in InP is attained following the secondary ion as $ZnCs^+$, using a Cs^+ primary beam. An additional complication is that the technique of SIMS suffers from a 'matrix effect' in which the probability of ion formation can change, in some cases by more than an order of magnitude, as one depth profiles from one material into another, as in a heterostructure.

In SIMS depth profiling, the data received is the signal intensity $I^q(\mathbf{A})$ as a function of the time required to expose the layer of specimen generating that signal. What is desired is the fractional concentration of **A**, $c(\mathbf{A}, z)$, as a function of depth. (Actually what is more desirable is the initial distribution of **A** as a function of depth in a specimen before sputtering artefacts have been created. This will be addressed in Section 8.4.) Depth scales are generally established by knowing the erosion rate of the matrix, and assuming a linear conversion of sputter time to depth. This can be determined routinely by measuring the depth of the crater left by the SIMS analysis with a profilometer and dividing the value obtained by the total time of primary ion beam bombardment.

While the preceding paragraphs have discussed factors involving the sensitivity of dynamic SIMS, without the potential of high depth resolution it would be of little assistance in the study of delta-doped structures. Returning to Eqn. (8.1), the factor '$c(\mathbf{A}, z)$' denotes the fractional concentration of impurity **A** at the surface of a specimen after a thickness z of material has been eroded away by the primary ion beam. More precisely, it corresponds to a larger volume of the order of three atomic layers thickness with an area matching that of the primary beam's projection on the specimen surface or of the greater area encompassed by a primary beam that is rastered over the surface, as will be discussed later. The thickness of this volume corresponds to the depth from which sputtered atoms and ions are capable of escaping. The difficulty arises because $c(\mathbf{A}, z)$ is not identical to $c_0(\mathbf{A}, z)$, the original distribution of **A** as a function of depth, and it is the difference between these functions that is responsible for depth resolution being generally greater than the escape length of sputtered particles. Only a small fraction of the energy imparted by a bombarding primary ion is used in the sputter removal of particles. The remainer is taken up in the redistribution of atoms within the target (Fig. 8.1). *Cascade mixing* refers to the general stirring up of the target atoms by the stochastic distribution of energy released by the primary projectile and subsequent (secondary, tertiary, etc.) collisions. Its effect is symmetric in that target atoms may be moved either closer to or further from the bombarded surface. The thickness of the damaged zone of the target will be sensitive to the target's density and the mass, kinetic energy and angle of incidence of the primary ion. *Recoil implantation* (or knock-on mixing) occurs when a primary ion has a nearly head-on collision with a target atom, which will it itself then become implanted deeper into the target. This is an asymmetric mixing process, the range of which will be governed by those conditions affecting cascade mixing as well as the mass of the 'knock-on' atom. Another source of degradation of depth resolution arises because the surface of the analysis region, after a period of sputter etching, may not correspond to a unique depth. Non-uniform sputtering may arise from non-uniformity of the primary ion beam, initial surface topography, or contaminants. The ion beam itself is unlikely to have a constant ion flux throughout its cross-section, resulting in sloped edges surrounding the SIMS crater, and thus a range of depth values instantaneously exposed to subsequent sputtering. Furthermore, for certain sputter conditions and target materials, topographic features such as ripples or cones may be induced onto the sputtered surface (Stevie *et al.*, 1988). We will return to a more detailed discussion of effects leading to loss of depth resolution and partial remedies following a discussion of SIMS instrumentation.

8.2 Equipment

A schematic of a generic SIMS spectrometer is shown in Fig. 8.2. The entire system is under ultra-high vacuum conditions, generally of the order of 10^{-7}–10^{-10} torr. This is necessary in order to permit good collimation of the primary ion beam, reduced scattering of secondary ions, reduced contamination of the specimen during sputtering from the ambient, and reduced background signal generated at the ion detector. Samples are generally introduced into the vacuum through a load-lock mechanism in commercially

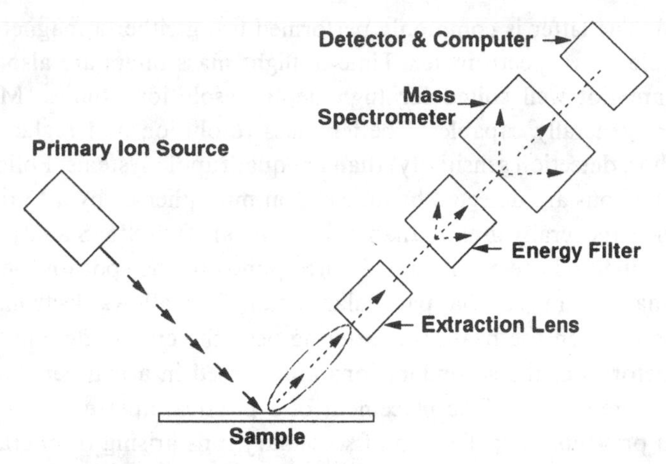

Fig. 8.2. Schematic of generic secondary ion mass spectrometer.

available instruments in order to allow reasonable throughput. As alluded to previously, while SIMS originally used noble gas ions for the primary beam in order to reduce chemical interactions while sputtering, in depth profiling typically O_2^+ or Cs^+ beams are used. The former is generated in a plasma while the latter is generally created by surface ionization of either evaporated Cs or a cesium-containing salt. High lateral resolution ion beams can be obtained using a liquid gallium source, but this is of limited use in delta-doping studies because of poor secondary ion yields and low primary ion flux. Once created, the primary ions are accelerated toward the target sample. The beam is focused along this path, attenuated by apertures or fields, and, in some instruments, is mass-filtered to remove unwanted contaminants. All commercial ion guns allow for this beam to be rastered along the X and Y directions over the sample surface. Proper beam focusing and rastering are necessary in order to create a crater on the specimen with a bottom that is flat enough to allow reasonable depth resolution.

Sample mounting varies among commercial instruments. In magnetic sector instruments such as Cameca IMS spectrometers, the mounted specimen can be moved only in the X and Y directions, with no capacity for tilting the sample relative to the beam. It is generally biased at ± 4.5 kV relative to ground, with polarity determined by the polarity of the secondary ions to be analyzed. The angle of incidence of the primary ion beam to the specimen is determined by the primary beam energy and the specimen bias. In quadrupole instruments, such as the Perkin Elmer 6000 series spectrometers, the specimen is unbiased and mounted on a sample positioner with three-axis motion and the ability to tilt the specimen relative to the primary ion beam, thereby allowing independent control of the primary beam energy and angle of incidence. Similar flexibility is available in other quadrupole instruments such as those from Atomika and EVA. Finally, modifications are becoming available for many instruments in order to allow sample rotation about the sample normal during analysis, a feature to be discussed further below.

Secondary ions are extracted from the sputtered region and are subsequently energy and mass filtered. The latter is commonly performed using either a magnetic sector or an electric quadrupole mass spectrometer. Time-of-flight mass filters are also used in SIMS equipment but are not well suited for high depth resolution studies. Magnetic sector spectrometers are generally capable of better mass resolution and higher ion collection efficiency (and, thus, detection sensitivity) than are quadrupole systems. Following the mass filter the secondary ions are detected by an electron multiplier or by a Faraday cup if the signal is too intense (generally greater than 10^6 counts/s). Two SIMS analysis modes exist. In the *microprobe* method the SIMS signal corresponds to the spot on the sample where the rastering primary beam is at that particular instant. This allows electronic signal gating to prevent counting when the beam is sputtering near the crater edge. In the *microscope* method the trajectories of the secondary ions are focused in a manner that preserves an image of the sputtered region. The placement of a physical aperture at a focal plane of these trajectories prevents the collection of secondary ions arising from crater edges.

As mentioned in the preceding section, generally less than 1% of the material sputtered from a specimen is ionic. There is a class of techniques related to SIMS, collectively referred to as sputtered neutral mass spectrometry (SNMS), in which the secondary ion signal is suppressed while an effort is made to ionize the neutral sputtered atoms. This is done with either an electron beam or laser excitation, resonant or non-resonant. While SNMS has the potential for greater sensitivity than SIMS, this has not yet been realized in routine depth profiling. Nevertheless, SNMS suffers significantly less from the matrix effects that influence SIMS quantization. Sputtering is performed with a primary ion beam, as in SIMS, and generally uses noble gases. Depth resolution constraints are the same as in SIMS. The use of SNMS is likely to increase within the near future (Downey *et al.*, 1990; Arlinghaus *et al.*, 1993).

8.3 Depth profiling

A variety of SIMS options exist such as obtaining lateral ion imaging or mass spectral analysis at designated depths. The depth profiling mode is of use in delta-doping studies. As sputtering ensues, the mass filter is multiplexed, allowing sequential monitoring of one or more preselected secondary ion masses. This results in a spectrum of signal versus time for each selected mass-to-charge ratio.

To change the time axis to depth requires a knowledge of the sputtering rate of the matrix material. When only a single matrix material is involved this is most easily obtained by measuring the depth of the crater created during the SIMS profiling with a profilometer and dividing the result by the total duration of sputtering time. For depths greater than 0.5 μm one can generally take such measurements with better than 10% accuracy as long as the unsputtered regions of the specimen surrounding the edge of the crater are reasonably flat. For shallower craters the effects of specimen curvature and the noise limitations of the profilometer become increasingly significant with a limiting accuracy of $\simeq 10$ nm. When studying very shallow features it is useful to make at least one 'deep' profiling measurement under identical primary beam conditions in order to establish a sputtering rate that can

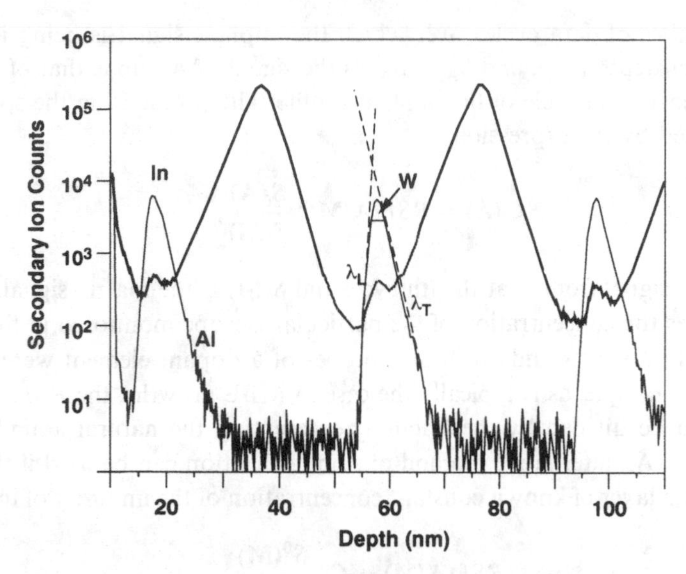

Fig. 8.3. SIMS profile of MBE-grown GaAs with AlAs monolayers at 200, 600, and 1000 Å, and InAs monolayers placed at 400, 800, and 1200 Å. Profile taken with 1 keV O_2^+ at 60° to the surface. Full width at half maximum (W) and leading and trailing exponential slopes (λ_L and λ_T).

be used for the shallow experiments. Commercial SIMS equipment can produce ion beams with current stability of better than 5% over several hours with sufficient warm-up time. In certain specimen structures, such as that represented in Fig. 8.3, separations between marker layers are already known by, for example, RHEED measurements during MBE growth or subsequent microscopic analysis of a cross-section of the specimen. This information can be used in place of profilometer measurements to determine sputtering rates. For those structures involving more than one matrix material, as in heterojunction structures, accurate depth scales require knowledge of the sputtering rates or relative sputtering rates of all the matrices involved. Relative sputtering rates are dependent upon primary beam identity, energy, and angle of incidence, so preexisting tables are only useful when these conditions are matched.

Conversion of signal intensity to local concentration requires the use of standards. An appropriate standard should consist of the same matrix material as the specimen, with a known amount of the same dopant in the matrix. Ion implants are often used for calibration. For example, to quantify a SIMS profile of delta-doped Si in GaAs an appropriate standard might be 50 keV ^{28}Si in GaAs at a dose measured by the implanter of 1×10^{15} atoms/cm^2. From this profile one calculates the relative sensitivity factor (RSF) of the implant **A** signal in relation to a signal from a matrix species (**M**):

$$RSF(\mathbf{A/M}) = \frac{N\phi S_{\mathrm{m}}^{0}}{d\sum_{\mathrm{i}}^{\mathrm{N}}(S_{\mathrm{A}_i}^{0} - S_{\mathrm{A}_B}^{0})}, \tag{8.2}$$

where ϕ is the implant dose (atoms/cm^2), S_{m}^{0} is the signal count rate of the matrix species

M, N is the number of data cycles over which the implant signal is being integrated, d is the SIMS crater depth (cm), and $S^0_{\mathbf{A}_i} - S^0_{\mathbf{A}_B}$ is the signal of **A** minus that of any existent background for each data cycle of the implant profile. The profile from the specimen then becomes quantified by the expression

$$C_i(\mathbf{A}) = RSF(\mathbf{A}/\mathbf{M})^* \frac{S_i(\mathbf{A})}{S(\mathbf{M})}, \tag{8.3}$$

where $S_i(\mathbf{A})$ is the signal from **A** at the ith cycle and $S(\mathbf{M})$ is the matrix signal. As written, this equation gives the concentration of the particular isotope monitored in the profiles of the specimen and implant standard. If all isotopes of a dopant element were used in the preparation of the sample, as is typically the case in MBE growths, the *RSF* calculated in Eqn. (8.2) should be divided by the fraction representing the natural abundance of the isotope monitored. An alternative method of standardization can be used if the specimen or a standard has a layer of known constant concentration of the impurity of interest. Then

$$RSF(\mathbf{A}/\mathbf{M}) = C^0_{\mathbf{A}} \frac{S^0(\mathbf{M})}{S^0(\mathbf{A})} \tag{8.4}$$

The concentration profile, $C^0_{\mathbf{A}}$, in standards such as this, generally includes all isotopes of **A**. This can then be used in Eqn. (8.3) to calculate the unknown concentrations of **A**.

The purpose of using relative sensitivity factors involving the signal of some matrix species is to remove errors in calculation that would arise from incomplete knowledge of primary beam density, detector collection efficiency, or aspects of specimen orientation or surface conditions that may affect sputter yield uniformly. In practice, these are measured fairly frequently, as some variation in impurity detectability does occur. Furthermore, the accuracy of an RSF is no better than the accuracy of the reported ion implant dose or the reported concentration of the impurity within a uniformly doped layer. Assuming that these are known to within 10%, SIMS quantification is generally accurate to within 20%. Lower error brackets are achievable with sufficient care. Relative concentration errors between samples or between regions within a single sample are less.

8.4 Sources and solutions to degraded depth resolution

The problem of SIMS depth resolution does not arise from an extended source of secondary ions from the bombarded sample. It is well established that the secondary ions originate from no deeper than the top three atomic planes, which would give a depth resolution of the order of only 1 nm. The limitation to resolution arises because after having destructively bombarded a specimen with primary ions to a depth z, the composition of that exposed surface is no longer the same as it was before sputtering began. Sources of depth resolution degradation can be classified under four categories: (i) crater curvature within the sampling region; (ii) data sampling rate versus the sputter etch rate; (iii) atomic mixing, including cascade mixing, recoil implantation, primary ion implantation, and diffusion enhancement, all caused by the sputtering process; and (iv) micro-topography, both that initially present and that induced by the primary ion bombardment. Before discussing these in greater detail, there are several observations to be made. Depth resolution degradation can be treated as the mathematical convolution (Ho and Lewis, 1976) of a SIMS response function $G(z')$ with the 'real' concentration profile $C_{\mathbf{A}}^{0}(z)$ of the impurity of interest, **A**:

$$C_{\mathbf{A}}(z) = \int_{0}^{\infty} C_{\mathbf{A}}^{0}(z - z')G(z')\,dz', \tag{8.5}$$

where $G(z')$ includes all the effects enumerated above. For the special case where the initial distribution **A** is a delta distribution centered at z_0, that is,

$$C_{\mathbf{A}}^{0}(z) = C^{0} * \delta(z - z_0),$$

then

$$\begin{aligned} C_{\mathbf{A}}(z) &= \int_{0}^{\infty} C^{0} * \delta(z - z_0 - z')G(z')\,dz' \\ &= C^{0}G(z - z_0). \end{aligned} \tag{8.6}$$

In other words, the measured distribution of **A** will be proportional to the response function centered at z_0. The ability to determine the response function is a major reason why SIMS experimentalists have become quite interested in analyzing delta structures, irrespective of the needs of material scientists and structure designers who desire to know what they have grown. Naturally, the function $G(z - z_0)$, is dependent upon the particular choice of SIMS conditions, the species **A**, and the surrounding matrix composition. The issue of a response function becomes more complex because, as will be seen, depth resolution generally degrades with increasing sputtered depth, so that the shape of $G(z - z_0)$ is also dependent on z_0. One might envision that measuring a series of delta distributions of **A** at different depths under a specific set of SIMS conditions would describe $G(z, z_0)$ sufficiently that one could deconvolve on a measured profile of **A** from an arbitrary initial distribution and be able to derive that distribution. Such efforts have been underway for the past few years (Clegg and Beall, 1989; Clegg and Gale, 1991). Furthermore, computer simulations of the basic causes of depth resolution degradation, discussed in detail below, have been performed and have given good comparison with experimental measurements of delta-doped structures (Badheka *et al.*, 1990). Limitations on using such simulations in deconvolving very narrow profiles to derive 'true' profiles, however, have been noted (Zalm and de Kruif, 1993).

It would be useful to define some mathematical values here which characterize response functions and actual SIMS profiles of delta distributions. Although the following treatments are fairly simplistic for true SIMS profiles from delta layers, they have been found useful when not using formal computer modeling. As a rough approximation the response function can be treated as a Gaussian distribution:

$$G(z, z_0) = \frac{1}{\sqrt{(2\pi)}\,\sigma_{\mathrm{R}}} \exp[-(z - z_0)^2/2\sigma_{\mathrm{R}}^2] \tag{8.7}$$

where σ_R itself may depend upon z_0. Such a distribution can be characterized by the width

$$\Delta z_R \equiv 2\sigma_R \approx 0.85 * W_R, \tag{8.8}$$

where W_R is the full width at half maximum of the response function (Fig. 8.3). One can further treat each of the individual components of depth resolution degradation (i.e. atomic mixing, initial surface roughness, induced micro-topography, individual crater shape effects, etc.) as having Gaussian responses with σ_i. It can then be shown that

$$\sigma_R^2 = \sum \sigma_i^2, \tag{8.9}$$

with similar relationships for Δz_R and W_R (Hofmann, 1991). However, all response functions are not Gaussian and some are not even symmetric. Generally in SIMS, the leading and trailing edges of a measured delta distribution are almost exponential, leading to cusped double exponentials with the approximate form

$$\begin{aligned} C_A(z) &\approx K * 10^{(z_0 - z)/\lambda_L} \qquad \text{for } z > z_0 \\ &\approx K * 10^{(z - z_0)/\lambda_T} \qquad \text{for } z < z_0. \end{aligned} \tag{8.10}$$

(Similar expressions and definitions may be obtained by replacing '10' by 'e'.) Values for the leading exponential (base 10) slopes, λ_L, can be measured as the difference between two measured depths corresponding to an increase in SIMS signal by a factor of 10. A similar calculation for the trailing slope, λ_T, can also be made.

The discussion of sources of depth resolution degradation now continues. The first two categories listed previously are the easiest to discuss. A non-rastering ion beam will form a crater with a curved surface, typically with a Gaussian profile, corresponding to the inhomogeneous ion flux density across the beam. This is unsuitable for good depth resolution as secondary ions are produced simultaneously from a range of depths (Fig. 8.4(a)). Rastering the ion beam over an area at least five times its diameter will produce a crater with a sufficiently large flat region for meaningful analysis. (In commercial spectrometers care is generally taken in the design of the raster algorithm to prevent the formation of grooves within the bottoms of craters from insufficient overlap of adjacent beam traces.) Non-flat domains will still extend from the borders of the craters. Their extent is minimized by optimizing the focus of the primary ion beam on the sample surface and by further increasing the raster area. Finally, the signal from the crater edge is avoided by either mechanical apertures or electrostatic gating, as previously described. Typically, secondary ions generated outside the central 10% of the crater area are rejected.

The issue of data sampling rate corresponds to the limitation of depth resolution to the amount of material sputtered away between adjacent measurements. For a sputter rate S, if one chooses to follow N secondary ion species by multiplexing, collecting the signal during each cycle for each species for time Δt, the depth between data points Δz_D is given by

$$\Delta z_D = SN\Delta t. \tag{8.11}$$

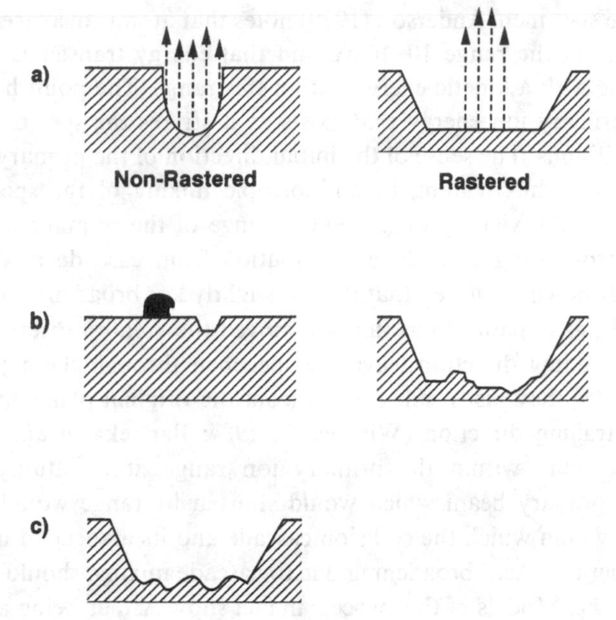

Fig. 8.4. (a) Crater shapes from non-rastered and rastered ion beams. Rays indicate from where secondary ion signal is received, demonstrating depth resolution loss in former case. (b) Effect on depth resolution of sputtering material with initial surface topography. (c) Ripples forming from initially smooth specimen. (Not to scale.)

To minimize Δz_D one would reduce these factors. Sputter rate S is decreased by reducing the ion beam current or increasing the raster size. Reducing S by too great an extent becomes impracticable, however. Signal intensity, and thus sensitivity, decreases with decreasing $S\Delta t$, resulting in a corresponding decrease of signal-to-noise. This in turn places limits on characterizing dopant distributions. Reducing S also increases analysis time. Neglecting resolution degradation effects related to beam roughening, it may be impracticable to sputter at 1 Å/s to allow reasonable sampling rates of a delta peak 5000 Å below a specimen's outer surface. This problem may be circumvented either by altering the specimen design to have the layer of interest initially closer to the surface (which will have other benefits, as discussed below) or by preceding the slow-sputter SIMS analysis by sputtering with a beam of greater current density, and stopping just prior to the depth of interest. In the latter case, of course, the final crater depth cannot be used to establish a depth scale for the SIMS profile. Multiplexing in SIMS is often used to monitor the signals of a matrix species for concentration calibration, species to denote composition changes as layer markers, contaminant species such as O or C, or other dopants. From Eqn. (8.11) it is clearly advisable to follow as many of these as possible in a profiling experiment separate from that requiring high depth resolution.

The dominant atomic mixing effect is that of *cascade mixing.* Typical atomic surface binding energies are 2–8 eV. However, sputtering yields are only 1–5 atoms per primary ion, with kinetic energies ranging from 1 to 10 keV. Thus most of the primary ion energy is deposited within the specimen. Anderson (1979) notes that atomic displacement energies for semiconductors are in the range 10–16 eV and that energy transfer is most efficient from a moving particle with a kinetic energy within this range. The point here is that the greatest part of the primary ion energy that exists as it strikes the specimen is lost in a process of multiple collisions. The sense of the initial direction of the primary ion is absent for most of these events, thus leading to an isotropic mixing of the specimen atoms, particularly at depths in the vicinity of the end-of-range of the primary ion. Given this isotropic effect, the broadening of a delta distribution from cascade mixing is largely symmetric. It has been shown, however, that there is slightly less broadening on the leading edge of a SIMS profile, as dopant atoms that are moved the greatest distances from their original plane in the leading direction leave the specimen through the exposed surface, while at the trailing edge they just reside further from the original plane, more deeply in the specimen in the trailing direction (Wittmaack, 1984; Badheka *et al.*, 1990). As the 'stirring-up' process occurs within the primary ion range, it is intuitive that those characteristics of the primary beam which would shorten its range would decrease the thickness of the layer within which the collision cascade and its associated mixing occurs. Thus one would predict that Δz_M, broadening due to cascade mixing, should decrease with primary beam energy, E_P. Models of this process in fact show Δz_M as being approximately proportional to $E_P^{1/2}$ (Anderson, 1979). Increasing the angle θ between the normal to the sample and the ion beam trajectory also reduces the range of mixing as it reduces the range of the primary ion implantation. And, for a given E_P, increasing the mass of the primary ion also decreases its implantation range and would be expected to decrease Δz_M. While this has been found to be experimentally true for most primary ion beams (including Ne^+, Ar^+, Xe^+, and Cs^+), there is an important exception in O_2^+ which will be mentioned again below.

A small fraction of collisions between the primary ions and the specimen atoms will result in the transfer of a large amount of the kinetic energy and momentum of the projectile to a target species. This target atom will now have an 'implant' range comparable to that of the primary ions. This effect, termed *recoil implantation*, would cause an asymmetric broadening of a delta profile in the direction away from the bombarded surface. Modeling indicates that this effect is significantly less than cascade mixing and is most pronounced for light target atoms and primary projectiles surrounded by a heavier atom matrix (Anderson, 1979).

At typical bombardment energies the primary ions are left implanted within the target until they are exposed and removed by subsequent sputtering. These lead to at least two potentially noticeable effects. First, to conserve atomic density within the specimen, those species originally residing above the implant range will be pushed outward toward the beam source, associated with an initial swelling of the target area. Using the standard method of profilometry to establish a profile's depth scale, this will lead to an underestimation of the original depth of a shallow delta layer, particularly for lighter primary ions (e.g. O_2^+) with greater ranges. Second, the incorporation of the implant reduces the density of the original species. If enough primary ions become incorporated, enhanced by factors such as lower sputtering yield and smaller angle of incidence, it has been argued that such a diluted delta distribution will show less effective cascade mixing than a sample with less incorporation. This is felt to be the reason that ions such as O_2^+, O^+, and N^+ offer better depth resolution than heavier ions at similar energies (Meuris *et al.*, 1991).

A final set of atomic mixing effects includes those where either a chemical interaction between the implanted primary species and target atoms affects the redistribution of atoms after bombardment, or the creation of defects within the target by bombardment measurably enhances the diffusion rate of a target species at the temperature of the analysis. Using O_2^+ under certain conditions, oxide-like layers can form, allowing the creation of electric potential gradients which may cause impurities existing in a charged state within a semiconductor to migrate. Generally these effects, which are not well modeled for SIMS profiling, are collectively referred to as *radiation enhanced diffusion*. Their importance is noted by Clegg and Gale (1991), who demonstrated that under similar sputtering conditions by O_2^+ of GaAs, Al delta spikes show significantly more broadening than do Si spikes, while their atomic masses, and thus anticipated kinematic effects, are quite similar. Redistribution of Sb in Si from O_2^+ sputtering has been indicated by Wittmaack (1994).

In all the forms of atomic mixing discussed above there exists a depth dependence on resolution until the surface plane of the sample recedes to the depth corresponding to the end-of-range of the primary ion within the sample as it was initially being sputtered. From that point and beyond, atomic mixing attains a steady-state condition as all subsequently exposed surfaces have experienced mixing effects of similar duration. Unfortunately, the pre-steady-state region is not particularly useful for SIMS analysis. The secondary ion yield improvement from the use of O_2^+ and Cs^+ does not become constant until the steady-state condition is achieved, so SIMS signal calibration is unreliable. Furthermore, ambient contamination in the outermost atomic layers of the pre-sputtered sample (particularly O, C, and H) may dominate the SIMS profile (by signal dilution, mass interferences, and matrix effects on ion yield) until the steady-state region is reached. Thus, the minimum depth from which useful profiling data is achievable corresponds closely to the range of the primary ion within the sample, which also largely controls the value of Δz_M, the broadening from cascade mixing.

The final category of sources of depth resolution degradation, surface roughening effects, does show significant dependence on the depth sputtered. Investigations by Werner (1982) and others have indicated that, to first order, depth resolution degrades linearly with depth, that is,

$$\Delta z \approx \alpha + \beta z, \tag{8.12}$$

with β, depending upon experimental conditions, having values in the range 0.00–0.03. This includes effects from crater shape degradation, mentioned above, and roughening. The value of β can be quite instrument-dependent (McPhail *et al.*, 1988). Depth independence of resolution has been achieved when special care has been taken to avoid non-linearities in primary beam rastering (Clegg and Beall, 1989; Dowsett *et al.*, 1992). Another source of roughening arises when a surface initially has topography, either from unevenness of the matrix surface or from particulate contamination on the surface. Generally, with subsequent conventional sputtering, the roughness of an exposed surface will be at least that of the initial surface (Fig. 8.4(b)). As particulate contamination often sputters more slowly than semiconductor matrix material, roughening will worsen as particles shadow matrix material from the primary ion beam. Metals are found to sputter quite unevenly, forming microscopic cones of recrystallized material as exposure to an ion beam increases. (This has also been observed with some semiconductor materials (Gries, 1991).) Interfaces between semiconductors and metal layers are often irregular, and, given differences in relative sputtering rates, this irregularity is propagated as profiling proceeds into the semiconductor. Thus metal contact (or silicide) layers over a delta layer will greatly reduce the ability to assess subtle distribution changes accurately. These problems can be avoided in the design of the specimen for SIMS analysis – avoid initial surface roughness, excess contamination, and metallic overlayers.

Even with only semiconductor matrix material, texture can be developed on the exposed surface with sputtering, a process that not only degrades depth resolution but can also change secondary ion yields (Stevie *et al.*, 1988). For example, conventional sputtering of GaAs with O_2^+ at incident angles between 32 and 58° produces quite periodic ripples (Cirlin *et al.*, 1990). This problem can be avoided by using Cs^+ as a primary ion where practical, sputtering near normal to the surface (Clegg and Beall, 1989), or at angles $\geqslant 60°$. Another solution is to have the specimen rotate while being sputtered by the primary beam (Cirlin *et al.*, 1990, 1991, 1993). While the source of the rippling effect is not fully understood, it typically forms normal to the projection of the primary ion beam trajectory on the sample surface. Sample rotation removes this directionality component and can nearly eliminate depth dependent resolution loss. Similar beneficial effects have been shown for metal and dielectric layers as well (Vajo and Cirlin, 1991; Stevie *et al.*, 1992). Sputtering with two ion beams from different directions will also reduce this effect. Finally, a fundamental source of surface roughening arises from the statistical process involved in sputtering itself. This has been estimated from atomic force microprobe measurements (Cirlin *et al.*, 1993) and theoretical considerations to be ≈ 1.6 nm (Hoffmann, 1991).

Before giving examples of how changing experimental conditions affect depth resolution, some 'competing' effects are noted. To reduce broadening from atomic mixing one is led to the use of low primary ion energies and oblique bombarding angles. However, in order to improve the data sampling rate, one chooses to reduce the sputtering rate by reducing the primary ion current density or by increasing the raster size. However, these choices, all ultimately reduce secondary ion yield and thus limit sensitivity. Furthermore, primary ion beams focus more poorly at low energies and the effect of crater shape on depth resolution increases in importance (Meuris *et al.*, 1989), and its effect increases with depth. Using an oblique angle can cause the rippling effect mentioned above and also leads to an increasingly inhomogeneous beam density profile that increases the crater shape effect. Thus, for the SIMS analyst, optimum profiling conditions for high depth resolution depend upon making trade-offs with sensitivity. The initial depth of the layers of interest should also affect the choice of operating conditions.

8.5 SIMS studies of delta-doped semiconductors

As has been previously pointed out, SIMS profiling of delta-doped structures gives direct information on the depth resolution capabilities of SIMS. Results are shown in Fig. 8.3 from measurements on a MBE-grown sample (grown at 500 °C, 150 Å/min) having an intended structure of GaAs with monolayers of AlAs inserted at depths of 20, 60, and 100 nm and of InAs at 40, 80, and 120 nm (Luftman *et al.*, 1990). The use of monolayers of Al and In as opposed to standard dopants was intended to permit the study to optimize depth resolution conditions without concern for substantial sensitivity loss. A typical profile using a 1 keV O_2^+ primary ion beam, 60° to the surface normal, is shown in Fig. 8.3. The secondary ion monitored for Al was AlO^+, as the signal from Al^+ was sufficiently intense to saturate the electron multiplier detector. Figures 8.5 and 8.6 show measured values from depth profiles of full width at half maximum, W, and the leading and trailing edge slopes, λ_L and λ_T (see equation (8.10)). This experiment was performed with a Physical Electronics 6300 SIMS spectrometer (a quadrupole-based ion microprobe) with O_2^+ or Cs^+ primary ion beams ranging in energy from 0.5 to 3.0 keV and an angle of incidence from 0 to 75°. Beams were rastered over 500 μm × 500 μm to 750 μm × 750 μm regions with secondary

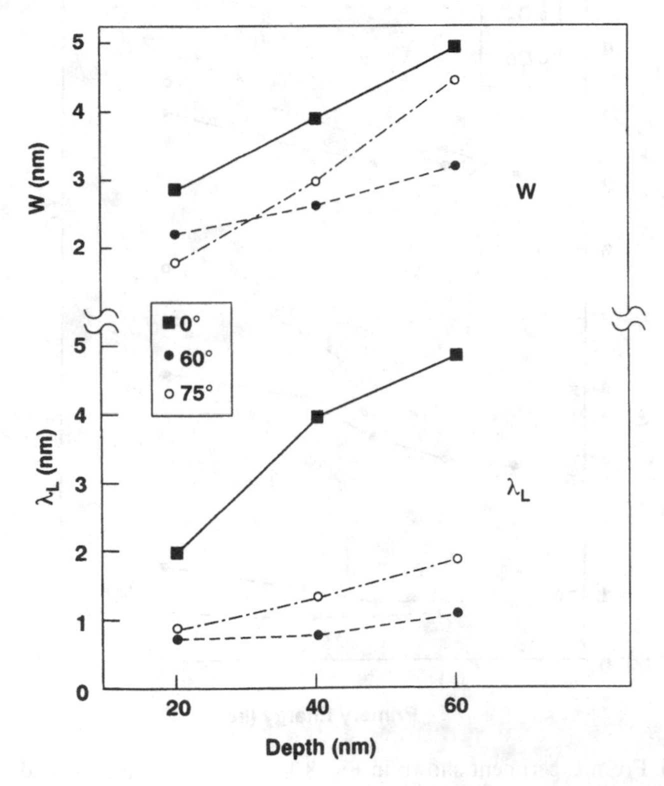

Fig. 8.5. From experiment shown in Fig. 8.3, depth and angle dependence of W and λ_L. All data taken with 1 keV O_2^+ rastered over 500 × 500 μm, monitoring AlO^+. O – 75°, ● – 60°, ■ – 0°.

ions extracted from only the centered 9% of the area. Figure 8.5 shows the dependence of W and λ_L on incidental beam angle and depth. At 20 nm the 75° beam produced the narrowest W observed in this experiment, but depth resolution degraded more quickly than when using a 60° beam, which was probably due to greater unevenness in the resulting crater. At 60°, W increased with depth while λ_L and λ_T (not shown) remained fairly constant. While depth dependence with a beam of normal incidence can be observed here, this might be avoidable with better instrumentation (Clegg and Beall, 1989). Resolutions for the 20 nm AlAs layer are shown in Fig. 8.6 as a function of primary beam energy using O_2^+ while monitoring AlO^+, and using Cs^+ while monitoring $AlCs^+$. The trailing edge slope, λ_T, is much more sensitive than λ_L to beam energy, as has been noted in other investigations (Clegg and Beall, 1989). Depth resolution for Al is better with an O_2^+ primary beam, mainly due to the improvement of λ_T with O_2^+ over Cs^+. In fact, λ_L increases with the use of O_2^+ following the mass effects of primary ions discussed by Wittmaack and Poker (1990).

Returning to Fig. 8.3, note the broader and nearly symmetric In^+ peaks. Both these aspects were unexpected for true delta distributions of an element heavier than Al. Subsequent examination with lattice imaging transmission electron microscopy (which is

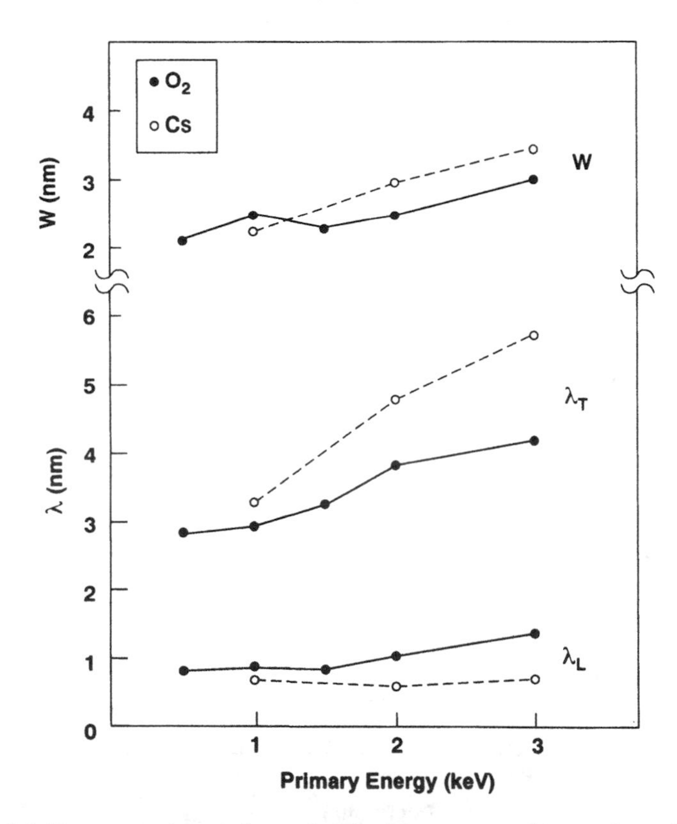

Fig. 8.6. From experiment shown in Fig. 8.3, energy and beam dependence of W and λ_L at the 20 nm AlAs layer. All data taken with a beam of 60° incidence rastered over 500 × 500 μm, at a depth of 20 nm. ● – O_2^+ primary ion, AlO^+ secondary ion; ○ – Cs^+ primary ion, $AlCs^+$ secondary ion.

of no use for dopant level delta layers) indicated that in fact the In was not confined to a monolayer in the original sample, as was the Al. This was attributed to the lattice mismatch and weaker bond strength of InAs versus GaAs, resulting in surface migration of the In during growth. Thus, the symmetry of the In^+ profile given in Fig. 8.3 was indicative of an initial In distribution skewed toward the specimen surface. Similar observations have been reported recently, coupled with simulations of the In^+ profiles (Dosanjh *et al.*, 1993).

Figure 8.7(a) shows Si delta distributions in GaAs analyzed with 2 keV O_2^+, following the secondary ion $^{28}Si^+$, and with 2 keV Cs^+, following $SiAs^-$. Both analyses were performed with a quadrupole SIMS using 60° incident beams. Depth resolution in this experiment is significantly better with O_2^+, but there is also significantly poorer sensitivity to Si in this mode. (Better sensitivity to Si with O_2^+ has been obtained by other researchers e.g. Clegg and Beall, 1989.) With best linear fits made to the data in Fig. 8.7(b), both primary ion beams give values of β (Eqn. (8.12)) of 0.075, indicating that in this case the depth dependence of depth resolution is independent of beam choice. Profiles of a similar sample are shown in Fig. 8.8 using 2 keV O_2^+ at 35° with and without sample rotation

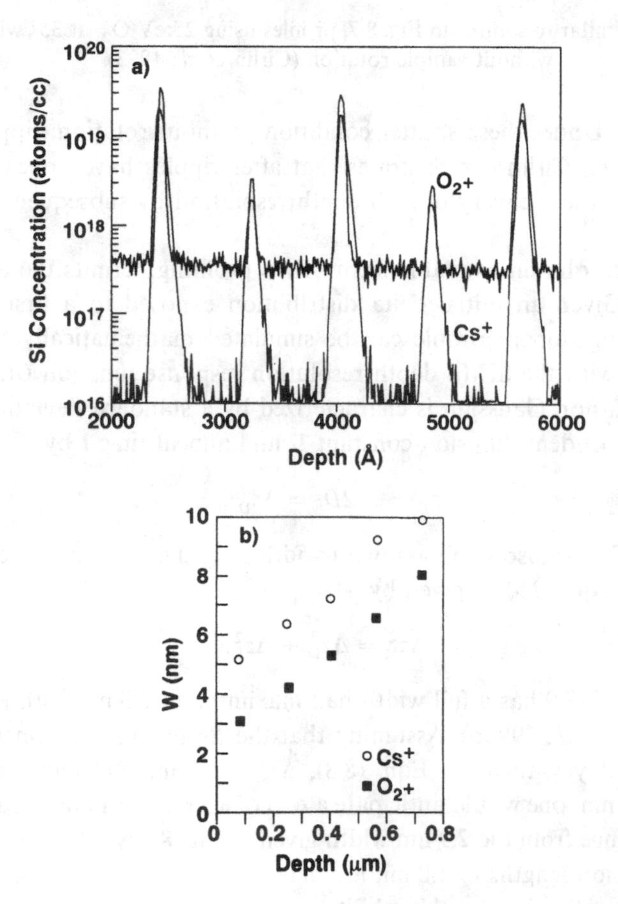

Fig. 8.7. (a) Profiles of δ-Si layers in GaAs using 2 keV Cs^+ and O_2^+ primary ion beams, (b) W versus depth for the 2×10^{19} cm^{-3} delta peaks.

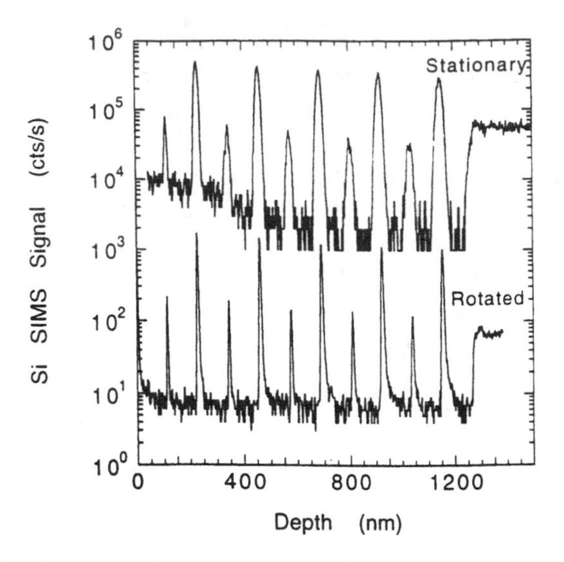

Fig. 8.8. Similar to sample in Fig. 8.7, profiles using 2 keV O_2^+ at 35° with and without sample rotation (Cirlin *et al.*, 1993).

(Cirlin *et al.*, 1993). Under these sputter conditions without rotation, ripples form on the GaAs surface. In fact, Cirlin demonstrates that after ripples have appeared they can be removed, leading to the recovery of high depth resolution by subsequent sputtering with rotation.

The high depth resolution available from SIMS profiling permits the analysis of short diffusion lengths. Given an initial delta distribution exposed to a first order diffusion process, the resulting dopant profile can be simulated mathematically by convolving a Gaussian function with the SIMS depth resolution response function $G(z)$ introduced in Eqn. (8.5). The diffusion Gaussian is characterized by a standard deviation σ_D related to the temperature-dependent diffusion constant D and anneal time t by

$$\sigma_D = \sqrt{2Dt} = \Delta z_D/2. \tag{8.13}$$

Assuming that $G(z)$ is also Gaussian with width Δz_R, then the measured profile width, as an extension to Eqn. (8.9), is given by

$$\Delta z_M^2 = \Delta z_D^2 + \Delta z_R^2. \tag{8.14}$$

The Be profile in Fig. 8.9 has a full width half maximum of 2.9 nm with a reproducibility of ±7% (Schubert *et al.*, 1990b). Assuming that the Be in this experiment was confined to a single atomic layer, then, by Eqn. (8.8), $\Delta z_R = 2.5$ nm. Referring to Eqn. (8.13), if one had a σ_D of 1.0 nm, one would anticipate a Δz_M of 3.2 nm, or a full width half maximum of 3.8 nm. This change from the 2.9 nm width given in Fig. 8.9 would be easily measurable, showing that diffusion lengths of 1.0 nm are detectable.

An interesting application of this ability of SIMS to detect the subtle broadening of near-delta B spikes in MBE-grown Si as a result of oxidation enhanced diffusion,

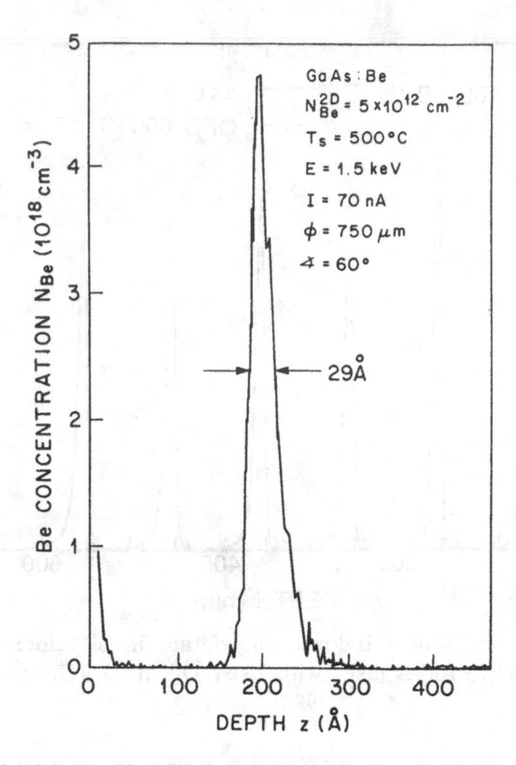

Fig. 8.9. Profile of δ-doped Be in GaAs using 1.5 keV O_2^+.

OED, was demonstrated by Gossmann *et al.* (1993). Figure 8.10 shows profiles of B^+ from samples before and after oxidation in dry oxygen at 800 °C for 15 min. Secondary ion mass spectrometry was performed with 2 keV O_2^+ at 60°. The degradation of SIMS depth resolution with depth is most evident in the decreasing peak heights with depth in the as-deposited profile. However, the profile following OED shows a reversed trend in the first four peaks. This indicates increased B diffusion for spikes closest to the surface, consistent with the model of Si self-interstitials injected at the surface promoting B diffusion. From this data it is possible to extract B diffusion constants as a function of depth from which the diffusion length of the self-interstitials can be deduced. Recent studies of delta-doped B and Sb in Si have been reported by Dowsett *et al.* (1992) and Wittmaack (1994).

Figure 8.11 corresponds to an experiment studying the surface segregation of Si in delta-doped GaAs as a function of doping density. The material was grown at a relatively high temperature of 660 °C to promote segregation. For all four profiles $\lambda_T \approx 9.0$ nm, while λ_L increases sublinearly with the Si areal density. As SIMS broadening processes create a skewness of true delta distributions away from the surface, the skewness toward the surface noted in this diagram indicates real segregation toward the surface. The shape of these profiles is consistent with segregation driven by Fermi level pinning as opposed to solubility-limited Si incorporation (Schubert *et al.*, 1990b). A different broadening process is evident in Fig. 8.12 of delta-doped Be in GaAs, grown at 500 °C (Schubert *et al.*, 1990a).

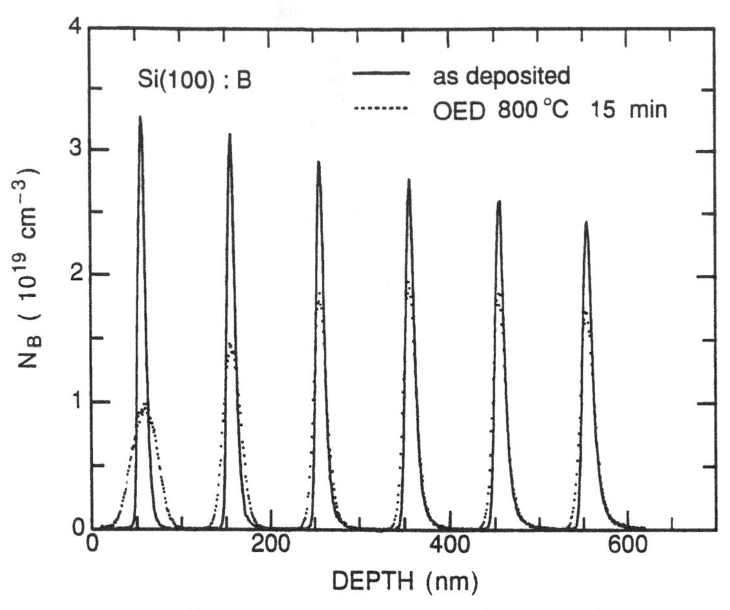

Fig. 8.10. SIMS profiles of a B-doped superlattice in Si before and after annealing in dry O_2. Profiles taken with 2 keV O_2^+ at 60° (Gossmann *et al.*, 1993).

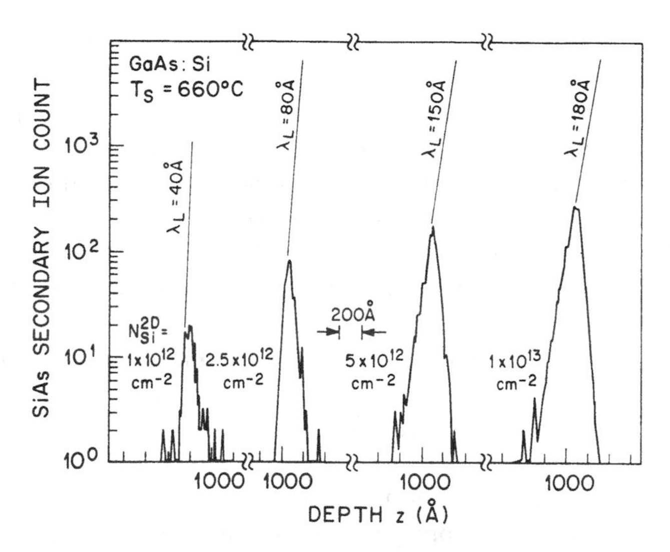

Fig. 8.11. Profiles of four specimens of δ-doped Si in GaAs with varying Si areal densities, all grown at 660 °C.

Significant broadening does not ensue until $1 \times 10^{14}\ cm^{-2}$ Be is introduced into the growth. Profiles are not skewed toward the surface here but broaden in both directions about the 100 nm depth where the Be was introduced. This result has been explained as a repulsive Coulombic interaction between Be ions within the GaAs matrix.

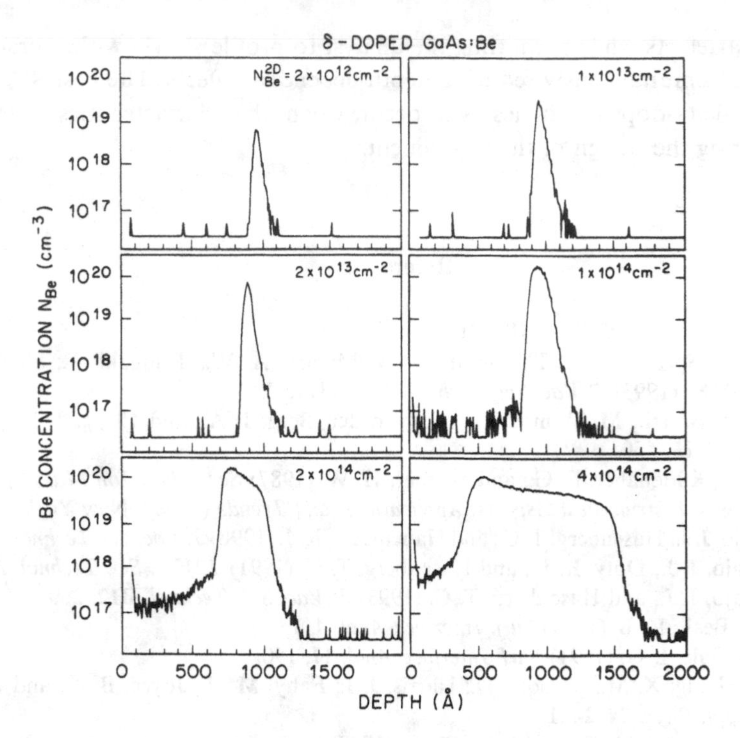

Fig. 8.12. Profiles of δ-doped Be in GaAs showing lack of Be confinement at higher areal densities.

8.6 Conclusion

Secondary ion mass spectrometry depth profiling is particularly well suited to characterizing actual dopant distributions in nominally delta-doped structures. Limitations to the depth resolving capabilities of SIMS are due to atomic mixing, inadequate data sampling rates, lack of crater flatness and sputter-induced micro-topography. The latter two classes of effect worsen resolution with increasing depth of analysis. The SIMS analyst has several means at his or her disposal for improving resolution. Primary ion beams should be at low energies and at fairly oblique angles of incidence. Heavier primary ions are preferable, with the exception of oxygen (and nitrogen) beams, which offer particularly good resolution characteristics. The beam should be well focused on the sample surface and rastered over an area significantly larger than its cross-sectional area. Secondary ions should be extracted from only a small region centered within the crater. Sputtering rates should be slow relative to data acquisition rates in order to allow sufficient sampling. Sample rotation during profiling can be particularly helpful when sufficiently oblique sputtering angles are impracticable. Finally, the designer of delta-doped specimens can assist analysis by preparing material with smooth surfaces, no metallic overlayers, and, where possible, features of interest should be placed at depths between 20 and 100 nm, avoiding potential

SIMS surface artefacts while permitting the analyst to profile slowly within practical time constraints, and simultaneously reducing depth-dependent effects. The best SIMS characterizations of delta-doped samples will occur when the characteristics of SIMS are considered during the design of the experiment.

References

Anderson, H. H. (1979) *Appl. Phys.* **18**, 131.

Arlinghaus, H. F.; Spear, M. T.; Thonnard, N.; McMahon, A. W.; Tanigaki, T.; Shichi, H.; and Holloway, P. H. (1993) *J. Vac. Sci. Technol.* **A11**, 2317.

Badheka, R.; Wadsworth, M.; Armour, D. G.; van den Berg, J. A.; and Clegg, J. B. (1990) *Surf. Interface Anal.* **15**, 550.

Benninghoven, A.; Rüdenauer, F. G.; and Werner, H. W. (1987) *Secondary Ion Mass Spectrometry: Basic Concepts, Instrumental Aspects, Applications, and Trends* (Wiley, New York).

Cirlin, E. H.; Vajo, J. J.; Hasenberg, T. C.; and Hauenstein, R. J. (1990) *J. Vac. Sci. Technol.* **A8**, 4101.

Cirlin, E. H.; Vajo, J. J.; Doty, R. E.; and Hasenberg, T. C. (1991) *J. Vac. Sci. Technol.* **A9**, 1395.

Cirlin, E. H.; Vajo, J. J., and Hasenberg, T. C. (1993) *J. Vac. Sci. Technol.* **B12**, 269.

Clegg, J. B. and Beall, R. B. (1989) *Surf. Interface Anal.* **4**, 1.

Clegg, J. B. and Gale, I. G. (1991) *Surf. Interface Anal.* **17**, 190.

Dosanjh, S. S.; Zhang, X. M.; Sansom, D.; Harris, J. J.; Fahy, M. R.; Joyce, B. A.; and Clegg, J. B. (1993) *J. Appl. Phys.* **74**, 2481.

Downey, S. W.; Emerson, A. B.; and Kopf, R. F. (1990) in *Resonance ionization spectroscopy 1990*, eds. Parks, J. E. and Omenetto, N., Inst. Phys. Conf. Ser. **114**, 401.

Dowsett, M. G.; Barlow, R. D.; Fox, H. S.; Kubiak, R. A. A.; and Collins, R. (1992) *J. Vac. Sci. Technol.* **B10**, 336.

Gossmann, H. J.; Vredenberg, A. M.; Rafferty, C. S.; Luftman, H. S.; Unterwald, F. C.; Jacobson, D. C.; Boone, T.; and Poate, J. M. (1993) *Appl. Phys. Lett.* **63**, 639.

Gries, W. H. (1991) *Surf. Interface Anal.* **17**, 719.

Ho, P. S. and Lewis, J. E. (1976) *Surface Sci.* **55**, 335.

Hofmann, S. (1991) *Prog. Surf. Sci.* **36**, 35.

Luftman, H. S.; Schubert, E. F.; and Kopf, R. F. (1990) *American Vacuum Society National Symposium* (Toronto), unpublished.

McPhail, D. S.; Clark, E. A.; Clegg, J. B.; Dowsett, M. G.; Gold, J. P.; Spiller, G. D. T.; and Sykes, D. (1988) *Scanning Microsc.* **2**, 639.

Meuris, M.; De Bisschop, P.; Leclair, J. F.; and Vanderworst, W.; (1989) *Surf. Interface Anal.* **14**, 739.

Meuris, M.; Vanderworst, W.; and Jackman, J. (1991) *J. Vac. Sci. Technol.* **A9**, 1482.

Schubert, E. F.; Kuo, J. M.; Kopf, R. F.; Luftman, H. S.; Hopkins, L. C.; and Sauer, N. J. (1990a) *J. Appl. Phys.* **67**, 1969.

Schubert, E. F.; Luftman, H. S.; Kopf, R. F.; Headrick, R. L.; and Kuo, J. M. (1990b) *Appl. Phys. Lett.* **57**, 1799.

Stevie, F. A.; Kahora, P. M.; Simons, D. S.; and Chi, P. (1988) *J. Vac. Sci. Technol.* **A6**, 76.

Stevie, F. A.; Kahora, P. M.; Moore, J. L.; and Wilson, R. G. (1992) in *Secondary Ion Mass Spectrometry SIMS VIII*, eds. Benninghoven, A. *et al.* (Wiley, W. Sussex) p. 327.

Vajo, J. J. and Cirlin, E.-H. (1991) *Surf. Interface Anal.* **17**, 786.

Werner, H. W. (1982) *Surf. Interface Anal.* **4**, 1.

Wilson, R. G.; Stevie, F. A.; and Magee, C. W. (1989) *Secondary Ion Mass Spectrometry: A Practical Handbook for Depth Profiling and Bulk Impurity Analysis* (Wiley, New York).

Wittmaack, K. (1984) *Vacuum* **34**, 119.
Wittmaack, K. and Poker, D. B. (1990) *Nucl. Instr. and Meth.* **B47**, 224.
Wittmaack, K. (1992) in *Practical Surface Analysis*, eds. D. Briggs and M. P. Seah (Wiley, Chichester), Vol. 2, p. 105.
Wittmaack, K. (1994) *J. Vac. Sci. Technol.* **B12**, 250.
Zalm, P. C. and de Kruif, R. C. M. (1993) *Appl. Surf. Sci.* **70–71**, 73.

9

Capacitance–voltage profiling

E. F. SCHUBERT

9.1 Introduction

Capacitance–voltage (C–V) profiling is an electrical characterization technique for spatial resolution of doping distributions in semiconductors. The technique is non-destructive and can be used at room temperature. The C–V technique provides highly accurate and precise values for the dopant concentration and its spatial distribution. The C–V measurement can be performed on Schottky contacts, on p^+n junctions, or on n^+p junctions. The capacitance of the junction is measured as a function of the bias which yields the $C(V)$ curve. The C–V concentration, N_{CV}, is inferred from the $C(V)$ curve using the relation

$$N_{CV} = \frac{-2}{e\varepsilon} \frac{\mathrm{d}V}{\mathrm{d}(1/C^2)} = \frac{C^3}{e\varepsilon} \frac{\mathrm{d}V}{\mathrm{d}C} \tag{9.1}$$

where e and ε are the elementary charge and the permittivity of the semiconductor, respectively, and $C(V)$ is the capacitance per unit area of the junction. The C–V depth, z_{CV}, is deduced from the reciprocal of the capacitance, that is,

$$z_{CV} = \varepsilon/C. \tag{9.2}$$

Equations (9.1) and (9.2) allow one to plot N_{CV} versus z_{CV}, which is called the *C–V profile*.

For homogeneously doped, non-compensated semiconductors in which all dopants are electrically active, the C–V concentration equals the donor concentration (N_D) or the acceptor concentration (N_A), that is,

$$N_{CV} = N_{D,A}. \tag{9.3}$$

Using the depletion approximation, one can easily show the validity of Eqns. (9.1), (9.2), and (9.3). The C–V concentration equals the *electrically active* impurity concentration if a fraction of the doping concentration is electrically inactive, that is,

$$N_{CV} = N_{D,A} - N_{\text{inactive}} \tag{9.4}$$

where $N_{D,A}$ is the total concentration of dopants and $N_{inactive}$ is the concentration of electrically inactive (neutral) dopants. For a partially compensated semiconductor containing acceptor and donor impurities the C–V concentration is given by

$$N_{CV} = N_D - N_A \quad \text{for} \quad N_D > N_A \tag{9.5a}$$

$$N_{CV} = N_A - N_D \quad \text{for} \quad N_D < N_A. \tag{9.5b}$$

The spatial resolution of the C–V profiling technique is given by the majority carrier screening length (Kennedy *et al.*, 1968 and 1969). In a semiconductor, abrupt spatial changes of the impurity concentration are not mirror-imaged by the free carrier concentration. Instead, the free carrier concentration is smeared out as compared to the dopant distribution. The characteristic smearing length is the majority carrier screening length. In non-degenerately doped semiconductors, the majority carrier screening length, that is, the Debye length, is given by

$$L_D = \sqrt{\left(\frac{\varepsilon kT}{e^2 n}\right)} \tag{9.6}$$

where n is the free carrier concentration. In degenerately doped semiconductors, the majority carrier screening length, that is, the Thomas–Fermi length, is given by

$$L_{TF} = \pi^{2/3}\sqrt{\left(\frac{\varepsilon h^2}{e^2 m^*(3n)^{1/3}}\right)}. \tag{9.7}$$

Note that the two screening lengths have different dependences on n and T. As a result of the screening length limitation, the C–V profile represents the free carrier distribution rather than the distribution of ionized impurities.

If the C–V technique is applied to semiconductors with quantum confinement of free carriers, its spatial resolution is not given by the screening length. Instead, the resolution has been shown to be limited by the spatial extent of the wave functions representing the quantum-mechanical free carrier system (Schubert *et al.*, 1990). The resolution of the C–V technique in semiconductors with quantum confinement will be discussed in detail in Section 9.2.

The C–V technique is not only used in the spatial resolution of doping distributions; it can also be used to determine the two-dimensional free carrier density (per cm^2) in thin films. This possibility is interesting in delta-doped semiconductors in which the two-dimensional (2D) carrier density is of interest. Assuming that the 2D free carrier system is located between z_1 and z_2, the 2D density, n^{2D}, of a free carrier system is given by

$$n^{2D} = \int_{z_1}^{z_2} n(z)\,dz = \int_{z_1}^{z_2} N_{CV}(z)\,dz. \tag{9.8}$$

Even though N_{CV} does not coincide with the free carrier concentration, that is, $N_{CV}(z) \neq n(z)$, the integral charge of the profile coincides exactly with the integral charge of the free carrier concentration (Kroemer *et al.*, 1980). Evaluation of a C–V profile using

Eqn. (9.8) thus allows one to determine the free carrier density and the electrical activity of dopants in δ-doped semiconductors.

9.2 Theory of C–V technique in semiconductors with quantum confinement

The *depletion approximation* forms the basis of the C–V profiling technique in semiconductors without quantum confinement of free carriers. It is assumed in this approximation that the semiconductor can be depleted of free carriers slice by slice, as the reverse voltage, applied to the Schottky contact or pn junction, is increased. Although the depletion approximation can be applied successfully to semiconductors with a classical carrier distribution, it does not hold for semiconductors with a quantized carrier distribution.

An example of a quantized carrier distribution is illustrated in Fig. 9.1(a), which shows the band diagram of a δ-doped semiconductor. The carrier distribution is represented by a wave function with a position expectation value of $\langle z \rangle$ and a spatial extent of z_0. The donors of δ-doped layer are located at a distance z_d below a Schottky contact which has a barrier height of $e\phi_B$. Figure 9.1(a) further shows that, for zero bias, the position expectation value coincides with the position of the donor plane. As the voltage on the metal–semiconductor is changed, the wave function is perturbed, including its amplitude, shape, and position expectation value. Inspection of Fig. 9.1(b) reveals that the position expectation value of the wave function is displaced from the donor plane. It is clear that the depletion approximation cannot be applied to a semiconductor quantum system, that

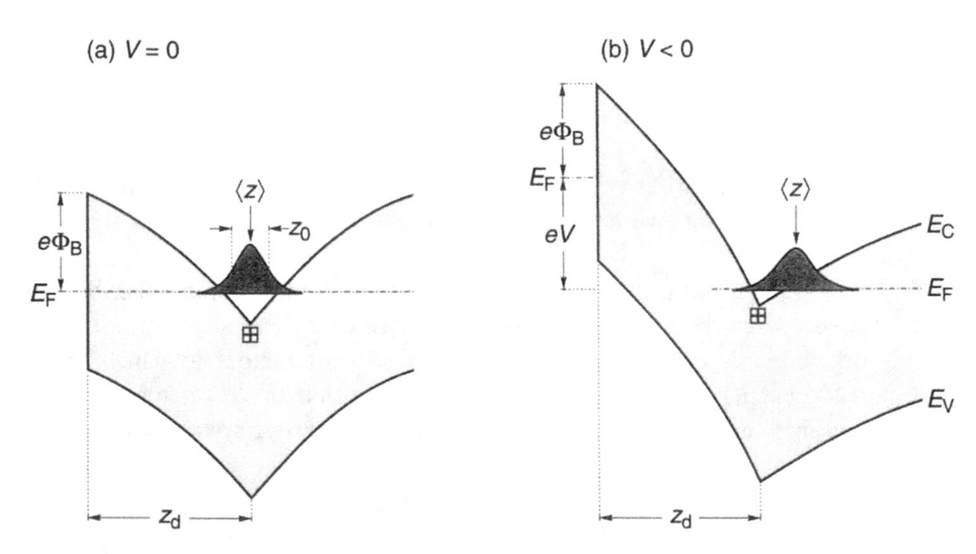

Fig. 9.1. Band diagram of a semiconductor with a δ-doped layer located at a distance z_d below a Schottky contact for (a) zero bias and (b) reverse bias. The Fermi level in the metal is located at $e\phi_B$ below the conduction band edge. The position expectation value of the electron system, $\langle z \rangle$, depends on the bias.

is, a wave function cannot be 'depleted' slice by slice. Instead, the entire wavefunction is perturbed due to a change in external bias. Below, we will quantify the perturbation and show that the spatial resolution of the C–V technique in quantum confined semiconductors is given by the spatial extent of the wave function.

The capacitance of a 2D quantum system is given by

$$C = \left(\frac{\mathrm{d}V}{\mathrm{d}Q}\right)^{-1} = \left(\frac{\mathrm{d}V_{\mathrm{el}}}{\mathrm{d}Q} + \frac{1}{e}\frac{\mathrm{d}(E_{\mathrm{F}} - E_0)}{\mathrm{d}Q} + \frac{1}{e}\frac{\mathrm{d}E_0}{\mathrm{d}Q}\right)^{-1}, \tag{9.9}$$

where Q is the charge per unit area and E_{F} and E_0 are the Fermi energy and the energy of the lowest subband respectively. The total capacitance is composed of three contributions which will be referred to as the *electrostatic*, the *quantum*, and the *Stark capacitance*, for the first, second, and third term respectively. The electrostatic capacitance can be inferred from Poisson's equation and is given by $C_{\mathrm{el}} = \varepsilon/\langle z \rangle$, where $\langle z \rangle$ is the position expectation value of the quantized carrier system. The quantum capacitance has its physical origin in the finite density of states of a 2D system. The Stark capacitance is due to the Stark effect of a 2D quantized electron gas.

The resolution of the C–V profiling technique in a quantized 2D system is given by the change in C–V profile depth, Δz_{CV}, during the depletion of the 2D electron gas. From Eqns. (9.2) and (9.9) one obtains

$$\Delta z_{CV} = \Delta\frac{\varepsilon}{C} = \Delta\frac{\mathrm{d}V_{\mathrm{el}}}{\mathrm{d}Q}\varepsilon + \Delta\frac{\mathrm{d}(E_{\mathrm{F}} - E_0)}{\mathrm{d}Q}\frac{\varepsilon}{e} + \Delta\frac{\mathrm{d}E_0}{\mathrm{d}Q}\frac{\varepsilon}{e}. \tag{9.10}$$

The three terms in Eqn. (9.10) constituting the resolution of the C–V profile can be calculated *analytically* in the one electron picture using variational wavefunctions. The quantum and the Stark capacitance of Eqn. (9.10) are much smaller than the first term for typical geometries. Neglecting the quantum and the Stark capacitance, Eqn. (9.10) simplifies to

$$\Delta z_{CV} \approx \Delta\langle z \rangle_0, \tag{9.11}$$

where $\Delta\langle z \rangle_0$ is the change in position expectation value of the ground state upon application of the depleting bias.

In the following, we show that the change in expectation value, $\Delta\langle z \rangle_0$, is comparable to the spatial extent of the ground state, z_0, that is, $\Delta z_{CV} \approx \Delta\langle z \rangle_0 \approx z_0$. The electrostatic potential resulting from a δ-function-like dopant distribution of density $N^{2\mathrm{D}}$ at $z = 0$ is given by

$$eV(z) = +e\mathscr{E}z \quad \text{for} \quad z \geqslant 0, \tag{9.12a}$$

$$eV(z) = -e\mathscr{E}z \quad \text{for} \quad z < 0, \tag{9.12b}$$

where the electric field is given by $\mathscr{E} = eN^{2\mathrm{D}}/2\varepsilon$. Equation (9.12) is based on the dopant charges only and the charges of the free carriers are neglected. This approximation is made in order to obtain an analytic result. For the ground state and the first and second excited states we use the variational wavefunctions (Schubert *et al.*, 1989):

$$\psi_0(z) = (\sqrt{(2\alpha_0)/5})(1 + \alpha_0 z)\, e^{-\alpha_0 z} \qquad \text{for} \qquad z > 0, \tag{9.13a}$$

$$\psi_1(z) = \sqrt{(2)}\alpha_1^{3/2} z\, e^{-\alpha_1 z} \qquad \text{for} \qquad z > 0, \tag{9.13b}$$

$$\psi_2(z) = \frac{2}{3}(\sqrt{(\alpha_2)/7})(\alpha_2^2 z^2 - 1)(1 + \alpha_2 z)\, e^{-\alpha_2 z} \qquad \text{for} \qquad z > 0. \tag{9.13c}$$

The wave functions can be symmetrized ($n = 0, 2$) and anti-symmetrized ($n = 1$) for $z < 0$. The trial parameters α_n are

$$\alpha_0 = \left(\frac{9}{2} e\mathscr{E} \frac{m^*}{\hbar^2}\right)^{1/3}; \quad \alpha_1 = \left(\frac{3}{2} e\mathscr{E} \frac{m^*}{\hbar^2}\right)^{1/3}; \quad \alpha_2 = \left(\frac{47}{6} e\mathscr{E} \frac{m^*}{\hbar^2}\right)^{1/3}. \tag{9.14}$$

The spatial extent of the ground-state wave function is then given by

$$z_0 = 2\sqrt{(\langle z^2\rangle - \langle z\rangle^2)} = 2\sqrt{(7/5)}(2\hbar^2/9e\mathscr{E}m^*)^{1/3}. \tag{9.15}$$

The position expectation value in the presence of a perturbing electric field can be calculated using second-order perturbation theory:

$$\langle z\rangle_0 = \langle z\rangle_0^0 + \sum_{n\neq 0} \frac{\langle\psi_0|z|\psi_n\rangle^2}{E_n - E_0} e\mathscr{E}, \tag{9.16}$$

where $\langle z\rangle_0^0 = 0$ is the unperturbed expectation value for the ground state. For the calculation we denote the perturbing field by $\mathscr{E} = \pm eN^{2D}/4\varepsilon$. The magnitude of this field depletes half of the free carriers as inferred from Gauss's equation. Calculating the sum of Eqn. (9.16) for $n = 1$ and $n = 2$ by using the variational wave functions of Eqn. (9.13) and estimating contributions to the sum for $n > 2$, one obtains

$$\Delta\langle z\rangle_0 \approx z_0. \tag{9.17}$$

Thus, the resolution limit of the C–V profiling technique on semiconductors with quantum confinement is given by the *spatial extent of the wave function* associated with the carrier system. Note that the result of Eqn. (9.17) is of general validity and applies to δ-doped semiconductors as well as to compositional quantum wells. The resolution of the C–V profiling technique in δ-doped semiconductors is obtained from Eqns. (9.11), (9.14), and (9.16) and is given by

$$\Delta z_{CV} = 2\sqrt{(7/5)}(2\hbar^2/9e\mathscr{E}m^*)^{1/3} = 2\sqrt{(7/5)}(4\varepsilon\hbar^2/9e^2N^{2D}m^*)^{1/3}. \tag{9.18}$$

The dependence of the resolution length on the doping concentration on the effective mass is different from the dependence of the screening length on these quantities.

Capacitance–voltage profiles on δ-doped semiconductors can also be calculated by solving Schrödinger's equation and Poisson's equation self-consistently in a numerical, iterative way (Schubert *et al.*, 1988). The calculation of the electron charge is carried out for different bias voltages, which yields the total charge as a function of the bias. Using

$C = \mathrm{d}Q/\mathrm{d}V$, the capacitance of the system can be calculated as a function of bias. Finally, the C–V profiles can be calculated from the $C(V)$ data. The widths of such numerically calculated C–V profiles are in good agreement with the analytically calculated profile widths given in Eqn. (9.18).

9.3 Experimental *C*–*V* profiles

Capacitance–voltage measurements are commonly performed on Schottky contacts with an area of 20–200 000 μm^2. The measurement can be carried out at room temperature. Capacitance bridges or parameter analyzers such as the Hewlett-Packard model HP 4194A can be used for the measurement. Analysis of the data and the conversion to C–V profiles are usually performed using a personal computer.

The unambiguous analysis of C–V data requires that the *capacitance* dominates the impedance of the sample. Modeling the sample as a junction capacitance, a (parallel) junction conductance (which describes the current leakage), and a resistor in series to the junction capacitance and conductance (contact resistance, etc.), the (serial) resistance and the (parallel) conductance must be much smaller than the reactance of the junction capacitance. This condition is usually met for 500 μm-diameter contacts if the measurement frequency is 1 MHz.

The dominance of the capacitance can be verified by measuring the phase angle between the small signal ac current and the ac voltage of the instrument. The phase angle and the capacitance of a δ-doped n-type GaAs sample is shown in Fig. 9.2. For voltages ranging

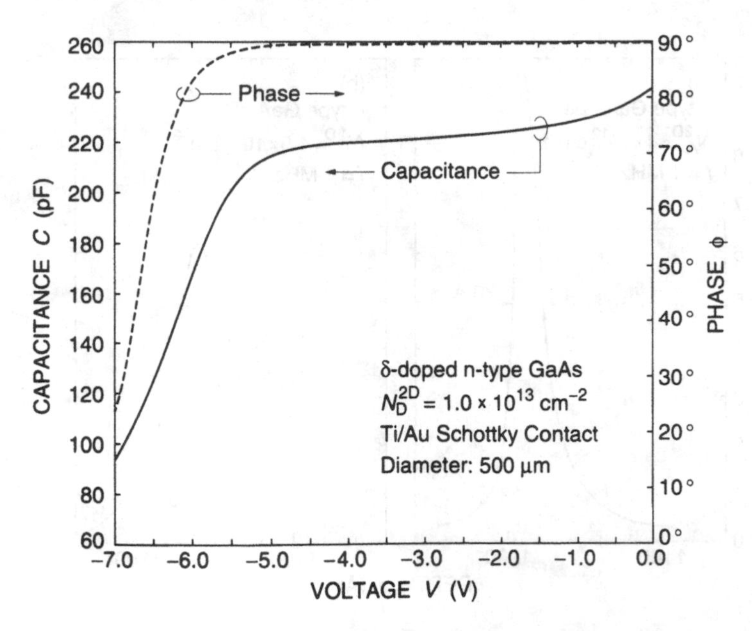

Fig. 9.2. Capacitance and phase of a δ-doped semiconductor–metal junction as a function of voltage.

from -6 V to 0 V, the phase angle is very close to 90°, indicating the dominance of the capacitance in the impedance of the sample. It is essential that the phase angle be close to 90° over the entire range of dc voltages and not just for zero dc voltage.

The C–V-curve of a δ-doped semiconductor is shown in Fig. 9.2. The capacitance is measured between a 500-μm-diameter circular Schottky contact and the highly conductive semiconductor substrate. As the bias voltage is decreased, the capacitance (i) initially decreases; (ii) remains nearly constant for a range of voltages; and (iii) drops rapidly for voltages < 5.0 V. The three distinct regions are interpreted as follows. In the first region, the top semiconductor layer is being depleted of free carriers. The capacitance remains nearly constant in the second region, during the depletion of the δ-doped layer. Finally, when the δ-doped layer is depleted, the capacitance drops rapidly. Note that the phase angle, which is also shown in Fig. 9.2, has a value $\approx 90°$ for the entire range of voltages of interest. This makes reliable capacitance measurements in the entire voltage range possible.

Experimental C–V-profiles on δ-doped GaAs and Si are shown in Figs. 9.3 and 9.4 respectively. The C–V-profiles of Be-doped and Si-doped GaAs are shown in Fig. 9.3(a) and (b) respectively. The profile on the p-type GaAs sample has a strikingly narrow width of only 20 Å. The peak concentration of the profile is in the high 10^{18} cm^{-3} range. This C–V-profile is the narrowest ever measured on a semiconductor. The profile on the n-type δ-doped GaAs sample has a width of 49 Å and a peak concentration of 6×10^{18} cm^{-3}. The integration of the C–V profile versus depth yields the electrically active impurity concentration. Concentrations of $N_A^{2D} = 4 \times 10^{12}$ cm^{-2} and $N_D^{2D} = 4.5 \times 10^{12}$ cm^{-2} are determined for the p-type and n-type sample respectively.

The widths of the C–V profiles compare favorably with theoretical profile widths.

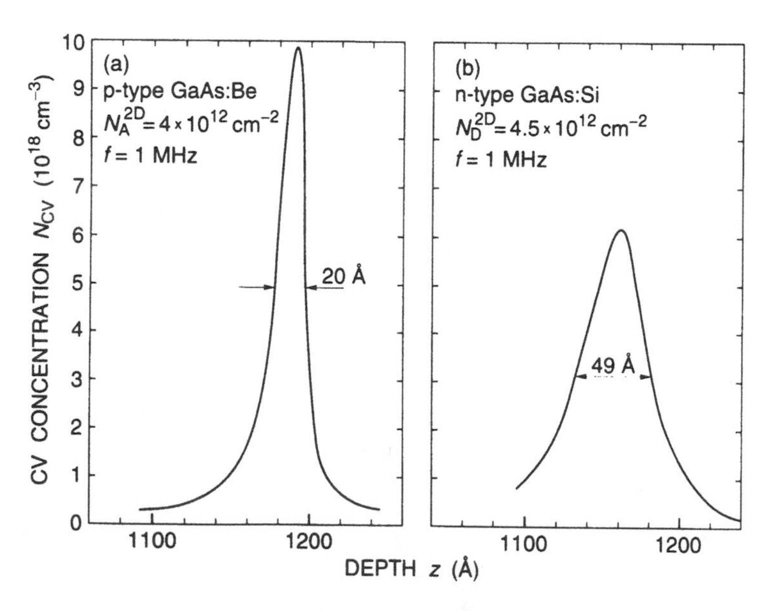

Fig. 9.3. Capacitance–voltage profiles on (a) p-type and (b) n-type GaAs measured at room temperature.

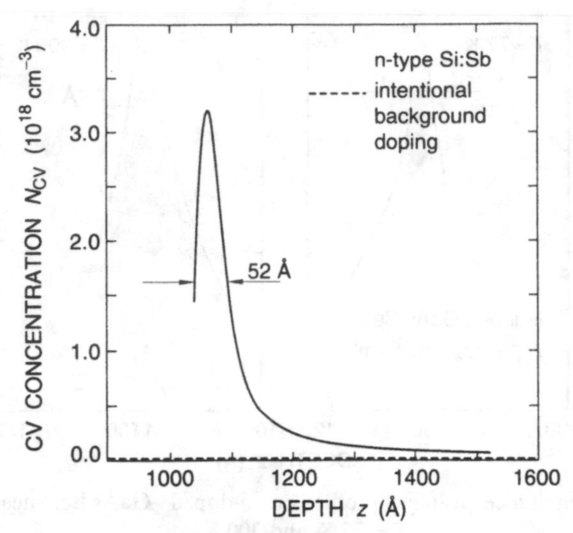

Fig. 9.4. Capacitance–voltage profile on Sb δ-doped Si measured at room temperature.

Using Eqn. (9.18) for the calculation of the profile widths, one obtains a theoretical width of 26 Å and 48 Å for the p-type ($m^*_{hh} = 0.45\, m_0$) and n-type ($m^*_e = 0.067\, m_0$) sample respectively. The free carrier concentration used for the calculation are the free carrier concentrations of the samples shown in Fig. 9.3. Comparison of the calculated and the experimental widths of the C–V profiles yields excellent agreement.

The C–V profile of a Sb δ-doped Si sample grown by molecular-beam epitaxy is shown in Fig. 9.4 (Gossmann and Schubert, 1993). The epitaxial layer was grown on a 0.01 Ω cm n-type (001)-oriented Si substrate. The epitaxial layer consists of a 1000-Å-thick buffer layer, the Sb δ-doped spike of density 4×10^{12} cm^{-2}, and a 1000-Å-thick Si cap layer. The buffer and cap layers are n-type with $N_D = 10^{16}$ cm^{-3}. The C–V profile exhibits a well-defined peak at a depth of 1070 Å, in good agreement with the value anticipated during growth. The profile measured at a frequency of 2 MHz has a width of 52 Å. For the experimental parameters of the sample, a C–V profile width of 20 Å is calculated using Eqn. (9.18), which is smaller than the experimental value. The difference could be due to a small amount of broadening of the δ-layer itself.

Narrower C–V profiles have been reported by van Gorkum *et al.* (1989) for concentrations exceeding 10^{13} cm^{-2}. However, in this range avalanche breakdown occurs which makes the interpretation of the C–V measurement questionable. Mattey *et al.* (1990) measured a 30 Å profile width on Si with a B-doping spike of concentration 2×10^{12} cm^{-2}. This compared favorably with the expected width of 30–40 Å.

The measurement temperature is of minor influence on the C–V profile characteristic. Figure 9.5 shows C–V profiles measured at $T = 77$ K and 300 K of a Be δ-doped GaAs sample. At room temperature, the profile is 21 Å wide. The width reduces to 17 Å at liquid nitrogen temperature. In addition to the narrowing of the profile, the peak concentration

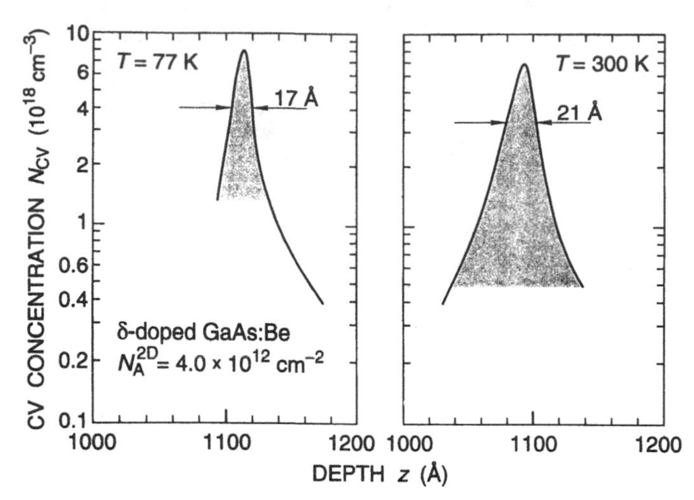

Fig. 9.5. Capacitance–voltage profile on δ-doped GaAs:Be measured at $T = 77$ K and 300 K.

increases from 7×10^{18} to 8×10^{18} cm^{-3} at low temperatures. Note that the changes are much smaller than the temperature dependence of the Debye length, which has a $\sqrt{T}$-dependence on temperature. The lack of strong temperature sensitivity of the C–V profiles can be attributed to the quantum nature of the free carrier gas. Consider the quantum limit, in which only the ground state is occupied. In this case, the distribution of carriers along the z-direction depends on the carrier wave function, which is, in the quantum limit, independent of temperature. Thus, no temperature dependence of the C–V profile is expected in the quantum limit. In a realistic 2D system, a weak temperature dependence is expected. Higher subbands become populated as the temperature increases. Such higher subbands have a larger spatial extent, that is, their contribution will broaden C–V profiles, which is indeed observed experimentally (see Fig. 9.5).

Capacitance–voltage profiles have proved to be very useful in assessing the redistribution of impurities during epitaxial growth or during post-growth annealing. If δ-doped semiconductors are grown at elevated substrate temperatures, the initially δ-function-like doping profiles may broaden into gaussian-function-like or other doping distributions (Schubert *et al.*, 1989). Likewise, if δ-doped semiconductors are subjected to post-growth annealing cycles, impurities diffuse and broaden the doping distribution (Schubert *et al.*, 1988). The broadening of the doping profiles can be assessed, in both cases, by the C–V profiling technique.

The C–V profiles of Si δ-doped $Al_xGa_{1-x}As$ samples grown by MBE at different temperatures are shown in Fig. 9.6. The growth temperatures are 500, 600, and 700 °C respectively. The Si δ-doped layers are located at a nominal depth of 1000 Å below the semiconductor surface. The nominal Si density is 5×10^{12} cm^{-2}. Figure 9.6 reveals a clear and drastic broadening of the C–V profile for epitaxial growth at temperatures of 600 and 700 °C. The profiles broaden from 51 Å for low-temperature growth to 305 Å for high-temperature growth. Note that the impurity redistribution may be more involved

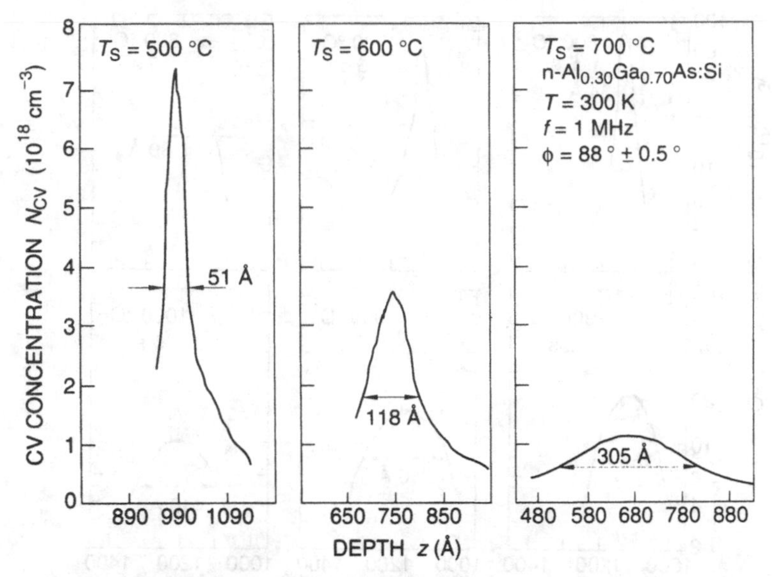

Fig. 9.6. Capacitance–voltage profiles of δ-doped n-type $Al_{0.30}Ga_{0.70}$As:Si grown by molecular-beam epitaxy at 500, 600, and 700 °C.

than just plain diffusion. Growth at different temperatures results in different crystal quality, that is, point defects are more abundant at either very low or very high growth temperatures. Since the diffusion of impurities is strongly influenced by point defects, a different diffusion mechanism may be required to explain the redistribution of impurities illustrated in Fig. 9.6.

Capacitance–voltage profiles of Be δ-doped GaAs annealed at different temperatures are shown in Fig. 9.7. The epitaxial layers were grown by MBE at a temperature of 500 °C. Subsequently, the sample was cleaved into pieces and the individual pieces were annealed for 5 s at temperatures ranging from 600 °C to 1000 °C in a rapid thermal annealing furnace heated by tungsten halogen lamps. During the anneals, the samples were covered by another GaAs wafer in order to reduce As out-diffusion from the sample. The profiles shown in Fig. 9.7 illustrate a progressive broadening with increasing annealing temperature. The magnitude of the broadening ranges from 34 Å for the 600 °C anneal to 440 Å for the 1000 °C anneal. Diffusion coefficients can be extracted from the impurity profile broadening, as is discussed further in Chapters 10 and 11 on impurity redistribution in this book.

The assessment of impurity distributions by the C–V profiling technique allows one to estimate the growth and processing conditions required to obtain or maintain δ-function-like doping distributions. Generally, low-growth temperatures < 550 °C are desirable. For rapidly diffusing problematic impurities such as Sb in Si, even lower temperatures (< 400 °C) may be necessary. High-temperature processing leads to impurity profile broadening. However, significant differences exist between different doping elements. Carbon in GaAs and B in Si are known to be very stable impurities which can withstand annealing at 800 °C without strong redistribution. Other impurities, such as Sn in GaAs, diffuse rapidly upon annealing.

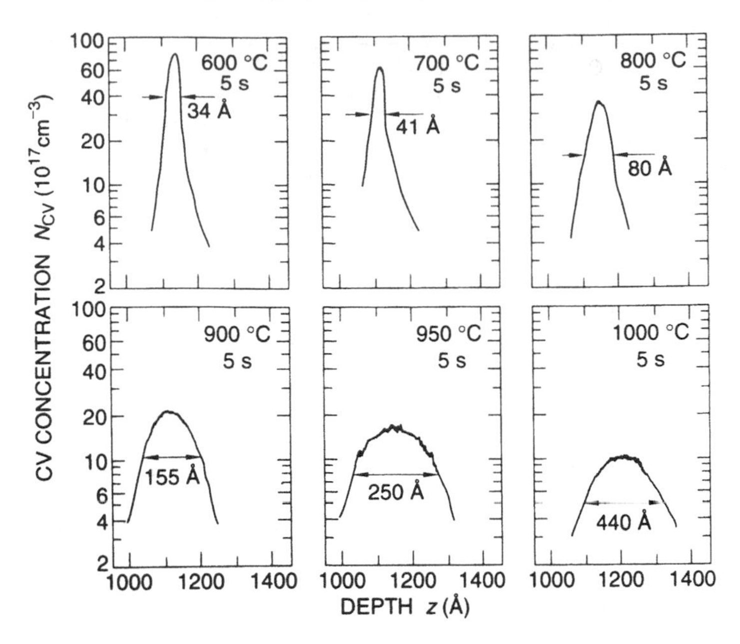

Fig. 9.7. Capacitance–voltage profiles of δ-doped p-type GaAs:Be subjected to post-growth annealing at temperatures ranging from 600 to 1000 °C for 5 s.

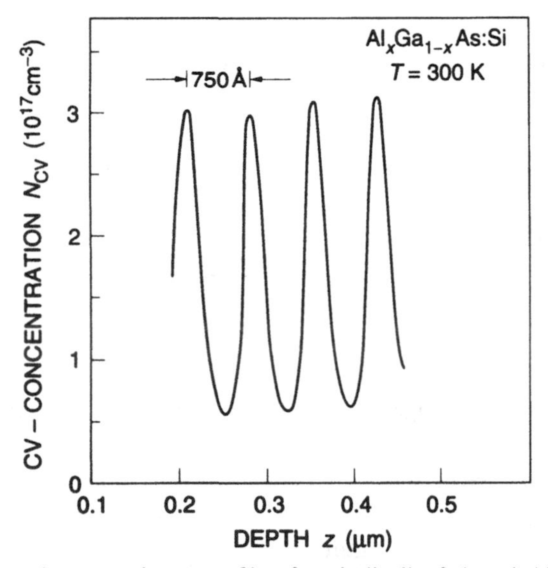

Fig. 9.8. Capacitance–voltage profile of periodically δ-doped $Al_xGa_{1-x}As$:Si measured at room temperature.

The C–V profiling technique can be used to characterize semiconductors with multiple δ-doped sheets. An example of a Si-doped $Al_xGa_{1-x}As$ sample containing several δ-doped sheets separated by 750 Å is shown in Fig. 9.8. The C–V profile provides several pieces of useful information. First, the profile yields the distance between the individual doping sheets and can thus be used as a thickness or growth rate calibration for epitaxial growth. Second, the free carrier density of the individual sheets can be obtained by integration of the profile.

Furthermore, C–V profiles on multiple δ-doped layers can be used to characterize dopant incorporation under different growth conditions. Consider an epitaxial crystal grown under conditions that differ for each δ-doped layer deposited. For example, the growth temperature is changed for each of the doped layers. Any change in the electrical activity can easily be detected by the C–V profile. The profile shown in Fig. 9.8 displays four doping spikes with the same peak concentration, as well as with the same full-width at half-maximum, which indicates that the electrical activity of the Si dopants is the same for all four doping spikes.

The electrical activity of dopants in δ-doped semiconductors is a topic of prime importance. During the δ-doping process, impurities are evaporated on the non-growing crystal surface. For unity electrical activity, the free carrier density is expected to be proportional to the duration of the impurity deposition (assuming that the impurity flux is constant). The C–V technique allows one to evaluate the free carrier density and thus the electrical activity of impurities in δ-doped structures.

The free carrier density of Si δ-doped GaAs is shown in Fig. 9.9 as a function of the Si deposition time. The free electron density is determined by the C–V technique as well as by using *van der Pauw* measurements. For Si deposition times <150 s the free carrier density is directly proportional to the deposition time, as indicated by the Hall measurement, as well as the C–V, results. However, for longer Si-deposition times ($t \geqslant 200$ s), the free carrier concentration actually *decreases*. The decrease is found for

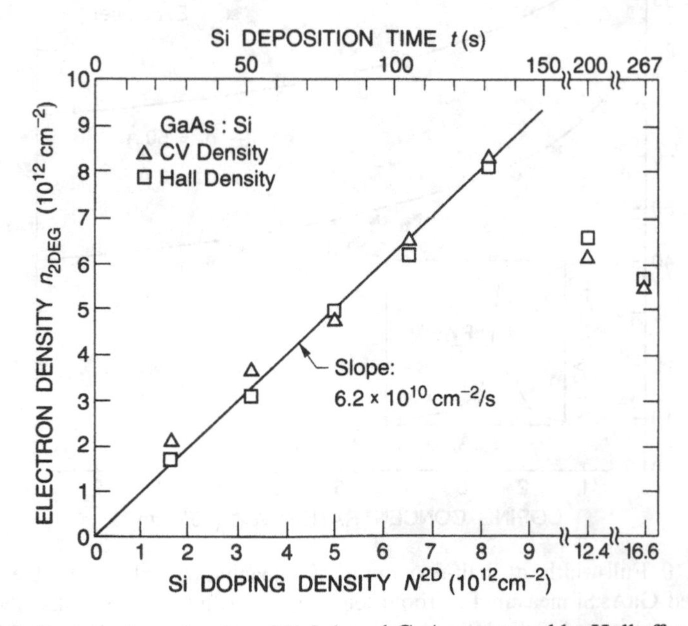

Fig. 9.9. Free electron density of Si δ-doped GaAs measured by Hall effect and capacitance–voltage profiling as a function of the Si deposition time.

both measurement techniques. This experimental result can be explained by assuming that the electrical activity of Si in GaAs decreases for densities exceeding 8×10^{12} cm^{-2}. The physical origin of the saturation was found to be either electrically inactive or compensating Si impurities (Schubert *et al.*, 1990; Beall *et al.*, 1989; Maguire *et al.*, 1987; Koenraad *et al.*, 1990). The exact density of the saturation of the electrical activity was investigated in detail by Köhler *et al.* (1993), who found that the density of saturation depends strongly on the growth temperature.

The C–V profiling technique is well suited to assessing the spatial localization of impurities in δ-doped semiconductors. The spatial resolution of the technique is of the order of several tens of angstroms and is thus just in the range of interest. In addition, the C–V technique probes *electrically active* impurities which are the species of interest. Electrically *inactive* impurities cannot be characterized by the C–V technique. The spatial localization of impurities in δ-doped semiconductors can be determined by the comparison of experimental and theoretical C–V profile widths. Resolution-limited C–V profile widths indicate a true δ-function-like distribution of impurities.

Experimental and theoretical C–V profile widths are shown in Fig. 9.10 as a function of doping density. Here the C–V profile *width* is the *full-width at half-maximum* of the profile, as illustrated in the inset of Fig. 9.10. Figure 9.10 reveals the general trend, that the profile width decreases with increasing doping concentration. This trend is consistent with the fact that the spatial extent of the ground-state wavefunction decreases with increasing doping density. Equation (9.18) predicts the proportionality $z_0 \propto (N^{2D})^{-1/3}$. The

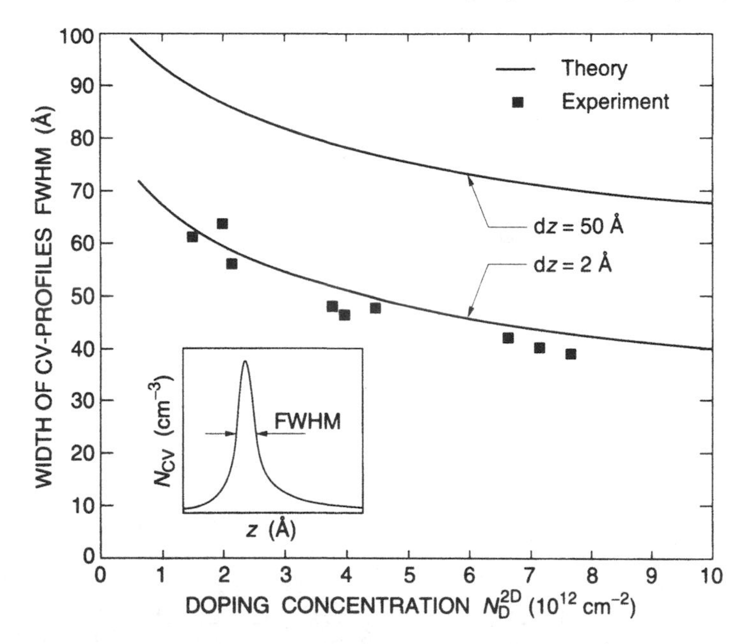

Fig. 9.10. Full-width at half-maximum of capacitance–voltage profiles on δ-doped GaAs:Si measured at room temperature (solid squares). Also shown are the calculated full-width at half-maximum for a top-hat-shaped donor distribution width dz = 2 Å and dz = 50 Å.

experimental points shown in Fig. 9.10 are from several Si δ-doped GaAs samples with different doping densities. The two theoretical curves are calculated C–V profile widths. The calculation was performed for two different doping distributions, namely a 2-Å-wide and a 50-Å-wide top-hat-shaped distribution. The theoretical curves are based on a set of fully self-consistent solutions of Schrödinger's and Poisson's equation.

The comparison of theoretical and experimental data shown in Fig. 9.10 shows that the experimental points are well described by theory if a strong spatial localization of impurities is assumed. The experimental points are well described by the theoretical curve representing the 2-Å-wide impurity distribution. The experimental points cannot be explained by the theoretical curve representing a 50-Å-wide impurity distribution. The comparison illustrates that the dopants are confined to a mathematical plane in δ-doped semiconductors.

Capacitance–voltage profiling has also been employed to characterize δ-doped *deep* centers in semiconductors (Piprek *et al.*, 1992a, 1992b; Piprek and Schenk, 1993). In these studies, the authors used Ti to introduce deep levels in a GaAs pn junction. Tunneling-assisted recombination via the Ti levels in the pn junction region was studied using a generalized Shockley–Read–Hall model. In addition, a characteristic signature of the deep centers created by Ti δ-doping of metal/n-GaSs Schottky structures was found in C–V measurements. The characteristic $C(V)$ curves were simulated using a numerical 1D model.

References

Beall R. B., Clegg J. B., Castagné J., Harris J. J., Murray R., and Newman R. C., *Semicond. Sci. Technol.* **4**, 1171 (1989).

Gossmann H. J. and Schubert E. F., *J. Mater. Res.* **18**, 1 (1993).

Kennedy D. P. and O'Brien, *IBM J. Res. Develop.* **13**, 212 (1969).

Kennedy D. P., Murley P. C., and Kleinfelder W., *IBM J. Res. Develop.* **12**, 399 (1968).

Köhler K., Granser, P., and Maier M., *J. Cryst. Growth* **127**, 720 (1993).

Koenraad P. M., de Lange W., Blom F. A. P., Leys M. R., Perenboom J. A. A. J., Singleton J., van der Vleuten W. C., and Wolter J., *Materials Science Forum* **65**, **66**, 461 (1990).

Kroemer H., Chien W. Y., Harris Jr J. S., and Edwall D. D., *Appl. Phys. Lett.* **36**, 295 (1980).

Maguire J., Murray R., Newman R. C., Beall R. B., and Harris J. J., *Appl. Phys. Lett.* **50**, 516 (1987).

Mattey N. L., Hopkinson M., Houghton R. F., Dowsett M. G., McPhail D. S., Whall T. E., Parker H. C., Booker G. R., and Whitehurst J., *Thin Solid Films*, **184**, 15 (1990).

Piprek J. and Schenk A., *J. Appl. Phys.* **73**, 456 (1993).

Piprek J., Kostial H., Krispin P., Lange C. H., and Böer K. W., *PIE Proc.* **1679**, 232 (1992a).

Piprek J., Krispin P., Kostial H., Lange C. H., and Boer K. W., *Phys. Stat. Sol.* **B173**, 661 (1992b).

Schubert E. F., Harris T. D., Cunningham J. E., and Jan W., *Phys. Rev.* B **39**, 11011 (1989).

Schubert E. F., Kopf R. F., Kuo J. M., Luftman H. S., and Garbinski P., *Appl. Phys. Lett.* **57**, 497 (1990).

Schubert E. F., Stark J. B., Chiu T. H., and Tell B., *Appl. Phys. Lett.* **53**, 293 (1988).

Schubert E. F., Stark J. B., Ullrich B., and Cunningham J. E., *Appl. Phys. Lett.* **52**, 1508 (1988).

Schubert E. F., Tu C. W., Kopf R. F., Kuo J. M., and Lunardi L. M., *Appl. Phys. Lett.* **54**, 2592 (1989).

van Gorkum, A. A., Nakagawa K., and Shiraki Y., *J. Appl. Phys.* **65**, 2485 (1989).

10

Redistribution of impurities in III–V semiconductors

E. F. SCHUBERT

During the δ-doping process, dopants are deposited on a non-growing semiconductor surface. Doping atoms on the surface are chemically bonded with at least one bond to atoms of the host semiconductor. Despite the chemical bonding, the mobility of doping atoms on the surface is considerable. At typical growth temperatures, surface atoms complete approximately 10^7 surface-diffusion hops per second (see, for example, Schubert, 1993). Assuming a diffusion hop distance of 3 Å (nearest neighbor distance), a diffusion length of $L_D = 3\ \text{Å}\sqrt{10^7} \approx 1\ \mu\text{m}$ can be estimated for a diffusion time of 1 s. Once growth is resumed and the doping atoms are covered by host atoms, the mobility of the atoms decreases drastically. Nevertheless, redistribution of the doping atoms of the δ-doped layer can certainly occur during subsequent overgrowth. The redistribution, which, in the simplest case is pure diffusion, will be especially pronounced at high growth temperatures and for long growth times. In addition to pure diffusion, other redistribution effects may play a role. (1) Impurity–surface interactions: after the termination of the dopant deposition and shortly after regular crystal growth has been resumed, the crystal surface is in close proximity to the doped layer. Interactions between doping atoms and the semiconductor surface can occur for such a spatial proximity. The interaction between the doped layer and the surface can be mediated by, for example, potential energy gradients in the near-surface region or by electric fields created by charged surface states. (2) Impurity–impurity interactions: interactions between doping atoms may also affect the redistribution of doping atoms. Closely spaced dopant atoms have a repulsive interaction due to their like coulombic charge. (3) Impurity–defect interactions: finally, the defect chemistry of the doped semiconductor crystal plays a pivotal role in the redistribution of impurities. Typical defects include vacancies, antisite defects, and interstitials. The concentration of such defects is known to depend on the doping concentration and to affect the redistribution of impurities considerably.

Redistribution of doping atoms can also occur during post-growth annealing or other thermal processing steps. In these cases, the doped layer is buried and interactions between the surface and doping atoms are reduced or even negligible. The redistribution can then be considered symmetrical with respect to the substrate and the surface direction. Diffusion, that is, the random motion of doping atoms within the semiconductor host, is the dominant redistribution process during post-growth annealing, even though other effects, such as coulombic interactions and strain interactions, may play a role.

In this chapter, we will first discuss conventional diffusion in δ-doped semiconductors. Both theoretical modeling and experimental results will be presented. Subsequently, surface related effects such as surface segregation and surface migration will be discussed. Finally, the magnitude of redistribution effects depends sensitivitly on the doping concentrations. The redistribution in the high-concentration regime will also be discussed.

Diffusion

The simplest impurity redistribution mechanism is atomic diffusion, that is, the thermally stimulated random movement of dopant atoms in the semiconductor lattice. This process is characterized by the diffusion constant D, which may depend on the doping concentration. In the present section we assume that D does not depend on doping concentration.

Consider a δ-function-like doping profile $N(z)$ with a total dopant density per unit area N^{2D}. The dopant profile is then given by

$$N(z) = N^{2D}\delta(z). \tag{10.1}$$

Consider, further, that this doping profile is subjected to an annealing cycle of duration t at a temperature T and that the impurity diffusion constant is D at this temperature. The initially δ-function-like impurity distribution broadens during the annealing cycle into a gaussian distribution given by

$$N(z) = \frac{N^{2D}}{\sqrt{(2\pi)}\sqrt{(2Dt)}} \exp\left(-\frac{z^2}{4Dt}\right). \tag{10.2}$$

The square root of the product Dt is called the diffusion length $L_D = \sqrt{(Dt)}$. The gaussian distribution given in Eqn. (10.2) is shown in Fig. 10.1 for different values of L_D. Note that the standard deviation of the gaussian distribution, σ, is related to the diffusion length by $\sigma = \sqrt{(2)}L_D = \sqrt{(2Dt)}$. Finally, the full-width at half-maximum (FWHM) of the gaussian distribution is related to the diffusion length by

$$FWHM = 2\sqrt{[2(\ln 2)\sigma]} = 4\sqrt{(\ln 2)}\sqrt{(Dt)} \approx 2.355\sigma.$$

Diffusive redistribution of dopants does occur during epitaxial growth as well as during post-growth annealing. The broadening of Be doping spikes in GaAs grown by molecular-beam epitaxy at temperatures of 500, 580, and 660 °C is shown in Fig. 10.2 (Schubert *et al.*, 1990a). Secondary ion mass spectrometry (SIMS) has been employed to measure the Be distributions shown in Fig. 10.2. At the low growth temperature of 500 °C, the SIMS

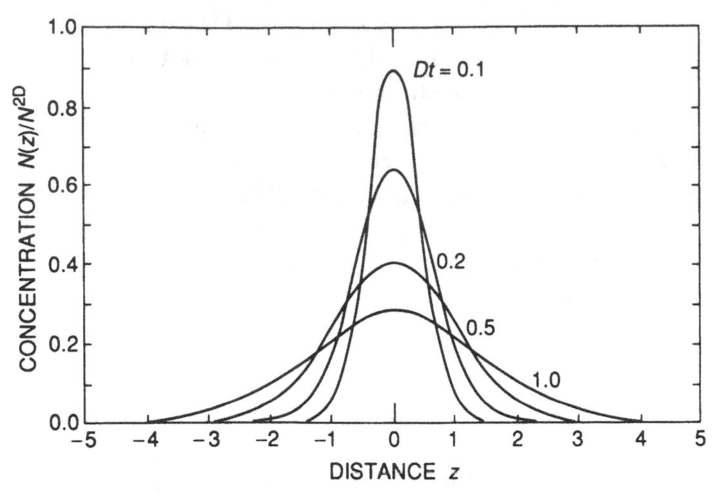

Fig. 10.1. Gaussian impurity profile for different values of Dt.

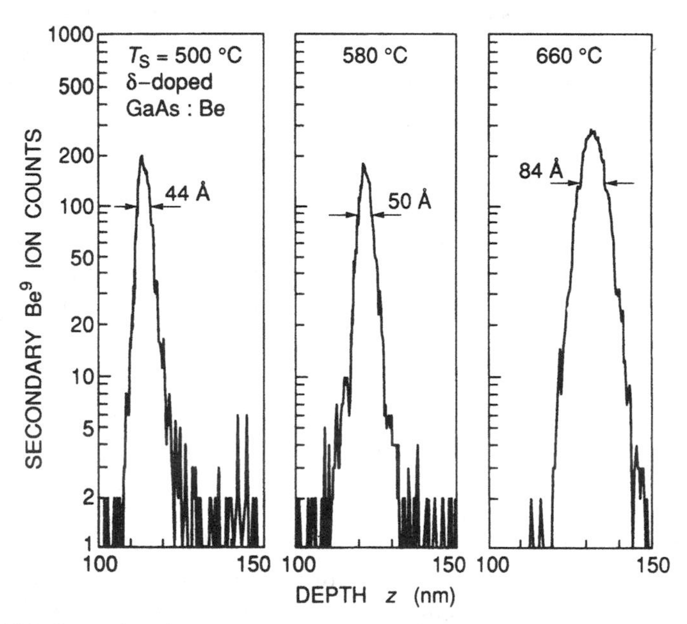

Fig. 10.2. Secondary ion mass spectrometry profiles of Be δ-doped GaAs grown at 500, 580, and 660 °C.

profile has a full-width at half-maximum (FWHM) of 44 Å. Noting the depth of the Be layer below the GaAs surface (1150 Å), the FWHM of 44 Å is resolution limited. However, as the growth temperature is increased to 580 °C and 660 °C, the SIMS profile broadens to 50 Å and 84 Å respectively. Knowing the resultion-limited FWHM, a diffusion length can be deduced from the broadening of the Be profiles grown at higher temperatures.

Assuming that the resolution function is gaussian with a standard deviation of σ_r, the diffusion length $L_D = \sqrt{(Dt)} = \sigma/\sqrt{2}$ can be determined using the relation

$$L_D = \sqrt{(\sigma_{ex}^2 - \sigma_r^2)}/\sqrt{2} \tag{10.3}$$

where σ_{ex} is the measured standard deviation of the experimental profiles shown in Fig. 10.2. Equation (10.3) is valid if the resolution broadening and the diffusion broadening are both gaussian and if the two processes can be considered to be independent. Using the profile width shown in Fig. 10.2 for $T = 580\,°C$ and using $\sigma_r = 44$ Å/2.355 = 19 Å, one obtains $\sigma_{ex} = 50$ Å/2.355 = 21 Å, and a diffusion length of $L_D = 6$ Å. At the growth temperature of $T = 660$ °C, one obtains a diffusion length of $L_D = 21$ Å. Diffusion over several tens of angstroms can thus be detected at elevated growth temperatures.

The redistribution of dopants during epitaxial growth may be complicated by the vicinity of a semiconductor surface. The formation energy of impurity incorporation is different on the surface than in the bulk, which results in a preferential movement of impurities either towards the surface or away from it. The influence of surface effects can be avoided by studying the diffusion of impurities at large distances from the surface during post-growth annealing. Consider a sample that has been grown at sufficiently low substrate temperatures that impurity redistribution can be neglected during epitaxial growth. If the δ-doped layer is far from the surface, then the surface will not influence the redistribution process. As a consequence, symmetric redistribution of dopants is expected in this case. If the redistribution is analyzed in terms of diffusion, which is by its very nature an isotropic redistribution process, then the diffusion constant can be deduced from the changes in profile widths incurred during post-growth annealing.

The redistribution of Si in $Al_xGa_{1-x}As$ for different post-growth annealing temperatures is shown in Fig. 10.3. The epitaxial growth of the sample was carried out at 500 °C. Subsequently, the sample was cleaved into several pieces. The individual pieces were annealed for 5 s at temperatures ranging from 600 °C to 1000 °C. Figure 10.3 displays a strong broadening of the impurity profiles with increasing annealing temperature. The full-width at half-maximum of the sample annealed at 1000 °C for 5 s is 410 Å, indicating significant diffusion. The activation energy E_a and the D_0 deduced from the data shown in Fig. 10.3 are $E_a = 1.3$ eV and $D_0 = 4 \times 10^{-8}$ cm^2/s (Schubert *et al.*, 1989).

The diffusion coefficients of Si, Be, and C in GaAs are shown in Fig. 10.4 (Schubert, 1990). The diffusion constants were deduced from the broadening of impurity profiles during post-growth annealing experiments using Eqns. (10.1)–(10.3). Carbon clearly has the lowest diffusion constant of the three elements. The other acceptor, beryllium, has a 1–2 orders of magnitude higher diffusion constant than does carbon.

The deduction of the diffusion constant from post-growth annealing of δ-doped samples probably provides the most reliable diffusion data obtainable today. Other methods, such as the diffusion of impurities from a thin film on the surface into the bulk of the semiconductor, pose several problems. First, the impurity concentration at the semiconductor surface is very high which can result, as will be seen later, in a completely different diffusive behavior. Second, the proximity of the surface further complicates the diffusion process. Native defects, such as As vacancies generated by As out-diffusion, can strongly

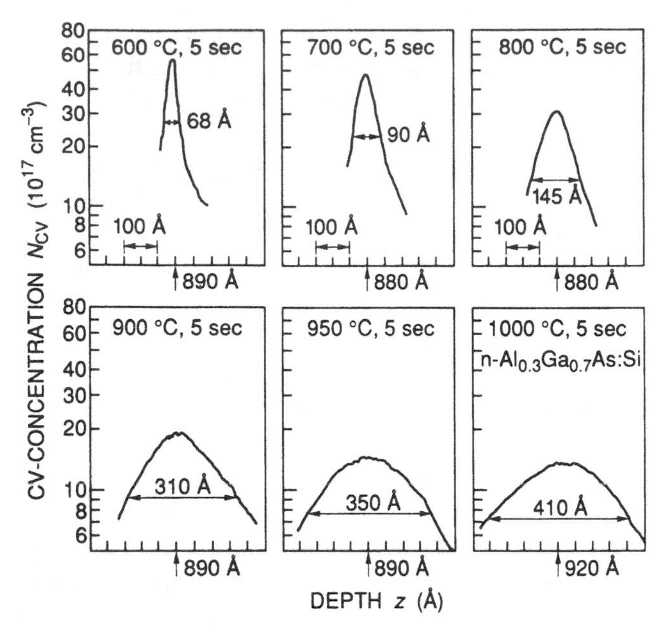

Fig. 10.3. Capacitance–voltage profiles of Si δ-doped $Al_{0.3}Ga_{0.7}As$ annealed at temperatures of 600–1000 °C for 5 s.

influence the diffusive characteristics of impurities. These problems are obviously avoided for δ-function-like impurity layers buried in the bulk of a semiconductor.

Nevertheless, caution is required when comparing and interpreting diffusion data of impurities in III–V semiconductors. The diffusion characteristics of impurities depend sensitivity on the *quality* of the host semiconductor material, that is, its stoichiometry and native defect concentration. The diffusion characteristics also depend on many experimental parameters such as the V/III flux ratio during growth and the growth temperature. Furthermore, the Fermi level is known to determine the concentration of charged native defects (see, for example, Schubert, 1993), and thereby the diffusion properties of impurities. Due to the strong sensitivity of the diffusion characteristics of impurities on many parameters, such as growth temperature and the quality of the host semiconductor, it is conceivable that different results may be obtained for D_0 and E_a, even for the *same* doping element in the *same* compound semiconductor. As an example, we consider Si in GaAs. It is generally agreed that the growth of spatially well-confined Si doping layers in GaAs by MBE requires substrate temperatures below 600 °C (Schubert *et al.*, 1988; Beall *et al.*, 1989; Lanzillotto *et al.*, 1989; Webb, 1989). These groups concluded that significant redistribution occurs at growth temperatures exceeding 600 °C. Quite to the contrary, Yang *et al.* (1992) studied Si δ-doped GaAs grown by oranometallic vapor-phase epitaxy (OMVPE) and demonstrated that very narrow doping profiles (32 Å) can be obtained at growth temperatures of 650 °C–700 °C. Even though this result clearly contradicts the results reported previously, there is no reason to believe that the results

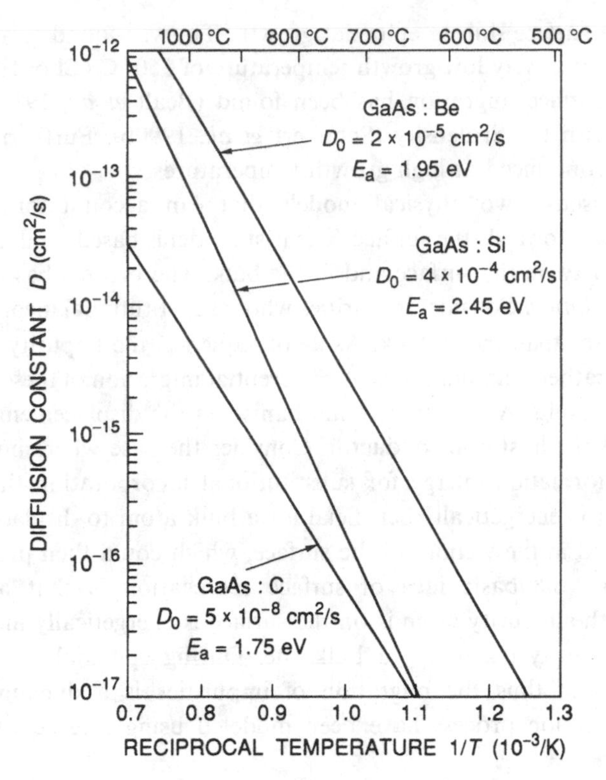

Fig. 10.4. Diffusion constants of Be, Si, and C in GaAs versus reciprocal temperature.

reported by Yang *et al.* are incorrect. It is, instead, very likely that the nature of the OMVPE growth process is sufficiently different from MBE growth that changes in the diffusion properties can result. It is, for example, well accepted that the average quality of OMVPE bulk material exceeds the quality of typical MBE-grown material. The point-defect concentration, which strongly influences the efficiency of radiative processes, is often lower in OMVPE material than in MBE-grown III–V semiconductors. Since the diffusion behavior of impurities depends sensitively on the point-defect concentration, it is conceivable that δ-doped GaAs can be grown at higher temperatures using OMVPE as compared to MBE. Accordingly, different diffusion constants will be obtained for materials grown by the two growth techniques.

d
Surface migration

While *diffusion* is an isotropic redistribution process, *migration* indicates the redistribution along a preferential direction. *Surface migration* thus denotes the preferential motion of impurities towards a crystal surface. Several impurities have been clearly identified to migrate towards the surface during epitaxial growth. Tin was found to migrate towards the surface even at a relatively low growth temperature of 550 °C (Cho, 1975). In δ-doped semiconductors, Si surface migration has been found (Beall *et al.*, 1988) and has been shown to increase with the Si density (Schubert *et al.*, 1990b). Furthermore, Si surface migration is more pronounced at high growth temperatures.

Below, we will discuss two physical models that can account for the preferential migration of impurities towards the surface. The first model is based on different formation energies of the impurity on the surface and in the bulk. The process based on this model is called *surface segregation.* Consider impurities whose substitutional incorporation energy is larger on the surface than in the bulk. As a consequence, the impurity atoms prefer to occupy surface sites rather than bulk sites. A preferential migration of these impurity atoms towards the surface results. An alternative mechanism is the displacement of an impurity atom by an atom of the host semiconductor. Consider the case where host atoms in the bulk have a larger formation energy for substitutional incorporation than do impurity atoms. As a result, it is energetically beneficial for a bulk atom to displace (kick out) the impurity atom located in the vicinity of the surface, which could then proceed to occupy a surface site. Thus, the basic idea of surface segregation is that an atomic configuration in which the impurity atom is on the surface is energetically more favorable as compared to the impurity occupying a bulk site. During epitaxial growth, the crystal surface propagates and thus the migration of impurities is a dynamic process. The kinetics of the segregation process have been modeled using rate equations by Harris *et al.* (1984).

The second model is based on the interaction of charged impurities with the surface electric field in semiconductors caused by the pinning of the Fermi level at the semiconductor surface (Schubert *et al.*, 1990c). In the most common III–V semiconductors such as GaAs, $Al_xGa_{1-x}As$, or InP, the Fermi level is pinned at an energy close to the middle of the forbidden gap. Consider a semiconductor with an n-type doping spike close to a pinned semiconductor surface. The band diagram of such a semiconductor is shown in Fig. 10.5(a). The doped layer is either fully or partially depleted of free carriers which have transferred to surface states. As a result of this transfer, an electric field is created between the surface and the donor layer, as shown in Fig. 10.5(a). If sufficiently mobile, donors drift under the influence of the electric field towards the surface. The gain in energy experienced by a donor impurity during the drift is of the order of one electron volt (i.e. the barrier height $e\phi_B$), which represents a considerable driving force for impurities to drift towards the surface.

Next we will analyze the surface drift of impurities quantitatively and present experimental data supporting the model. To calculate the impurity drift, we assume that the semiconductor surface is in the xy-plane and that the growing crystal surface is moving along the z-axis with a velocity v_s. For a donor sheet located at z_d with a two-dimensional donor density N_D^{2D}, the electric field of the dipole is given by

$$\mathscr{E} = \begin{cases} eN_D^{2D}/\varepsilon & (10.4a) \\ \phi_B/(z_s - z_d) & (10.4b) \end{cases}$$

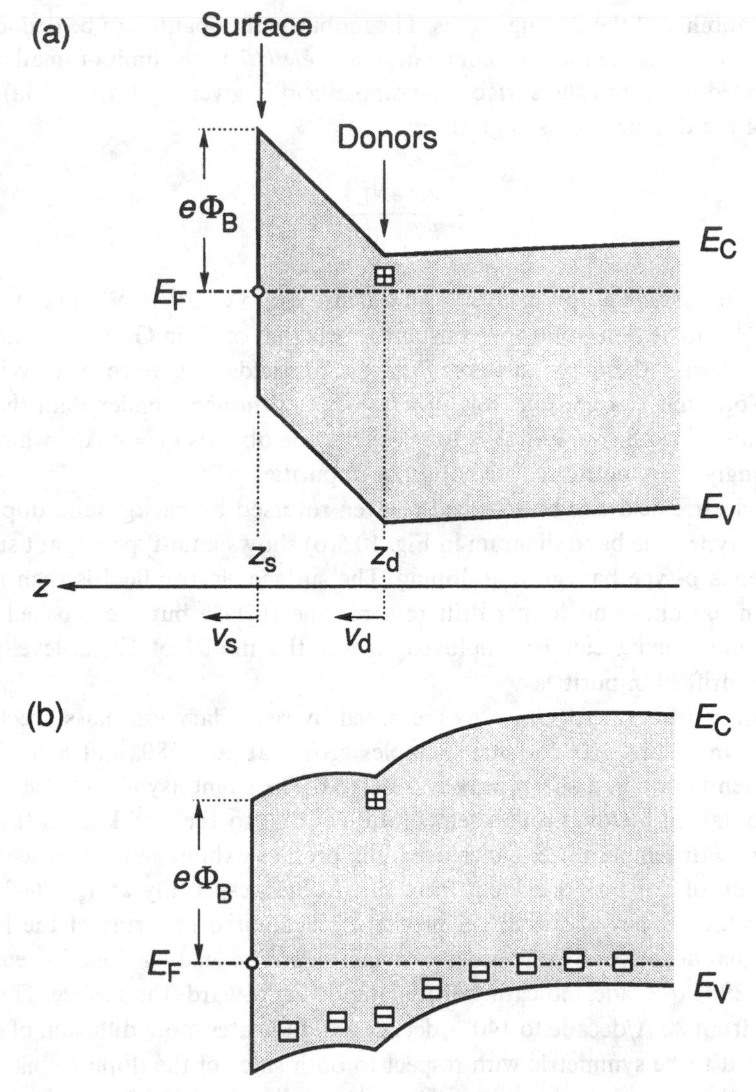

Fig. 10.5. (a) Band diagram of a δ-doped n-type semiconductor whose Fermi level is pinned at the surface at an energy $e\Phi_B$ below the conduction band edge. (b) Band diagram of the semiconductor shown in (a) with additional p-type background doping.

where z_s is the position of the surface and ϕ_B is the position of the Fermi level at the surface with respect to the conduction band edge. Equation (10.4a) applies if the doped layer is entirely depleted of free carriers, that is, if $z_s - z_d < \phi_B \varepsilon / e N_D^{2D}$, whereas Eqn. (10.4b) is valid if the doped layer is only partially depleted, that is, if $z_s - z_d > \phi_B \varepsilon / e N_D^{2D}$. The drift velocity of impurities towards the surface is then given by

$$v_d = \frac{dz_d}{dt} = \mu \mathscr{E} \tag{10.5}$$

where μ is the mobility of the doping atoms. The mobility of dopants can be deduced from their diffusion constant using the Einstein relation $\mu = eD/kT$. In the limit of small distances between the doped layer and the surface, the dipole field is given by Eqn. (10.4a) and the drift velocity of the doping atoms is given by

$$v_d = \frac{eD}{kT} \frac{eN_D^{2D}}{\varepsilon}. \tag{10.6}$$

As an example we assume a donor layer with a density of $5 \times 10^{12}\ \text{cm}^{-2}$ and a diffusion constant of $10^{-15}\ \text{cm}^2/\text{s}$. This diffusion constant equals that of Be in GaAs at about 680 °C (see Fig. 10.4). Using these parameters, Eqn. (10.6) yields a surface drift velocity of $v_d = 0.8$ Å/s. Note that is a considerable drift velocity, although smaller than the typical MBE growth rate of ≈2.8 Å/s. For $D = 10^{-14}\ \text{cm}^2/\text{s}$ one obtains $v_d = 8$ Å/s, which would result in a strongly asymmetric redistribution of impurities.

The surface electric field can be screened or even reversed by background doping with another dopant type. The band diagram in Fig. 10.5(b) shows an n-type dopant spike, but also homogeneous p-type background doping. The surface electric field is then reversed. As a consequence, donors no longer drift towards the surface but are repelled from it. Thus, background doping can be employed to test the model of Fermi-level-pinning-induced surface drift of impurities.

Silicon-dopant profiles in $Al_xGa_{1-x}As$ measured by secondary ion mass spectrometry (SIMS) are shown in Fig. 10.6 for three samples grown at 500, 580, and 660 °C. At the lowest growth temperature, a sharp peak is observed. The slight asymmetric shape of the SIMS profile obtained at low growth temperature is due to the well-known 'knock-on' effect. As the growth temperature is increased, the profiles exhibit significant asymmetry. Surface migration of dopants is evident from the profiles, especially at $T_s = 660$ °C. The leading and trailing slopes of the SIMS profile are evaluated in terms of the length at which the Si signal decreases by one order of magnitude. The leading slope increases from 35 Å/decade to 390 Å/decade, indicating drift of impurities towards the surface. The trailing slope increases from 80 Å/decade to 140 Å/decade and indicates more diffusion of dopants, which is expected to be symmetric with respect to both sides of the doping spike.

Qualitatively the same, but quantitatively weaker redistribution of impurities towards the surface has been observed for Si in GaAs (Schubert *et al.*, 1990c). The smaller redistribution of impurities in GaAs is consistent with a smaller diffusion constant found in GaAs as compared with $Al_xGa_{1-x}As$.

Fermi-level pinning at the semiconductor surface causes the doped layer to be depleted of free carriers and localization of electrons in surface states. If the resulting dipole field were the driving force for surface migration, then this process could be reduced by screening the dipole field. Figure 10.7 shows two SIMS profiles on samples which have a p-type Be background doping. The concentration of Be is chosen to be $N_A = 4 \times 10^{18}\ \text{cm}^{-3}$ in order to compensate for the Si dopants within 50 Å. Figure 10.7 reveals that the segregation length is drastically reduced from 150 Å/decade to 80 Å/decade at a growth temperature of 580 °C. The decrease of the segregation length is attributed to the screening of the surface

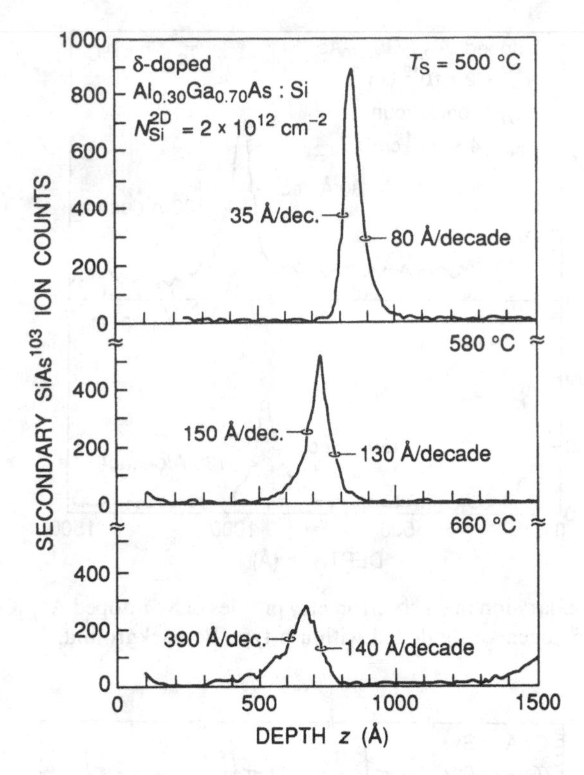

Fig. 10.6. Secondary ion mass spectrometry profiles of Si δ-doped $Al_{0.3}Ga_{0.7}As$ grown by MBE at 500, 580, and 660 °C.

dipole. The trailing slope of the SIMS profile is changed insignificantly, indicating that diffusion and the 'knock-on' effect are not influenced by background doping. The SIMS profile for the sample grown at $T = 660$ °C shows the same qualitative trend as the sample grown at the intermediate temperature (see Figs. 10.6 and 10.7). The surface-segregation length is reduced and the SIMS profile has a more symmetric shape, indicating that diffusion dominates the dopant redistribution. The experimental results show that the electronic property of Fermi-level pinning causes significant redistribution of dopants during growth and that this interaction can be screened by appropriate background doping.

A net shift of the entire Si peak towards the surface is found at higher growth temperatures, as shown in Figs. 10.6 and 10.7. While the net shift is expected without background doping (Fig. 10.6), it is unexpected with background doping (Fig. 10.7). We therefore attribute the shallower depth of the Si peak at high growth temperatures to a thinner $Al_xGa_{1-x}As$ top layer. At the high growth temperature of 660 °C, reevaporation of $Al_xGa_{1-x}As$ is known to occur, which results in a smaller thickness of the top $Al_xGa_{1-x}As$ cap layer.

The surface migration of impurities has also been observed for Be in GaAs (Schubert *et al.* 1990c). While the surface migration is not detectable at a growth temperature of

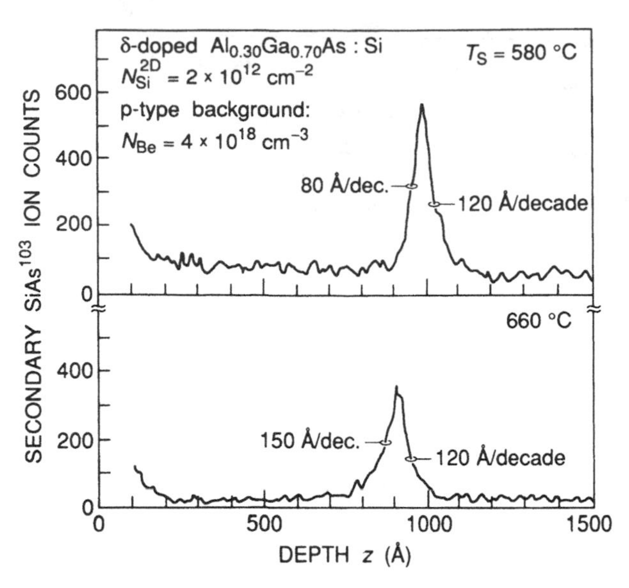

Fig. 10.7. Secondary ion mass spectrometry profiles of Si δ-doped $Al_{0.3}Ga_{0.7}As$ homogeneously doped with a p-type Be background.

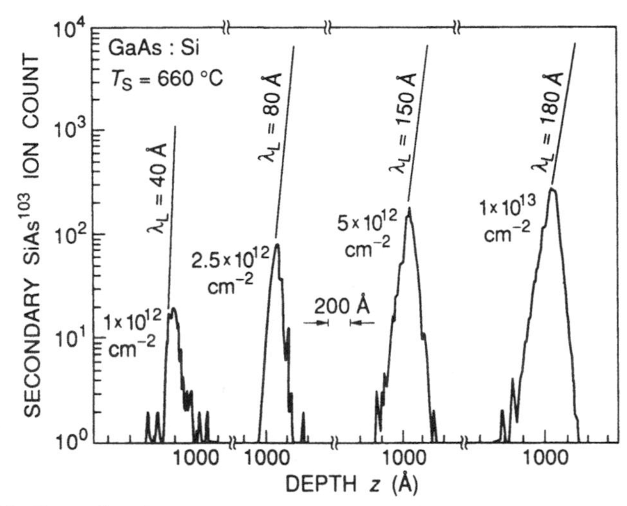

Fig. 10.8. Secondary ion mass spectrometry profiles of Si δ-doped GaAs doped at doping densities 1×10^{12}, 2.5×10^{12}, 5×10^{12}, and 1×10^{13} cm^{-2}.

500 °C, it is pronounced at growth temperatures exceeding 650 °C. The surface migration of Be can be reduced by appropriate background doping, similar to the case of Si in GaAs.

The surface migration of Si in GaAs increases with the Si density. The SIMS profiles of four Si δ-doped GaAs samples with different Si densities are shown in Fig. 10.8 (Schubert *et al.*, 1990b). The Si densities range between 1×10^{12} and 1×10^{13} cm^{-2}. The growth temperature of 660 °C is unusually high for the MBE growth of GaAs. However, segregation effects are more prominent at such temperatures. The SIMS profiles shown in Fig. 10.8 are skewed towards the surface, indicating surface migration. The migration is especially pronounced at the highest Si densities. We evaluate the segregation quantitatively by defining a migration length of the leading slope of the SIMS profiles, λ_L, in which the secondary-ion signal decreases by one order of magnitude. The trailing slope of the SIMS profiles is $\lambda_T = 90$ Å per decade of the secondary-ion count. The trailing slope remains approximately constant for the four samples. Note that λ_T is partly determined by the 'knock-on' effect. Both λ_L and λ_T are broadened due to the SIMS-induced 'cascade mixing' effect and due to conventional diffusion (Wilson *et al.*, 1989).

The surface migration length evaluated in Fig. 10.8 increases substantially with the Si dose. The migration length increases from $\lambda_L = 40$ Å to 180 Å for Si densities of 1×10^{12} cm^{-2} and 1×10^{13} cm^{-2} respectively. The increase is in agreement with the model of Fermi-level-pinning-induced surface migration, since the surface electric field is proportional to the doping density (see Eqn. (10.6)).

High concentration regime

Impurity redistribution processes at low and medium densities ($<10^{13}$ cm^{-2}) are well described by diffusion and surface migration, which have been discussed in the preceding sections. Redistribution in the high concentration regime, that is, for two-dimensional doping densities exceeding 10^{13} cm^{-2}, can be fundamentally different in terms of magnitude and nature. As a general rule, redistribution processes become more pronounced as the doping density increases.

Several models of impurity redistribution processes predict a diffusion constant which increases with concentration (for an overview of different diffusion models, see Schubert, 1993). For example, the well-known substitutional-interstitial diffusion model predicts that the diffusion constant increases with the impurity concentration according to a power law. Another example of redistribution processes increasing with concentration is the Fermi-level-pinning-induced drift model discussed in the preceding section.

However, the magnitude of redistribution processes of impurities may go much beyond the gradual increase mentioned above. The drastic redistribution of Be in GaAs occurring at high Be densities is illustrated in Fig. 10.9, which shows the SIMS profiles of six δ-doped GaAs samples with Be densities ranging from 2×10^{12} cm^{-2} to 4×10^{14} cm^{-2}. All samples were grown by MBE at a growth temperature of 500 °C using a growth rate of 1.0 μm/hr. After the growth of a buffer layer, the epitaxial growth was interrupted for the deposition of the Be. A 1000-Å-thick GaAs layer was grown to cap the δ-doped layer.

At the three lowest Be densities shown in Fig. 10.9 ($N_{Be}^{2D} = 2 \times 10^{12}$, 1×10^{13}, and 2×10^{13} cm^{-2}), the SIMS profiles indicate a well-confined Be distribution, as indicated by the narrow width. However, a drastic Be redistribution occurs for Be densities in the 10^{14} cm^{-2} range. The profile width increases and becomes top-hat shaped. The Be profile with density 4×10^{14} cm^{-2} has a width of approximately 1000 Å. The 3D concentration of the profiles reaches a maximum of 10^{20} cm^{-3}, which is not exceeded in any of the

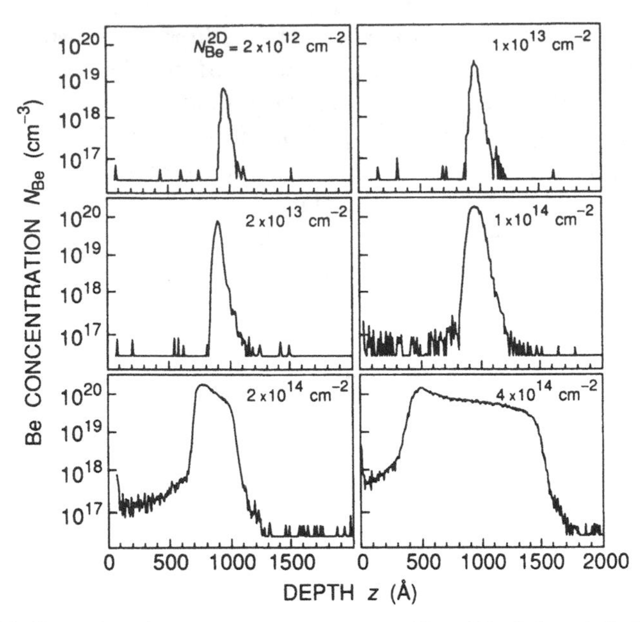

Fig. 10.9. Secondary ion mass spectrometry profiles of Be δ-doped GaAs doped at different densities.

profiles. The width of the Be profiles increases whereas the peak concentration remains constant. The product of profile width and peak concentration yields the 2D Be density.

The maximum 3D Be concentration depends on the growth temperature. The Be profiles in GaAs for the two growth temperatures 400 °C and 500 °C are shown in Fig. 10.10. Both samples have a 2D Be density of 4×10^{14} cm^{-2}. The sample grown at 400 °C displays a narrower width and a higher peak concentration than does the sample grown at 500 °C. The peak concentration of the 400 °C growth is a factor of two higher than the 500 °C growth. Furthermore, the profile width is just half of the 500 °C growth. The product of peak concentration and width is the same for the two samples shown in Fig. 10.10. These results indicate that the redistribution is reduced at low growth temperature, but is similar in nature at both temperatures.

A model explaining the upper limit of the Be concentration in GaAs has been published by Schubert *et al.* (1992). The authors showed that a repulsive interaction of charged impurities due to (i) the coulombic charge of the impurities and (ii) the increase in electronic energy at high doping levels can limit the maximum achievable doping concentration. We will first discuss the coulombic interaction between impurities and then the increase in electronic energy.

Consider a semiconductor whose Fermi level is pinned at the surface due to surface states (Bardeen states). The pinning of the Fermi level results in free-carrier depletion of the near-surface region. The force exerted on charged doping atoms due to coulombic interactions is directed towards the semiconductor surface and results in a movement of impurities towards the semiconductor surface. The electric field at the semiconductor

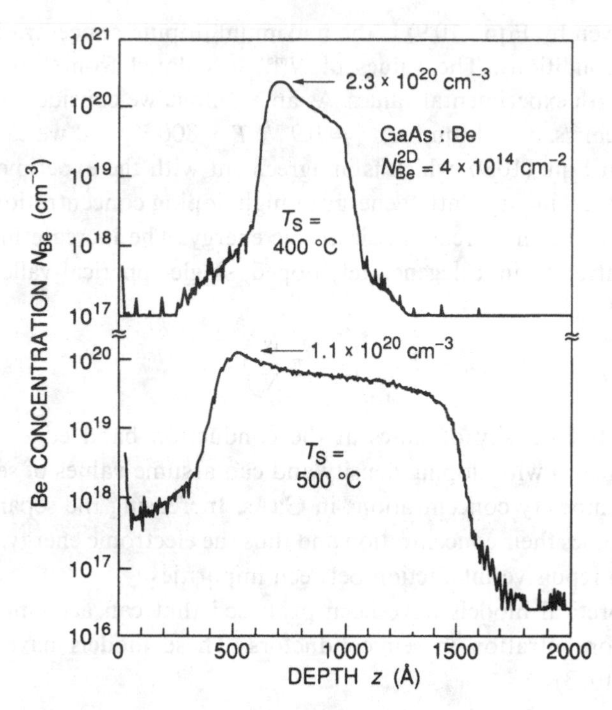

Fig. 10.10. Secondary ion mass spectrometry profile of Be δ-doped GaAs grown at 400 and 500 °C by MBE.

surface due to charged impurities in the depletion layer can be obtained from Poisson's equation and the electric field is given by

$$\mathscr{E} = \sqrt{(2eN\phi_B/\varepsilon)}, \tag{10.7}$$

where N is the impurity concentration and ϕ_B is the barrier height of the pinned surface. Charged impurities experience a driving force towards the surface. The drift velocity of impurities is given by $v_d = \mu\mathscr{E}$. Using the Einstein relation $\mu = eD/kT$, one obtains the drift velocity of charged doping atoms:

$$v_d = \frac{eD}{kT}\sqrt{(2eN\phi_B/\varepsilon)} \tag{10.8}$$

where D and ε are the impurity diffusion constant and the semiconductor permittivity respectively. Note that Eqn. (10.8) is valid for homogeneously doped semiconductors, whereas the previously derived Eqn. (10.6) applies to δ-doped semiconductors.

Thus, the coulombic repulsion between impurities results in a redistribution of impurities that tends to reduce the doping concentration. The maximum doping concentration is obtained when the epitaxial growth rate v_g equals the drift velocity of doping atoms towards the surface, that is, $v_d = v_g$. Using Eqn. (10.8), this condition yields

$$N^{\max} = \left(\frac{v_g kT}{eD}\right)^2 \frac{\varepsilon}{2e\phi_B}. \tag{10.9}$$

The concentration given by Eqn. (10.9) is the maximum doping concentration achievable under the assumed conditions. The values of $N^{\max}$ calculated from Eqn. (10.9) are in excellent agreement with experimental values. As an example, we consider Be-doped GaAs with $D \approx 3 \times 10^{-16}\ \mathrm{cm^2/s}$, $v_g = 1\ \mu\mathrm{m/hr}$, $\phi_B = 0.9$ V, $T = 800$ K, and we calculate $N_{\mathrm{Be}}^{\max} \approx 1.6 \times 10^{20}\ \mathrm{cm^{-3}}$ from Eqn. (10.9), which is in agreement with the experimental value.

An increase in the total incorporation energy at high doping concentrations results from phase-space filling, that is, an increase in electronic energy. The increase in Fermi energy with impurity concentration in a degenerately doped, single-spherical-valley n-type semiconductor is given by

$$E_F - E_C = \left(\frac{3}{4}\sqrt{\pi}\,\frac{N}{N_c}\right)^{2/3}, \qquad (10.10)$$

where N_c is the effective density of states at the conduction band edge. The electronic energy, $E_F - E_C$, increases with doping density and can assume values of several hundred meV at high n-type impurity concentrations in GaAs. Increasing the separation between impurity atoms, decreases their concentration and thus the electronic energy. The electronic energy thus acts as a repulsive interaction between impurities.

Several other theoretical models have been proposed that can account for maximum achievable doping concentration in semiconductors. These models have recently been reviewed (Schubert, 1993).

References

Beall R. B., Clegg J. B., and Harris J. J. (1988) *Semicond. Sci. Technol.* **3**, 612.

Beall R. B., Clegg J. B., Castagné J., Harris J. J., Murray R., and Newman R. C. (1989) *Semicond. Sci. Technol.* **4**, 1171.

Cho A. Y. (1975) *J. Appl. Phys.* **46**, 1733.

Harris J. J., Ashenford D. E., Foxon C. T., Dobson P. J., and Joyce B. A. (1984) *Appl. Phys.* A **33**, 87.

Lanzillotto A.-M., Santos M., and Shayegan M. (1989) *Appl. Phys. Lett.* **55**, 1445.

Schubert E. F. (1990) *J. Vac. Sci. Technol.* A **8**, 2980.

Schubert E. F. (1993) *Doping in III–V Semiconductors* (Cambridge University Press).

Schubert E. F., Gilmer G. H., Kopf R. F., and Luftman H. S. (1992) *Phys. Rev.* **46**, 15 078.

Schubert E. F., Kuo J. M., Kopf R. F., Jordan A. S., Luftman H. S., and Hopkins L. C. (1990c) *Phys. Rev.* B **42**, 1364.

Schubert E. F., Kuo J. M., Kopf R. F., Luftman H. S., Hopkins L. C., and Sauer N. J. (1990a) *J. Appl. Phys.* **67**, 1969.

Schubert E. F., Luftman H. S., Kopf R. F., Headrick R. L., and Kuo J. M. (1990b) *Appl. Phys. Lett.* **57**, 1799.

Schubert E. F., Stark J. B., Ullrich B., and Cunningham J. E. (1988) *Appl. Phys. Lett.* **54**, 2091.

Schubert E. F., Tu C. W., Kopf R. F., Kuo J. M., and Lunardi L. M. (1989) *Appl. Phys. Lett.* **54**, 2592.

Webb C. (1989) *Appl. Phys. Lett.* **54**, 2091.

Wilson R. G., Stevie F. A., and Magee C. W. *Secondary Ion Mass Spectrometry* (Wiley, New York, 1989).

Yang G. M., Park S. G., Seo K. S., and Choe B. D. (1992) *Appl. Phys. Lett.* **60**, 2380.

11

Dopant diffusion and segregation in delta-doped silicon films

H.-J. GOSSMANN

Delta (δ-)doping makes available very large concentration gradients as well as very-high-volume concentrations of substitutional dopant atoms that may exceed the equilibrium solubility by many orders of magnitude. These two fundamental properties open up a variety of new directions into the study of dopant diffusion and the interactions between dopant atoms and intrinsic Si point-defects (self-interstitials and vacancies). In this chapter we will explore some of those directions and show that δ-doping offers unique opportunities for designing experiments in the area of the interaction of dopants with native point-defects. We will also discuss some new phenomena that occur during segregation of dopant atoms at very high concentrations.

11.1 Diffusion of dopants and their interaction with point-defects

11.1.1 The standard model of dopant diffusion in Si

It is generally accepted that diffusion of substitutional dopants in Si is mediated by intrinsic Si point-defects, X, that is, diffusion of the dopant, A, occurs only when mediated by Si self-interstitials and/or vacancies (Fahey *et al.*, 1989). In the context of δ-doping we will restrict ourselves here to electrically active dopants, that is, A = B, Ga, In, P, As, or Sb. In general, point-defects exist in various charge states, i, so that for a substitutional donor atom $A^{(+1)}$

$$A^{(+1)} + X^{(i-1)} \leftrightarrow AX^{(i)} \tag{11.1a}$$

and for a substitutional acceptor atom $A^{(-1)}$

$$A^{(-1)} + X^{(i+1)} \leftrightarrow AX^{(i)}. \tag{11.1b}$$

For compactness we can rewrite Eqns. (11.1a) and (11.1b) as

$$A^{(\pm 1)} + X^{(i\mp 1)} \leftrightarrow AX^{(i)}, \tag{11.2}$$

where the top of the double sign applies to n-type material (donors) and the bottom to p-type (acceptors). Note that the standard model of dopant diffusion in Si does not make any assumptions about the precise mechanism of diffusion except that the interaction can be written in the form of the rate equation, Eqn. (11.2). In particular, the entity $AX^{(i)}$ does not necessarily imply a pair. The standard model as formulated here does exclude a dissociative mechanism, in which a substitutional dopant atom becomes interstitial, leaving a Si vacancy behind. However, extensive experimental evidence indicates that the dissociative mechanism does not play a role for those dopants in Si that we consider here (Fahey *et al.*, 1989).

Along with Eqn. (11.2), the standard model assumes that Fick's law is valid, that is, that the flux of each complex $AX^{(i)}$ is proportional to the gradient of its concentration, $\nabla C_{AX^{(i)}}$. The proportionality constant is the diffusivity, $d_{AX^{(i)}}$. The total flux of complexes is then

$$\mathbf{J} = \sum_{X,i} -d_{AX^{(i)}} \nabla C_{AX^{(i)}} + i\mu_{AX^{(i)}} C_{AX^{(i)}} \mathbf{E}. \tag{11.3}$$

The summation extends over all defects and charge states. In addition to Fickian diffusion a second term has been added in Eqn. (11.3) that represents the drift of the (charged) dopant-defect complexes in the electric field **E**; $\mu_{AX^{(i)}}$ is their mobility.

The third and final assumption of the standard model is the validity of Boltzmann statistics. The internal field is related to the Fermi-level, ε_F, and the carrier concentration, m, via

$$\mathbf{E} = -\frac{1}{q}\nabla\varepsilon_F = \mp\frac{kT}{q}\nabla\ln\frac{m}{n_i}. \tag{11.4}$$

Here q is the elementary charge and $m = n$ for donors and $m = p$ for acceptors, n and p being the electron and hole concentration respectively; k represents Boltzmann's constant, T is the absolute temperature. The intrinsic electron concentration is denoted by n_i and is shown in Fig. 11.1 for Si (Morin and Maita, 1954a, 1954b). Further, Einstein's relation holds:

$$\mu_{AX^{(i)}} = \frac{q}{kT} d_{AX^{(i)}}. \tag{11.5}$$

Inserting Eqns. (11.4) and (11.5) into Eqn. (11.3) yields

$$\mathbf{J} = -\sum_{X,i}\left[d_{AX^{(i)}} \nabla C_{AX^{(i)}} \pm i d_{AX^{(i)}} C_{AX^{(i)}} \nabla \ln\frac{m}{n_i}\right]. \tag{11.6}$$

Experimentally, one can only measure the total concentration of dopant atoms, C_A, and the corresponding diffusivity, D_A. In terms of these accessible parameters,

$$\mathbf{J} = -D_A \nabla C_A. \tag{11.7}$$

If all dopant atoms are electrically active, that is, no precipitation or other losses have occurred, charge neutrality requires that

$$C_A + p = n \tag{11.8a}$$

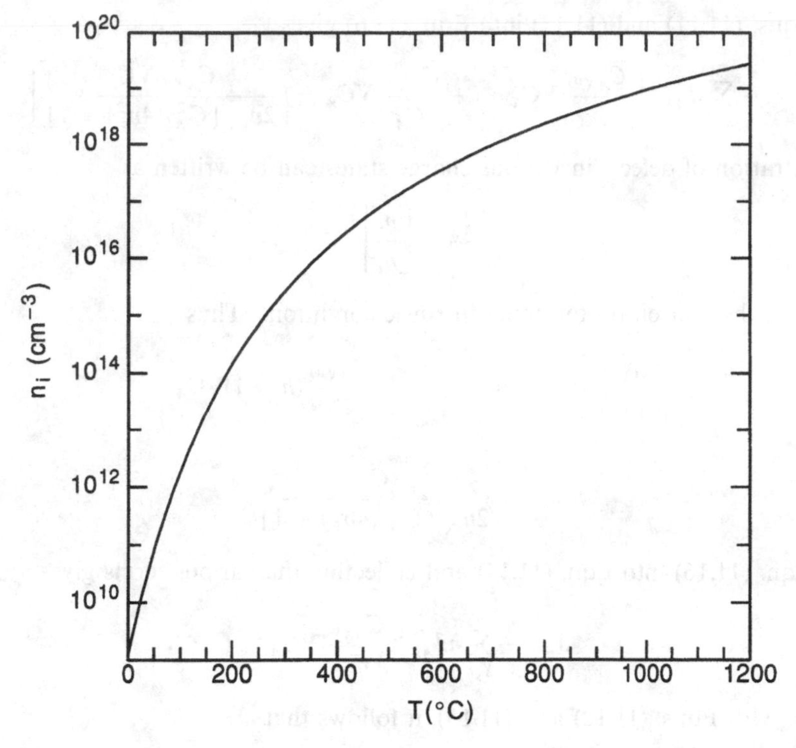

Fig. 11.1. Intrinsic carrier concentration in Si as a function of temperature.

for the donors and

$$C_A + n = p \tag{11.8b}$$

for the acceptors. Replacing p and n in Eqns. (11.8a) and (11.8b), respectively, by the use of

$$pn = n_i^2 \tag{11.9}$$

yields a quadratic equation for n and p, respectively, with the solution

$$\frac{m}{n_i} = \frac{C_A}{2n_i} + \left[\frac{C_A^2}{4n_i^2} + 1\right]^{1/2}. \tag{11.10}$$

Then

$$\nabla \ln \frac{m}{n_i} = \frac{\nabla(m/n_i)}{m/n_i} = \frac{1}{n_i} \frac{\nabla C_A}{2\sqrt{[C_A^2/(4n_i^2) + 1]}}. \tag{11.11}$$

From Eqn. (11.2), the mass action law yields

$$C_A C_{X^{(i\mp1)}} = k_{X,i} C_{AX^{(i)}}, \tag{11.12}$$

where $k_{X,i}$ is a constant, and thus

$$C_A \nabla C_{X^{(i\mp1)}} + C_{X^{(i\mp1)}} \nabla C_A = k_{X,i} \nabla C_{AX^{(i)}} = \frac{C_A C_{X^{(i\mp1)}}}{C_{AX^{(i)}}} \nabla C_{AX^{(i)}}. \tag{11.13}$$

Inserting Eqns. (11.11) and (11.13) into Eqn. (11.6) gives

$$\mathbf{J} = -\sum_{X,i} d_{AX^{(i)}} \left[\frac{C_{AX^{(i)}}}{C_{X^{(i\mp1)}}} \nabla C_{X^{(i\mp1)}} + \frac{C_{AX^{(i)}}}{C_A} \nabla C_A \pm i \frac{C_{AX^{(i)}} \nabla C_A}{2n_i\sqrt{[C_A^2/(4n_i^2)+1]}} \right]. \quad (11.14)$$

The concentration of defects in various charge states can be written as

$$\frac{C_X^{(i)}}{C_{X_{\mathrm{INT}}^{(i)}}} = \left[\frac{m}{n_i}\right]^{\mp i}. \quad (11.15)$$

where $C_{X_{\mathrm{INT}}^{(i)}}$ is the concentration under intrinsic conditions. Thus

$$\frac{C_{AX^{(i)}}}{C_{X^{(i\mp1)}}} \nabla C_{X^{(i\mp1)}} = \mp(i \mp 1) \frac{C_{AX^{(i)}}}{C_A} (h-1) \nabla C_A, \quad (11.16)$$

where

$$h = 1 + \frac{C_A}{2n_i\sqrt{[C_A^2/(4n_i^2)+1]}}. \quad (11.17)$$

Inserting Eqn. (11.16) into Eqn. (11.14) and collecting the various terms gives

$$\mathbf{J} = -\sum_{X,i} h d_{AX^{(i)}} \frac{C_{AX^{(i)}}}{C_A} \nabla C_A, \quad (11.18)$$

from which, with Eqns. (11.12) and (11.15), it follows that

$$\mathbf{J} = -\sum_{X,i} h \frac{d_{AX^{(i)}}}{k_{X,i}} C_{X_{\mathrm{INT}}^{(i)}} \left[\frac{m}{n_i}\right]^{\mp i} \nabla C_A. \quad (11.19)$$

Defining

$$D_{AX}^{(i)} = \frac{d_{AX^{(i)}} C_{X_{\mathrm{INT}}^{(i)}}}{k_{X,i}} \quad (11.20a)$$

and

$$D_A^{(i)} = \sum_X D_{AX}^{(i)}, \quad (11.20b)$$

we obtain by comparison of Eqn. (11.19) with Eqn. (11.7) the familiar form for the diffusivity of a dopant A under extrinsic conditions:

$$D_A = h \sum_i D_A^{(i)} \left[\frac{m}{n_i}\right]^{\mp i}; \quad (11.21)$$

for p-type material, $m = p$ and the bottom of the double sign applies; for n-type material, $m = n$ and the top is applicable. The prefactor h takes values between 1.00 for $C_A \ll n_i$ and 2.00 for $C_A \gg n_i$. Each pair diffusivity $D_A^{(i)}$ can be written in Arrhenius form:

$$D_A^{(i)} = D_0^{(i)} \mathrm{e}^{E_a^{(i)}/kT}. \quad (11.22)$$

Intrinsic diffusivities are given by

$$D = h \sum_i D_{AX}^{(i)}. \quad (11.23)$$

Table 11.1. *Prefactors $D_0^{(i)}$ in units of $cm^2\ s^{-1}$ and activation energies $E_a^{(i)}$ in units of eV for bulk diffusion in Si under equilibrium conditions (from Fair 1981)*

Element		$D_0^{(0)}$	$E_a^{(0)}$	$D_0^{(1)}$	$E_a^{(1)}$	$D_0^{(-1)}$	$E_a^{(-1)}$	$D_0^{(-2)}$	$E_a^{(-2)}$
p-type	B	0.037	3.46	0.76	3.46				
	Al	1.385	3.41	2480	4.20				
	Ga	0.374	3.39	28.5	3.92				
	In	0.785	3.63	415	4.28				
n-type	P	3.85	3.66			4.44	4.00	44.2	4.37
	As	0.066	3.44			12.0	4.05		
	Sb	0.214	3.65			15.0	4.08		

In Si experiments show that for the common dopants $|i| \leqslant 2$ in Eqn. (11.22); values under equilibrium conditions are listed in Table 11.1 (Fair, 1981). Intrinsic diffusivities under equilibrium conditions are plotted in Fig. 11.2 for the Si dopants B, Sb, and As.

Dopant diffusion in Si is mediated by two point-defects, vacancies ($X = V$) and self-interstitials ($X = I$). Under intrinsic conditions Eqn. (11.18) becomes

$$\mathbf{J} = -\left(d_{AV}\frac{C_{AV}}{C_A} + d_{AI}\frac{C_{AI}}{C_A}\right)\nabla C_A. \tag{11.24}$$

Defining a fractional interstitial component of diffusion under equilibrium conditions as (Fahey *et al.*, 1989)

$$f_{AI} = \frac{D^*_{AI}}{D^*_{AI} + D^*_{AV}}, \tag{11.25}$$

where the '*' denotes equilibrium, we can write

$$\frac{D_A}{D^*_A} = (1 - f_{AI})\frac{C_{AV}}{C^*_{AV}} + f_{AI}\frac{C_{AI}}{C^*_{AI}}. \tag{11.26}$$

From Eqn. (11.12) we have

$$\frac{C_{AX}}{C^*_{AX}} = \frac{C_X}{C^*_X}, \tag{11.27}$$

so that Eqn. (11.26) becomes

$$\frac{D_A}{D^*_A} = (1 - f_{AI})\frac{C_V}{C^*_V} + f_{AI}\frac{C_I}{C^*_I}. \tag{11.28}$$

For Si dopants such as Sb ($f_{\mathrm{SbI}} \approx 0$) and B ($f_{\mathrm{BI}} \approx 1$) (Fahey *et al.*, 1989), Eqn. (11.28) simplifies further to

$$\frac{D_{\mathrm{Sb}}}{D^*_{\mathrm{Sb}}} = \frac{C_V}{C^*_V} \quad \text{and} \quad \frac{D_{\mathrm{B}}}{D^*_{\mathrm{B}}} = \frac{C_I}{C^*_I}. \tag{11.29}$$

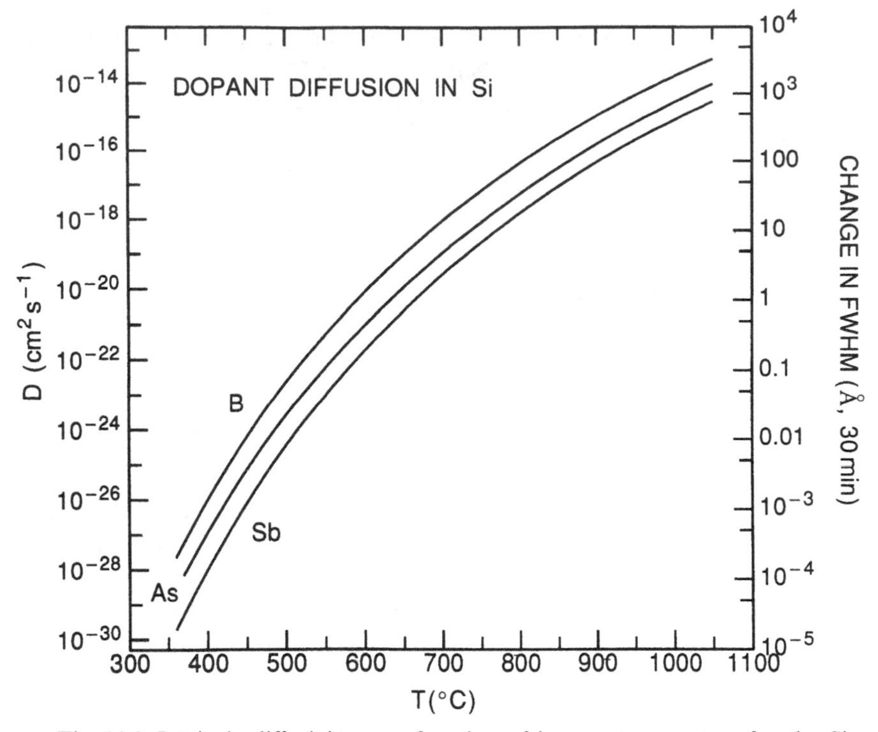

Fig. 11.2. Intrinsic diffusivity as a function of inverse temperature for the Si dopants B, As, and Sb. Also shown on the right ordinate is the change in full-width-at-half-maximum of a δ-doped spike after a 30 min anneal.

This is an important result and we will refer to it frequently in the following. Essentially, it implies that a dopant can be used as a tracer of point-defects, since its diffusivity is proportional to the concentration of a particular point-defect. The diffusivity of B is proportional to the concentration of Si-self-interstitials, and thus B traces interstitials; the diffusivity of Sb is proportional to the concentration of Si vacancies, and thus Sb can be used to trace vacancies in Si.

11.1.2 Extraction of diffusivities

Combining the continuity equation

$$\frac{\partial C_A}{\partial t} + \nabla \mathbf{J} = 0 \tag{11.30}$$

with Eqn. (11.7) yields the diffusion equation

$$\frac{\partial C_A}{\partial t} = \nabla(D_A \nabla C_A), \tag{11.31}$$

which, together with an appropriate initial profile $C_A(\mathbf{x}, 0)$, completely describes the evolution of $C_A(\mathbf{x}, t)$ as a function of position $\mathbf{x}$ and time t. In the present context of a diffusing two-dimensional dopant sheet, C_A will only depend on one spatial coordinate, the depth z into the sample, that is, $C_A(\mathbf{x}, t) \equiv c(z, t)$. Then Eqn. (11.31) becomes

$$\frac{\mathrm{d}c}{\mathrm{d}t} = \frac{\mathrm{d}}{\mathrm{d}z}\left[D\frac{\mathrm{d}c}{\mathrm{d}z}\right], \tag{11.32}$$

where for convenience we have dropped the subscript 'A'.

In the intrinsic case, where D does not depend on the dopant concentration, Eqn. (11.32) has an analytic solution:

$$c(z, t) = \frac{1}{2\sqrt{(\pi D t)}} \int_{-\infty}^{\infty} c(\zeta, 0) \mathrm{e}^{-(z-\zeta)^2/4Dt}\, \mathrm{d}\zeta, \tag{11.33}$$

where $c(z, 0)$ is the concentration at depth z before diffusion started, and $c(z, t)$ represents the concentration after diffusion for a time interval t at a fixed temperature T. The experimental concentrations $\tilde{c}(z, 0)$ and $\tilde{c}(z, t)$, determined, for example, by secondary ion mass spectrometry (SIMS), come in the form of n discrete data points, $\tilde{c}_i(z_i, 0)$ and $\tilde{c}_i(z_i, t)$, $i = 1, 2, \ldots, n$ respectively. Equation (11.33) can easily be solved numerically. An optimization shell calculates the residual vector, $\mathbf{r}$:

$$r_i = c_i(z_i, t) - \tilde{c}_i(z_i - z_0, t), \tag{11.34}$$

and minimizes the norm, $\|r\|$, by varying D and z_0. The parameter z_0 serves to line up both depth profiles at exactly the same depth in order to prevent errors arising from slightly varying depth scales.

The above procedure has the advantage that it is exact for any kind of instrument response function and initial dopant concentration profile. This can easily be seen by considering Eqn. (11.33) as a convolution between the initial dopant profile c_0 and an exponential function, Δ:

$$c = c_0 ** \mathrm{e}^{(z^2/4Dt)} \equiv c_0 ** \Delta, \tag{11.35}$$

where the ** symbolizes a convolution. The experimental data is, of course, influenced by the instrument response function Φ, that is,

$$\tilde{c} = c ** \Phi \quad \text{and} \quad \tilde{c}_0 = c_0 ** \Phi. \tag{11.36}$$

The convolution operator is associative and commutative so that

$$\tilde{c} = (c_0 ** \Delta) ** \Phi = (c_0 ** \Phi) ** \Delta = \tilde{c}_0 ** \Delta. \tag{11.37}$$

For δ-doping, peak concentrations will often exceed the intrinsic carrier concentration and the diffusivity will be concentration-dependent. Ghezzo (1972) has proposed a modification of the Boltzmann (1894)–Matano (1933) analysis that yields the diffusion coefficient as

$$D(c(z, t)) = -\frac{zc(z, t)}{2t\, \mathrm{d}c/\mathrm{d}z}. \tag{11.38}$$

While in theory this approach provides all the desired information it poses two problems. (1) The implicit assumption of a δ-function for the as-deposited profile requires deconvolution of the instrument response function from the actual data after diffusion. This is an inherently unstable numerical procedure for real data with noise. (2) The algorithm requires division by the first derivative of the concentration with respect to depth. This again is a numerically unstable procedure in the presence of noise.

A better alternative is to solve Eqn. (11.31) numerically, with the initial condition $\tilde{c}(z, 0)$, and vary D in the same optimization shell as above. While this introduces an error due to the instrument response function, the error for profiling techniques such as SIMS is usually small compared with other experimental uncertainties. Numerical solution of the diffusion equation by a process simulator, such as PROPHET (Pinto *et al.*, 1992), provides additional benefits such as the correct treatment of the effect of internal fields.

The statistical error in D can be estimated by a Monte-Carlo procedure. There are essentially two main error sources: fluctuations in the concentrations determined, and errors in the depth scale, for example due to varying SIMS sputter beam current, measurement errors in the determination of the crater depth, set-up errors, or variations in the Si deposition rate during growth. For each pair of data-sets $\tilde{c}_i(z_i, 0)$ and $\tilde{c}_i(z_i, t)$ a certain number of new pairs $\tilde{c}_i^{MC}(z_i^{MC}, 0)$ and $\tilde{c}_i^{MC}(z_i^{MC}, t)$ are generated numerically. Here

$$\tilde{c}_i^{MC} = \tilde{c}_i + G\sqrt{(\tilde{c}_i\gamma)} \tag{11.39a}$$

and

$$z^{MC} = z(1 + G\chi), \tag{11.39b}$$

where G is a Gaussian distributed random variable of mean $E(G) = 0$ and standard deviation $E(G^2) = 1$, γ is the concentration corresponding to a count of 1 in the experimental apparatus and χ represents one standard deviation in the relative depth scale error. Equation (11.39a) adds noise to each data point, while Eqn. (11.39b) stretches or contracts the depth scale as a whole. From each of the new pairs of data a diffusion coefficient can be extracted as described above; from the whole set a mean D and its standard deviation can then be calculated. Note that estimates for *systematic* errors, such as errors in the annealing temperature, still have to be added.

11.1.3 The influence of growth defects on diffusion

Due to the difficulties associated with doping in Si (see Chapter 7 by Gossmann on low-temperature MBE of Si) the first attempts to investigate diffusion in δ-doped Si were all performed on films grown and doped by solid phase epitaxy (SPE). Fukatsu *et al.*

(1991) studied the diffusion of Sb at concentrations of $N_{Sb}^{2D} = 4 - 9.5 \times 10^{13}$ cm^{-2} in a flowing $N_2:H_2$ environment. No change in the concentration profile was observed up to a temperature of 550 °C. Above 750 °C the authors observed anomalous diffusion in the form of a kink in the profiles; above 900 °C significant surface accumulation occurred. Using Ghezzo's (1972) algorithm for the analysis of their data, Fukatsu *et al.* found an enhancement of the Sb diffusivity by one to two orders of magnitude over the corresponding bulk values. Above a concentration of $\approx 5 \times 10^{17}$ cm^{-3} the diffusion constant became concentration-dependent, rising approximately as $c^{1/3}$. The authors attribute the anomalous behavior to electric field drift of the ionized donors, although they could not definitely rule out other mechanisms such as oxide generated during post-growth annealing.

Slijkerman *et al.* (1991) have studied even higher two-dimensional concentrations of Sb, again grown by SPE. They found enhanced diffusion at the c-Si/α-Si interface during recrystallization between 550 and 650 °C, at least two orders of magnitude larger than the bulk-diffusion coefficient in α-Si. Van IJzendoorn *et al.* (1992) reported high lateral diffusion, that is, in the plane of the dopant, in SPE films doped with Sb at a concentration of 3.1×10^{14} cm^{-2}. Using transmission electron microscopy (TEM) and ion scattering the authors detected precipitation. From the density of precipitates, $N_p^{2D} = 9 \times 10^{10}$ cm^{-2} after annealing for a time $t = 3600$ s at 650 °C it is possible to estimate the Sb diffusivity, $D_{Sb} \approx 1/(8N_p^{2D}t) \approx 4 \times 10^{-16}$ cm^2/s. This is five orders of magnitude more than the intrinsic diffusivity of Sb at 650 °C. Van Opdorp *et al.* (1992) have interpreted this result within the framework of Mathiot and Pfister's (1989) percolation theory. Essentially, this theory predicts a sudden drastic increase in diffusion when the dopant concentration exceeds 2.5×10^{20} cm^{-3}. As a consequence, the theory predicts the formation of precipitates in the plane of the doping-spike, even at temperatures at which ordinary diffusion is negligible. If we demand that the mean distance between dopant atoms in a uniformly doped film of concentration $N_{A,D}$ be the same as in the plane of the dopant sheet in a δ-doped film of two-dimensional concentration $N_{A,D}^{2D}$, then $N_{A,D}^{2D} = (N_{A,D})^{2/3}$ (Schubert *et al.*, 1987). Thus Mathiot and Pfister's critical dopant concentration corresponds to a sheet concentration of 4×10^{13} cm^{-2} in a δ-doped film.

It is important to note here that there is extensive evidence that SPE films contain a large number of defects. Bean and Poate (1980) reported the existence of voids based on epitaxial crystallization measurements; electrical measurements show incomplete activation of the dopants and a reduced carrier mobility (Vescan *et al.*, 1985; Casel *et al.*, 1986; Casel *et al.*, 1990); positron annihilation spectroscopy indicates open-volume defects with concentrations of the order of 10^{18}–10^{19} cm^{-3} (Schut *et al.*, 1991; Asoka-Kumar *et al.*, 1993). By contrast, material grown and doped by molecular beam epitaxy is of excellent quality (see Chapter 7 by Gossmann on low-temperature MBE of Si). A recent systematic study of B- and Sb-δ-doped films found that the diffusive behavior of dopants in films grown by SPE differs markedly from that in films produced by low-temperature MBE (LT-MBE) and attributed this to the vacancy-like defects that are intrinsic to growth by SPE but not to growth by LT-MBE (Gossmann *et al.*, 1993a). Figure 11.3(a) shows, in the upper panel, a depth profile of a Sb-δ-doped film after annealing in a vacuum for 16 hr (squares). The solid line represents the best fit result of a numerical solution to the

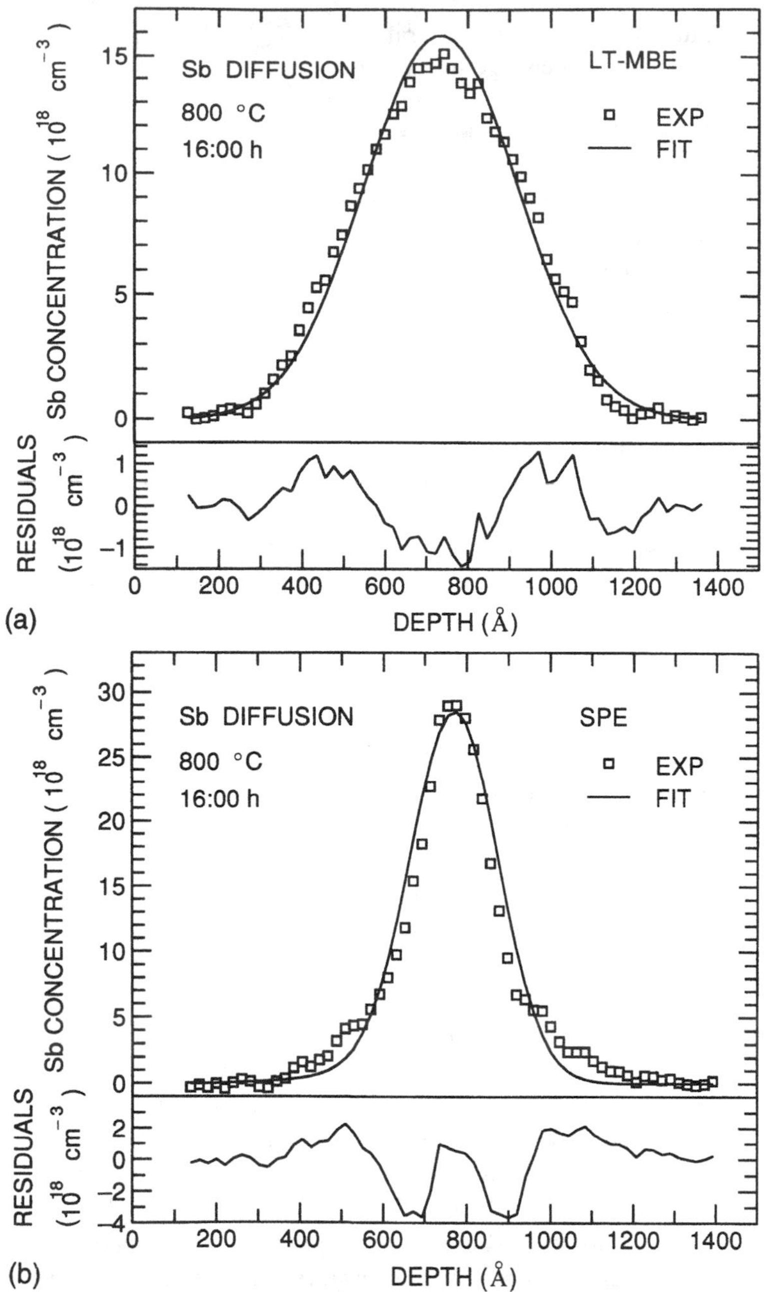

Fig. 11.3. Concentration vs. depth profile of a Sb doping spike, annealed for 16 hr at 800 °C. Shown are the results for a film grown by (a) LT-MBE and (b) SPE. In each case the two-dimensional concentration of the spike is $N_{\mathrm{Sb}}^{\mathrm{2D}} = 0.7 \times 10^{14}\ \mathrm{cm}^{-2}$. Experimental data are symbolized by squares; the solid lines are best fits to the data, solving Eqn. (11.1) with a concentration-independent diffusion coefficient. The difference between experiment and calculation, the residuals, are displayed in the bottom panel of each diagram.

one-dimensional diffusion equation with a concentration-independent diffusion coefficient, using the actual as-deposited profile as the starting point. Since the peak dopant concentrations reach to $\approx 10^{20}\,\text{cm}^{-3}$, the assumption of a concentration-independent diffusion coefficient leads to a slight disagreement between fit and actual data. The experimental profile has a slightly flatter top and somewhat steeper sides than does the calculated profile. The differences between calculated and experimental datapoints, the residuals, are shown in the bottom panel of Fig. 11.3(a). They exhibit a characteristic shape, with two peaks and one valley.

By contrast, for the film grown by SPE (Fig. 11.3(b)) the center of the profile is sharper and the shoulders flatter than the calculation, leading to residuals that have one peak and two valleys. The profile appears to consist of two components, one central peak and a low concentration background. Figure 11.4 shows that dopant-spikes in SPE material are (after regrowth) always wider than corresponding ones in films grown by LT-MBE. As Slijkerman *et al.* (1991) have pointed out for the case of Sb as dopant, this enhancement of diffusion during recrystallization of the amorphous film cannot be explained on the basis of bulk diffusion, but is due to enhanced diffusion at the α-Si/*c*-Si interface. It is thus inherent to growth by SPE and limits the minimum achievable width in δ-doped layers grown by this technique. The broadening is significant and amounts to 60–70 Å for the above data (regrowth took place at 550 °C and took $\approx$15 min).

Overall, the diffusivity in SPE grown films is much smaller than that in LT-MBE grown films, as Fig. 11.4 illustrates. There, depth profiles are plotted, obtained before and after vacuum annealing; each diagram compares samples grown by LT-MBE and by SPE that are otherwise identical. This retardation of diffusion in SPE films is not only true for the anneals described in Fig. 11.4, but is true in general: Fig. 11.5 shows the Sb- (in the top right part of the diagram) and B- (in the bottom left part) diffusion coefficient in films grown by SPE, normalized to the corresponding diffusion coefficient in films grown by LT-MBE, as a function of annealing time at the indicated annealing temperature. All ratios show a value below one and decrease steeply with increasing annealing time. As was discussed above, the Sb diffusion coefficient is proportional to the vacancy concentration and the B diffusion coefficient is proportional to the interstitial concentration. A reduction of the vacancy concentration in SPE films from the equilibrium value could explain the observed suppression of the Sb diffusion coefficient. However, if such a suppression existed at small times, thermal generation of interstitial-vacancy pairs and indiffusion of vacancies from the bulk substrate should lead to an increase in the vacancy concentration with time, contrary to the observed decrease in diffusion coefficient. Further, a vacancy-deficit does not lead simultaneously to a deficit in interstitial concentration, as evidenced by the reduced B diffusion coefficient. Deviations of the Si self-interstitial and vacancy concentrations alone cannot therefore be responsible for the experimental observations.

Analyzing the central spike only, while ignoring the shoulders, gives the results shown in Fig. 11.6. There, we plot the change in width of the central spike in SPE-grown films, compared to the original width before *ex-situ* annealing, as a function of annealing time, for various annealing temperatures. After some transient diffusion that is not resolved but

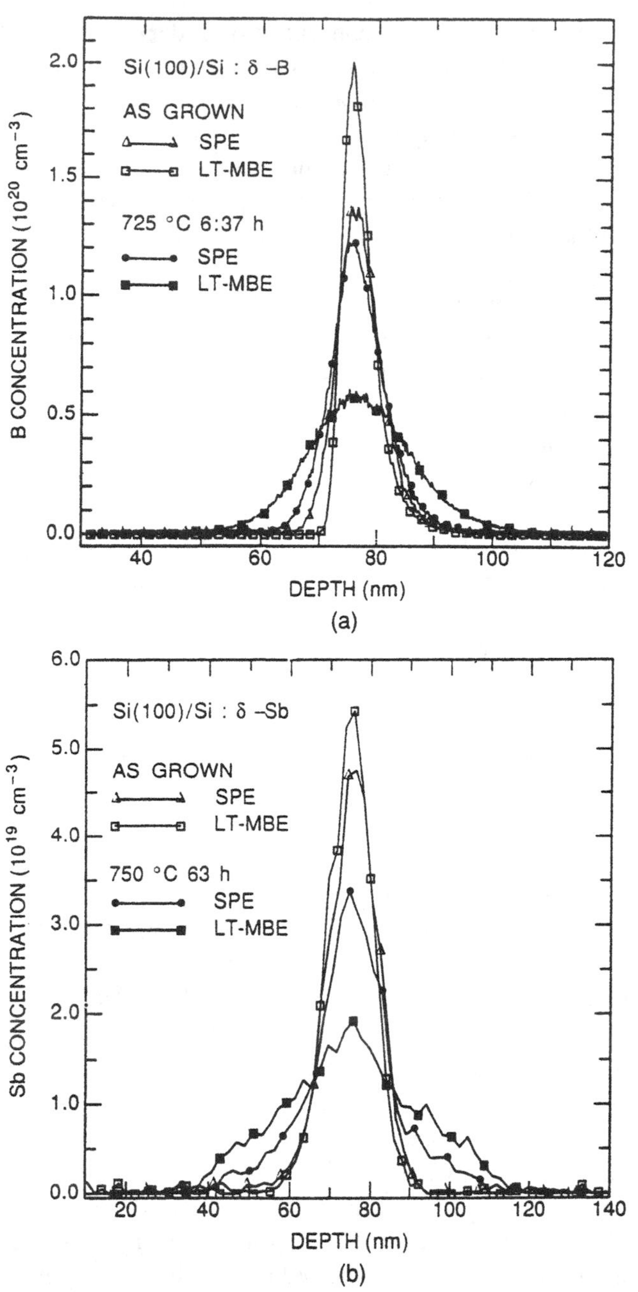

Fig. 11.4. Depth profiles for a (a) B- and (b) Sb-δ-doped layer, before ('as grown') and after annealing in vacuum at (a) 725 °C, 6:37 hr, and (b) 750 °C, 63 hr. Each diagram compares a film grown by LT-MBE with a structure produced by SPE. Two-dimensional concentrations are $N_{B}^{2D} = 1.30 \times 10^{14}$ cm^{-2} and $N_{Sb}^{2D} = 0.73 \times 10^{14}$ cm^{-2}, respectively; they are identical among the four samples in each diagram. The B profiles were obtained by SIMS, the Sb profiles by RBS.

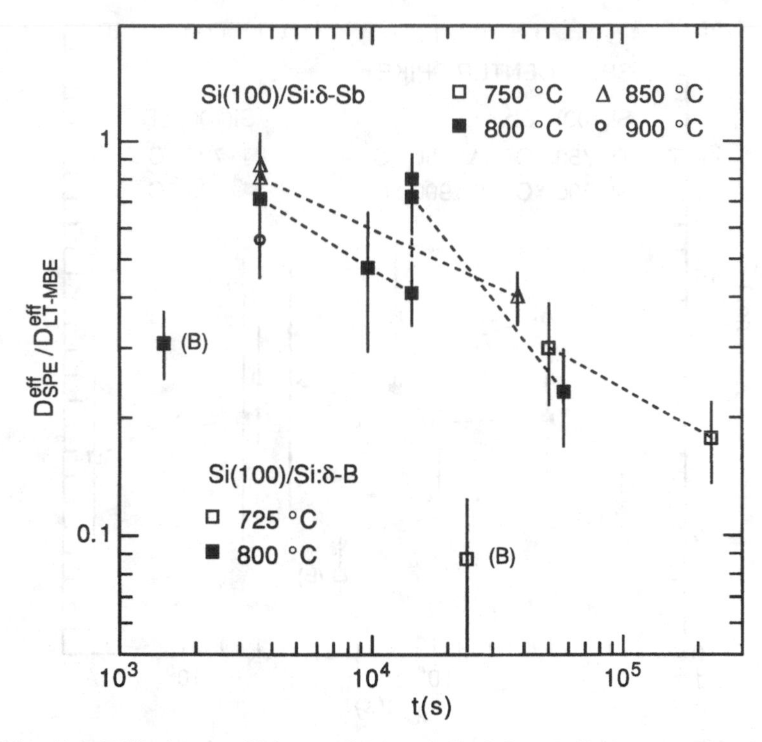

Fig. 11.5. Diffusion coefficients of B- and Sb-δ-doped films grown by SPE, normalized to the corresponding diffusion coefficient obtained from films grown by LT-MBE, as a function of annealing time. Sb-data are clustered in the upper right of the plot, B-data in the lower left. Annealing temperatures are indicated in the diagram. The doping spikes have sheet concentrations of $N_B^{2D} = 1.3 \times 10^{14}\ \text{cm}^{-2}$ and $N_{Sb}^{2D} = 0.7 \times 10^{14}\ \text{cm}^{-2}$ for B and Sb respectively. The dashed lines serve as a guide only.

adds ≈ 10 nm to the width, the central spike does not broaden further within the experimental uncertainty under any of the annealing conditions.

The behavior of the dopant in the films grown by SPE indicates that traps exist in the SPE film that retard the diffusive motion of B- and Sb-atoms during annealing. Besides the retardation in diffusion, the recrystallization and/or annealing must also lead to a coalescence of the majority of the dopant in the SPE films into a form that is practically invariant with time and does not diffuse. Precipitation would account for such behavior and has indeed been observed, as discussed above. Nevertheless, the results given in Figs. 11.3–11.4 do not show any evidence for Mathiot and Pfister's (1989) percolation theory. Instead, the diffusive effects observed in SPE films appear to be peculiar to the material and are not a manifestation of fundamental diffusion mechanisms operating at high concentrations. This indicates that any conclusions with regard to fundamental diffusion processes that are derived from structures grown by SPE need to be evaluated carefully.

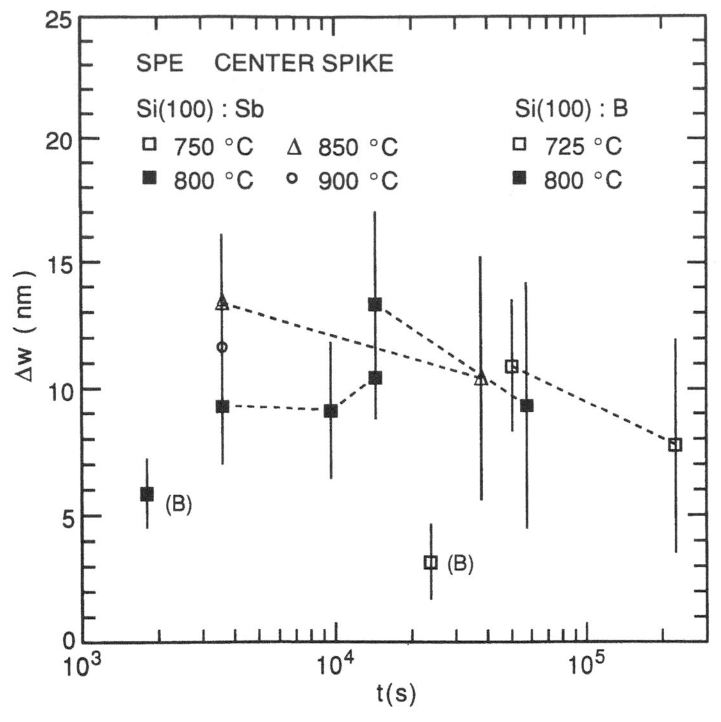

Fig. 11.6. Change in width of the central spike in B- and Sb-δ-doped films grown by SPE, as a function of annealing time at the temperatures indicated in the diagram. The doping spikes have sheet concentrations of $N_B^{2D} = 1.3 \times 10^{14}$ cm^{-2} and $N_{Sb}^{2D} = 0.7 \times 10^{14}$ cm^{-2} for B and Sb respectively. The dashed lines serve as a guide only.

11.1.4 Dopants as tracers of point-defect behavior

Quantitative information about the fundamental parameters of point-defect diffusion and recombination in Si is crucial to a detailed understanding and modeling of diffusion and diffusion transients during Si device processing. Unfortunately, reliable and consistent data is scarce. Point-defects cannot be observed directly but only through their effects on some physical quantity of the sample, such as the growth of dislocation loops (Hu, 1974). This fundamental experimental difficulty represents a formidable obstacle and is largely responsible for the controversies existing in the literature of silicon point-defects (Fahey *et al.*, 1989). On the other hand, advanced Si device technologies make understanding and control of non-equilibrium effects more and more important, as characteristic dimensions reach 0.1 μm (Mazuré and Orlowski, 1989; Rafferty *et al.* 1993). Figure 11.7 shows an example of the phenomenon known as transient diffusion (Angelucci *et al.*, 1987). A B-doped Si layer was grown on a n-doped Si wafer leading to a junction depth of $\approx$0.3 μm. The wafer was masked with SiO_2 and patterned to form alternating stripes of oxide and bare Si. Thus, subsequent Si implantation created damaged and undamaged regions.

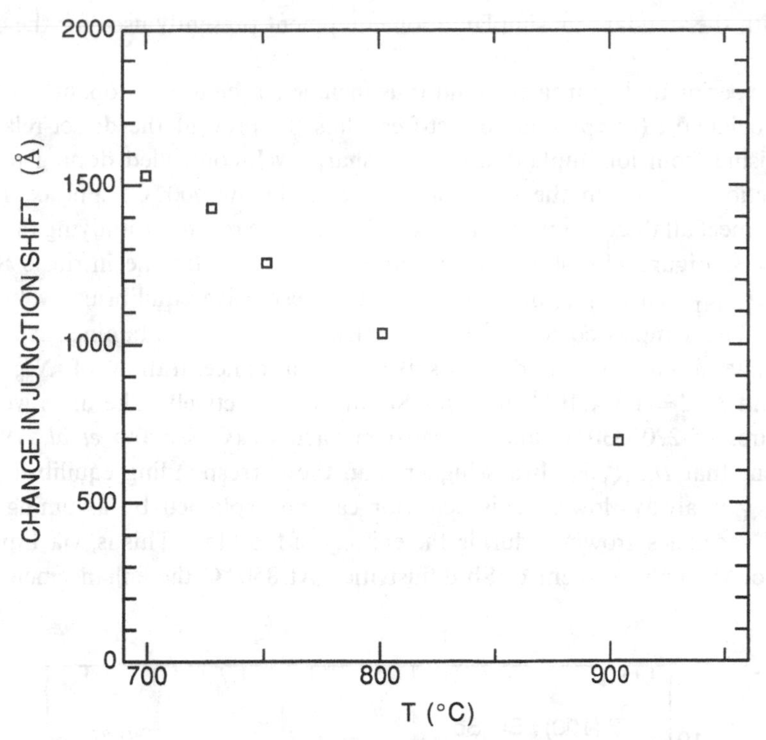

Fig. 11.7. Difference in junction shift between an ion-implanted and an undamaged junction (after Angelucci *et al.*, 1987).

During annealing the dopant moves in both regions, leading to a junction shift. The difference in junction shift between damaged and undamaged regions is plotted in Fig. 11.7 and shows quite an unintuitive *increase* with *decreasing* temperature. This phenomenon is of course due to the point-defects that the ion-implantation has created in large numbers in the unmasked regions and that enhance diffusion of the dopant according to Eqn. (11.29) until they have been annealed out. As Fig. 11.7 shows, this enhancement can be quite large, in particular at the lower temperatures that a process might wish to use in order to prevent bulk diffusion elsewhere in a device structure. It is also quite obvious that, in order to predict and model this behavior, a thorough understanding of point-defects and their interactions both between themselves and with dopants is indispensable.

Modern Si device processing technologies employ temperatures significantly smaller than 900 °C, for example for the gate-oxide growth of metal-oxide–semiconductor (MOS) devices (Yan *et al.*, 1992). Diffusion distances under relevant processing conditions may be as small as 100 Å, but, with the characteristic lengthscales of devices approaching 0.1 μm, such diffusion lengths are not negligible. For example, in modulation-doped field effect transistors (MODFETs) the maximum tolerable diffusion distance of the dopant is set by the set-back of the dopant-spike from the well, which is of the order $\leqslant$100 Å (People *et al.*, 1985; Monroe *et al.*, 1992). The study of dopant diffusion phenomena under these conditions requires initial dopant profiles with typical lengthscales of $\ll$100 Å, impossible to achieve with the standard ion-implantation equipment presently used for the doping of devices.

The ideal experimental approach would thus include (i) the use of dopants as tracers of point-defect behavior; (ii) epitaxial, defect-free films (to prevent the defect-related complications arising from ion-implantation); (iii) sharp, well-controlled doping profiles (to enable extraction of data in the temperature regime below 900 °C). Epitaxially grown δ-doped films meet all these requirements and offer a unique means of studying the behavior of point-defects. Figure 11.8 shows an example. There, we plot the intrinsic Sb and B diffusivities, $D_{Sb,INT}$ and $D_{B,INT}$, normalized to their respective equilibrium values (Table 11.1, p. 257). The samples consist of 1500-Å-thick films of Si containing, at a depth of 750 Å, a single δ-function-like dopant-spike with a concentration of $N_{Sb}^{2D} = 0.73 \times 10^{14}\ cm^{-2}$ and $N_B^{2D} = 1.3 \times 10^{14}\ cm^{-2}$ for Sb and B respectively. The films were grown at temperatures of 220–330 °C and annealed in vacuum (Gossmann *et al.*, 1994). It is quite apparent that $D_{Sb,INT}$ is always higher than the corresponding equilibrium values, whereas $D_{B,INT}$ is always lower. This behavior can be explained by assuming a supersaturation of vacancies grown-in during the epitaxy of the films. This is, via Eqn. (11.29), responsible for the enhancement of Sb diffusivities. At 850 °C the enhancement is about

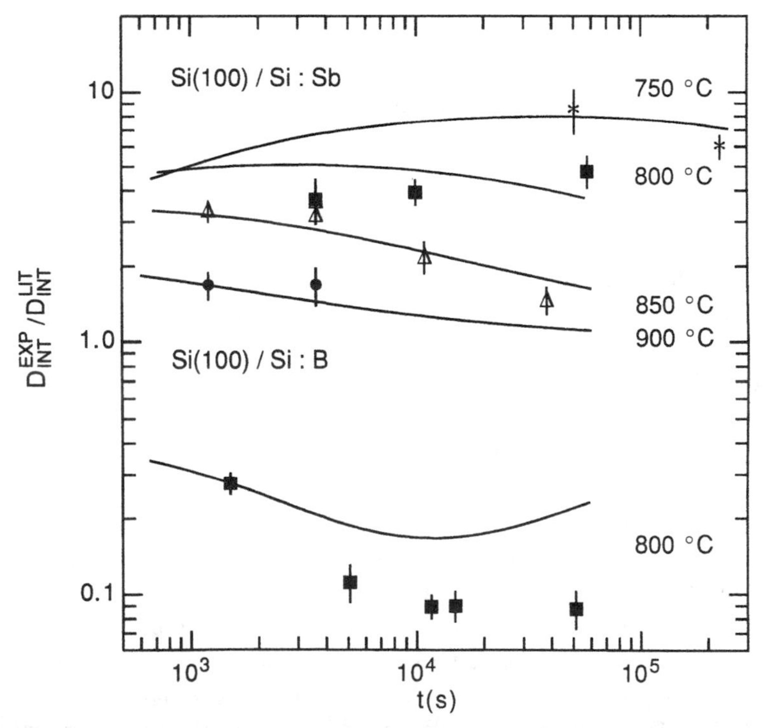

Fig. 11.8. Sb (top half of diagram) and B (bottom half) intrinsic diffusion coefficients, normalized to their respective equilibrium values, as a function of annealing time. Different symbols represent annealing temperatures as indicated in the diagram. Solid lines are the result of a simulation.

a factor of 3, that is, $C_V \approx 3C_V^*$. The equilibrium concentration of vacancies at that temperature is $\approx 1 \times 10^{14}$ cm^{-3} (Zimmermann and Ryssel, 1992), so that the number of grown-in vacancies is $\approx 3 \times 10^{14}$ cm^{-3}. The retardation of B diffusion is caused by recombination of interstitials with the excess vacancies. Qualitatively, this explains all the observations: the slow decline of the antimony enhancement with time is due to out-diffusion of vacancies from the film to the bulk and to the surface. The lower enhancement of antimony diffusion at higher temperatures is due to the increase in C_V^* with temperature. Bulk recombination of interstitials with vacancies is responsible for the decrease in $D_{B,INT}$. The model has been examined quantitatively (Gossmann *et al.*, 1994), resulting in an activation energy of thermal vacancy generation of 1.4 eV for the temperature range 750–900 °C, a bulk recombination lifetime of 7×10^3 s, and a vacancy diffusivity of $D_V = 2.5 \times 10^{-13}$ cm^2/s at 850 °C.

11.1.5 δ-doping-superlattices and the Si-self-interstitial diffusivity

Equation (11.29) indicates that the diffusivity of B at a particular point in a sample is proportional to the interstitial concentration at that point. The same is true for Sb and vacancies. This property can be utilized to map the point-defect concentration as a function of depth during processing. An epitaxially grown doping superlattice made from B or Sb dopant-spikes, spaced at regular intervals, is exposed to a particular processing step. For example, an essentially infinite source of interstitials could be placed at the surface of the sample by annealing in dry oxygen. After processing, the diffusivity of the dopant can be extracted in each spike individually. For each spike the diffusivity is proportional to the average concentration of interstitials (for B) or vacancies (for Sb) at the depth of the particular spike. The superlattice as a whole thus gives a depth profile of the average concentration of point-defects. In the case of oxide growth and with a B-doping-superlattice (DSL), this technique has been demonstrated as a tool for the detection of defects in epitaxial material (van Oostrum, 1992) and to yield the interstitial diffusivity (Gossmann *et al.*, 1993b). Although all applications of this technique have focused on the behavior of Si-self-interstitials, the procedure itself is generic and is suitable for a wide variety of processing steps (e.g. oxidation, nitridation, ion-implantation, etc.) and materials (e.g. Si, Ge, GaAs, etc.).

Figure 11.9 illustrates the concept. It shows the dopant depth profile of a B doping superlattice as deposited (solid line) and after dry oxidation at 800 °C for 15 min (dotted line). The interstitials created at the surface diffuse into the bulk of the sample, as symbolized by the arrows. The height of the individual dopant peaks before oxidation appears to decrease with increasing depth. This is purely an artefact of the degrading SIMS depth resolution that makes peaks at successive depths appear successively wider. After oxidation, this peak envelope is reversed: now the peak at the surface is the widest, with the amount of B diffusion decreasing as the distance to the interstitial source increases. Extracting B diffusion coefficients results in Fig. 11.10. There, for each dopant-spike the intrinsic B diffusivity is plotted as a function of the spike's depth. Figure 11.10 is also a

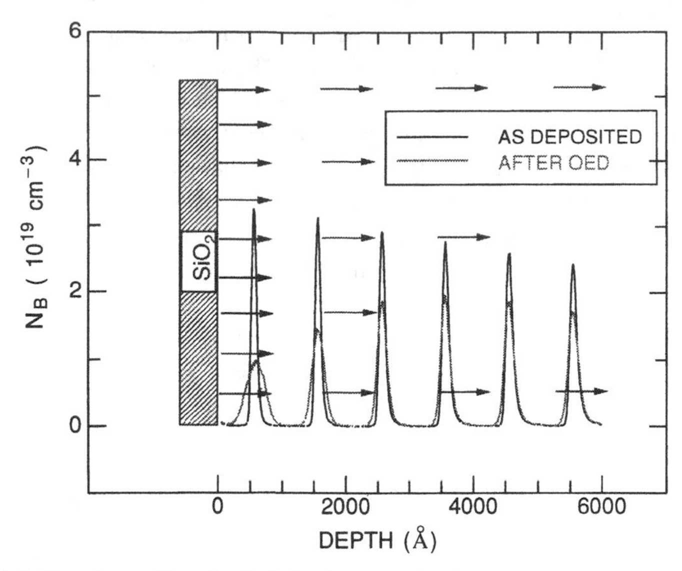

Fig. 11.9. Depth profile of a B-δ-doping-superlattice as-deposited (solid line) and after annealing at 800 °C for 15 min in oxygen (dotted line). Oxidation of Si generates Si self-interstitials at the oxide/Si interface. The interstitials diffuse into the bulk of the sample, as symbolized by the arrows.

depth profile of the interstitial concentration, averaged over the time of the oxidation. Assuming that the concentration of interstitials at the surface is constant during the oxidation, the interstitial concentration at a particular time, t, is proportional to a complementary error-function,

$$C_\mathrm{I} \propto erfc(a) \tag{11.40}$$

where

$$a = \frac{z}{\sqrt{(4D_\mathrm{I}t)}} \tag{11.41}$$

and D_I is the interstitial diffusivity. However, the experimentally determined B diffusivity is proportional to the time averaged C_I. Thus, fitting a complementary error-function (solid line) to the data in Fig. 11.10 yields an average interstitial diffusivity, $D_\mathrm{I}^{(\mathrm{av})}$, that is related to D_I by

$$erfc(a_0^{(\mathrm{av})}) = erfc(a_0) + 2a_0^2 erfc(a_0) - \frac{2a_0}{\sqrt{\pi}} \mathrm{e}^{-a_0^2}, \tag{11.42}$$

with

$$a_0^{(\mathrm{av})} = \frac{z}{\sqrt{(4D_\mathrm{I}^{(\mathrm{av})}t_0)}}, \quad a_0 = \frac{z}{\sqrt{(4D_\mathrm{I}t_0)}} \tag{11.43}$$

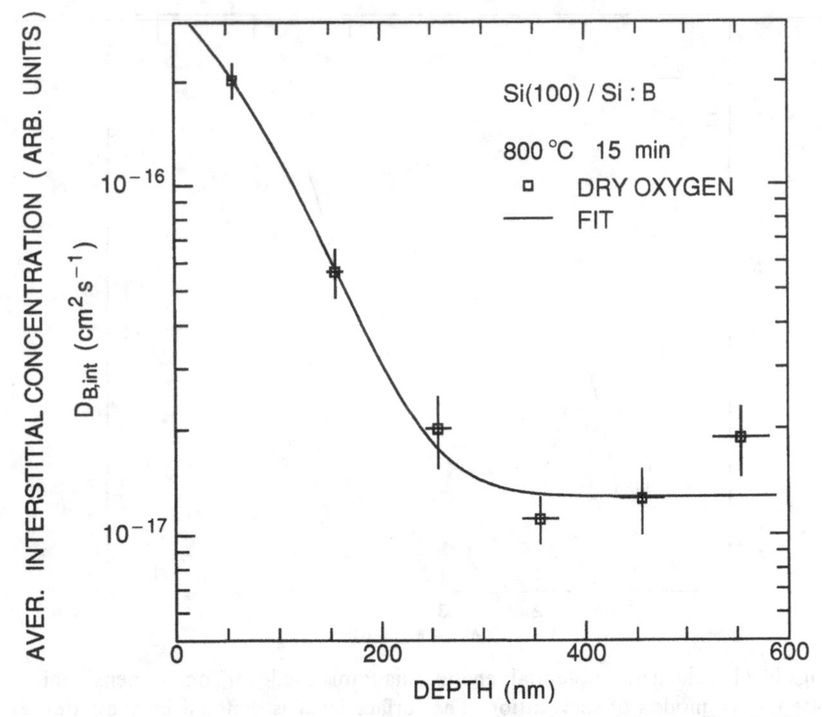

Fig. 11.10. B diffusivities, extracted from a comparison of the profiles in Fig. 11.9, as a function of depth; alternatively, average Si- self-interstitial concentration as a function of depth. The solid line is the result of a fit to this average interstitial concentration profile that yields, assuming a constant surface concentration, the interstitial diffusivity.

where t_0 is the oxidation time. The right-hand side can be approximated by *erfc*$(1.4a_0)$, so that $D_I \approx 2D_I^{(av)}$. Thus one obtains from Fig. 11.10 the interstitial diffusivity as $(1.4 \pm 0.4) \times 10^{-13}\ \mathrm{cm^2\ s^{-1}}$ at 800 °C.

11.2 Dopant segregation

11.2.1 Rate theory of segregation

Segregation of a component X to the surface is often favored by a reduction of the enthalpy of the system (Buck, 1982). The element X could represent a dopant in a semiconductor, e.g. Sb in Si. All common dopants employed in Si molecular beam epitaxy (Sb, Ga, In, B) segregate to the surface under certain conditions, which makes δ-doping difficult (see Chapter 7 by Gossmann on low-temperature MBE of Si). Several theoretical models, based on rate theory, have been developed that allow prediction of the amount of segregation and of the resulting dopant profile (Tabe and Kajiyama, 1983; Metzger and Allen, 1984a, 1984b; Barnett and Greene, 1985; Barnett *et al.*, 1986; Jorke, 1988; Ni *et al.*, 1989). They assume

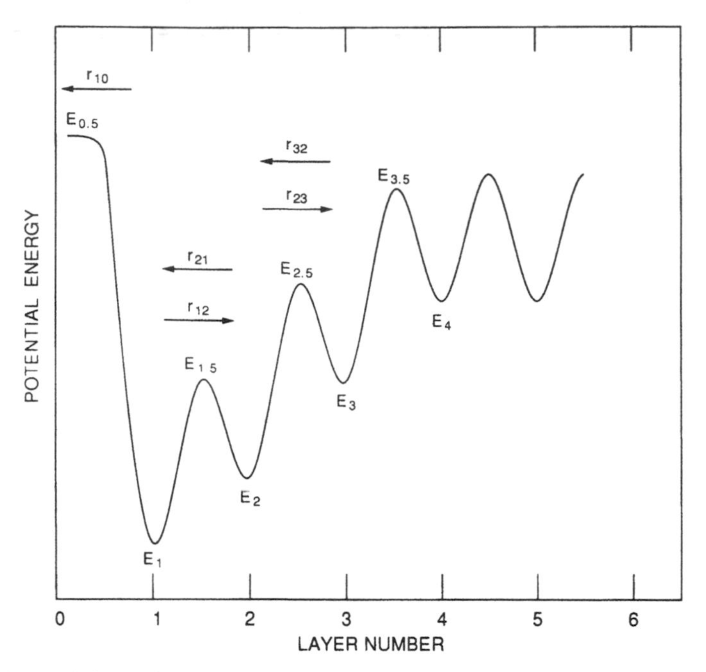

Fig. 11.11. Schematic potential energy diagram used for one-dimensional rate theory models of segregation. The surface layer is denoted by the index '1'. Also shown are selected energy levels E_i and rate constants r_{ij}.

a potential energy diagram as shown in Fig. 11.11. The surface layer is denoted by the index '1'. The change in concentration of dopant, C_i, in each layer is assumed to have the form

$$\frac{dC_i}{dt} = \sum_{j=i-1,i+1} [r_{ji}C_j - r_{ij}C_i], \tag{11.44}$$

where the r_{ij} are rate constants of the form

$$r_{ij} = \nu e^{E_{ij}/kT}, \tag{11.45}$$

with $E_{ij} = E_{(i+j)/2} - E_i$ and $\nu \approx 2 \times 10^{12}\ s^{-1}$, the attempt frequency. Usually, deposition is treated as a series of discrete steps. In each step a certain amount of dopant is deposited together with a monolayer of Si. This layer becomes the new surface layer and the indices of the layers in the preceding step are increased by one. Transitions of dopant atoms between layers are then allowed for a time $\tau = N^{2D}_{(klm)}/J$, where J is the deposition rate of the film and $N^{2D}_{(klm)}$ is the number of atoms per unit area on the surface of orientation (klm) (for Si $N^{2D}_{(100)} = 6.78 \times 10^{14}\ cm^{-2}$). For a δ-doped Si film growing at a sufficiently high temperature most of the dopant atoms will segregate to the surface layer during the time τ that it takes to cap the dopant-spike with one monolayer of Si. Only a fraction $\beta \ll 1$ of the originally deposited $N_{A,D}$ number of dopant atoms will remain in the layer where they were originally placed. After completion of the growth, the concentration in the nth

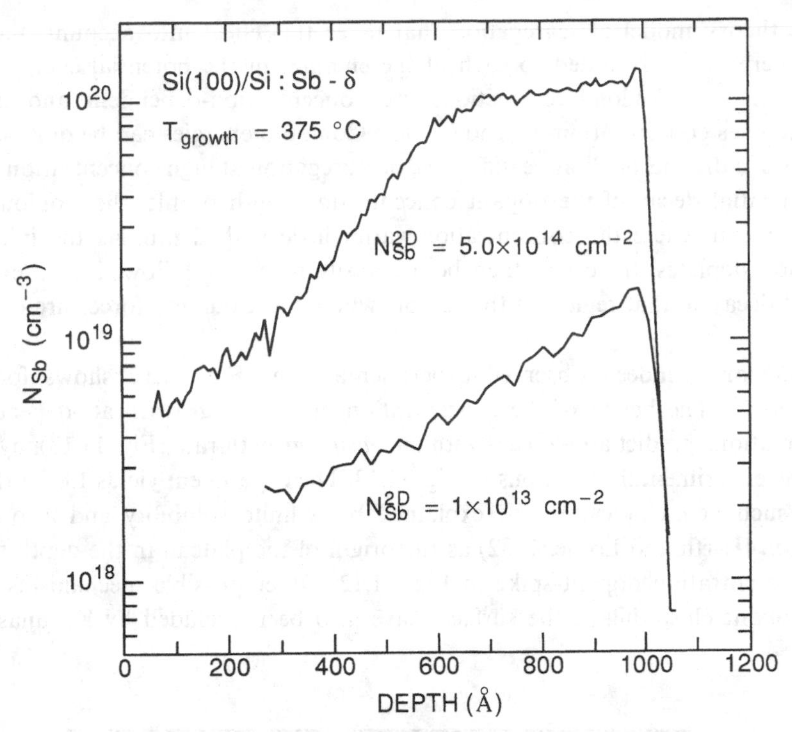

Fig. 11.12. Depth profile of Sb-δ-doped films grown at $T_{\mathrm{growth}} = 375$ °C and a growth rate of 1.2×10^{14} cm^{-2}. The dopant spikes were placed at a depth of 1000 Å and had a concentration of $N_{\mathrm{Sb}}^{2\mathrm{D}} = 5.0 \times 10^{14}$ cm^{-2} and $N_{\mathrm{Sb}}^{2\mathrm{D}} = 1.0 \times 10^{13}$ cm^{-2}.

layer is $(1-\beta)^n \beta N_{\mathrm{A,D}}$. Such dependence on thickness is equivalent to an exponential decay of the dopant concentration towards the surface. This is a characteristic feature of all the standard rate theory models and is quite independent of the details of implementation. For initial sheet concentrations $N_{\mathrm{A,D}}^{2\mathrm{D}} \geqslant 10^{13}$ cm^{-2} an exponential decay is indeed observed experimentally (Fig. 11.12). Any general initial dopant concentration can always be viewed as a superposition of appropriate δ-spikes and, in particular for Sb, rate theory is quite successful in predicting experimental doping profiles.

11.2.2 Segregation at very high concentrations: Coulomb repulsion

Implicit in the standard rate theory model of segregation is the assumption that the energy levels E_i are independent of the concentration of the dopant. For typical dopant concentrations this is a correct assumption. However, δ-doping can lead to extremely high peak volume concentrations and the dopant atoms in such a doping-spike may only be a few interatomic distances away from each other. In this case the Coulomb field of a charged impurity will no longer be completely screened, and thus dopant atoms will repel each other (Schubert *et al.*, 1992). Karunasiri *et al.* (1994) have recently proposed an extension of the rate theory model of segregation that takes this effect into account. Basically a Coulomb-energy E_i^{Coul} is added to each of the energies in the potential energy diagram of Fig. 11.11. The Coulomb corrections are concentration-dependent and thus segregation becomes concentration-dependent. The Coulomb energies can be quite large and thus we expect a drastic, qualitative difference of segregation at high concentrations: instead of an exponential decay of the dopant concentration-depth profile, the Coulomb terms lead to a plateau where the concentration varies little with depth. As the initial sheet concentration depletes, there will then be a transition region, followed by the classical, exponential decay in that region of the sample where the Coulomb forces are completely screened.

This behavior is indeed observed experimentally, as Fig. 11.12 shows for $N_{Sb}^{2D} = 5.0 \times 10^{14}$ cm^{-2}. The height of the concentration plateau is growth-rate-dependent; the model calculations predict an *increase* with *increasing* growth rate (Fig. 11.13), by a factor of 2.0 for the experimental conditions of Fig. 11.13. The experiment yields 1.8 for this ratio. Note that such behavior cannot be explained by a finite solubility and also excludes pipe-diffusion (Hirth and Lothe, 1982) as the origin of the plateau in the depth profile of the high concentration dopant-spike in Fig. 11.12. Other possible mechanisms, such as strain or dopant clustering at the surface, have also been excluded by Karunasiri *et al.* (1994).

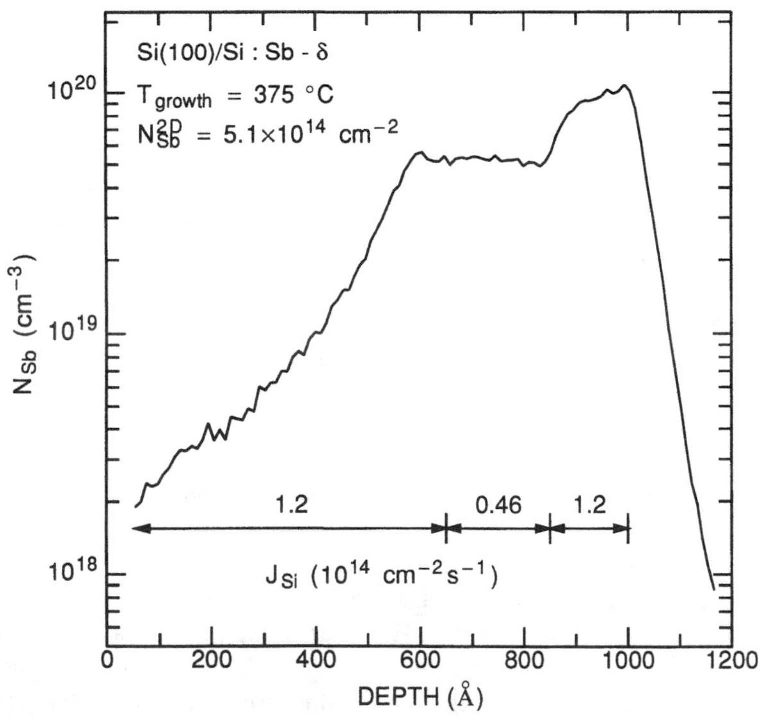

Fig. 11.13. Depth profile of a Sb-δ-doped film grown at $T_{growth} = 375$ °C. The Si growth rate, J_{Si}, varied as indicated in the diagram.

11.2.3 *Fermi-level-pinning-induced segregation*

The breaking of the three-dimensional symmetry of a crystal due to the formation of a surface or an interface may lead to the introduction of states into the band gap. These surface or interface states are able to pin the Fermi-level, as shown schematically in Fig. 11.14(a). There, we have assumed that the bulk of the semiconductor is n-doped, that the surface states are all acceptor-like and located in the middle of the band gap at an energy E_{SS} away from the bottom of the conduction band. Charge will flow into the empty surface states until equilibrium is established and the Fermi-level is flat. As a consequence, a space charge of ionized donors builds up (symbolized by the '+'-signs in Fig. 11.14(a)) and the bands bend upward. The magnitude of the internal field is given by

$$|\mathbf{E}| = \frac{eN_{D,A}}{\varepsilon}(W - z), \tag{11.46}$$

where e is the elementary charge, ε is the permittivity of the semiconductor, W is the width of the depletion region, and z measures the depth into the sample. For a δ-doped sample the situation is very similar (Fig. 11.14(b)), with the exception that now the field is independent of depth and is given by

$$|\mathbf{E}| = \begin{cases} \dfrac{eN_{A,D}^{2D}}{\varepsilon}, & z_\delta \leqslant \dfrac{E_{SS}\varepsilon}{eN_{A,D}^{2D}} \\ \dfrac{E_{SS}}{z_\delta}, & z_\delta > \dfrac{E_{SS}\varepsilon}{eN_{A,D}^{2D}} \end{cases}, \tag{11.47}$$

where the doping-spike with a two-dimensional concentration of $N_{A,D}^{2D}$ is located at a depth z_δ. A similar relation can be written for a p-type semiconductor and donor-like surface states. The internal field that is set up in this way will lead to a drift of dopant atoms towards the surface (Schubert *et al.*, 1990). This Fermi-level-pinning-induced segregation must be added to the amount of segregation that may be present from thermodynamical causes (Section 11.2.1). An order of magnitude estimate of the amount of Fermi-level-pinning-induced segregation can be obtained by considering sufficiently small segregation times. Then the velocity of a dopant atom is given by

$$v_{A,D} = \mu|\mathbf{E}|, \tag{11.48}$$

which becomes, with Eqn. (11.5), at sufficiently small distances from the surface

$$v_{A,D} = \frac{D_{A,D}e^2N_{A,D}^{2D}}{\varepsilon kT}. \tag{11.49}$$

To render the effect observable requires values of $v_{A,D}$ that are of the order of the growth-rate, that is, $D_{A,D} \approx 10^{-16}$ cm^2/s, as well as a sufficiently small thermodynamical driving force for segregation. It is therefore not surprising that Fermi-level-pinning-induced

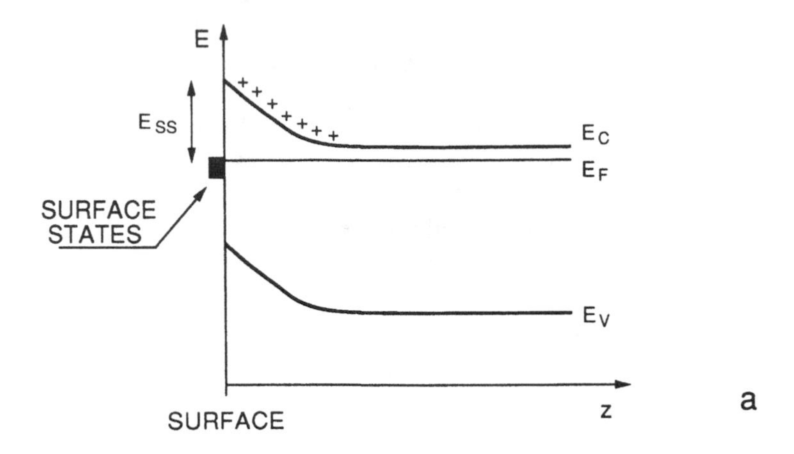

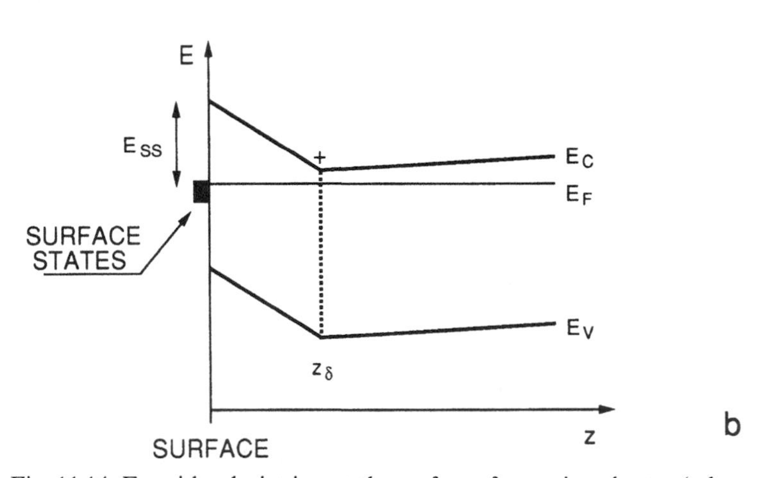

Fig. 11.14. Fermi-level pinning at the surface of a semiconductor (schematic). It is assumed that only empty surface states exist in the middle of the band gap and that the semiconductor is n-doped, uniformly in panel (a), with a δ-function-like dopant spike at depth z_δ in (b).

segregation has not been reported for any dopants in Si: at growth temperatures above $\approx$400 °C, dopant incorporation in Si is dominated by segregation (see Chapter 7 by Gossmann on low-temperature MBE of Si), whereas below that temperature $D_{A,D} \ll 10^{-16}$ cm^2/s. However, for Si as an n-type dopant in $Al_{0.30}Ga_{0.70}As$, segregation is unlikely (Beall *et al.*, 1988) and $D_{Si} = 6 \times 10^{-15}$ cm^2/s at 660 °C (Schubert *et al.*, 1990). Thus in this system Fermi-level-pinning-induced segregation should be observable and has indeed been reported by Schubert *et al.* (1990).

References

Angelucci, R., Cembali, F., Negrini, P., Servidori, M., and Solmi, S., *J. Electrochem. Soc.* **134**, 3130 (1987).

Asoka-Kumar, P., Gossmann, H.-J., Leung, T. C., Talyanski, V., Nielsen, B., Lynn, K. G., Unterwald F. C., and Feldman, L. C., *Phys. Rev.* B **48**, 5345 (1993).

Barnett, S. A and Greene, J. E., *Surf. Sci.* **151**, 67 (1985).

Barnett, S. A., Winters, H. F., and Greene, J. E., *Surf. Sci.* **165**, 303 (1986).

Beall, R. B., Clegg, J. B., and Harris, J. J., *Semicond. Sci. Technol.* **3**, 612 (1988).

Bean, J. C. and Poate, J. M., *Appl. Phys. Lett.* **36**, 59 (1980).

Boltzmann, L., *Ann. Phys. Chem.* **53**, 959 (1894).

Buck, T. M., in *Chemistry and Physics of Solid Surfaces IV*, Vanselow, R., and Howe, R., eds. (Springer, New York, 1982), p. 435.

Casel, A., Jorke, H., Kasper, E., and Kibbel, H., *Appl. Phys. Lett.* **48**, 922 (1986).

Casel, A., Kibbel, H., and Schäffler, F., *Thin Solid Films* **183**, 351 (1990).

Fahey, P. M., Griffin, P. B., and Plummer, J. D., *Rev. Mod. Phys.* **61**, 289 (1989).

Fair, R. B., in *Impurity Doping Processes in Silicon*, Wang, F. F. Y., ed. (North Holland, Amsterdam, The Netherlands, 1981), ch. 7.

Fukatsu, S., Kubo, S., Shiraki, Y., and Ito, R., *Appl. Phys. Lett.* **58**, 1152 (1991).

Ghezzo, M., *J. Electrochem. Soc.* **119**, 977 (1972).

Gossmann, H.-J., Vredenberg, A. M., Rafferty, C. S., Luftman, H. S., Unterwald, F. C., Jacobson, D. C., Boone, T., and Poate, J. M., *J. Appl. Phys.* **74**, 3150 (1993a).

Gossmann, H.-J., Rafferty, C. S., Luftman, H. S., Unterwald, F. C., Boone, T., and Poate, J. M., *Appl. Phys. Lett.* **63**, 639 (1993b).

Gossmann, H.-J., Rafferty, C. S., Vredenberg, A. M., Luftman, H. S., Unterwald, F. C., Eaglesham, D. C., Jacobson, D. C., Boone, T., and Poate, J. M., *Appl. Phys. Lett.* **64**, 312 (1994).

Hirth, J. P. and Lothe, J., *Theory of Dislocations* (Wiley, New York 1982), pp. 506.

Hu, S. M., *J. Appl. Phys.* **45**, 1567 (1974).

van IJzendoorn, L. J., Fredriksz, C. W., van Opdrop, C., Gravesteijn, D. J., Vandenhoudt, D. E. W., van de Walle, G. F. A., and Bulle-Lieuwma, C. W. T., *Nucl. Instrum. Methods Phys. Res.* B **64**, 120 (1992).

Jorke, H., *Surf. Sci.* **193**, 569 (1988).

Karunasiri, R. P. U., Gilmer, G. H., and Gossmann, H.-J., *Surf. Sci.*, **317**, 361 (1994).

Matano, C., *Jpn J. Phys.* **8**, 109 (1933).

Mathiot, D. and Pfister, J. C., *J. Appl. Phys.* **66**, 970 (1989).

Mazuré, C. and Orlowski, M., *IEEE Electron Dev. Lett.* **10**, 556 (1989).

Metzger, R. A. and Allen, F. G., *J. Appl. Phys.* **55**, 931 (1984a).

Metzger, R. A. and Allen, F. G., *Surf. Sci.* **137**, 397 (1984b).

Monroe, D., Xie, Y. H., Fitzgerald, E. A., and Silverman, P. J., *Phys. Rev.* B **46**, 7935 (1992).

Morin, F. J. and Maita, J. P., *Phys. Rev.* **94**, 1525 (1954a).

Morin, F. J. and Maita, J. P., *Phys. Rev.*, **96**, 28 (1954b).

Ni, W.-X., Knall, J., Hasan, M. A., Hansson, G. V., Sundgren, J.-E., Barnett, S. A., Markert, L. C., and Greene, J. E., *Phys. Rev.* **B 40**, 10449 (1989).

van Oostrum, K. J., Zalm, P. C., de Boer, W. B., Gravesteijn, D. J., and Maes, J. W. F., *Appl. Phys.* **72**, 1513 (1992).

van Opdorp, C., van IJzendoorn, L. J., Fredriksz, C. W., and Gravesteijn, D. J., *J. Appl. Phys.* **72**, 4047 (1992).

People, R., Bean, J. C., and Lang, D. V., *J. Vac. Sci. Technol* A. **3**, 846 (1985).

Pinto, M. R., Boulin, D. M., Rafferty, C. S., Smith, R. K., Coughran, W. M. Jr, Kizilyalli, I. C., and Thoma, M. J., *Proc. IEDM*-**92**, 923 (1992).

Rafferty, C. S., Giles, M. D., Vuong, H.-H., Eshraghi, S. A., Pinto, M. R., and Hillenius, S. J., *VPAD*, Nara, Japan (1993).
Schubert, E. F., Cunningham, J. E., and Tsang, W. T., *Solid State Communications* **63**, 591 (1987).
Schubert, E. F., Kuo, J. M., Kopf, R. F., Jordan, A. S., Luftman, H. S., and Hopkins, L. C., *Phys. Rev.* B **42**, 1364 (1990).
Schubert, E. F., Gilmer, G. H., Kopf, R. F., and Luftman, H. S., *Phys. Rev.* B **46**, 15078 (1992).
Schut, H., van Veen, A., van de Walle, G. F. A., and van Gorkum, A. A., *J. Appl. Phys.* **70**, 3003 (1991).
Slijkerman, W. F. J., Zagwijn, P. M., van der Veen, J. F., van de Walle, G. F. A., and Gravesteijn, D. J., *J. Appl. Phys.* **70**, 2111 (1991).
Tabe, M. and Kajiyama, K., *Jpn J. Appl. Phys.* **22**, 423 (1983).
Vescan, L., Kasper, E., Meyer, O., and Maier, M., *J. Cryst. Growth* **73**, 482 (1985).
Yan, R. H., Lee, K. F., Jeon, D. Y., Kim, Y. O. B., Park, G., Pinto, M. R., Rafferty, C. S., Tennant, D. M., Westerwick, E. H., Chin, G. M., Morris, M. D., Early, K., Mulgrew, P., Mansfield, W. M., Watts, R. K., Voshchenkov, A. M., J. Bokor, J., Swartz, R. G., and Ourmazd, A., *IEEE Electron Dev. Lett.* **13**, 256 (1992).
Zimmermann, H. and Ryssel, H., *Appl. Phys.* A **55**, 121 (1992).

12

Characterisation of silicon and aluminium delta-doped structures in GaAs: local mode spectroscopy

R. C. NEWMAN

12.1 Introduction

Characterisation of a δ-layer requires the determination of the areal concentration of the dopant, the extent of spreading of the impurities by diffusion and surface segregation during growth, the electrical and optical properties of the structure, the local concentrations of intrinsic defects and, finally, the lattice sites that are occupied by the dopant atoms. Analytical techniques which yield this information have been, or will be, described elsewhere in this volume, except for the last two requirements. In this chapter it will be shown how information can be obtained about impurity lattice sites, together with some limited information about lattice defects, from local vibrational mode (LVM) spectroscopy.

The substitution of a host lattice atom by an impurity with a significantly lower mass will lead to a vibrational mode that has a frequency greater than the maximum lattice frequency ω_{max}, provided that the local force constants are comparable with, or greater than, those of the host. Such a mode cannot propagate and only the impurity and its nearest neighbours will have significant vibrational displacements: thus the mode is spatially *localised.* Because the force constants depend on the local bonding, it follows that a silicon donor atom in GaAs occupying a Ga-lattice site (written henceforth as Si_{Ga}) and bonded to As-neighbours will have a different vibrational frequency from that of a Si_{As} acceptor bonded to Ga-neighbours (Spitzer, 1971). Such modes may be detected as sharp lines in infrared (IR) absorption or Raman scattering spectra. Since the assignments of most lines in GaAs have now been established, the sites occupied by the impurities can be determined in particular samples. It will be shown that the IR and Raman techniques are complementary, but both require calibrations to be established before a line 'strength' can be converted into an impurity concentration. If an impurity atom forms a complex with a second impurity or an intrinsic defect there will be modifications of the local masses and force constants so that two or three new LVM lines may be produced depending upon the reduction in the local tetrahedral symmetry.

These ideas will be expanded and illustrated in Section 12.2, mainly by a discussion of silicon impurities in GaAs, since δ-doping with silicon has been studied more extensively than other systems. It is necessary first to discuss homogeneously doped material so that a comparison can be made of the distribution of occupied lattice sites with that found in regions adjacent to a δ-plane when some diffusion or surface segregation has occurred. The development of LVM, IR and Raman absorption spectroscopies will be outlined in Sections 12.3.1 and 12.3.2, respectively, with emphasis being placed on the assignments of lines and the available calibration data.

12.2 Local vibrational modes

A bonded substitutional impurity in a compound semiconductor, such as GaAs, is illustrated in Fig. 12.1. A simple two-parameter Keating model (Keating, 1966), involving only host lattice bond stretching and bending force constants, written as α and β, respectively, gives a reasonable fit to the phonon dispersion and the density of phonon modes, although there is no splitting of the LO and TO branches at the Γ-point in the Brillouin zone due to Coulombic interactions. Procedures for determining values of α and β together with numerical data for various hosts have been given by Sangster *et al.* (1992). The bonds centred on the impurity have modified constants α' and β' (at the apex of the impurity) which can be determined from measured LVM frequencies. The modified parameters have similar values to α and β for isoelectronic impurities, such as Al_{Ga} or P_{As}

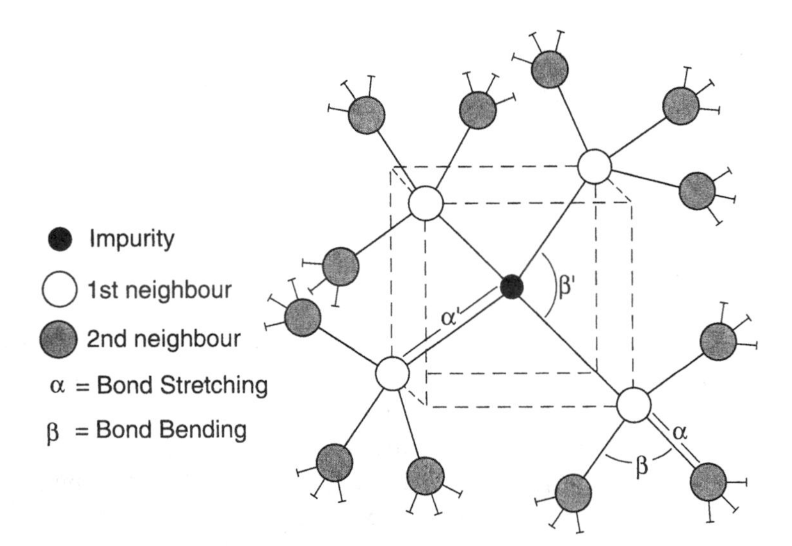

Fig. 12.1. Keating cluster model used to study the LVM of a light impurity. The lines depict bonds with a stretching force constant α and there are angle bending constants β between adjacent bonds. The bonds around the impurity α' and β' can take modified values. A displacement of the impurity to the right clearly requires displacements of the first neighbours to the left so that there is no displacement of the centre of mass.

in GaAs, increased values ($\approx 20\%$) for group IV atoms acting either as donors (Si_{Ga}) or acceptors (Si_{As}) but reduced values ($\approx 20\%$) for group II (Be_{Ga}) acceptors and group VI (S_{As}) donors. This pattern can be understood in terms of changes in the bond lengths arising from local Coulombic interactions when the charge on the impurity and the ionicity of the III–V compound (Kunc, 1984) are taken into account. For example, the bonds around a Si_{Ga}^{+} donor are shortened because of the attraction to the nearest neighbour As^{-} atoms.

When the substitutional impurity (Fig. 12.1) is displaced along the positive x-axis in a high frequency LVM, the neighbouring atoms must have components of their displacements in the opposite direction so that there is no movement of the centre of mass. Since the displacements of second neighbours are negligible, as shown by computer simulations (Sangster *et al.*, 1992), the LVM angular frequency ω_{LVM} may be written as:

$$(\omega_{LVM})^2 = k\left(\frac{1}{M_{imp}} + \frac{1}{\chi M_{nn}}\right) \tag{12.1}$$

where M_{imp} is the mass of the impurity, M_{nn} is the mass of the nearest neighbours and χ is a parameter that usually has a value close to 2 and depends on the ratio β'/α'; $k = 4\alpha' + 2\beta' + 3\beta$ is the restoring isotropic force constant for the impurity in the so-called static well approximation (Leigh and Newman, 1988), where the displacements of nearest neighbours are identically zero, corresponding to $M_{nn}/M_{imp} \to \infty$. The identification of a measured IR absorption line at wavenumber $v = \omega/2\pi c$ (cm^{-1}), where c is the velocity of light (in $cm\ s^{-1}$), is straightforward if the impurity has two or more naturally occurring isotopes with reasonably comparable abundances. Silicon just falls into this category since there are three isotopes, namely ^{28}Si (92.3%), ^{29}Si (4.7%) and ^{30}Si (3%), leading to three well resolved LVMs (Fig. 12.2) with separations of $\approx 5\ cm^{-1}$. For studies of other impurities, specially grown crystals containing enriched isotopes such as ^{13}C (Newman *et al.*, 1972) and ^{18}O (Schneider *et al.*, 1989) to replace ^{12}C and ^{16}O have been used to allow positive identification of LVM lines. Alternatively, for Raman studies, ion implanted isotopes such as ^{29}Si or ^{30}Si may be detected (Ramsteiner *et al.*, 1988). For the remaining impurities of interest, such as ^{9}Be and ^{27}Al, there are no other isotopes and the assignment of LVM lines has had to rely on doping studies, analyses of crystals using secondary ion mass spectrometry (SIMS) and processes of elimination of other known lines. A list of LVM frequencies for selected substitutional impurities in GaAs is given in Table 12.1.

From the discussion thus far it is not clear how the LVM lines from Si_{Ga} and Si_{As} might be distinguished. Even if the electrical properties of the GaAs are known, there is no *a priori* reason to suppose that the associated dipole moments or Raman scattering cross-sections should be equal, so that the relative concentrations of donors and acceptors cannot be determined. For GaAs hosts, the solution to the problem is implicit in Eqn. (12.1) since arsenic has only one isotope ^{75}As (100%), whereas there are two naturally occurring gallium isotopes, namely ^{69}Ga (60%) and ^{71}Ga (40%). It follows that the configuration of a silicon donor $^{28}Si_{Ga}{}^{75}As_4$ is unique so that there is only one LVM frequency, which is revealed as a sharp line with a full width at half power $\Delta \approx 0.4\ cm^{-1}$ in high resolution

Table 12.1. *Vibrational modes of substitutional impurities in GaAs and related infrared data* (*Newman* 1993)

Impurity	LVM frequency (cm^{-1})	Apparent charge η (units of e)	Concentration to give $IA = 1\ cm^{-2}$ ($cm^{-3} \times 10^{16}$)
$^{9}Be_{Ga}$	482	1.1 ± 0.1	5.5
$^{12}C_{As}$	582	3.0	0.95 ± 0.29
$^{13}C_{As}$	561	—	—
$^{27}Al_{Ga}$	362	2.5	3.1 ± 0.4
$^{28}Si_{Ga}$	384	2.0	5.0 ± 0.4
$^{29}Si_{Ga}$	379	—	—
$^{30}Si_{Ga}$	373	—	—
$^{28}Si_{As}$	399	1.9	5.5
$^{30}Si_{As}$	389	—	—

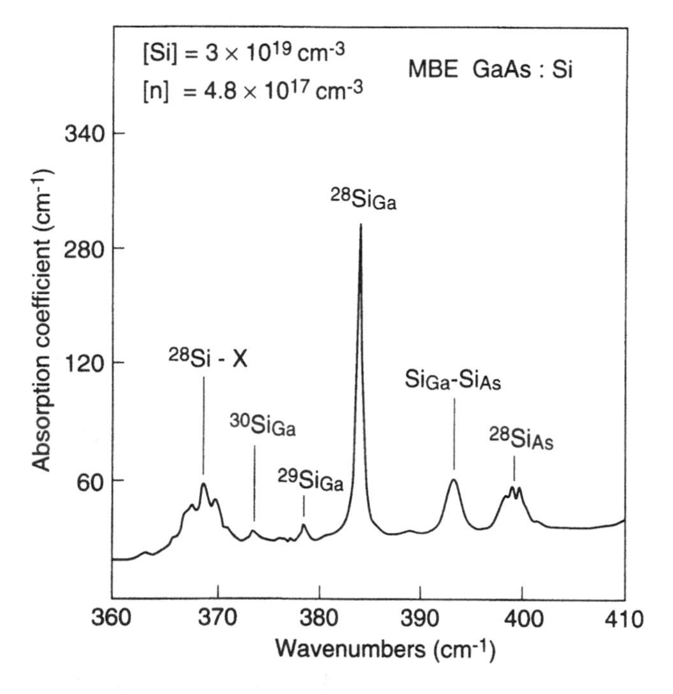

Fig. 12.2. Infrared absorption spectrum of a homogeneously Si doped MBE (001) layer after subtraction of the intrinsic two-phonon absorption features of GaAs showing LVM lines from Si impurities in various sites. The doping level was [Si] = $3 \times 10^{19}\ cm^{-3}$ but the as-grown carrier concentration was only $4.8 \times 10^{17}\ cm^{-3}$ due to the presence of the electron trap labelled Si-X. The sample had been irradiated with 2 MeV electrons to remove free-carrier absorption.

IR spectra. On the other hand, there are five combinations of the four Ga isotopes surrounding a Si_{As} acceptor and different clusters have different symmetries (Theis and Spitzer, 1984; see also Laithwaite and Newman, 1977; Theis *et al.*, 1982). The complexes Si_{As}–$^{69}Ga_4$ (T_d), Si_{As}–$^{69}Ga_3{}^{71}Ga$ (C_{3v}), Si_{As}–$^{69}Ga_2{}^{71}Ga_2$ (C_{2v}), Si_{As}–$^{69}Ga{}^{71}Ga_3$ (C_{3v}) and Si_{As}–$^{71}Ga_4$ (T_d) lead to a total of nine closely spaced LVM frequencies which can be calculated from the Keating cluster model, or by perturbation methods (Leigh and Newman, 1982). The disposition of these modes is shown in Fig. 12.3: the overall spread in the structure is reduced as the ratio β'/α' is increased and is smaller for $^{12}C_{As}$ than for $^{28}Si_{As}$ or $^{11}B_{As}$ (Gledhill *et al.*, 1984). For Si_{As}, the fine structure is partially resolved (Fig. 12.2) and the measured value of Δ is $\approx 1.5\ cm^{-1}$, which is significantly larger than $\Delta \approx 0.4\ cm^{-1}$ found for Si_{Ga} donors.

The limiting width of an LVM line also depends upon the lifetime of the oscillator in its first excited state, the degree of broadening due to strains produced by dislocations and point defects including the impurities themselves and also upon electron–phonon interactions. By definition, an LVM has an energy $\hbar\omega$ that can only be lost by a second, third or higher order interaction with lattice phonons. The LVM frequencies of Si in its various

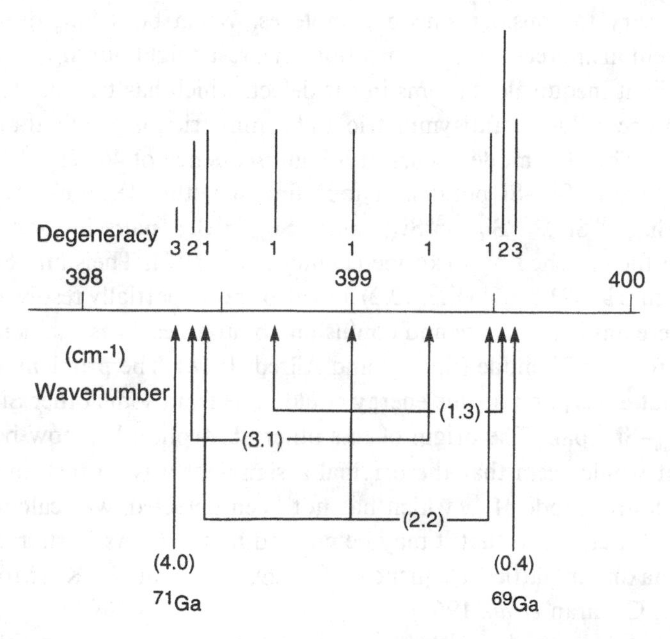

Fig. 12.3. The calculated disposition of the fine structure components of the LVM from Si_{As} resulting from the presence of mixed ^{69}Ga and ^{71}Ga nearest neighbour isotopes: lines arising from the combination $^{71}Ga_3{}^{69}Ga$ are shown by the arrows labelled (3.1), etc. The heights of the component lines are calculated from the relative isotopic abundance of the combinations and the degeneracy factor of the vibrational mode. (NB: the strengths of the three highest frequency components for the corresponding diagram for C_{As} in GaAs shown in Figure 3 of Newman, 1993, are too small in relation to the other components.)

lattice sites fall between ω_{max} and $2\omega_{max}$ of GaAs and so decay by a two-phonon process is possible. To obtain the low values of Δ quoted above, it is necessary to reduce the sample temperature, T, below 77 K, preferably to a value close to 4.2 K. Residual strains in GaAs are usually low, leading to small values of Δ, unless the doping levels are in excess of $\approx 10^{18}$ cm^{-3}. However, other epitaxial layers, such as InAs grown heteroepitaxially on GaAs, are significantly strained and only relatively broad LVM lines have been found (Addinall *et al.*, 1991, 1992).

When electrically active impurities are present, LVM lines will be superposed on the continuum of free-carrier absorption in IR spectra. Since there is degeneracy of the vibrational and electronic states at the LVM frequency, mixing of the two wavefunctions can lead to a redistribution of electronic absorption round the spectral position of the LVM. The line may then appear with an asymmetric derivative shape or even a dip in the continuum and there could also be a small shift from the usual line position. These features, called Fano profiles (Fano, 1961), have been observed in Be, C and Si doped p-type GaAs (Ashwin *et al.*, 1993a): such effects are weaker in Si-doped n-type GaAs. When these electron–phonon interactions occur, all the fine structure of the LVM line due to nearest neighbour isotopic interactions is lost.

Next it is necessary to consider silicon complexes. When both Si_{Ga} donors and Si_{As} acceptors are present in appreciable concentrations, nearest neighbour Si_{Ga}–Si_{As} pairs form. As there are two light inequivalent atoms in the defect, which has trigonal C_{3v} symmetry, four LVMs could occur due to antisymmetric and symmetric longitudinal and transverse modes respectively. The $^1\Gamma^-$ mode occurs at a high frequency of 464 cm^{-1}, as it involves stretching of the strong Si—Si bond and four fine structure features are found from $^{28}Si_{Ga}$–$^{28}Si_{As}$, $^{28}Si_{Ga}$–$^{30}Si_{As}$, $^{30}Si_{Ga}$–$^{28}Si_{As}$ and $^{30}Si_{Ga}$–$^{30}Si_{As}$ pairs in GaAs doped with silicon containing the enriched ^{30}Si isotope (Leung *et al.*, 1974; Theis and Spitzer, 1984). The $^2\Gamma^-$ mode occurs at 393 cm^{-1} (Fig. 12.2) and also shows partially resolved Ga isotopic fine structure. There has been debate and confusion about an early assignment of an LVM line at 367 cm^{-1} to the $^2\Gamma^+$ mode (Spitzer and Allred, 1968). The problem arose because it became clear that absorption at this energy could arise from some other Si complex not related to the Si_{Ga}–Si_{As} pair. The origin of this other absorption has now been identified (see below) and it would seem that the original assignment was correct, in spite of later questioning. The fourth mode $^1\Gamma^+$, which has not been detected, was calculated to have a low energy of ≈ 327 cm^{-1}, so that it may be masked by the GaAs Reststrahl absorption just below the maximum lattice frequency of ≈ 295 cm^{-1} at 4.2 K (Mooradian and McWhorter, 1967; Cochran *et al.*, 1961).

It was postulated from early photoluminescence (PL) measurements that Si_{Ga}–V_{Ga} second neighbour pairs should be present in GaAs (Williams and Bebb, 1972). Local vibrational mode measurements made on (a) plastically deformed and annealed GaAs (Ono and Newman, 1989) and (b) material grown by molecular beam epitaxy (MBE) at a low temperature of 200 °C or 250 °C (McQuaid *et al.*, 1992, 1993) have confirmed directly that these defects exist. The defect, originally called a Si–Y defect, has C_s symmetry so that the triply degenerate mode of isolated Si_{Ga} donors splits into three non-degenerate modes at 366.8 cm^{-1}, 367.5 cm^{-1} and 397.8 cm^{-1}. The lines are sharp, indicating the presence of a Si_{Ga} atom, while the disposition of the lines is similar to those arising from donor–acceptor pairs of the type Si_{Ga}–Cu_{Ga}, Si_{Ga}–Zn_{Ga}, Si_{Ga}–Li_{Ga}, etc. (Spitzer, 1971). There would appear to be no ambiguity about the assignment. It is apparent that the two lines at 366.8 cm^{-1} and 367.5 cm^{-1} are close to and overlap the line at 367 cm^{-1} originally attributed to Si_{Ga}–Si_{As} pairs. The Si_{Ga}–V_{Ga} pair is expected to be a net electron trap as V_{Ga} defects are thought to exist as V_{Ga}^{-}, V_{Ga}^{2-} or V_{Ga}^{3-}. The observation of the paired vacancy complex is informative because there is the implication that isolated V_{Ga} defects may also be present. These defects are likely to lead to rapid Si diffusion, similar to the enhancement of the diffusion of the group III elements, leading to intermixing of GaAs/AlAs quantum well structures (Pavesi *et al.*, 1992).

The final defect to be considered, called Si–X, not to be confused with a DX-centre, gives rise to absorption, with unique structure, close to 369 cm^{-1}, again immediately adjacent to the lines at 367 cm^{-1} (Fig. 12.2) (Murray *et al.*, 1989) and (Fig. 12.4) (Wagner and Ramsteiner, 1989: Wagner *et al.*, 1989a). This defect becomes dominant in highly Si-doped GaAs grown by the Bridgman technique or by MBE on a (001) surface at the normal growth temperature of 580 °C (Maguire *et al.*, 1987). It acts as a deep electron trap so that the carrier concentration first saturates and then decreases as the silicon concentration is increased in homogeneously doped material. This electrical behaviour has now also been reported for δ-doped structures (Köhler *et al.*, 1993). The important point is that a limiting value of n does not result simply from Si site-switching from Ga- to As-lattice sites. The Si–X defect is only found in as-grown material when Si_{As} acceptors are present and is destroyed by the in-diffusion of interstitial Li which combines with V_{Ga} to form Li_{Ga} acceptors (Spitzer *et al.*, 1969). Since, in addition, it is not present in GaAs grown by liquid phase epitaxy (LPE) it has been suggested that the defect incorporates a Si_{As} acceptor and a V_{Ga} defect (Murray *et al.*, 1989). Such a pairing of two negatively charged centres would not be expected, although the absence of a bond which normally links the Si_{As} atom to one of its neighbours would explain why no associated LVM occurs in the region of 399 cm^{-1}. Recently, it has been found that low temperature anneals ($\approx$350 °C) of MBE material grown at 200 °C or 250 °C convert Si_{Ga}–V_{Ga} defects to Si–X centres, implying that long-range migration is not involved (McQuaid *et al.*, 1993). It has been speculated that successive diffusion jumps, first of the bridging As-atom into the V_{Ga} and then of the Si_{Ga} atom to the resulting V_{As}, lead to a planar V_{Ga}–Si_{As}–As_{Ga} complex. This defect has a donor–acceptor characteristic and could explain deep PL emission ($\approx$1.0 eV) found in highly Si doped n-type GaAs. It is also clear that it could be a net electron trap, even though the As_{Ga} antisite defect is a double donor. The presence of Si–X defects is therefore another indicator that V_{Ga} defects might be present in the GaAs crystal.

LVM frequencies for Si complexes in GaAs are summarised in Table 12.2 and a more detailed discussion has been given by Newman (1994).

In the above discussion it has been implied that IR measurements might always be made on the as-received material. However, the presence of free-carrier absorption in *thick* ($\geqslant$1–2 μm) highly doped material including MBE layers causes them to be opaque and LVM measurements can only be made if electrical compensation can be effected. The usual procedure is to irradiate the material with 2 MeV, electrons (Spitzer *et al.*, 1969) in order

Table 12.2. *Vibrational modes of substitutional complexes in GaAs and related infrared data* (*Newman* 1993)

Impurity complex	LVM frequencies (cm^{-1})	Electrical behaviour	Concentration to give $IA = 1\ cm^{-2}$ ($cm^{-3} \times 10^{16}$)
$^{28}Si_{Ga}$–$^{28}Si_{As}$	367, 393, 464	Neutral	ND
$^{28}Si_{Ga}$–V_{Ga}	366.8, 367.5, 397.8	Electron trap	ND
Si–X	369	Electron trap	2.7 ± 1.0†

ND = not determined.
† Murray *et al.* (1989).

to produce randomly distributed intrinsic defects (vacancies, self-interstitials and Frenkel pairs (Murray *et al.*, 1987)) that act as electron/hole traps, at sites remote from the Si impurities. The dose ($e^-\ cm^{-2}$) required to remove the electronic absorption is numerically close to the free-carrier concentration n (cm^{-3}). In order to avoid confusion in the analysis of thin epitaxial films, the much thicker substrate should be undoped and semi-insulating but grown by the liquid encapsulated Czochralski (LEC) technique to eliminate the possibility of inadvertent silicon incorporation originating from SiO_2 components used during the growth of Bridgman GaAs.

It is known that some intrinsic defects introduced during room temperature irradiation are mobile and may be selectively trapped by impurities, although such processes are inhibited in strongly n-type GaAs (Murray *et al.*, 1987). Nevertheless, it was natural to ask whether the silicon site distribution is modified by the irradiation, especially as at least one, and possibly two, of the silicon complexes listed above incorporate gallium vacancies. The use of Raman spectroscopy applied to samples, first in their as-grown state and then after irradiation, has demonstrated that no detectable changes occur in n-type material (Murray *et al.*, 1989). On the other hand, evidence has been obtained from IR spectra that changes can occur in p-type Si-doped MBE GaAs grown on a (111)A face (Ashwin *et al.*, 1993a).

12.3 Measurement techniques

12.3.1 IR absorption

Traditionally, LVM studies have been made by IR absorption using dispersive grating spectrometers. Fortunately, the current requirement to examine thin epitaxial layers coincided with the commercial availability of Fourier transform spectrometers, with their much higher speed and their very high spectral resolution. Much-improved infrared detectors, including bolometers operating at 4.2 K and electronic detectors using mercury cadmium telluride (77 K), also became readily available. In practice, the LVM absorption has to be distinguished from much stronger underlying two-phonon absorption which has to be subtracted from the composite spectrum of the substrate with the overgrown epitaxial layer. The two-phonon absorption sharpens and decreases in strength when the sample temperature is lowered to $T \leqslant 77$ K (Cochran *et al.*, 1961) and thus samples should be cooled to 4.2 K.

The strength, defined as the integrated absorption coefficient (IA), of an LVM line is proportional to the concentration $[C]$ of impurities present in the particular lattice site giving rise to that line according to the relation:

$$IA = \int \alpha(\nu)\,d\nu = \frac{\pi[C]\eta^2}{nc^2 M_{imp}} \quad (\text{cm}^{-2}\text{: c.g.s. units}), \tag{12.2}$$

where $\alpha(\nu)$ is the absorption coefficient due to the LVM, M_{imp} is the mass of the impurity, n is the refractive index and η, an effective charge, is the dipole moment per unit displacement in the normal co-ordinate of the mode. To find values of η, measurements have to be made on samples for which $[C]$ is known by some other technique, or a combination of techniques, such as SIMS and electrical measurements. This process is relatively straightforward for a grown-in impurity which occupies only one type of lattice site, such as Al_{Ga} or Be_{Ga}, but the problem is complicated for an amphoteric impurity such as silicon, and is compounded by the formation of impurity complexes. The interpretation of electrical measurements may also be uncertain since high concentrations of active native defects may lower the measured electron or hole concentrations by significant factors. Values of η for substitutional impurities in GaAs are listed in Table 12.1, together with the impurity concentrations that give rise to an integrated absorption coefficient of 1 cm^{-2}. A more comprehensive tabulation has been given by Newman (1993).

The sensitivity of the IR measurements is such that $^{28}Si_{Ga}$ donors present in a concentration of 3×10^{17} cm^{-3} in a homogeneously doped epitaxial layer 1 μm in thickness can just be detected at a spectral resolution of 0.1 cm^{-1}. Thus, the detection limit is $\approx 3 \times 10^{13}$ atom cm^{-2}, corresponding to 0.05 of a monolayer (ML) on a (001) surface of GaAs (6.2×10^{14} Ga-atom sites cm^{-2}) (Murray *et al.*, 1989). The sensitivity for detecting Si_{As} acceptors is lower by a factor of about three ($\approx 10^{14}$ atoms cm^{-2}) because of the greater linewidth of the LVM, and because the frequency lies in a less favourable position in relation to the underlying two-phonon features (Ashwin *et al.*, 1993b).

12.3.2 Raman scattering

Studies of the vibrational modes of crystals and molecules using Raman scattering are well known and have been extended to impurity-induced LVMs in semiconductors. There was a question as to whether or not there would be adequate sensitivity since there was no *a priori* reason to suppose that the scattering cross-section, σ, of an impurity should be significantly different from that of a host lattice atom. For low impurity concentrations $[C]$, the LVM scattering was therefore expected to be weak and superposed on intrinsic multiphonon scattering, which has to be measured for undoped crystals and then subtracted from the composite spectra of the Raman shift. There is again an advantage in lowering the sample temperature to 77 K as the background scattering becomes sharper and less intense, while the LVM lines become sharper (Ramsteiner *et al.*, 1988; Wagner and Ramsteiner, 1989). The back scattering geometry $z(x, y)\bar{z}$ is commonly used in which light, plane polarised along the [100] direction, is incident on the sample (001) surface so that the refracted beam is almost parallel to the [001] direction and x, y and z denote [100], [010] and [001] crystallographic directions respectively. Raman scattering is then allowed for vibrational modes with Γ_{15} symmetry (corresponding to an LVM of an impurity with T_d symmetry (Nakamura and Katoda, 1985)) and Γ_{25} symmetry but the component of the intrinsic two-phonon background scattering with Γ_1 symmetry is suppressed (Cardona, 1982).

For homogeneously doped material there is an advantage in using a large scattering volume, V, since the intensity of the scattered light per unit solid angle can be written as $\partial S/\partial\Omega \approx \sigma[C]V$. If the incident laser energy is smaller than the bandgap E_g of the semiconductor, V will be proportional to the sample thickness. The LVMs of $^{11}B_{Ga}$ and $^{10}B_{Ga}$ in GaP crystals were examined in this way by Hon *et al.*, (1970). For semiconductors with a smaller value of E_g, there is a disadvantage in using sub-bandgap irradiation since the scattered intensity varies as the fourth power of the laser frequency according to classical analysis. However, when the photon energy is greater than E_g, the scattering can occur only from a thin surface layer of depth $d_L \approx 1/2\alpha_L$, where α_L is the absorption coefficient of the incident radiation. For GaAs, d_L varies from ≈ 1400 Å to ≈ 80 Å as $h\nu$ increases from 1.92 to 3 eV, and is usually taken as $d_L \approx 100$ Å for the 3 eV line of a Kr-ion laser (see Cardona and Harbeke, 1962; Aspnes and Studna, 1983; Ramsteiner *et al.*, 1988). The scattered light is filtered and dispersed, either by a double or, now more usually, by a triple monochromator, and then detected, either with a single element detector with many multiple scans, or, more efficiently, with a detector array to provide parallel processing. In the latter case, the limiting spectral resolution ($\approx$2–5 cm^{-1}) is determined by the spacing of the elements in the array and the dispersion of the third stage of the monochromator.

The scattering cross-sections for Si_{As} and Si_{Ga} are dependent on the energy of the incident photons. For Si_{As} acceptors, there is a smooth increase by a factor of ≈ 3 as $h\nu$ is increased from 1.9 eV to 2.7 eV (Ramsteiner *et al.*, 1988), similar to that found for deformation potential scattering by LO phonons. There is a further increase in the strength of the intrinsic scattering as $h\nu$ approaches 3 eV, corresponding to a resonance with the E_1-bandgap, and it was assumed that this increase also occurred for the impurity LVM scattering of Si_{As}. The behaviour of Be_{Ga} acceptors appears to be similar to that of Si_{As} acceptors in highly p-type GaAs, but excitation above 2.5 eV is required to eliminate a contribution to the background from a ‘hot’ photoluminescence band extending up to this energy (Wagner *et al.*, 1991a): at the resonance condition ($T = 77$ K, $h\nu = 3.0$ eV) $\partial S/\partial\Omega = (1.1 \pm 0.6) \times 10^{-24}$ Sr^{-1} cm^{-2}. By contrast, the scattering from Si_{Ga} donors showed no increase over the range 1.9 eV $\leqslant h\nu \leqslant$ 2.7 eV, and for slightly higher energies the LVM line was not detectable in some of the samples examined by Ramsteiner *et al.* (1988). However, at the resonance condition ($T = 77$ K, $h\nu = 3.0$ eV), the Si_{Ga} LVM becomes very strong with $\partial S/\partial\Omega = 3.2 \pm 0.6 \times 10^{-24}\,\mathrm{Sr}^{-1}\,\mathrm{cm}^{-2}$ (Wagner, 1990), some 30 times greater than that for deformation potential LO phonon scattering per lattice atom. Under these particular conditions the Raman spectrum, after subtraction of the background multiphonon scattering, closely resembles the IR absorption spectrum (Fig. 12.4). Similar resonance effects for silicon donors have been reported for InAs and InSb hosts (Wagner *et al.*, 1991b). Thus the overall appearance of the Raman spectrum of GaAs: Si is very dependent on the experimental conditions used (see, for example, Uematsu and Koichi, 1990), and it is not always easy to make direct comparisons of measurements taken by different research groups.

To determine Si_{Ga} impurity concentrations experimentally, the Raman LVM line first has to be 'normalised'. Initially, Wagner *et al.* (1989b) compared its height $I(Si_{Ga})$ above the local background with the height $I(460\,\mathrm{cm}^{-1})$ of the Γ_{15} component of the intrinsic phonon background at $460\,\mathrm{cm}^{-1}$ above the minimum at $485\,\mathrm{cm}^{-1}$. The ratio $R = I(Si_{Ga})/I(460\,\mathrm{cm}^{-1})$ is then independent of variations in the laser intensity and the precise geometry used. A value of $R = 1$ corresponds to $[Si]_{Ga} = 5.6 \pm 0.5 \times 10^{18}\,\mathrm{cm}^{-3}$ (resolution $5\,\mathrm{cm}^{-1}$, $T = 77$ K, $h\nu = 3$ eV) and a similar calibration has been found for the Si–X

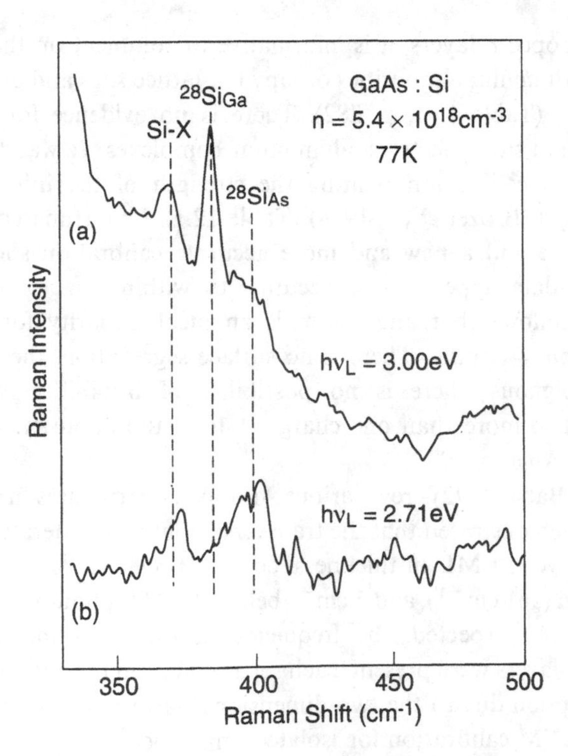

Fig. 12.4. Raman scattering from a Si doped MBE (001) layer with [Si] = $(4 \pm 0.6) \times 10^{19}\,\mathrm{cm}^{-3}$ (SIMS) after the subtraction of the intrinsic background phonon scattering showing the LVMs from Si impurities in various sites. Note the huge change in the strength of the line from Si_{Ga} when the incident photon energy is increased from 2.71 eV to 3.00 eV to obtain resonance with the E_1 bandgap (after Wagner and Ramsteiner, 1989).

complex. The scattering from Si_{As} at 399 cm^{-1} overlaps that from Si_{Ga}–Si_{As} pairs at 393 cm^{-1}, but it has been inferred that the calibrations for these two centres are similar. Measurements have not been obtained for Si_{Ga}–V_{Ga} complexes. In later work, Wagner (1990, 1991) used a modified ratio $I(Si_{Ga})/I(540\ cm^{-1})$ in order to avoid problems of electronic scattering near 485 cm^{-1}. A value of $R = 1$ of the original ratio transforms to $R' = 0.22$ for the new ratio, according to the published diagrams.

The smallest concentrations of $[Si_{Ga}]$ and $[Be_{Ga}]$ that can be detected by Raman scattering in uniformly doped GaAs are ≈ 2–3×10^{18} cm^{-3}, under optimum conditions and with a spectral resolution of 5 cm^{-1} (Wagner, 1990). Since the probing depth is only ≈ 100 Å, the detection limit is $\approx 2 \times 10^{12}$ atoms cm^{-2} ($\approx 3 \times 10^{-3}$ ML).

12.4 Analysis of δ-layers in GaAs

12.4.1 LVM studies of aluminium δ-layers

Before discussing Si-doped δ-layers, it is informative to comment on the available results for Al-doped layers. Aluminium impurities occupy Ga-lattice sites and give rise to an LVM at 361.8 cm^{-1} (4.2 K) (Table 12.1, p. 282). There is no evidence for the formation of aluminium impurity antisite defects or aluminium complexes. It was therefore straightforward to establish a calibration relating the strength of the infrared LVM to the concentration of $[Al_{Ga}]$ (Brozel *et al.*, 1978) (Table 12.1). Unfortunately, this latter work was not comprehensive and a new and more accurate calibration should be obtained: nevertheless, the old data appear to be accurate to within a factor of two (Ono and Furuhata, 1991). It follows that aluminium is an ideal impurity for the study of IR absorption arising from δ-doping. There is no surface segregation, the lattice sites of the impurity are unambiguous, there is no possibility of bistability and there are no complications related to more than one charge state of the impurity, or to free carriers (see also Tanino and Amano, 1992).

Recently, Ono and Baba (1992) grew various Al-doped δ-structures in GaAs by MBE on (001) substrates and demonstrated that the transverse doubly degenerate vibrational mode of two-dimensional layers, 1 ML in thickness, occurs at 358 cm^{-1}, 3 cm^{-1} below that of the isolated Al_{Ga} LVM (361 cm^{-1}), and 5 cm^{-1} below the TO–phonon mode at the Γ-point of AlAs (363 cm^{-1}). As expected, the frequency of this two-dimensional mode was unchanged when 20 δ-layers were present, each separated from the next by 20 ML of GaAs. The integrated absorption due to the two-dimensional layer appeared to be close to that calculated from the LVM calibration for isolated impurities.

These experiments have been repeated and extended (Ashwin *et al.*, 1994). Aluminium δ-layer superlattice structures were grown at 400 °C by MBE (001) with a spacing of 500 Å between adjacent layers, and the final layer was capped with 50 Å of GaAs. The Al planar concentrations examined (per layer) were 0.1 ML (20 layers), 0.5 ML (20 layers) and 1.0 ML (100 layers). X-ray measurements of the satellite structure around the 002 reflection were used to verify these sample specifications and indicated negligible spreading of the Al atoms. Infrared absorption spectra with the samples at 10 K showed the Al_{Ga} LVM at 361 cm^{-1} for the lowest doping level and a shift to 358 cm^{-1} for the two higher doping levels. These results are therefore in agreement with Ono and Baba's (1992) data. Raman scattering measurements using a Kr-ion laser (3 eV) were then made in the back scattering configurations $z(x, y)\bar{z}$, with the samples at 77 K. Subtraction of the scattering from undoped GaAs from the composite spectra yielded the spectra shown in Fig. 12.5. The sample with the lowest doping showed a line at 360 cm^{-1}, close to that observed by IR absorption. There was then development of scattering at ≈ 390 cm^{-1}, about 10 cm^{-1} below the LO frequency of AlAs for the two more highly doped layers. This latter shift is consistent with that expected for very thin layers of AlAs, as discussed by Tanino and Amano (1992). At the same time, there was a progressive reduction in the strength of the TO-like mode, as expected, because this mode is not Raman active for bulk AlAs for the scattering configuration used. The important points to note are the observation of the early stages of the transition from two-dimensional to three-dimensional behaviour in the Raman

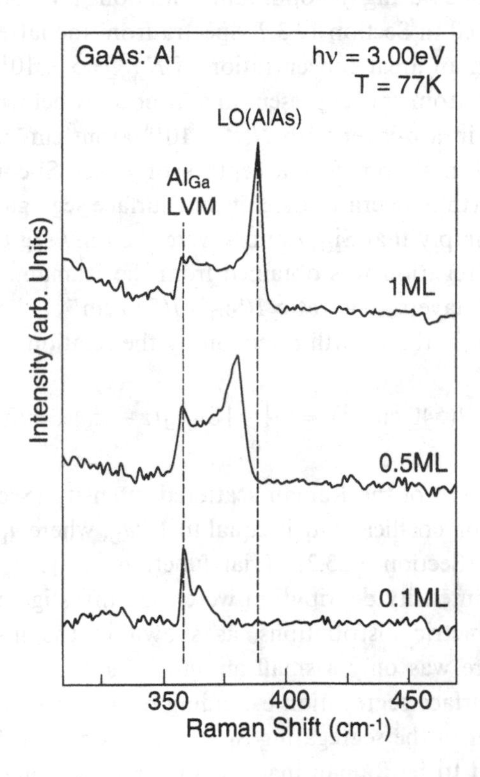

Fig. 12.5. Raman scattering for the $z(x, y)\bar{z}$ configuration from Al δ-layers of thickness 0.1 ML, 0.4 ML and 1 ML, respectively, in GaAs, after subtraction of the scattering from pure GaAs. At the lowest doping level the principal feature is the Al_{Ga} LVM but there is a progressive transition towards LO(AlAs) scattering as the Al concentration is increased (after Ashwin *et al.*, 1994).

spectra, the fact that the Raman scattering was strong and the fact that IR absorption increased as the total areal concentration of Al was increased.

It will be necessary to recall these results when Si-doped δ-layers are discussed, as it will be shown that the strength of the Si LVM absorption tends to zero as the thickness of each layer is increased close to about 0.5 ML and there is no detectable Raman scattering.

12.4.2 Silicon doping

(*a*) *As-grown GaAs*

The first LVM measurements of silicon and beryllium δ-layers in GaAs were made on samples grown by molecular beam epitaxy on (001) substrates held at 580 °C (Wagner *et al.*, 1989b). The structures contained either a single Si δ-layer, with an areal concentration of $N_D^{2D} = 8 \times 10^{12}$ atom cm^{-2} located at various depths, z_0 (50–500 Å) from the uppermost surface, or alternating Be and Si δ-layers to produce a 'saw-tooth' doping superlattice. Raman spectra were obtained using the optimum conditions for resonant scattering of the Si_{Ga} impurities, as described in Section 12.3.2. Spectra from the latter samples showed the LVM line from Be_{Ga} with an areal concentration of $N_A^{2D} = 5 \times 10^{12}$ atom cm^{-2} (deeper layers were doped to 10^{13} atom cm^{-2}) present in a layer 53 Å below the surface. An LVM line due to Si_{Ga}, present in a δ-layer with $N_D^{2D} = 10^{13}$ atom cm^{-2}, 105 Å below the Be layer was also present. Lines from Si_{As} acceptors or other Si-complexes were absent. Because of the 'high' growth temperature, significant surface segregation was expected and the measurements could imply that Si_{Ga} donors were the only segregating species.

Direct evidence for segregation was obtained from the Raman spectra of the samples containing the single Si δ-layers. The ratio $I(Si_{Ga})/I(540\ \text{cm}^{-1})$ should be related to the distribution of Si_{Ga} atoms in the growth direction by the relation:

$$I(Si_{Ga})/I(540\ \text{cm}^{-1}) = k \int_0^{\infty} [Si_{Ga}](z - z_0)e^{-2\alpha_L z}\, dz, \tag{12.3}$$

using the later normalisation of the Raman scattered intensity (Section 12.3.2) (Wagner, 1990, 1991). The absorption coefficient α_L is equal to $1/2d_L$, where $d_L = 100$ Å for photons with an energy of 3 eV (Section 12.3.2). Trial functions of $[Si_{Ga}](z - z_0)$ that would reproduce the measured intensity distribution were then investigated. Fits could only be obtained for very asymmetric distributions, as shown in the inset of Fig. 12.6. The implication was that there was only a small amount of spreading of the δ-layer due to diffusion but there was surface segregation extending over a distance of some 200 Å. For $z_0 < 200$ Å, the remainder of the segregating Si atoms was assumed to be located on the surface of the sample and to be Raman inactive. Other measurements made on samples grown in a similar way, but at a lower temperature of 500 °C, showed much narrower δ-layers with a width of <50 Å (Fig. 12.7) (Wagner, 1991, 1992a). The procedure outlined is indirect but is the best that can be achieved at the present time.

Supporting evidence for the surface segregation of silicon impurities over a distance

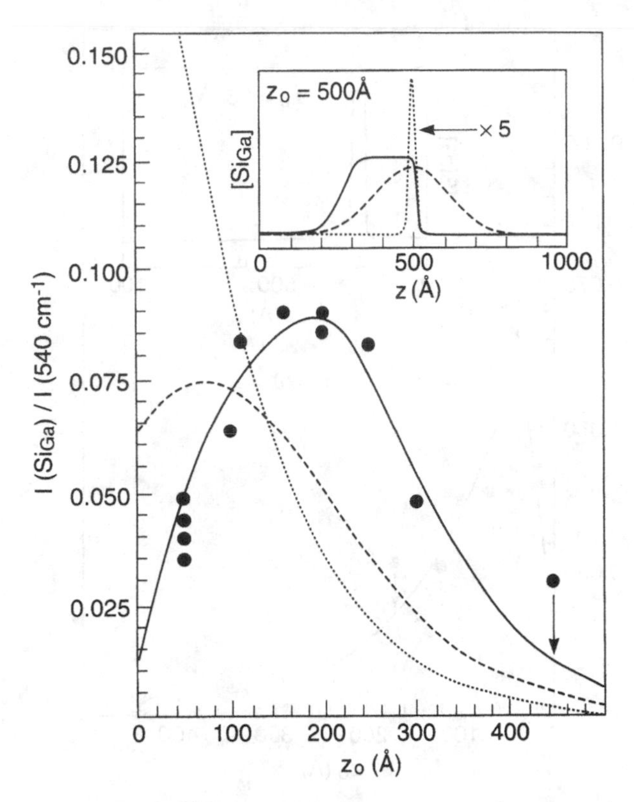

Fig. 12.6. Normalised $^{28}Si_{Ga}$ LVM Raman scattering intensity $I(Si_{Ga})/I(540\ cm^{-1})$ vs depth z_0 for Si doping spikes with $[Si] = 8 \times 10^{12}\ cm^{-2}$ located at various depths z_0 in MBE (001) layers grown at 580 °C. The inset shows different assumed profiles of $[Si_{Ga}]$ used to evaluate the integral in Eqn. (12.3). Neither the sharp spike (dotted lines) nor the symmetrically broadened spike (dashed lines) agree with the measurements. Good agreement is obtained only with an asymmetric $[Si_{Ga}]$ distribution produced by surface segregation (after Wagner 1990).

of ≈20 ML (=56 Å) was obtained in the work of Fahy *et al.* (1992) from studies of the progressive changes of the surface reconstruction from 3 × 1 to 2 × 4 revealed by reflection high energy diffraction patterns as GaAs was deposited on the Si δ-layers at 580 °C. Secondary ion mass spectrometry measurements made on a δ-doped (100 periods) superlattice grown at 550 °C also showed significant spreading but another result of interest was that the only centres observed by IR LVM measurements were Si_{Ga} donors (Ashwin *et al.*, 1993b). There is, therefore, a second independent observation that segregating Si atoms appear to occupy mainly Ga-lattice sites.

Raman measurements (Brandt *et al.*, 1991) were made on MBE (001) samples for which the Si deposition was carried out in an As-free environment at 450 °C in a separate chamber from that used for the growth of the GaAs. The capping layer of GaAs was grown at 580 °C after an initial deposition at 300 °C (Crook *et al.*, 1992). The thickness of the silicon

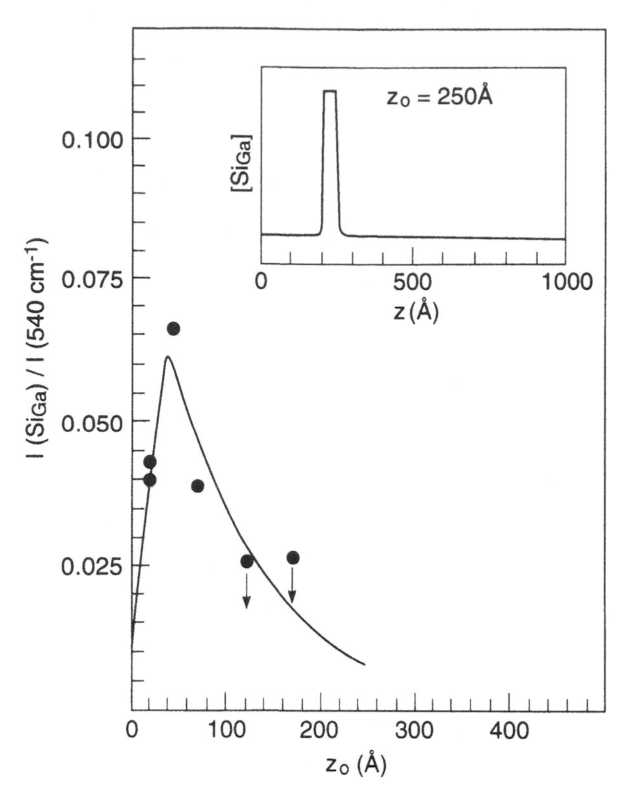

Fig. 12.7. Normalised $^{28}Si_{Ga}$ LVM Raman scattering intensity $I(Si_{Ga})/I(540\ cm^{-1})$ for Si doping spikes with $[Si] = 2.8 \times 10^{12}\ cm^{-2}$ located at various depths z_0 in MBE (001) layers grown at a temperature below 500 °C. Agreement with Eqn. (12.3) is obtained by assuming that the Si is confined to a layer less than 5 ML (<14 Å) in width (after Wagner 1992a).

layers was either 0.33 ML or 0.74 ML and a range of capping layer thicknesses from 50 to 2000 Å was investigated. Secondary ion mass spectrometry measurements revealed a spike close to the position of the deposited Si-layer with extended Si segregation extending over ≈ 400 Å at a level of $[Si_{Ga}] \approx 10^{19}\ cm^{-3}$. The LVM spectra showed mainly Si_{Ga} donors at large distances from the δ-layer (thick capping layers), but closer to the layer LVM lines from Si_{Ga}, Si_{As}, Si_{Ga}–Si_{As} pairs and Si–X were all detectable (Fig. 12.8). Thus the overall spectrum resembled that found for highly Si-doped bulk MBE GaAs grown at 580 °C (Fig. 12.2) and the presence of Si–X implied that V_{Ga} defects might have been present. There was in addition a band, labelled A, at 480 cm^{-1}, which was assigned to the vibrations of Si–Si bonds (Fig. 12.8): these bonds must have been distinct from those of Si_{Ga}–Si_{As} pairs which have a longitudinal frequency of 464 cm^{-1}. There was therefore an implication that three-dimensional clustering of Si atoms had occurred, as 480 cm^{-1} is the spectral position observed for vibrations of Si–Si bonds in amorphous silicon (Bermejo and Cardona, 1979; Hayashi, 1989). Clustering of Si for thicker deposits was confirmed by the transmission electron microscopy (TEM) studies of Brandt *et al.*, (1993), while

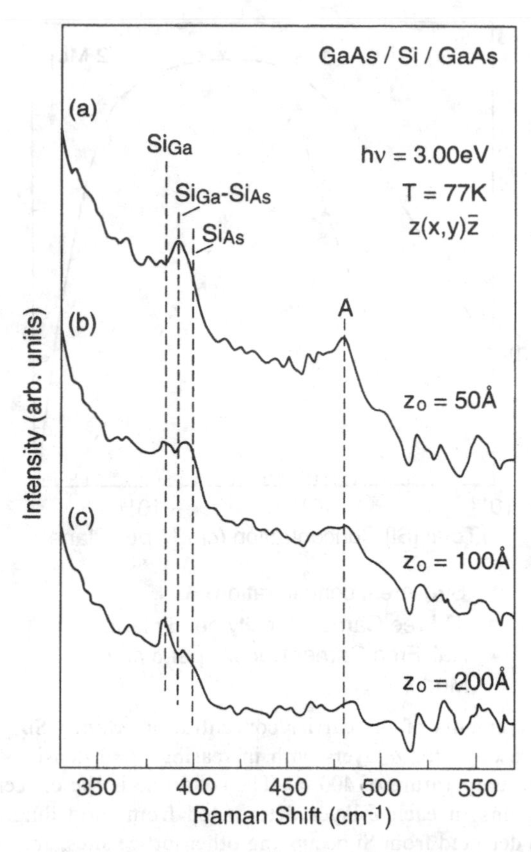

Fig. 12.8. Raman spectra after the subtraction of the two-phonon scattering of GaAs samples with Si δ-layers at depths of 50, 100 and 200 Å below the top surface of the capping layer. The Si layers, each with a thickness of 0.74 ML (4.8×10^{14} cm^{-2}), were grown in the absence of an As-flux at $T < 500$ °C. The spectra indicate that the concentrations of $[Si_{As}]$, $[Si_{Ga}-Si_{As}]$ pairs and the defect giving the line A are greatest in the region of the deposited δ-layer. It has been suggested that band A was due to Si–Si vibrations (after Brandt *et al.* 1991).

Raman scattering near 480 cm^{-1} has also been found for such silicon layers (Tanino *et al.*, 1991; Scamarcio *et al.*, 1992; Sorba *et al.*, 1992).

An alternative approach was used in parallel IR LVM measurements. Here, the aim was to inhibit the segregation and diffusion of the Si δ-layers by effecting growth at the lower temperature of 400 °C with the As flux maintained (Ashwin *et al.*, 1993b). Secondary ion mass spectrometry data indicated that the total spreading should not exceed $\approx$1–2 ML (Clegg and Beall, 1989; Harris *et al.*, 1991): temperatures lower than 400 °C were not used because it had been shown that lattice defects are introduced into the layers (Blood and Harris, 1984). Triple axis X-ray analyses of as-grown superlattices have shown that the spreading was no greater than $\approx$1.5 ML and it was inferred that the Si atoms were located

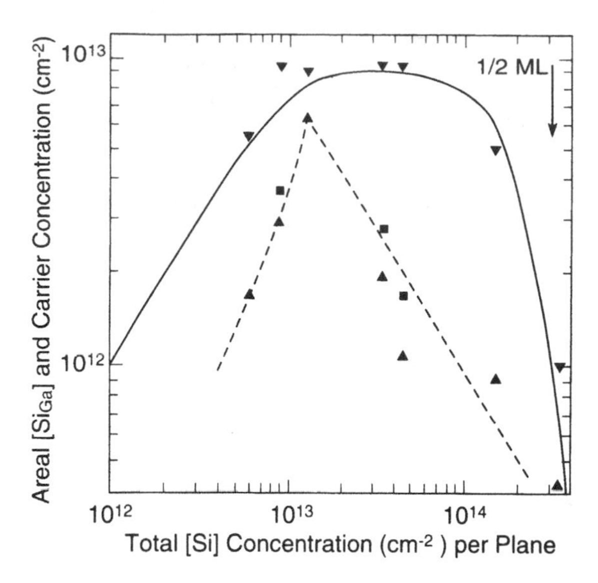

Fig. 12.9. Measurements of the carrier concentration n and [Si_{Ga}] made by LVM IR spectroscopy for δ-layers with increasing areal densities grown by MBE (001) at a temperature of 400 °C. The solid line is the concentration of isolated Si_{Ga} atoms in each δ-layer determined from modelling. No LVM absorption was detected from Si occupying other lattice sites, and line A (Fig. 12.8) was absent (after Ashwin *et al.*, 1993b).

on Ga-lattice sites for layer thicknesses up to 0.5 ML (Hart *et al.*, 1993). The LVM spectra showed only the line from Si_{Ga}, the strength of which first increased with increasing areal concentrations of silicon, followed by a plateau region and a final decay until *no* absorption was detectable (Fig. 12.9). Modelling the occupation of surface Ga sites by Si atoms during its deposition led to estimates of the numbers of 'isolated' Si atoms which followed the strength of the LVM line. Once two-dimensional clusters and/or pairs became dominant (according to the calculations), the IR absorption was lost. This result could be understood if three-dimensional Si crystallites were produced but there was no evidence for their existence from X-ray measurements, TEM or reflection high energy electron diffraction (RHEED). At first sight, a two-dimensional sheet of Si_{Ga} atoms would have been expected to produce strong absorption by analogy with the results for Al_{Ga} (Section 12.4.1). Even if a double layer of Si_{Ga}–Si_{As} formed, lines near 393 cm^{-1} and 464 cm^{-1} would have been expected. Limited Raman measurements made on Ashwin *et al.*'s (1993b) samples showed only a very weak LVM line from Si_{Ga}, while line A at 480 cm^{-1} (Si–Si pairs) was not observed (Wagner, 1992b). It was speculated that pairs of silicon atoms formed a two-dimensional ordered array of Si_{Ga}^{+}–Si_{Ga}^{-} (DX) centres, possibly leading to a low dipole moment and a low polarisability.

Another puzzle was that the electrical conductivity of these layers first increased as the silicon concentration increased, but it also passed through a maximum and fell to a value close to zero for layers 0.5 ML in thickness (Ashwin *et al.*, 1993b). Such electrical behaviour has also been observed by Köhler *et al.*, (1993). In our first study (Beall *et al.*, 1989), this sequence had not been established and so samples were unnecessarily irradiated with 2 MeV electrons to remove the expected free-carrier absorption. However, later samples for LVM studies were not given this post-growth treatment.

(*b*) *Effects of a post-growth anneal*

Samples grown at 400 °C and not subjected to a post-growth irradiation were heat treated at 600 °C for 30 min, leading to an increase in the number of free carriers (Beall *et al.*, 1989). After electron irradiation to remove the electronic absorption, IR LVM spectra

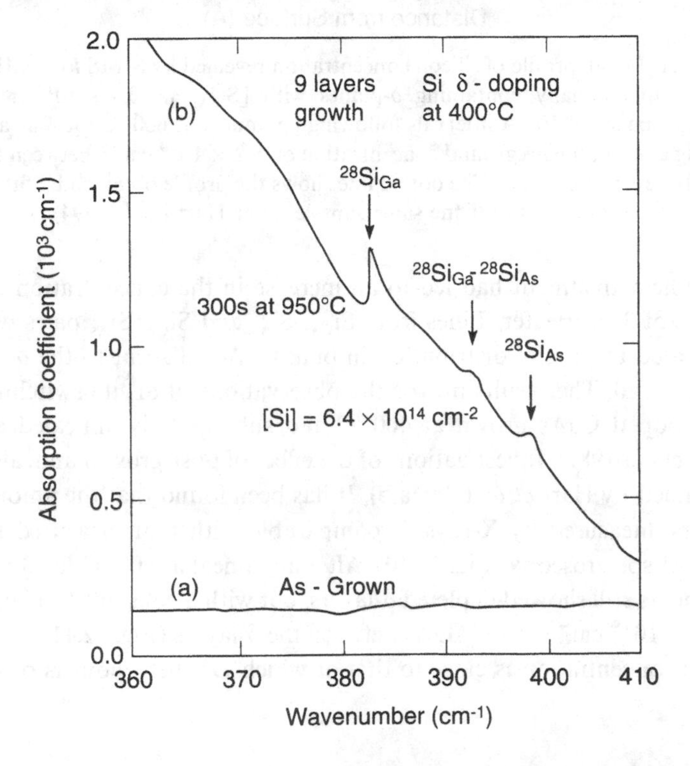

Fig. 12.10. IR LVM spectra from GaAs MBE (001) samples grown at 400 °C containing nine Si δ-layers, each with an areal concentration of $[Si]_A = 6.4 \times 10^{14}$ cm^{-2} (1 ML) after subtraction of the intrinsic two-phonon absorption. Line (a) for as-grown sample showing zero detectable LVM absorption and no free-carrier absorption (zero slope) and (b) the same sample after heating at 950 °C for 300 s leading to diffusion of Si out of the δ-layers and the presence of free carriers (sloping background).

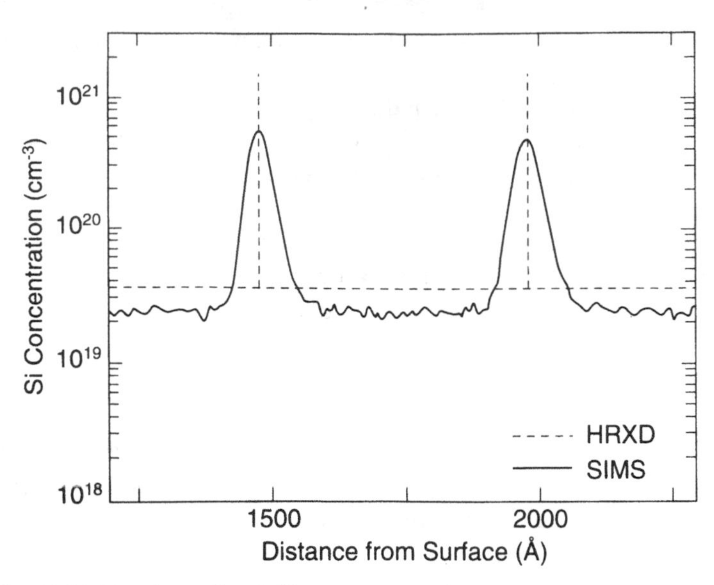

Fig. 12.11. Depth profile of silicon concentration revealed by SIMS for a MBE (001) sample initially containing δ-planes with $[\mathrm{Si}]_{\mathrm{Areal}} = 3.4 \times 10^{14}\ \mathrm{cm}^{-2}$ (0.5 ML) spaced at 500 Å intervals following an anneal at 600 °C for 3 hr and showing a uniform background concentration of $\approx 2 \times 10^{19}\ \mathrm{cm}^{-3}$ between the partially depleted δ-layers. The dotted line shows the profile obtained by fitting to the X-ray data for the same sample (after Hart *et al.*, 1994).

showed that the heat treatment had led to an increase in the concentration of detectable silicon by factors of 3 or greater. Lines from $\mathrm{Si_{Ga}}$, $\mathrm{Si_{As}}$ and $\mathrm{Si_{Ga}}$–$\mathrm{Si_{As}}$ pairs were detected and could be related to the site distribution in bulk GaAs adjacent to the δ-layers, where diffusion had occurred. The results mirror the observations of Si site-switching found for homogeneously doped GaAs grown at 400 °C and subsequently annealed at 500 °C or 600 °C (Murray *et al.*, 1989). Investigations of the effect of post-growth anneals on δ-layers have been continued by Hart *et al.* (1994a,b). It has been found that the amount of Si lost from the δ-layers, measured by X-rays, is comparable with that measured in the GaAs lattice by IR LVM spectroscopy (Fig. 12.10). After an anneal at 600 °C for 3 hr, SIMS and X-ray measurements still showed depleted δ-layers, but with a constant total concentration of silicon of $\approx 2 \times 10^{19}\ \mathrm{cm}^{-3}$ in the GaAs between the δ-layers (Fig. 12.11). It is interesting to note that this concentration is close to that at which DX behaviour is observed.

12.5 Conclusions

In Section 12.2, the basic theory of localised vibrational modes was outlined and illustrated mainly by the available information about the sites occupied by silicon impurities in homogeneously doped MBE layers of GaAs. Five defects, namely $\mathrm{Si_{Ga}}$ donors, $\mathrm{Si_{As}}$ acceptors, $\mathrm{Si_{Ga}}$–$\mathrm{Si_{As}}$ pairs, $\mathrm{Si_{Ga}}$–$\mathrm{V_{Ga}}$ pairs and Si–X defects have been identified in high resolution IR spectra, and the first four have also been reported in Raman scattering spectra. Calibrations have been obtained relating the concentrations of silicon in those particular sites to the strengths of the LVMs for both IR (Section 12.3.1) and Raman (Section 12.3.2) spectra. The latter calibrations are valid only for specific experimental conditions with excitation at 3.0 eV and with a sample temperature of 77 K; very different calibrations would apply for even small changes in the incident photon energy and great care is required in making intercomparisons of measurements made by different researchers. It should be noted that corresponding IR and Raman calibrations have also been obtained for Be_{Ga} in p-type MBE GaAs. The IR LVM line appears with a derivative profile but the problem of determining the integrated absorption coefficient is overcome by subjecting samples to electron irradiation to effect electrical compensation, when the LVM assumes a normal symmetrical absorption shape.

It appears that the systems GaAs:Si and GaAs:Be grown on a (001) substrate at a temperature $T_g \geqslant 400$ °C are reasonably well understood. Perhaps the most important result to emerge for homogeneously doped n-type material (GaAs:Si) is that at the highest doping levels the carrier concentration *falls*, not because of site switching of Si_{Ga} to Si_{As}, as is commonly assumed, but due to the presence of a rapidly increasing concentration of Si–X defects. As T_g is reduced from 580 °C to 400 °C, [Si–X] decreases, allowing n_{max} to increase: however further reductions in T_g again lead to low values of n due to the presence of excess Ga vacancies. A possible structure for the Si–X defect incorporating a V_{Ga} defect was discussed, but there is a need to verify or reject this model so that attention can be directed towards understanding the mechanism of its formation. Apart from Si_{Ga}–Si_{As} pairs, LVM spectroscopy has provided no evidence for the presence of other small silicon clusters.

The amount of information obtained for δ-doped samples is extremely limited and even that is not well focused. For Si-doped layers in GaAs the results depend critically upon the growth conditions which control the surface segregation and the silicon diffusion during growth. When there is significant spreading, the resulting doping pattern appears to be similar to that in highly homogeneously doped GaAs. In addition, the carrier concentration reaches a maximum value and then falls with further increases in the doping level. X-ray data have confirmed that any spreading was $\leqslant 1.5$ ML for δ-layers grown at 400 °C under an As-flux and it was concluded that all the Si atoms were present as Si_{Ga} to explain the measured tetragonality of doping superlattice structures. These structures are probably as close as it is possible to get to true planar or δ-doping when account is taken of the presence of steps and terraces on the surface, but there is currently no explanation for the progressive loss of all IR LVM absorption, as the areal concentration of Si was increased. Transmission electron microscopy measurements have not revealed the presence of three-dimensional silicon precipitate particles in such layers. It is interesting to ask whether we should consider deposits less than 1–2 ML in thickness to be silicon arsenide (Si_3As_4) rather than silicon since the loss of optical activity is not found for Al δ-layers. It could then be speculated that ordered Ga-vacancies are present: alternatively, Si DX-behaviour may be involved. The latter possibility is not radically different from the former since 'vacancies' would be formed when Si_{Ga} atoms moved off their lattice sites, and there is recent evidence for this process from positron annihilation measurements made on AlGaAs (Mäkinen *et al.*, 1993). There is clearly an intriguing puzzle to be resolved.

The relative strengths and weaknesses of the IR and Raman techniques for the study of impurity LVMs in δ-layers can be inferred from the foregoing text but it is helpful to present an overview here.

1. The spectral resolution of IR measurement (0.05 cm^{-1}) is superior to that of current Raman measurements by a factor of ≈ 50, which is advantageous when there are closely spaced LVM lines from different defect centres such as Si–X (369 cm^{-1}) and Si_{Ga}–V_{Ga} pairs (366.8 cm^{-1} and 367.5 cm^{-1}).
2. The problem of obtaining adequate IR transparency of highly conducting GaAs can be overcome by subjecting samples to high energy (1–2 MeV) electron irradiation. This treatment is not required for Raman measurements because only a very thin layer is probed.
3. The sensitivity of the Raman technique, in terms of the minimum areal concentration of Si_{Ga} that can be detected ($\approx 2 \times 10^{12}$ cm^{-2}), is better than that of the IR method ($\approx 3 \times 10^{13}$ cm^{-2}) by an order of magnitude.
4. The Raman technique has good depth resolution, limited only by the penetration depth of the incident laser radiation. Consequently, it is possible to obtain information about impurity profiles throughout layers that are only ≈ 200 Å in thickness but a range of samples with grown-in δ-layers at different depths z_0 from the surfaces has to be examined. Such measurements complement SIMS measurements but it has to be assumed that the growth of individual layers is completely reproducible. Care is also necessary in the interpretation because of possible site-switching of silicon, due to Fermi level effects resulting from the presence of surface states, and silicon surface segregation effects, as deduced from scanning tunnelling microscopy measurements (Pashley and Haberern, 1991) when the δ-layer is close to the surface.
5. The IR method has the advantage of having, in effect, no depth limitation. Thus, many (≈ 100) δ-layers can be grown into MBE GaAs at regular intervals to form a doping superlattice and, in principle, there is no limit to the number of layers. There is a great advantage in using such structures rather than a single δ-layer for the determination of the site distribution and the spreading effects of Si impurities when the LVM (and SIMS) measurements are complemented by high resolution X-ray analyses (Hart *et al.*, 1993, 1994).
6. The Raman technique yields information about the point symmetry of isolated impurities when different polarisations of the incident and scattered light are used. Only limited data are available for B_{Ga} in GaP (Hon *et al.*, 1970), Si_{As} in GaAs (Nakamura and Katoda, 1985) and H–N_{Se} pairs in Zn_{Se} (Wolk *et al.*, 1993). Such information can be obtained from IR measurements only if samples are subjected to a uni-axial stress, but, nevertheless, this procedure has been used quite extensively for LVM lines at higher frequencies relating to paired hydrogen

complexes (see, for example, Pajot *et al.*, 1991). However, the author is unaware of such measurements being applied to δ-layers.

The use of LVM spectroscopy has added another technique to those available for the study of δ-layers and has allowed the sites occupied by Si atoms to be determined, information which is not available from chemical techniques such as SIMS. There is clearly scope for comprehensive studies of Be-doping, further work involving Si and then investigations of other dopants. Such LVM studies need not be limited to GaAs hosts but could be made on InAs and InSb, for which both IR and Raman measurements have already been made for Si and Be in homogeneously doped samples, allowing limited calibration data to be obtained (Addinall *et al.*, 1992).

Note added in proof

Since this chapter was prepared, new Raman (Wagner *et al.*, 1995) and IR (Newman *et al.*, 1995) LVM spectra with improved signal/noise ratios have been obtained from samples containing silicon δ-layers by using the most recently available equipment. In addition to Si_{Ga} (Fig. 12.9), Si_{As}, Si_{Ga}–Si_{As} pairs and Si–X are present for layers grown at 400 °C with silicon areal concentrations, $[Si]_A$, less than 0.5 ML but the concentrations of all threse defects fall to zero for $[Si]_A > 0.5$ ML. Line A (Fig. 12.8), observed only in Raman spectra, then evolves in the manner described by Tanino *et al.* (1991). New X-ray simulations that agree with the measurements of Hart *et al.* (1993) and take these new LVM observations into account have been developed. It is now clear that such simulations do not provide sufficient information to determine a unique structure of a silicon δ-layer.

Acknowledgements

The author would like to thank Dr J. J. Harris and Dr J. Wagner for making critical comments and suggesting improvements to the manuscript. Miss D. Pullar-MacMillan is thanked for the preparation of the manuscript and Mr N. Powell for preparing the illustrations. The Science and Engineering Research Council, UK, are thanked for their financial support of this work.

References

Addinall R., Murray R., Newman R. C., Wagner J., Parker S. D., Williams R. L., Droopad R., de Oliveira A. G., Ferguson I. F. and Stradling R. A. (1991) *Semicond. Sci. & Technol.* **6**, 147.

Addinall R., Newman R. C., Ferguson I. T., Mohades-Kassai A., Brozel M. R., Sharma V. K. M., McPhail D. and Sangster M. J. L. (1992) *Mater. Sci. Forum* **83–87**, 1027.

Ashwin M. J., Fahy M. R. and Newman R. C. (1993a) *J. Appl. Phys.* **73**, 3574.

Ashwin M. J., Fahy M. R., Harris J. J., Newman R. C., Sansom D. A., Addinall R., McPhail D. S. and Sharma V. K. M. (1993b) *J. Appl. Phys.* **73**, 633.

Ashwin M. J., Fahy M. R., Hart L., Newman R. C. and Wagner J. (1994) *J. Appl. Phys.* **76**, 7627.
Aspnes D. E. and Studna A. A. (1983) *Phys. Rev.* B **27**, 985.
Beall R. B., Clegg J. P., Castagné J., Harris J. J., Murray R. and Newman R. C., (1989) *Semicond. Sci. & Technol.* **4**, 1171.
Bermejo D. and Cardona M. (1979) *J. Non-Cryst. Solids* **32**, 405.
Blood P. and Harris J. J. (1984) *J. Appl. Phys.* **56**, 933.
Brandt O., Crook G. E., Ploog K., Wagner J. and Maier M. (1991) *Appl. Phys. Lett.* **59**, 2730.
Brandt O., Crook G. E., Ploog K., Bierworf R., Hohenstein M., Maier M. and Wagner J. (1993), *Jpn J. Appl. Phys.* **32**, L24.
Brozel M. R., Clegg J. B. and Newman R. C. (1978) *J. Phys. D.* **11**, 1331.
Cardona M. (1982) *Light Scattering in Solids II*, edited by M. Cardona and G. Güntherodt (New York: Springer), p. 19.
Cardona M. and Harbeke G. (1962) *J. Appl. Phys.* **34**, 816.
Clegg J. B. and Beall R. B. (1989) *Surf. Int. Anal.* **14**, 308.
Cochran W., Fray S. J., Johnson F. A., Quarrington J. E. and Williams N. (1961) *J. Appl. Phys.* **32**, 2102.
Crook G. E., Brandt O., Tapfer L. and Ploog K. (1992) *J. Vac. Sci. Technol.* **10**, 841.
Fahy M. R., Ashwin M. J., Harris J. J., Newman R. C. and Joyce B. A. (1992) *Appl. Phys. Lett.* **61**, 1805.
Fano U. (1961) *Phys. Rev.* **124**, 1866.
Gledhill G. A., Newman R. C. and Woodhead J. (1984) *J. Phys. C: Solid St. Phys.* **17**, L301.
Harris J. J., Clegg J. B., Castangné J., Woodbridge K. and Roberts C. (1991) *J. Cryst. Growth* **111**, 239.
Hart L., Fahy M. R., Newman R. C. and Fewster P. F. (1993) *Appl. Phys. Lett.* **62**, 2218.
Hart L., Ashwin M. J., Fewster P. F. Zhang X., Fahy M. F. and Newman R. C. (1994a) *Semicond. Sci. and Technol.* **10**, 32.
Hart L., Fewster P. F., Fahy M. R., Ashwin M. J. and Newman R. C. (1994b), *Mater. Sci. Forum* **143–147**, 647.
Hayashi S. (1989) *Jpn J. Appl. Phys.* **23**, 665.
Hon D. T., Faust W. L., Spitzer W. G. and Williams P. F. (1970) *Phys. Rev. Lett.* **25**, 1184.
Keating P. N. (1966) *Phys. Rev.* **145**, 637.
Köhler K., Granzer P. and Maier M. (1993) *J. Cryst. Growth* **127**, 720.
Kunc K., (1984) *Electronic Structure, Dynamics and Quantum Structural Properties of Condensed Matter*, NATO Series B, Vol 121, eds. J. T. Devreese and P. van Camp, pp. 227–310.
Laithwaite K. and Newman R. C. (1977) *Phil. Mag.* **35**, 1689.
Leigh R. S. and Newman R. C. (1982) *J. Phys. C.: Solid State Phys.* **15**, L1045.
Leigh R. S. and Newman R. C. (1988) *Semicond. Sci. & Technol.* **3**, 84.
Leung P. C., Fredrickson J., Spitzer W. G., Kahan A. and Bouthilette L. (1974) *J. Appl. Phys,* **45**. 1009.
Maguire J., Murray R., Newman R. C., Beall R. B. and Harris J. J. (1987) *Appl. Phys. Lett.* **50**, 516.
Mäkinen J., Laine T., Saarinen K., Hautojärvi P., Corbel C., Airaksinen V. M. and Gilbart P. (1993) *Phys. Rev. Lett.* **71**, 3154.
McQuaid S. A., Newman R. C., Missous M. and O'Hagan S. (1992) *Appl. Phys. Lett.* **61**, 3008.
McQuaid S. A., Newman R. C., Missous M. and O'Hagan S. (1993) *J. Cryst. Growth* **127**, 515.
Mooradian A. and McWhorter A. L. (1967) *Phys. Rev. Lett.* **19**, 849.
Murray R., Newman R. C. and Woodhead J. (1987) *Semicond. Sci. & Technol.* **2**, 399.
Murray R., Newman R. C., Sangster M. J. L., Beall R. B., Harris J. J., Wright P. J., Wagner J. and Ramsteiner M. (1989) *J. Appl. Phys.* **66**, 2589.
Nakamura T. and Katoda T. (1985) *J. Appl. Phys.* **57**, 1084.
Newman R. C. (1993) *Semiconductors and Semimetals*, Vol 38, ed. E. E. Weber (Academic Press, NY), p. 117.
Newman R. C. (1994) *Semicond. Sci. Technol.* **9**, 1749.

Hewman R. C., Ashwin M. J., Wagner J., Fahy M. R., Hart L., Holmes S. N. and Roberts C. (1995) *Mater. Res. Soc. Symp. Proc.*, eds. S. Ashok, J. Chevallier, N. M. Johnson, I. Akasaki and B. L. Sopori, Spring Meeting 1995, in press.
Newman R. C., Thompson F., Hyliands M. and Peart R. F. (1972) *Solid St. Commun.* **10**, 505.
Ono H. and Baba T. (1992) *Mater. Sci. Forum* **83–87**, 1409.
Ono H. and Furuhata N. (1991) *Appl. Phys. Lett.* **59**, 1881.
Ono H. and Newman R. C. (1989) *J. Appl. Phys.* **66**, 141.
Pajot B., Clerjaud B. and Chevallier J. (1991) *Physica* B **170**, 371.
Pashley M. D. and Haberern K. W. (1991) *Phys. Rev. Lett.* **67**, 2697.
Pavesi L., Nguyen Hong Ky, Ganière J. D., Reinhart F. K., Baba-Ali N., Harrison I., Tuck B. and Henini M. (1992) *J. Appl. Phys.* **71**, 2225.
Ramsteiner M., Wagner J., Ennen H. and Maier M. (1988) *Phys. Rev.* B **38**, 10669.
Sangster M. J. L., Newman R. C., Gledhill G. A. and Upadhyay S. B. (1992) *Semicond. Sci. and Technol.* **7**, 1295.
Scamarcio G., Spagnolo V., Molinari E., Tapfer L., Sorba L., Bratina G. and Franciosi A. (1992) *Phys. Rev.* B **46**, 7296.
Schneider J., Dischler B., Seelewind H., Mooney P. M., Lagowski J., Matsui M., Beard D. R. and Newman R. C. (1989) *Appl. Phys. Lett.* **54**, 1442.
Sorba L., Bratina G., Franciosi A., Tapfer L., Scamarcio G., Spagnolo V. and Molinari E. (1992) *Appl. Phys. Lett.* **61**, 1570.
Spitzer W. G. (1971) *Festkörperprobleme XI*, ed. O. Madelung (Pergamon: Vieweg), p. 1.
Spitzer W. G. and Allred W. (1968) *J. Appl. Phys.* **39**, 4999.
Spitzer W. G., Kahan A. and Bouthillette L. (1969) *J. Appl. Phys.* **40**, 3398.
Tanino H. and Amano S. (1992) *Surface Science* **267**, 422.
Tanino H., Amano S., Kawanami H. and Matsuhata H. (1991) *J. Appl. Phys.* **70**, 7068.
Theis W. M. and Spitzer W. G. (1984) *J. Appl. Phys.* **56**, 890.
Theis W. M., Bajaj K. K., Litton C. W. and Spitzer W. G. (1982) *Appl. Phys. Lett.* **41**, 70.
Uematsu M. and Koichi M. (1990) *Jpn J. Appl. Phys.* **29**, 301.
Wagner J. (1990) *Mater. Sci. Forum* **65–66**, 1.
Wagner J. (1991) *Light Scattering in Semiconductor Structures and Superlattices*, eds. D. J. Lockwood and J. F. Young (Plenum Press, New York), p. 275.
Wagner J. (1992a), SPIE Conf. on Spectroscopic Characterisation Techniques for Semiconductor Technology, *SPIE* **110**, 1678.
Wagner J. (1992b), Private communication.
Wagner J. and Ramsteiner M. (1989) *IEEE J. Quantum Electron* QE-**25**, 993.
Wagner J., Newman R. C. and Roberts C. (1995) *J. Appl. Phys.*, in press.
Wagner J., Koidl P. and Newman R. C. (1991b) *Appl. Phys. Lett.* **59**, 1729.
Wagner J., Maier M., Murray R., Newman R. C., Beall R. B. and Harris J. J. (1991a) *J. Appl. Phys.* **69**, 971.
Wagner J., Ramsteiner M., Murray R. and Newman R. C. (1989a) *Mater. Sci. Forum* **38–41**, 815.
Wagner J., Ramsteiner M., Stolz W., Hauser M. and Ploog K. (1989b) *Appl. Phys. Lett.* **55**, 978.
Williams E. W. and Bebb H. B. (1972) in *Semiconductors and Semimetals*, eds. R. K. Willardson and A. C. Beer (Academic Press, New York), vol 8, p. 321.
Wolk J. A., Ager J. W. III, Duxstad K. J., Haller E. E., Taskar N. T., Dorman D. R. and Olego D. J. (1993) *Appl. Phys. Lett.* **63**, 2756.

13

The DX-center in silicon delta-doped GaAs and $Al_xGa_{1-x}As$

P. M. KOENRAAD

Introduction

The electronic properties of Si-doped GaAs and $Al_xGa_{1-x}As$ are controlled by the co-existence of shallow effective mass donor states and a deep donor state, the so-called DX center. This deep center, which is dominant for $0.25 < x < 0.6$, is responsible for the effect of persistent photo-conductivity (PPC), which is observed at temperatures below 100 K. Far infrared measurements by Theis *et al.* [1] showed that all deep DX-states in $Al_xGa_{1-x}As$ act as shallow effective mass states after photo-ionization. The nature of the deep donor state is a long-standing problem. Lang *et al.* [2, 3], who were the first to observe the deep donor state, proposed a large lattice relaxation model to explain the features of the deep center, whereas Saxena [4] used a band structure model or small lattice relaxation model to explain the same features.

Hydrostatic pressure experiments on GaAs by Tachikawa *et al.* [5] and Lifshitz *et al.* [6] showed that the deep state can be induced by pressure. From these experiments it has been proposed that the deep state is tied to the L conduction band minimum and that the predominance of either shallow or deep states depends on the relative positions of their levels, see Fig. 13.1 and Fig. 13.2. The total number of populated deep and shallow states was proved to be nearly equal to the amount of Si doping [7]. Therefore, it has been concluded that both states are induced by the same donor. Li *et al.* [8] showed that the photo-ionization cross-section and the thermal capture and emission energies are nearly the same for the pressure-induced deep state in GaAs and for the deep state in $Al_xGa_{1-x}As$. The same conclusion can be drawn from deep level transient spectroscopy (DLTS) measurements [9, 10].

In GaAs structures with a low doping concentration a hydrostatic pressure of 20 kbar is needed to populate the DX-center. At this pressure the Fermi energy is resonant with the DX-level. Maude *et al.* [11, 12] showed that the hydrostatic pressure needed to populate DX-centers is smaller in samples with a higher doping concentration due to the higher Fermi energy in these samples. In their sample with the highest doping concentration the

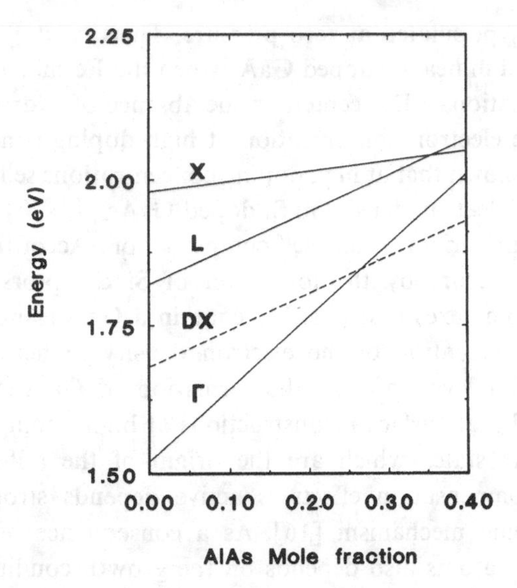

Fig. 13.1. Schematic diagram showing the energy position of the shallow Γ state (dotted line) and the DX-level (dashed line) as a function of the AlAs mole fraction. At a mole fraction of 0.25 the DX-level is resonant with the shallow Γ state.

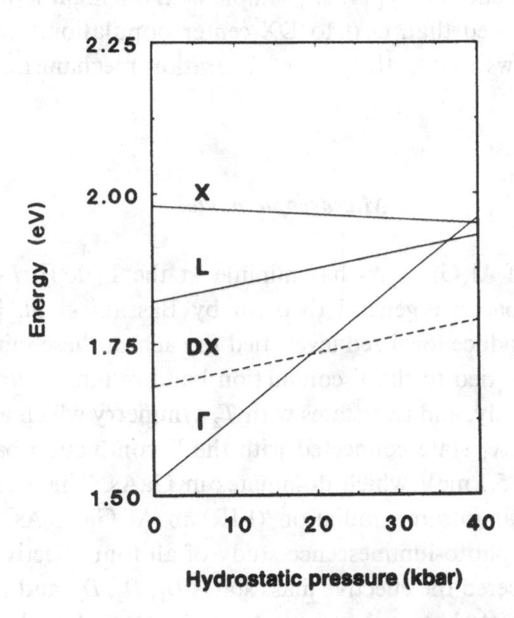

Fig. 13.2. Schematic diagram showing the energy position of the shallow Γ state (dotted line) and the DX-level (dashed line) as a function of applied hydrostatic pressure. At 20 kbar the DX-level is resonant with the shallow Γ state.

DX-centers were already populated at zero pressure. Theis *et al.* [13] also showed that DX-centers are populated in heavily doped GaAs when the Fermi energy is resonant with the DX-level. This population of DX-centers in the absence of hydrostatic pressure limits the maximum attainable electron concentration at high doping concentrations. In other publications it has been shown that at high doping concentrations self-compensation limits the maximum attainable electron density in Si-doped GaAs [14–20]. Several mechanisms have been proposed as the origin of this self-compensation. According to Maguire *et al.* [15], self-compensation occurs by the formation of Si acceptors, neutral Si donor–acceptor pairs, and Si complexes that possibly contain a Ga vacancy [21]. Walukiewicz [17, 18] explained the saturation of the electron density in heavily doped GaAs by stabilization of the Fermi level through the formation of Ga vacancies. Pashley and Haberern [19] proposed that surface reconstructions at high doping concentrations give rise to surface acceptors states which are the origin of the self-compensation effect. Which of these auto-compensation effects is active depends strongly on the growth conditions and the doping mechanism [16]. As a consequence of this, the maximum number of active dopant atoms also depends on the growth conditions and the doping mechanism.

Thus there are at least two basically different types of saturation mechanism which limit the electron density at high doping concentrations, that is, DX-center population and self-compensation. In Maude *et al.*'s [11, 12] sample with the highest doping concentration, LVM measurements showed that, next to DX-center population, self-compensation was also occurring. This shows that both types of saturation mechanism can be active in the same structure.

Microscopic model

The conduction band of $Al_xGa_{1-x}As$ has minima at the $\Gamma[000]$, $L\langle 111\rangle$, and $X\langle 100\rangle$ points in momentum space. A general theorem by Bassani *et al.* [22, 23] states that donor-like impurities produce localized levels tied to each of these minima. There are two states with A_1 symmetry, tied to the Γ conduction band minimum and the L conduction band minimum, respectively, and two states with T_2 symmetry which are states with mixed L and X character. The A_1 state connected with the Γ conduction band minimum is the normal shallow state of 5.7 meV which dominates in GaAs. This shallow state has been studied extensively by far infrared radiation (FIR) in $Al_xGa_{1-x}As$ [24]. Henning [25] have made an extensive photo-luminescence study of all four effective mass-like states in $Al_xGa_{1-x}As$. They numbered the effective mass states D_1, D_2, D_3, and D_4 according to their binding energy. The D_4 state has an A_1 symmetry, is connected to the L conduction band minimum and is positioned at about 200 meV below the L conduction band minimum. According to Saxena's [26] band structure model, Henning *et al.* proposed that the D_4 state is the DX-state. However, Dmochowski *et al.* [27] have shown that the D_4 line in Hennings PL spectra is due to Ge contamination and not to intentional Si doping. They showed that the deep A_1 state of Si is resonant with the shallow A_1 state of 5.7 meV

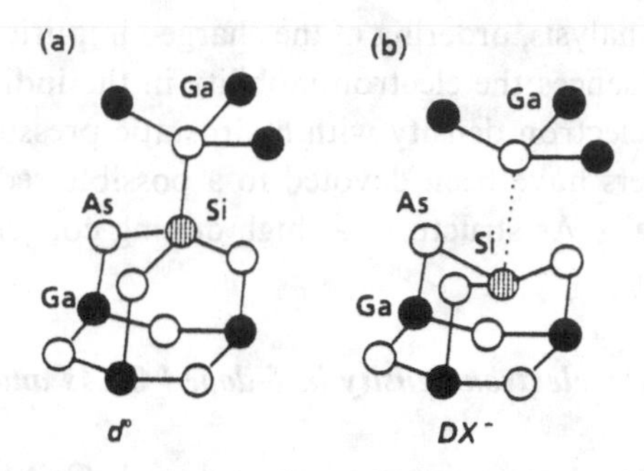

Fig. 13.3. (a) Si donor in a substitutional position. The Si donor behaves as an effective mass-like donor. (b) Si donor in the relaxed position. The Si donor is in the DX-configuration. In $Al_xGa_{1-x}As$ there are three nearest neighbor positions that can be filled with either Ga atoms or Al atoms [after 29].

at a hydrostatic pressure of 30 kbar. This deep A_1 state is not the DX-center because it does not show any persistent illumination effect.

Chadi and Chang [28, 29] proposed that another deeper state can be reached by large lattice relaxation of the Si atom in the ⟨111⟩ direction, that is, the Si atom moves from a substitutional site to an interstitial site, see Fig. 13.3. Recently, this has been corroborated by positron annihilation experiments [30]. Deep level transient spectroscopy measurements have shown that, in agreement with Chadi and Chang's calculations, a number of distinct deep levels exist in Si-doped $Al_xGa_{1-x}As$ depending on the number of Al atoms in second nearest neighbor positions [31].

The large lattice relaxation is induced by the breaking of a bond between the Si atom and a neighboring As atom. The electron needed for this bond breaking is obtained from another Si donor, that is, $2D^0 \rightarrow d^+ + DX^-$, where d^+ is the shallow donor state. Fujisawa *et al.* [32] have shown in GaAs co-doped with Si and Ge that, under hydrostatic pressure, the Si donors indeed take up two electrons. The donor thus exhibits a negative effective Hubbard correlation energy, or so-called negative U.

Recent and more extensive reviews on the DX-center are given by P. M. Mooney [33] and by several authors in [34] and [35].

DX-centers in δ-doped GaAs and $Al_xGa_{1-x}As$

The study of DX-centers in δ-doped GaAs and $Al_xGa_{1-x}As$ has mainly been devoted to the saturation effect of DX-centers on the electron density in structures with high doping concentrations. Usually, magneto-transport experiments under hydrostatic pressure are performed in order to discriminate between saturation effects on the electron density by either self-compensation or population of deep states.

In a smaller number of papers, of which only one deals with experimental work, the influence of ordering on the spatial distribution of charged DX-centers is discussed. In the negative U model the charge distribution consists of positive and negative charged centers.

According to theoretical analysis, ordering of the charged impurities in both cases, through Coulomb interaction, influences the electron mobility in the individual subbands and the depopulation rate of the electron density with hydrostatic pressure.

Also, a number of papers have been devoted to a possible reduction in the DX-center activity in δ-doped $Al_xGa_{1-x}As$ structures at high doping concentrations.

Saturation of the electron density in δ-doped GaAs and $Al_xGa_{1-x}As$

The maximum attainable electron density in δ-doped GaAs structures varies from 8×10^{12} cm^{-2} to 2.5×10^{13} cm^{-2} depending on the growth temperature [36, 37], see Fig. 13.4, and growth conditions [36–39]. Different mechanisms can be responsible for the saturation of the electron density at increasing doping concentrations. The first one is self-compensation in the largest sense. Under this term we consider the following self-compensating effects: Si atoms incorporated on As sites instead of Ga sites, Si–Si pair formation and/or Si cluster formation, formation of Ga vacancies, formation of Si_{As}–V_{Ga} complexes, etc. A second possibility for saturation of the electron density occurs when the DX-level, or possibly the deep A_1 level, becomes resonant with the Fermi level. At high doping concentrations the Fermi level easily rises to 200 meV or more above the Γ conduction band minimum. At these high Fermi energies the DX-center becomes populated, see Fig. 13.1. A third possibility for saturation of the electron density at the Γ conduction band minimum occurs when the Fermi level is resonant with the L conduction band minimum. Up to now there are no experimental observations of this effect. Therefore, we will not discuss this possibility in further detail.

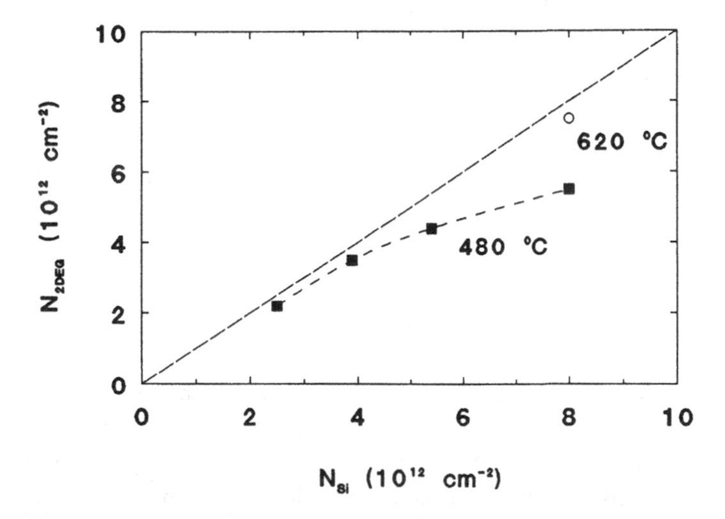

Fig. 13.4. The electron density in δ-doped GaAs as a function of the Si concentration in samples grown at 480 °C and 620 °C. The electron density was obtained by summing up the electron densities of the individual subbands as determined by Schubnikov–de Haas measurements.

One is able to discriminate between self-compensation and population of DX-states or deep A_1 states as the origin of the saturation of the electron density by studying the electron density as a function of hydrostatic pressure, and to look for persistent photo-conductivity. When hydrostatic pressure is applied, the energy separation between the Γ conduction band minimum and the DX-level or the deep A_1 level, which are both more or less tied to the L conduction band minimum, decreases [5, 6, 11, 12, 40]. This implies that more DX-centers or deep A_1 states become populated when hydrostatic pressure is applied. In the case where a decreasing electron density with increasing hydrostatic pressure and a concurrent PPC effect are observed, one has to conclude that population of DX-states limits the electron density. In the case of a pressure dependent electron concentration without a concurrent PPC effect, population of the A_1 states has occurred. And, finally, when no pressure dependence of the electron density is observed, self-compensation is the origin of the saturation of the electron concentration.

In GaAs structures we have grown at 480 °C and doped at 8×10^{12} cm^{-2}, we find an electron density of only 5.5×10^{12} cm^{-2}. The absence of both a dependence of the electron density on hydrostatic pressure and a strong PPC effect shows that in these structures the electron density is limited by self-compensation [41]. Self-compensation at high doping concentrations is also observed in bulk-doped samples when the doping concentration is higher than about 10^{19} cm^{-3} [14–20]. In δ-doped structures where the doping layer is very narrow the local Si density can easily exceed 10^{19} cm^{-3} [42]. Thus in samples with a narrow doping profile and an increasing doping concentration, self-compensation can occur prior to the population of the DX-level.

Self-compensation in Si δ-doping layers has been observed by several authors [38, 43, 44]. Secondary ion mass spectrometry studies have shown that at high δ-doping concentrations two types of Si distribution are found: a non-diffusing distribution of Si atoms which are electrically neutral and a fast diffusing distribution of charged donors [43]. The authors propose that the non-diffusing neutral distribution consists of Si clusters. X-ray and LVM spectroscopy also show that in samples with high doping concentrations part of the silicon is built in as clusters [44, 45]. It has been observed that Si atoms only occupy As sites up to very high δ-doping concentrations [46], see Fig. 13.5. In our self-compensated sample, grown at 480 °C and doped at 8×10^{12} cm^{-2}, we observed that, after annealing, the number of electrons in the two-dimensional electron gas (2DEG) increases [46] and that the diffusion rate is considerably higher than in samples with a lower doping concentration [37, 46]. Dissolving of Si clusters might explain the increase in the electron density after annealing but this seems improbable according to the SIMS results of Beall *et al.* [43]. Possibly the presence of Ga vacancies, as proposed by Walukiewicz [17, 18], must be considered because this might explain the self-compensation and increased diffusion rate in the samples with high δ-doping concentrations.

Self-compensation in samples with a high δ-doping concentration and a narrow doping profile can be circumvented by distributing the dopant atoms over a few separate δ-doping layers, all within a 20 Å region. In this way the electronic properties are still very similar to a true monolayer-thick δ-doped structure because the wavefunctions and subband energies are not sensitive to the thickness of the doping layer when the width of the doping

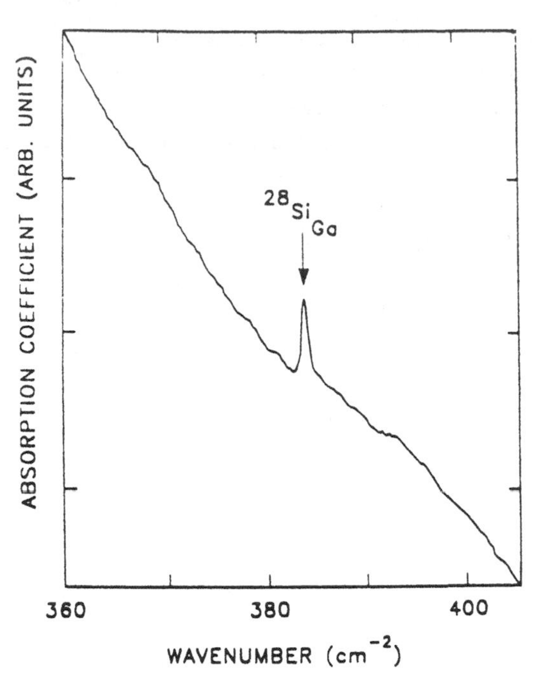

Fig. 13.5. LVM spectrum for a GaAs sample δ-doped at 3.45×10^{13} cm^{-2} and grown at a temperature close to 400 °C. The measurement shows that no acceptor sites (Si_{As}) are occupied by the Si atoms up to this very high doping concentration [after 36].

layer is smaller than 20 Å. This idea of distributing the dopant atoms over a few separate δ-doping layers in order to circumvent self-compensation has been used by Skuras *et al.* [47]. In their structures a hydrostatic pressure dependent electron concentration and a PPC effect are clearly observed. This proves that in their sample, population of DX-states is at least one of the origins of the saturation of the electron density.

At high growth temperatures, $T > 500$ °C, the dopant atoms are spread over a larger region due to diffusion and surface segregation [48–50]. This spreading of the Si dopant atoms prevents the local density from reaching the critical self-compensation limit. Thus in samples grown at high temperature, DX-centers can be the sole origin of the saturation of the electron density. However, due to stronger diffusion and segregation of the dopant atoms at higher growth temperatures, the potential well also becomes wider. This leads to a reduction in the Fermi energy in samples with equal electron densities but thicker doping layers, as shown in Fig. 13.6. Thus only at a higher electron density or hydrostatic pressure does the Fermi level becomes resonant with the DX-level.

This situation occurs in the doping layers studied by Zrenner *et al.* [41] and in the structures of Arscott *et al.* [51], who observe populated DX-states at a high doping concentration. Zrenner *et al.* studied the subband population by magneto-transport measurements under hydrostatic pressure. Arscott *et al.* show that a deep center appears in the DLTS spectra only for structures with the highest δ-doping concentration of

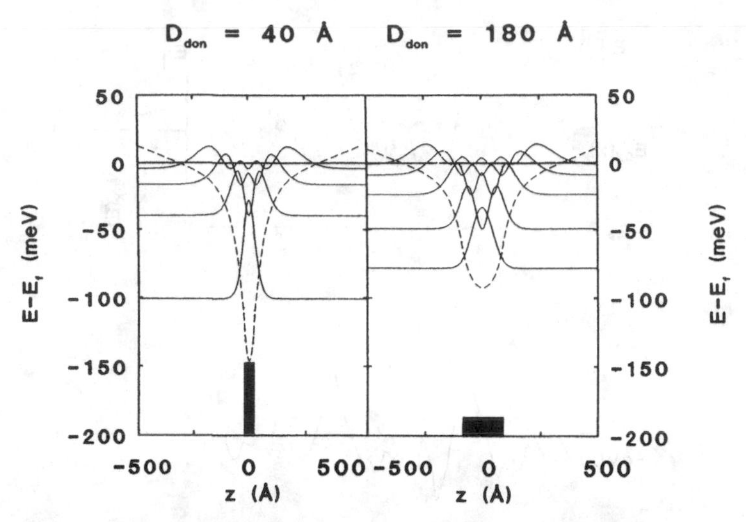

Fig. 13.6. The calculated energy and probability distribution, $|\varphi_i(z)|^2$, of each subband (solid lines) and the electrostatic potential (dashed line) for a δ-doped structure. The 4.5×10^{12} cm^{-2} donors are distributed over 40 Å and 180 Å. Note that the Fermi energy in the structure with the thick doping layer is lower than in the structure with the narrow doping layer.

1×10^{13} cm^{-2}. This deep center in the DLTS spectrum has the same thermal activation energy as the DX-center in bulk doped GaAs [9, 52].

Harris *et al.* [53] have studied the saturation of the electron density in δ-doped quantum wells of varying width and have observed a Fermi level pinning at about 190 meV above the Γ conduction band minimum. Because they did not observe any PPC effect they proposed that in this case the deep A_1 state might be responsible for the saturation of the electron density.

In conclusion, in Si δ-doped GaAs samples grown at low temperature the saturation of the electron density is due to self-compensation unless precautions are taken to prevent a very high local concentration of Si atoms. In single δ-doping layers grown at 480 °C we find that partial self-compensation is active at doping concentrations above 4×10^{12} cm^{-2}. At higher growth temperatures, self-compensation is less important due to spreading of the dopant atoms, in which case population of DX-centers can occur and thus limit the maximum attainable electron density.

Spatial correlations in the distribution of charged DX-centers

We have calculated the position of the Fermi level in structures where the saturation of the electron density is due to population of DX-centers. By assuming that the Fermi level in these structures is pinned by the DX-level we find a relation between the thickness of the δ-doping layer and the energy position of the DX-state relative to the Γ conduction band minimum [42]. We find that the energy position of the DX-level is higher in samples

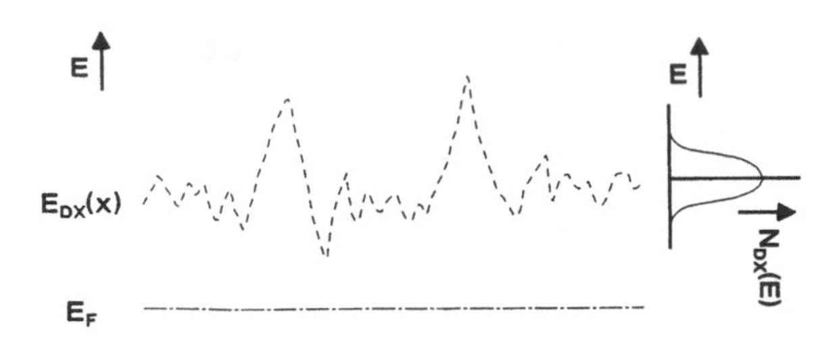

Fig. 13.7. Schematic representation of the local potential energy and local energetic position of the DX-level. The Fermi energy is indicated by the dashed line. The energetic distance of individual DX-centers to the Fermi energy forms a broadened distribution function due to local potential fluctuations.

with a narrower doping profile. Maude *et al.* [11, 12] observed a similar shift of the DX-level in bulk-doped GaAs in samples with increased doping concentration. Recently, Wilamowski *et al.* [54] proposed that the energy position of the DX-states relative to the Fermi level forms a distribution which is broadend due to local potential fluctuations, see Fig. 13.7. Thus the width of the energy distribution function of DX-states depends on the strength of the fluctuations in the local Coulomb potential. The wider the distribution, the lower is the electron density or pressure needed to populate DX-states. By calculating the strength of the potential fluctuations, with the inclusion of the spatial correlation effects of the charged DX-centers and ionized shallow donor states, Wilamowski *et al.* were able to explain the apparent shift of the DX-state with doping concentration that was observed by Maude *et al.* Probably the same arguments hold for the apparent shift of the DX-state with the thickness of the δ-doping layer. If we assume that in narrow δ-doping layers the fluctuations in the local potential are smaller, due, for instance, to stronger correlation effects, this results in a smaller width of the energy distribution function of the DX-states. This implies an apparent shift of the DX-level in structures with a narrow δ-doping profile. Brunthaler *et al.* [55] have studied the influence of these correlation effects on the saturation of the electron density in δ-doped $GaAs/Al_xGa_{1-x}As$ quantum wells, both experimentally and theoretically, and find these correlations to be important.

A theoretical study of the spatial correlation effects of charged centers in δ-doping layers under pressure has been presented in two papers by Sobkowicz *et al.* [56, 57]. They propose that these correlation effects will influence the mobility and the subband population dependence on hydrostatic pressure. In an extensive study of the subband population and mobility in δ-doped GaAs under hydrostatic pressure, Stradling *et al.* [58] looked for the influence of correlation effects on the subband mobility and the pressure dependent subband mobility and the pressure dependent subband population. Although they state that little evidence was found for any correlation effects, their electron mobility in the lowest subband for the structure under 15 kbar hydrostatic pressure is at least a factor of two higher than the mobility in a δ-doped GaAs structure with a similar electron density under zero pressure. In the sample under hydrostatic pressure a large number of DX-centers are populated and thus correlation effects might be important in contrast to the sample under zero pressure where no DX-centers are populated [59].

Persistent photo-conductivity in δ-doped GaAs and $Al_xGa_{1-x}As$

Arscott *et al.* [60] state that they have seen persistent photo-conductivity in δ-doped GaAs due to the depopulation of DX-centers. They studied the subband population and subband mobility before, under, and after illumination. Surprisingly, they did not find a higher total electron density after illumination but only a considerable change in the shape of the confining potential well. Normally, a higher electron density has to be expected when DX-centers are depopulated. The different shape of the potential well was attributed to an asymmetrical photo-ionization of the DX-centers. Possibly the effects observed had a different origin. The subband population that was found prior to illumination can only be fitted to a calculated subband population when a high background impurity concentration is assumed. An asymmetrical photo-ionization of this high concentration of background impurities could explain the change in the potential well just as efficiently.

Sánchez-Dehesa *et al.* [61] have studied the effect of illumination on δ-doped $Al_{0.2}Ga_{0.8}As$. They found an increasing total electron density after illumination and also a pressure dependent total electron density, as would be expected from the normal DX-center characteristics. We have also found a strong PPC effect on the Hall electron density in Si δ-doped $Al_{0.33}Ga_{0.67}As$ [42].

The PPC effect on the Hall electron density in δ-doped $Al_{0.3}Ga_{0.7}As$ has been shown to depend strongly on the doping concentration [62]. These results are discussed in more detail in the next paragraph.

Influence of δ-doping on DX-center characteristics

It has been reported by several authors that the concentration of DX-centers and/or the behavior of the DX-center is changed at high δ-doping concentrations. Hall electron density measurements, photo-luminescence (PL) measurements and deep level transient spectroscopy (DLTS) measurements have been reported in respect of this problem.

A study of the Hall electron density as a function of temperature in either δ-doped $Al_{0.3}Ga_{0.7}As$ [62] or δ-doped $Al_{0.26}Ga_{0.74}As$ [63] has been reported in two papers. Both groups [62, 63] observe a reduction in the temperature dependence of the Hall electron density when the doping concentration is increased. In both papers the authors argue that this is an indication of reduced DX-center activity. However, this reduction of the temperature dependence at high doping concentrations is also observed in δ-doped GaAs [36, 38] where no DX-centers are active.

One has to be careful in drawing conclusions from Hall electron density measurements in δ-doped samples while the Hall electron density is an averaged quantity over the population of the individual subbands and the subband mobilities [59]. The reduced temperature dependence of the Hall electron density in δ-doped samples at higher doping concentrations is due to both reduced freeze out of the electrons at low temperatues and reduced temperature dependence of the subband mobility. The freeze out of electrons at low temperature disappears because the binding energy of the shallow donor state decreases to zero when the electron wavefunctions overlap at high doping concentrations. The subband mobilities become less temperature dependent due to higher Fermi velocity in samples with higher electron concentration.

Etienne and Thierry-Mieg [62] define the concentration of DX-centers as $(n_{\mathrm{Hall}}^{\mathrm{ill}} - n_{\mathrm{Hall}}^{\mathrm{dark}})/n_{\mathrm{Hall}}^{\mathrm{ill}}$, where $n_{\mathrm{Hall}}^{\mathrm{ill}}$ and $n_{\mathrm{Hall}}^{\mathrm{dark}}$ are the Hall electron density measured at 77 K after illumination and in the dark respectively. They observe only a small PPC effect on the Hall electron density at high doping concentrations. According to their definition of the DX-center concentration, this implies a strong reduction of the percentage of DX-centers at high doping concentrations. Again one must be careful with the interpretation of the Hall electron density measurements. Measurements in δ-doped GaAs have shown that the Hall electron density decreases after illumination, whereas the true total electron density, the sum of electron concentrations in the individual subbands, increases [36]. Furthermore, it remains unclear whether the reduced illumination dependence of the Hall electron density in the samples from Etienne and Thierry-Mieg is influenced by self-compensation or by other effects occurring at the very high doping concentrations that they have used (up to 3% of a monolayer in the sample with the highest doping concentration).

From PL measurements on the same set of samples, Mejri *et al.* [64, 65] also conclude that the concentration of DX-centers decreases at high doping concentrations. Four effective mass states exist in Bassani's theory. Henning has studied these states in Si bulk-doped $Al_xGa_{1-x}As$ [25]. Mejri *et al.* identify four peaks in their PL spectra which they attribute to the effective mass states of the Si donors. Because the $D_{4\mathrm{h}}$ line, which they attribute to the DX-center, disappears at high doping concentrations, they conclude that the DX-center concentration drops at high doping concentrations. However, when one compares their identification of the PL peak with the work of Henning *et al.*, one must label their PL peaks differently. Their $D_{1\mathrm{h}}$ peaks corresponds to the combined D_1–D_3 peak in the paper by Henning *et al.* (see Fig. 4 in [25]). Their $D_{2\mathrm{h}}$ peak is most probably a phonon replica of the bound exciton peak (see the illumination intensity dependence in Fig. 3 of [65]). Their $D_{3\mathrm{h}}$ peak corresponds to the D_4 peak in the paper by Henning, and their $D_{4\mathrm{h}}$ is of unknown origin. Henning proposed that this D_4 peak is due to the DX level. However, Dmochowski *et al.* [27] showed that this D_4 peak is due to Ge contamination of the samples. Therefore, the conclusions that Mejri *et al.* draw from the $D_{4\mathrm{h}}$ line concerning the DX-centers are probably flawed.

Another indication of the possible influence of the doping concentration on the DX-center behavior in δ-doped $Al_{0.26}Ga_{0.74}As$ is shown by the DLTS work of Solomon *et al.* [63]. They observe that the DLTS peak-height increases and the DLTS peak-width decreases in samples with increased doping concentration. They state that the captured cross-section of the deep state is six orders of magnitude smaller in the sample with the highest doping concentration as compared with the sample with the lowest doping concentration. A disadvantage of the DLTS technique is the fact that the interpretation of the measurements is notoriously difficult and can easily lead to misinterpretations. Solomon *et al.* propose, just as do Etienne and Thierry-Mieg, that the DX-center characteristics are changed by the local stress induced by the Si-doping layer on the GaAs matrix. These local stresses have been measured by X-ray diffraction in δ-doped GaAs samples with a very high doping concentration of 3.4×10^{14} cm^{-2} [66].

At the moment there is no strong evidence for any influence of the doping concentration on the DX-center activity. Further and more detailed experiments are needed in order to resolve this problem.

Conclusions

The presence and characteristics of the well-known DX-center in Si-δ-doped GaAs has been the subject of many papers. Experiments have shown that the maximum attainable electron density at high doping concentrations can be limited by both self-compensation and/or DX-center population. Self-compensation is effective in δ-doping layers where the local concentration of Si exceeds a critical value. This is very likely to occur in narrow doping layers with a high doping concentration.

It has been proposed that at the high doping concentrations in a δ-doping layer, correlations due to Coulomb interaction are important. At the moment there is little evidence for this correlation effect. Finally, there is also no strong evidence for the influence of the high doping concentrations in δ-doping layers on the characteristics of the DX-center.

References

[1] T. N. Theis, T. F. Kuech, L. F. Palmateer, and P. M. Mooney, *Inst. Phys. Conf. Ser.* **74**, 241 (1984).
[2] D. V. Lang and R. A. Logan, *Phys. Rev. Lett.* **39**, 635 (1977).
[3] D. V. Lang, R. A. Logan, and M. Jaros, *Phys. Rev.* B **19**, 1015 (1979).
[4] A. K. Saxena, *Solid State Electron.* **25**, 127 (1982).
[5] M. Tachikawa, M. Mizuta, H. Kukimoto, and S. Minomura, *Jpn J. Appl. Phys.* **24**, L281 (1985).
[6] N. Lifshitz, A. Jayaraman, R. A. Logan, *Phys. Rev.* B **21**, 670 (1980).
[7] M. O. Watanabe, M. Morizuka, M. Mashita, Y. Ashizaway, and Y. Zohte, *Jpn J. Appl. Phys.* **23**, 1103 (1984).
[8] M. F. Li, P. Y. Yu, E. R. Weber, and W. Hansen, *Phys. Rev.* B **36**, 4531 (1987).

[9] M. Mizuta, M. Tachikawa, H. Kukimoto, and S. Minomura, *Jpn J. Appl. Phys.* **24**, L143 (1985).
[10] P. M. Mooney, N. S. Caswell, S. L. Wright, *J. Appl. Phys.* **62**, 4786 (1989).
[11] D. K. Maude, J. C. Portal, L. Dmowski, T. Foster, L. Eaves, M. Nathan, M. Heiblum, J. J. Harris, and R. B. Beall, *Phys. Rev. Lett.* **59**, 815 (1987).
[12] L. Eaves, T. J. Foster, D. K. Maude, J. C. Portal, R. Murray, R. C. Newman, L. Dmowski, R. B. Beall, J. J. Harris, M. I. Nathan, and M. Heiblum, *Inst. Phys. Conf. Ser.* **95**, chapter 5, 315 (1988) [*Inst. Conf. Shallow Impurities in Semiconductors*].
[13] T. N. Theis, P. M. Mooney, and S. L. Wright, *Phys. Rev. Lett.* **60**, 361 (1988).
[14] T. Makimoto and Y. Horikoschi, *Jpn J. Appl. Phys.* **29**, 12250 (1990).
[15] J. Maguire, R. Murray, R. C. Newman, R. B. Beall, and J. J. Harris, *Appl. Phys. Lett.* **50**, 516 (1987).
[16] E. F. Schubert, *Doping in III–V Semiconductors*, Cambridge University Press (1993).
[17] W. Walukiewicz, *Appl. Phys. Lett.* **54**, 2094 (1989).
[18] W. Walukiewicz, *Materials Science Forum* **143–147**, 519 (1994) [*Int. Conf. on Defects in Semiconductors, ICDS-17*].
[19] M. D. Pashley and K. W. Haberern, *Phys. Rev. Lett.* **67**, 2697 (1991).
[20] Y. G. Chai, R. Chow, and C. E. C. Wood, *Appl. Phys. Lett.* **39**, 800 (1981).
[21] R. Murray, R. C. Newmann, J. L. Sangster, R. B. Beall, J. J. Harris, P. J. Wright, J. Wagner, and M. Ramsteiner, *J. Appl. Phys.* **66**, 2589 (1989).
[22] F. Bassani, G. Iadonisi, and B. Preziosi, *Phys. Rev.* **186**, 735 (1969).
[23] F. Bassani, G. Iadonisi, and B. Preziosi, *Rep. Prog. Phys.* **37**, 1099 (1974).
[24] T. N. Theis, *Inst. Phys. Conf. Ser.* **95**, chapter 5, 307 (1988) [*Int. Conf. Shallow Impurities in Semiconductors*].
[25] J. C. M. Henning, *Solid State Phenomena* **10**, 145 (1989).
[26] A. K. Saxena, *J. Phys. C.: Solid St. Phys.* **13**, 4323 (1980).
[27] J. E. Dmochowski, Z. Wasilewski, and R. A. Stradling, *Materials Science Forum* **65–66**, 449 (1990) [*Int. Conf. Shallow Impurities in Semiconductors*].
[28] D. J. Chadi and K. J. Chang, *Phys. Rev. Lett.* **61**, 873 (1988).
[29] D. J. Chadi and K. J. Chang, *Phys. Rev.* B. **39**, 10063 (1989).
[30] K. Saarinen, J. Mäkinen, P. Hautojärvi, S. Kuisma, T. Laine, C. Corbel, and C. le Berre, *Materials Science Forum* **143–147**, 983 (1994) [*Int. Conf. on Defects in Semiconductors, ICDS-17*].
[31] P. M. Mooney, T. N. Theis, and E. Calleja, *Journ. of Elec. Mat.* **20**, 23 (1991).
[32] T. Fujisawa, J. Yoshino, and H. Kukimoto, *Jpn J. Appl. Phys.* **29**, L388 (1990).
[33] P. M. Mooney, *J. Appl. Phys.* **67**, R1 (1990).
[34] *Journal of Electronic Materials* **20** (1991).
[35] *Semicond. Sci. Technol.* B **6** (1991).
[36] P. M. Koenraad, F. A. P. Blom, C. J. G. M. Langerak, M. R. Leys, J. A. A. J. Perenboom, J. Singleton, S. J. R. M. Spermon, W. C. van der Vleuten, A. P. J. Voncken, and J. H. Wolter, *Semicond. Sci. Technol.* **5**, 861 (1990).
[37] P. M. Koenraad, I. Bársony, J. C. M. Henning, J. A. A. J. Perenboom, and J. H. Wolter, NATO Advance Research Workshop *Semiconductor Interfaces at the Sub-nanometer Scale*, eds. H. W. M. Salemink and M. Pashley, *NATO ASI series* E **243**, 35 (1993).
[38] G. Gillman, B. Vinter, E. Barbier, and A. Tardella, *Appl. Phys. Lett.* **52**, 972 (1988).
[39] E. F. Schubert, R. F. Kopf, J. M. Kuo, H. S. Luftman, P. A. Garbinski, *Appl. Phys. Lett.* **57**, 497 (1990).
[40] R. B. Beall, J. J. Harris, R. J. Clegg, J. P. Gowers, B. A. Joyce, J. Castagnè, and V. Welch, *Int. Conf. Gallium Arsenide and Related Compounds 1988*, ed. J. J. Harris, IOP publishing Ltd, Bristol.
[41] A. Zrenner, F. Koch, R. L. Williams, R. A. Stradling, K. Ploog, and G. Weimann, *Semicond. Sci. Technol.* **3**, 1203 (1988).

[42] P. M. Koenraad, W. de Lange, F. A. P. Blom, M. R. Leys, J. A. A. J. Perenboom, J. Singleton, and J. H. Wolter, *Semicond. Sci. Technol.* **6**, B143 (1991).
[43] R. B. Beall, J. B. Clegg, J. Castagné, J. J. Harris, R. Murray, and R. C. Newman, *Semicond. Sci. Technol.* **4**, 1171 (1989).
[44] M. J. Ashwin, M. Fahy, J. J. Harris, R. C. Newman, D. A. Sansom, R. Addinall, D. S. McPhail, and V. K. M. Sharma, *J. Appl. Phys.* **73**, 633 (1993).
[45] L. Hart, P. F. Fewster, M. J. Ashwin, M. R. Fahy, and R. C. Newman, *Materials Science Forum* **143–147**, 647 (1994) [*Int. Conf. on Defects in Semiconductors, ICDS-17*].
[46] P. M. Koenraad, I. Bársony, A. F. W. van de Stadt, J. A. A. J. Perenboom, and J. H. Wolter, *Materials Science Forum* **143–147**, 663 (1994) [*Int. Conf. on Defects in Semiconductors, ICDS-17*].
[47] E. Skuras, R. Kumar, R. L. Williams, R. A. Stradling, J. E. Dmochowski, E. A. Johnson, A. Mackinnon, J. J. Harris, R. B. Beall, C. Skierbeszeswki, J. Singleton, P. J. van der Wel, and P. Wisniewski, *Semicond. Sci. Technol.* **6**, 535 (1991).
[48] R. B. Beall, J. B. Clegg, and J. J. Harris, *Semicond. Sci. Technol.* **3**, 612 (1988).
[49] E. F. Schubert, H. S. Luftman, R. F. Kopf, R. L. Headrick, and J. M. Kuo, *Appl. Phys. Lett.* **57**, 1799 (1990).
[50] Ph. Jansen, M. Meuris, M. van Rossum, and G. Borghs, *J. Appl. Phys.* **68**, 3766 (1990).
[51] S. Arscott, M. Missous, and L. Dobaczewski, *Semicond. Sci. Technol.* **7**, 620 (1992).
[52] P. M. Mooney, T. N. Theis, S. L. Wright, *Inst. Phys. Conf. Ser.* **91**, 359 (1988). [*Int. Conf. Gallium Arsenide and Related Compounds 1987*].
[53] J. J. Harris, R. Murray, and C. T. Foxon, *Semicond. Sci. Technol.* **8**, 31 (1993).
[54] Z. Wilamowski, J. Kossut, W. Jantsch, and G. Ostermayer, *Semicond. Sci. Technol.* **6**, B38 (1991).
[55] G. Brunthaler, M. Seto, G. Stöger, G. Ostermayer, and K. Kökler, *Materials Science Forum* **143–147**, 641 (1994) [*Int. Conf. on Defects in Semiconductors, ICDS-17*].
[56] P. Sobkowicz, Z. Wilamowski, and J. Kossut, *Semicond. Sci. Technol.* **7**, 1155 (1992).
[57] P. Sobkowicz, Z. Wilamowski, and J. Kossut, *Acta Fysica Polonia* **82**, 645 (1992).
[58] R. A. Stradling, E. A. Johnson, A. Mackinnon, R. Kumar, E. Skuras, and J. J. Harris, *Semicond. Sci. Technol.* **6**, B137 (1991).
[59] P. M. Koenraad, chapter 17 of this book, Electron mobility in δ-doped Layers.
[60] S. Arscott, M. Missous, L. Dobaczewski, P. C. Harness, D. K. Maude, and J. C. Portal, MRS meeting, Dec. 1992, Boston, *MRS Res. Soc. Symp. Proc.* **281**, 19 (1993).
[61] J. Sánchez-Dehesa, D. Lavielle, E. Ranz, B. Goutiers, J. C. Portal, E. Barbier, A. . Cho, and D. L. Sivco, *Semicond. Sci. Technol.* **6**, 445 (1991).
[62] B. Etienne and V. Thierry-Mieg, *Appl. Phys. Lett.* **52**, 1237 (1988).
[63] G. S. Solomon, G. Roos, and J. S. Harris, *Journ. Crys. Growth* **127**, 737 (1993).
[64] H. Mejri, S. Alaya, H. Maaref, J. C. Bourgoin, and B. Etienne, *Semicond. Sci. Technol.* **5**, 900 (1990).
[65] H. Mejri, A. Selmi, H. Maaref, J. C. Bourgoin, *J. Appl. Phys.* **69**, 4060 (1991).
[66] L. Hart, M. R. Fahy, R. C. Newman, P. F. Fewster, *Appl. Phys. Lett.* **62**, 2218 (1993).

PART FIVE

Physical characteristics

14

Luminescence and ellipsometry spectroscopy

HUADE YAO AND E. F. SCHUBERT

The optical properties of δ-doped doping superlattices and multilayer δ-doped semiconductors are reviewed in this chapter. First the electronic structure of sawtooth superlattices is introduced and the homogeneously doped and δ-doped doping superlattices are compared. Then the electronic sub-band structures of V-shaped quantum wells are calculated by the variational method. The improved optical properties of δ-doped doping superlattices are illustrated in terms of photoluminescence, absorption, and transmission spectroscopy. The dielectric functions of multilayer δ-doped GaAs are measured by spectroscopic ellipsometry and compared with homogeneously doped and undoped GaAs. The applications of doping superlattices in optoelectronic devices will be reviewed in a separate part of the book (see Chapters 24 and 25).

14.1 Delta-doped doping superlattices

Doping superlattices form a class of modulated semiconductor structures which have many unique properties not found in other modulated semiconductor structures. The modulation of the band-edge potential in doping superlattices is achieved by means of impurity charges. A periodic modulation results from the alternating placement of positively charged and negatively charged impurities along the crystal growth direction.

The doping profile of the doping superlattices discussed here consists of a train of alternating n- and p-type δ-functions. Such a doping profile is shown in Fig. 14.1(a), along with the mathematical expression for the doping profile. If the period of the superlattice, z_p, is sufficiently small, electrons originating from donors recombine with holes originating from their parent acceptors. If the two-dimensional densities of donors (N_D^{2D}) and acceptors (N_A^{2D}) have the same value (and all of them are ionzied) there are no excess carriers of extrinsic origin. That is, the superlattice is depleted of free carriers.

The only charges in a doping superlattice under thermal equilibrium conditions are impurity charges. A schematic band diagram of a doping superlattice is shown in

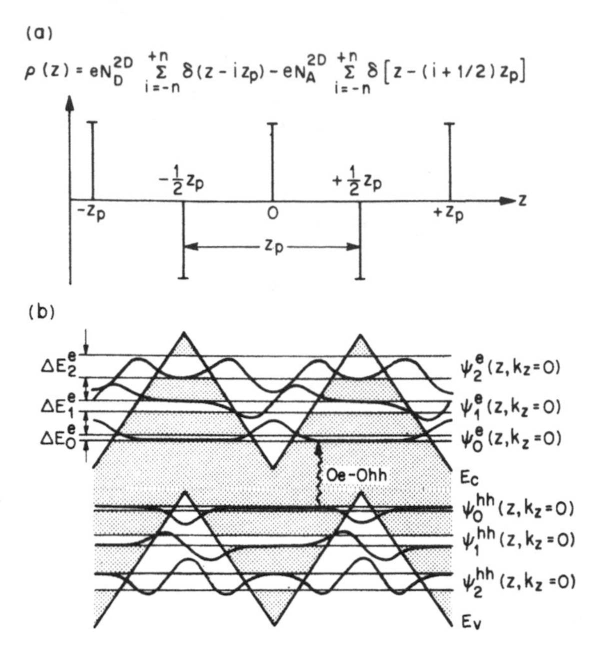

Fig. 14.1. (a) Doping profile of a δ-doped doping superlattice consisting of a train of alternating n- and p-type δ-functions. (b) Schematic band diagram and wavefunctions of the δ-doped doping superlattice. The wavefunctions $\psi_n(z)$ are shown for $k_z = 0$. Minibands have a width of ΔE_n.

Fig. 14.1(b), along with the electron- and hole-wavefunctions of the superlattice. Also shown is an absorption transition from the top heavy-hole sub-band to the lowest conduction sub-band. Inspection of the transition energy reveals that it is smaller than the fundamental gap energy of the host semiconductor. That is, doping superlattices allow the gap of any semiconductor to be extended to lower energies, which is an intriguing feature. For example, a GaAs doping superlattice could be used as a light-detecting or light-emitting device at $\lambda = 1.3$ μm, that is, below the fundamental gap of the GaAs host lattice.

It is appropriate to define the energy gap of a doping superlattice as the energy separation of the lowest conduction sub-band and highest valence sub-band. Assuming a degenerate valence band at $k = 0$ of the host semiconductor, heavy- and light-hole sub-bands are formed. The superlattice energy gap, E_g^{SL}, is thus defined as

$$E_g^{SL} = E_g - eV_z + E_0^e + E_0^{hh}, \tag{14.1}$$

where the amplitude of the superlattice modulation is given by

$$V_z = \frac{1}{4}\frac{eN^{2D}}{\varepsilon} z_p. \tag{14.2}$$

The lowest sub-band energies for electrons and heavy holes, E_0^e and E_0^{hh}, are yet to be determined.

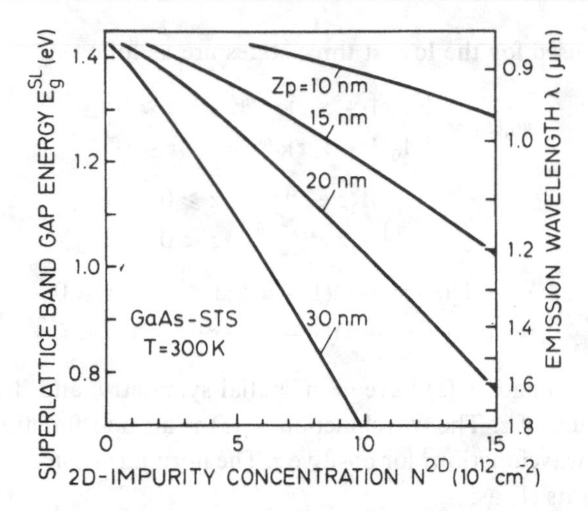

Fig. 14.2. Energy gap and emission wavelength of a GaAs sawtooth superlattice (STS) at 300 K as a function of the doping density and period. The energy gap has values smaller than the gap of the GaAs host lattice.

The superlattice energy gap is shown in Fig. 14.2 as a function of impurity density for different periods of the superlattice.[1] The superlattice gap decreases with the period and the doping density of the superlattice. Figure 14.2 reveals that the wavelength range of 0.9–1.6 μm, which is important for optical communication devices, is accessible to GaAs doping superlattices. The results obtained from the variational calculations are presented below, including the eigenstate energies and wavefunctions of a V-shaped quantum well in δ-doped semiconductors.

14.2 Variational solutions of the V-shaped quantum well

The variational method yields eigenstate energies and wavefunctions of good accuracy. The variational method has been applied to the V-shaped quantum well[2] and yields wavefunctions of simple analytic form.

A thin δ-function-like doping profile results in an electric field of magnitude:

$$\mathbf{E} = eN^{2D}/2\varepsilon, \tag{14.3}$$

Consequently, the potential energy of a one-dimensional V-shaped potential well is given by

$$V(z) = \begin{cases} e\mathbf{E}z & z \geqslant 0 \\ -e\mathbf{E}z & z < 0 \end{cases}, \tag{14.4}$$

The trial functions used for the lowest three states are as follows:

$$\psi_0(z) = \begin{cases} A_0(1 + \alpha_0 z)e^{-\alpha_0 z} & z \geqslant 0 \\ A_0(1 - \alpha_0 z)e^{\alpha_0 z} & z < 0 \end{cases}, \tag{14.5}$$

$$\psi_1(z) = \begin{cases} A_1 z\, e^{-\alpha_1 z} & z \geqslant 0 \\ A_1 z\, e^{\alpha_1 z} & z < 0 \end{cases}, \tag{14.6}$$

$$\psi_2(z) = \begin{cases} A_2(\alpha_2^2 z^2 - 1)(1 + \alpha_2 z)e^{-\alpha_2 z} & z \geqslant 0 \\ A_2(\alpha_2^2 z^2 - 1)(1 - \alpha_2 z)e^{\alpha_2 z} & z < 0 \end{cases}. \tag{14.7}$$

The wavefunctions $\psi_0(z)$ and $\psi_2(z)$ have even spatial symmetry, and decay exponentially for large absolute values of z. The wavefunction $\psi_1(z)$ is an odd function and is identical to the Fang–Howard wavefunction[3] for positive z. The normalization condition $\langle\psi|\psi\rangle = 1$ determines the constants A_n as:

$$A_0^2 = \tfrac{2}{5}\alpha_0, \quad A_1^2 = 2\alpha_1^3, \quad A_2^2 = \tfrac{4}{63}\alpha_2. \tag{14.8}$$

Minimization of the energy expectation values yields the trial parameters α_n:

$$\alpha_0 = \left(\frac{9}{4}\right)^{1/3}\left(e\mathbf{E}\frac{2m^*}{\hbar^2}\right)^{1/3}, \tag{14.9}$$

$$\alpha_1 = \left(\frac{3}{4}\right)^{1/3}\left(e\mathbf{E}\frac{2m^*}{\hbar^2}\right)^{1/3}, \tag{14.10}$$

$$\alpha_2 = \left(\frac{47}{12}\right)^{1/3}\left(\frac{e\mathbf{E}2m^*}{\hbar^2}\right)^{1/3}, \tag{14.11}$$

and the energies:

$$E_0 = \frac{3}{10}\left(\frac{9^2}{2}\right)^{1/3}\left(\frac{e^2\hbar^2\mathbf{E}^2}{2m^*}\right)^{1/3}, \tag{14.12}$$

$$E_1 = \frac{3}{2}\left(\frac{9}{2}\right)^{1/3}\left(\frac{e^2\hbar^2\mathbf{E}^2}{2m^*}\right)^{1/3}, \tag{14.13}$$

$$E_2 = \frac{9}{7}\left(\frac{47}{12}\right)^{2/3}\left(\frac{e^2\hbar^2\mathbf{E}^2}{2m^*}\right)^{1/3}. \tag{14.14}$$

Comparison of this variational result with the mathematically exact solution of the V-shaped potential well using Airy functions[4] yields that both methods agree to better than 2%.

14.3 Optical properties of doping superlattices

The concept of generating a periodic potential by means of ionized impurities was proposed in 1970 by Esaki and Tsu.[5] The doping superlattice structure that they proposed is

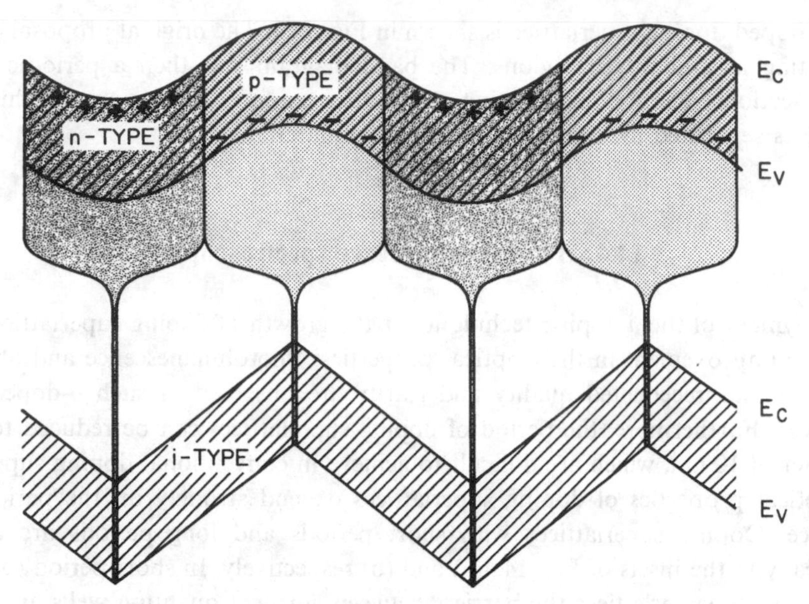

Fig. 14.3. Comparison of the (top) original proposal for doping superlattices and of the δ-doped doping superlattice. The δ-doped structure is superior due to (1) a larger potential modulation, (2) reduced potential fluctuations and (3) the feasibility of smaller superlattice periods.

shown in the top part of Fig. 14.3 and consists of alternating n- and p-type regions, which are homogeneously doped. Novel properties such as negative differential conductivity (NDC) have been postulated for such doping superlattices.

Shortly after the initial proposal of doping superlattices, published work was exclusively of a theoretical nature.[6–9] First experimental work began in the 1980s and included the observation of the tunability of the energy gap in doping superlattices.[10] Extended reviews on homogeneously doped doping superlattices have now become available.[11]

Quantum-confined interband transitions were readily observed in the compositional $Al_xGa_{1-x}As/GaAs$ superlattices. However, in doping superlattices, quantum-confined interband transitions were not observed initially. Doping superlattices which were grown using the δ-doping technique resulted in a significant improvement in their optical properties. Quantum-confined optical interband transitions were observed for the first time in δ-doped doping superlattices in absorption spectroscopy[4] as well as in photoluminescence spectroscopy.[2]

The improved δ-doped doping superlattice structure consists of a train of alternating n- and p-type δ-doping sheets separated by intrinsic (undoped) layers. A periodic n–i–p–i sequence results, whose band diagram consists of linear sections. The δ-doped superlattice has several important advantages over the homogeneously doped structure, including larger superlattice modulation, the feasibility of shorter periods and the minimization of potential fluctuations.[12]

A comparison between the structure of the originally proposed doping superlattice and the δ-doped doping superlattice is shown in Fig. 14.3. The original proposal consisted of alternating n- and p-type regions. The band diagram was then a periodic series of parabolic sections. The δ-doped doping superlattice confines the dopants in thin, highly doped sheets separated by undoped (i = intrinsic) material.

14.4 Photoluminescence spectroscopy

The employment of the δ-doping technique for the growth of doping superlattices results in a drastic improvement in their optical properties. Photoluminescence and absorption transitions of unprecedented quality and clarity are observed in such δ-doped doping superlattices. Furthermore, the period of doping superlattices can be reduced to lengths of the order of 10 nm, which are difficult to achieve in conventional doping superlattices.

The optical properties of doping superlattices depend strongly on the period of the superlattice. Doping superlattices with short periods and long periods are displayed schematically in the insets of Fig. 14.4(a) and (b) respectively. In short-period (or type A) sawtooth doping superlattices the barriers between adjacent quantum wells are thin (e.g. 10 nm) and coupling between adjacent wells becomes significant. Furthermore, the overlap

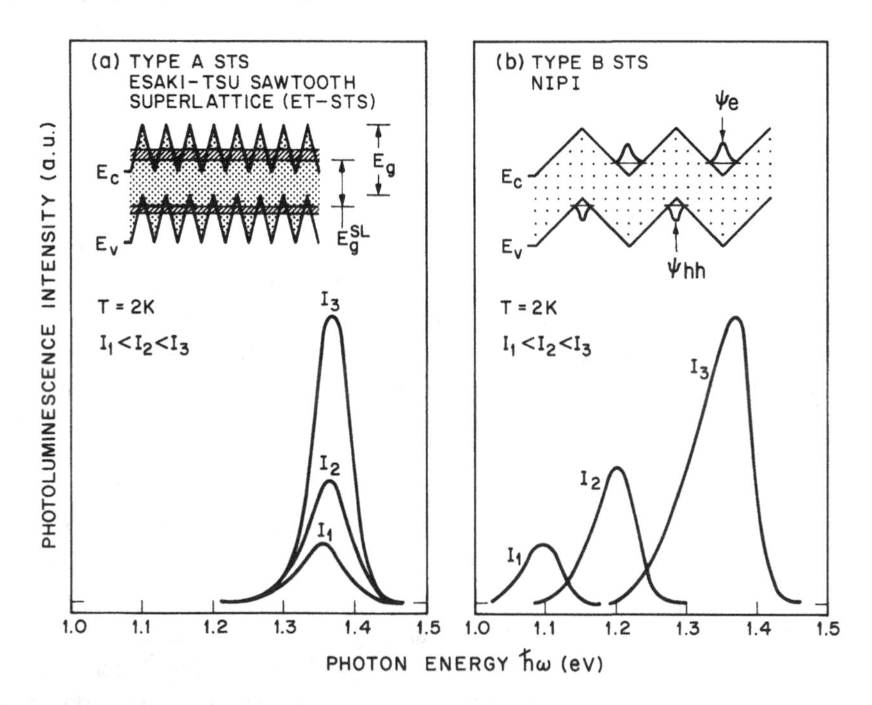

Fig. 14.4. Low-temperature photoluminescence spectra on (a) short-period (type A) sawtooth superlattices and (b) long-period (type B) sawtooth superlattices at 2 K for three different excitation intensities ($I_1 < I_2 < I_3$). The emission wavelength is approximately constant for the type A superlattice, whereas it is clearly tunable for the type B superlattice.

between the electron- and hole-wavefunctions is largest for short-period doping superlattices. Long-period (or type B) doping superlattices with thick barriers between adjacent quantum wells (e.g. 60 nm) result in negligible tunneling between adjacent quantum wells. They also result in negligible overlap between electron- and hole-wavefunctions and long recombination lifetimes of free carriers.

Drastic differences in the optical properties of long- and short-period doping superlattices were, indeed, found in low-temperature photoluminescence experiments.[13] The photoluminescence spectra of a short-period (type A) and a long-period doping superlattice are shown in Fig. 14.4(a) and (b) respectively. The periods of the superlattices are 15 and 60 nm. The doping density is 1×10^{13} cm^{-2}. The photoluminescence spectrum shown in Fig. 14.4(a) reveals an emission peak energy at 1.37 eV. Note that the emission energy is smaller than the energy gap of GaAs, which is $E_g = 1.512$ eV at low temperatures. Three different excitation intensities $I_1 < I_2 < I_3$ are used for excitation. The peak energy of the spectra shifts slightly to higher energies with increasing excitation intensity. However, the shift is quite small as compared with the width of the spectra. Thus, small tunability of the peak energy as a function of excitation intensity is found in short-period doping superlattices.

Strong tunability is found in long-period doping superlattices, as shown in Fig. 14.4(b). The wavelength is continuously tunable from 1.1 to 1.4 eV by changing the excitation intensity. The tuning range is larger than the full-width-half-maximum (FWHM) of the luminescence line, and confirms earlier observations of tunability in doping superlattices.

Clearly resolved quantum-confined interband transitions in photoluminescence spectra are shown in Fig. 14.5. The samples have a carefully balanced donor and acceptor density of 1.25×10^{13} cm^{-2} and a period of 15 nm. Three photoluminescence peaks of unprecedented clarity are observed at $\lambda \approx 0.98$, 1.02, and 1.09 μm. Furthermore, a shoulder is observed on the high-energy side of the spectrum at $\lambda \approx 0.93$ μm. We attribute the luminescence peaks to transitions between quantum-confined conduction- and valence-band states. The assignment of luminescence peaks is confirmed by calculation of transition energies and comparison with experimentally observed peak energies. A very good fit between experimental and calculated peak energies is obtained by using $N^{2D} = 1.3 \times 10^{13}$ cm^{-2} and $z_p = 14.2$ nm. Five transitions can be identified, namely the 0e → 0hh, 0e → 0lh, 0e → 1hh, 0e → 2hh, 1e → 0hh transitions.

Most strikingly, however, the photoluminescence spectrum of Fig. 14.5 displays not only the lowest transition, but also transitions via excited states, for example the 0e → 1hh transition. Furthermore, excited-state transitions (e.g. 0e → 0lh or 0e → 1hh) are more intense than the ground-state (0e → 0hh) transition. The light-hole transition is stronger than the heavy-hole transition, even though the density of states of the light-hole sub-band is much smaller (by approximately a factor of $m^*_{hh}/m^*_{lh} \approx 7$). We will now show that the specific characteristics of the photoluminescence spectrum can be explained consistently in terms of the unique energy dependence of the oscillator strength of the sawtooth structure.

For completeness, we note that multisub-band transitions have been observed in compositional quantum-well structures at very high excitation intensities.[14] However, the

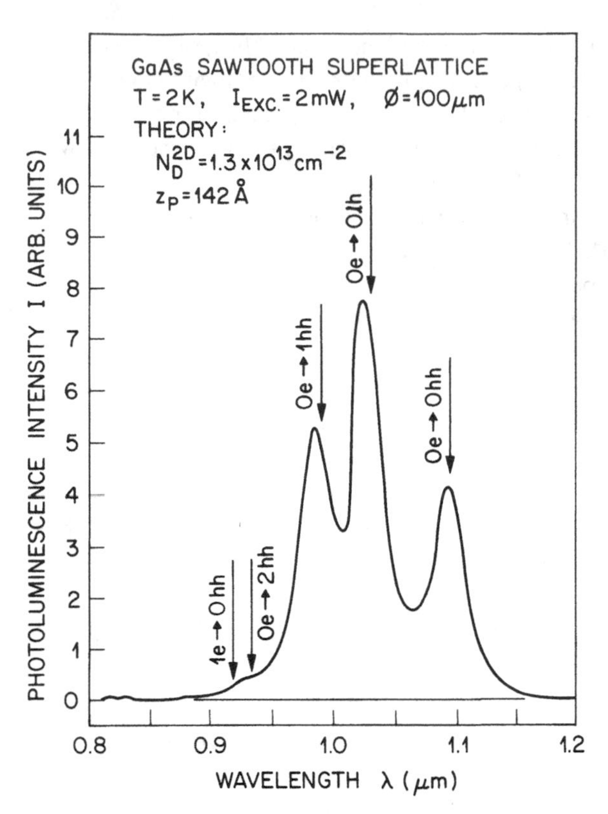

Fig. 14.5. Low-temperature photoluminescence spectrum on a 100-μm-diameter GaAs sawtooth doping superlattice at 2 K ($I_{EXC} = 2$ mW). The arrows indicate theoretical transition energies calculated for a superlattice with $N^{2D} = 1.3 \times 10^{13}$ cm^{-2} and $z_p = 142$ Å.

intensities of those transitions have been found to decrease exponentially with energy. Multisub-band photoluminescence spectra were not observed in conventional homogeneously doped n–i–p–i structures.

The photoluminescence lineshape in semiconductors and semiconductor quantum-well structures can usually be determined by the product of the joint density of states and the thermal distribution of carriers. The latter can be modeled in terms of a Boltzmann distribution and a carrier temperature. The carrier temperature depends on the photoluminescence excitation intensity and is typically in the range of $10 \leqslant T_c \leqslant 50$ K at a lattice temperature of $T_1 = 2$ K. The transition-matrix element depends weakly on the energy in the homogeneous semiconductors or compositional semiconductor quantum wells. By contrast, the oscillator strength increases exponentially with the energy in the sawtooth superlattice. Thus, the oscillator strength has an opposite dependence on energy as compared with the thermal distribution. Consequently, transitions via excited states are more likely to be observed in the sawtooth structure, even though they may be sparsely populated.

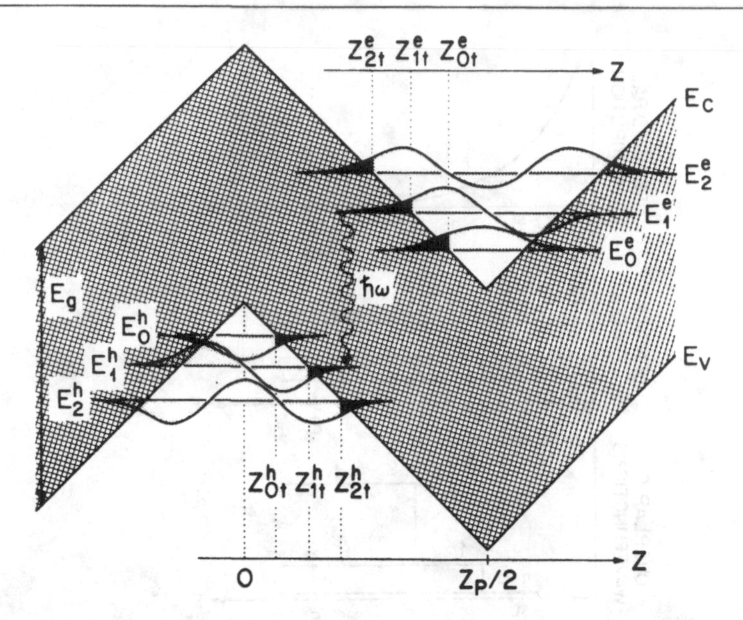

Fig. 14.6. Schematic band diagram of a sawtooth superlattice. The overlap of electron and hole wavefunctions occurs in the classically forbidden region (cross-hatched), that is, beyond the classical turning points, and increases exponentially with sub-band index.

The energy-dependence of the oscillator strength can be visualized with the aid of Fig. 14.6, which shows an overlap of wavefunctions which increases with energy. The matrix element of a quantum-confined transition in a sawtooth superlattice mainly involves the exponentially decaying parts of the wavefunctions. Therefore, the limits of integration are chosen in such a way that integration is limited to the region beyond the classical turning points z_{it} and z_{ft} of the initial (electron) and final (hole) state, as shown in Fig. 14.6. Evaluation of the overlap integral yields[2]

$$\int_{z_{ft}}^{z_p/2 - z_{it}} \psi_i(z)\psi_f(z)\,dz \approx \frac{\psi_i(z_{it})\psi_f(z_{ft})}{\alpha_f - \alpha_i}\,e^{-\alpha_i(z_p/2 - z_{it} - z_{ft})}, \tag{14.15}$$

where α_i and α_f are the decay constants of the initial and final state wavefunction respectively. For the calculation of this result we use the fact that the electron mass is lighter than either the heavy- or light-hole mass, that is, $\alpha_i \leqslant \alpha_f$. Since the turning points depend on the eigenstate energies according to $z_{it} = E_i/e\mathbf{E}$ and $z_{ft} = E_f/e\mathbf{E}$, it is obvious that the oscillator strength depends exponentially on the eigenstate energy due to an increasing overlap of wavefunctions.

The interplay between the thermal distribution of carriers, the density of states, and the oscillator strength of the transitions is illustrated in Fig. 14.7. The thermal distribution of carriers and the oscillator strength have an opposite (exponential) dependence on energy. The transition probability is then a function with multiple peaks.

Next, a quantitative analysis is provided in order to quantify the shift of the quantum-confined photoluminescence transitions as a function of excitation intensity. In this

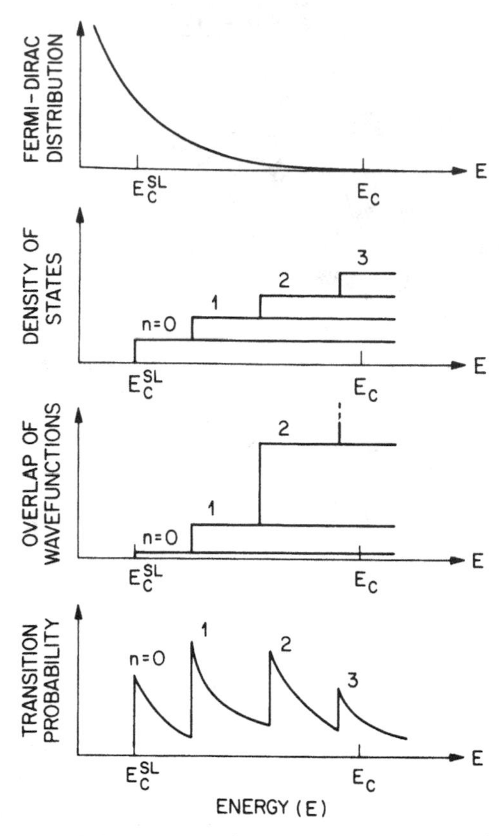

Fig. 14.7. Schematic illustration of thermal carrier distribution, density of states and the overlap of wavefunctions as a function of energy. The lowest diagram shows the transition probability for optical transitions, which is the product of the top three diagrams. Note that the thermal carrier distribution and the overlap of wavefunctions have an opposite exponential dependence on energy.

analysis we assume an exponential decay of the radiative recombination. In CW (continuous wave) photoluminescence experiments the generation rate and recombination rate coincide:

$$\frac{dn}{dt} = \frac{n}{\tau}. \tag{14.16}$$

The radiative lifetime τ can be determined from our experiments, since the generation rate dn/dt is known at a given excitation intensity and the free-carrier concentration n can be determined from screening caused by the free carriers. To determine n we consider the shift of quantum-confined transitions in Fig. 14.8 which is further illustrated by the broken lines. For an increase in the excitation intensity from 2 mW (25 W cm^{-2}) to 20 mW (250 W cm^{-2}), the $0e \rightarrow 0lh$ transition, which is the strongest transition, increases in peak energy from 1.208 to 1.240 eV. We attribute this change of peak energy to screening of the

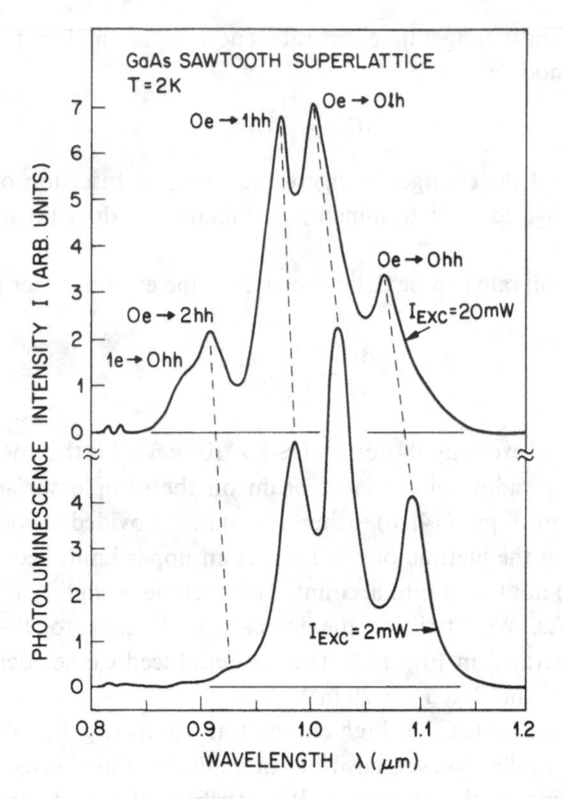

Fig. 14.8. Low-temperature photoluminescence spectrum of a GaAs sawtooth superlattice at 2 K for two different excitation intensities of 2 mW (25 W cm^{-2}) and 20 mW (250 W cm^{-2}). The broken lines indicate the shift of quantum-confined transitions with excitation intensity. At high excitation intensity a new transition arises, namely the 1e → 0hh and 0e → 2hh transitions.

dopant charge by the photogenerated carrier. The result of screening is easily illustrated by considering the photoluminescence energy, given by

$$E = E_{\mathrm{g}}^{\mathrm{GaAs}} = \tfrac{1}{2}e\mathbf{E}z_{\mathrm{p}} + E_n^{\mathrm{e}} + E_n^{\mathrm{h}}, \tag{14.17}$$

where $E_{\mathrm{g}}^{\mathrm{GaAs}}$ is the gap energy of GaAs, $\frac{1}{2}e\mathbf{E}z_{\mathrm{p}}$ is the modulation of the superlattice potential, and E_n^{e} and E_n^{h} are the electron and hole eigenstate energies respectively. The electric field $\mathbf{E}$ is screened under photoexcitation and changes by an amount

$$\Delta\mathbf{E} = \frac{e}{2\varepsilon}\,\Delta n_{\mathrm{2D}}, \tag{14.18}$$

where Δn_{2D} is the excitation-induced density of carriers per quantum well. Thus, the recombination energy increases due to the reduction in potential modulation, as shown in Eqn. (14.17). Upon screening, the potential modulation is reduced, and the individual V-shaped quantum wells become shallower. The eigenstate energies of the confined states reduce accordingly. The change in eigenstate energies is smaller than the change in band-edge potential modulation:

$$\Delta E/e = \Delta \mathbf{E}\tfrac{1}{2}z_{\mathrm{p}}, \tag{14.19}$$

and we therefore neglect the changes in eigenstate energies. Insertion of Eqn. (14.18) into Eqn. (14.17) allows $\Delta n_{2\mathrm{D}}$ to be determined as a function of the shift of the luminescence energy ΔE.

Finally, the generation rate can be determined from the exciting laser power P according to

$$\frac{\mathrm{d}n}{\mathrm{d}t} = \frac{P}{\hbar\omega}\frac{\alpha}{r^2\pi}, \tag{14.20}$$

where $\hbar\omega$ is the energy of exciting photons, α ($\approx 4 \times 10^4\ \mathrm{cm}^{-1}$) is the absorption coefficient, and r ($=50\ \mu\mathrm{m}$) is the radius of the laser beam on the sample surface. The lifetime of carriers obtained from Eqn. (14.16) using the data provided above is $\tau \approx 10$ ns at $P_{\mathrm{L}} = 20$ mW. Note that the lifetime of $\tau \approx 10$ ns is an upper limit, because changes in the eigenstate energies are not taken into account. This lifetime is slightly longer than lifetimes in homogeneous GaAs. We attribute the increase in lifetime to a smaller overlap of wavefunctions, as illustrated in Fig. 14.6. The photoinduced carrier density inferred from this lifetime is $4 \times 10^{17}\ \mathrm{cm}^{-3}$ at $P_{\mathrm{L}} = 20$ mW.

Figure 14.8 further illustrates that high-energy transitions (e.g. 0e → 1hh and 0e → 0lh) gain intensity relative to the lowest transition. In addition, a new peak and shoulder arise on the high-energy side of the spectrum. We attribute the peak and shoulder to the 0e → 2hh and 1e → 0hh transitions, in agreement with absorption measurements on the same sample. The occurrence of the new transitions cannot be understood solely on the basis of screening. The new high-energy transitions can be explained by band filling and a higher effective carrier temperature at increased excitation intensity.[2]

14.5 Quantum-confined absorption

Optical transitions in δ-doping superlattices are not governed by the conventional selection rules which apply to compositional quantum wells and state that interband transitions are allowed only if $\Delta n = 0$, for example an $n = 0$ conduction sub-band electron can recombine with an $n = 0$ sub-band hole. In doping superlattices, the selection rules are fundamentally different, that is, the optical dipole matrix elements are finite and non-zero for all transitions. The matrix element involves an initial and a final state, which have an exponentially decaying part and a spatially oscillating part, as shown in Fig. 14.1, yielding a finite non-zero transition probability for all transitions. This property of the sawtooth superlattice is in contrast to compositional superlattices, where conventional selection rules do apply (e.g. the selection rule $\Delta n = 0$ for (Al, Ga)As/GaAs superlattices with no electric field present).

The GaAs epitaxial layers used for this study were grown by gas-source MBE on undoped semi-insulating GaAs substrates. The growth temperature was kept below $T = 550\,°C$ to avoid diffusion of n-type (silicon) and p-type (beryllium) impurities. The design parameters of the superlattice include a period of $z_p = 15.0$ nm and a two-dimensional doping density of $N_D^{2D} = N_A^{2D} = 1.25 \times 10^{13}\ cm^{-2}$. The samples have ten periods of 20 dopant sheets separated by $\frac{1}{2}z_p = 7.5$ nm. The samples have a closely balanced impurity concentration: $N_D^{2D} \approx N_A^{2D}$. Such a balance is essential, because its absence would blue-shift the absorption edge according to the Burstein–Moss shift. Absorption measurements were performed on polished $0.25\ cm^2$ samples. A dual-beam Perkin–Elmer model 330 spectrophotometer and a variable-temperature cold-finger cryostat were used.

The results of absorption measurements on GaAs sawtooth superlattices measured at $T = 6$ K are shown in Fig. 14.9. The gap energy of the undoped GaAs substrate corresponds to a wavelength of $\lambda = 820$ nm and is shown by a double arrow. The substrate absorbs light at energies slightly below the fundamental gap; this absorption of bulk material is known at the Urbach tail. We determined the corresponding Urbach-tail energy to be $E_U = 6$ meV for our undoped GaAs samples. A typical absorption spectrum of an undoped GaAs sample is shown as a broken curve in Fig. 14.9.

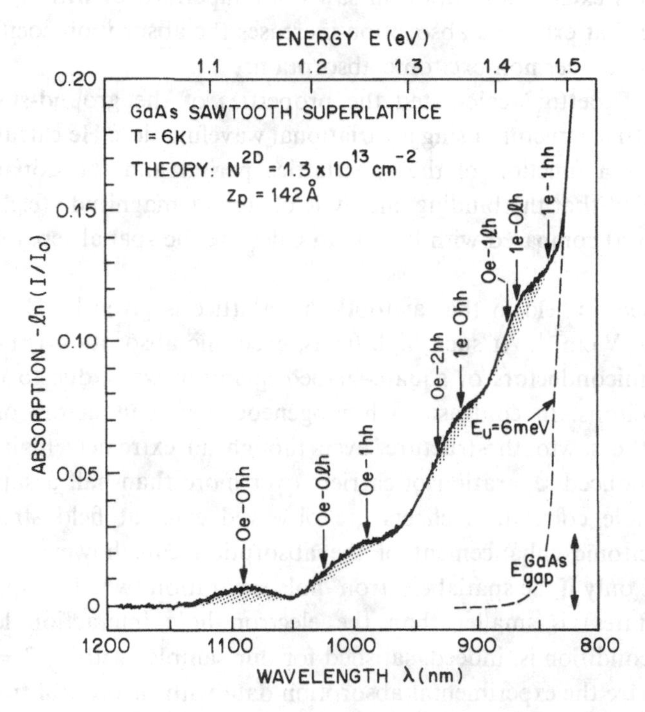

Fig. 14.9. Optical interband absorption spectrum of a GaAs sawtooth superlattice at $T = 6$ K. Theoretical transition energies are indicated by arrows. The lowest electron to lowest heavy-hole transition is referred to as $0e \rightarrow 0hh$. The parameters used for the calculation are a period of 142 Å and a dopant concentration of $1.3 \times 10^{13}\ cm^{-2}$. The energy gap of the substrate is marked by a double arrow. The absorption tail of the substrate is characterized by an Urbach-tail energy of $E_U = 6$ meV.

The absorption spectrum given in Fig. 14.9 shows strong absorption below the fundamental gap of GaAs in a range of 400 meV below the bandgap of the GaAs host lattice. The absorption below the gap of GaAs is the result of the high built-in electric fields of doping superlattices, and is known as the Franz–Keldysh effect. The most striking aspects of the absorption spectrum are four distinct features: an absorption maximum (peak) at $\lambda = 1090$ nm and three shoulders at wavelengths of 1000, 920, and 865 nm. This structure is attributed to transitions between quantum-confined states in the valence and conduction bands. Such quantum-confined interband transitions have not been observed in homogeneously doped doping superlattices. Furthermore, the absorption does not increase monotonically with energy, but has a clear peak at $\lambda = 1090$ nm. Unlike the absorption spectrum shown in Fig. 14.9, the joint density of states does increase monotonically with energy. The occurrence of such an absorption peak thus shows the presence of excitonic or electron–hole correlation effects.

The formation of excitons necessitates an extended understanding of the physical properties of doping superlattices. According to previous beliefs, electron–hole separation naturally objects to the formation of excitons. However, the absorption peak shown in Fig. 14.9 elucidates exciton formation in sawtooth superlattices with appropriate design parameters. Note that excitonic absorption increases the absorption coefficient by several orders of magnitude over non-excitonic absorption.

Very recently, Proetto[15] calculated the properties of the ground-state exciton in a sawtooth superlattice structure using a variational wavefunction. He calculated the exciton binding energy as a function of the superlattice period and the dopant density. This calculation revealed that the binding energy is of similar magnitude (e.g. $E_b = 6$ meV for a period of 10.0 nm) compared with bulk GaAs, despite the spatial separation of electrons and holes.

The built-in electric field in the sawtooth superlattice is given by $\mathbf{E} = eN^{2D}/2\varepsilon$, which gives $\mathbf{E} \geqslant 5 \times 10^5$ V cm^{-1}. At such high fields, excitonic absorption is not observable in homogeneous semiconductors or square-shaped quantum wells, due to the field-induced ionization of excitons. In contrast to homogeneous semiconductors or square-shaped quantum wells, the sawtooth structure, even though an extremely high field is present, opposes a field-induced separation of carriers over more than half a superlattice period. Thus, electron–hole correlation effects are observed even at field strengths exceeding 10^5 V cm^{-1}. Excitonic enhancement of the absorption can, however, occur in doping superlattices, but only if the spatial electron–hole separation (which is approximately $z_p/2$ for the $n = 0$ states) is smaller than the electron–hole interaction length (excitonic diameter). This condition is, indeed, satisfied for our samples with $z_p/2 = 7.5$ nm.

We now compare the experimental absorption data with theoretical transition energies. The arrows shown in Fig. 14.9 are calculated energies of quantum-confined transitions. The lowest electron ($n = 0$) to lowest heavy-hole ($n = 0$) transition is referred to as the $0e \rightarrow 0hh$ transition. Agreement between calculated quantum-confined transition energies and experimental ones is observed over a wide range of energies. For our calculation, a period of $z_p = 14.2$ nm and a doping density of $N_D^{2D} = N_A^{2D} = 1.3 \times 10^{13}$ cm^{-2} have been used.

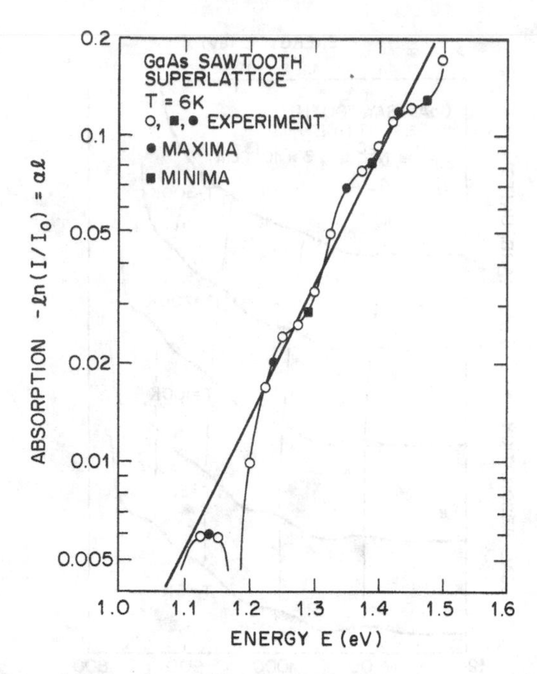

Fig. 14.10. Logarithmic absorption of a GaAs sawtooth superlattice at $t = 6$ K plotted against photon energy. The quasi-linear functional dependence (solid line) indicates that the transition probability increases exponentially with energy. Solid circles and solid squares represent maxima and minima of the absorption spectrum.

The absorption is plotted in Fig. 14.10 on a logarithmic scale against energy. A linear relationship (straight line) between the envelope of $\ln(\alpha l)$ and E is obtained. Thus, an exponentially increasing transition probability with energy is inferred from Fig. 14.10. The envelope of the absorption is exponentially increasing due to an increasing overlap of wavefunctions and, accordingly, an exponentially increasing matrix element.

Additional absorption spectra for temperatures $6 \leqslant T \leqslant 300$ K are shown in Fig. 14.11. The absorption edge of the substrate shifts from 820 nm at $T = 6$ K to 875 nm at $T = 300$ K. The quantum-confined interband transitions of the sawtooth superlattice exhibit an identical qualitative shift to those of longer wavelengths, as shown in Fig. 14.11. Such a shift of peak energies of the superlattice absorption is expected to track the shift of the band edge of the host material (see Eqn. (14.1)).

14.6 Transmission spectroscopy

Transmission spectroscopy on δ-doped doping superlattices, using semiconductor p–i–n diodes to detect the transmitted light, is advantageous due to the much better signal-to-noise ratio compared with (spectrophotometer) absorption measurements. Transmission experiments were carried out with a cooled germanium photodiode which detects the light

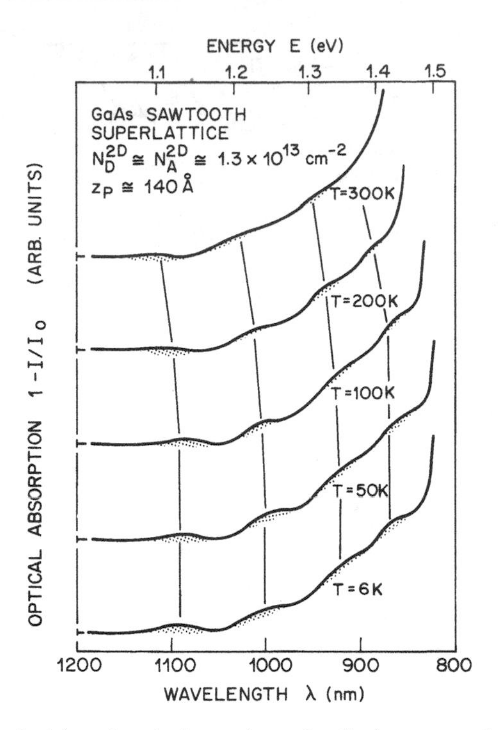

Fig. 14.11. Optical interband absorption of a GaAs sawtooth superlattice for sample temperatures $6 \leqslant T \leqslant 300$ K. $N_D^{2D} \approx N_A^{2D} \approx 1.3 \times 10^{13}$ cm^{-2} and $z_p \approx$ 140 Å. A shift of the fundamental gap as well as the quantum-confined transitions is found with increasing temperature.

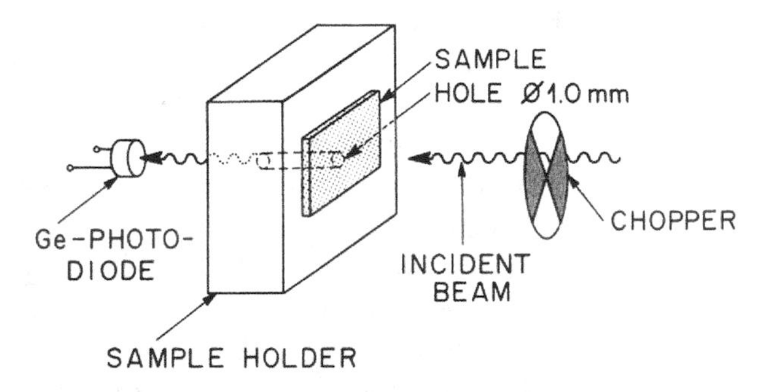

Fig. 14.12. Schematic illustration of the transmission experiment comprising a sample illustrated by chopped monochromatic light. The transmitted light is detected with a germanium p–i–n detector, and calibrated through the 1.0 mm hole.

transmitted through the doping superlattice. When using a phase-sensitive lock-in technique for the detection, the transmission signal carries very little noise.

For the transmission measurements, a sample holder with a 1.0-mm-diameter hole is used, as illustrated in Fig. 14.12. The germanium photodiode detector is thermally decoupled from the sample holder. For monochromatic illumination of the sample, we used a 250 W halogen lamp with a double monochromator (HRD 600 Jobin Yvon) incorporating 1200 and 600 line mm^{-1} gratings. The holographic 1200 line mm^{-1} grating was used for the measurements because of its smooth optical response. In addition, optical filters with cutoff wavelengths of 665 and 780 nm were employed.

The measured spectra were normalized to a reference measurement performed on a semi-insulating substrate with a thickness of 300 μm. In this way, transmission spectra were obtained without detailed knowledge of the response of the measuring set-up. From these corrected curves, the first derivative was obtained numerically. The transmission spectra and its derivative of a δ-doped doping superlattice are shown in Fig. 14.13. The transmission spectrum displays four shoulders which are identified as quantum-confined interband transitions. The structure is stronger in the derivative of the transmission spectrum. The arrows indicate calculated transition energies using the exact Airy-function solutions[4] of the V-shaped potential well. The agreement between calculated and measured data is closest if a doping density of 1.3×10^{13} cm^{-2} and a period of 14.2 nm are used in the calculation.

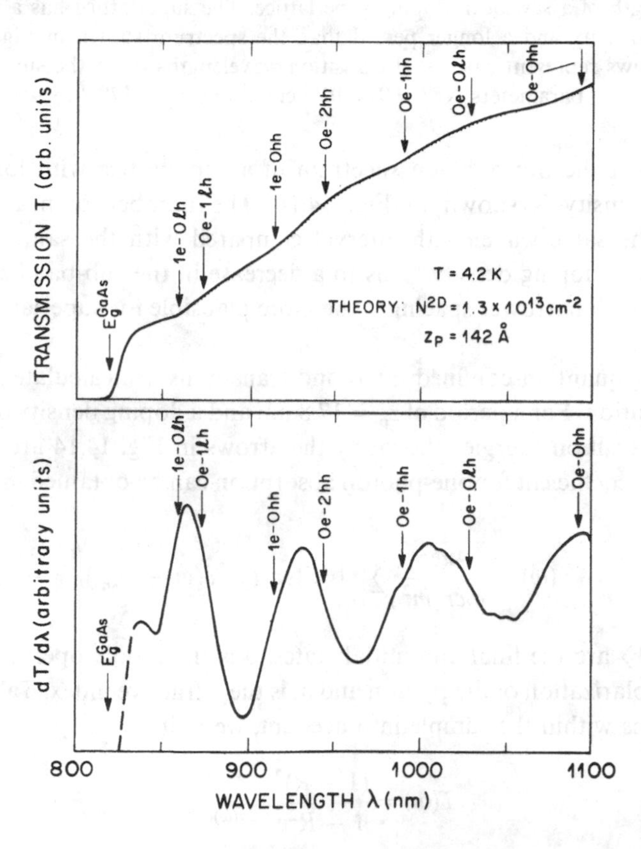

Fig. 14.13. Transmission and derivative of transmission plotted against wavelength of a sawtooth doping superlattice at 4.2 K. The arrows represent calculated transition wavelengths using the superlattice parameters $N^{2D} = 1.3 \times 10^{13}$ cm^{-2} and $z_p = 142$ Å.

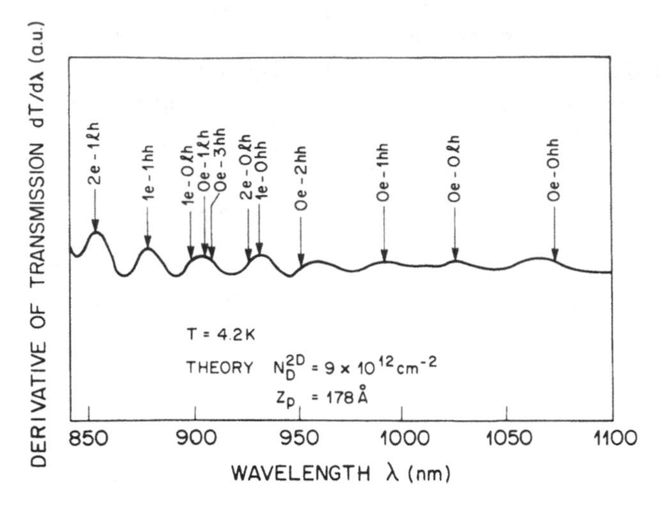

Fig. 14.14. Low-temperature (4.2 K) derivative of transmission plotted against wavelength of a sawtooth doping superlattice. The superlattice has a smaller doping density and a longer period than the spectrum shown in Fig. 14.13. The arrows represent calculated transition wavelengths using the superlattice parameters $N^{2D} = 9 \times 10^{12}$ cm^{-2} and $z_p = 178$ Å.

The derivative of the transmission spectrum of a superlattice with longer period and smaller doping density is shown in Fig. 14.14. The number of peaks has increased significantly in the same wavelength interval compared with the sample shown in Fig. 14.13. A decrease in doping density leads to a decrease in the sub-band spacing. A larger number of peaks with narrower spacing is therefore plausible for superlattices with smaller doping density.

The energies of quantum-confined interband transitions are calculated using the exact Airy-function solution. For a period of $z_p = 17.8$ nm and a doping density of 9×10^{12} cm^{-2} the theoretical transition energies shown by the arrows in Fig. 14.14 are obtained.

The absorption coefficient for one-photon absorption can be obtained in terms of Fermi's golden rule:

$$\alpha(\omega) = \frac{4\pi e^2}{\omega c n_r m^2} \sum_{n \neq n'} |\langle F|p\varepsilon|I\rangle|^2 \delta(\omega - \omega_{nn'}), \tag{14.21}$$

where $\langle F|$ and $|I\rangle$ are the final and initial states, p is the dipole operator, c is the speed of light, ε is the polarization of the photon and n_r is the refractive index. Taking the multiple internal reflections within the sample into account, we write

$$T(\omega) = \frac{(1 - R)^2 e^{-\alpha(\omega)t}}{1 - R^2 e^{-2\alpha(\omega)t}}, \tag{14.22}$$

where R is the reflectivity given by $R = (n_r - 1)^2/(n_r + 1)^2$ in the long-wavelength limit and t is the thickness of the sample. The estimated order of magnitude of the absorption coefficient for our sample is < 1000 cm^{-1} and the denominator can be treated as unity.

This yields $T(\omega) = (1 - R)^2 e^{-\alpha(\omega)t}$ and we can write its derivative with respect to the wavelength as

$$\frac{dT(\omega)}{d\lambda} = \frac{2\pi tc}{\lambda^2}\frac{d\alpha(\omega)}{d\omega}(1 - R)^2 e^{-\alpha(\omega)t}. \quad (14.23)$$

The absorption coefficient for sub-band transition is a broadened step function. Therefore, its derivative has a peak when the photon energy is equal to any transition energy, and has a dip when the photon energy lies in the middle of two transition energies.

14.7 Ellipsometry spectroscopy

In δ-doped semiconductors, the energy band structure is modified by the dopants. This will inevitably affect the material's dielectric optical response, which is very important for optical devices. The dielectric functions of GaAs (100) samples with multiple-δ-doping layers were measured by spectroscopic ellipsometry (SE) at room temperature.

Ellipsometry measures the change in the state of polarization of light reflected from a sample surface. As shown in Fig. 14.15, a linearly polarized, collimated light beam is incident on the sample at an external angle of incidence Φ. After reflection from the sample, the light is in general elliptically polarized. The optical properties of the sample can be studied by analyzing the changes in the reflected wave polarization as a function of wavelength and angle of incidence.

Ellipsometric measurements accurately determine the ratio of complex reflectance R_p to R_s, where R_p and R_s are the reflection coefficients of light polarized parallel to (p) or perpendicular to (s) the plane of incidence. In the simplest case of the two-phase model: ambient (air or vacuum)/substrate with no overlayers, $R_{p,s}$ are simply the Fresnel reflection

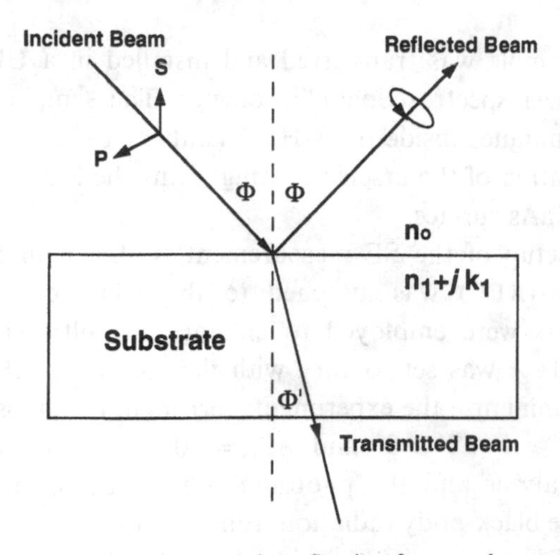

Fig. 14.15. Schematic of polarized light reflection from a plane sample surface.

coefficients. The ratio has been traditionally defined as:

$$\rho = R_p/R_s = \tan(\psi)e^{i\Delta}, \tag{14.24}$$

where the values of $\tan(\psi)$ and Δ are the amplitude and phase of the complex ratio. The results of the SE experimental measurements can be expressed as $\psi(h\nu_i, \Phi_j)$ and $\Delta(h\nu_i, \Phi_j)$, where $h\nu_i$ is the photon energy and Φ_j is the external angle of incidence. The ψ and Δ are sensitive to changes in the surface conditions, overlayer thicknesses, dielectric functions, and other parameters of the sample.[16–19]

The pseudodielectric function $\langle\varepsilon\rangle$ can be obtained from the ellipsometrically measured values of ρ, based on a two-phase model (ambient/substrate):[16]

$$\langle\varepsilon\rangle = \langle\varepsilon_1\rangle + i\langle\varepsilon_2\rangle = \varepsilon_a\left[\left(\frac{1-\rho}{1+\rho}\right)^2 \sin^2\Phi\tan^2\Phi + \sin^2\Phi\right], \tag{14.25}$$

regardless of the possible presence of surface overlayers. The ε_a in Eqn. (14.25) represents the ambient dielectric function (i.e. $\varepsilon_a = 1$ in a vacuum). Therefore, pseudodielectric functions $\langle\varepsilon\rangle$ measured from a clean and smooth sample surface (with no oxide overlayers and surface roughness), by spectroscopic ellipsometry (SE), represent the intrinsic bulk dielectric response ε of the material more accurately.

Delta-doped GaAs samples were grown by growth-interrupted impurity deposition during MBE at $\approx$580 °C, on semi-insulating GaAs (100). Samples no. 1 and no. 2 are constructed with nine Si δ-doped layers with a dopant density of $N^{2D} = \approx 2 \times 10^{13}/\text{cm}^2$, with layer-spacings of 100 and 25 Å, and with surface-GaAs-layers of 250 and 25 Å respectively. The number of δ-doping layers (nine) was selected to obtain adequate light probing sensitivities, from SE measurements, to the δ-doping effects. After finishing the MBE growth, the samples remained in the chamber and were allowed to cool down. The GaAs top layer was then capped with arsenic ($\approx$1000 Å), which prevented surface oxidation in air.

After growth, the sample was transferred and installed in a UHV chamber equipped with a rotating-polarizer spectroscopic ellipsometer. The sample was heated to $\approx$350–400 °C for about ten minutes inside the UHV chamber. A clean and smooth surface was obtained after evaporation of the arsenic coating. Thus the SE measurements were taken from an unoxidized GaAs surface.

The experimental setup of the SE measurement is shown in Fig. 14.16. A rotating-polarizer ellipsometer (RPE) was attached to the UHV chamber. Two air spaced Glan–Taylor polarizers were employed to enhance the ultra-violet transmission. The analyzer azimuth angle A was set to vary with the changes of the measured parameter $\psi(h\nu_i, \Phi_j)$ in order to minimize the experimental errors in the ellipsometry system.[20–22] It was set such that $5° \geqslant A - \Psi \geqslant 0°$ and $A_{min} = 10°$. The monochromator was placed between the fixed analyzer and the photomultiplier tube for improved ambient light rejection (including the black-body radiation from the hot sample and holder). The sample was clamped on a resistor–heater plate inside the UHV chamber, which could be rotated

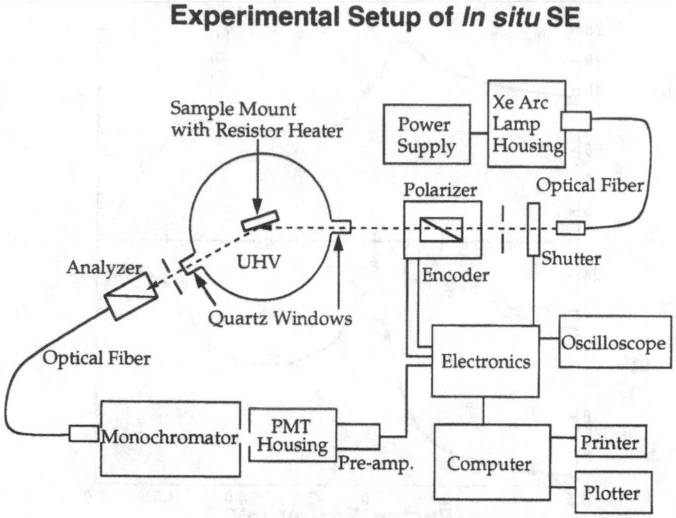

Fig. 14.16. Schematic of the experimental setup of the SE measurement.

and tilted by a rotary drive. The angle of incidence Φ was determined by measuring a known GaAs sample at room temperature. The Φ was treated as a fitting parameter, and the value $\Phi = 74.7° \pm 0.02°$ was obtained. Repeatable angle setting was ensured by two apertures positioned in the entrance and reflected beam paths. Two low-strain, fused-quartz windows[23] were employed. The base pressure of the UHV chamber was typically 1×10^{-9} torr, and the SE measurements were carried out without arsenic overpressure.

Dielectric functions of multilayer-δ-doped GaAs $\varepsilon = \varepsilon_1 + i\varepsilon_2$ were measured by SE at room temperature (RT), in the range 1.5–5.0 eV, in a UHV chamber from an unoxidized surface, as described above. Figure 14.17(a) shows the SE measured ε_1 and ε_2 of δ-doped GaAs sample no. 1, with a layer-spacing of 100 Å. Also shown in Fig. 14.17(b) and (c) for comparison are dielectric functions ε of an undoped GaAs and a uniformly doped GaAs, measured by the same SE system. A comparison between the undoped and the uniformly doped GaAs shows that the E_1 peak ($\approx$2.92 eV at RT)[24] in the ε_2 spectrum of uniformly doped n-GaAs is diminished and broadened. This is known as broadened optical transition at the E_1 energy, induced by dopants.[25] From the ε_2 spectrum of the multilayer-δ-doped n-GaAs, it is obvious that the broadening of optical transitions at the E_1 energy is much less as compared with that of uniformly doped GaAs, although the averaged 3D dopant concentration of the δ-doped GaAs, estimated from the 2D concentration N^{2D}, is higher by a factor of 3–4. We believe this to be due to the strong localization of photoexcited electrons confined in the space charge potential of multilayer-δ-doped GaAs, since the optical beam penetrating into the GaAs passes through large portions of undoped regions between the δ-doped layers. This is consistent with evidence observed by other optical techniques, such as Raman scattering[26] and photoluminescence.[27,28]

The dielectric function of the multilayer-δ-doped GaAs with similar dopant con-

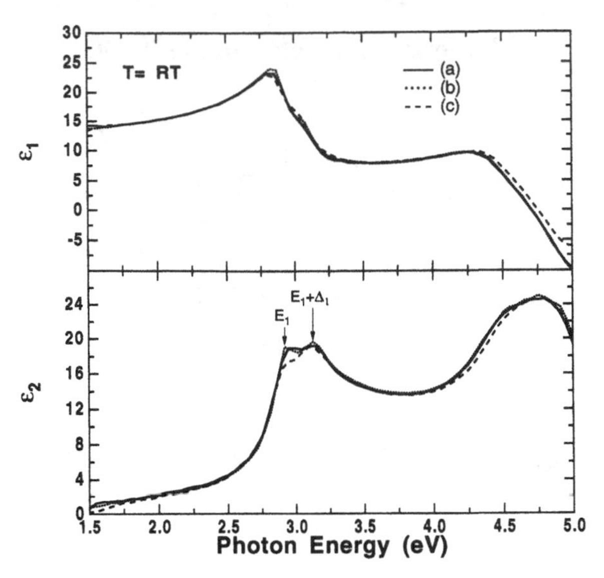

Fig. 14.17. Room temperature dielectric functions ε of GaAs: (a) nine-layer Si δ-doped at dopant concentration $N^{2D} = \approx 2 \times 10^{13}/\text{cm}^2$, with a layer-spacing of 100 Å and top GaAs cap 250 Å thick; (b) undoped; (c) Si uniformly doped at $N = \approx 5.5 \times 10^{18}/\text{cm}^3$.

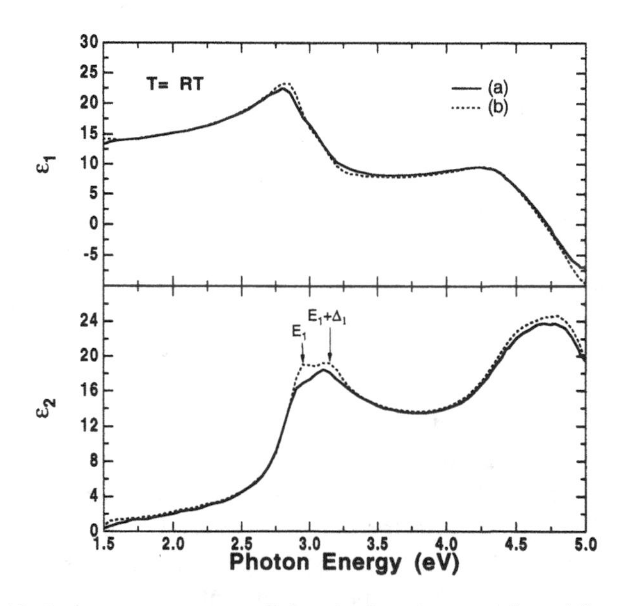

Fig. 14.18. Room temperature dielectric functions ε of multilayer-δ-doped GaAs: (a) nine-layer Si δ-doped at dopant concentration $N^{2D} = \approx 2 \times 10^{13}/\text{cm}^2$, with a layer spacing of 25 Å and top GaAs cap 25 Å thick; (b) nine-layer Si δ-doped at dopant concentration $N^{2D} = \approx 2 \times 10^{13}/\text{cm}^2$, with a layer spacing of 100 Å and top GaAs cap 250 Å thick.

centration but smaller layer-spacing (25 Å) is shown in Fig. 14.18(a). For comparison, ε_1 and ε_2 of δ-doped GaAs sample no. 1, with a layer-spacing of 100 Å, are also shown in Fig. 14.18(b). It is obvious that, due to the smaller layer-spacing, both the E_1 and the $E_1 + \Delta_1$ peaks are greatly diminished and broadened. This comparison further confirms the quantized spatial dopant distribution in δ-doped semiconductors. It should be pointed out that the electron activation of the sample, with layer-spacing of 25 Å, is low ($\approx 12\%$), characterized by the Hall measurement. The low activity is probably related to the extremely high average-3D-dopant concentration of this sample.

14.8 Summary

The optical properties of doping superlattices have been significantly improved by the employment of the δ-doping technique due to the minimization of potential fluctuations resulting from random impurity distribution. The improved optical characteristics are impressively illustrated by observation of quantum-confined interband transitions in doping superlattices. The quantum-confined interband transitions can be observed in several spectroscopic methods, including photoluminescence spectroscopy, absorption spectroscopy, and transmission spectroscopy. The optical transitions have been clearly resolved and their discrete energies can be attributed to calculated transition energies.

The dielectric functions of multilayer-δ-doped GaAs are measured by spectroscopic ellipsometry (SE). Compared with uniformly doped GaAs, the SE measurements indicate reduced broadening of the optical transitions between the E_1 and $E_1 + \Delta_1$ critical energies due to the δ-function-like dopant profile at a layer-spacing of 100 Å.

Acknowledgements

For the ellipsometry and other optical characterization studies of δ-doped GaAs, performed at the University of Nebraska-Lincoln (UNL), the author, H. Yao, acknowledges the partial support by the Research Grant-In-Aid from the Research Council of UNL, partial support by NASA-Lewis Grant NAG-3-154, and the Center for Microelectronic and Optical Materials Research of UNL.

References

1. E. F. Schubert, Y. Horikoshi, and K. Ploog, *Phys. Rev.* B **32**, 1085 (1985).
2. E. F. Schubert, T. D. Harris, J. E. Cunningham, and W. Jan, *Phys. Rev.* B **39**, 11011 (1989).
3. F. F. Fang and W. E. Howard, *Phys. Rev. Lett.* **16**, 797 (1966).
4. E. F. Schubert, B. Ullrich, T. D. Harris, and J. E. Cunningham, *Phys. Rev.* B **38**, 8305 (1988).
5. L. Esaki and R. Tsu, *IBM J. Res. Dev.* **14**, 61 (1970).
6. Y. A. Romanov, *Soviet Phys. Semicond.* **5**, 1256 (1972).

7. Y. A. Romanov and L. K. Orlov, *Soviet Phys. Semicond.* **7**, 182 (1973).
8. G. H. Döhler, *Phys. Status Solidi* **52**, 79 (1972).
9. G. H. Döhler, *Phys. Status Solidi* **52**, 533 (1972).
10. G. H. Döhler, H. Künzel, D. Olego, K. Ploog, P. Ruden, H. J. Stolz, and G. Abstreiter, *Phys. Rev. Lett.* **47**, 864 (1981).
11. G. H. Döhler, *IEEE J. Quantum Electron.* **QE-32**, 1682 (1986).
12. E. F. Schubert, T. D. Harris, and J. E. Cunningham, *Appl. Phys. Lett.* **53**, 2208 (1988).
13. E. F. Schubert, J. E. Cunningham, and W. T. Tsang, *Phys. Rev.* B **36**, 1348 (1987).
14. R. C. Miller, D. A. Kleinman, O. Munteanu, and W. T. Tsang, *Appl. Phys. Lett.* **39**, 1 (1981).
15. C. R. Proetto, *Phys. Rev.* B **41**, 6036 (1990).
16. R. M. A. Azzam and N. M. Bashara, *Ellipsometry and Polarized Light* (North-Holland, Amsterdam, 1977).
17. D. E. Aspnes, in *Handbook of Optical Constants of Solids*, ed. E. D. Palik (Academic, New York, 1985), p. 89.
18. H. Yao and P. G. Snyder, *Thin Solid Films* **206**, 283 (1991).
19. H. Yao, P. G. Snyder, and J. A. Woollam, *J. Appl. Phys.* **70**, 3261 (1991).
20. W. Budde, *Appl. Opt.* **1**, 201 (1962).
21. P. S. Hauge and F. H. Dill, *IBM J. Res. Dev.* **17**, 472 (1973).
22. D. E. Aspnes, *J. Opt. Soc. Am.* **64**, 639 (1974).
23. A. A. Studna, D. E. Aspnes, L. T. Florez, B. J. Wilkens, J. P. Harbison, and R. E. Ryan, *J. Vac. Sci. Technol.* B **7**, 3291 (1989).
24. P. Lautenschlager, M. Garriga, S. Logothetidis, and M. Cardona, *Phys. Rev.* B **35**, 9174 (1987).
25. M. Erman, J. B. Theeten, N. Vodsdani, and Y. Demay, *J. Vac. Sci. Technol.* B **1**, 328 (1983).
26. H. Yao, E. F. Schubert, and R. F. Kopf, *Mat. Res. Soc. Symp. Proc.* **261**, 57 (1992).
27. A. C. Maciel, M. Tatham, J. F. Ryan, J. M. Worlock, R. E. Nahory, J. P. Harbison, and L. T. Florez, *Surf. Sci.* **228**, 251 (1990).
28. S. M. Shibli, L. M. R. Scolfaro, and J. R. Leite, *Appl. Phys. Lett.* **60**, 2895 (1992).

15

Photoluminescence and Raman spectroscopy of single delta-doped III–V semiconductor heterostructures

J. WAGNER AND D. RICHARDS

In this chapter we will focus on the photoluminescence (PL) and Raman spectroscopy of *single* δ-doped structures in which the doping spike is placed at the centre of a III–V semiconductor quantum well or double-heterostructure. It will be shown how, from a combination of these two experimental techniques, detailed information is obtained on the electronic structure of the two-dimensional electron-gas (2DEG) or hole-gas (2DHG) located at the n- or p-type δ-doping spike. In the first section, n-type δ-doped structures will be discussed. It includes the introduction of heterojunction barriers for the confinement of photogenerated minority carriers, the effect of the surface Fermi level position on the subband structure and occupancy, and the study of δ-doped strained quantum wells. The second section deals with p-type δ-doped double-heterostructures, highlighting the dependence of the hole-subband structure on the dopant density, the identification of heavy- and light-hole levels by circularly polarised PL spectroscopy, and the spin-relaxation kinetics of photogenerated electrons. The third section will be devoted to the so-called Fermi edge singularity as an example of many-body effects in the optical spectra of the 2DEG (2DHG) formed in δ-doped GaAs:Si(Be)/$Al_xGa_{1-x}As$ double-heterostructures.

15.1 N-type δ-doped structures

15.1.1 Confinement of minority carriers for efficient emission from the 2DEG

Spectroscopy of luminescence, arising from the radiative recombination of photogenerated holes with electrons from an electron-gas, has proved to be a powerful experimental technique for the investigation of two-dimensional electron systems. Similar to the situation found in n-modulation-doped heterojunctions, PL spectroscopy of single n-type δ-doping spikes is complicated by the fact that the potential confining the electrons in the 2DEG is repulsive for the holes. This repulsion reduces the overlap between electron and hole wavefunctions considerably and, consequently, the probability for radiative recombination is too low to detect emission from the 2DEG (Perry *et al.*, 1988; Henning *et al.*, 1991). For the PL spectroscopy of the 2DEG formed in modulation-doped $GaAs/Al_xGa_{1-x}As$ heterojunctions a p-type lightly doped δ-spike (Kukushkin *et al.*, 1988, 1989) or a second heterointerface (Delalande *et al.*, 1987) have been placed sufficiently close to the 2DEG to confine the photocreated holes.

In n-type δ-doped GaAs some confinement of the photogenerated holes at the surface side could be achieved by placing the doping spike a few hundred Å below the sample surface (Wagner *et al.*, 1990). This is shown in Fig. 15.1(a), where the low-temperature emission spectrum is plotted for Si δ-doped GaAs with the doping spike placed at $z_0 = 300$ Å beneath the GaAs surface. Figure 15.1(b) displays the corresponding emission spectrum of an undoped reference sample. Both spectra were excited at 3.00 eV in order to minimise the penetration depth of the exciting light to about 200 Å (Cardona and Harbeke, 1962). The spectra are dominated by bound exciton recombination (BX) and band-to-carbon acceptor emission (C_{As}) from the undoped GaAs buffer layer. The difference

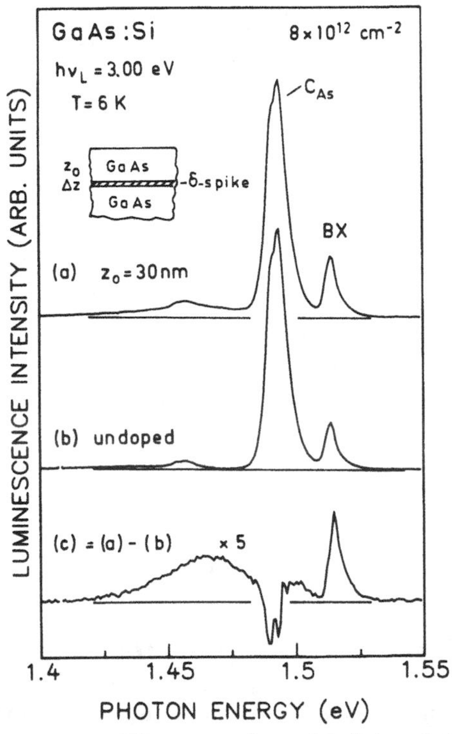

Fig. 15.1. Low-temperature PL spectra from (a) δ-doped GaAs:Si with the doping spike placed at $z_0 = 300$ Å beneath the surface and from (b) an undoped reference sample. In (c) the difference spectrum (a)–(b) is displayed magnified by a factor of 5. The spectra were excited at 3.00 eV with a power density of ≈ 1 kW/cm^2. The sample structure is shown schematically in the inset (Wagner *et al.*, 1990).

spectrum (a)–(b) in Fig. 15.1(c) shows a broad band centred around 1.47 eV. This band has been identified as recombination from the 2DEG at the Si doping spike (Wagner *et al.*, 1990). The features at ≈1.49 and ≈1.52 eV are artefacts due to the subtraction procedure. However, as can be seen from Fig. 15.1, the PL intensity from the 2DEG is still weak because of competing surface recombination. The emission intensity can be significantly enhanced by replacing the bare GaAs surface with an $Al_xGa_{1-x}As/GaAs$ heterointerface (Wagner *et al.*, 1990), as shown in Figure 15.2. There, the same sequence of PL spectra is plotted for an $Al_{0.33}Ga_{0.67}As/GaAs$ heterostructure with the Si doping spike placed in the GaAs layer 300 Å below the heterointerface. The thickness of the $Al_{0.33}Ga_{0.67}As$ top-barrier is 100 Å. The overall luminescence intensity is strongly increased because of the much lower interface recombination rate as compared with that of the free GaAs surface (Dawson and Woodbridge, 1984). The PL spectra in Fig. 15.2(a) and (b) are dominated by band-to-band recombination (*eh*), which indicates a much higher density of photocreated carriers compared with the preceding example, even though the

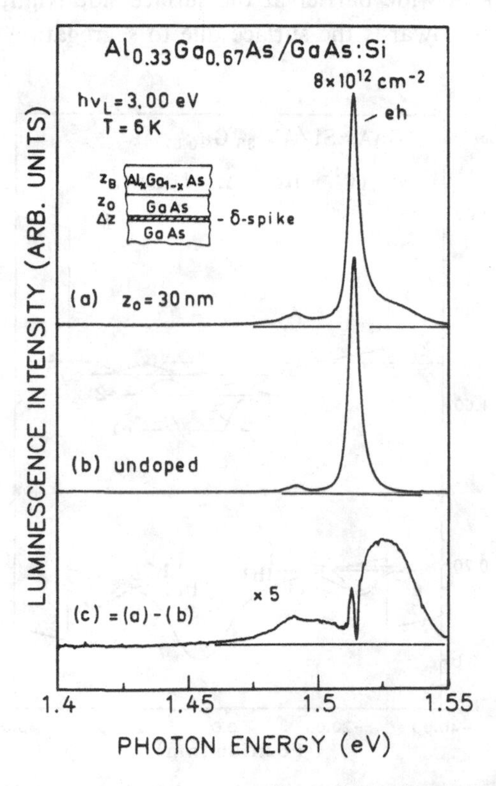

Fig. 15.2. Low-temperature PL spectra from (a) a δ-doped $Al_{0.33}Ga_{0.67}As$/GaAs:Si heterostructure with the doping spike placed at $z_0 = 300$ Å beneath the heterointerface and from (b) an undoped reference sample. In (c) the difference spectrum (a)–(b) is displayed, magnified by a factor of 5. The spectra were excited at 3.00 eV with a power density of ≈1 kW/cm^2. The sample structure is shown schematically in the inset (Wagner *et al.*, 1990).

optical excitation intensity was kept constant. This is a consequence of the enhanced carrier lifetime in the heterostructure. The difference spectrum in Fig. 15.2(c) shows a well-resolved maximum at 1.525 eV accompanied by a somewhat weaker shoulder at $\approx$1.49 eV. These features have again been assigned to radiative recombination from the 2DEG formed at the doping spike (Wagner *et al.*, 1990). The difference in 2DEG emission energy, compared to that observed in the δ-doped structure without the $Al_xGa_{1-x}As$ top-barrier (Fig. 15.1), can be explained by the different densities of photogenerated carriers, even though the optical excitation intensity was the same, and by changes in the self-consistent potential caused by the different depths of the doping spike underneath the surface.

By placing a second $Al_xGa_{1-x}As$ barrier at the substrate side of the doping spike a further drastic increase in the 2DEG emission intensity can be achieved (Wagner *et al.*, 1991a). The self-consistent potentials for electrons and holes in such a δ-doped GaAs/$Al_xGa_{1-x}As$ double-heterostructure are plotted in Fig. 15.3 (Richards *et al.*, 1992). The n-type doping spike of 8×10^{12} donors/cm^2 is placed in the centre of a 600-Å-wide GaAs layer sandwiched between a 200-Å-thick $Al_{0.33}Ga_{0.67}As$ barrier on the substrate side (left) and a corresponding 100-Å-wide barrier at the surface side (right). The doping layer is assumed to be spread out towards the surface due to segregation effects (Wagner *et al.*,

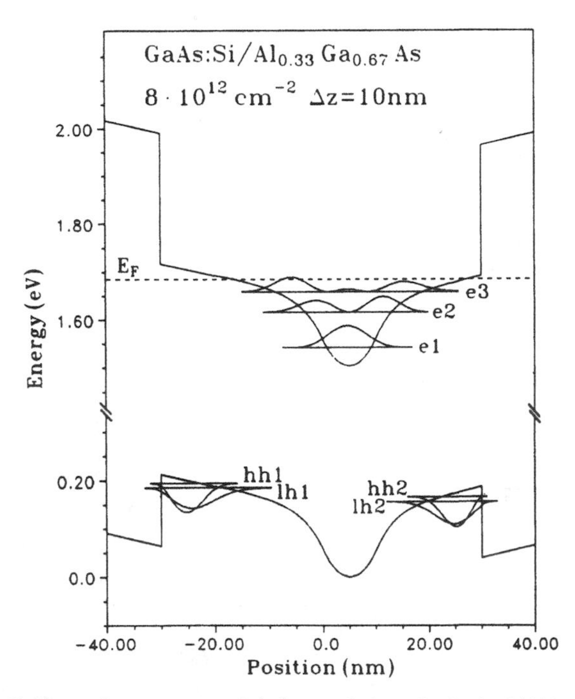

Fig. 15.3. Self-consistent potential in a δ-doped GaAs:Si/$Al_{0.33}Ga_{0.67}As$ double-heterostructure. The surface is on the right at 420 Å. A Si concentration of 8×10^{12} cm^{-2} is taken with a 100 Å dopant spread towards the surface. Energies and probability densities are shown for electron subbands *ei* and for the first heavy- (*hh*1) and light-hole (*lh*1) subbands for an in-plane wavevector $k_{\parallel} = 0$ (Richards *et al.*, 1992).

1989), resulting in a $\Delta z = 100$-Å-wide homogeneously doped layer. Electron- and hole-subband energies and probability densities are also indicated. This potential diagram shows that there is a clear confinement of photogenerated holes at either side of the potential well confining the 2DEG, which leads to a finite electron–hole wavefunction overlap. On the other hand, the heterointerfaces are sufficiently remote that the 2DEG is entirely confined by the space-charge-induced potential well and not by the band discontinuities.

Low-temperature (6 K) PL spectra of such a δ-doped GaAs:Si/$Al_xGa_{1-x}As$ double-heterostructure are plotted in Fig. 15.4. For excitation at 3.00 eV (upper curve in Fig. 15.4), as also used for recording the PL spectra shown in Figs. 15.1 and 15.2, the present spectrum is entirely dominated by the luminescence from the 2DEG centred at about 1.52 eV. Band-to-acceptor emission from the GaAs buffer layer is just resolved as a weak shoulder at ≈ 1.49 eV. This is in striking contrast to the relative emission intensities observed for samples without heterointerfaces or with just one placed at the surface side of the doping spike (Figs. 15.1 and 15.2 respectively). The dominance of the 2DEG emission is also preserved for excitation at a photon energy of, for example, 1.65 eV, which is much closer to the fundamental band-gap energy of GaAs and therefore has a much larger penetration

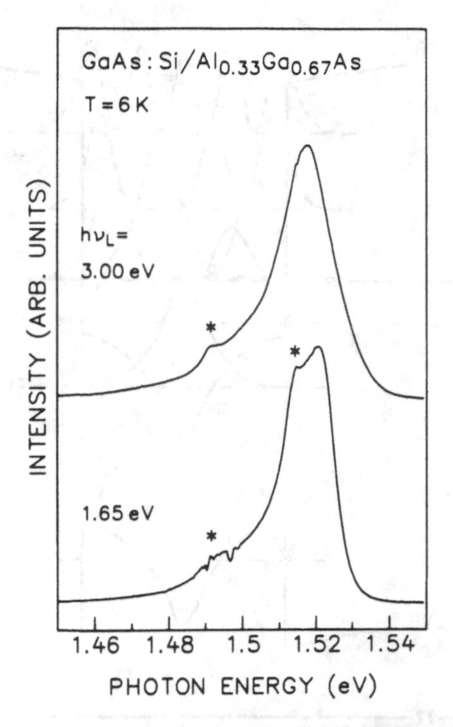

Fig. 15.4. Low-temperature PL spectra of a δ-doped GaAs:Si/$Al_{0.33}Ga_{0.67}As$ double-heterostructure with a dopant concentration of 8×10^{12} cm^{-2}. The spectra were excited at 3.00 eV (top) and 1.65 eV (bottom), respectively, with a power density of ≈ 200 W/cm^2. The asterisks mark band-to-acceptor emission (at ≈ 1.49 eV) and bound exciton emission (at 1.513 eV) from the undoped GaAs buffer layer.

depth of about 0.8 μm (Aspnes *et al.*, 1986). As shown in the lower spectrum in Fig. 15.4, the band-to-acceptor recombination, as well as the bound exciton emission from the GaAs buffer, are also significantly weaker than the luminescence signal arising from the 2DEG. This proves the usefulness of the above sample structure for PL studies of δ-doping layers. As can be seen from Fig. 15.3, there are three occupied electron subbands. Electrons residing in the second and third subbands contribute dominantly to the observed 2DEG emission because of the much larger wavefunction overlap with the photogenerated holes (Richards *et al.*, 1992). Similar effects were seen previously in doping superlattice structures (Schubert *et al.*, 1989).

If one reduces the barrier separation, one progresses from a double-heterostructure to a δ-doped quantum well, where at least the higher-lying occupied subbands are confined by the heterojunction barriers rather than by the space-charge-induced potential. As pointed out by Harris *et al.* (1993), this additional heterojunction confinement provides an extra degree of freedom, as compared with the double-heterostructure, to tune the subband spacings and occupancies. Such n-type δ-doped quantum wells have been studied by several groups combining capacitance–voltage profiling and PL spectroscopy (Shih and Streetman, 1991; Kim *et al.*, 1993), PL and PL excitation (PLE) spectroscopy (Ke *et al.*,

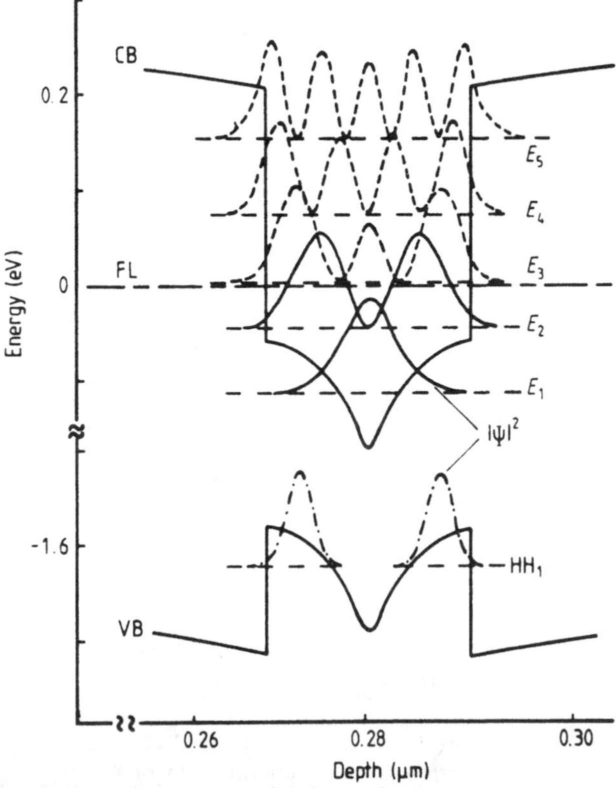

Fig. 15.5. Self-consistent potential in a δ-doped GaAs:Si/$Al_{0.33}Ga_{0.67}As$ quantum well. The well width is 200 Å and the dopant concentration 5×10^{12} cm^{-2}. Energies and probability densities are shown for electron subbands E_i and for the first heavy-hole (HH_1) subband (Harris *et al.*, 1993).

1992; Ke and Hamilton, 1993), and PL and PLE with magneto-transport measurements (Harris *et al.*, 1993).

Figure 15.5 shows the self-consistent potential profiles for electrons and holes for a GaAs/$Al_{0.33}Ga_{0.67}As$ quantum well of width 200 Å, n-type δ-doped at its centre (Harris *et al.*, 1993). For the given Si dopant density of 5×10^{12} cm^{-2} the Fermi level lies very close to the third electron subband, resulting in a significant population of the first two subbands. It can clearly be seen that electrons occupying the second subband are confined by the heterojunction barriers and that there is considerable electron–hole wavefunction overlap for electrons occupying the second and higher subbands. Photoluminescence and PLE spectra are shown in Fig. 15.6 for 100-Å- and 300-Å-wide Si δ-doped GaAs quantum

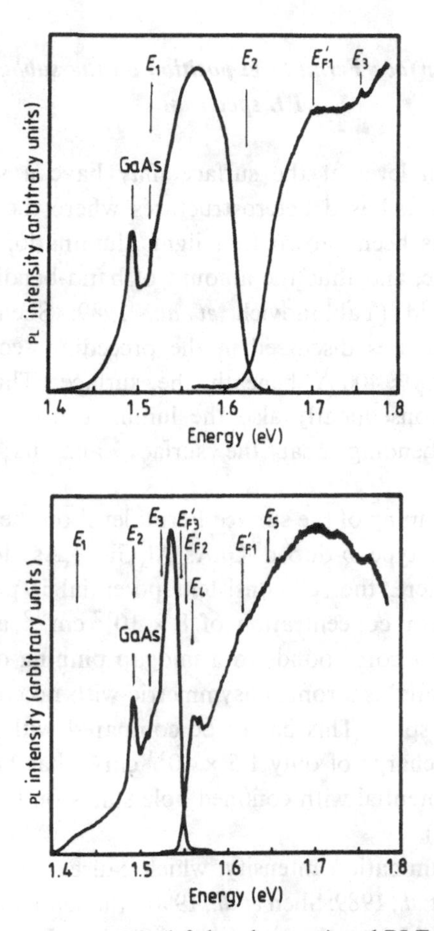

Fig. 15.6. Low-temperature PL (left-hand curves) and PLE spectra (right-hand curves) of centre-δ-doped GaAs:Si/$Al_{0.33}Ga_{0.67}As$ quantum wells. The well width is 100 Å (top) and 300 Å (bottom) respectively. The Si concentration is kept constant at 5×10^{12} cm^{-2}. Calculated interband transition energies are marked by E_i. 'GaAs' denotes emission from the undoped GaAs buffer. E'_{Fi} denotes the onset of in k-space direct absorption from the valence band into states above the Fermi level in the ith electron subband (Harris *et al.*, 1993).

wells (Harris *et al.*, 1993). As can be seen from the calculated subband positions marked in the diagram, there is an increase in subband spacing with decreasing well width. With the dopant concentration kept constant at $5 \times 10^{12}\ cm^{-2}$ the number of occupied electron subbands decreases from three in the 300-Å-wide well to just one for the 100-Å-wide quantum well. As a consequence of this the Fermi energy relative to the bottom of the conduction band rises, for a given electron concentration, with decreasing well width because of the reduction in the available density of states due to the decreasing number of occupied subbands. Thus an increase in the spectral width of the 2DEG emission band is expected for a decrease in well width, as is indeed observed in the PL spectra shown in Fig. 15.6.

15.1.2 Effect of the surface Fermi level position on the subband structure and the PL spectrum

The position of the Fermi level at the surface may have a significant effect on the electronic properties of GaAs-based heterostructures where the doped region is placed close to the surface. It has been shown that light illumination can change the Fermi level position at the surface, and thus the amount of band-bending, that is, the strength of the surface electric field (Yablonovich *et al.*, 1989; Shen *et al.*, 1990). In the δ-doped double-heterostructures discussed in the preceding section (see Fig. 15.3) the Si doping spike lies only ≈ 400 Å beneath the surface. Thus we expect the self-consistent potential, and consequently also the luminescence spectrum, to be strongly influenced by the band-bending near the surface, and its possible change upon illumination.

The effect of a midgap pinning of the surface Fermi level on the self-consistent potential in the above-mentioned n-type δ-doped $GaAs/Al_xGa_{1-x}As$ double-heterostructure is illustrated in Fig. 15.7. There, the self-consistent potential is plotted for electrons and holes calculated for a donor concentration of $8 \times 10^{12}\ cm^{-2}$ and a surface charge of $1.85 \times 10^{12}\ cm^{-2}$. The latter corresponds to a midgap pinning of the Fermi level at the surface. The resulting potential is strongly asymmetric with no confined hole states at the surface side of the doping spike. This has to be compared with the potential plotted in Fig. 15.3, where a surface charge of only $1.5 \times 10^{11}\ cm^{-2}$ has been assumed, leading to an essentially symmetric potential with confined hole states on both sides of the δ-doping layer (Richards *et al.*, 1992).

Upon increasing the illumination intensity, which causes an unpinning of the surface Fermi level (Yablonovich *et al.*, 1989; Shen *et al.*, 1990), the self-consistent potential changes from that depicted in Fig. 15.7 to the situation plotted in Fig. 15.3. This change has a strong effect on the luminescence spectrum of the 2DEG, arising from the recombination of photogenerated holes localised at the heterointerface on the surface side of the doping spike. When the surface Fermi level is pinned midgap, these holes are not localised at that heterointerface and drift towards the surface where they recombine mostly non-radiatively. With increasing illumination intensity, the energy of the conduction band edge at the

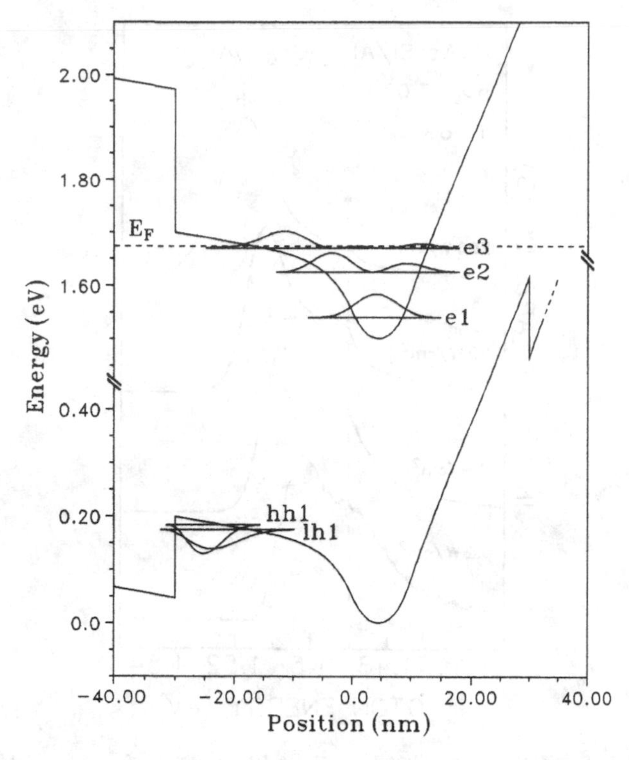

Fig. 15.7. Self-consistent potential in a δ-doped GaAs:Si/$Al_{0.33}Ga_{0.67}As$ double-heterostructure. The surface is on the right at 420 Å. A Si concentration of 8×10^{12} cm^{-2} is taken with a 100 Å dopant spread towards the surface. The surface Fermi level is assumed to be pinned midgap corresponding to a surface charge of 1.85×10^{12} cm^{-2}. Energies and probability densities are shown for electron subbands *ei* and for the first heavy- (*hh*1) and light-hole (*lh*1) subbands for an in-plane wavevector $k_{\parallel} = 0$ (Richards *et al.*, 1992).

surface moves towards the Fermi level, which results in bound states for the photogenerated holes at the topmost heterointerface, and radiative recombination of these holes with electrons in the 2DEG should be observed. A further increase in illumination intensity leads to a further decrease in the surface electric field and thus to a high-energy shift of the 2DEG emission arising from the spatially indirect recombination, with holes localised at the top heterointerface. In the calculation of the potential shown in Fig. 15.3, the effect of photogenerated carriers on the self-consistent potential has been neglected for simplicity (Richards *et al.*, 1992). However, even though the presence of photogenerated holes will create band-bending near the surface, the general scheme discussed above should still be valid.

In order to test this experimentally, PL spectra have been recorded for optical excitation at 3.00 eV. Because of the small penetration depth of $1/\alpha \approx 200$ Å (Cardona and Harbeke, 1962), where α denotes the absorption coefficient, this light is strongly absorbed in the near-surface region and thus only generates holes localised at the topmost heterointerface.

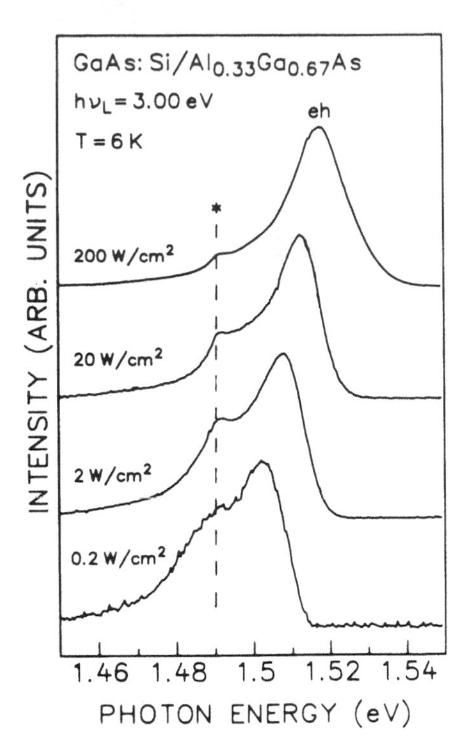

Fig. 15.8. Excitation-intensity-dependent PL spectra of a δ-doped GaAs:Si/$Al_{0.33}Ga_{0.67}As$ double-heterostructure with a dopant concentration of 8×10^{12} cm^{-2}. The spectra were recorded at 6 K with excitation at 3.00 eV. *eh* denotes band-to-band recombination from the 2DEG. The asterisk marks band-to-acceptor emission from the undoped GaAs buffer layer (Richards *et al.*, 1992).

Such excitation-intensity-dependent PL spectra, plotted in Fig. 15.8, show a low-energy shift of the 2DEG band-to-band (*eh*) emission with decreasing excitation intensity, while the band-to-acceptor emission from the GaAs buffer remains fixed in energy (Richards *et al.*, 1992). This is exactly what one expects from the above considerations if one progresses from a symmetric potential with the surface Fermi level unpinned (see Fig. 15.3) to a situation where this Fermi level is closer to the pinned midgap position.

For excitation with light which penetrates much deeper, and therefore generates holes at both sides of the doping spike, two contributions to the 2DEG emission can be resolved corresponding to the recombination of holes localised at the front and the back heterointerface respectively. This is illustrated in Fig. 15.9, which shows excitation-intensity-dependent PL spectra excited with 1.65 eV photons with a penetration depth $1/\alpha \approx 0.8$ μm (Aspnes *et al.*, 1986). For high excitation intensities the two contributions coincide in emission energy (topmost curve). With decreasing excitation energy the portion of the emission spectrum arising from the recombination of holes at the top

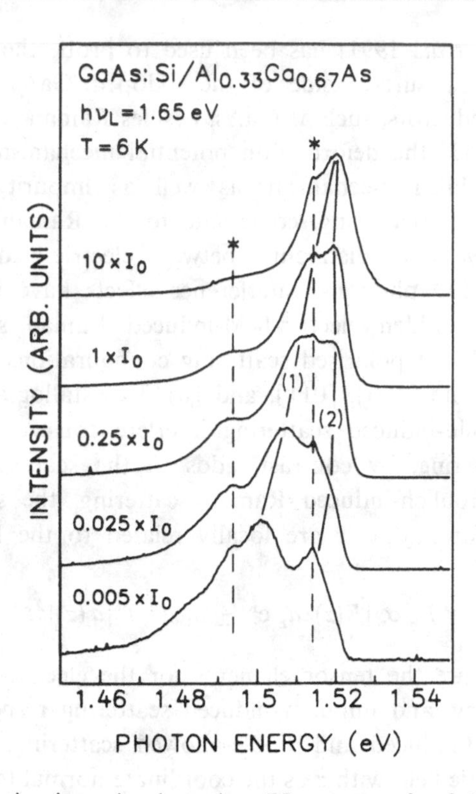

Fig. 15.9. Excitation-intensity-dependent PL spectra of a δ-doped GaAs:Si/$Al_{0.33}Ga_{0.67}As$ double-heterostructure with a dopant concentration of 8×10^{12} cm^{-2}. The low-temperature (6 K) spectra were excited at 1.65 eV. I_0 corresponds to a power density of 20 W/cm^2. The excitation-intensity-dependent peak positions of 2DEG band-to-band recombination lines, involving photogenerated holes at the top (1) and bottom heterointerface (2), are connected by full lines as a guide to the eye. The asterisks mark band-to-acceptor emission (at $\approx$1.49 eV) and bound excition emission (at 1.513 eV) from the undoped GaAs buffer layer; both transitions occur at a constant energy.

hetero-interface (1) shows a much larger low-energy shift than does the portion due to recombination of holes localised at the back interface (2). As can be seen from a comparison of Figs. 15.3 and 15.7, the transition energy for the latter recombination process is expected to be essentially independent of the surface Fermi level position, and thus the excitation energy, because of the screening of the surface electric field by the 2DEG. Part of the observed small-energy shift of the luminescence involving holes at the back heterointerface probably arises from changes in the self-consistent potential caused by photogenerated holes accumulating at the heterointerfaces. This effect is not included in the present potential calculations (Richards *et al.*, 1992).

As a further independent proof of the effect of photoexcitation on the surface Fermi level position, electric-field-induced Raman scattering (Schäffler and Abstreiter, 1986; Shen *et al.*, 1987; Pletschen *et al.*, 1991) has been used to probe the strength of the surface electric field on the near-surface side of the δ-doped GaAs layer (Richards *et al.*, 1992). In polar semiconductors, such as GaAs, besides Raman scattering by longitudinal optical (LO) phonons via the deformation potential mechanism, intrinsic LO phonon scattering by the Fröhlich mechanism, as well as impurity-induced and electric-field-induced LO phonon scattering, contribute to the Raman spectrum for resonant excitation (Cardona, 1982). To distinguish between electric-field-induced and impurity-induced scattering by LO phonons, interference effects have been exploited between dipole-allowed and -forbidden electric-field-induced Raman scattering (Shen *et al.*, 1987). For the two different polarised scattering configurations $x(z', z')\bar{x}$ and $x(y', y')\bar{x}$, where x, y', and z' denote [100], [0$\bar{1}$1], and [011] crystallographic directions, dipole-allowed and electric-field-induced scattering interfere constructively and destructively Impurity-induced scattering, by contrast, adds to the scattering signal incoherently. Neglecting intrinsic Fröhlich-induced Raman scattering, the scattering intensities I_+ for $x(z', z')\bar{x}$ and I_- for $x(y', y')\bar{x}$ are locally related to the Raman tensor elements by:

$$I_\pm \propto |E(z) \cdot a_E \, e^{i\phi} \pm a_{DP}|^2 + |a_I(z)|^2. \qquad (15.1)$$

where a_E, a_{DP}, and a_I are the tensor elements for the electric-field-induced scattering, dipole-allowed scattering, and impurity-induced scattering respectively; ϕ is the phase difference between field-induced and dipole-allowed scattering, and $E(z)$ is the depth-dependent surface electric field with z as the coordinate normal to the surface. Taking the difference between the scattering intensities for the two different scattering configurations we can define a quantity $\Delta G \propto (E^*/\alpha) \cdot |a_{DP}| \cdot |a_E| \cdot \cos\phi$. This quantity is proportional to E^*, the electric field averaged over the probing depth (Pletschen *et al.*, 1991), which is $1/(2\alpha) \approx 100$ Å for excitation at 3.00 eV in resonance with the E_1 band gap of GaAs (Cardona and Harbeke, 1962).

Figure 15.10 displays polarised Raman spectra excited at 3.00 eV in resonance with the GaAs E_1 band gap and recorded for the two different scattering configurations $x(y', y')\bar{x}$ and $x(z', z')\bar{x}$. For low excitation intensities there is a marked difference in the LO phonon scattering intensity due to electric-field-induced Raman scattering, whereas for high intensities this difference is much less pronounced. This is more clearly seen in the insert in Fig. 15.10, where the difference in LO phonon intensity ΔG, normalised to the strength of purely intrinsic two-LO phonon scattering I(2LO), is plotted versus the excitation power density. There is a decrease in the strength of the average electric field E^* ($\Delta G \propto E^*$) by one order of magnitude over the measured range of excitation intensities. This is clear proof of a photoinduced decrease in the electric field strength in the near-surface region of the δ-doped GaAs layer. We note in passing that for the resonance conditions chosen, we only probe the electric field on the near-surface side of the GaAs layer and *not* on the topmost $Al_xGa_{1-x}As$ barrier, which can be seen from the absence of the AlAs-like $Al_xGa_{1-x}As$ LO phonon in the spectra shown in Fig. 15.10. Thus the results of the electric-field-induced Raman scattering experiment complement the view obtained from the above PL measurements nicely.

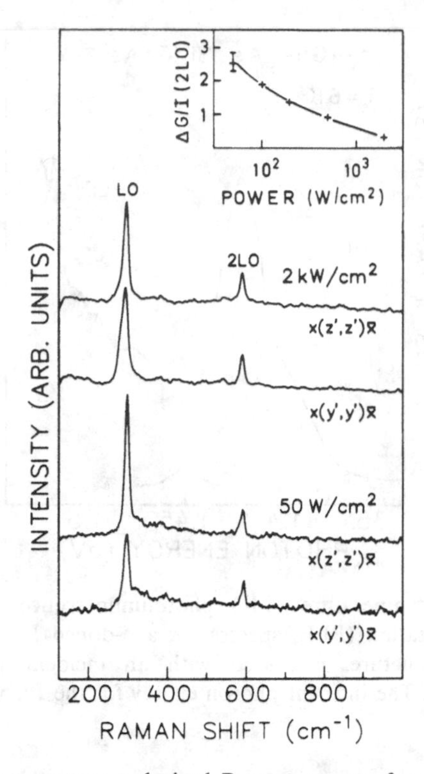

Fig. 15.10. Low-temperature polarised Raman spectra for polarisation along [011] $[x(z', z')\bar{x}]$ and $[0\bar{1}1]$ directions $[x(y', y')\bar{x}]$. The inset shows the difference $\Delta G \propto E^*$ between the LO phonon scattering intensities for these two different configurations, normalised to the 2LO scattering intensity I(2LO) and plotted versus incident power density (Richards *et al.*, 1992).

15.1.3 N-type δ-doped $In_xGa_{1-x}As/GaAs$ double-heterostructures: combined PL and Raman spectroscopic study

One problem with spectroscopic investigations of GaAs δ-doped systems can be the presence of luminescence from the undoped GaAs buffer layer (see, for example, Figs. 15.4, 15.6, 15.8, and 15.9) which can complicate the analysis of spectra. However, this problem may be avoided by replacing the active GaAs layer, in which the δ-doping spike is placed, by another material with a lower band gap than GaAs. In this way, luminescence from the undoped GaAs buffer and the 2DEG in the active δ-doped layer are spectrally separated. One suitable material system for such a structure is that of a pseudomorphically strained $In_xGa_{1-x}As/GaAs$ quantum well (Richards *et al.*, 1993b). In what follows we illustrate how complementary information on the electron-subband structure can be obtained from a combination of PL and intersubband Raman spectroscopy.

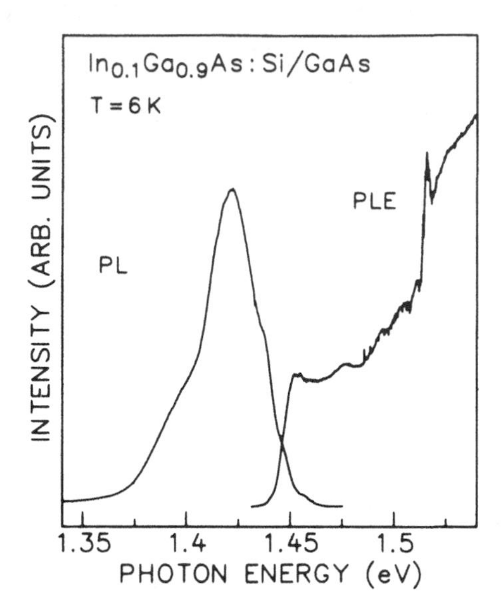

Fig. 15.11. Low-temperature (6 K) photoluminescence (PL) and photoluminescence excitation (PLE) spectra for a δ-doped $In_{0.1}Ga_{0.9}As$:Si/GaAs quantum well structure, measured with an incident power density of $10\ W\ cm^{-2}$. The incident photon energy for the PL was 1.65 eV.

The PL and PLE spectra are shown in Fig. 15.11, for a 300 Å $GaAs/In_{0.1}Ga_{0.9}As/GaAs$ strained quantum well with a Si δ-doping spike placed at its centre. The intended doping density for this structure is $6 \times 10^{12}\ cm^{-2}$, with the doping spike located 500 Å beneath the sample surface. The PLE shows absorption in the $In_{0.1}Ga_{0.9}As$ quantum well from 1.45 eV, displaying the characteristic step-like structure of a two-dimensional density of states. There is a large increase in the PLE intensity at 1.52 eV, marking the onset of absorption in the GaAs cap and buffer layer. The PL spectrum is due solely to emission from carriers in the $In_{0.1}Ga_{0.9}As$ layer.

Figure 15.12 shows the self-consistent potential profile and electron-subband energy levels and probability densities for this structure (Richards *et al.*, 1993b). The Si doping density is $6 \times 10^{12}\ cm^{-2}$ and a rectangular doping profile with a width of 10 Å has been assumed. Following on from Section 15.1.2, it has been assumed, that, under the present illumination conditions, the Fermi level is unpinned at the sample surface. There are three occupied subbands, with only the first two (labelled $c1$ and $c2$) experiencing the characteristic 'V'-shaped potential of a δ-doping layer. The third occupied subband ($c3$) and the unoccupied fourth subband ($c4$) experience the quantum well potential to a much greater extent (i.e. the confining potential of the GaAs barriers). For an isolated δ-doping layer the energy levels come closer together with increasing energy, whereas for a square well the energy levels move further apart. This leads to the $c2 \rightarrow c3$ and $c3 \rightarrow c4$ energy spacing being equal, due to the transition from one potential type to the other.

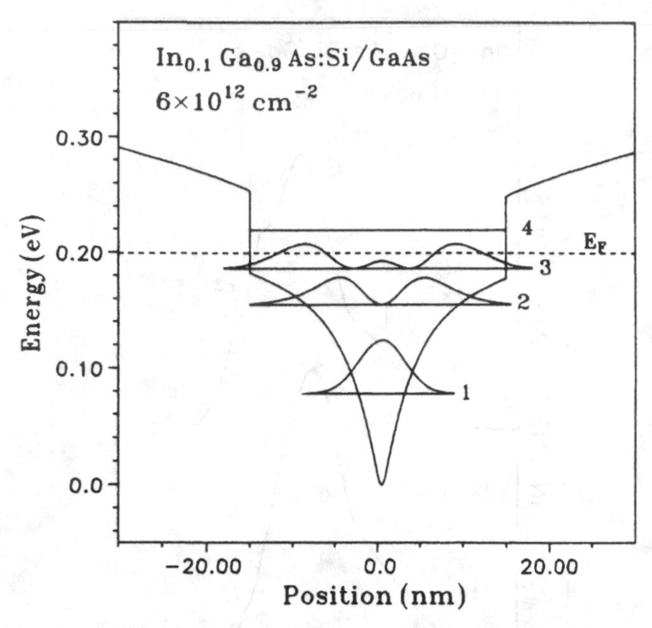

Fig. 15.12. Potential profile and subband ci energies (labelled i) and envelope probability densities $|\psi(z)|^2$, determined from self-consistent subband calculations. There is a 10-Å-wide 6×10^{12} cm^{-2} Si-dopant layer positioned in the centre of a 300-Å-wide $In_{0.1}Ga_{0.9}As/GaAs$ quantum well. A 1.5×10^{11} cm^{-2} transfer of electrons to both the substrate and surface is assumed.

Photoluminescence spectra for two different incident power densities are shown in Fig. 15.13. For the lower power density I_0 of 1 W cm^{-2} there is a peak centred at 1.42 and a shoulder at 1.39 eV, which can be assigned to luminescence from the second ($c2$) and first ($c1$) subband respectively. As the majority of electrons reside in these two subbands, since the Fermi level is close to the minimum of the $c3$ subband, emission from the $c1$ and $c2$ subbands will dominate the PL. Although the density of electrons in subband $c1$ is much greater than that in $c2$, there is a much smaller overlap of the $c1$ electron wavefunction, with the hole states localised at the heterointerfaces. Hence, the PL intensity is greater for emission from $c2$. With increase in power density to $100 \times I_0 \approx 100$ W cm^{-2}, further features appear on the high-energy side of the $c2$ emission band at 1.44 and 1.45 eV which are due to luminescence from photoexcited electrons in the third ($c3$) and fourth ($c4$) subband respectively. In particular, the $c2 \rightarrow c3$ and $c3 \rightarrow c4$ energy spacings appear to be equal from the PL spectra, in agreement with self-consistent subband calculations.

Subband energy spacings can also be measured by electronic Raman scattering, which is a commonly used technique in the study of the properties of the 2DEG present in n-modulation-doped quantum wells and heterojunctions (Pinczuk and Abstreiter, 1989). Intersubband transitions are usually observed via inelastic light scattering by collective spin-density or charge-density (plasmon) excitations, which are seen in depolarised and polarised Raman spectra respectively (Abstreiter *et al.*, 1984; Pinczuk and Abstreiter, 1989). For depolarised scattering, in a backscattering geometry, incident and scattered

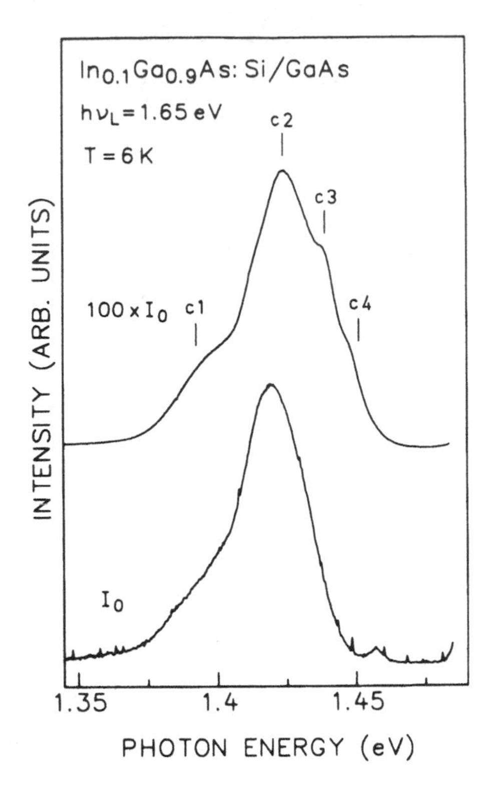

Fig. 15.13. Low-temperature (6 K) PL spectra at two incident power densities for a δ-doped $In_{0.1}Ga_{0.9}As$:Si/GaAs structure. $I_0 = 1$ W cm^{-2}. Luminescence from the different conduction subbands *ci* are identified (see Fig. 15.12). Upon increasing power density, and hence photoexcited carrier density, the higher subbands become occupied.

polarisations are perpendicular and for the polarised case they are parallel. Spin-density excitation energies are shifted slightly down in energy from the bare subband spacings, due to exchange–correlation effects (Pinczuk *et al.*, 1989; Gammon *et al.*, 1992), whereas charge-density excitations are shifted to higher energies by the so-called depolarisation shift, which depends on the electron density (Pinczuk and Abstreiter, 1989). Raman scattering by single-particle excitations is also possible in both polarised and depolarised spectra (Pinczuk *et al.*, 1989) and has been observed to have comparable strength to that from collective spin- and charge-density excitations in a high-density n-modulation-doped GaAs/Al_xGa_{1-x}As quantum well (Peric *et al.*, 1993).

Intersubband Raman scattering has been measured from single doping layers embedded in plain GaAs (Abstreiter *et al.*, 1986; Wagner *et al.*, 1991c) and from multiple δ-doped n–i–n–i GaAs structures (Maciel *et al.*, 1990). For the former, measurements were made around the $E_0 + \Delta_0$ band gap of GaAs, where in n-type doped structures electronic Raman scattering could be expected to be strongly resonant. Although this facilitated the observation of intersubband excitations, the Raman signals were always superimposed on

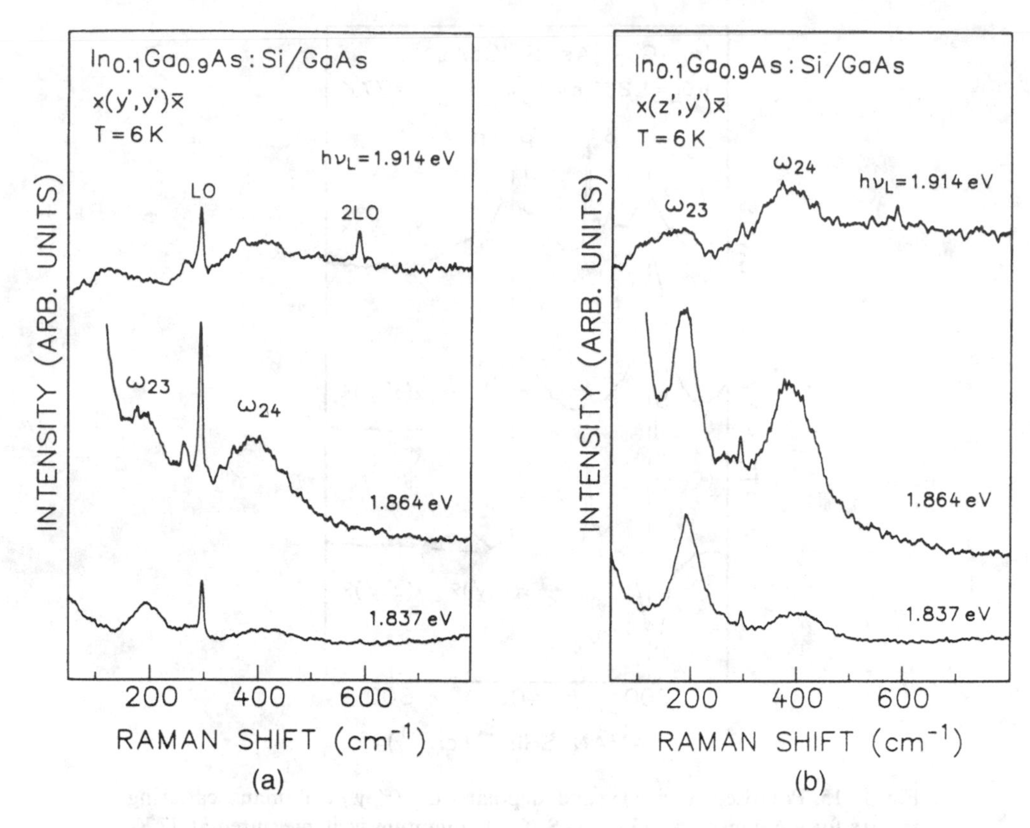

Fig. 15.14. (a) Polarised $x(y', y')\bar{x}$ and (b) depolarised $x(z', y')\bar{x}$ low-temperature (6 K) Raman scattering spectra for a δ-doped $In_{0.1}Ga_{0.9}As$:Si/GaAs quantum well. The spectra were measured with different incident photon energies around the $E_0 + \Delta_0$ band gap of the $In_{0.1}Ga_{0.9}As$ quantum well, for a power density of 1 kW cm^{-2}. In the $x(y', y')\bar{x}$ spectra LO and 2LO scattering are observed. ω_{23} and ω_{24} mark $c2 \rightarrow c3$ and $c2 \rightarrow c4$ intersubband single-particle excitations respectively (see text).

the much stronger $E_0 + \Delta_0$ photoluminescence from the bulk GaAs. This problem, however, is again not present in an $In_xGa_{1-x}As$ structure as the $E_0 + \Delta_0$ interband energy of the δ-doped quantum well is significantly less than that of the surrounding GaAs layers.

Figure 15.14 shows polarised $x(y', y')\bar{x}$ (Fig. 15.14(a)) and depolarised $x(z', y')\bar{x}$ (Fig. 15.14(b)) spectra for the above δ-doped $In_{0.1}Ga_{0.9}As$ quantum well structure, recorded at 6 K for three different incident photon energies around the $In_{0.1}Ga_{0.9}As$ $E_0 + \Delta_0$ band-gap energy. There are two Raman modes, at 200 cm^{-1} (25 meV) and 410 cm^{-1} (51 meV), which are present in both scattering configurations. Polarised and depolarised spectra recorded at 77 K are shown in Fig. 15.15, where it is clear that these excitations occur at the same frequency for the two cases. Hence, these peaks can be assigned to scattering from intersubband single-particle excitations which, from comparison with self-consistent subband calculations, are thought to be associated with $c2 \rightarrow c3$ and $c2 \rightarrow c4$

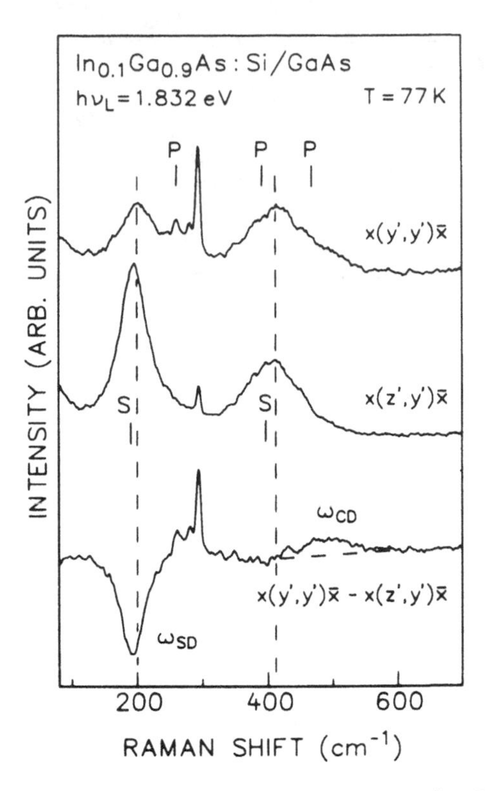

Fig. 15.15. Polarised $x(y', y')\bar{x}$ and depolarised $x(z', y')\bar{x}$ Raman scattering spectra for a δ-doped $In_{0.1}Ga_{0.9}As$:Si/GaAs quantum well, measured at 77 K. The subtracted spectrum is also shown, where charge-density ω_{CD} and spin-density ω_{SD} scattering peaks become apparent. *P* and *S* mark calculated intersubband plasmon and spin-density energies respectively.

transitions (see Fig. 15.12), labelled ω_{23} and ω_{24} in Fig. 15.14. It can also be seen from Fig. 15.14 that these modes, as well as the unscreened $In_{0.1}Ga_{0.9}As$ LO phonon, are strongly resonant with the $In_{0.1}Ga_{0.9}As$ quantum well $E_0 + \Delta_0$ band gap.

If the 77 K spectra in Fig. 15.15 are subtracted from one another, other structures become apparent, although it should be noted that results obtained from the subtraction of spectra taken under different conditions can sometimes be misleading. Nevertheless, it becomes clear from the subtracted spectrum shown in Fig. 15.15 that there is polarised scattering in $x(y', y')\bar{x}$ from excitation at about 490 cm^{-1} (60 meV), which can be assigned to a coupled intersubband plasmon–phonon mode. The LO phonon at 295 cm^{-1} and an intersubband plasmon–phonon mode at 260 cm^{-1} (32 meV) are also present in the $x(y', y')\bar{x}$ spectrum. Intersubband–plasmon–LO–phonon energies can be obtained from calculations within the random phase approximation (RPA) of the dielectric response, employing self-consistently determined wavefunctions (Fasol *et al.*, 1989). Using a total electron density of 5.7×10^{12} cm^{-2} and the experimentally determined intersubband energy spacings of 25 meV for $c1 \rightarrow c3$, and 26 meV, for $c3 \rightarrow c4$, good agreement with experiments is obtained, as shown in Fig. 15.15, where the calculated energies are marked. Note that as the plasmon energies are consistent with a density of $\approx 6 \times 10^{12}$ cm^{-2}, plasma frequencies being strongly dependent on carrier concentration, this confirms the assumption of no transfer of electrons to surface states for the conditions of the optical experiments.

It is possible to calculate approximate spin-density energies in a similar way to plasmon energies, by replacing the direct Coulomb interaction between electrons in the calculation by an exchange–correlation interaction, given by the local spin-density approximation (Gammon *et al.*, 1992). These energies, which are just 1–2 meV less than the bare intersubband energies, are also marked in Fig. 15.15. It is possible that the larger intensity of the 200 cm^{-1} line in the $x(z', y')\bar{x}$ spectrum is due to a superposition of the spin-density and single-particle excitation peaks. From the subtracted spectrum it appears that the spin-density mode is shifted to a slightly lower energy than the single-particle peak at 200 cm^{-1}, with the calculated $c2 \rightarrow c3$ spin-density energy being in agreement with the energy of the peak in the subtracted spectrum. Slightly below 400 cm^{-1}, there is almost a coincidence of calculated spin-density and charge-density excitations which apparently leads to a cancellation of both contributions in the difference spectrum shown in Fig. 15.15.

The assignment of the single-particle excitations ω_{23} and ω_{24} can be confirmed by consideration of the resonance profiles of these Raman peaks (Richards *et al.*, 1933b). The Raman scattering intensity versus the incident photon energy for these excitations is plotted in Fig. 15.16, where the strong resonance is readily apparent. In addition, it can be seen that the ω_{24} resonance curve is displaced to a higher energy than the ω_{23} resonance curve by approximately 25 meV, which happens to be the $c3 \rightarrow c4$ transition energy. In a

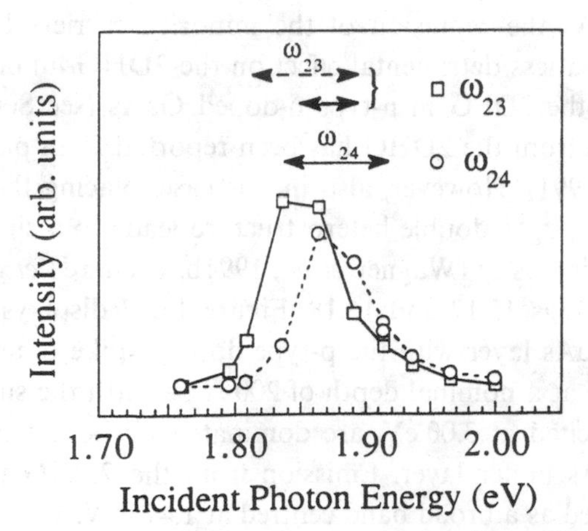

Fig. 15.16. Resonance Raman profiles for the ω_{23} and ω_{24} excitations of a δ-doped $In_{0.1}Ga_{0.9}As$:Si/GaAs quantum well (see Fig. 15.14). Lines are to guide the eye. The calculated ranges for resonance are marked by double-headed arrows above the profiles. The dashed double-headed arrow marks the resonance range for ω_{23} if the occupancy of the third electron subband $c3$ is ignored.

simple model of the resonant Raman mechanism for intersubband single-particle excitations, scattering is described by a two-stage process in which an electron in the initial conduction subband and a hole in the split-off valence band are created, followed coherently by the annihilation of an electron in the final conduction subband with the hole (Abstreiter *et al.*, 1984). So, for the excitation $i \rightarrow f$, all wavevector-conserving processes involving occupied states in the final conduction subband f and unoccupied states in the initial subband i will contribute to the Raman scattering cross-section. Thus a range of resonant incident energies is expected for each excitation; the expected ranges for $c2 \rightarrow c3$ and $c2 \rightarrow c4$ single-particle excitations are marked in Fig. 15.16. These have been determined using the same subband parameters as for the calculation of plasmon frequencies (above) and with $E_0 + \Delta_0$ energies obtained from self-consistent calculations (Fig. 15.12). The agreement is good for the ω_{24} and becomes so for the ω_{23} excitation by either relaxing the condition that the initial states are unoccupied, or by assuming a vanishing electron density in subband $c3$. From inspection of Fig. 15.12, it can be seen that other possible candidates for the Raman peaks at 200 cm^{-1} and 410 cm^{-1} are $c3 \rightarrow c4$ and $c1 \rightarrow c2$ transitions respectively. However, the expected resonance range for the first is very much smaller than that observed and, for the second, it is very much greater. Thus, the resonance behaviour of these two peaks is consistent with their assignment to $c2 \rightarrow c3$ and $c2 \rightarrow c4$ intersubband single-particle excitations.

15.2 P-type δ-doped double-heterostructures

15.2.1 Dopant density dependence of the PL spectrum

In p-type δ-doped GaAs the repulsion of the minority carriers by the hole-confining potential seems to have a less detrimental effect on the 2DHG luminescence efficiency, as compared with that of the 2DEG in n-type δ-doped GaAs (see Section 15.1.1), because radiative recombination from the 2DHG has been reported from plain Be δ-doped GaAs layers (Gilinsky *et al.*, 1991). However, also in this case, placing the doping spike at the centre of a $GaAs/Al_xGa_{1-x}As$ double-heterostructure leads to a significant enhancement of the 2DHG emission intensity (Wagner *et al.*, 1991b; Richards *et al.*, 1993a).

This is illustrated in Figs 15.17 and 15.18. Figure 15.17 displays low-temperature PL spectra of a δ-doped GaAs layer with the p-type doping spike of acceptor concentration 8×10^{12} Be/cm^2 placed at a nominal depth of 200 Å beneath the surface (Richards *et al.*, 1993a). The spectra, excited at 3.00 eV, are dominated by bound exciton emission (*BE*) from the undoped GaAs buffer layer. Emission from the 2DHG introduced by the Be δ-doping spike is resolved as a broad band centred at 1.475 eV. With decreasing excitation intensity, the relative strength of the 2DHG emission decreases and band-to-acceptor emission (*eA*) from the buffer layer appears at 1.493 eV. The spectral shape and position of the present 2DHG emission band are very similar to those reported by Gilinsky *et al.* (1991) for Be δ-doping spikes placed 0.5 μm below the surface. A similar series of spectra is shown in Fig. 15.18 for a sample in which the doping spike of 8×10^{12} Be/cm^2 is

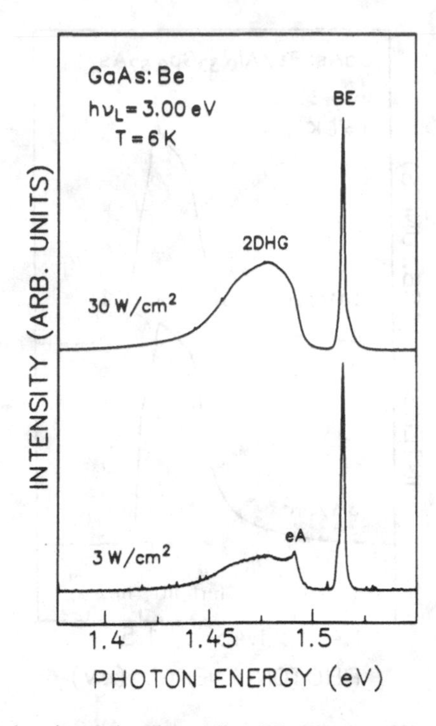

Fig. 15.17. Excitation-intensity-dependent PL spectra of p-type δ-doped GaAs. The doping spike with a dopant concentration of 8×10^{12} Be/cm^2 was placed at a nominal depth of 200 Å below the surface. The spectra excited at 3.00 eV were recorded with the sample cooled to 6 K (Richards *et al.*, 1993a).

positioned at the centre of a 600-Å-wide GaAs layer sandwiched between $Al_xGa_{1-x}As$ barriers (Richards *et al.*, 1993a). Under identical experimental conditions the 2DHG emission intensity is enhanced by a factor of 50 as compared with the sample without $Al_xGa_{1-x}As$ barriers, and no bound exciton luminescence from the GaAs buffer layer is observed.

Figure 15.19 shows the self-consistent potential for electrons and holes in such a p-type δ-doped GaAs:Be/$Al_xGa_{1-x}As$ double-heterostructure. The structure parameters and the doping density of 8×10^{12} cm^{-2} are identical to those of the sample used to record the spectra plotted in Fig. 15.18. A spread of the dopant atoms of $\Delta z = 20$ Å was assumed for the calculations (Richards *et al.*, 1993a; Wagner *et al.*, 1993). For the given doping density two subbands are occupied, namely the first heavy-hole ($hh1$) and the first light-hole ($lh1$) subbands. Compared with n-type δ-doping with the same dopant concentration (see Fig. 15.3), the width of the space-charge-induced potential well is significantly reduced because of the larger effective hole mass (see Section 15.2.2).

Optical spectroscopy with circularly polarised light can be used to distinguish experimentally between interband transitions involving light-hole and heavy-hole (sub)bands (Meier and Zakhachenya, 1984). In the present p-type δ-doped double-heterostructure the incident light of photon energy 1.65 eV is essentially absorbed in the almost bulk-like

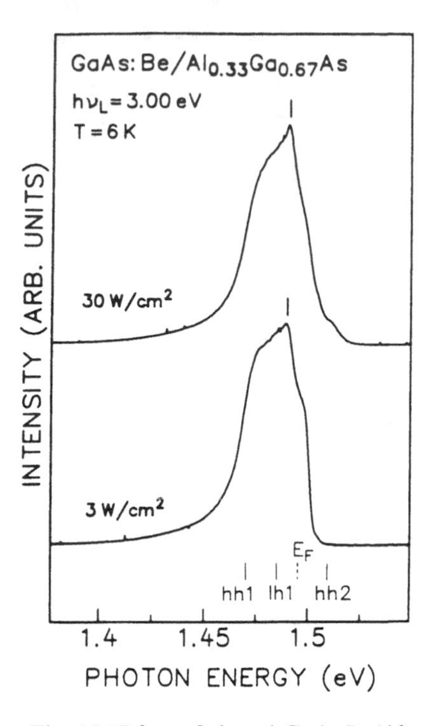

Fig. 15.18. Same as Fig. 15.17 for a δ-doped GaAs:Be/$Al_{0.33}Ga_{0.67}As$ double-heterostructure. Calculated $k_{\|} = 0$ subband energies are marked at the bottom of the diagram (see Section 15.2.2). Calculations are for an 8×10^{12} cm^{-2} Be density and a dopant spread of $\Delta z = 20$ Å. Heavy- and light-hole character is denoted by *hh* and *lh* respectively. E_F marks the Fermi energy (Richards *et al.*, 1993a).

intrinsic regions of the 600-Å-wide GaAs layer above and below the doping spike (Richards *et al.*, 1993a; Wagner *et al.*, 1993). Therefore, excitation with, for example, right circularly polarised light generates heavy-holes with spin $-\frac{3}{2}$ and spin $-\frac{1}{2}$ electrons as well as spin $-\frac{1}{2}$ light-holes and spin $+\frac{1}{2}$ electrons. The probabilities for these transitions are 3:1, leading to a majority of spin $-\frac{1}{2}$ electrons with a maximum degree of polarisation of 0.5 (Parsons, 1969; D'yakonov and Perel', 1971). The number of photogenerated holes is small compared to the doping level ($\leqslant 1\%$) and the spin polarisation of these holes is neglected. The radiative recombination of the spin-polarised electrons involves quantised hole states with the heavy-hole–light-hole degeneracy removed (see Fig. 15.19). Thus the maximum degree of polarisation for the emitted light is also 0.5. Recombination with light-holes results in the emission of photons with circular polarisation opposite to that of photons from heavy-hole recombination.

Figure 15.20 shows low-temperature PL spectra of the above p-type δ-doped GaAs:Be/$Al_xGa_{1-x}As$ double-heterostructure excited at 1.65 eV. Polarised (I^+) and depolarised (I^-) luminescence spectra were recorded, with the detected light having the same and opposite circular polarisation of the pump light respectively. In the total intensity spectrum

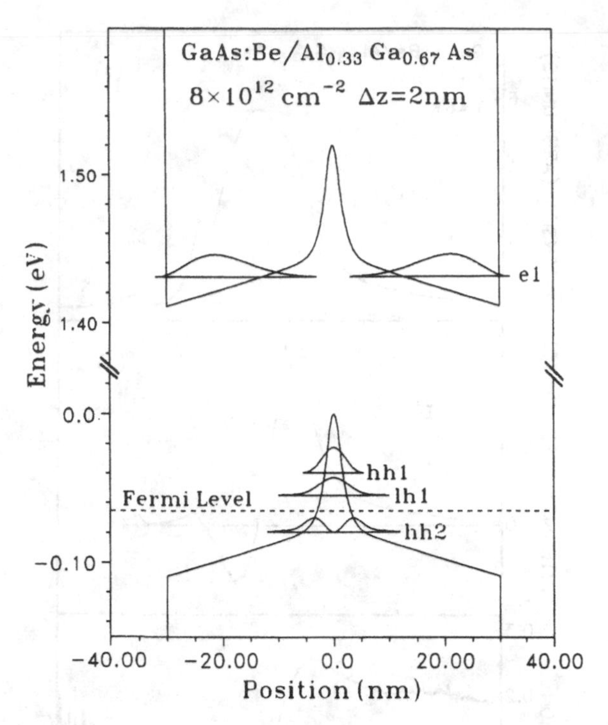

Fig. 15.19. Self-consistent potential profile for the 2DHG in a δ-doped GaAs:Be/$Al_{0.33}Ga_{0.67}As$ double-heterostructure. An 8×10^{12} cm^{-2} acceptor concentration is taken with a 20 Å dopant spread. Subband energies and probability densities at $k_{\parallel} = 0$ are also shown. *hh* and *lh* denote heavy-hole and light-hole subbands, respectively; denotes the first electron subband (Wagner *et al.*, 1993).

$(I^+ + I^-)$, shown in Fig. 15.20(a), the emission from the 2DHG appears as a broad band centred around 1.47 eV. The peaks marked with asterisks at 1.513 and 1.493 eV arise from bound exciton and band-to-acceptor recombination, respectively, in the GaAs buffer layer. We note in passing that, here, optical excitation has to be performed at photon energies between the E_0 and $E_0 + \Delta_0$ band-gap energies of GaAs in order to generate spin polarisation of the photogenerated electrons (D'yakonov and Perel', 1971; Wagner *et al.*, 1993). The difference spectrum $(I^+ - I^-)$ (Fig. 15.20(b)) and the spectral dependence of the degree of circular polarisation $P = (I^+ - I^-)/(I^+ + I^-)$ (Fig. 15.20(c)) show a maximum at 1.46 eV and a minimum at 1.48 eV. Based on the polarisation selection rules discussed above, these extrema are assigned to recombination involving the first heavy-hole and the first light-hole subband respectively.

In Fig. 15.21 a series of difference spectra $(I^+ - I^-)$ is displayed, recorded from samples with different doping levels (Richards *et al.*, 1933a). The Be concentrations were 3×10^{12} cm^{-2} (bottom spectrum), 8×10^{12} cm^{-2} (middle spectrum), and 3×10^{13} cm^{-2} (top spectrum) respectively. For the lowest doping level there is one pronounced maximum corresponding to recombination from the first heavy-hole subband and a slight minimum

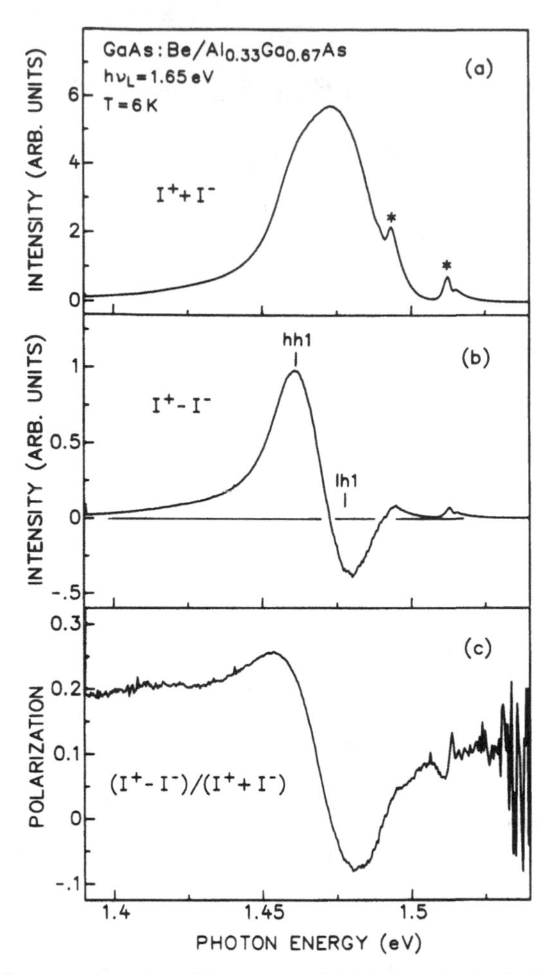

Fig. 15.20. Low-temperature PL spectra of a δ-doped GaAs:Be/$Al_{0.33}Ga_{0.67}As$ double-heterostructure with a dopant concentration of $8 \times 10^{12}\,cm^{-2}$. The spectra were recorded at 6 K and excited by right-hand circularly polarised light of 1.65 eV photon energy. (a) Total intensity spectrum $I^+ + I^-$, (b) difference spectrum $I^+ - I^-$ and (c) polarisation spectrum $(I^+ - I^-)/(I^+ + I^-)$. Subband energies, calculated for a dopant spread of 20 Å, are marked in (b) with the position of the first heavy-hole level fixed at the peak of the *hh*1 emission band. Emission from the undoped GaAs buffer layer is marked by asterisks (Wagner *et al.*, 1993).

at the high-energy side arising from recombination out of the first light-hole subband, which is just populated at this doping level. With increasing dopant concentration the minimum corresponding to the recombination of light-holes becomes more pronounced, which reflects the increase in the population of the *lh*1 subband, and for the highest doping level a second maximum appears arising from the recombination of holes residing in the *hh*2 subband. For a more detailed discussion of the dopant density dependence of the

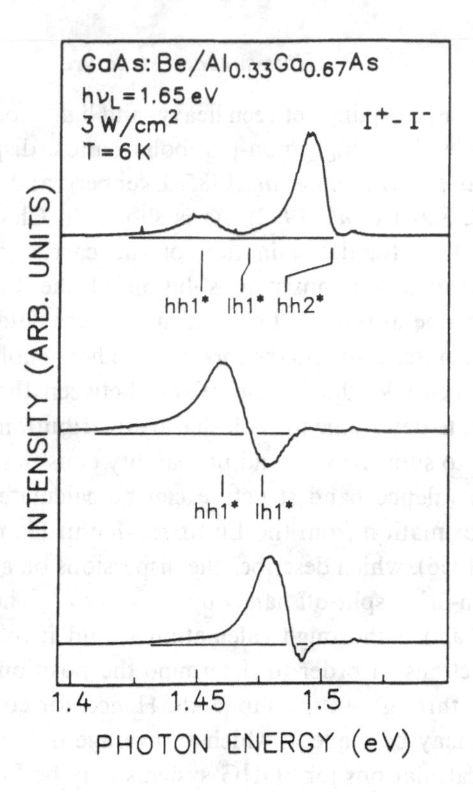

Fig. 15.21. Low-temperature (6 K) PL spectra of δ-doped GaAs:Be/$Al_{0.33}Ga_{0.67}As$ double-heterostructures with dopant concentrations of 3×10^{12} cm^{-2} (bottom), 8×10^{12} cm^{-2} (middle), and 3×10^{13} cm^{-2} (top spectrum). The spectra were excited by right-hand circularly polarised light and the difference spectra $I^+ - I^-$ between right and left-hand circularly polarised emission are displayed. Emission from a heavy-hole band results in a positive signal, that from a light-hole band results in a negative signal. Subband energies, calculated for dopant spreads Δz of 20 Å (middle spectrum) and 60 Å (top spectrum, upper set of vertical lines) or 75 Å (top spectrum, lower set of vertical lines), are marked for the two most heavily doped samples. The position of the first heavy-hole level *hh*1 is fixed at the peak of the lowest subband. Asterisks (*) indicate that the absolute energetic positions are not given but, rather, are an indication of the expected peak positions (Richards *et al.*, 1993a).

subband structure, see the following section, Section 15.2.2. The overall width of the emission spectrum increases with increasing doping level which mirrors the increase in Fermi energy. The low-energy edge of the 2DHG emission shifts to lower energies, reflecting an increase in the modulation depth of the space-charge-induced potential well and possibly a renormalisation of the band-gap energy due to many-body effects (Sernelius, 1986; Delalande *et al.*, 1987; Borghs *et al.*, 1989).

15.2.2 Hole-subband structure

In two-dimensional systems, coupling between heavy- and light-holes results in complicated valence band structures with strongly non-parabolic bands, dispersions sometimes displaying electron-like masses (Altarelli *et al.*, 1985; Ekenberg and Altarelli, 1985; O'Reilly, 1989; Eaves *et al.*, 1992; Kash *et al.*, 1992). As is shown in Chapter 2 on the theory of δ-doped structures, by Proetto, determination of the carrier confinement energies of a δ-doped structure requires self-consistent solution of the Poisson and Schrödinger equations within the Hartree approximation, which involves determination of the carrier density distribution at each iteration. As the forms of the hole-subband wavefunctions vary with the in-plane wavevector $\mathbf{k}_{\parallel}$ due to the mixing between the bands (O'Reilly, 1989; Ekenberg, 1990), in order to determine the hole density distribution for an acceptor δ-doped structure, it is necessary to sum the subband probability densities explicitly over states up to the Fermi level. The valence band structure can be calculated within the multiband envelope-function approximation from the Luttinger–Kohn Hamiltonian (Luttinger and Kohn, 1955; Luttinger, 1956), which describes the dispersions of, and interactions between, the heavy, light and spin-orbit split-off bands up to order k^2. The valence bands at large $\mathbf{k}_{\parallel}$ are very anisotropic and a thorough calculation would involve calculating the band dispersions in all k-directions in order to determine the position of the Fermi level and the distribution of holes throughout the subbands. Hence, for computational expediency, an axial approximation may be made in which an average over the in-plane directions is taken. Details of these calculations for 2DHG systems may be found elsewhere (O'Reilly, 1989; Cohen and Marques, 1990; Ekenberg, 1988; Richards *et al.*, 1993a).

Following on from the arguments presented in Section 15.1.2, the potential of a δ-doped double-hererostructure can be considered to be symmetric with respect to the doping layer. Self-consistent potential profiles are shown in Fig. 15.22 for sheet acceptor concentrations of (a) $3 \times 10^{13}\ \mathrm{cm}^{-2}$ and (b) $8 \times 10^{12}\ \mathrm{cm}^{-2}$. Dopant spreads of (a) $\Delta z = 60$ Å and (b) $\Delta z = 20$ Å, respectively, have been assumed, these values being consistent with secondary ion mass spectroscopy (SIMS) data for samples with such Be concentrations (Richards *et al.*, 1993a). Also shown are the zone centre ($\mathbf{k}_{\parallel} = 0$) energy levels and probability densities $|\psi|^2$ (where the $\psi(z)$ are the envelope wavefunctions). What is immediately apparent is the very strong localisation of the hole density around the dopant spike, with a confinement length of only ≈ 100 Å. The intercepts of the subband energies of the unoccupied levels (those below the Fermi level) with the confining potential give an indication of the onset of the decay of the wavefunctions. From Fig. 15.22 it can be seen that the wavefunction spread of these levels is significantly larger than that of the occupied levels, with a spread of ≈ 220 Å for the higher-density structure (Fig. 15.22(a)).

The valence subband structure for a system with a δ-doping of $3 \times 10^{13}\ \mathrm{Be/cm}^2$ is shown in Fig. 15.23, where the strongly non-parabolic nature and anti-crossings between the bands are readily apparent. These hole dispersions have been calculated within the axial approximation (see above). The labelling of the bands as the different quantised (i) heavy-hole (hhi) or light-hole (lhi) levels is only relevant at $\mathbf{k}_{\parallel} = 0$, as the states become strong mixtures of the different angular momentum states away from the zone centre

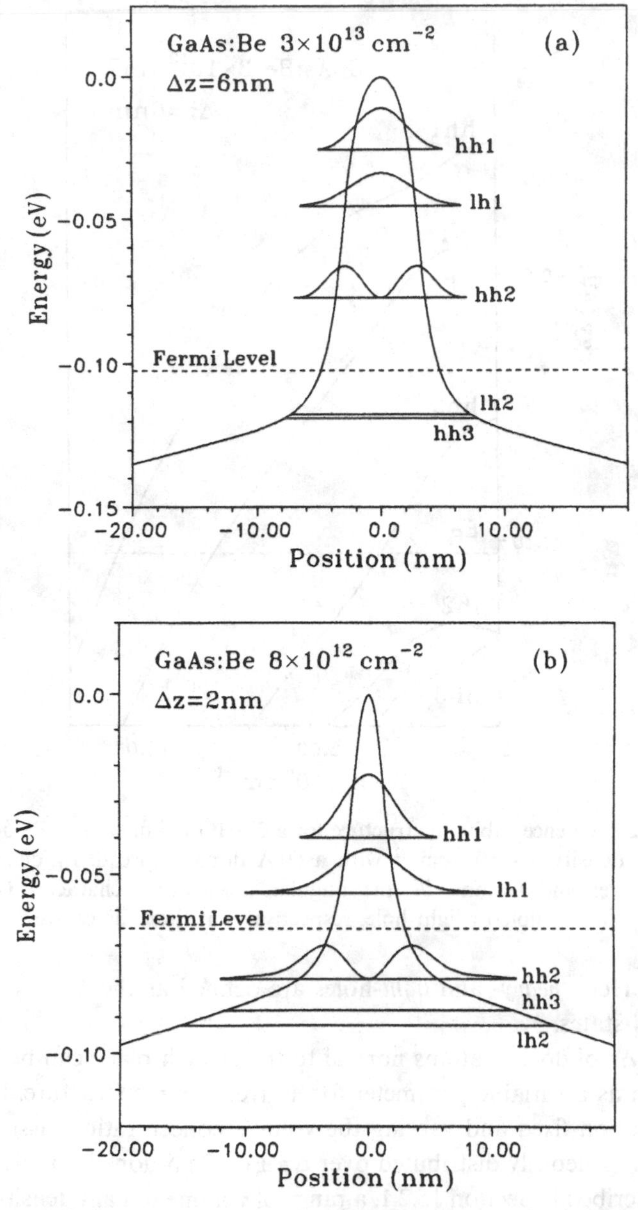

Fig. 15.22. Self-consistent potential profiles for the quasi-two-dimensional-hole-gas at a Be δ-doped layer in GaAs. (a) 3×10^{13} cm^{-2} Be density with 60 Å dopant spread. (b) 8×10^{12} cm^{-2} Be with 20 Å dopant spread. Subband energies and probability densities $|\psi(z)|^2$ at $k_\parallel = 0$ are also shown. *hh* and *lh* denote heavy-hole and light-hole, respectively. Note that the vertical scales for (a) and (b) are different.

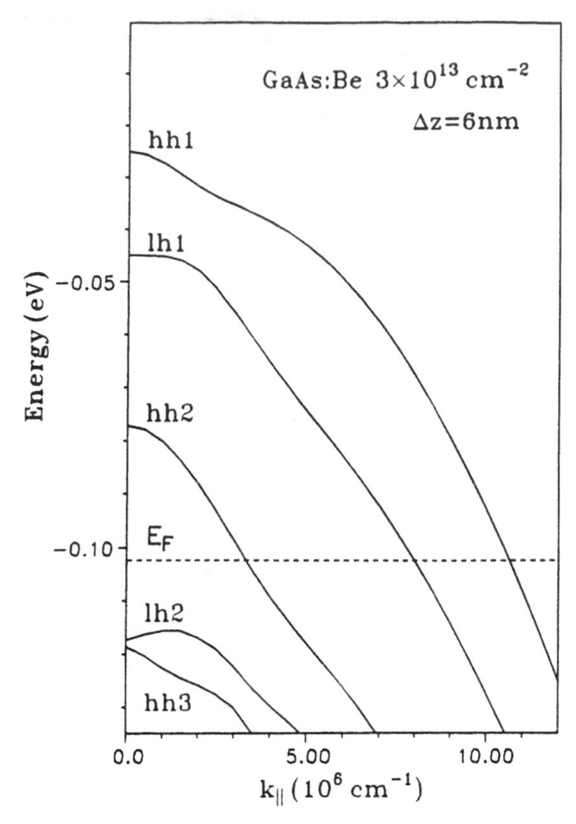

Fig. 15.23. Valence subband structure for a 2DHG confined at a Be δ-doping layer of density 3×10^{13} cm^{-2} with a 60 Å dopant spread. E_F denotes the Fermi level and *hh* and *lh*, the angular momentum character (whether heavy-hole or light-hole, respectively) at the zone centre.

(O'Reilly, 1989). Here, *heavy*- and *light*-holes are defined as the $M_J = \frac{3}{2}$ and $M_J = \frac{1}{2}$ zone centre spin-$J = \frac{3}{2}$ states.

As the spread Δz of dopant atoms normal to the growth plane can be rather uncertain, this can be taken as a variable parameter for a given sample structure, keeping the sheet dopant concentration fixed and varying the volume concentration, assuming the dopant atoms to be homogeneously distributed over Δz. For Be δ-doped double-heterostructures such as those described in Section 15.2.1, a range of volume dopant densities from 2×10^{19} to 10^{20} Be/cm^3 may be realistically considered. Thus, for a structure doped with 8×10^{12} Be/cm^2, this leads to a possible dopant spread between 8 Å and 40 Å. Over this range of Δz the wavefunction extent of the hole states is greater than that of the dopant spread, causing the spacing of the energy levels to be fairly insensitive to the form of the potential around the dopants. Contrary to this, for the structure doped with 3×10^{13} Be/cm^2, the extent of the hole density matches that of the dopants for spreads $\Delta z \geqslant 70$ Å very closely. So there is a progression from a parabolic to a square potential profile as the dopant spread increases, and the volume concentration decreases

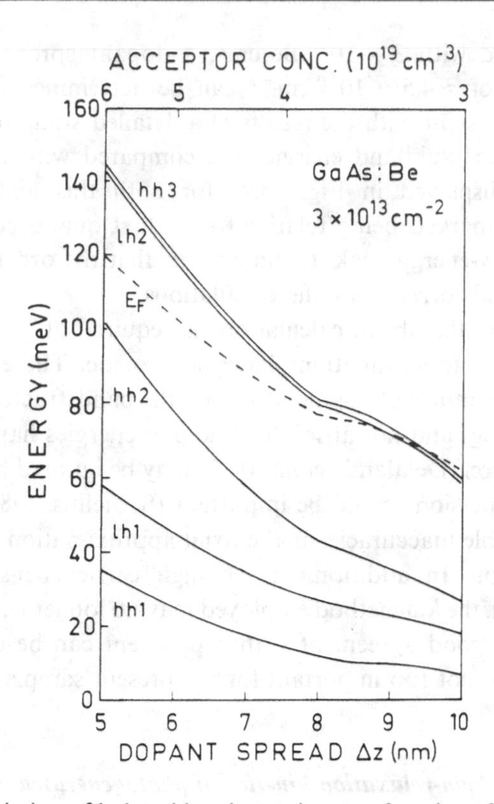

Fig. 15.24. Variation of hole subband energies, as a function of dopant spread Δz, for p-type δ-doped GaAs with 3×10^{13} Be/cm^2. Subband energies are for $k_{\parallel} = 0$ and are with respect to the valence band maximum at the centre of the dopant spike. For convenience the negative of the energies are in fact plotted, that is, they are given as positive. For a comparison, the horizontal scale at the top gives the volume dopant concentration.

correspondingly and so comes into line with that of the hole density. This leads to a large variation in the confinement energies and subband spacings with dopant spread, as illustrated in Fig. 15.24, where the variation with dopant spread Δz of the $\mathbf{k}_{\parallel} = 0$ hole confinement energies and the Fermi energy are plotted. Energies are with respect to the valence band maximum at the centre of the dopant spike. At a dopant spread of $\Delta z = 80$ Å for this high-density structure there is a kink in the curves of Fig. 15.24, indicating occupation of a new subband. Although the Fermi level is not above the *lh*2 level at $\mathbf{k}_{\parallel} = 0$, hole occupation of this band commences at this point at non-zero $\mathbf{k}_{\parallel}$: from Fig. 15.23 it can be seen that there exists a maximum in the *lh*2 band dispersion where a pocket of holes in *k*-space will initially form. This emphasises the need to consider the full band structure when considering such quasi-2DHG systems.

The strong variation of subband spacings with dopant spread Δz, described above for a structure δ-doped with 3×10^{13} Be/cm^2, enables determination of Δz in such a structure from a comparison of the subband energies with PL measurements. Hence, from a comparison with the top spectrum shown in Fig. 15.21, recorded from a double-heterostructure δ-doped with 3×10^{13} Be/cm^2, a dopant spread of ≈ 67 Å with a Be volume concentration of $\approx 4.5 \times 10^{19}$ cm^{-3} can be determined for this structure. This magnitude of Δz is consistent with the results of a detailed study reported by Schubert *et al.* (1990). The calculated subband energies are compared with the circularly polarised luminescence results displayed in Fig. 15.21 for both this and the 8×10^{12} Be/cm^2 structure, the energies marked being relative to the first quantised heavy-hole level *hh*1, which is fixed at the low-energy peak. It can be seen that the ordering and spacing of the levels has been predicted correctly by the calculations.

A word of caution on the above calculations is required at this point as it should be noted that a number of approximations have been made. The effects of the high hole density and high concentration of dopant atoms on the bandstructure have been neglected. As only subband spacings and not absolute band-gap energies have been addressed here, band-gap renormalisation (Delalande *et al.*, 1987) may be ignored but second-order effects on the actual band dispersions could be important (Sernelius, 1986; Borghs *et al.*, 1989). Another cause of possible inaccuracies is the axial approximation used in calculating the valence band dispersion. In addition, at the high carrier densities, and hence large wavevectors considered, the **k.p** method employed may no longer be a good approximation. However, the fact that good agreement with experiment can be obtained indicates that these considerations are not too important for the present sample structures.

15.2.3 Spin-relaxation kinetics of photogenerated electrons

Time-resolved polarisation-dependent PL measurements have been proved to be a powerful method for the study of the spin-relaxation kinetics of, for example, photogenerated electrons in p-type semiconductors (Seymour and Alfano, 1980; Damen *et al.*, 1991; Viña *et al.*, 1992). Low-temperature spin-relaxation (or spin-flip) times in p-type bulk GaAs are $\leqslant 200$ ps for acceptor concentrations $\geqslant 10^{19}$ cm^{-3} (Zerrouati *et al.*, 1988). Based on measured hole density and temperature dependencies, it was suggested that electron–hole scattering with a simultaneous exchange interaction (the so-called Bir–Aronov–Pikus (BAP) mechanism (Bir *et al.*, 1976)) dominates electron spin-relaxation at low temperatures and high doping levels. A reduction in the spin-relaxation time by a factor of $\frac{1}{3}$–$\frac{1}{4}$, as compared with bulk material with comparable hole densities, has been observed in p-modulation-doped quantum wells (Damen *et al.*, 1991). This has been attributed to the enhanced electron–hole wavefunction overlap in quantum wells and has been taken as additional evidence for the BAP mechanism being the dominant mechanism at low temperatures. In p-type δ-doped double-heterostructures (see Sections 15.2.1 and 15.2.2), by contrast, the electron–hole wavefunction overlap is drastically reduced. Therefore, such structures are expected to show a significant enhancement in electron spin-relaxation time.

In what follows we review the results of time-resolved circularly polarised PL experiments on a δ-doped GaAs:Be/Al$_x$Ga$_{1-x}$As double-heterostructure. In the particular structure investigated (see Fig. 15.19), the photogenerated electrons are spatially separated from the holes, while still maintaining sufficient wavefunction overlap for efficient radiative recombination to probe the spin polarisation of the electrons (Wagner *et al.*, 1993).

Time-resolved PL spectra provide direct information on electron recombination and spin dynamics. The rate equations, which describe the population change of spin-up electrons (N^+) and spin-down electrons (N^-), that is, electrons with spin aligned parallel and antiparallel to the propagation direction of the exciting light, are given by (Seymour and Alfano, 1980)

$$dN^+/dt = -N^+/\tau_{rec} - N^+/\tau_{sp} + N^-/\tau_{sp} \quad (15.2)$$

$$dN^-/dt = -N^-/\tau_{rec} - N^-/\tau_{sp} + N^+/\tau_{sp}. \quad (15.3)$$

Here a δ-shaped excitation pulse is assumed, generating the initial spin-up and spin-down electron populations $N^+(0)$ and $N^-(0)$ respectively. The τ_{rec} and τ_{sp} are the recombination time and electron spin-relaxation time. The change in the total population $N(t) = N^+(t) + N^-(t)$ is given by a single exponentially decaying function

$$N(t) = [N^+(0) + N^-(0)] \cdot \exp(-t/\tau_{rec}). \quad (15.4)$$

The temporal evolution of the degree of spin polarisation P defined by $P = (N^+ - N^-)/(N^+ + N^-)$ is given by

$$P(t) = P(0) \cdot \exp(-2t/\tau_{sp}), \quad (15.5)$$

where $P(0) = [N^+(0) - N^-(0)]/[N^+(0) + N^-(0)]$.

Thus, with the emission intensities I^+ and I^- being proportional to the electron populations N^+ and N^-, the recombination time τ_{rec} and the electron spin-relaxation time τ_{sp} can be extracted from the time-resolved total intensity spectra $(I^+ + I^-)$ (Fig. 15.25(a)) and polarisation spectra $(I^+ - I^-)/(I^+ + I^-)$ (Fig. 15.25(b)). At 6 K the recombination shows biexponential behaviour with an initial time constant $\tau_{rec} = 2$ ns, followed by a somewhat slower decay time of 3.5 ns. For the spin-relaxation time a significantly larger value of $\tau_{sp} = 20 \pm 4$ ns is obtained. With increasing time delay the 2DHG emission shifts to lower energies. This is explained as follows. After the exciting laser pulse, photogenerated carriers reduce the spatial modulation of the band-edge energies. With the concentration of these carriers decreasing with increasing time delay, the spatial modulation increases and causes a red shift of the emission from the recombination of spatially separated photocreated electrons and the 2DHG.

The initial degree of polarisation P_0 for the *hh*1 transition is 0.3. This value is lower than the maximum achievable polarisation of 0.5, which indicates that some spin-relaxation takes place while the photogenerated electrons relax in energy and become spatially separated from the photoexcited holes. Taking into account the reduction factor (Parsons, 1969) $\tau_{sp}/(\tau_{rec} + \tau_{sp})$, which relates the initial polarisation P_0 to the time-averaged one, P_{av}, observed in the cw measurement, one thus obtains an expected averaged value of $P_{av} = 0.27$, in good agreement with the measured cw polarisation of $P_{av} = 0.26$ (see Fig. 15.20(c)).

The present low-temperature spin-relaxation time $\tau_{sp} = 20$ ns is two orders of magnitude longer than the relaxation time of $\simeq 200$ ps found in homogeneously p-type doped GaAs

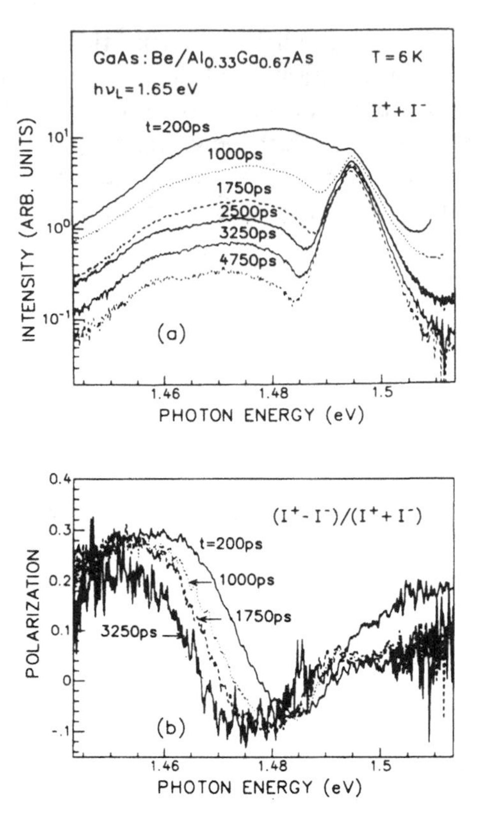

Fig. 15.25. Time-resolved PL spectra of a δ-doped GaAs:Be/$Al_{0.33}Ga_{0.67}As$ double-heterostructure with a dopant concentration of $8 \times 10^{12}\,cm^{-2}$ for various time delays t. The spectra were recorded at 6 K with excitation at 1.65 eV. (a) Total intensity spectra $I^+ + I^-$ and (b) polarisation spectra $(I^+ - I^-)/(I^+ + I^-)$. The long-lived emission at 1.49 eV in (a) is due to donor-to-acceptor recombination in the GaAs buffer layer (Wagner *et al.*, 1993).

with a comparable acceptor concentration of $2.8 \times 10^{19}\,cm^{-3}$ (Zerrouati *et al.*, 1988). As described above, it has been concluded from work on homogeneously p-type doped samples, as well as on p-modulation doped quantum wells, that the BAP mechanism is relevant for electron spin-relaxation in degenerate p-type GaAs at low temperatures. For this mechanism the spin-relaxation rate $1/\tau_{sp}$ in a degenerate semiconductor is proportional to $(\Delta_{exch})^2$ (Bir *et al.*, 1976; Aronov *et al.*, 1983; Zerrouati *et al.*, 1988); Δ_{exch} is the exchange splitting of the exciton ground state and is given by (Chen *et al.*, 1988; Rössler *et al.*, 1990)

$$\Delta_{exch} = \Delta_{exch}(3D)|\Phi^{2D}(0)|^2/|\Phi^{3D}(0)|^2 \cdot \int (X_e(z) \cdot X_h(z))^2\,dz, \qquad (15.6)$$

where $\Delta_{exch}(3D)$ is the exchange splitting of the 3D exciton, $\Phi^{2D}(0)$ and $\Phi^{3D}(0)$ are, respectively, the 2D and 3D exciton wavefunction amplitude at zero electron–hole relative coordinate, and $X_e(z)$ and $X_h(z)$ are the electron and hole subband envelope wavefunctions.

Ignoring quantum size effects, we can write $|\Phi^{2D}(0)|^2/|\Phi^{3D}(0)|^2 = 8a_0$, where a_0 is the 3D exciton radius (150 Å for GaAs). Based on the scheme presented in Fig. 15.19, one can calculate the electron–hole wavefunction overlap in Eqn. (15.6). These calculations yield a two orders of magnitude decrease in $1/\tau_{sp}$, in agreement with the above-described experiment. This agreement confirms that electron–hole scattering with simultaneous exchange interaction (the BAP mechanism) is the dominant electron spin-relaxation process in highly p-type doped GaAs (Wagner *et al.*, 1993).

When neglecting excitonic effects, electron spin-relaxation times $\geqslant 10$ ns are expected only for hole concentrations below 10^{16} cm^{-3} (Bir *et al.*, 1976; Zerrouati *et al.*, 1988). However, it has been shown that exciton formation leads to a drastic reduction in the spin-relaxation time (Damen *et al.*, 1991; Viña *et al.*, 1992). Taking a hole concentration of about 2×10^{16} cm^{-3} as the limit where exciton formation becomes important, it follows that spin-relaxation times are always $\leqslant 4$ ns in homogeneously doped GaAs or in p-type doped quantum wells. In the p-type δ-doped double-heterostructures, by contrast, where exciton formation is suppressed by the built-in electric fields (Fig. 15.19), at least one order of magnitude longer relaxation times can be realised.

15.3 Fermi edge singularities in the optical spectra of δ-doped double-heterostructures

The optical absorption and emission spectra of high-density two-dimensional electron- and hole-systems are strongly influenced by many-body interactions (see e.g. Schmitt-Rink *et al.*, 1986). A widely studied manifestation of such many-body effects is the enhanced probability of optical interband transitions involving electron states close to the Fermi energy E_F, the so-called Fermi edge singularity (FES). This effect, originally proposed by Mahan (1981), is suppressed for states well below E_F because of the exclusion principle.

Most of the experimental studies on the FES have been carried out on n-modulation-doped quantum wells, where a Fermi edge enhancement has been observed both in the PL spectrum (Skolnick *et al.*, 1987; Chen *et al.*, 1990; Skolnick *et al.*, 1991; Chen *et al.*, 1992; Hawrylak *et al.*, 1992) and the absorption (PLE) spectrum (Livescu *et al.*, 1988). In optical absorption (PLE), the experimental observation of a Fermi edge enhancement is straightforward because electron–hole pairs can be generated by momentun-conserving transitions at wavevector equal to or larger than the Fermi wavevector $\mathbf{k}_F$. To observe a Fermi edge enhancement in emission, by contrast, electrons with energies close to E_F, and consequently a wavevector close to $\mathbf{k}_F$, have to recombine with photogenerated holes relaxed to the top of the valence band. At low temperatures these holes have a wavevector that is essentially zero. If only momentum-conserving transitions were allowed, this recombination would be forbidden and, therefore, the FES should be strongly suppressed. However, a low-dispersion hole band or hole localisation, and hence a spread in k-space, may permit optical emission (Skolnick *et al.*, 1987).

Another possible mechanism for the observation of a FES in emission, despite the mismatch in electron and hole wavevectors, has been suggested by Chen *et al.* (1990, 1992).

Based on magneto-PL experiments on n-modulation-doped quantum wells with the Fermi level just below the second electron subband, these authors concluded that the near resonance of E_F with an unoccupied subband leads to a hybridisation of occupied states at the Fermi edge with virtual excitons involving the unoccupied band. This hybridisation can cause a large enhancement of the optical matrix element. The whole process becomes analogous to an edge singularity occurring at zero wavevector and therefore the need for localisation of the minority carriers (holes) is relaxed. The problem of a resonant hybridisation of occupied states at the Fermi edge with unoccupied states in a higher-lying subband has been treated theoretically by Mueller (1990), within the approximation of a rigid Fermi surface, and by Hawrylak (1991), taking into account the dynamic response of the Fermi sea. Both authors arrive at the conclusion that such a resonance should indeed favour the observation of a FES in the PL spectrum.

A Fermi edge enhancement has also been observed in the absorption (PLE) spectrum of n-type δ-doped GaAs:Si/$Al_xGa_{1-x}As$ double-heterostructures (Wagner *et al.*, 1991a). In this system there is a significant overlap between the electron wavefunctions and the ionised donors in the doping layer. This is in contrast to n-modulation-doped heterostructures, discussed above, where this overlap is drastically reduced by the introduction of an undoped spacer layer. In spite of the low electron mobility due to ionized impurity scattering in such δ-doped heterostructures, there is a well-resolved Fermi edge enhancement in the low-temperature PLE spectrum (Fig. 15.26). The strength of this enhancement decreases with increasing temperature, which is a characteristic fingerprint of a FES (Schmitt-Rink *et al.*, 1986; Skolnick *et al.*, 1987). This is due to the increasing energy spread of the electrons around E_F with increasing temperature, which becomes comparable with the binding energy of the Mahan exciton.

So far, the experimental investigation of Fermi edge singularities has been restricted to n-type doped structures. Recently a FES has also been observed in the emission spectrum of a high-density 2DHG formed in p-type δ-doped GaAs:Be/$Al_xGa_{1-x}As$ double-heterostructures (Wagner *et al.*, 1991b; Richards *et al.*, 1993a). The observation of a Fermi edge enhancement in the PL spectrum of a 2DHG is greatly facilitated by the use of such a δ-doped double-heterostructure, where two-dimensional hole concentrations well above 10^{13} cm^{-2} can be achieved. Such high hole concentrations, which have to exceed the critical density for the formation of classical free excitons significantly, might be difficult to achieve, for example in p-modulation-doped GaAs/$Al_xGa_{1-x}As$ heterojunctions and quantum wells because of the comparatively small valence band offset in this material system.

Figure 15.27 shows temperature-dependent PL spectra of a δ-doped GaAs:Be/$Al_xGa_{1-x}As$ double-heterostructure with a dopant concentration of 3×10^{13} cm^{-2}. For optical excitation at 3.00 eV, above the band-gap energy of the $Al_{0.33}Ga_{0.67}As$ barriers, a well-resolved FES is observed in the luminescence spectrum for temperatures $\leqslant 15$ K (Wagner *et al.*, 1991b). Unlike the case of a FES in the emission spectrum of a 2DEG, where hole localisation is one possible way of achieving an effectively large mass of the photoinduced minority carriers (Skolnick *et al.*, 1987), in the present case of a 2DHG the photogenerated electrons are unlikely to be localised because of their much smaller mass. Thus, to explain the occurrence of a FES in the 2DHG emission spectrum, one has to

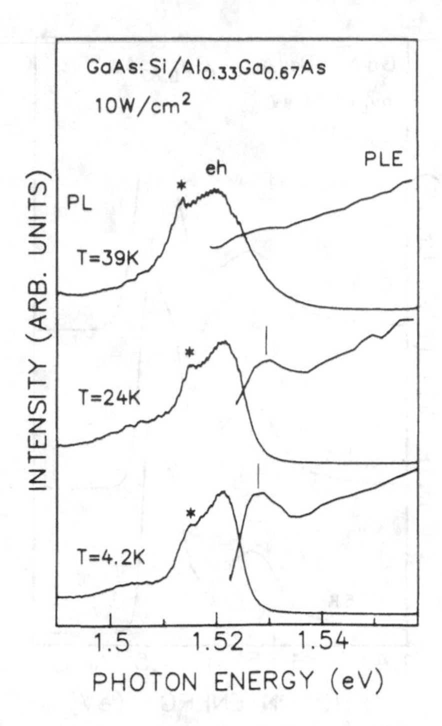

Fig. 15.26. Temperature-dependent PL (left) and PLE spectra (right) of a δ-doped GaAs:Si/$Al_{0.33}Ga_{0.67}As$ double-heterostructure with a dopant concentration of 8×10^{12} cm^{-2}. When recording the PLE spectra, detection was set to the 2DEG band-to-band (eh) recombination. The PL spectra were excited at 1.559 eV. The Fermi edge enhancement is marked in the PLE spectra by vertical lines. The asterisks mark emission from the undoped GaAs buffer layer (Wagner *et al.*, 1991a).

assume that in this case the observation of a Fermi edge enhancement is brought about by the hybridisation between occupied states at the Fermi edge and an unoccupied hole subband lying close to the hole Fermi energy. That, for the present doping level, unoccupied subbands are indeed expected to lie close to the Fermi level can be seen from the self-consistent hole potential shown in Fig. 15.22(a) (Richards *et al.*, 1993a). Two unoccupied subbands, namely the third heavy-hole (*hh*3) and the second light-hole (*lh*2) band, lie only 10–20 meV away from E_F. The existence and energy positions of these subbands have been verified experimentally by temperature-dependent PL measurements where, for temperatures $\geqslant 40$ K, the *hh*3 and *lh*2 subbands become thermally populated and exhibit strong luminescence (Richards *et al.*, 1993a).

One possible way of tuning the energy positions of higher-lying subbands in particular, in such double-heterostructures is the variation of the barrier separation. As discussed in Section 15.2.1 for p-type δ-doped quantum wells, placing the $Al_xGa_{1-x}As$ barriers closer to the doping spike increases the energy spacing of these subbands. Therefore, varying the

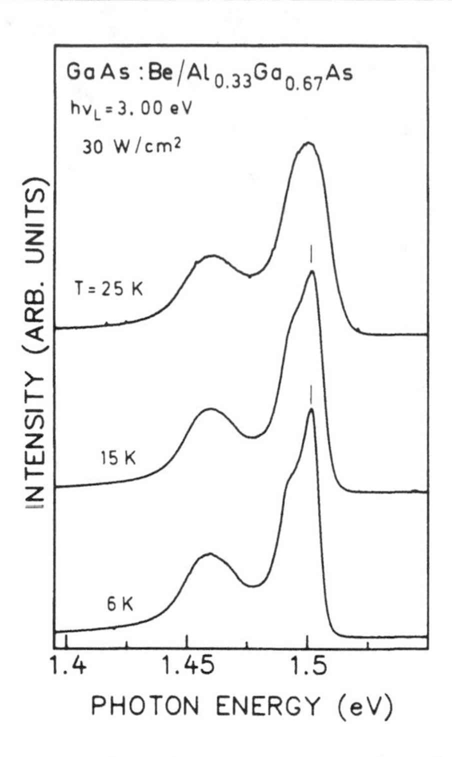

Fig. 15.27. Temperature-dependent PL spectra of a δ-doped GaAs:Be/$Al_{0.33}Ga_{0.67}As$ double-heterostructure with a dopant concentration of $3 \times 10^{13}\,cm^{-2}$. The spectra were excited at 3.00 eV. The vertical lines mark the enhancement in luminescence intensity at the Fermi edge (Wagner *et al.*, 1991b).

width of the p-type δ-doped GaAs layer allows one to tune the energy separation between E_F and nearby unoccupied hole subbands and thus to test the above hypothesis of a resonant hybridisation. Figure 15.28 illustrates a sequence of low-temperature PL spectra of δ-doped GaAs:Be/$Al_xGa_{1-x}As$ double-heterostructures where the width d of the δ-doped GaAs layer was varied between 200 and 600 Å. The two-dimensional carrier concentration was kept fixed at $4 \times 10^{13}\,cm^{-2}$ (Wagner *et al.*, 1993). For $d = 600$ Å there is a well-resolved Fermi edge enhancement at the high-energy side of the emission spectrum. When the layer width is decreased to $d = 400$ Å the strength of the FES is reduced considerably and for $d = 200$ Å no detectable enhancement remains. The energy spacing between E_F and the nearest unoccupied hole subband has been measured by temperature-dependent PL to increase from ≈ 15 meV for $d = 600$ Å to $\simeq 20$ meV for $d = 400$ Å. For $d = 200$ Å this spacing increases further, such that no recombination from thermally excited holes could be detected out of that subband. Thus these experimental findings give full support to the above model that the observation of a FES in the PL spectrum of a 2DHG is due to a resonant hybridisation between occupied states at the Fermi edge and unoccupied levels (Wagner *et al.*, 1993).

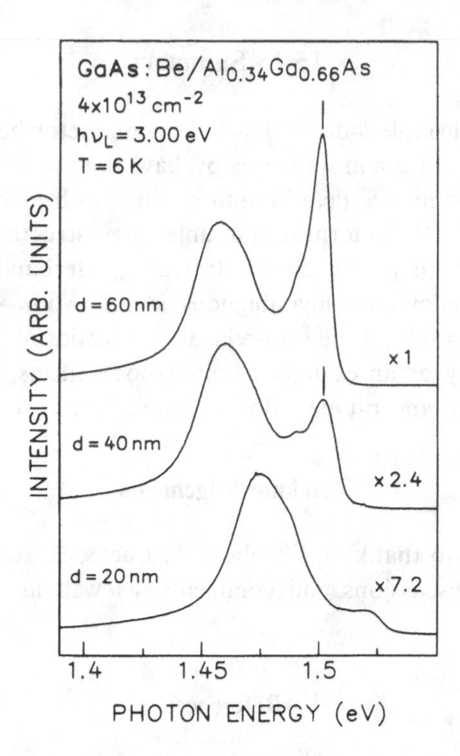

Fig. 15.28. Low-temperature PL spectra of p-type δ-doped GaAs:Be/$Al_{0.34}Ga_{0.67}As$ double-heterostructures as a function of the width d of the GaAs layer in the centre of which the doping spike is placed. The spectra were excited at 3.00 eV. Vertical lines mark the enhancement in luminescence intensity at the Fermi edge (Wagner *et al.*, 1993).

It has been concluded by Rodriguez and Tejedor (1993) that a hybridisation between states from adjacent subbands is not possible for carriers confined in a symmetric potential. The present δ-doped GaAs:Be/$Al_xGa_{1-x}As$ double-heterostructure is indeed symmetric by design. However, there are a number of possible causes for the introduction of some asymmetry, such as the effect of the position of the Fermi level at the nearby surface (see also Section 15.1.2) and a possible segregation of the dopant atoms not detectable within the depth resolution of SIMS (Richards *et al.*, 1993a).

For the narrowest layer width of $d = 200$ Å the whole luminescence spectrum (Fig. 15.28) is shifted to higher energies, primarily due to the reduction in the electron–hole separation. This reduced separation leads, for a constant electric field above and below the doping spike (see Fig. 15.19), to a blue-shift of the 2DHG emission. A secondary effect may be an increase in confinement energy of the photogenerated electrons. As can be seen from Fig. 15.19, these electrons are localised in a triangular potential well formed by the $Al_xGa_{1-x}As$ barriers and by the repulsive space-charge-induced potential from the p-type doping spike. Thus, in principle, moving these barriers towards the doping spike reduces the effective well width and consequently increases the electron confinement energy.

15.4 Summary

The optical properties of single δ-doped III–V semiconductor heterostructures, as studied by photoluminescence and Raman spectroscopy, have been reviewed. From a combination of these spectroscopic techniques, detailed information has been obtained on the electronic structure of the 2DEG (2DHG) formed in double-heterostructures, the centre of which is n-type (p-type) δ-doped. In p-type doped structures, interband spectroscopy using circularly polarised light allows an unambiguous identification of heavy- and light-hole subbands as well as the study of the spin-relaxation kinetics of photogenerated electrons. A Fermi edge singularity, as an example of many-body effects, has been observed in the optical spectra of both n- and p-type δ-doped double-heterostructures.

Acknowledgements

The authors would like to thank U. Ekenberg, F. Fuchs, B. Jusserand, P. Koidl, and H. Schneider for helpful discussions and comments, as well as H. S. Rupprecht for his continuous support.

References

[Abstreiter *et al.*, 1984]: G. Abstreiter, M. Cardona, and A. Pinczuk, *Light Scattering in Solids* IV, eds. M. Cardona and G. Güntherodt (Springer, Heidelberg, 1984), p. 5.

[Abstreiter *et al.*, 1986]: G. Abstreiter, R. Merlin, and A. Pinczuk, *IEEE J. Quantum Electron.* **22**, 1771 (1986).

[Altarelli *et al.*, 1985]: M. Altarelli, U. Ekenberg, and A. Fasolino, *Phys. Rev.* B **32**, 5138 (1985).

[Aronov *et al.*, 1983]: A. G. Aronov, G. E. Pikus, and A. N. Titkov, *Zh. Eksp. Teor. Fiz.* **84**, 1170 (1983) [*Sov. Phys. JETP* **57**, 680 (1983)].

[Aspnes *et al.*, 1986]: D. E. Aspnes, S. M. Kelso, R. A. Logan, and R. Bhat, *J. Appl. Phys.* **60**, 754 (1986).

[Bir *et al.*, 1976]: G. L. Bir, A. G. Aronov, and G. E. Pikus, *Zh. Eksp. Teor. Fiz.* **69**, 1382 (1975) [*Sov. Phys. JETP* **42**, 705 (1976)].

[Borghs *et al.*, 1989]: G. Borghs, K. Bhattacharyya, K. Deneffem, P. van Mieghem, and R. Mertens, *J. Appl. Phys.* **66**, 4381 (1989).

[Cardona and Harbeke, 1962]: M. Cardona and G. Harbeke, *J. Appl. Phys.* **34**, 813 (1962).

[Cardona, 1982]: M. Cardona, in *Light Scattering in Solids II*, eds. M. Cardona and G. Güntherodt (Springer, Heidelberg, 1982), p. 19.

[Chen *et al.*, 1988]: Y. Chen, B. Gil, P. Lefebre, and H. Mathieu, *Phys. Rev.* B **37**, 6429 (1988).

[Chen *et al.*, 1990]: W. Chen, M. Fritze, A. V. Nurmikko, D. Ackley, C. Colvard, and H. Lee, *Phys. Rev. Lett.* **64**, 2434 (1990).

[Chen *et al.*, 1992]: W. Chen, M. Fritze, W. Walecki, A. V. Nurmikko, D. Ackley, M. Hong, and L. L. Chang, *Phys. Rev.* B **45**, 8464 (1992).

[Cohen and Marques, 1990]: A. M. Cohen and G. E. Marques, *Phys. Rev.* B **41**, 10608 (1990).

[D'yakonov and Perel', 1971]: M. I. D'yakonov and V. I. Perel', *Zh. Eksp. Teor. Fiz.* **60**, 1954 (1971) [*Sov. Phys. JETP* **33**, 1053 (1971)].

[Damen *et al.*, 1991]: T. C. Damen, L. Viña, J. E. Cunningham, J. Shah, and L. J. Sham, *Phys. Rev. Lett.* **67**, 3432 (1991).

[Dawson and Woodbridge, 1984]: P. Dawson and K. Woodbridge, *Appl. Phys. Lett.* **45**, 1227 (1984).

[Delalande *et al.*, 1987]: C. Delalande, G. Bastard, J. Organasi, J. A. Brum, H. W. Liu, M. Voos, G. Weimann, and W. Schlapp, *Phys. Rev. Lett.* **59**, 2690 (1987).

[Eaves *et al.*, 1992]: L. Eaves, R. K. Hayden, M. L. Leadbeater, D. K. Maude, E. C. Valadares, M. Henini, F. W. Sheard, O. H. Hughes, J. C. Portal, and L. Cury, *Surf. Sci.* **263**, 199 (1992).

[Ekenberg and Altarelli, 1985]: U. Ekenberg, and M. Altarelli, *Phys. Rev.* B **32**, 3712 (1985).

[Ekenberg, 1988]: U. Ekenberg, *Phys. Rev.* B **38**, 12664 (1988).

[Ekenberg, 1990]: U. Ekenberg, *Surf. Sci.* **229**, 419 (1990).

[Eppenga *et al.*, 1987]: R. Eppenga, M. F. H. Schuurmans, and S. Colak, *Phys. Rev.* B **36**, 1554 (1987).

[Fasol *et al.*, 1989]: G. Fasol, R. D. King-Smith, D. Richards, U. Ekenberg, N. Mestres, and K. Ploog, *Phys. Rev.* B **39** 12695 (1989).

[Gammon *et al.*, 1992]: D. Gammon, B. V. Shanabrook, J. C. Ryan, D. S. Katzer, and M. J. Yang, *Phys. Rev. Lett.* **68**, 1884 (1992).

[Gilinsky *et al.*, 1991]: A. M. Gilinsky, K. S. Zhuravelev, D. I. Lubyshev, V. P. Migal, V. V. Preobrazhenskii, and B. R. Semiagin, *Superlattices and Microstructures* **10**, 399 (1991).

[Harris *et al.*, 1993]: J. J. Harris, R. Murray, C. T. Foxon, *Semicon. Sci. Technol.* **8**, 31 (1993).

[Hawrylak, 1991]: P. Hawrylak, *Phys. Rev.* B **44**, 3821 (1991); *Phys. Rev.* B **44**, 6262 (1991); *Phys. Rev.* B **44**, 11236 (1991).

[Hawrylak *et al.*, 1992]: P. Hawrylak, N. Pulsford, and K. Ploog, *Phys. Rev.* B **46**, 15193 (1992).

[Henning *et al.*, 1991]: J. C. M. Henning, Y. A. A. R. Kessner, P. M. Koenraad, M. R. Leys, W. van der Vleuten, J. H. Wolter, and A. M. Frens, *Semicond. Sci. Technol.* **6**, 1079 (1991).

[Kash *et al.*, 1992]: J. A. Kash, M. Zachau, M. A. Tischler, and U. Ekenberg, *Phys. Rev. Lett.* **69**, 2260 (1992).

[Ke and Hamilton, 1993]: M.-L. Ke and B. Hamilton, *Phys. Rev.* B **47**, 4970 (1993).

[Ke *et al.*, 1992]: M.-L. Ke, J. S. Rimmer, B. Hamilton, J. H. Evans, M. Missous, K. E. Singer, and P. Zalm, *Phys. Rev.* B **45**, 14114 (1992).

[Kim *et al.*, 1993]: Y. Kim, M.-S. Kim, and S.-K. Min, *Appl. Phys. Lett.* **62**, 741 (1993).

[Kukushkin *et al.*, 1988]: I. V. Kukushkin, K. von Klitzing, and K. Ploog, *Phys. Rev.* B **37**, 8509 (1988).

[Kukushkin *et al.*, 1989]: I. V. Kukushkin, K. von Klitzing, K. Ploog, and V. B. Timofeev, *Phys. Rev.* B **40**, 7788 (1989).

[Livescu *et al.*, 1988]: G. Livescu, D. A. B. Miller, D. S. Chemla, M. Ramaswamy, T. Y. Chang, N. Sauer, A. C. Gossard, and J. H. English, *IEEE J. Quantum Electron.* **QE-24**, 1677 (1988).

[Luttinger and Kohn, 1955]: J. M. Luttinger, and W. Kohn, *Phys. Rev.* **97**, 869 (1955).

[Luttinger, 1956]: J. M. Luttinger, *Phys. Rev.* **102**, 1030 (1956).

[Maciel *et al.*, 1990]: A. C. Maciel, M. Tatham, J. F. Ryan, J. M. Worlock, R. E. Nahory, J. Harbison, and L. T. Florez, *Surf. Sci.* **228**, 251 (1990).

[Mahan, 1981]: G. D. Mahan, *Many-Particle Physics* (Plenum, New York, 1981).

[Meier and Zakhachenya, 1984]: *See*, e.g., *Optical Orientation*, eds. F. Meier and B. P. Zakhachenya (North-Holland, Amsterdam, 1984).

[Mueller, 1990]: J. F. Mueller, *Phys. Rev.* B **42**, 11189 (1990).

[O'Reilly, 1989]: E. P. O'Reilly, *Semicond. Sci. Technol.* **4**, 121 (1989).

[Parsons, 1969]: R. R. Parsons, *Phys. Rev. Lett.* **23**, 1152 (1969).

[Peric *et al.*, 1993]: H. Peric, B. Jusserand, D. Richards, and B. Etienne, *Phys. Rev.* B **47**, 12722 (1993).

[Perry *et al.*, 1988]: C. H. Perry, K. S. Lee, W. Zhou, J. M. Worlock, A. Zrenner, F. Koch, and K. Ploog, *Surf. Sci.* **196**, 677 (1988).

[Pinczuk and Abstreiter, 1989]: *See* e.g., A. Pinczuk, and G. Abstreiter, *Light Scattering in Solids V*, eds. M. Cardona and G. Güntherodt (Springer, Berlin, 1989), p. 153.

[Pinczuk *et al.*, 1989]: A. Pinczuk, S. Schmitt-Rink, G. Damen, J. P. Valadares, L. N. Pfeiffer, and K. West, *Phys. Rev. Lett.* **63**, 1633 (1989).

[Pletschen *et al.*, 1991]: W. Pletschen, J. Wagner, G. Kaufel, and K. Köhler, *Appl. Phys. Lett.* **59**, 2299 (1991).

[Richards *et al.*, 1992]: D. Richards, J. Wagner, A. Fischer, and K. Ploog, *Appl. Phys. Lett.* **61**, 2685 (1992).

[Richards *et al.*, 1933a]: D. Richards, J. Wagner, H. Schneider, G. Hendorfer, M. Maier, A. Fischer, and K. Ploog, *Phys. Rev.* B **47**, 9629 (1993).

[Richards *et al.*, 1993b]: D. Richards, J. Wagner, M. Maier, and K. Köhler, *Semicond. Sci. Technol.* **8**, 1412 (1993).

[Rodriguez and Tejedor, 1993]: F. J. Rodriguez and C. Tejedor, *Phys. Rev.* B **47**, 1506 (1993).

[Rössler *et al.*, 1990]: U. Rössler, S. Jorda, and D. Brodio, *Solid State Commun.* **73**, 209 (1990).

[Schäffler and Abstreiter, 1986]: F. Schäffler and G. Abstreiter, *Phys. Rev.* B **34**, 4017 (1986).

[Schmitt-Rink *et al.*, 1986]: S. Schmitt-Rink, C. Ell, and H. Haug, *Phys. Rev.* B **33**, 1183 (1986).

[Schubert *et al.*, 1989]: E. F. Schubert, T. D. Harris, J. E. Cunningham, and W. Jan, *Phys. Rev.* B **39**, 11011 (1989).

[Schubert *et al.*, 1990]: E. F. Schubert, J. M. Kuo, R. F. Kopf, H. S. Luftman, L. C. Hopkins, and N. J. Sauer, *J. Appl. Phys.* **67**, 1969 (1990).

[Sernelius, 1986]: B. E. Sernelius, *Phys. Rev.* B **33**, 8582 (1986).

[Seymour and Alfano, 1980]: R. J. Seymour and R. R. Alfano, *Appl. Phys. Lett.* **37**, 231 (1980).

[Shen *et al.*, 1987]: H. Shen, P. Parayanthai, F. H. Pollack, R. N. Sacks, and G. Hickman, *Solid State Commun.* **63**, 357 (1987).

[Shen *et al.*, 1990]: H. Shen, M. Dutta, L. Fotiadis, P. G. Newman, R. P. Moerkirk, W. H. Chang, and R. N. Sacks, *Appl. Phys. Lett.* **57**, 2118 (1990).

[Shih and Streetman, 1991]: Y. C. Shih and B. G. Streetman, *Appl. Phys. Lett.* **59**, 1344 (1991).

[Skolnick *et al.*, 1987]: M. S. Skolnick, J. M. Rorison, K. J. Nash, D. J. Mowbray, P. R. Tapster, S. J. Bass, and A. D. Pitt, *Phys. Rev. Lett.* **58**, 2130 (1987).

[Skolnick *et al.*, 1991]: M. S. Skolnick, D. M. Whittaker, P. E. Simmonds, T. A. Fisher, M. K. Saker, J. M. Rorison, R. S. Smith, P. B. Kirby, and C. R. H. White, *Phys. Rev.* B **43**, 7354 (1991).

[Viña *et al.*, 1992]: L. Viña, T. C. Damen, J. E. Cunningham, J. Shah, and L. J. Sham, *Superlattices and Microstructures* **12**, 379 (1992).

[Wagner *et al.*, 1989]: J. Wagner, M. Ramsteiner, W. Stolz, M. Hauser, and K. Ploog, *Appl. Phys. Lett.* **55**, 978 (1989).

[Wagner *et al.*, 1990]: J. Wagner, A Fischer, and K. Ploog, *Phys. Rev.* B **42**, 7280 (1990).

[Wagner *et al.*, 1991a]: J. Wagner, A. Fischer, and K. Ploog, *Appl. Phys. Lett.* **59**, 428 (1991).

[Wagner *et al.*, 1991b]: J. Wagner, A. Ruiz, and K. Ploog, *Phys. Rev.* B **43**, 12134 (1991).

[Wagner *et al.*, 1991c]: J. Wagner, M. Ramsteiner, D. Richards, G. Fasol, and K. Ploog, *Appl. Phys. Lett.* **58**, 143 (1991).

[Wagner *et al.*, 1993]: J. Wagner, H. Schneider, D. Richards, A. Fischer, and K. Ploog, *Phys. Rev.* B **47**, 4786 (1993).

[Wagner *et al.*, 1994]: J. Wagner, D. Richards, H. Schneider, A. Fischer, and K. Ploog, *Solid State Electron*, **37**, 1871 (1994).

[Yablonovich *et al.*, 1989]: E. Yablonovich, B. J. Skromme, R. Bhat, J. P. Harbison, and T. J. Gmitter, *Appl. Phys. Lett.* **54**, 555 (1989).

[Zerrouati *et al.*, 1988]: K. Zerrouati, F. Fabre, G. Bacquet, J. Bandet, J. Frandon, G. Lampel, and D. Paget, *Phys. Rev.* B **37**, 1334 (1988).

16

Electron transport in delta-doped quantum wells

W. TED MASSELINK

16.1 Introduction

Delta-doping of semiconductor quantum wells makes simultaneous use of the two fundamental ways in which to confine electrons or holes in a quasi-two-dimensional (Q2D) sheet. These are through the electric fields which form in response to non-uniform doping profiles within the structure and through the use of heterojunctions between materials with differing electron affinities. The simultaneous use of confinement due to band bending and to heterojunctions is also employed in the modulation-doped heterostructure, the difference being in the location of the dopant atoms. Electron transport in a Q2D system has provided both a basis for much of today's electronic device technology, and a fertile ground for fundamental physical studies. The principal electronic device using Q2D electron transport are the metal-xide–semiconductor field effect transfer (MOSFET) and the modulation-doped field-effect transistor (MODFET). With current gain cutoff frequencies (f_T) predicted to reach 400 GHz [1], the MODFET is the fastest three-terminal device in existence. The speed advantage of MODFETs based on InAlAs/InGaAs/InP, AlGaAs/InGaAs/GaAs, and AlGaAs/GaAs is derived from a combination of high sheet electron density, high electron velocity, and, to a lesser extent, high electron mobility, and is realized through the use of extremely short gate-lengths [1, 2]. Other than the high mobilities achieved in modulation-doped structures, these same characteristics can largely be achieved using delta-doping directly into the channel [3]. From the point of view of f_T, when very short gate-lengths are employed, transistors using a shallow dopant profile and no heterostructures can successfully compete against MODFETs in the same material system [4]. On the other hand, to maintain a very shallow profile and to decrease leakage through the transistor gate, putting the dopant sheet into a quantum well is advantageous [2].

Electron transport in Q2D semiconductor systems has also proved worthy of study on more fundamental grounds. Already in 1966, Fowler, Fang, Howard, and Stiles demonstrated that the silicon inversion layer in a MOS system is a two-dimensional electron gas (2DEG) [5]. With the invention of modulation-doping, high-mobility 2DEG was demonstrated in 1979 [6]. Today, the modulation-doped AlGaAs/GaAs system has been improved to the point that low-field mobilities exceeding 10^7 cm^2/Vs have been realized [7]. Such extremely high mobilities allow investigations of fundamental questions in physics, such as the quantum Hall effect [8].

The emphasis in this chapter, however, is on transport as it is relevant to electronic devices such as field-effect transistors (FETs). Transistor-like devices are not usually used at the low temperature and low electric fields which are necessary to achieve the aforementioned extremely high mobilities. Since the first reports of FETs based on the modulation-doping structure [9], it has become ever clearer that FET speed performance results from a combination of electron mobility and electron velocity. Even though the FETs reported in Ref. [9] had extremely long gates (400 μm), the improvement in transconductance (g_m) was not as great as that in mobility. Since then, many authors have noted that, especially as gate-length is decreased, the relative importance of mobility decreases and that of velocity increases. Fischetti and Laux [10] point out that with extremely short gates and relatively large drain-source voltages, so much of the Brillouin zone is sampled by the electron gas that differences between different semiconductor materials become rather small.

Between the regimes of mobility-dominated FET characteristics and the 'universal FET' of Ref. [10] is a regime in which the entire velocity versus electric-field characteristic of the material is important. In this chapter, we will review measurements of electron velocity in heterostructures and attempt to relate these measurements to FET performance.

16.2 High-field transport in modulation-doped heterostructures

This section is intended to serve as a motivation for the study of transport in doped quantum wells. Like doped quantum wells, modulation-doped heterostructures use a combination of band bending and heterojunctions to confine the Q2D sheet of electronic charge. The difference, of course, is that in the case of modulation-doped quantum wells, the separation of the donors from the channel results in extremely high mobilities [7]. Structures which have been optimized for extremely high 2DEG mobilities have a resulting sheet electron density, n_{2d}, which is too low to be of use in FETs. There is always a trade-off in these structures, that is, higher mobility requires a design which results in lower n_{2d}. Even when the structure is optimized for maximum n_{2d}, however, the conduction band discontinuity limits the carrier concentration; in GaAs, this limit is about 1×10^{12} cm^{-2}. The highest mobilities that one finds in modulation-doped heterostructures with sufficient 2D electron concentration to be useful for FETs are (at 77 K) in the range of 7×10^4 cm^2/Vs. Of course, such a high mobility is only observed at very low electric fields. Once a higher electric field is applied across the channel, this mobility quickly degrades. The geometrical magnetoresistance (GMR) technique is well suited for measuring electron mobility and sheet concentration at low and moderate electric fields in FET structures.

Such measurements are described in Ref. [11]; Fig. 16.1 reproduces some data from

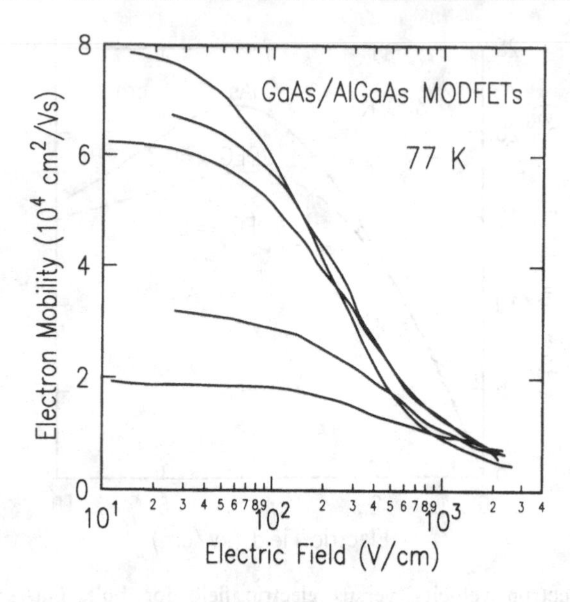

Fig. 16.1. Measured mobilities as functions of electron field for GaAs/AlGaAs modulation-doped structures with a variety of spacer thicknesses and low-field mobilities.

Ref. [12]. By varying the thickness of the undoped AlGaAs spacer between the doped AlGaAs and the GaAs channel in various MODFET structures, the resulting mobility varies between 2×10^4 and 7×10^4 cm^2/Vs. The structure with the lowest low-field mobility has no spacer at all. If one desires a current density of 200 mA/mm and the maximum sheet electron concentration is 1×10^{12} cm^{-2}, then, even in the parasitic source and drain regions, where the electric field is lowest, none of these structures will contain electric fields below about 1000 V/cm. What is significant about the data in Fig. 16.1 is that already at 1000 V/cm there is very little difference in mobility between the high-mobility and low-mobility structures. Furthermore, because the low-mobility structures have a higher electron concentration, they are actually better suited for FETs than are the high-mobility structures.

On the other hand, if we compared structures with electron mobilities typically found at room temperature instead of at 77 K, even at a field of 1000 V/cm, higher mobilities would still be beneficial, especially in the parasitic source region, allowing the extrinsic transistor to perform closer to the corresponding intrinsic transistor.

Electron velocity has also been measured for modulation-doped heterostructures [13, 14]. Figure 16.2 depicts the results of this measurement. Electron velocity in $Al_xGa_{1-x}As$/GaAs modulation-doped heterostructures with $x = 0.3$ and 0.5 are compared to the electron velocity in undoped GaAs. These data indicate that although all three samples have similar low-field mobilities, the peak velocity of the 2DEG is lower than that of electrons in lightly doped bulk GaAs and occurs at a somewhat lower electric field. This behavior is explained through two mechanisms. First, because the electrons are spatially

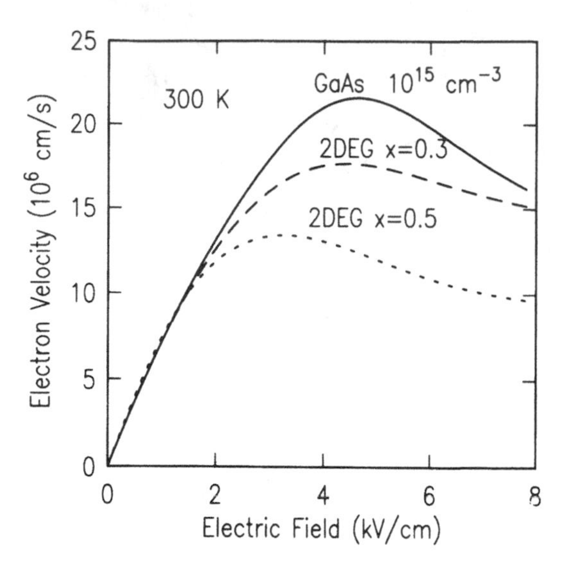

Fig. 16.2. Electron velocity versus electric field for bulk GaAs ($n = 1 \times 10^{15}$ cm^{-3}) and $Al_xGa_{1-x}As/GaAs$ modulation-doped heterostructures with $x = 0.3$ and 0.5.

confined, the density of states is step-like, with zero states at the bottom of the Γ valley conduction band. The lowest energy available for electron states coincides with the lowest Γ valley subband, with an energy of about 40 meV above the bottom of the band [15]. The lack of electron states below this subband results in the effective Γ–L energy separation being reduced from about 0.31 eV to about 0.27 eV. Since intervalley transfer and, therefore, peak velocity followed by negative differential mobility depends on this energy difference in an exponential manner, this reduction will be important. The second effect contributing to the lower peak velocity of the 2DEG as compared with electrons in bulk undoped GaAs is real-space transfer. In the heterostructure, besides being able to transfer from the Γ valley into the L valleys in GaAs, the electrons may also transfer into the Γ valley in the $Al_xGa_{1-x}As$ and also into the L and X valleys in the $Al_xGa_{1-x}As$. These valleys are characterized by higher electron effective mass and also by the addition of alloy scattering, which means that electrons scattering into these valleys will have much reduced velocities compared with those in the Γ valley of GaAs. For a higher AlAs mole fraction $Al_xGa_{1-x}As$, the X valleys become more important. This increased scattering into the X valleys may account for the measured lower peak electron velocity in $Al_{0.5}Ga_{0.5}As$ as compared with $Al_{0.30}Ga_{0.70}As$. This measurement implies that allowing electrons to scatter into $Al_xGa_{1-x}As$ with high x can reduce the maximum electron velocity significantly and thereby limit FET performance.

At both 300 and 77 K, the 2DEG samples exhibited lower peak electron velocity than the lightly doped bulk GaAs. The comparison of the velocity-field characteristic of the 2DEG in a GaAs/AlGaAs heterstructure to that of three-dimensional electrons in very lightly doped bulk GaAs demonstrates the additional scattering mechanism resulting from

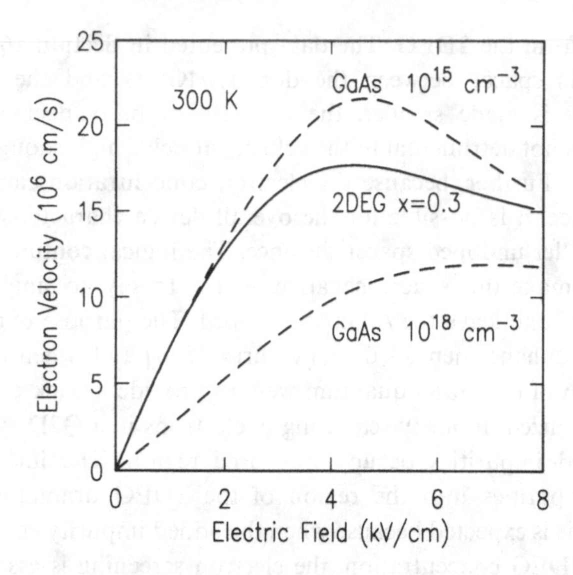

Fig. 16.3. A comparison of electron velocity in undoped GaAs, the two-dimensional electron gas, and doped GaAs at 300 K.

the heterostructure and the 2D confinement. For analysis appropriate to FET performance, however, it is more useful to compare the modulation-doped samples with bulk GaAs doped $\approx 10^{18}$ cm^{-3}. Only with such heavy doping can one obtain the gate capacitance (i.e. charge control) and shallow channel (necessary for short gate FETs) similar to what one can obtain with the modulation-doped structures. Comparing the heavily doped GaAs results with the 2DEG results, as shown in Fig. 16.3, one can easily see that the modulation-doped heterostructures have both superior low-field mobility and higher-field velocity than the heavily doped bulk GaAs.

To summarize, the extremely high mobility that one observes in MODFET 2DEG structures at 77 K does not contribute to the performance of the FETs any more than do much more modest mobilities at 77 K. In fact, the sacrifice of electron concentration necessary to achieve enhanced mobilities may even be detrimental in balance. On the other hand, the very low mobilities that one finds in heavily doped GaAs suited to metal–semiconductor field-effect transistors (MESFETs) also leads to lower velocities than in the case of the modulation-doped heterostructure. Thus, for comparable electron concentration, one expects the MODFET to outperform the MESFET; this expectation is ultimately derived from the higher mobility. In the next section we will discuss the lower-mobility structures based on transport in heavily doped material further.

16.3 Electron mobility in doped quantum wells

In Section 16.2, we summarized measurements of electron velocity in the modulation-doped heterojunction system; there, the ionized-impurity scattering is kept very small by locating the impurities away from the 2DEG. The data presented in Section 16.2 indicate that as the undoped AlGaAs spacer between the doped AlGaAs and the GaAs channel in MODFET structures is made smaller, the low-field mobility decreases, but that this decrease in mobility is not detrimental to the velocity at fields high enough to be interesting in an operating FET. Further, because the electron concentration can be higher with a smaller spacer distance, it is possible for the overall device characteristics to actually be improved with a smaller undoped spacer distance. The logical continuation of that work would have been to make this spacer negative, that is to say, to simply put the dopant into the GaAs channel and leave the AlGaAs undoped. The purpose of the AlGaAs in this case is to provide an enhancement of the gate barrier [16] and, when the AlGaAs is also present at the bottom of the GaAs quantum well, to provide a back confinement. In this section we describe ionized-impurity scattering of electrons in a Q2D system in which the electrons and ionized impurities occupy the same region. The introduction of small concentrations of impurities into the region of the 2DEG dramatically degrades the mobility [17–19]. This is expected because when the added impurity concentration is small compared with the 2DEG concentration, the electron screening is essentially unaffected, but the effective number of scattering centers is increased.

In the case of a bulk semiconductor, the total wavefunction of the electrons overlaps with the ionized impurities, which are the same in number as the electron density. If such a uniformly doped bulk semiconductor is confined in one dimension, the overlap of the electronic wavefunction with the ionized impurities (an effective impurity concentration) will be altered because the electronic wavefunction will be peaked in the center, but vanishing at the edges of the resulting quantum well.

Here we describe in detail how ionized-impurity scattering in a Q2D system differs from that in bulk. Specifically, we show that the ionized-impurity scattering in uniformly doped AlGaAs/GaAs quantum wells is greater than in a similarly doped GaAs. This result is explained through a detailed calculation of the ionized-impurity scattering rates in a Q2D system [20].

To compare ionized-impurity scattering in three dimensions with that in two dimensions, we compare Si-doped bulk GaAs with uniformly doped GaAs quantum wells with AlGaAs barriers. All samples were grown by molecular beam epitaxy (MBE) under identical conditions on (100)-oriented undoped GaAs substrates. All sources were solid and a low substrate temperature of about 560 °C was used. The bulk GaAs was 0.25 μm thick and was doped with Si to a level of $6 \times 10^{17}\ \mathrm{cm}^{-3}$. The quantum wells, which were in mulitple-quantum-well samples with 25–50 periods, were either 100 or 50 Å wide with, respectively, 34- or 17-Å-wide barriers of $Al_{0.4}Ga_{0.6}As$. These barriers were narrow enough to ensure that the 2D concentration of traps in the AlGaAs was small compared with the 2D electron and dopant concentrations, and yet were thick enough to contain the wavefuction in the GaAs quite well. All quantum wells were also doped to an average 3D concentration $6 \times 10^{17}\ \mathrm{cm}^{-3}$. Both 100 Å wells and 50 Å wells were prepared with progressively narrower doping profiles, maintaining the same 2D concentration (i.e. the same average doping concentration if one divides the sheet concentration by the well width). The widest doping profile comprised 84% of the well width; the entire well was

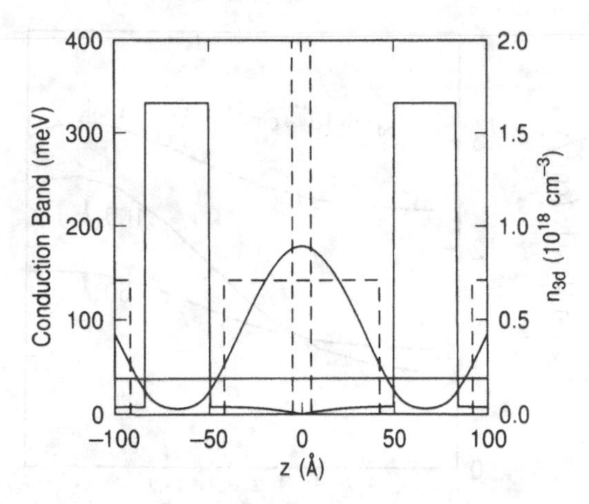

Fig. 16.4. Conduction band profile for the doped quantum well samples discussed in the text. The dashed lines are Si doping profiles. Also depicted are the Fermi level and the electron density.

not doped in order to prevent D–X center formation near the interfaces. The narrow-profile limits of this sample series included samples that were delta-doped [21, 22] with sheet densities of 6×10^{11} cm^{-2} in the 100 Å wells and 2.8×10^{11} cm^{-2} in the 50 Å wells. Wells 100 Å wide, delta-doped with the same sheet density (6×10^{11} cm^{-2}), but doped away from the center, were also studied for comparison. Figure 16.4 shows the conduction band edge and dopant profiles for representative samples.

Some studies of Si delta-doping diffusion in GaAs indicate that the Si atoms deposited in one atomic plane diffuse less than two lattice constants during MBE growth [23, 24]. Other studies [25–27] show evidence of much wider final Si atom profiles. Reference [26] indicates that the 3D concentration of active Si in the delta-doping spike is limited by the D–X center to the maximum Si concentration in bulk GaAs, which results in a maximum sheet concentration of about 7×10^{12} cm^{-2}. In any case, the Si concentrations in this study are much lower than in Refs. [23–27] and we expect negligible diffusion. Additionally, some diffusion will not affect our results because the electron wavefunction is quite flat at the well centers and thus spreading of the doping spike in the center of the well has little effect on the scattering.

Van der Pauw–Hall measurements have been made on the samples between 10 and 300 K using low electric and magnetic fields. In heavily doped semiconductors, such as the samples studied here, the Hall factor is very close to unity, irrespective of which scattering mechanisms dominate. The curves in Fig. 16.5 show the temperature dependence of the mobilities of three uniformly doped samples: bulk GaAs, 100 Å wells, 50 Å wells. The mobility of the bulk sample is what is typically measured at this doping level and is well understood theoretically [28]. Mobilities for the doped quantum wells are significantly lower than that observed in bulk GaAs for all temperatures. Previous data [29, 30] indicate that interface roughness scattering of a 2DEG in 100-Å-wide wells is quite small. These same data, on the other hand, also indicate that in 50 Å wells it can be significant.

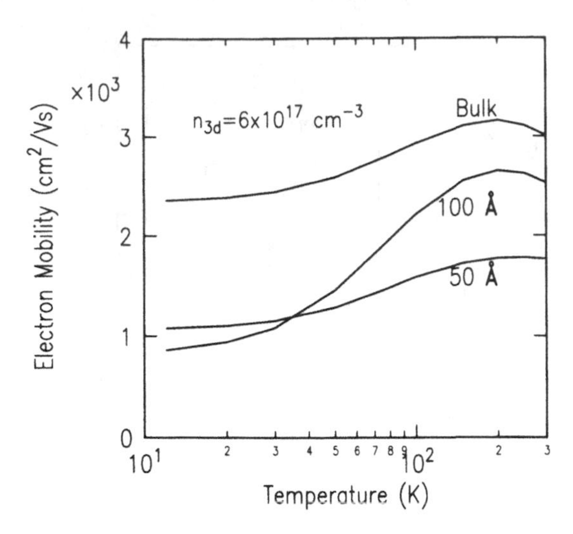

Fig. 16.5. Measured Hall mobilities as functions of temperature for bulk GaAs, 100-Å-wide quantum wells, and 50-Å-wide quantum wells.

Structures with 50 Å wells and AlAs barriers have mobilities limited to about 1000 cm^2/Vs [29]; when the barrier alloy is $Al_xGa_{1-x}As$ with x between 0.30 and 0.35, mobilities of about 10^4 cm^2/Vs have been measured [30]. Electrons in 100 Å wells achieve mobilities of 10^5 cm^2/Vs even with the rougher AlAs barriers [29]. From this we conclude that the interface roughness scattering in our 100 Å samples with $Al_{0.4}Ga_{0.6}As$ barriers is not important compared with the ionized-impurity scattering, but almost certainly needs to be accounted for in the 50 Å well samples.

The confinement of the electronic wavefunction affects the ionized-impurity scattering in two ways: first, it changes the way the wavefunction overlaps with the impurities; and, second, it changes the electron screening term in the dielectric function. Following Ando [31], the relaxation time for electrons in one subband scattering from ionized impurities is given by

$$\frac{\hbar}{\tau_c(k)} = 2\pi \int dz\, N(z) \sum_{\mathbf{q}} \left[\frac{2\pi e^2}{q\varepsilon(q)}\right] |F(q, z)|^2 (1 - \cos\theta)\delta(\varepsilon_{\mathbf{k}} - \varepsilon_{\mathbf{k}-\mathbf{q}}), \tag{16.1}$$

where $N(z)$ is the doping profile, $q = 2\,k \sin(\theta/2)$ and $\varepsilon_{\mathbf{k}} = \hbar^2k^2/2m$ where $\mathbf{k}$ is the wave vector, and θ is the scattering angle. The static dielectric function is given by

$$q\varepsilon(q) = q + \frac{2\pi e^2}{\kappa}\frac{2m}{2\pi\hbar^2}F(q), \tag{16.2}$$

where κ is the dielectric constant and m is the electron effective mass. Here we use the $T = 0$ formula for the polarization term which introduces only small errors away from $T = 0$ in the case of these degenerate samples. In 2D systems, degenerate statistics result in no explicit n_{2d} dependence of the polarization. The form factor appearing in Eqn. (16.1),

$F(q, z)$, is a measure of the overlap of the electron density with charge centers at location z and is given by

$$F(q, z) = \int dz' |\zeta(z')|^2 \exp(-q|z - z'|), \tag{16.3}$$

where $\zeta(z)$ is the z component of the total electronic envelope wavefunction $\Psi(\mathbf{r}, z) = \zeta(z) \exp(i\mathbf{k} \cdot \mathbf{r})$. The other form factor $F(q)$ which appears in Eqn. (16.2) may be thought of as the 'screening of the screening' [32] and is given by

$$F(q) = \int dz \int dz' |\zeta(z)|^2 |\zeta(z')|^2 \exp(-q|z - z'|); \tag{16.4}$$

because it never exceeds unity; it only decreases the screening. The sum over q in Eqn. (16.1) may be replaced by an integral over θ so that Eqn. (16.1) is written

$$\frac{\hbar}{\tau_c(k)} = \frac{4\pi m e^4}{\hbar^2} \int_0^{\pi} d\theta (1 - \cos\theta) \frac{1}{(q\varepsilon(q))^2} \int dz |F(q, z)|^2 N(z). \tag{16.5}$$

The mobility is finally given by

$$\mu(T) = \frac{e\langle \tau_c \rangle}{m}. \tag{16.6}$$

The temperature dependence of $\mu(T)$ enters in the averaging of $\tau_c(k)$ over k. We choose a simple wavefunction,

$$\zeta(z) = \sqrt{\left(\frac{2}{W}\right)} \cos\left(\frac{\pi z}{W}\right), \tag{16.7}$$

where W is the width of the quantum well. This wavefunction, along with the one-subband approximation, is justified by noting that even with such narrow barriers, especially in the 100 Å case, the wavefunction squared is close to zero at the interface, and that the second subband is nearly unpopulated. The self-consistent solution of Schrödinger's and Poisson's equations used to generate Fig. 16.4 shows that, compared with the conduction band offset, the band bending with these doping levels is small. The doping profile $N(z)$ for symmetrically doped samples is given by

$$N(z) = \begin{cases} \dfrac{1}{uW} n_{2d} & |z| < \dfrac{uW}{2} \\ 0 & |z| > \dfrac{uW}{2} \end{cases}, \tag{16.8a}$$

where u is the fraction of the well which is doped. In the case of the delta-doped wells,

$$N(z) = n_{2d} \delta(z - z_\delta), \tag{16.8b}$$

where z_δ is the position of the delta-doping spike. It is then straightforward to evaluate the integrals of Eqns. (16.3) and (16.4) analytically and of Eqns. (16.5) and (16.6) numerically to obtain the theoretical mobilities as functions of dopant distribution, temperature, and well width. We ignore certain corrections which have been incorporated into more detailed calculations of ionized-impurity scattering in bulk semiconductors [33] such as multiple scattering, as well as the effect of well-to-well screening which we expect to be small.

The data displayed by the curves in Fig. 16.5 do not really provide a definitive comparison of ionized-impurity scattering in two dimensions with ionized-impurity scattering in three dimensions because although the bulk sample is unfirmly doped, the wells are doped in the center 84%, which will result in lower mobility than if they were uniformly doped. (Uniform doping over 100% of the wells was avoided in order to prevent D–X center formation at the interfaces.) The quantum well mobilities may be corrected, however, by using Matthiessen's rule to subtract the extra scattering rate resulting from the slightly more concentrated dopant profile. In the case of the 50 Å wells, an additional (assumed temperature-independent) scattering rate equivalent to a mobility of 6000 cm^2/Vs was also subtracted. These corrected data of mobility as a function of temperature are shown in Fig. 16.6. These data are slightly different from the raw data of the solid curves and demonstrate unambiguously that Q2D confinement results in a decrease in mobility and therefore an increase in ionized-impurity scattering over a broad temperature range.

Figure 16.7 depicts the dependence of low-field mobility in doped quantum wells on well width. Quantum well samples with well widths ranging from 50 Å to 400 Å are compared with each other and with bulk GaAs. In all cases, the average 3D dopant and carrier concentration is approximately $6 \times 10^{17}\ cm^{-3}$. Figure 16.7 illustrates that progressively wider wells lead to progressively higher mobilities, asymptotically approaching that for bulk GaAs. Only the sample with 50 Å wells appears to deviate from a fairly

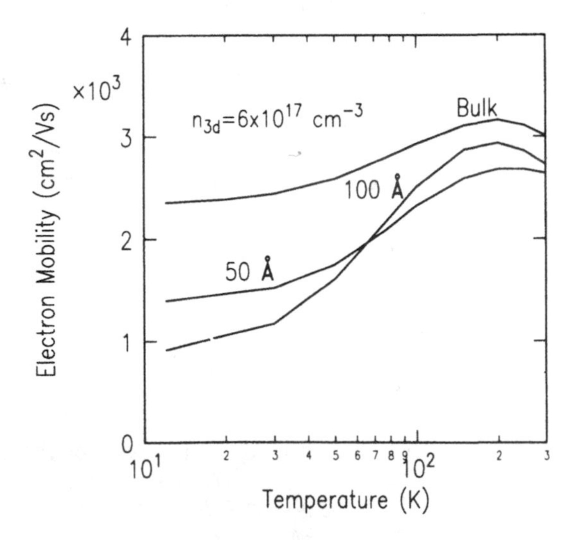

Fig. 16.6. Hall mobilities as functions of temperature for bulk GaAs, 100 Å wide quantum wells, and 50 Å wide quantum wells corrected as described in the text to show uniform doping profiles.

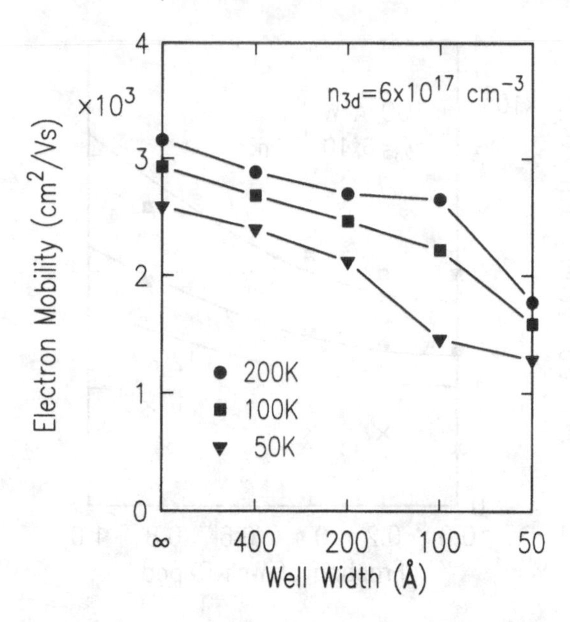

Fig. 16.7. Electron mobility versus quantum well width in doped AlGaAs/GaAs quantum wells.

constant decrease in mobility with well width; the additional decrease is consistent with the addition of interface roughness scattering in that sample.

From Figs. 16.5–16.7 it is clear that uniformly doped quantum wells have a lower mobility than does identically doped bulk GaAs and, of course, studies of modulation-doped quantum wells show that when the dopants are largely moved out of the quantum well, the mobility is higher. Figure 16.8 shows the effect of concentrating the dopants into the center of the well. In each case, the areal concentration is about 6×10^{11} cm^{-2}, but the distributions vary. When the fraction of the well which is doped reaches 0, the well is delta-doped in the center; when the fraction is 1.0, it is uniformly doped. The solid curves in Fig. 16.8 are calculated as described above with no corrections or adjustable parameters. (Phonon scattering is not included in the calculations.) The symbols are experimental data. We see that the doping profile has a significant impact on the ionized-impurity mobility and that (except at low temperatures) the model described above describes the physics adequately. At higher temperatures (and especially for larger fractions of the well which is doped), phonon scattering, which we have omitted, becomes important. A structure was also prepared which was otherwise identical to the delta-doped quantum well sample except that it contained no AlGaAs barriers. This sample consisted of delta-like spikes of 6×10^{11} cm^{-2} located every 100 Å; the mobility $\mu(T)$ for the sample was nearly identical to that of the bulk GaAs of identical doping for all temperatures. The chapter by Koenraad, Chapter 17, and that by Schubert, Chapter 1, describe the properties of delta-doped bulk GaAs in detail.

Concentrating the dopant profile to a delta function in the center of the well serves to decrease the mobility further. This latter effect is quite easy to visualize because,

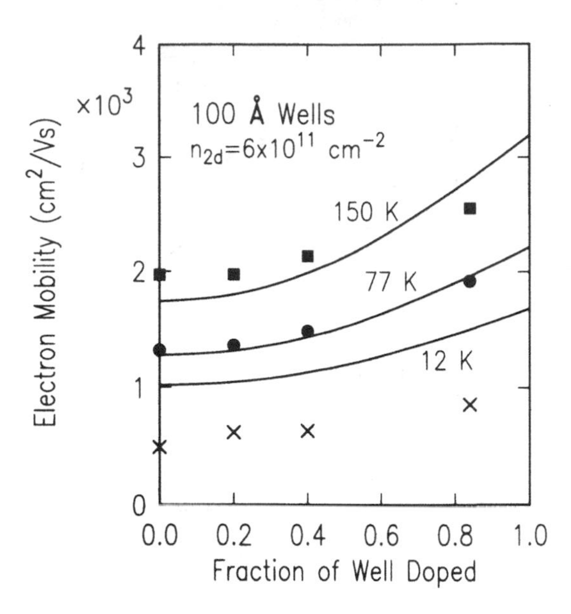

Fig. 16.8. Electron mobility versus the fraction of the quantum well which is doped for doped GaAs/AlGaAs quantum wells. The sheet doping concentration is the same in all samples; a fraction of '0.0' is a δ-function doping spike in the center of the well. The solid lines are calculations of Ref. [20].

in these samples, only the ground-state wavefunction is occupied. Thus, the electron population is strongly peaked in the center of the well. If the ionized impurities are also concentrated in the center of the well, then the scattering of electrons from ionized impurities will be enhanced and the mobility will decrease. The effect can be observed experimentally, as shown in Fig. 16.8, which shows mobility versus the fraction of the well which is doped. This effect may be contrasted with what one observes when higher subbands are occupied. For a related, but quite distinct, situation in which the semiconductor is doped in a single delta spike [34, 35], mobility enhancement is reported, which is in part caused by the odd-parity subband electron wavefunction having a vanishing overlap with the dopant spike. This same effect observed only for hot electrons is discussed in Section 16.4.

Mathematically, there is an equivalence between the two effects of a more concentrated doping profile, resulting in an increase in ionized-impurity scattering, and that of a more peaked electron distribution profile, resulting in an increase in ionized-impurity scattering. If we write the scattering rate as in Eqn. (16.1), then we can define an effective doping profile, $N_{\rm eff}(q)$, in terms of the actual doping profile, $N(z)$, and the form factor measuring the overlap of the electronic wavefunction and a given dopant ion, $F(q, z)$. The effective doping profile is given by

$$N_{\rm eff}(q) = \frac{1}{W}\int {\rm d}z|F(q, z)|^2 N(z), \qquad (16.9)$$

where the form factor, $F(q, z)$, is

$$F(q, z) = \int dz' |\zeta(z')|^2 \exp(-q|z - z'|). \tag{16.10}$$

The function $\zeta(z)$ is the z component of the total electronic envelope wavefunction $\Psi(\mathbf{r}, z) = \zeta(z) \exp(i\mathbf{k} \cdot \mathbf{r})$. The electron concentration is then simply given by

$$n(z') = n_{2d} |\zeta(z')|^2. \tag{16.11}$$

Thus, the effective dopant concentration profile may be written as

$$N_{\text{eff}}(q) = \frac{1}{W n_{2d}} \iint dz\, dz'\, N(z) n(z') \exp(-q|z - z'|). \tag{16.12}$$

What is significant about this form is that it is immediately clear that $N(z)$ and $n(z)$ (the actual dopant profile and the electron concentration profile) appear symmetrically. Thus the two-dimensionality of the electronic wavefunction decreases the mobility in the same way as delta-doping of the quantum well. Of course, when both are centered on the center of the well, this effect is compounded.

From Fig. 16.8 we see that the lowest mobility is observed in the wells with delta-doping in the center. This effect results from all the ionized impurities being localized where the electronic wavefunction is maximum (i.e. maximizing N_{eff}). By moving the delta-doping away from the center, we can increase the mobility. Figure 16.9 shows the mobility versus

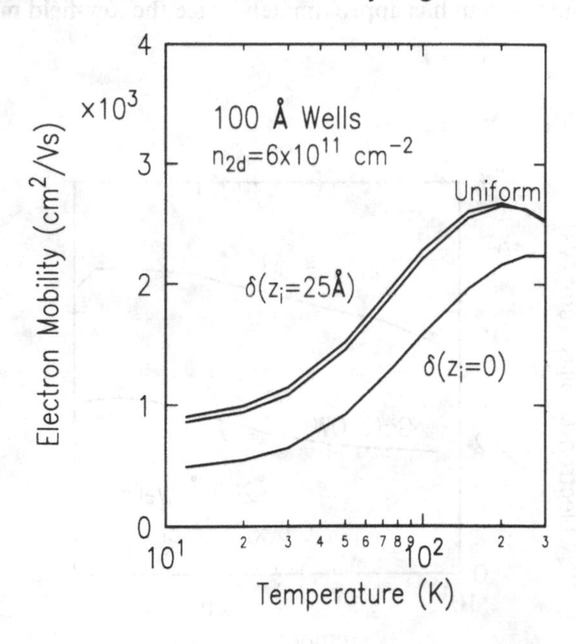

Fig.16.9. Electron mobility of doped 100-Å GaAs/AlGaAs quantum wells, comparing uniformly doped wells, δ-doped wells doped in the well centers ($z_i = 0$), and δ-doped wells doped half way between the bottom interface and the center ($z_i = 25$ Å).

temperature for three samples. The solid curves are experimental. The one labeled 'Uniform' is uniformly doped in the center 84% of the well. The lowest curve is for the center delta-doped wells. The third curve is from a sample which is also delta-doped, but with the doping midway between the center and the edge of the well. From Fig. 16.9, we see that the mobility is higher than when the dopants are all at the center of the well. If the dopants were moved farther from the center, the mobility would be even higher; by moving them entirely out of the wells, we are back to the modulation-doped example.

Thus, the electron mobility in doped quantum wells is lower than for electrons in similarly doped bulk GaAs, indicating an enhancement of the ionized-impurity scattering rate of Q2D carriers. Our calculations of electron mobility in doped quantum wells show that this enhanced scattering rate is characteristic of the confinement in one dimension and is due to a decrease in screening, an increase in the effective overlap of electronic wavefunction with the dopant atoms, and an increase in large-angle scattering in confined systems.

For transistor applications, the InGaAs/InAlAs lattice matched to InP has advantages over the GaAs/AlGaAs system because of higher electron mobility and larger Γ–L separation in InGaAs than in GaAs. As a result, MODFETs using InGaAs/InAlAs hold most FET performance records today [1]. Doped quantum wells in InGaAs/InAlAs also show higher mobilities than in GaAs/AlGaAs. Figure 16.10 compares these two material systems. The structures were, other than the materials used, identical. For all measured temperatures, the InGaAs system has approximately twice the low-field mobility than does the GaAs system.

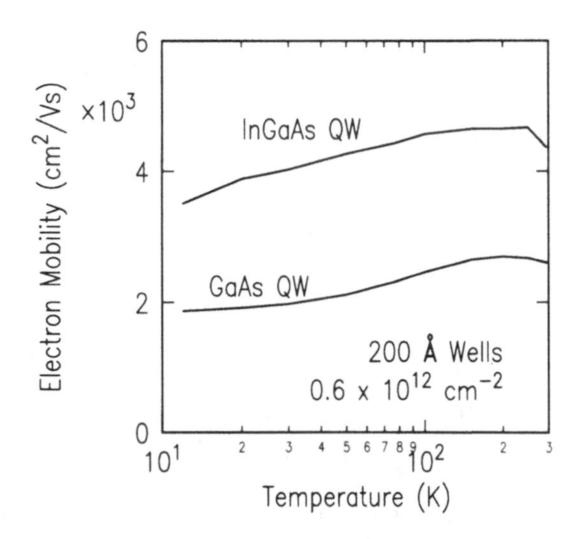

Fig. 16.10. Mobilities of electrons in 200 Å quantum wells, comparing GaAs and InGaAs.

16.4 Electron velocity in doped quantum wells

The transport properties of the electron gas were measured at high fields using a sinusoidally varying electric field with a frequency of 35 GHz [14] to avoid the formation of charge and field domains in the samples. This technique has been shown to be reliable in the measurement of electron velocity in bulk GaAs [13, 36] and has also been used to obtain the data given in Section 16.2. Figure 16.11 shows the measured electron velocity as a function of the electric field for three samples, all with a 3D doping density of 6×10^{17} cm^{-3}: the bulk sample of total thickness 0.25 μm, a sample with 100 Å quantum wells, and a sample with 50 Å quantum wells.

From Fig. 16.11 we see that even with the same average 3D electron density, the electron gas confined to narrower quantum wells has a lower peak velocity than in bulk GaAs. This result is qualitatively similar to the lower peak velocity found in the 2DEG in undoped GaAs when compared to that of 3D electrons in bulk undoped GaAs, as described in Section 16.2 and depicted in Fig. 16.2. Comparing Figs. 16.11 and 16.2, however, we see that while in Fig. 16.2, the mobility at low electric fields of the 2D case is essentially the same as in the 3D case, this is not true in Fig. 16.11. In doped bulk GaAs, as the donor concentration is increased, both the peak electron velocity *and* the low-field mobility decrease. The decrease in both these values is due to an increase in ionized-impurity scattering resulting from a higher ionized-impurity concentration. When comparing 2D and 3D samples, the low-field mobility in already lower in the 2D case because of the increased scattering with ionized impurities which was discussed in Section 16.3. A fairer comparison to make regarding the data in Fig. 16.11, then, is to compare the peak velocities

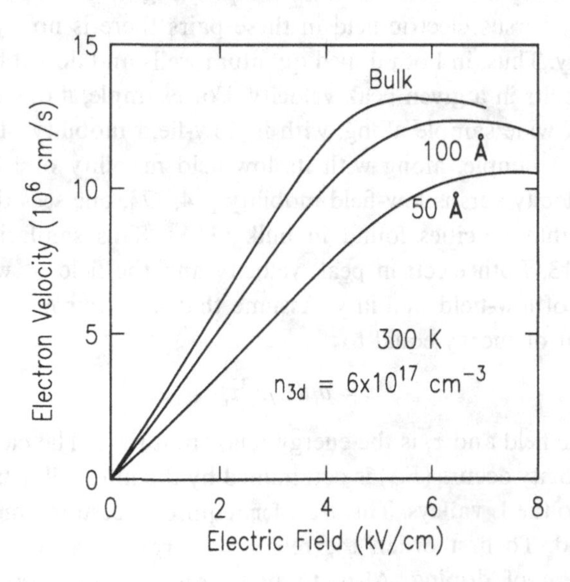

Fig. 16.11. Measured electron velocities as functions of electric field for bulk GaAs, 100 Å doped quantum wells, and 50-Å doped quantum wells, all with the identical volume doping density.

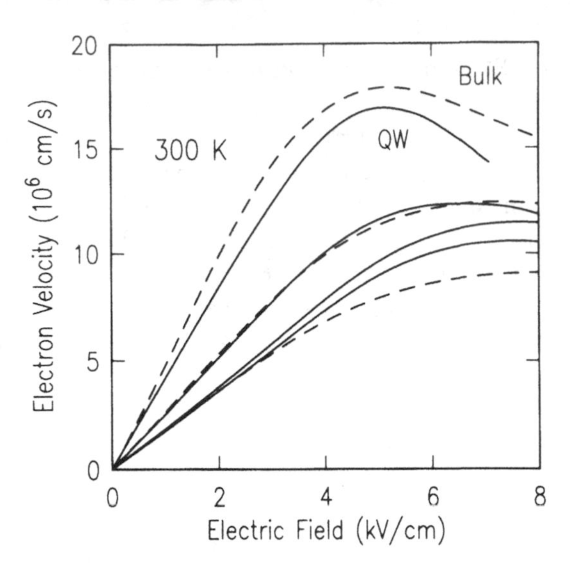

Fig. 16.12. A comparison of electron velocity in bulk GaAs and in doped GaAs quantum wells. In each of the three cases, the low-field mobilities are very similar; in order to have similar low-field mobilities, however, the volume doping density must be higher in the bulk GaAs samples.

as functions of low-field mobility. Consider several sample pairs, each with one doped quantum well sample and one bulk GaAs sample, and comparable low-field mobilities. (In order that each pair will have comparable mobilities, the doped quantum well sample must have a lower doping level than the bulk sample.) Figure 16.12 shows that when one measures the velocity versus electric field in these pairs there is no significant difference in higher field velocity. Thus, in both doped quantum wells and doped bulk GaAs, a given low-field mobility results in a given peak velocity. For example, if one compares the peak velocity of the 100-Å-wide sample along, with its low-field mobility of 2536 cm^2/Vs, and that of the 50-Å-wide sample, along with its low-field mobility of 1767 cm^2/Vs, with a summary of peak velocity versus low-field mobility [14, 37], one sees that these velocities are consistent with the velocities found in bulk GaAs. This similarity is shown more generally in Fig. 16.13. Both electron peak velocity and the field at which it occurs are plotted as functions of low-field mobility. Assume that an electron in a uniform electric field gains an amount of energy equal to

$$\varepsilon = q\mu(E)E^2\tau_\varepsilon \tag{16.13}$$

where E is the electric field and τ_ε is the energy relaxation time. The electric field at which the peak electron velocity occurs (E_{th}) is determined by the intervalley transfer of electrons from the Γ valley into the L valleys. This transfer requires a certain amount of energy and is thermally activated. To first order, the relevant energy is the Γ–L energy difference and is not a function of doping. Also, to first order we may take the mobility to be constant below E_{th}. This is the commonly used two-piece approximation. Finally, we will also assume here that the energy relaxation time is a constant. Assuming that

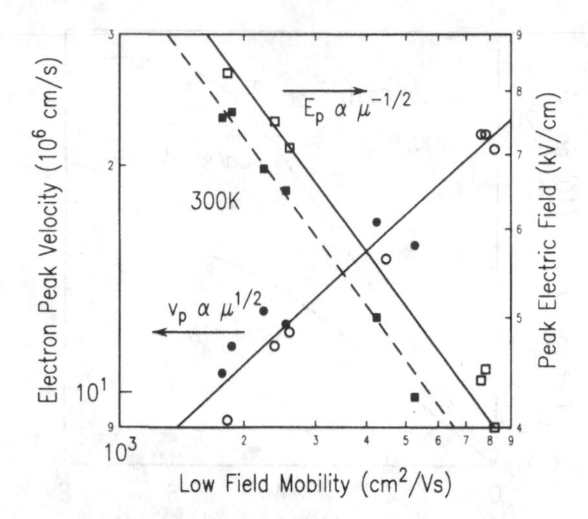

Fig. 16.13. Summary of comparisons of electron velocity in bulk GaAs and in doped GaAs quantum wells. The open circles are bulk GaAs and the solid circles doped quantum wells.

1. the Γ–L energy difference is a constant,
2. the mobility, $\mu(E)$, is about constant until the velocity is near the peak velocity,
3. and the energy relaxation time, τ_ε, is also a constant, Eqn. (16.13) implies that

$$\mu_0(E)E^2 = \text{constant}. \tag{16.14}$$

From this simple expression, one derives the fact that the peak velocity, v_p, and the threshold electric field, E_{th}, should depend on low-field mobility as

$$v_p \propto \mu_0^{1/2} \tag{16.15}$$

$$E_{th} \propto \mu_0^{-1/2} \tag{16.16}$$

The open circles shown in Fig. 16.13 are peak velocities in bulk GaAs and the open squares are peak electric fields. The solid straight lines are drawn to conform to Eqns. (16.15) and (16.16). The solid circles and squares are peak velocities and peak electric fields from doped 100 Å GaAs/AlGaAs quantum wells. As can also be seen in Fig. 16.12, the peak velocity does not differ significantly from that found in bulk GaAs *with the same low-field mobility*. (That the peak velocity occurs at a lower electric field will be explained later in this chapter.) We therefore conclude that what limits the high-field velocity in these structures is *not* enhanced transfer of electrons out of the Γ valley into the GaAs L valleys and (real space transfer) into the AlGaAs as occurs in modulation-doped heterostructures, but, rather, simply an enhancement of the ionized-impurity scattering.

Higher low-field mobility in InGaAs also helps electrons in InGaAs to achieve higher peak velocity. Figure 16.14 shows electron velocity as a function of electric field in similar InGaAs and GaAs quantum wells. Both have approximately the same sheet electron concentration, but in addition to the InGaAs having a low-field mobility almost twice that of GaAs (see Fig. 16.10), the peak velocity is also almost twice as great.

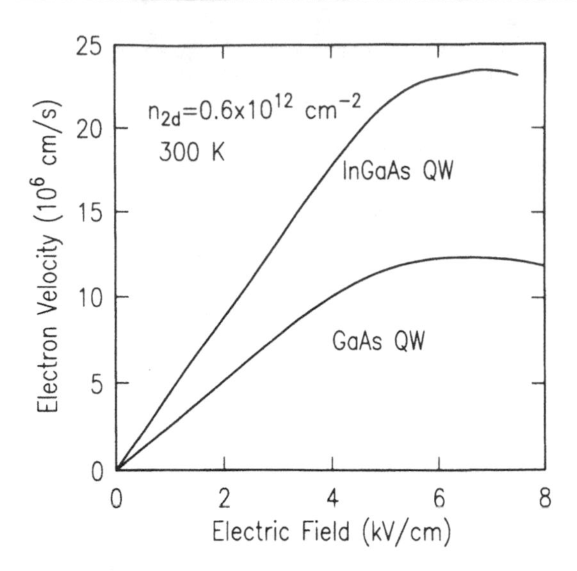

Fig. 16.14. Electron velocities as functions of electric field, comparing GaAs and InGaAs.

From the analysis and data presented in Section 16.3, it is clear that uniformly doped quantum wells have a lower mobility than does identically doped bulk GaAs. Figure 16.8 in Section 16.3 shows the effect of concentrating the dopants into the center of the well. Electron velocity at high electric fields was also measured for the delta-doped samples [38, 39]. Figure 16.15 depicts the measured electron velocity as a function of the electric field for the three samples. Although the on-center delta-doped quantum well sample had the lowest mobility at low electric fields, the peak velocity of electrons in this sample was higher than in either the uniformly doped wells or in the off-center delta-doped wells. This higher velocity was possible because the differential mobility increases at electric fields between 0 and ε_p, the electric field at which the peak velocity occurs.

Particularly noticeable about Fig. 16.15 is that the velocity of the lowest mobility sample – the on-center delta doped sample – is the highest. This is because of a superlinear increase of velocity with increasing electric field over some range of electric field. This effect is more clearly seen in Fig. 16.16, which depicts the derivatives of the velocity–field curves of Fig. 16.15. From Fig. 16.16 we see that the differential mobility of the electrons in the on-center δ-doped quantum wells increases much more than that of the electrons in uniformly doped quantum wells or in off-center delta-doped quantum wells. We expect that the ionized-impurity scattering rate of the on-center delta-doped wells would decrease relatively more than that of the uniformly doped wells simply because the ionized-impurity scattering rate is so much greater at low electric fields and, therefore, has much farther to fall. From Fig. 16.16, however, we see that the differential mobility of the electrons in the on-center delta-doped wells actually increases sufficiently to be significantly greater than that for the electrons in the uniformly doped wells. The large increase in differential mobility in the

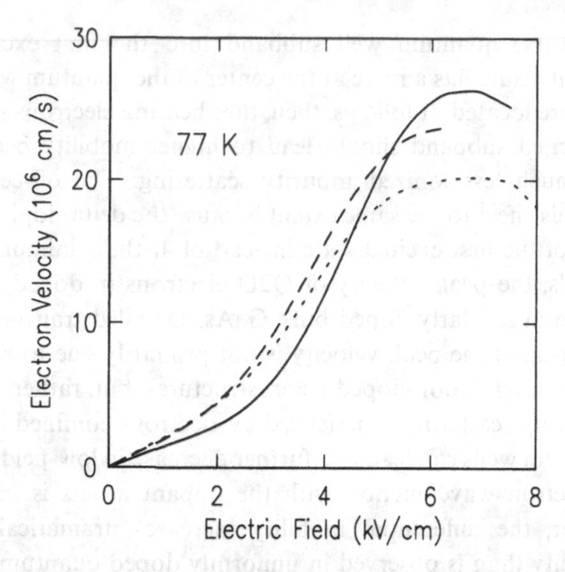

Fig. 16.15. Electron velocities as functions of electric field for uniformly doped quantum wells (dashed line), center delta-doped quantum wells (solid line), and off-center delta-doped quantum wells (dotted line). All well widths are 100 Å.

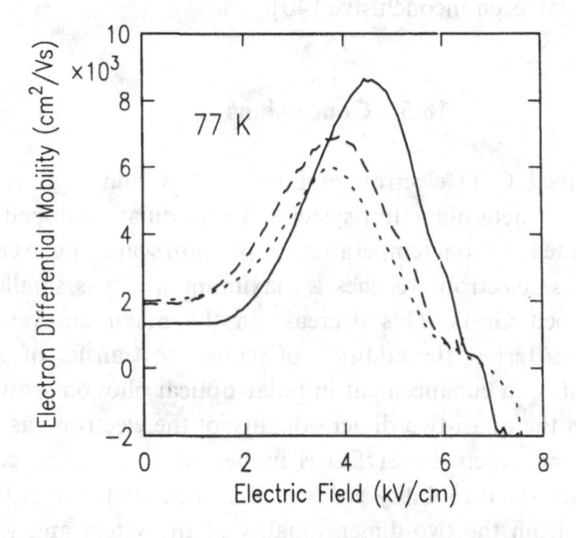

Fig. 16.16. Electron differential mobilities as functions of electric field for uniformly doped quantum wells (dashed line), center delta-doped quantum wells (solid line), and off-center delta-doped quantum wells (dotted line). All well widths are 100 Å.

on-center delta-doped sample may also be contrasted with the smaller increase in the off-center delta-doped sample's mobility.

These data may be explained as follows. The dramatic increase in the differential mobility of the on-center delta-doped samples is due to the heating of the electrons from the ground-state (even parity) quantum well subband into the first excited (odd-parity) subband. This odd-parity state has a node at the center of the quantum well, exactly where the ionized impurities are located. It follows, then, that heating electrons so that they begin to occupy the first excited subband should lead to higher mobility because the excited electrons experience much less ionized-impurity scattering. The off-center delta-doped sample will not show this effect to the same extent because the delta-doping spike is located at one of the maxima of the first excited state instead of at the minimum.

At high electric fields, the peak velocity of Q2D electrons in doped quantum wells is lower than we measure in similarly doped bulk GaAs. Detailed transport measurements indicate that this lowering of the peak velocity is not primarily due to real-space transfer as is the similar effect in modulation-doped heterostructures, but, rather, is also due to the enhanced ionized-impurity scattering experienced by electrons confined in one dimension. Delta-doping the quantum wells can lead to a further decrease in low-field mobility because the overlap of the electron wavefunction with the dopant atoms is increased. At high electric fields, however, the differential mobility increases dramatically, leading to a somewhat higher velocity than is observed in uniformly doped quantum wells, in spite of the lower low-field mobility. This large increase in differential mobility may be the result of heating the electrons out of the symmetric ground state into the anti-symmetric first excited state. Attempts to use this effect to enhance the performance of FETs with high channel doping have so far been inconclusive [40].

16.5 Conclusions

This chapter has described Q2D electron mobility at low and high electric fields for electrons in AlGaAs/GaAs heterojunction systems. In modulation-doped structures, the mobility is very high, limited at room temperature by phonon scattering. At elevated electric fields the velocity of these electrons reaches a maximum which is smaller than that of electrons in bulk undoped GaAs. This decrease in the maximum velocity is due to enhancements in Γ–L scattering, the addition of real-space transfer of electrons to the AlGaAs, and, at 77 K, also an enhancement in polar-optical phonon scattering. All these mechanisms follow from the quasi-two-dimensionality of the electron gas.

At the opposite extreme, when the 2DEG is immersed in an equal concentration of ionized impurities, the low-field mobility is lower than one finds for a 3D electron gas. This result also follows from the two-dimensionality of the sytem and will be observed generally in systems such as doped quantum wells. The higher field velocity of these structures is also lower than that of 3D electrons in the same doping concentration, again because of enhanced ionized-impurity scattering.

When a quantum well is delta-doped in the center of the well, the mobility is lower than when the same 2D electron concentration is uniformly doped throughout the well because the maximum of the electronic wavefunction lies in the same region as the dopant spike. At higher electric fields, however, these center delta-doped structures have higher electron velocity; this is probably due to electrons becoming heated out of the ground-state wavefunction into the first excited state and, therefore, becoming spatially separated from the dopant spike.

References

[1] L. D. Nguyen, L. E. Larson and U. K. Mishra, *Proceedings of the IEEE* **80**, 494–518 (1992).
[2] W. T. Masselink, *Thin Solid Films* **231**, 86–94 (1993).
[3] E. F. Schubert, A. Fischer and K. Ploog, *IEEE Trans. on Electron Devices* **33**, 625–32 (1986).
[4] M. Feng, C. L. Lau, V. Eu and C. Ito, *Appl. Phys. Lett.* **57**, 1233–5 (1990).
[5] A. B. Fowler, F. F. Fang, W. E. Howard and P. J. Stiles, *Phys. Rev. Lett.* **16**, 901–3 (1966).
[6] H. L. Störmer, R. Dingle, A. C. Gossard, W. Wiegmann and M. D. Sturge, *J. Vac. Sci. Technol.* **16**, 1517–19 (1979).
[7] L. Pfeiffer, K. W. West, H. L. Störmer and K. W. Baldwin, *Appl. Phys. Lett.* **55**, 1888–90 (1989).
[8] J. P. Eisenstein and H. L. Störmer, *Science* **248**, 1510–16 (1990).
[9] T. Mimura, S. Hiyamizu, T. Fujii and K. Nanbu, *Jpn J. Appl. Phys. Lett.* **19**, L225–7 (1980).
[10] M. V. Fischetti and S. E. Laux, *IEEE Trans. Electron Devices* **38**, 650–60 (1991).
[11] W. T. Masselink, W. Kopp, T. Henderson and H. Morkoç, *IEEE Electron Device Lett.* **EDL-6**, 539–41 (1985).
[12] W. T. Masselink, T. Henderson, J. Klem, W. F. Kopp and H. Morkoç, *IEEE Trans. Electron Devices* **33**, 639–45 (1986).
[13] W. T. Masselink, N. Braslau, W. I. Wang and S. L. Wright, *Appl. Phys. Lett.* **51**, 1533–5 (1987).
[14] W. T. Masselink, *Semicond. Sci. Technol.* **4**, 503–12 (1989).
[15] F. Stern and S. Das Sarma, *Phys. Rev.* B **30**, 840–8 (1984).
[16] H. Morkoç, S. G. Bandy, R. Sankaran, G. A. Antypas and R. L. Bell, *IEEE Trans. Electron Devices* **25**, 619–27 (1978).
[17] R. J. Haug, K. von Klitzing and K. Ploog, *Proceedings of 19th International Conference on the Physics of Semiconductors, Warsaw, 1988*, ed. W. Zawadzki (Inst. of Physics, Polish Academy of Sciences, Warsaw, 1988), pp. 307–10.
[18] S. Mori and T. Ando, *J. Phys. Soc. Jpn* **48**, 865–73 (1980).
[19] A. Gold, *J. Phys. (Paris) Colloque* **C5**, 255–8 (1987).
[20] W. T. Masselink, *Phys. Rev. Lett.* **66**, 1513–16 (1991).
[21] C. E. C. Wood, G. Metze, J. Berry and L. F. Eastman, *J. Appl. Phys.* **51**, 383–7 (1980).
[22] E. F. Schubert and K. Ploog, *Jpn J. Appl. Phys. Lett.* **24**, L608–10 (1985).
[23] E. F. Schubert, J. B. Stark, B. Ullrich and J. E. Cunningham, *Appl. Phys. Lett.* **52**, 1508–10 (1988).
[24] E. F. Schubert, J. B. Stark, T. H. Chiu and B. Tell, *Appl. Phys. Lett.* **53**, 293–5 (1988).
[25] A. Zrenner, F. Koch and K. Ploog, *Proceedings of Int. Symp. GaAs and Related Compounds, Heraklion, Greece, 1987*, ed. A. Christou and H. S. Rupprecht (Inst. of Physics Publishing Ltd, Bristol, 1988), pp. 171–4.
[26] A. Zrenner, F. Koch, R. L. Williams, R. A. Stradling, K. Ploog and G. Weimann, *Semicond. Sci. Technol.* **3**, 1203–9 (1988).
[27] J. Wagner, M. Ramsteiner, W. Stolz, M. Hauser and K. Ploog, *Appl. Phys. Lett.* **55**, 978–80 (1989).
[28] J. R. Meyer and F. J. Bartoli, *Phys. Rev.* B **36**, 5989–6000 (1987).
[29] H. Sakaki, T. Noda, H. Hirakawa, M. Tanaka and T. Matsusue, *Appl. Phys. Lett.* **51**, 1934–6 (1987).
[30] R. Gottinger, A. Gold, G. Abstreiter, G. Weimann and W. Schlapp, *Europhys. Lett.* **6**, 183–8 (1988).
[31] T. Ando, *J. Phys. Soc. Jpn* **51**, 3900–07 (1982).

[32] P. J. Price, *J. Vac. Sci. Technol.* **19**, 599–603 (1981).
[33] D. Chattopadhayay and H. J. Queisser, *Rev. Mod. Phys.* **53**, 745–68 (1981).
[34] E. F. Schubert, J. E. Cunningham and W. T. Tsang, *Solid State Commun.* **63**, 591–4 (1987).
[35] H.-J. Gossmann and F. C. Unterwald, *Phys. Rev.* **B 47**, 12618–24 (1993).
[36] M. V. Fischetti, *IEEE Trans. Electron Devices* **38**, 634–49 (1991).
[37] W. T. Masselink and T. F. Kuech, *J. Electronic Materials* **18**, 579–84 (1989).
[38] W. T. Masselink, *Appl. Phys. Lett.* **59**, 694–6 (1991).
[39] W. T. Masselink, *Proceedings of Int. Symp. GaAs and Related Compounds, Seattle, USA, 1991*, edited by G. B. Stringfellow (Inst. of Physics Publishing Ltd, Bristol, 1992), pp. 425–30.
[40] J. K. Zahurak, A. A. Iliadis, S. A. Rishton and W. T. Masselink *Proceedings: IEEE/Cornell Conference on Advanced Concepts in High Speed Semiconductor Devices and Circuits* (IEEE, New York, 1993), pp. 270–8.

17

Electron mobility in delta-doped layers

P. M. KOENRAAD

Introduction

In a perfect δ-doped layer the dopant atoms should be confined to a single atomic layer and the doping atoms should be distributed on a perfect regular lattice in the plane of the doping layer. If the dopant atoms are arranged on a perfect lattice in the doping plane, the scattering of the free carriers on the ionized dopant doping atoms will be absent. However, at the moment it is still impossible to have such complete control over the positioning of the dopant atoms. For instance, images of Be δ-doped planes obtained using cross-sectional scanning tunneling microscopy [1] show that, although the thickness of the doping plane is only a few atomic layers, the arrangement of the Be atoms in the doping plane is far from perfect, see Fig. 17.1. When the ionized scattering centring centres are not ordered on a perfect lattice, then ionized impurity scattering in δ-doping layers is by far the most important carrier scattering mechanism due to the strong overlap of the ionized impurities and the free carriers.

Up until recently the study of electrical transport in δ-doped structures has been limited to III–V materials, in particular to Si-δ-doped GaAs. Due to the low mobility of the electrons in Si-δ-doped GaAs structures, high magnetic fields of at least 20 T are needed for detailed study of electrical transport properties. The study of electron transport in the group IV and II–VI materials is more difficult due to an even lower electron mobility. The mobility in these materials is lower due to a larger electron effective mass. The study of electric transport in p-type δ-doped layers is very complex due to strong non-parabolicity of the hole dispersion relations, coupling of the light and heavy hole dispersion relations and the heavy effective masses. In this chapter electron transport in Si-δ-doped GaAs will mainly be discussed.

The study of electric transport in δ-doped samples can roughly be divided into two parts, that is, the study of δ-doped structures in the high and the low doping concentration regimes. Structures in the low doping concentration regime have a doping concentration below the critical Mott density, in which case the electron wavefunctions of the individual

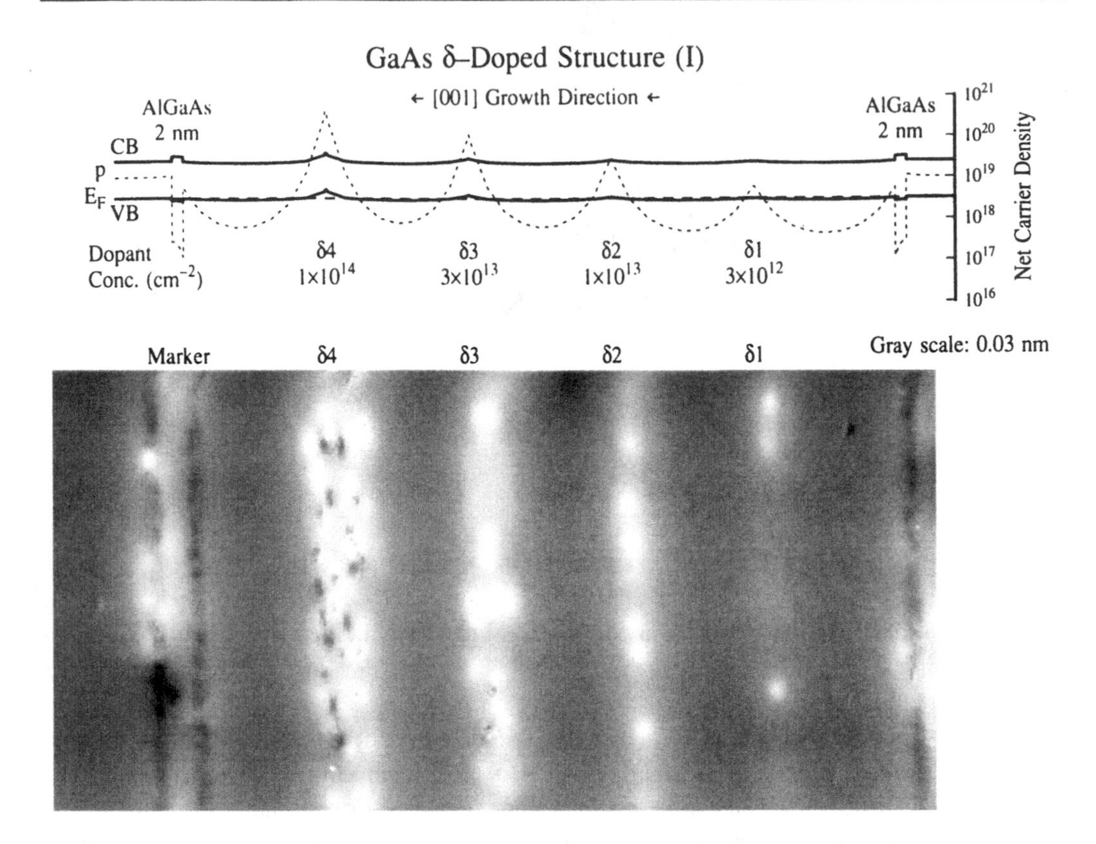

Fig. 17.1. STM image of a set of Be δ-doped layers in GaAs. The upper part gives a band diagram showing the position and intended concentration of the δ-doped layers equally spaced between AlGaAs marker layers. The total lateral extent between the marker layers is about 1200 Å. The lower part shows a large-scale (1500 Å*500 Å), As-related image of a set of δ-doped layers aligned with the upper part. Tunneling conditions: sample bias -1.0 V and tunneling current 20 pA. Electrically active Be-dopants appear as white hillocks about 25 Å in diameter.

Si-donors do not have an important overlap with one another. No impurity band is formed and the conduction at low temperatures takes place by electrons that hop from one donor site to another [2]. This is the regime where it is not possible to split off a homogeneous part in the distribution of ionized impurities. I will only briefly discuss the results obtained on δ-doped structures in this low concentration regime.

In GaAs the critical n-type doping concentration is about 3×10^{11} cm^{-2}. Above this doping concentration, the donors no longer act as isolated trapping centres and an impurity band is formed. The electrons are free to move in the plane of the doping atoms and at low temperatures they do not freeze-out on the donors. In the direction perpendicular to the doping layer, the electrons are confined in the Coulomb potential of the sheet of positively charged Si-donors. The dimension of the V-shaped Coulomb potential is smaller

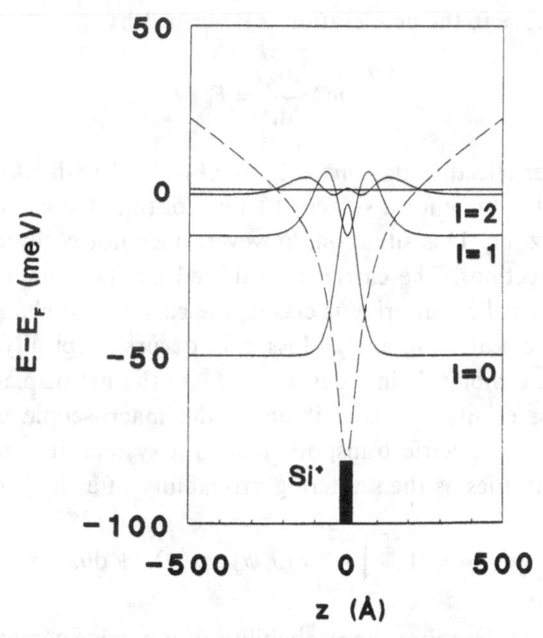

Fig. 17.2. Calculated potential energy distribution and probability distribution of the electrons for a GaAs structure δ-doped at 2×10^{12} cm^{-2}. E_F is the Fermi energy. The black box indicates the position of the ionized donors, the dashed line the electrostatic potential and the solid lines the probability distribution, $|\psi_i(z)|^2$, and energy position of each subband. The width of the doping profile is 20 Å.

than the de Broglie wavelength of the electrons and thus quantum states (subbands) are formed in the well. Due to the high electron density in δ-doped structures, a two-dimensional electron gas (2DEG) is formed which normally has more than one populated subband.

The wavefunctions and energy positions of the subbands can be found by solving the coupled Poisson and Schrödinger equations describing the δ-doped structure. A typical result for a GaAs structure with an n-type doping concentration of 2×10^{12} cm^{-2} and a 20-Å-thick doping layer is shown in Fig. 17.2. Only states below the Fermi energy are occupied. Three subbands are populated in this example. In δ-doped structures one expects a fairly low mobility because there is a strong overlap of the electron wavefunctions and the ionized scattering centres in the δ-doping layer.

Electrical transport theory

Scattering times

In this chapter we will consider the electrical transport in δ-doped semiconductors under the influence of electric and magnetic fields. If we apply a force on a carrier in a semiconductor with $k_x \approx 0$, the acceleration is described by

$$m^* \cdot \frac{\mathrm{d}v_x}{\mathrm{d}t} = F_x \tag{17.1}$$

where m^* is the carrier effective mass $m^* = 2\hbar^2(\partial\varepsilon(k)/\partial k)^{-1}$ with $\varepsilon(k)$ being the dispersion relation. If we apply the force for a sufficient length of time the carrier will reach the end of the first Brillouin zone. This situation, however, does not easily occur in real crystals due to crystal imperfections. The carrier is scattered by these imperfections, which also include phonons. When the scattering is elastic, the carrier will change its direction only; in the case of inelastic scattering energy loss/gain occurs. Typically, a carrier will move through a crystal like a molecule in a dense gas. Only the net displacement of the carrier after many scattering events will contribute to the macroscopic charge transport. To describe the macroscopic electric transport in such a system, it is usual to make use of such microscopic quantities as the scattering probability of a single carrier:

$$\sigma = \int_0^{2\pi} \int_{-\pi/2}^{\pi/2} \sigma(\theta, \psi) \cos(\psi)\, \mathrm{d}\psi\, \mathrm{d}\theta. \tag{17.2}$$

In this equation $\sigma(\theta, \psi)$ describes the probability of a carrier coming from the direction $\theta = 0$ to scatter elastically in the direction (θ, ψ). A second microscopic quantity connected to the scattering probability is τ, the averaged time between two scattering events. We have to discriminate between various scattering times. The first is the averaged elastic scattering time, which is normally called the quantum lifetime or single-particle relaxation time. In a two-dimensional structure the quantum lifetime is calculated from the scattering probability:

$$\frac{1}{\tau_\mathrm{q}} = \int_0^{2\pi} \sigma(\theta) 2\pi\, \mathrm{d}\theta. \tag{17.3}$$

When we study the macroscopic charge transport, only scattering events away from $\theta = 0$ are important. Each scattering event is weighted according to its projection of the outgoing wavevector in the $\theta = 0$ direction. This so-called transport scattering time is defined by

$$\frac{1}{\tau_\mathrm{t}} = \int_0^{2\pi} \sigma(\theta)\{1 - \cos(\theta)\} 2\pi\, \mathrm{d}\theta. \tag{17.4}$$

Because of weighting by $1 - \cos(\theta)$ of the scattering probability $\sigma(\theta)$, the transport and quantum scattering times can differ. For isotropic scattering, such as phonon or alloy scattering, the scattering times are equal. However, for ionized impurity scattering the transport scattering time will be longer than the quantum scattering time while the scattering distribution is strongly peaked in the forward direction. In GaAs/AlGaAs heterostructures one normally finds values of 5–20 for the ratio of the transport time to the quantum scattering time [3].

In δ-doped structures we expect ionized impurity scattering of the electrons on the ionized parent donors to be the main scattering mechanism because of the very strong overlap of the electron wavefunctions with the ionized scattering centres. At low temperatures the ionized impurity scattering mechanism is much stronger than acoustic and optical phonon scattering. Other scattering mechanisms, such as alloy scattering, only occur in ternary III/V and II/VI materials, and interface scattering is certainly not present in normal δ-doped structures.

Finally, we define the inelastic scattering time τ_{in}, which is normally much longer than the elastic scattering time. Thus an electron scatters elastically many times before it encounters an inelastic scattering event. During the elastic scattering sequence the quantum mechanical phase of the carrier is conserved. There is always a non-zero probability that the carrier will scatter back to its original position. This gives rise to weak localization in the absence of a magnetic field [4]. The most important inelastic scattering processes are direct electron–electron scattering and Nyquist scattering. Nyquist scattering is the scattering of a single electron on the fluctuating potential of all other moving electrons in the 2DEG.

Mobility analysis

The mobility of a carrier is defined by

$$\mu = (e/m^*)\cdot\langle\tau\rangle \tag{17.5}$$

where

$$\langle\tau\rangle = \frac{\int_0^\infty \tau(\varepsilon)\cdot\varepsilon\cdot(-\partial f(\varepsilon-\varepsilon_F)/\partial\varepsilon)\,d\varepsilon}{\int_0^\infty \varepsilon\cdot(-\partial f(\varepsilon-\varepsilon_F)/\partial\varepsilon)\,d\varepsilon} \tag{17.6}$$

with $f(\varepsilon-\varepsilon_F)$ being the Fermi–Dirac distribution function, where ε_F is the Fermi energy. At low temperatures in highly degenerate systems $\langle\tau\rangle = \tau(\varepsilon_F)$. The total conductivity of a 2DEG in a magnetic field is

$$\sigma_{xx} = \frac{en\mu_t}{(1+\mu_t^2B^2)} = \sigma_{yy} \quad \text{and} \quad \sigma_{xy} = \frac{en\mu_t\cdot\mu_t B}{(1+\mu_t^2B^2)} = -\sigma_{yx} \tag{17.7}$$

where n is the 2D carrier density, B is the magnetic field and μ_t is the transport mobility. The resistivity tensor elements are obtained from the inverse of the conductivity tensor:

$$\rho_{xx} = \frac{1}{en\mu_t} = \rho_{yy} \quad \text{and} \quad \rho_{xy} = -\frac{B}{en} = -\rho_{yx}. \tag{17.8}$$

The preceding equations apply for the case where there is only one type of carrier. In the case where the conductivity is determined by more than one type of carrier, the total conductivity is the sum of the individual conductivity contributions of each type of carrier. This is the situation in δ-doped structures where more than one subband is occupied. Thus, the total conductivity is the sum of the conductivities of the individual subbands, $\sigma_{xx} = \sum_i \sigma_{xx_i}$ and $\sigma_{xy} = \sum_i \sigma_{xy_i}$. The resistivity tensor elements in the case of multi-subband transport are

$$\rho_{xx}(B) = \frac{\sum_i \sigma_{xx_i}(B)}{\left(\left(\sum_i \sigma_{xx_i}(B)\right)^2 + \left(\sum_i \sigma_{xy_i}(B)\right)^2\right)}$$
$$\rho_{xy}(B) = -\frac{\sum_i \sigma_{xy_i}(B)}{\left(\left(\sum_i \sigma_{xx_i}(B)\right)^2 + \left(\sum_i \sigma_{xy_i}(B)\right)^2\right)}. \quad (17.9)$$

By analysing the magnetic field dependence of either $\rho_{xx}(B)$ and $\rho_{xy}(B)$ or $\sigma_{xx}(B)$ and $\sigma_{xy}(B)$, one is able to determine the carrier density and carrier transport mobility in each subband.

In simple characterization measurements, such as van der Pauw measurements [5], one assumes that the conductivity is described by a single type of carrier. In this case the resistivity is written as

$$\rho_{xx} = \frac{1}{en_{\mathrm{H}}\mu_{\mathrm{H}}} \quad \text{and} \quad \rho_{xy} = \frac{B}{en_{\mathrm{H}}}. \quad (17.10)$$

Thus in structures with multiple occupied subbands, as in δ-doped structures, this leads to a magnetic-field-dependent Hall mobility and Hall density. In the case where only two subbands are populated one finds, in the limit of a disappearing magnetic field, that the expressions for the Hall density and Hall mobility can be reduced to

$$n_{\mathrm{H}} = \frac{(n_0\mu_0 + n_1\mu_1)^2}{(n_0\mu_0^2 + n_1\mu_1^2)} \quad \text{and} \quad \mu_{\mathrm{H}} = \frac{(n_0\mu_0 + n_1\mu_1)}{(n_0 + n_1)}. \quad (17.11)$$

In these equations n_0 and n_1 are the carrier densities and μ_0 and μ_1 are the transport mobilities in the individual subbands.

Up until now we have described the electrical transport by the classical Drude expressions, see Eqns. (17.7). However, in high and low magnetic fields quantum effects occur. In the low magnetic field region, the destruction of the weak localization by the magnetic field will give rise to negative magneto-resistance. In high magnetic fields the carrier motion is quantized into cyclotron orbits. The number of cyclotron orbits that a carrier can perform is $\tau_{\mathrm{q}} \cdot \omega_{\mathrm{c}}$, where $\omega_{\mathrm{c}} = eB/m^*$ is the cyclotron frequency. Schubnikov–de Haas oscillations observed in high magnetic fields, $\tau_{\mathrm{q}} \cdot \omega_{\mathrm{c}} = \mu_{\mathrm{q}} B > 1$, are due to Landau quantization of the carriers.

The correction to the magneto-resistivity in high magnetic fields [6] is

$$\Delta\rho_{xx}(B) = 4\rho_{xx}(B=0) \sum_{s=1}^{\infty} D(sX) \cdot \exp\left(-\frac{\pi s}{\mu_{\mathrm{q}} B}\right) \cdot \cos\left(\frac{2\pi^2 \hbar s n}{eB} - \pi s\right) \quad (17.12)$$

where

$$D(sX) = sX/\sinh(sX)$$

with $X = 2\pi m^* k_B T/\hbar eB$. The first harmonic, $s = 1$, is the strongest and Schubnikov–de Haas oscillations are normally analysed by only accounting for this first harmonic. The amplitude dependence of the Schubnikov–de Haas oscillations on the magnetic field can be used to determine the quantum mobility.

According to Wittmann and Schmid [7] the magnetic field dependence of the weak localization correction to the Drude expressions is

$$\Delta\sigma_{xx}(B) = -\frac{e^2}{2\pi^2\hbar}\cdot\frac{b}{(1+\gamma)^2}\cdot\sum_n^{\infty}\frac{\Psi_n^3(b)}{1+\gamma-\Psi_n(b)} \tag{17.13}$$

with

$$\Psi_n(b) = \int_0^{\infty} \exp(-(\xi + b\xi^2/4))L_n(b\xi^2/2)\,\mathrm{d}\xi$$

and

$$L_n(y) = \sum_m^n (1/m!)\binom{n}{m}(-y)^m.$$

In these equations $\gamma = \tau_0/\tau_\Phi$ is the ratio of the elastic transport scattering time to the inelastic scattering time, $b = 2el_0^2B/(1+\gamma^2)\hbar$ and $l_0 = \sqrt{(2E_F/m^*)}\cdot\tau_0$ is the mean free path. The negative magneto-resistance due to destruction of the weak localization is fully determined by the Fermi energy, the elastic scattering time and the inelastic scattering time. The elastic transport time and the Fermi energy can be determined from other experiments. Thus the inelastic scattering time can be determined by analysing the negative magneto-resistance. In low magnetic fields Eqn. (17.13) reduces to the well-known digamma function [8, 9].

Experimental results

Hall mobility

Several authors have reported Hall-mobility measurements on Si-δ-doped GaAs samples as a function of temperature [10, 11, 12, 13] or as a function of electron density [12, 13, 14, 15]. Values for the Hall electron density in Si-δ-doped samples range from 1×10^{11} cm^{-2} to 2×10^{13} cm^{-2}. The lower limit is determined by the Mott-criterion and the higher limit is determined by self-compensation of the dopant material [12, 16, 17].

Hall mobilities have also been measured in other materials: Si-δ-doped AlGaAs [18], Si-δ-doped InGaAs [19], S-δ-doped InP [20], Si-δ-doped InP [21], Si-δ-doped InSb [22] and B- or Sb-δ-doped Si [23, 24].

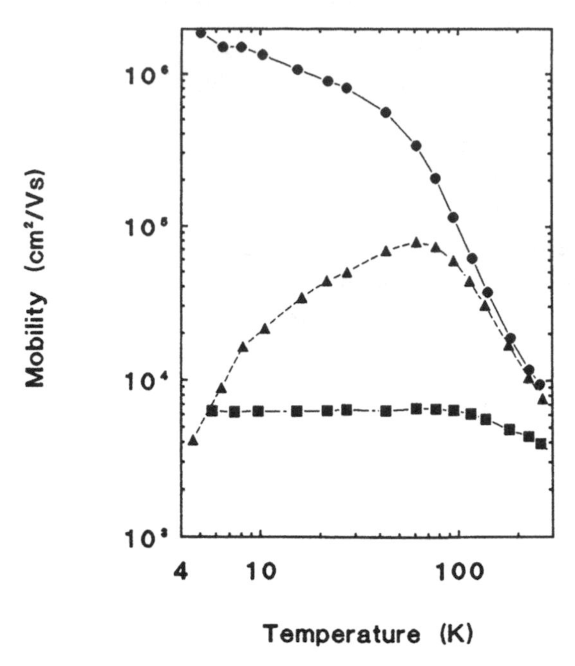

Fig. 17.3. Hall mobility in a bulk doped GaAs sample (▲), a GaAs/AlGaAs heterostructure (●) and a δ-doped GaAs structure (■).

Temperature dependence

In Fig. 17.3 we show the Hall mobility in a bulk doped GaAs sample, a GaAs/AlGaAs heterostructure and a δ-doped GaAs structure. In the heterostructure very high mobilities are obtained due to the spatial separation of the electrons and their ionized parent donors. The mobility in the heterostructure decreases with temperature because of acoustic and optical phonon scattering. At low temperatures the mobility in the bulk doped GaAs samples increases with temperature. There are two reasons for this. Firstly, in a non-degenerate 3DEG screening increases with temperature. Secondly, in a non-degenerate electron gas we can use the Boltzmann distribution function, see Eqn. (17.6), instead of the Fermi–Dirac distribution function. Thus the averaged energy of the carriers increases rapidly with temperature, leading to a reduced ionized impurity scattering rate. At high temperatures the mobility decreases because of phonon scattering. In the δ-doped structure, the averaged mobility is very low and almost independent of temperature. The low mobility is of course due to the overlap of the electron wavefunctions with the ionized donors. The slight temperature dependence is due to the fact that the electron gas in δ-doped structures is highly degenerate. Because of the high degeneracy the screening is temperature-independent. Also, Eqn. (17.6) depends solely on temperature when k_BT is comparable to E_F. At a typical electron density in a Si-δ-doped structure of 2×10^{12} cm^{-2}, the Fermi energy is 65 meV while k_BT is 8.63 meV at 100 K. Schubert *et al.* [14] also reported that the temperature dependence of the Hall mobility was small.

The temperature dependence of the Hall mobility has been studied in more detail by Gillman *et al.* [12]. In samples with low doping concentration, $n_H = 1 \times 10^{12}$ cm^{-2}, they find temperature-dependent behaviour of the Hall mobility as in bulk doped non-degenerate samples. At higher doping concentrations they also observe only minimal temperature dependence.

Makimoto *et al.* [10] have analysed the temperature dependence of the Hall electron density and Hall mobility by assuming that electric conduction in the δ-layer is determined by two populated subbands. The electrons in the lowest, most heavily populated, subband have lower mobility compared with the electrons in the highest subband. Furthermore, Makimoto *et al.* assume a temperature-independent ratio of the mobility in these two subbands, $\mu_0 = b\mu_1$, and a temperature-activated population in the second subband. They obtain a reasonable fit to the measurements with a typical activation energy of 10 meV and b close to 3. The problem with this analysis is that at their doping concentration of 1.3×10^{13} cm^{-2}, it is certain that more than two subbands are populated. Most probably the activated behaviour occurs in higher subbands. The electrons gain a factor of three in mobility when they are transferred from a lower to a higher subband which is roughly 10 meV away.

Electron density dependence

There are several ways of studying the dependence of the Hall mobility on the electron density, which can be varied either by changing the doping concentration, by applying a gate on the structure or by illumination of the sample.

The density dependence of the Hall mobility at 300 K, as observed by Schubert *et al.* [14], showed a reduction of the mobility by a factor of 3 when the doping concentration was increased from 10^{11} cm^{-2} to 10^{13} cm^{-2}. At low doping concentrations the mobility agreed well with the mobility measured in bulk doped samples, when a 3D concentration was calculated from the equation $n_{3D} = (n_{2D})^{3/2}$. However, in the high doping concentration regime the measured Hall mobility was considerable higher than the 3D Hall mobility measured by Hilsum [25] in bulk doped samples. Schubert *et al.* attributed this to the high degeneracy of the electron gas, the increased screening and the spatial separation of the electrons in subbands $i = 1, 3, 5$, etc., from the ionized scattering centres in the doping layer.

Gillman *et al.* [12] also observed a strong decrease in the Hall mobility at 300 K with increasing doping concentration. At 4.2 K they again observed a reduction in the Hall mobility with the doping concentration. However, at their lowest doping concentration, $n_H = 1 \times 10^{12}$ cm^{-2}, the mobility was reduced compared with samples with a higher doping concentration.

Our results [13] showed that the decrease in the Hall mobility at 4.2 K with doping concentration depends on the growth temperature. We observed that the Hall mobility in samples grown at low temperature, $T_{growth} = 480$ °C, decreases with the doping concentration. In samples grown at a higher growth temperature, $T_{growth} = 640$ °C, the Hall mobility is almost independent of the doping concentration. The thickness of the doping layer increases with the growth temperature and, thus, the overlap of the electrons with scatterers increases in samples grown at higher temperatures. It seems reasonable to assume that the mobility in the higher subbands decreases with this increased overlap. This leads to a different doping concentration dependence of the Hall mobility.

Kotelnikov *et al.* [15] studied the electron density dependence of the Hall mobility over a much smaller electron density range by using gated δ-doped structures. A problem with gating is that the δ-doping layer must be fairly close to the gate. Typically, a depth of 50 nm is needed in order to obtain a capacitance that is large enough to ascertain a large change in the electron density with gate voltage. This requirement is incompatible with the minimum δ-layer depth of about 5000 nm needed to ensure an undisturbed, symmetrical potential. Although the electron density range was quite small they clearly observed an increasing Hall mobility with increasing electron density. The increase in the Hall mobility was explained by the filling of the second subband, which has a higher mobility than the lowest subband.

After illumination it has also been observed that the Hall mobility increases [13, 18, 21]. In the case of illumination the electron density increases persistently due to photo-excited electrons that originate from the depletion regions, whereas the concentration of ionized donors remains constant.

Other materials

Lavielle *et al.* [18] report a rather low Hall mobility of about 400 cm^2/Vs in two δ-doped AlGaAs samples doped at 1×10^{12} cm^{-2}. In Si-δ-doped InP a Hall mobility of 1346 cm^2/Vs is found in a structure with a doping concentration of 2.15×10^{12} cm^{-2} [21]. For Si-δ-doped InGaAs a Hall mobility of 5000 cm^2/Vs is reported [19] in a structure doped at 6.25×10^{12} cm^{-2}. In a sample with a ten times smaller doping concentration the Hall mobility increased to only 5800 cm^2/Vs. In InSb a Hall mobility of 100 000 cm^2/Vs has been reported [22]. This very high value of the Hall mobility is mainly due to the low effective mass of $0.0136 \times m_e$ in InSb. In Sb-δ-doped Si, a Hall mobility as low as 65 cm^2/Vs was reported [23]. Gossmann and Unterwald [24] have shown that in Sb-δ-doped Si the Hall mobility is considerably higher than in bulk Sb-doped Si.

Conclusion

In conclusion, by increasing the doping concentration the Hall mobility decreases, but by increasing the electron density in the 2DEG and leaving the number of ionized donors constant, the Hall mobility increases. Thus by adding a donor to the system the effect of the extra ionized scattering centre is more important than the improved screening and higher Fermi velocity from the extra electron.

Transport mobility

The magnetic field dependence of ρ_{xx} and ρ_{xy} can be used to determine the transport mobilities and the electron density in the individual subbands. Zrenner [26] was the first to analyse the $\rho_{xx}(B)$ and $\rho_{xy}(B)$ curves in order to determine the transport mobilities in the individual subbands using Eqn. (17.9). The electron densities in the individual subbands were determined from Schubnikov–de Haas measurements. Zrenner obtained the following results in a structure doped at 3.8×10^{12} cm^{-2}: $\mu_0 = 1370$ cm^2/Vs, $\mu_1 = 2220$ cm^2/Vs and $\mu_2 = 6000$ cm^2/Vs. These results show that the mobility in higher subbands is larger than that in the lower subbands. A problem with this analysis is that only a few subbands may be populated. It is difficult enough to obtain accurate parameters when one has to fit to either $\rho_{xx}(B)$ and $\rho_{xy}(B)$ or $\sigma_{xx}(B)$ and $\sigma_{xy}(B)$ with three subbands.

Skuras *et al.* [27, 28] used a two-subband model to analyse measured ρ_{xx} and ρ_{xy} curves. In three different samples they found a transport mobility of about 1200 cm^2/Vs for the lowest subband and a transport mobility of 3000 cm^2/Vs in the second subband. The doping concentration in all samples was close to 9×10^{12} cm^2/Vs, but the width varied between 20 Å and 100 Å. In the samples with a narrow doping profile, the subband population obtained from this analysis was in agreement with the subband population obtained from the Schubnikov–de Haas oscillations. In the sample with a 100-Å-thick doping layer, there was no agreement with the Schubnikov–de Haas results. This discrepancy was probably caused by the fact that in thick doping layers at least five subbands are populated at these doping concentrations and thus a two-subband analysis is not correct.

Gusev *et al.* [29] found a transport mobility of 2200 cm^2/Vs in the lowest subband and a transport mobility of about 5300 cm^2/Vs in the two higher subbands in a structure doped at 4×10^{12} cm^{-2}.

We have also determined the transport mobility in δ-doped samples [30], using a method developed by Beck and Anderson [31], by which a measurement of either $\sigma_{xx}(B)$ and $\sigma_{xy}(B)$ or $\rho_{xx}(B)$ and $\rho_{xy}(B)$ can be transformed into a so-called 'mobility spectrum'. In the magnetic field range, where the quantum corrections to the Drude expressions are negligible, the magneto-conductivity can be expressed in a very general way by

$$\sigma_{xx} = \int_{-\infty}^{\infty} \frac{en(\mu_t)\mu_t}{(1 + (\mu_t B)^2)} d\mu_t$$
$$\sigma_{xy} = \int_{-\infty}^{\infty} \frac{en(\mu_t)\mu_t^2 B}{(1 + (\mu_t B)^2)} d\mu_t \quad (17.14)$$

In these equations one can adopt the convention that electrons have negative mobility and holes have positive mobility. Based on these equations, Beck and Anderson developed a technique for obtaining a so-called mobility spectrum, $n^*(\mu_t)$. This mobility spectrum gives the maximum number of carriers present in the conduction layer with mobility μ_t.

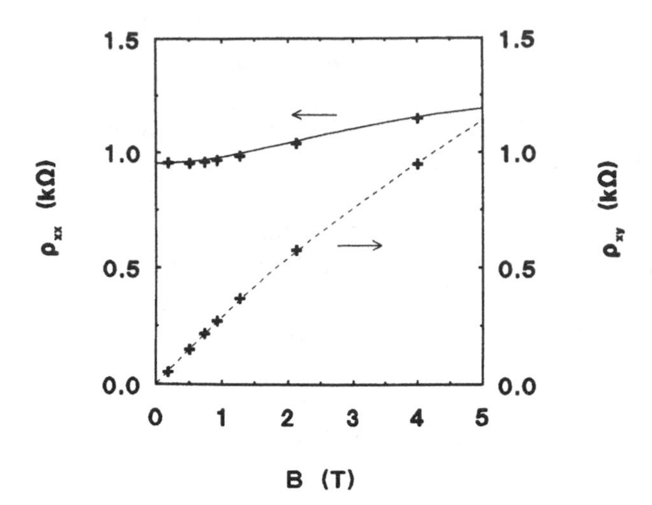

Fig. 17.4. $\rho_{xx}(B)$ and $\rho_{xy}(B)$ measured (+) in a sample doped at 3.5×10^{12} cm^{-2} and grown at 480 °C. The calculated $\rho_{xx}(B)$ and $\rho_{xy}(B)$ curves using a simple two-subband model and the values from the mobility spectrum analysis are indicated by the solid and dashed lines respectively.

Thus, the real distribution function $n(\mu_t)$ is always smaller than or equal to the mobility spectrum $n^*(\mu_t)$. In order to obtain the mobility spectrum the values of $\sigma_{xx}(B)$ and $\sigma_{xy}(B)$ at up to ten magnetic field positions must be known. We used this technique to determine the transport mobility in each subband.

In Fig. 17.4 the crosses indicate the $\rho_{xx}(B)$ and $\rho_{xy}(B)$ data measured on a sample doped at 3.5×10^{12} cm^{-2} and grown at 480 °C. These values were used in the mobility spectrum analysis. The mobility spectrum, shown in Fig. 17.5, clearly shows two peaks corresponding to two populated subbands. The structure observed below 300 cm^2/Vs is a spurious effect of the analysis technique. The electron density and transport mobility in an individual subband are obtained from the peak height and peak position respectively. The curve in Fig. 17.4 is calculated with a simple two-subband model, as in Eqn. (17.9), using the peak positions and peak heights in the mobility spectrum. The agreement with the original measurement proves that our mobility analysis method works well. A second proof of the correctness of the mobility spectrum analysis is the fact that the subband population obtained from the mobility spectrum analysis agrees very well with the subband population obtained from Schubnikov–de Haas oscillations (see Table 17.1). In this table we also give the calculated subband population for a δ-doped structure with a 20-Å-thick doping layer. The table shows that the transport mobility in the $i = 0$ subband is smaller by almost a factor of 4 than the transport mobility in the $i = 1$ subband. This difference is due to the stronger overlap of the electrons in the lowest subband with the ionized scatterers. In Fig. 17.6 we show the transport mobility as a function of temperature. At the moment we have no explanation for the slight increase in transport mobility with temperature in the $i = 1$ subband.

Table 17.1. *Electron density and transport mobility in the individual subbands before and after illumination, as determined from the mobility spectra (mobspec) and the Schubnikov–de Haas measurements (FFT). The sample was doped at* 3.5×10^{12} cm^{-2}. *In the calculations, we used the same doping concentration and a 20-Å-thick doping layer. After illumination, we assumed a total neutralization of all depletion charges in the calculations*

	Electron density (10^{12} cm^{-2})			
i	mobspec	FFT	calc	μ_t (cm^2/Vs)
		Dark		
0	2.5	2.53	2.57	1250
1	0.7	0.66	0.62	4800
tot	3.2	3.19	3.19	
		Illuminated		
0	2.4	2.47	2.45	1600
1	0.9	0.77	0.92	7800
2		0.25	0.12	–
tot	3.3	3.49	3.49	

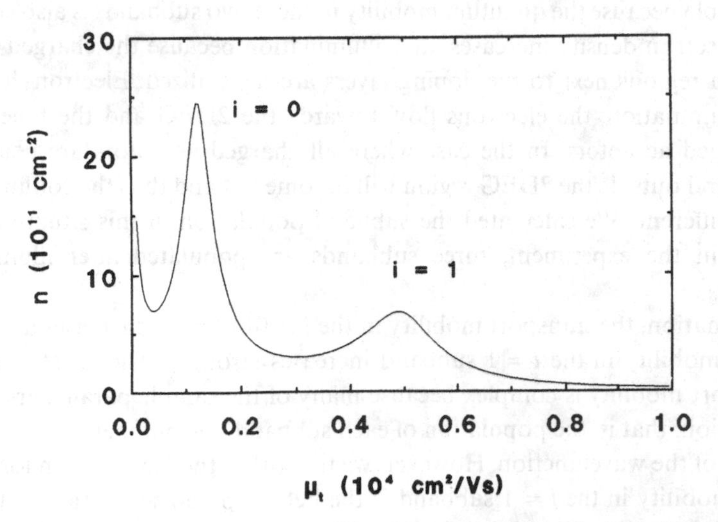

Fig. 17.5. The mobility spectrum obtained from ρ_{xx} and ρ_{xy} shown in Fig. 17.4.

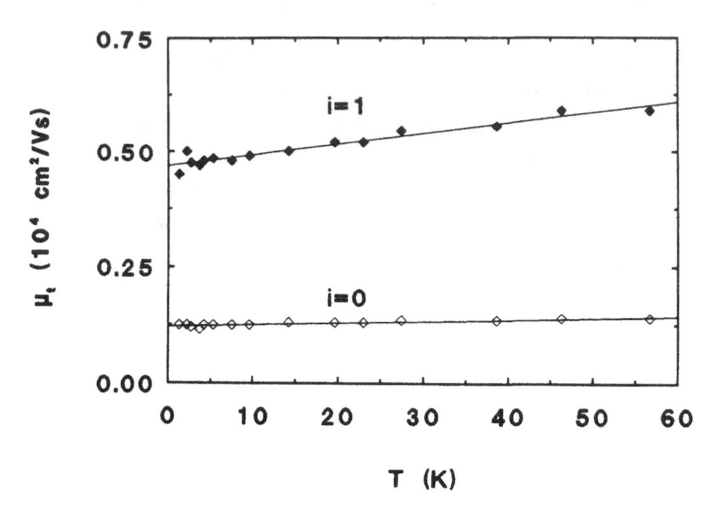

Fig. 17.6. Measured transport mobility of the $i = 0$ and $i = 1$ subband as a function of temperature.

The influence of illumination

After illumination, a third subband ($i = 2$) was observed in the Schubnikov–de Haas measurements. This subband was not found in the mobility spectrum obtained after illumination (see Table 17.1). It is possible that the transport mobility in the $i = 1$ and $i = 2$ subbands is equal. In this case the electron density obtained from the peak at 7800 cm^2/Vs in the mobility spectrum is the sum of the electron densities in the $i = 1$ and $i = 2$ subband obtained from the Schubnikov–de Haas measurements. This assumption seems reasonable because the quantum mobility in these two subbands is also equal [30].

The total electron density increases after illumination because the charged acceptors in both depletion regions next to the doping layers are neutralized. Electron–hole pairs are created by illumination; the electrons flow towards the 2DEG and the holes recombine with the charged acceptors. In the case where all charged acceptors are neutralized, the conduction band outside the 2DEG region will become flat and thus the confining potential will become different. We calculated the subband population in this situation and found that, just as in the experiment, three subbands are populated after illumination, see Table 17.1.

After illumination, the transport mobility in the $i = 0$ subband increases a little, whereas the quantum mobility in the $i = 1$ subband increases strongly. The effect of illumination on the transport mobility is complex because many of the sample parameters are changed after illumination; that is, the population of each subband, the number of depletion charges and the shape of the wavefunction. However, we think that the main reason for the increase in transport mobility in the $i = 1$ subband is the better spatial separation of the electrons in this subband and the ionized scatterers. The wavefunction increases in width due

Table 17.2. *Subband transport mobility in structures δ-doped at 2×10^{12} cm^{-2} and grown at different temperatures. The width of the doping layers was determined from a fit of the measured subband population to calculated subband populations*

T_{growth} (°C)	480	530	620
D_{don} (Å)	20	30	60
μ_0 (cm^2/Vs)	1500	1875	2300
μ_1 (cm^2/Vs)	7100	7650	2300

to the fact that the potential well becomes wider after neutralization of the depletion charges. In particular, electrons in the higher subbands 'feel' this change in the depletion charge.

The influence of growth temperature

We used the mobility spectrum technique to study the transport mobility in each subband as a function of the growth temperature. In Table 17.2 we show the subband mobility in samples doped at 2×10^{12} cm^{-2}, but grown at various temperatures. It is well known that the width of a δ-doping layer increases with the growth temperature. Our results also show an increase in the doping-layer thickness at higher growth temperatures. At low growth temperatures there is a large difference in the transport mobility in the two subbands, but at the highest growth temperature the transport mobility is equal in both subbands. This is mainly due to a decrease in the mobility in the highest subband. The stronger overlap of the electrons in the highest subband with the ionized Si-donors seems to be the most reasonable explanation. The increase in mobility in the lowest subband is due to weaker confinement of the electrons in a structure with a thicker doping layer. Moreover, the ionized scattering centres have moved away from the central maximum in the probability distribution of the electrons in the lowest subband.

Gated δ-doped structures

Kotelnikov *et al.* [15] studied the subband transport mobilities in a gated δ-doped structure when the second subband becomes populated. The transport mobility in the lowest subband varied from 600 to 700 cm^2/Vs. The transport mobility in the second subband increased strongly from 700 to 1600 cm^2/Vs upon filling of this subband. At the highest gate voltages Kotelnikov *et al.* observed a saturation of the transport mobilities in both subbands. This can be explained by an increase in the intersubband scattering when the third subband gets close to the Fermi energy. The transport mobility in the lowest subband is half of that observed by other authors. This is probably due the close proximity of the surface while the doping layer is only 50 nm below the gated surface.

δ-doped quantum well structures

The overlap of the electron wavefunction with the ionized scatters is different in centre- and edge-doped δ-doped quantum well structures. Another way of changing the overlap is to grow δ-doped quantum wells with different well thicknesses. Masselink [32] has grown δ-doped structures with different positions of the δ-layer in order to study the influence of the overlap of the electron wavefunction with the scatterers. In his structures only a single subband was populated, $n_0 = 0.5 \times 10^{12}$ cm^{-2}. He observed lower transport mobility in the centre δ-doped quantum well structure. In the sample that was δ-doped midway between the centre and the edge, the transport mobility was 900 cm^2/Vs. In the centre δ-doped quantum well structures, the transport mobility decreased to 500 cm^2/Vs. This shows clearly that the overlap is an important factor for the mobility.

The transport mobility in these centre-doped quantum wells is considerably lower than the mobility in the pure δ-doped structures with an even higher doping concentration. This reduction is probably not due solely to a change in the overlap. A sharp reduction in the Hall mobility at low doping concentrations has been observed by Gillman *et al.* [12] in the same density region. A reduction in the mobility has to be expected when we reach the critical Mott density, as was shown in detail by the theoretical work of Gold *et al.* [33].

It is interesting to note that Masselink *et al.* observed better conduction in the homogeneously doped quantum well as compared with the centre δ-doped quantum well. This agreed well with his theoretical work. However, Bhode *et al.* [34] calculated the opposite in the case where the scattering potential is determined by the correlated sum of the individual ionized scattering potentials. In the case where uncorrelated scattering potentials are used, there is little difference in the conductivity of a homogeneously doped quantum well and a centre δ-doped quantum well.

Harris *et al.* [35] also studied δ-doped quantum wells. However, he used a much higher doping concentration of 5×10^{12} cm^{-2}. Table 17.3 shows his results. To understand these results in a simple picture is difficult because the energy and wavefunction of the subbands change with quantum well thickness. This means that the electron density in the subbands also varies with the width of the quantum well. Using the same doping concentration as was used in the study by Harris *et al.*, maximum conductivity in a δ-doped quantum well 150 Å wide was reported by Shih and Streetman [36].

Table 17.3. *Transport mobility in centre δ-doped quantum wells as measured by Harris et al.* [35]. *The doping concentration in these structures was* 5×10^{12} cm^{-2}. D_{qw} *is the thickness of the quantum well*

D_{qw} (Å)	100	200	300
μ_0 (cm^2/Vs)	1500	1100	1000
μ_1 (cm^2/Vs)	–	5800	7200

Other materials

The subband transport mobility results in non-Si δ-doped GaAs can be found in the work of Wenchao Cheng *et al.* [20] on S-δ-doped InP. The authors report transport mobilities in samples with different doping concentrations and different widths of the doping layer. In the sample doped at 8.4×10^{12} cm^{-2} with a doping layer width of 50 Å, they found the following transport mobilities: $\mu_0 = 940$ cm^2/Vs, $\mu_1 = 4300$ cm^2/Vs, $\mu_2 = 4600$ cm^2/Vs. In a sample doped at 2×10^{12} cm^{-2} with the same width the transport mobility in the lowest subband increased to 1440 cm^2/Vs, whereas the transport mobility in the second subband was nearly the same as in the sample with the higher doping concentration. In a third sample with a doping concentration of 2.7×10^{12} cm^{-2} but a width of 100 Å, only a small decrease was observed in the transport mobility in the second subband, compared with the sample doped at 2×10^{12} cm^{-2}. It seems that qualitatively, as well as quantitatively, the results on S-δ-doped InP are comparable with Si-δ-doped GaAs.

Conclusion

In conclusion, all the results seem to agree on the fact that in structures containing a single undisturbed δ-doping layer the transport mobility in the lowest subband is about 1200 cm^2/Vs and is not very dependent on the broadening of the doping layer or the doping concentration. The transport mobility increases with the subband number in δ-doped structures with a narrow doping profile. Upon broadening of the doping layer the transport mobility in the higher subbands falls to the value in the lowest subband. In samples well above the critical Mott density, the transport mobilities are independent of temperature up to 60 K. The transport mobility can be signficantly influenced by changing the overlap of the electron wavefunction with the scatterers in quantum well structures. Finally, the density range over which the transport mobility can be determined is limited, while only measurements on structures with two or three populated subbands can be analysed.

Quantum mobility

As was shown in Eqn. (17.12) the envelope of the Schubnikov–de Haas oscillations depends exponentially on the quantum mobility. For a detailed analysis of the quantum mobility in the individual subbands, the contribution of each individual subband must be separated. The most suitable method is to take the Fourier transform of either a ρ_{xx} or a σ_{xx} curve, as shown in Fig. 17.7. The measurement shown in this diagram was obtained on the same sample as the one from Table 17.1. The diagram clearly shows that one begins to observe the oscillations with the fastest period above 10 T. These oscillations correspond to the most heavily populated subband. Thus, in order to observe all populated subbands, one has to perform Schubnikov–de Haas measurements up to magnetic fields well above 10 T. The Fourier transform of the measurement in Fig. 17.7 is shown in Fig. 17.8.

The simplest way of obtaining the quantum mobility was first proposed by Skuras *et al.* [27]. They showed that the Fourier transform of the first harmonic of the Schubnikov–de Haas oscillations, see Eqn. (17.12), can be approximated by

$$A_i(n) \propto \frac{1}{[\alpha^2 + (n - n_i)^2]^{1/2}} \tag{17.15}$$

where $\alpha = (2\pi e/\mu_{q,i}h)$ where $\mu_{q,i}$ is the quantum subband mobility and n_i is the subband electron density. Skuras *et al.* assumed that the term $D(X)$ in Eqn. (17.12) can be approximated by 1. The half-width of this Fourier peak is

$$\Delta n = \frac{e\sqrt{3}}{h\mu_{q,i}}. \tag{17.16}$$

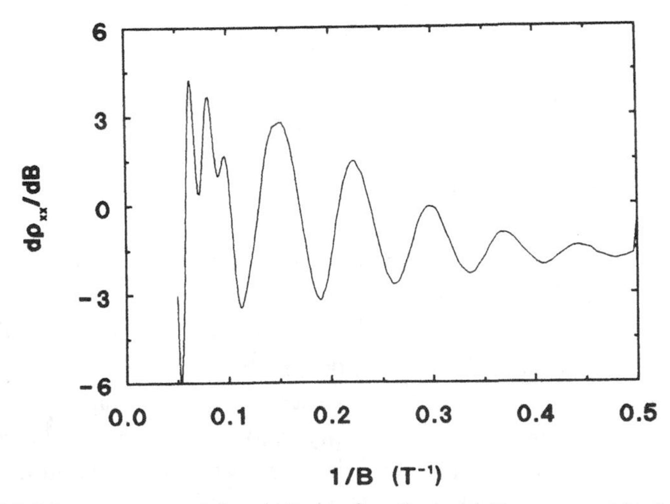

Fig. 17.7. Measurement of $d\rho_{xx}/dB$ as a function of $1/B$ on a sample δ-doped at 3.5×10^{12} cm^{-2} and grown at 480 °C. A modulation field of 18 mT was used.

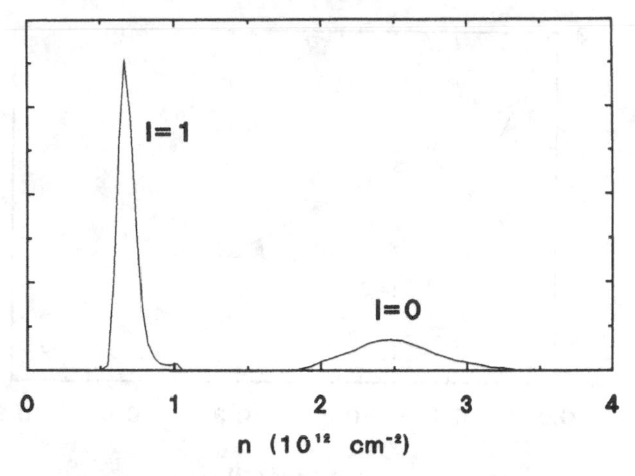

Fig. 17.8. Fourier transform of the measurement shown in Fig. 17.7.

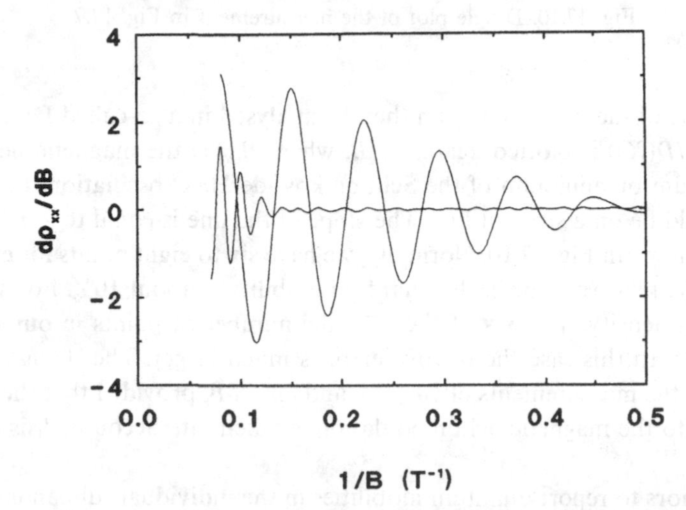

Fig. 17.9. The separated Schubnikov–de Haas oscillations of the measurement in Fig. 17.7.

Thus, the half-width of the Fourier peak can be used to find the quantum mobility in the individual subbands. The analysis method must be used with great caution, while the equation is obtained by integration over all Schubnikov–de Haas oscillations in the magnetic field domain. In an experiment, however, the magnetic field range is always limited. Also, the approximation of $D(X)$ by 1 is not always applicable. The mobility obtained in this way can differ considerably from the actual mobility.

The other method of obtaining the quantum mobility in the individual subbands uses a Fourier filtering technique. With this technqiue one filters out a peak in the frequency (or density) domain corresponding to a single subband. Subsequently, an inverse Fourier transform on this filtered peak is performed. In this way we obtain the separate Schubnikov–de Haas oscillations of a single subband (see Fig. 17.9).

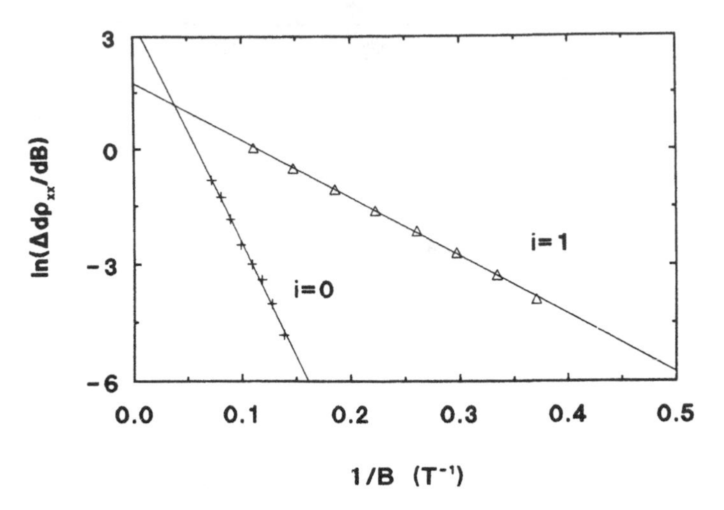

Fig. 17.10. Dingle plot of the measurement in Fig. 17.7.

The amplitude of the oscillations can then be analysed in a so-called Dingle plot, where $\log(|\Delta\rho_{xx}(B_{\text{ext}})|/D(X))$ is plotted against $1/B$, where B_{ext} is the magnetic field position of either a maximum or minimum of the Schubnikov–de Haas oscillation. In the ideal case the points should lie on a straight line. The slope of the line is equal to $-\pi/\mu_{q,i}$. A typical Dingle plot is shown in Fig. 17.10. Normally, we have six to eight points for every subband, in which case the relative error in the quantum mobility is about 10%. For subbands with a small electron density, $n_i < 3 \times 10^{11}$ cm^{-2}, the number of points in our Dingle plot is reduced to three. In this case the relative error is much larger. The 'Dingle' analysis can also be used for the measurements of $\partial\rho_{xx}/\partial B$ and $\partial\rho_{xy}/\partial B$, provided that the scaling factor $J_1(B^2_{\text{mod}}/B)$ due to the magnetic field modulation is taken into account; J_1 is the first-order Bessel function.

The first authors to report quantum mobilities in the individual subbands were Yamada and Makimoto [37]. They used the Fourier filtering technique together with Dingle plots for the individual subbands. For a structure doped at 7×10^{12} cm^{-2} they found that $\mu_0 = 440$ cm^2/Vs, $\mu_1 = 990$ cm^2/Vs, $\mu_2 = 2560$ cm^2/Vs, and $\mu_3 = 9570$ cm^2/Vs. The results for the lowest subband were probably misinterpreted because their maximum field of only 10 T was much too small to observe enough clear Schubnikov–de Haas oscillations from the $i=1$ subband. We assume that they interpreted the second harmonic of the $i = 1$ subband as the $i = 0$ subband, in which case one finds a population that is twice the population of the $i = 1$ subband and a mobility that is half the mobility in the $i = 1$ subband. Both the measured electron density and the mobility agree with these arguments. But when we study the quantum mobilities in the $i = 1$, $i = 2$ and $i = 3$ subbands, we notice that the quantum mobility, just as the transport mobility, increases with the subband number.

In the same sample as was used for the determination of the transport mobility in the individual subbands, we also determined the quantum mobility in the individual subbands.

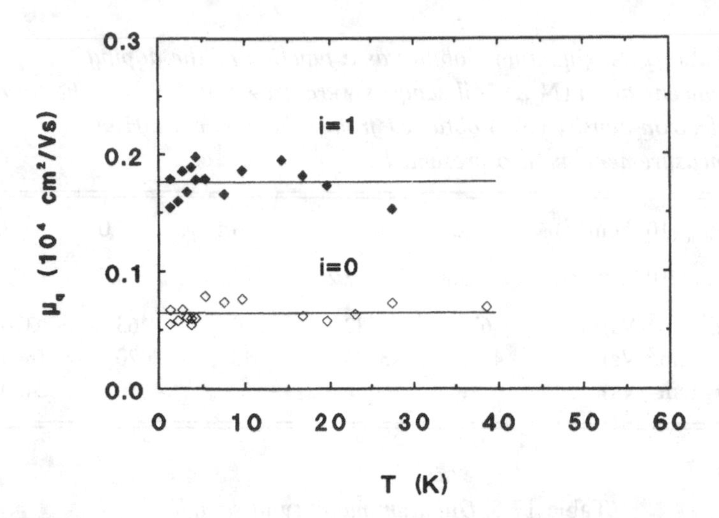

Fig. 17.11. The quantum mobility in the individual subbands as a function of temperature. The corresponding transport mobility in each subband of the same sample is shown in Fig. 17.6.

The quantum mobility in each subband was determined by Fourier filtering and a Dingle plot. The Schubnikov–de Haas measurements were carried out in magnetic fields up to 20 T. The weak oscillations in ρ_{xx} and ρ_{xy} were resolved by measuring $d\rho_{xx}/dB$ and $d\rho_{xy}/dB$ with a modulation field of 18 mT at a frequency of 21.3 Hz. The subband population was determined from the Fourier transform of $d\rho_{xx}/dB$ and $d\rho_{xy}/dB$. The quantum mobility was also determined as a function of temperature (see Fig. 17.11). Analysis was possible up to 40 K. Above this temperature the Schubnikov–de Haas oscillations were too weak to perform the Dingle analysis. Apart from better experimental accuracy in the case where we measured $d\rho_{xx}/dB$ and $d\rho_{xy}/dB$, we found the same results from analysis of either the ρ_{xx} and ρ_{xy} or the $d\rho_{xx}/dB$ and $d\rho_{xy}/dB$ measurements.

The results presented in Fig. 17.11 show that the quantum mobility in the $i = 0$ subband is about three times smaller than that in the $i = 1$ subband. A similar ratio was observed for the transport mobility in the $i = 0$ and $i = 1$ subbands. Temperature independence of the quantum mobility is to be expected when ionized impurity scattering is the dominant scattering mechanism in a degenerate 2DEG.

Influence of the doping concentration

We studied the quantum mobility in samples grown with different doping concentrations (see Table 17.4). All the samples were grown at 480 °C and the width of the doping layer was 20 Å or less. At high doping concentrations, the total electron density is much smaller than the doping concentration. This is due to self-compensation of the Si-dopant atoms. At low doping concentrations there is also a small difference between the doping concentration and the total electron density, but this is due to the formation of depletion

Table 17.4. *Quantum mobility as a function of the doping concentration* (N_{don}). *All samples were grown at* 480 °C. *The total electron density* (N_{tot}) *obtained from Schubnikov–de Haas measurements is also presented*

N_{don} (10^{12} cm^{-2})	2.0	2.5	3.5	5.0	8.0
N_{tot} (10^{12} cm^{-2})	1.8	2.2	3.2	4.5	5.6
μ_0 (cm^2/Vs)	670	582	500	463	370
μ_1 (cm^2/Vs)	4400	3390	2040	2020	1640
μ_2 (cm^2/Vs)	–	–	–	4080	2450

Table 17.5. *Quantum mobility in samples grown at different temperatures. All the samples were doped at* 2×10^{12} cm^{-2}. *The width of the doping layer, as determined from the comparison of measured and calculated subband populations, is also given*

T_{growth} (°C)	480	540	620
D_{don} (Å)	20	30	60
μ_0 (cm^2/Vs)	670	690	790
μ_1 (cm^2/Vs)	4400	3350	1830

layers next to the δ-doping layer. We observed that the quantum mobility in the lowest subband shows a clear but small decrease with increasing doping concentration. One of the reasons for the drop in mobility with doping concentration is the compression of the wavefunction of the lowest subband at higher doping concentrations. The quantum mobility in the higher subbands is more sensitive to compression of the wavefunction at increasing doping concentrations.

Influence of growth temperature

We studied the quantum mobility in samples grown at different temperatures (see Table 17.5). The thickness of the layer is determined from a comparison of the subband population measured and the calculated subband population, and is also given in Table 17.5. At higher growth temperatures the thickness of the doping increases. Just as the results on the transport mobility showed, we observed a small increase of the quantum mobility in the lowest subband and a very strong reduction of the mobility in the second subband.

Influence of δ-doping layer thickness

To analyse the thickness dependence in more detail and to exclude different growth conditions, we studied the quantum mobilities in samples that were grown at 480 °C and annealed afterwards at 800 °C for times ranging from 3 s to 300 s in a Rapid thermal annealer [38].

In Fig. 17.12(a) we show the subband population in the annealed samples as a function of the thickness of the doping layer, which was determined from a fit of the measured subband population to the calculated subband population. In the calculations we used the doping profile thickness as the free parameter. The thickness we give is the width of a square doping profile. In Fig. 17.12(b) the quantum mobility is shown as a function of the thickness of the doping layer. It is obvious that the mobility in all subbands decreases rapidly to the value of the quantum mobility in the $i = 0$ subband. In the samples with the thickest doping profile the quantum mobility is the same in all subbands.

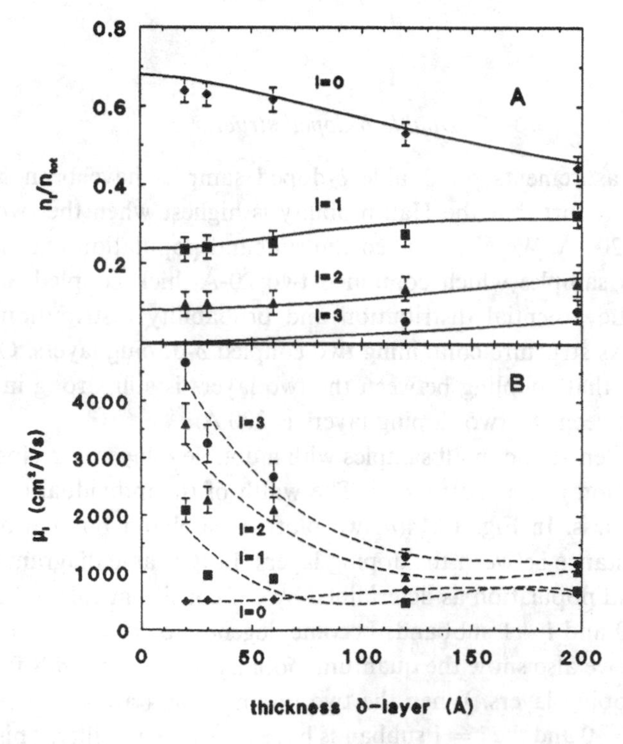

Fig. 17.12. (a) Subband population in an annealed structure containing a single doping layer. The solid lines represent the calculated subband population. (b) Quantum subband mobility in the annealed structure containing a single doping layer. The dashed lines are only given to guide the eye.

Influence of hydrostatic presssure

Stradling *et al.* [28] determined the quantum mobility in Si-δ-doped GaAs structures under hydrostatic pressure. They determined the quantum mobility from the Fourier peak half-width. Subbands are depopulated and the quantum mobility in most levels increases with increasing pressure. The depopulation is due to localized states which are pushed below the Fermi energy. The localized states are mainly DX-centres. Stradling *et al.* observed an increase in quantum mobilities by a factor of two when the electron density is decreased by a factor of two. If the DX centre is neutral in the ground state, then this increase is partially due to the fact that with every electron that is removed from the 2DEG one ionized Si-donor is neutralized. However, such a large change in quantum mobility is not seen in narrow doping layers, even when the doping concentration is reduced by a factor of three (see Table 17.4). Moreover, it is well accepted by now that the DX centre in the ground state is negatively charged [39]. This means that although the number of free electrons decreases after population of the DX state, the number of scatterers remains constant. Stradling *et al.* argue that correlation effects in the charge distribution, as calculated by Sobkowicz *et al.* [40], might play a role.

Double δ-doped structures

Hall mobility measurements on double δ-doped samples have been reported recently. Zheng *et al.* [41] report that the Hall mobility is highest when the two δ-doping layers are separated by 200 Å. We also studied the subband population and quantum subband mobility in GaAs samples which contained two 20-Å-thick coupled Si-δ-layers. In Fig. 17.13 we show the potential distribution and probability distribution of the subband electrons in a GaAs structure containing two coupled δ-doping layers. Our self-consistent calculations show that coupling between the two layers is still strong in structures where the separation between the two doping layers is 200 Å.

The doping concentration in all samples with a double δ-layer was close to the intended doping concentration of 5×10^{12} cm^{-2}. The width of the individual doping profiles was equal to 20 Å or less. In Fig. 17.14(*a*) we plot the subband population measured as a function of the distance between the doping layers. In the same diagram we also show the calculated subband population as determined by self-consistent calculations. At a distance of 150 Å the $i = 0$ and $i = 1$ subbands become degenerate.

In Fig. 17.14(*b*) we also show the quantum mobility in the subbands for a structure with two adjacent δ-doping layers. When the two doping layers are sufficiently separated, we observe that the $i = 0$ and the $i = 1$ subbands have the same mobility. This is to be expected because the two levels are degenerate at this separation. The quantum mobility in the $i = 2$ and, to some extent (though not very accurate), in the $i = 3$ subband, show decreasing mobility when the doping layers are separated.

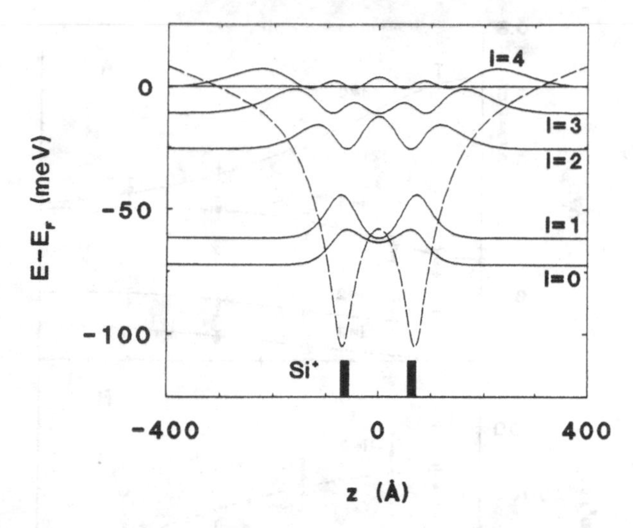

Fig. 17.13. The probability distribution of the electrons in the various subbands (solid line) and the electrostatic potential (dashed line) in a GaAs structure containing two adjacent Si-δ-layers. The total electron density is 2.5×10^{12} cm^{-2} and the donor concentration is equal in the two doping layers that are separated by 120 Å.

Measurements of the Hall mobility in these structures show decreasing Hall mobility with increasing separation. This result is in agreement with the observed fall in the quantum mobility in the two lowest subbands. However, observe that the gradual fall in the Hall mobility at increasing separation contradicts the results of Zheng *et al.* [41].

Comparison of the double-doped and annealed structures

When we compare the subband population in a double δ-doped structure with the subband population in an annealed sample, see Fig. 17.15, we observe that the relative population in the $i = 1$, $i = 2$ and $i = 3$ subbands is almost the same in both types of structure. The absolute population also has to be the same, while all structures were doped at 5×10^{12} cm^{-2}. Thus, we produced structures with similar subband populations, but with different distributions of ionized donors.

When we compare the mobilities we see that in the annealed samples the mobility in all subbands falls to the same low value in the $i = 0$ subband. In the double δ-doped structure, however, the quantum mobility in the $i = 2$ and $i = 3$ subbands does not decrease to the low value of the $i = 0$ subband. We think there are two possible explanations for this observation. Firstly, the separation of the scatterers and electrons in the $i = 2$ and $i = 3$ subbands is better in the double δ-doped sample than in the annealed samples, where the scatterers are spread over a large part of the well region. Secondly, in the double δ-doped structure the local electron density, in particular due to the $i = 0$ and $i = 1$

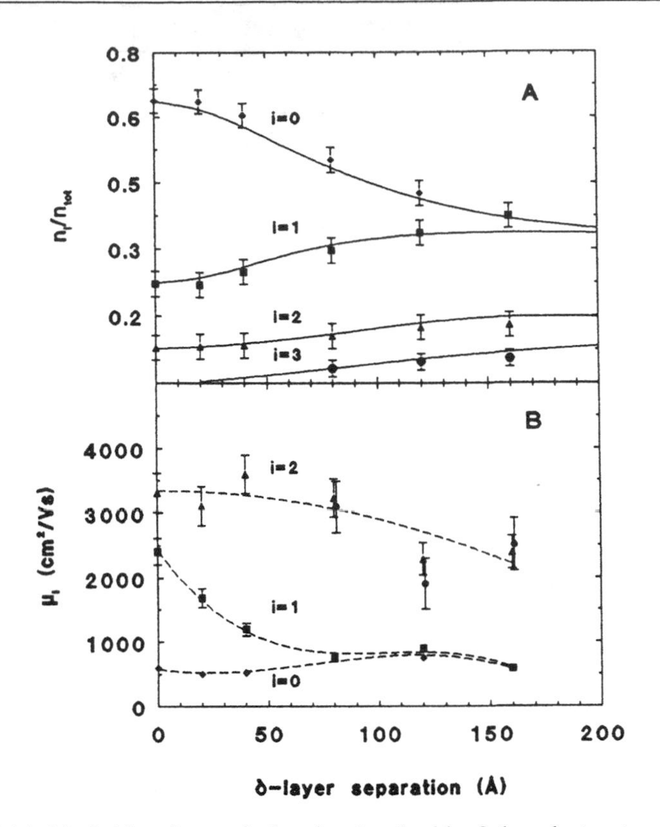

Fig. 17.14. (a) Subband population in the double δ-doped structures as a function of the separation of doping layers. The solid line represents the calculated subband population. (b) Quantum subband mobility in the double δ-doped structure as a function of the separation of doping layers. The dashed lines are only to guide the eye.

subbands, at the position of the scatterers is higher and thus leads to better screening of the scatterers for the electrons in the $i = 2$ and $i = 3$ subbands.

Quantum mobility to transport mobility ratio

Measurement of both transport and quantum electron mobility provides information on the dominant scattering mechanism, see Eqns. (17.3) and (17.4). In δ-doped structures, scattering on the ionized donors in the doping layer is expected to be the dominant scattering mechanism. Also, in heterostructures with a small spacer layer, ionized impurity scattering is the main scattering mechanism. The ratio μ_t/μ_q is typically 5–20 in these heterostructures. When the separation between the ionized donors and the electrons becomes smaller, the ratio μ_t/μ_q also decreases [42]. In Table 17.6 we show the ratio μ_t/μ_q obtained in our samples. Only samples with a doping concentration below 5×10^{12} cm^{-2} are reported, because in the samples with higher doping concentrations we were not able

Table 17.6. *Ratio of transport mobility to quantum mobility in δ-doped structures grown at different temperatures and doping concentrations*

N_{don} (10^{12} cm^{-2})	2.0	2.0	2.0	3.5
D_{don} (Å)	20.0	30.0	60.0	20.0
$i = 0$ subband	2.2	2.7	2.9	2.5
$i = 1$ subband	1.6	1.9	1.3	2.5

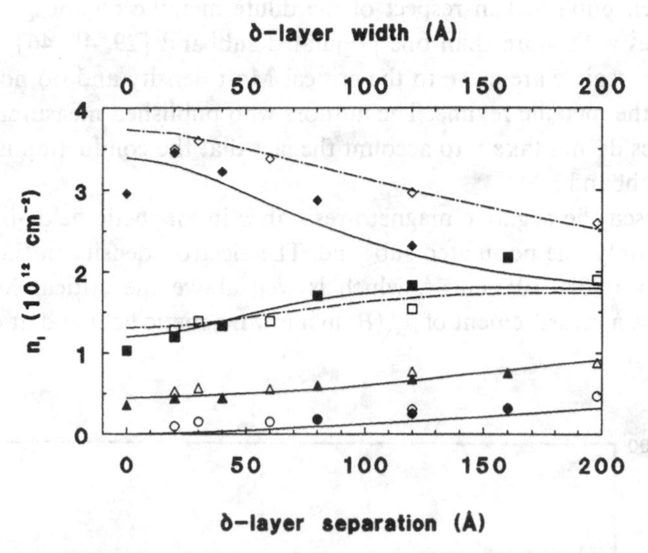

Fig. 17.15. Subband population in annealed single δ-doped structures as a function of the doping layer width (open symbols) and subband population in structures containing two interacting δ-doping layers as a function of the separation between the two doping layers (filled symbols).

to determine the transport mobility. From Table 17.6 it is clear that the ratio μ_t/μ_q in δ-doped samples is typically 2–3. These results are in agreement with results of Harris *et al.* [35] on quantum well δ-doped structures.

Conclusion

In conclusion, the quantum mobility in the individual subbands in δ-doped structures is far more accessible than is the transport mobility. Just as the transport mobility, the quantum mobility increases with subband number and is independent of temperature up to 40 K. The quantum mobility in each subband decreases with doping concentration, mainly due to compression of the wavefunctions at higher doping concentrations. The quantum mobility in the lowest subband increases slightly with the thickness of the doping layer, whereas the quantum mobility in the higher subbands shows a large fall towards the mobility in the lowest subband.

Inelastic scattering time

It is interesting to determine the dominant inelastic scattering process in δ-doped structures. The inelastic scattering time can be determined from the negative magneto-resistance due to weak localization. Up until now, measurements of inelastic scattering time in δ-doped samples have been published in respect of the dilute metallic regime [43, 44] or highly δ-doped structures with more than one populated subband [29, 45, 46]. The samples in the dilute metallic regime are close to the critical Mott density and do not compare with the structures in the metallic regime. The authors who published measurements on highly δ-doped structures do not take into account the fact that the conduction is determined by more than one subband.

We have analysed the negative magneto-resistance in magnetic fields below 0.5 T [47] in a sample with only one populated subband. The electron density in this quantum well δ-doped sample was 3×10^{12} cm^{-2}, which is well above the critical Mott density. In Fig. 17.16 we show a measurement of $\rho_{xx}(B)$ in a low magnetic field and at 4.2 K. The solid

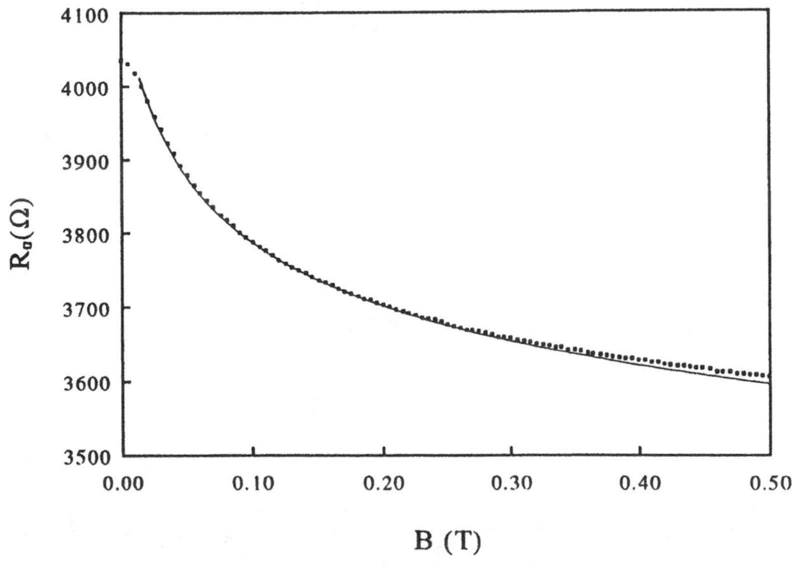

Fig. 17.16. Measurement of the negative magneto-resistance in a sample containing only one populated subband (dotted curve). The solid line is a fit with Eqn. (17.13).

curve gives a fit to the measurement with Eqn. (17.13) from Wittmann and Schmid. We used τ_0 and τ_ϕ as the fitting parameters. E_F was calculated from the electron density as determined from Schubnikov–de Haas measurements. At 4.2 K we found the best fit with $\tau_0 = 3.87 \times 10^{-14}$ s and $\tau_\phi = 1.95 \times 10^{-11}$ s. This shows that an electron in a δ-layer can scatter elastically some 500 times before it encounters an inelastic scattering event. Using the well-known digamma dependence of the weak localization on the magnetic field we were unable to obtain such a good a fit as is shown in Fig. 17.16.

In the temperature range from 1.7 K to 4.2 K we determined τ_0 and τ_ϕ as a function of temperature; τ_0 showed a very weak dependence on the temperature, but τ_ϕ decreased by a factor of two in the temperature range. We have compared the measured inelastic scattering time with the direct electron–electron scattering time and the Nyquist scattering time. The direct electron–electron inelastic scattering time [8, 48] is

$$1/\tau_{ee} = F^2 \cdot \frac{\pi(k_B T)^2}{2\hbar E_F} \cdot \ln(E_F \tau_0/\hbar) \tag{17.17}$$

where

$$F = \int_0^\pi \frac{1}{\pi(1 + 2(K_F/\varepsilon_0\varepsilon_r)\sin(\theta/2))}\, d\theta.$$

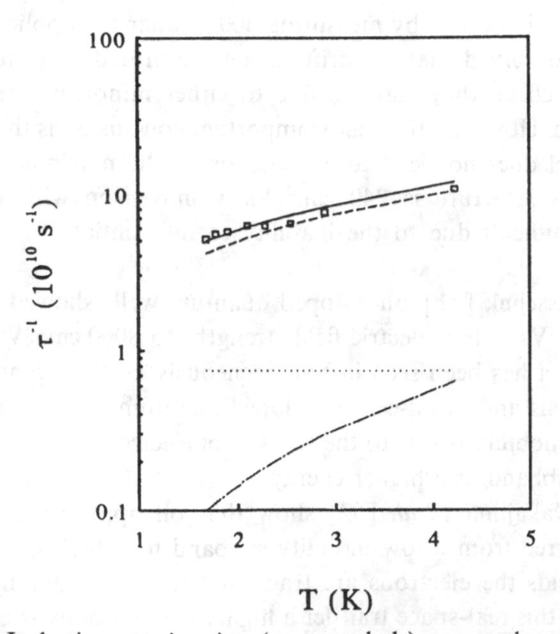

Fig. 17.17. Inelastic scattering time (open symbols) measured as a function of temperature. The Nyquist scattering time (dashed line), direct electron–electron scattering time (dash–dotted line) and a temperature-independent background scattering time (dotted line) are indicated. The solid line represents the total scattering time due to the separate scattering processes.

The Nyquist (many-body) electron scattering time [49] is

$$1/\tau_{\rm N} = \frac{k_{\rm B}T}{2E_{\rm F}\tau_0}\cdot\ln(E_{\rm F}\tau_0/\hbar). \tag{17.18}$$

In Fig. 17.17 we have plotted the measured inelastic scattering time τ_ϕ as a function of temperature. The direct electron–electron inelastic scattering time and the Nyquist inelastic scattering time are also indicated. The solid line gives the total scattering time due to direct electron–electron, Nyquist and temperature-independent background scattering. The results clearly show that Nyquist scattering, that is, the scattering of an electron on the total fluctuating potential of all moving electrons, is the dominant scattering mechanism.

Drift mobility under hot electron conditions

The first hot electron work on δ-doped samples was reported by Zrenner [26], who observed a real-space transfer of electrons from the δ-well to the depletion regions.

Balynas *et al.* [50] studied the drift velocity, $v_{\rm D} = \mu_{\rm D}E$, in δ-doped GaAs structures as a function of time after application of an electric pulse. When they measured the drift velocity 30 ps after they applied the electric pulse, they observed that the drift velocity saturated at 1.2×10^7 cm/s. This value is slightly above the maximum theoretical drift velocity in pure GaAs. However, by measuring 300 ps after the application of the electric pulse, Balynas *et al.* observed that the drift velocity saturated at a much higher value of 2.8×10^7 cm/s. This effect, they state, is due to either minority carrier injection or to avalanche breakdown. However, their most important conclusion is that the use of doping in the 2DEG channel does not lead to a reduction in the maximum drift velocity. The drift mobility in this structure is 2400 cm^2/Vs. Comparison with the other subband mobilities is rather difficult due to the inability to differentiate between the individual subband effects.

Later work by Masselink [51], on δ-doped quantum wells showed an increase in drift mobility of 2000 cm^2/Vs at low electric field strengths to 8000 cm^2/Vs at a field strength of 4 kV/cm. This effect has been seen in homogeneously doped quantum wells, in centre δ-doped quantum wells and in off-centre δ-doped quantum wells. Masselink argues that this increase in drift mobility is due to the transfer of electrons in a low mobility subband to a high mobility subband at a higher energy.

Experiments by Nakajima *et al.* [52] show that on application of an electric field electrons are transferred from a low mobility subband to a high mobility subband first. At higher electric fields the electrons are tranferred to the depletion layers next to the doping layer. Due to this real-space transfer a higher drift mobility is observed in δ-doped samples as compared with the drift mobility in bulk doped samples.

Recently, Kostial *et al.* [53] showed that the differential conductivity is determined by both electron heating and real-space transfer in GaAs structures with multiple parallel δ-doping layers.

Theory

Up until now theoretical work on mobility calculation in δ-doped samples has been rather limited, although δ-doped structures are ideal model structures. No interfaces are present in δ-doped structures and ionized impurity is by far the most important scattering mechanism.

Low doping concentration regime

Gold *et al.* [33] studied structures for which the doping concentration changes from the hopping regime to the impurity band regime. In the impurity band regime the calculations are valid for structures with only one filled subband. Their theory includes the influence of disorder in the doping layer on the density of states and the screening. The calculated mobilities are in good agreement with experimental data.

The influence of the two-dimensionality of scattering has been discussed by Masselink [32] for homogeneously and δ-doped quantum wells. His calculations are applicable for structures with only a single populated subband. Masselink arrives at the conclusion that scattering in two dimensions is more effective than scattering in three dimensions due to less effective screening and an increased possibility for back-scattering in two-dimensional systems. Comparison with experiments shows qualitative agreement. Masselink's theoretical as well as experimental results show that the conductivity in a centre δ-doped quantum well is worse as compared with a homogeneously doped quantum well. This contradicts the work of Bhode *et al.* [34], who conclude the opposite.

High doping concentration regime

The Hall mobility in δ-doped samples with more than one populated subband was calculated by Gillman *et al.* [12]. They found qualitative agreement with their experiments. It is a great pity that only a few details are given in this paper about the calculations and that only the Hall mobility is calculated.

Multi-subband conduction was studied in more detail theoretically by Mezrin and Shik [54, 55]. They calculated the transport subband mobility in the individual subbands of δ-doped structures in the absence of depletion charges and with true mathematical δ-distribution of the ionized donors. They used an analytical expression for the confining potential instead of a self-consistently calculated potential. The results of their calculations are shown in Fig. 17.18. The transport mobility in all subbands decreases with doping concentration and increases with subband number. The fall in mobility with doping concentration is mainly due to the stronger confinement of electrons on the doping layer. The available experimental results are compared with these calculations in a recent review article by Shik [56]. However, in this paper quantum and transport mobility results were

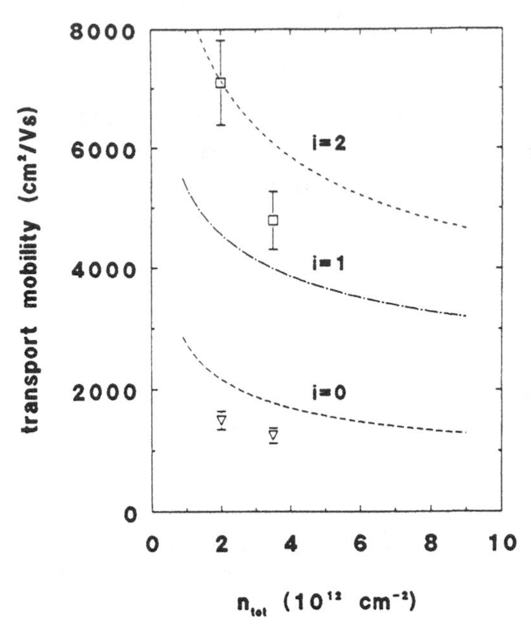

Fig. 17.18. Transport mobility in the individual subbands as a function of the total electron density in Si-δ-doped GaAs, as calculated by Mezrin and Shik [54]. Measured transport mobilities in the $i = 0$ (∇) and $i = 1$ subband ($\square$) as obtained from our analysis are indicated.

mixed up. Also, results were used that had been obtained on samples with thick doping layers. We showed that the mobility in the higher subbands is very sensitive to the broadening of the doping layer. Comparison of the calculated transport mobility with the measured transport mobilities shows that in the $i = 0$ subband the measured values are below the calculated mobility. However, in the $i = 1$ subband the measured transport mobility is higher than the calculated mobility. It is difficult to attribute the observed discrepancy to the calculations, while the experimental accuracy is rather limited. Also, the electron density range over which the transport mobility can be measured is rather small.

The quantum mobility in the lowest subband is also presented in the paper by Mezrin and Shik [54]. In Fig. 17.19 we have compared their calculations with our measurements. We obtain a perfect fit between theory and experiment. However, no results are presented concerning the quantum mobility in the higher subbands.

We have used Mezrin and Shik's formalism to calculate the transport and quantum mobility in every populated subband. The only difference from their calculations is the fact that we used wavefunctions which were obtained from the coupled Poisson and Schrödinger equations. We have calculated the transport and quantum mobilities as a function of the doping concentration for all populated subbands [57]. The results are presented in Figs 17.20 and 17.21. Our results compare well with those of Mezrin and Shik. We also observe a good fit to experimental quantum mobility in the lowest subband.

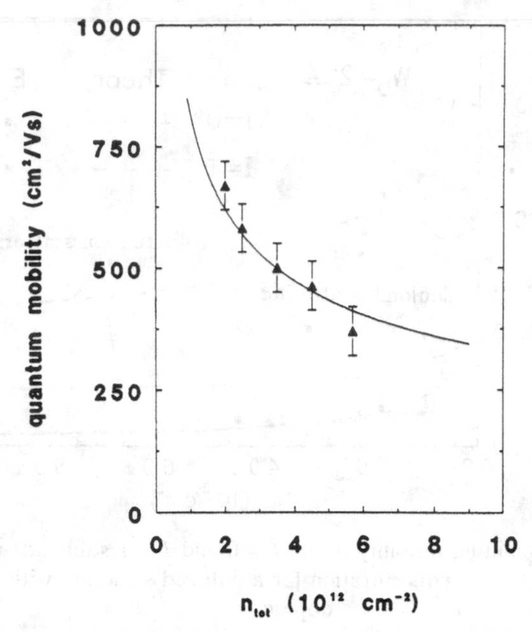

Fig. 17.19. Quantum mobility in the $i = 0$ subband for Si-δ-doped GaAs structures (solid line) as calculated by Mezrin and Shik [54]. The quantum mobilities measured in the $i = 0$ subband are also plotted (▲).

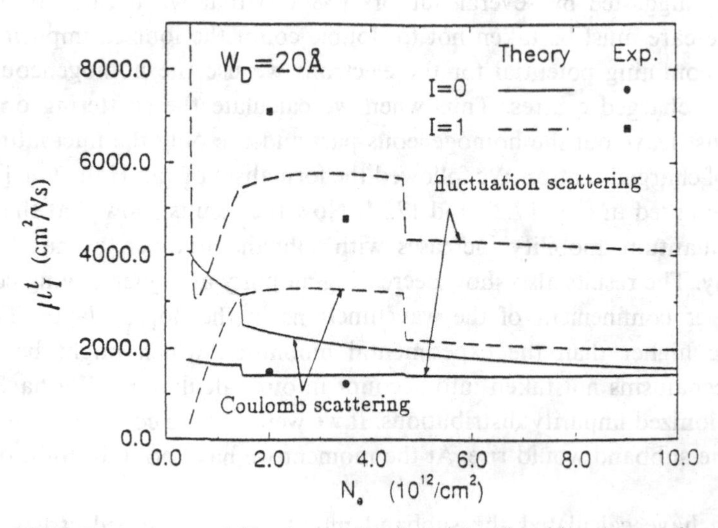

Fig. 17.20. Transport mobility in the $I = 0$ and $I = 1$ subbands as a function of the total electron concentration for a δ-doped structure with a 20-Å-thick doping layer.

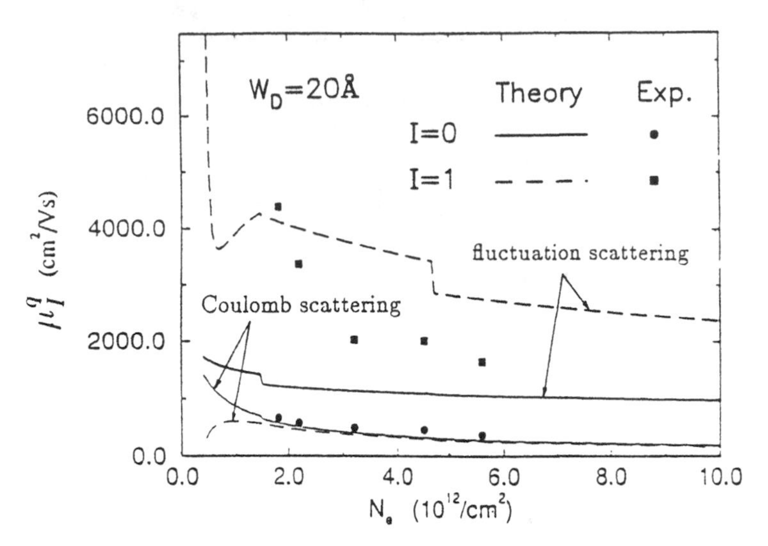

Fig. 17.21. Quantum mobility in the $I = 0$ and $I = 1$ subbands as a function of the total electron concentration for a δ-doped structure with a 20-Å-thick doping layer.

However, the quantum mobility in the $i = 1$ subband is equal to the quantum mobility in the $i = 0$ subband, whereas our experimental results have shown that the mobility increases with subband number. Also, the calculated mobility in the $i = 1$ subband is lower than the mobility found experimentally. This is physically incorrect. The experimental mobility has to be equal to or lower than the calculated mobility because not all scattering mechanisms might be accounted for in the calculations.

It has been suggested by several authors [58–60] that when calculating the ionized scattering rate care must be taken not to double count the ionized impurities. When we calculate the confining potential for the electrons we use the homogeneous part of the distribution of charged centres. Thus when we calculate the scattering on the charged centers we must leave out the homogeneous part and use only the fluctuating part of the distribution of charged centres. We followed the formalism of van Hall *et al.* [58] to obtain the results presented in Figs 17.20 and 17.21. Now the results show that the transport, as well as the quantum, mobility increases with subband number, as has been observed experimentally. The results also show decreasing mobility at higher doping concentrations due to stronger confinement of the wavefunctions in the doping layer. The calculated mobilities are higher than the experimental mobilities, which might be due to extra scattering mechanisms not taken into account in our calculations. We have used totally uncorrelated ionized impurity distributions. If we were to take correlated distributions the mobility in the subband would rise. At the moment we have no indication that this is the case.

Finally, we have calculated the subband mobility in a δ-doped structure where the thickness of the doping layer was increased, see Fig. 17.22. Our calculations show that, just as in the experiment, the mobility in all subbands falls to the quantum mobility in the

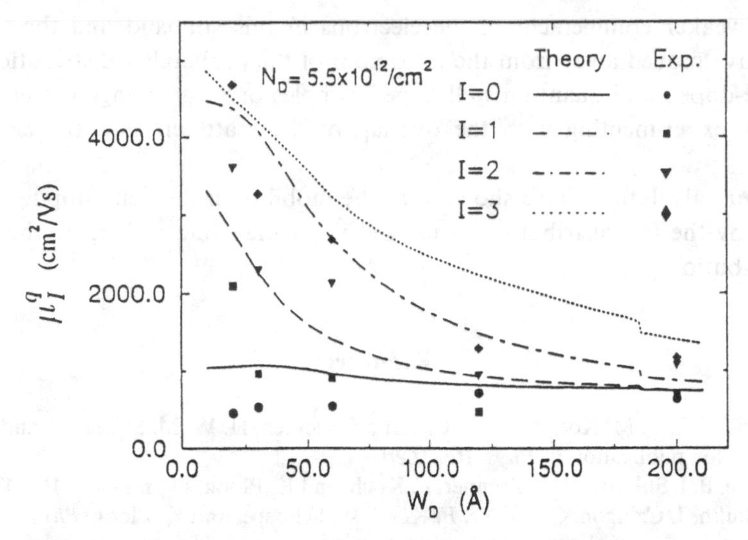

Fig. 17.22. Quantum mobility in the individual subbands as a function of the thickness of the doping layer for a structure doped at 5.5×10^{12} cm^{-2}.

$i = 0$ subband. Also, in agreement with our experiments, we find that the mobility in the $i = 0$ subband is roughly independent of the layer thickness. We obtained these results by calculating the scattering on the fluctuations in the charge distribution. We were not able to obtain an agreement between theory and experiments when we calculated the scattering on the full distribution. This again shows that the scattering is due to fluctuations in the charge distribution.

Conclusions

In δ-doped samples with a low doping concentration, $n_{\text{dop}} < 0.5 \times 10^{12}$ cm^{-2}, the mobility increases with the doping concentration because the conduction changes from hopping conduction to metallic conduction. At higher doping concentrations the scattering is fully dominated by ionized impurity scattering.

In samples with a narrow doping profile ($D_{\text{dop}} < 20$ Å), at a doping concentration just above the critical Mott density, the mobility in the lowest subband is about 1200 cm^2/Vs. Due to stronger confinement at higher doping concentrations, the mobility in the $i = 0$ subband decreases slightly. By increasing the doping concentration more subbands become populated. Every new subband that is populated has a higher mobility than the preceding filled subband due to the spatial separation of the electrons and scatterers. Within each higher subband the mobility decreases with the doping concentration due to the stronger confinement of the electrons at higher doping concentrations.

When the thickness of the doping layer is increased the mobility in all subbands except the $i = 0$ subband decreases. This is due to a stronger overlap of electron wavefunction with the ionized scattering centres. The mobility in the $i = 0$ subband increases slightly due to the weaker confinement of the electrons in this subband and the fact that the scatterers have moved away from the maximum of the probability distribution.

In double-doped and quantum well doped samples one can change the mobility of the electrons by experimenting with the overlap of the scatterers and the electron wave-functions.

Finally our calculations have shown that the mobility in the delta-doping layers is not determined by the full distribution of ionized impurities, but, rather, by the fluctuations in this distribution.

References

[1] M. B. Johnson, P. M. Koenraad, W. C. van der Vleuten, H. W. M. Salemink, and J. H. Wolter, Accepted for publication in *Phys. Rev. Lett.*

[2] Qiu-yi Ye, B. I. Shklovskii, A. Zrenner, F. Koch, and K. Ploog, *Phys. Rev.* B **41**, 8477 (1990), and M. E. Raikh, J. Czingon, Qiu-yi Ye, F. Koch, W. Schoepe, and K. Ploog, *Phys. Rev.* B **45**, 6015 (1992).

[3] J. P. Harrang, R. J. Higgins, R. K. Goodall, P. R. Jay, M. Laviron, and P. Delescluse, *Phys. Rev.* B **32**, 8126 (1985).

[4] G. Bergmann, *Phys. Rev.* B **28**, 2914 (1983).

[5] L. J. van der Pauw, *Philips Res. Repts.* **13**, 1 (1958).

[6] A. Isihara and L. Smrcka, *J. Phys.* C **19**, 6777 (1986).

[7] H. P. Wittmann and A. Schmid, *Journ. of Low Temp. Phys.* **69**, 131 (1987).

[8] B. L. Altshuler, A. G. Aronov, P. A. Lee, *Phys. Rev. Lett.* **44**, 1288 (1980).

[9] S. Hakami, A. Larkin, and Y. Nagaoka, *Prog. Theor. Phys.* **63**, 707 (1980).

[10] T. Makimoto, N. Kobayashi, and Y. Horikoshi, *J. Appl. Phys.* **63**, 5023 (1988).

[11] T. Makimoto, N. Kobayashi, and Y. Horikoshi, *Jpn Journ. Appl. Phys.* **27**, L770 (1988).

[12] G. Gillman, B. Vinter, E. Barbier, and A. Tardella, *Appl. Phys. Lett.* **52**, 972 (1988).

[13] P. M. Koenraad, F. A. P. Blom, C. J. G. M. Langerak, M. R. Leys, J. A. A. J. Perenboom, J. Singleton, S. J. R. M. Spermon, W. C. van der Vleuten, A. P. J. Voncken, and J. H. Wolter, *Semi. Sci. Tech.* **5**, 861 (1990).

[14] E. F. Schubert, J. E. Cunningham, W. T. Tsang, *Solid State Comm.* **63**, 591 (1987).

[15] I. N. Kotelnikov, V. A. Kokin, B. K. Medvedev, V. G. Mokerov, Yu, A. Rzhanov, and S. P. Anokhina, *Sov. Phys. Semicond.* **26**, 821 (1992).

[16] T. Makimoto and Y. Horikoshi, *Jpn Journ. Appl. Phys.* **29**, L2250 (1990).

[17] P. M. Koenraad, I. Barsony, J. C. M. Henning, J. A. A. J. Perenboom, and J. H. Wolter, NATO Advanced Research Workshop *Physical Properties of Semiconductor Interfaces at Sub-Nanometer Scale*, Riva del Garda, Italy, 1992, NATO ASI series E243, p. 35.

[18] D. Lavielle, E. Ranz, and J. C. Portal, *Inst. Phys. Conf. Ser.* **106**, 297 (1990).

[19] G. Nachtwei, S. Heide, H. Künzel, and W. Passenberg, *J. Phys. Condes. Matter* **5**, 1091 (1993).

[20] Wenchao Cheng, A. Zrenner, Qiu-yi Ye, F. Koch, D. Grützmacher, and P. Balk, *Semi. Sci. Tech.* **4**, 16 (1989).

[21] M. A. di Forte-Poisson, C. Brylinski, E. Blondeau, D. Lavielle, and J. C. Portal, *J. Appl. Phys.* **66**, 867 (1989).

[22] M. J. Yang, W. J. Moore, R. J. Wagner, J. R. Waterman, C. H. Yang, P. E. Thompson, and J. L. Davis, *Surf. Sci.* **263**, 595 (1992).

[23] H. P. Zeindl, T. Wegehaupt, I. Eisele, H. Oppolzer, H. Reisinger, G. Tempel, and F. Koch, *Appl. Phys. Lett.* **50**, 1164 (1987).

[24] H. J. Gossmann and F. C. Unterwald, *Phys. Rev.* B **47**, 12618 (1993).

[25] C. Hilsum, *Elec. Lett.* **10**, 259 (1974).
[26] A. Zrenner, thesis Munich University (1987).
[27] E. Skuras, R. Kumar, R. L. Williams, R. A. Stradling, J. E. Dmochowski, E. A. Johnson, A. Mackinnon, J. J. Harris, R. B. Beall, C. Skierbeszeswki, J. Singleton, P. J. van der Wel, and P. Wisniewski, *Semi. Sci. Tech.* **6**, 535 (1991).
[28] R. A. Stradling, E. A. Johnson, A. Mackinnon, R. Kumar, E. Skuras, and J. J. Harris, *Semi. Sci. Tech.* **6**, B137 (1991).
[29] G. M. Gusev, Z. D. Kvon, D. I. Lubyshev, V. P. Migal, and A. G. Pogosov, *Sov. Phys. Semicond.* **25**, 364 (1991).
[30] P. M. Koenraad, B. F. A. van Hest, F. A. P. Blom, R. van Dalen, M. Leys, J. A. A. J. Perenboom, and J. H. Wolter, *Physica* B **177**, 485 (1992).
[31] W. A. Beck and J. R. Anderson, *J. Appl. Phys.* **62**, 541 (1987).
[32] W. T. Masselink, *Phys. Rev. Lett.* **66**, 1513 (1991).
[33] A. Gold, A. Ghazall, and J. Serre, *Semi. Sci. Tech.* **7**, 972 (1992).
[34] S. Bhode, W. Porod, and S. Bandyopadhyay, *Solid-state Elec.* **32**, 1651 (1989), and S. Bhode, W. Porod, and S. Bandyopadhyay, *Phys. Stat. Sol.* A **125**, 375 (1991).
[35] J. J. Harris, R. Murray, and C. T. Foxon, *Semi. Sci. Tech.* **8**, 31 (1993).
[36] Y. C. Shih and B. G. Streetman, *Appl. Phys. Lett.* **59**, 1344 (1991).
[37] S. Yamada and T. Makimoto, *Inst. Phys. Conf. Ser.* **106**, 429 (1990).
[38] P. M. Koenraad, A. C. L. Heessels, F. A. P. Blom, J. A. A. J. Perenboom, and J. H. Wolter, *Physica* B **184**, 221 (1993).
[39] P. M. Mooney, *Journ. App. Phys.* **67**, R1 (1990).
[40] P. Sobkowicz, Z. Wilamowski, and J. Kossut, *Semi. Sci. Tech.* **7**, 1155 (1992).
[41] X. Zheng, T. K. Carns, K. L. Wang, and B. Wu, *Appl. Phys. Lett.* **62**, 504 (1993).
[42] S. Das Sarma and F. Stern, *Phys. Rev.* B **32**, 8442 (1985).
[43] M. Asche, K. J. Friedland, P. Kleinert, H. Kostial, J. Herzog, and R. Hey, *Superlatt. Microstruc.* **10**, 425 (1991).
[44] M. Asche, K. J. Friedland, P. Kleinert, and H. Kostial, *Semi. Sci. Tech.* **7**, 923 (1992).
[45] M. V. Budantsev, Z. D. Kvon, and A. G. Pogosov, *Sov. Phys. Semicond.* **26**, 879 (1992).
[46] H. Taniguchi, J. Takahara, Y. Takagaki, K. Gamo, S. Namba, S. Takaoka, and K. Murase, *Jpn Journ. Appl. Phys.* **30**, 2808 (1991).
[47] P. M. Koenraad, T. P. C. van Noije, J. A. Melsen, J. H. Wolter, to be published.
[48] H. Fukuyama and E. Abrahams, *Phys. Rev.* B **27**, 5976 (1983).
[49] S. Chakaravarty and A. Schimd, *Phys. Rep.* **140**, 193 (1986).
[50] Y. Balynas, A. Krotkus, T. Lideikis, A. Stalnionis, and G. Treideris, *Electron Lett.* **27**, 2 (1991).
[51] W. T. Masselink, *Appl. Phys. Lett.* **59**, 694 (1991).
[52] S. Nakajima, N. Kuwata, N. Nischiyama, N. Shiga, and H. Hayashi, *Semi. Sci. Tech.* **7**, B372 (1992).
[53] H. Kostial, Th. Ihn, P. Kleinert, R. Hey, M. Asche, and F. Koch, *Phys. Rev.* B **47**, 4485 (1993).
[54] O. A. Mezrin and A. Y. Shik, *Superlatt. Microstruc.* **10**, 107 (1990).
[55] O. A. Mezrin, A. Y. Shik, and V. O. Mezrin, *Semi. Sci. Tech.* **7**, 664 (1992).
[56] A. Y. Shik, *Sov. Phys. Semicond.* **26**, 649 (1992).
[57] G. Q. Hai, F. M. Peeters, J. T. Devreese, P. M. Koenraad, A. F. W. van der Stadt, and J. H. Wolter, accepted at 22nd *Int. Conf. on the Physics of Semiconductors, Vancouver, Canada, August 15–19, 1994.*
[58] P. J. van Hall, T. Klaver, and J. H. Wolter, *Semicond. Sci. Technol.* **3**, 120 (1988).
[59] R. Lassnig, *Solid State Comm.* **65**, 765 (1988).
[60] E. F. Schubert, L. Pfeiffer, K. W. West, and A. Izabelle, *Appl. Phys. Lett.* **54**, 1350 (1989).

18

Hot electrons in delta-doped GaAs

M. ASCHE

Introduction

In semiconductor physics, hot electrons, that is, carriers far from thermodynamical equilibrium, can be excited by laser irradiation or created by energy gain in an electric field. They form an example of a situation which is determined by the nonlinear relationship between cause and effect. In this chapter we are concerned with the nonlinear properties of electric conductivity, which are an important feature for device aspects.

In the classical concept of solid state physics, phonons as well as lattice defects are regarded as scattering centers for electrons. Between two such scattering events the electron is considered to move under the influence of an applied electric field, in accordance with the common law for the dynamics of a free electron, assuming, of course, inhibition by potential confinement leading to reduced dimensionality. Consequently, during this collision free propagation in the electric field, the energy of a carrier grows. On the other hand, the electron can change its energy and momentum in a collision process. In the present context, the influence of an electric field can be neglected during this scattering event because the field strengths concerned are not extremely high. The time between collisions of electrons and phonons by deformation potential and piezoelectric interaction shortens with increasing carrier energy and with respect to polar optical phonons, too, as long as the carrier energies do not exceed the energy of the phonons by several times. In our semiconductor system, consequently, the energy transferred to the lattice per unit time rises with carrier heating. In this way a stationary state is obtained with a mean carrier energy, which is higher than in the absence of the electric field. If the lattice is kept at constant temperature T by means of a bath, the electrons are designated as hot in comparison to T. The increase in their mean energy is determined by the balance of energy gained by the field and the net loss by the collision processes, which is given by the difference in the scattering processes out of the relevant electron state and those events from other states leading into this one; both are types of scattering probability with the assistance of emission as well as absorption of phonons. As interference effects are assumed to play no role under the conditions assumed here, the probabilities of all scattering processes simply sum up.

When there are several minima of allowed carrier energies (i.e. subbands or different conduction bands), scattering processes between states belonging to these band edges also have to be taken into account. When the energies and/or the wave vectors of these minima differ, attention must be paid to this fact in terms of energy and momentum conservation. Firstly, there are electron–phonon interaction processes which lead to a final state at a different band edge than the initial state of the electron; furthermore such transitions can result from electron–electron collisions. Additionally, the latter interaction processes can also be phonon assisted. Carrier heating enhances the probabilities of such transitions and a stationary state is achieved when the energy dependent probabilities of these scattering events match the population of the relevant bands.

Field induced electron transfer into the interlayer space

When the donor concentration in δ-doped multilayers is high, the electron wavefunctions overlap and lead to the formation of states that are extended in the two dimensions of the potential wells along the δ-planes. Self-consistent potential and electron density calculations by lhn *et al.* (1992) for a δ-doped multilayer GaAs structure with 5×10^{11} Si atoms per cm^2 in each sheet and 100 nm between the layers are shown in Fig. 18.1, exhibiting three subbands and the distribution of the electrons (Ψ^2) in each of them. As can be seen in the upper subband, the electron density is already concentrated in the space between the layers and approximates to a free electron state above the barriers, that is, a three dimensional electron gas (3DEG).

At low lattice temperatures and weak electric fields the diffusive motion of the electrons confined to the potential wells[1] in the two energetically lower subbands determines the conductivity. Increasing the lattice temperature and/or enhancing the carrier energy by an electric field or infrared radiation renders the contribution of the extended states important. This situation resembles real space transfer of hot electrons from the wells into the states above the barrier in heterostructures, as reviewed by Gribnikov *et al.* (1995). Due to the lower barrier height in δ-doped GaAs as compared with GaAs/Al$_x$Ga$_{1-x}$As with $x \approx 0.3$, the transfer of the electrons into the space between the layers for a given carrier excitation is enhanced in this case. Consequently, a change in the contributions to the transport from the confined to the extended states is expected at much lower electric fields in comparison with the deeper rectangular quantum wells. An important qualitative difference when comparing real space transfer in hetero- and δ-doped structures concerns the relevant changes in mobility due to the different role played by the scattering of electrons on ionized impurities. In the first case, due to intentional doping of the barriers, the carrier transfer from the GaAs wells into the states above the Al$_x$Ga$_{1-x}$As barriers leads to a decrease in mobility due to enhanced momentum scattering rates. This may even lead to a more

[1] Denoted as a two dimensional electron gas (2DEG).

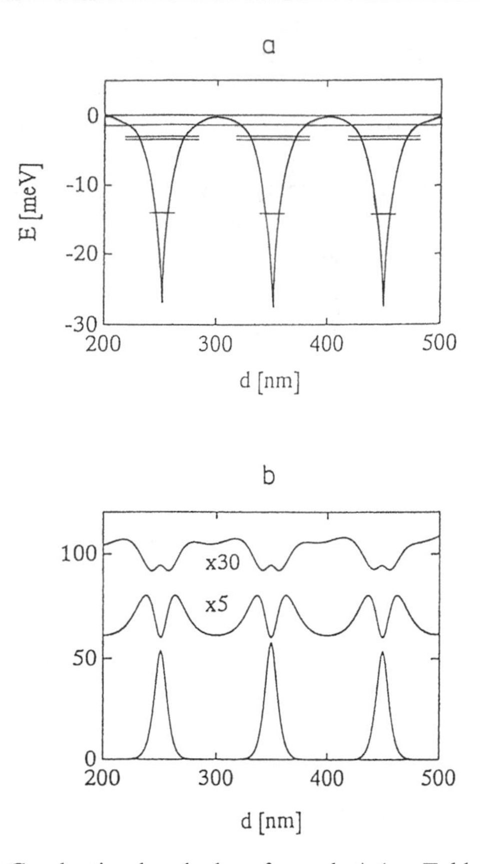

Fig. 18.1. (a) Conduction band edge of sample A (see Table 18.1). (b) Electron densities in the zeroth, first, and second subband.

pronounced negative differential conductivity than that due to a transition of the electrons from the Γ- to the L-band edges alone, as shown by Brennan and Park (1989) and Moško and Novák (1990), for instance. By contrast, in δ-doped multilayers the real space transfer results in carriers with higher mobility in the interlayer space, which is not intentionally doped, while the wells are formed by the impurity potential. This effect was clearly demonstrated by far infrared radiation (FIR) cyclotron responance for δ-doped InP by Mares *et al.* (1993). As the line width is determined by the scattering processes, the cyclotron resonance exhibits a very narrow line for orientation of the magnetic field **B** parallel to the layers, thereby demonstrating the low collision rate of carriers bunched in the interlayer space in contrast to perpendicular orientation of **B**.

Conductivity measurements in δ-multilayers and their results

The conductivity as a function of the electric field applied along the layers was investigated by Kostial *et al.* (1993) in respect of dependence on sheet doping concentration and layer

Table 18.1

	Sample			
	A	B	C	D
Number of δ-layers	8	10	10	12
Sheet doping concentration (10^{11} cm^{-2})	5	3.4	1.7	1.35
Layer spacing d [nm]	100	17.1	24.3	27.1
Equivalent volume doping (10^{16} cm^{-3})	5	20	7	5
Average distance of dopants in a sheet $\langle a \rangle$ (nm)	14.1	17.1	24.3	27.1
Aspect ratio $r = \langle a \rangle / d$	0.14	1	1	1

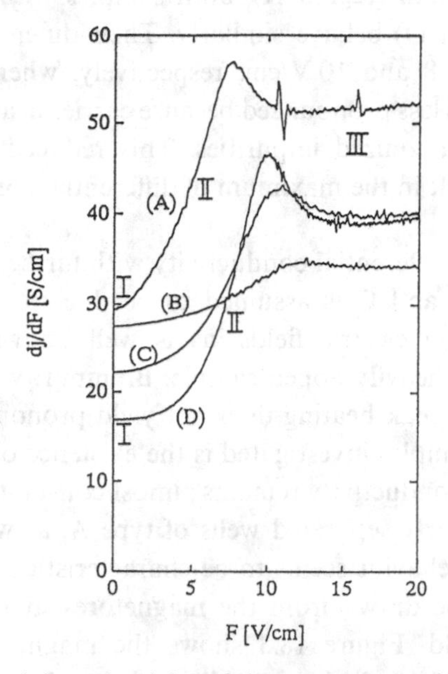

Fig. 18.2. Differential conductivity of the samples as a function of the electric field strength at 4.2 K.

spacing. The samples were grown by molecular beam epitaxy (MBE) at 520 °C, with growth interruption being employed for δ-doping. The unintentional background doping was p-type in the 10^{14} cm^{-3} range. The data from the samples are presented in Table 18.1. The current–voltage characteristics were measured using Hall bar samples immersed in liquid helium. The differential conductivity was obtained numerically from the measured current–voltage characteristic. The results are shown in Fig. 18.2 and illustrate the influence of the donor density on the carrier density and on the mobility, including the effect of partial ordering of the dopants.

Regarding the zero field limit, it can be seen that in the multilayers with aspect ratio 1 (i.e. the average impurity distance within one δ-sheet equals the distance between the layers), lowering of the dopant concentration (sample B–D) leads to lower differential conductivity. This is caused by the dominating decrease in carrier density as compared with that of reduced impurity scattering. However, the differential conductivities of samples A and B exhibit nearly the same value in spite of the higher volume doping density in B in comparison with the equivalent volume doping density in A. It is assumed that this effect can be explained by enhanced mobility due to the partial ordering of dopants in sample A.

In the weak field region up to 2 V/cm (region I), the differential conductivity of sample A increases, whereas the curves of the samples with aspect ratio 1 remain constant. This is clear evidence that the partial ordering of the dopants in clearly spatially separated layers has a significant influence on carrier heating. This is assumed to be connected with the transfer of the carriers into the interlayer space.

With increasing field strength (region II), both samples with the same weak effective volume doping (types A and D) behave similarly. Their differential conductivities show a significant increase up to 8 and 10 V/cm, respectively, whereas at higher equivalent volume doping the increase is less pronounced because carrier heating becomes less effective with increasing scattering at ionized impurities. This reduced heating at a given field strength also explains the shift in the maximum of differential conductivity towards higher fields.

The subsequent fall in the differential conductivity with further increase in field strength observed for samples A, D, and C is assumed to be due to the increasing interaction with phonons on heating by electric fields, as is well known from hot electrons in bulk semiconductors. In the heavily doped sample, B, impurity scattering still dominates behavior and, therefore, the weak heating does not yield pronounced peculiarities.

A specific feature of the samples investigated is the existence of an extended field region III, in which the differential conductivity remains almost constant. As the saturation occurs in the δ-multilayers with clearly separated wells of type A, as well as in samples of type D with aspect ratio 1, this behavior seems to be characteristic for a 3DEG.

Similar conclusions can be drawn from the magnetoresistivity and the Hall effect as functions of the electric field. Figure 18.3 shows the magnetoresistivity measured for sample A in a magnetic field applied perpendicularly to the layers and to the current direction. In the weak electric field limit a negative magnetoresistivity is observed, which is a characteristic feature of quantum interference effects. The quasidiffusive motion of a 2DEG in a disturbed system leads to a high back scattering probability known as weak localization, which becomes disturbed by a magnetic field due to the consequent change of phase of the electron wave. For δ-doped systems this effect was investigated at lower lattice temperatures by Asche *et al.* (1992). On increasing the applied electric field the value of the magnetoresistivity is diminished and exhibits saturation behavior. It even changes its sign with increasing carrier heating at higher magnetic fields. The dependence on electric field strength can be explained in past by a reduction in the quantum interference effects due to carrier heating, as obtained by Davies *et al.* (1981) in Si/SiO_2 inversion channels, due to the increasing probability of phase breaking scattering processes.

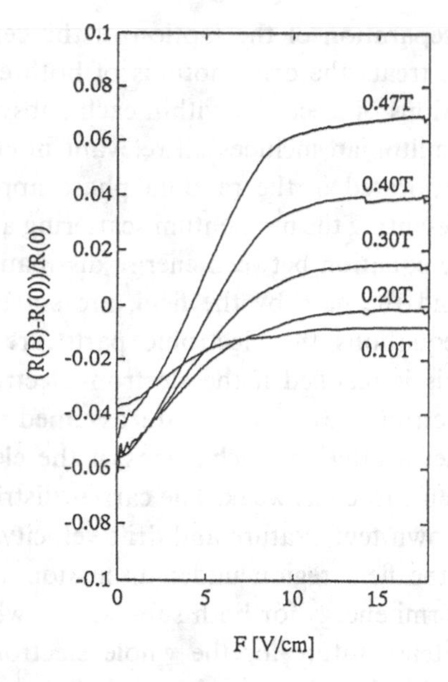

Fig. 18.3. Magnetoresistance of the sample A as a function of the electric field strength at 4.2 K and various magnetic fields.

However, in addition, Fig. 18.3 indicates the remarkable contribution of positive magnetoresistivity, as is well known for a 3DEG, cancelling the negative magnetoresistivity of the 2DEG.

Theoretical model and results of the simulation

In order to elucidate the influences of the field induced real space transfer from the confined to the extended states, and of carrier heating on the electron–phonon interaction processes and the intrasubband drift velocities as functions of the electric field, the experimental investigations were accompanied by numerical simulations. For this purpose the multiple δ-layers (Fig. 18.1) were approximated by a rigid superlattice of 10-nm-wide wells with one quantized electron level 18 meV below, and the free electron states just above, the flat barriers. This simplification assumes electrons with energies lower than the barrier as reflected specularly, while those exceeding the barrier energy are assumed to suffer no reflections from quantum mechanical effects. The potential wells are treated as constant and are not calculated self-consistently in respect of their dependence on real space transfer. Nevertheless, much of the essential physics of the transport should be pointed out.

Kleinert (see Kostial *et al.*, 1993) addressed this nonlinear transport problem using the balance equation approach published by Lei and Ting (1985), who developed a perturbation theory about a displaced electron state in 'thermal equilibrium' at enhanced electron temperature. This implies that the time evolution from the initial to the final state is a rapid microscopic thermalization due to strong intercarrier coupling. In our case, this method proceeds from the separation of the motion of the centers of mass of the two carrier subsystems, that is, it treats the drift motions of both electron gases as separate entities from the relative motions of electrons within each subsystem.

The second quantized Hamiltonian includes all relevant interaction processes and the dynamical screening is incorporated in the random phase approximation. The balance equation for the forces – representing the momentum scattering and the force of the electric field – as well as the balance equation between energy dissipation of the total electronic and phononic excitations, and the gain by the field, are solved as functions of carrier heating. In these balance equations the electronic parts are calculated using shifted Fermi-type distributions. This is justified if the electron–electron collisions within both subsystems (2D and 3D electrons, respectively) are assumed to play the main role in comparison with all the other scattering mechanisms, if the electron–electron collisions between both subsystems are regarded as weak. The carrier distribution in each subsystem can then be described by its own temperature and drift velocity.

For moderate heating in the field region under discussion, a further simplification is introduced via a common Fermi energy for both subsystems, which, strictly, is only valid in the case of one electron temperature for the whole electron gas. However, such an approximation permits one to neglect dynamical particle balance between the subsystems and to take the conservation of the number of particles into account. This may be justified for a low transfer rate into the 3D states in order to describe the experimentally observed features qualitatively.

The calculations were produced using the constants used by Lei *et al.* (1985). With respect to the deformation potential constant it should be noted that the bulk value of 7 eV used for heterostructures by Vickers (1992) could also be used instead of 8.5 eV because the simulation was performed to obtain a qualitative picture only. On the other hand, Manion *et al.* (1987) claimed to require a much higher value for their fit of Schubnikov–de Haas oscillations on account of a weak piezoelectric interaction. With respect to the strength of piezoelectric interaction, different statements have been published (compare, for instance, the statements by Lax and Narayanamurti, 1981; Karl *et al.*, 1988; and Hawker *et al.*, 1989).

The results of the frictional forces are shown in Fig. 18.4. The impurity scattering obviously dominates all other scattering processes of the 2D electrons by orders of magnitude, whereas in the 3DEG this contribution is less pronounced. The numerical ratio of the results for 3DEG to 2DEG depends on the value of background doping, assumed here to be relatively high ($N_D + N_A \approx 5 \times 10^{15}$ cm^{-3}). For $\approx 10^{14}$ cm^{-3}, only the interaction with optical and acoustical phonons rapidly gains influence. Figure 18.5 demonstrates the contributions to the energy loss rates per volume as functions of the electric field. The following features should be recalled: the dissipation of the energy per volume to acoustical phonons (including deformation potential and piezoelectric coupling) is an order of magnitude lower for the 3DEG than for the 2D carriers due to the different electron densities, whereas the energy loss per carrier at a given electric field does not exhibit such a remarkable difference. The contributions by deformation potential, as well as by piezoelectric interaction, to the energy loss of the 2DEG are also shown separately in

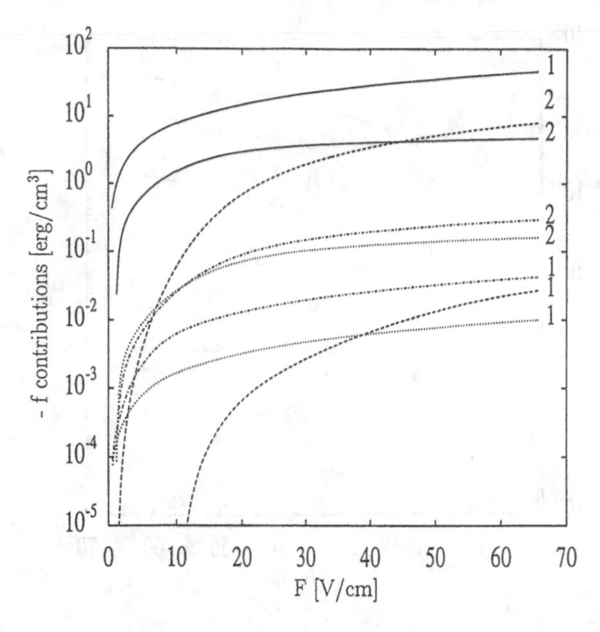

Fig. 18.4. Friction forces as functions of electric field strength for momentum relaxation on ionized impurities (solid lines), acoustic phonons by deformation potential (dashed–dotted lines), as well as by piezoelectric scattering contributions (dotted lines) and on polar optic phonons (dashed lines), contribution of piezoelectric interaction alone depicted by a dotted line. The data for the superlattice approximation are given in the text. (1) and (2) are the 2D and 3D parts respectively.

order to demonstrate the role of both processes with respect to the total scattering on acoustical phonons in the context of our treatment. Since the stronger heating in the interlayer space (see Fig. 18.1) leads to a high probability of phonon emission, the energy dissipation to optical phonons dominates in the average loss rate per 3D electron as well as per volume. This mechanism balances the input power at fields above some V/cm.

The numerical values of the carrier temperatures in both subsystems are depicted in Fig. 18.6. The strongly pronounced rise in temperature at comparatively weak fields is very remarkable, especially for the 3DEG. Of course, lower background doping in the interlayer space would again be reflected by a still more dramatic rise in the 3D electron temperature. A second important feature is the subsequent low increase in carrier temperatures at higher fields. For the 3D electrons the mean carrier energy remains almost constant for the field range investigated due to the high emission of optical phonons. Both effects demonstrate the nonlinear character of the system under investigation in a very distinct way.

The drift velocities for both subsystems are shown in Fig. 18.7 as functions of the applied field. While for the 2DEG the drift velocity remains lower than the sound velocity and increases almost linearly with applied electric field, for the 3DEG the drift velocity already exceeds the sound velocity at some V/cm and its increase with increasing field strength

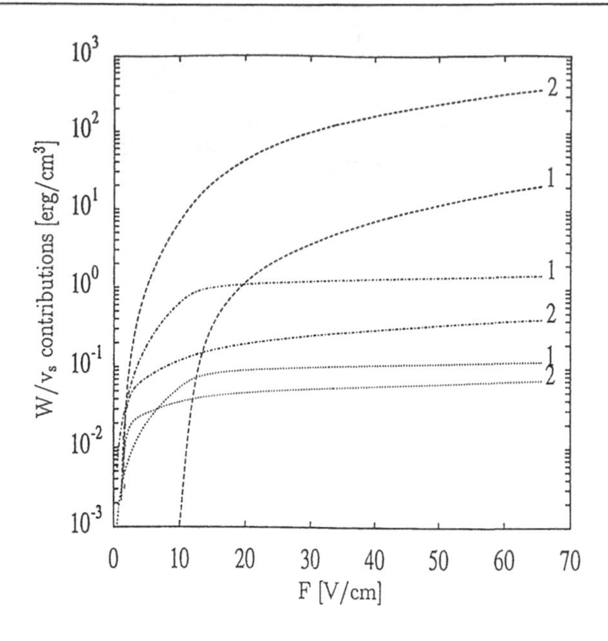

Fig. 18.5. Energy losses as a function of the electric field strength (with the same denotation as in Fig. 18.4).

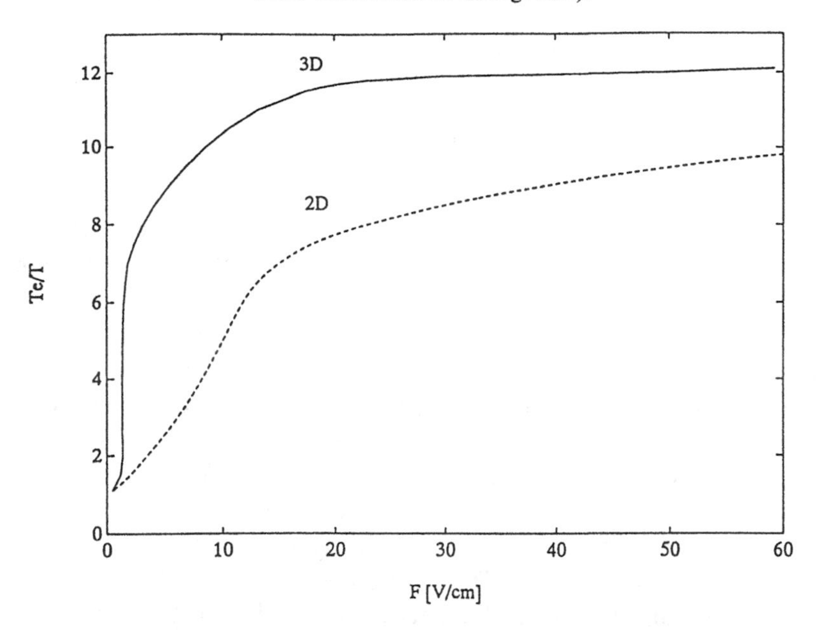

Fig. 18.6. Electron temperatures versus electric field strength.

can best be judged by the change in its slope (Fig. 18.7(b)). On account of the high ratios v^{3D}/v^{2D} a comparatively small percentage of 3D carriers already yields a dominating current contribution at fields of some V/cm and is about five times higher at about 10 V/cm for the parameters chosen in the present simulation (for details see Kostial, 1993). The

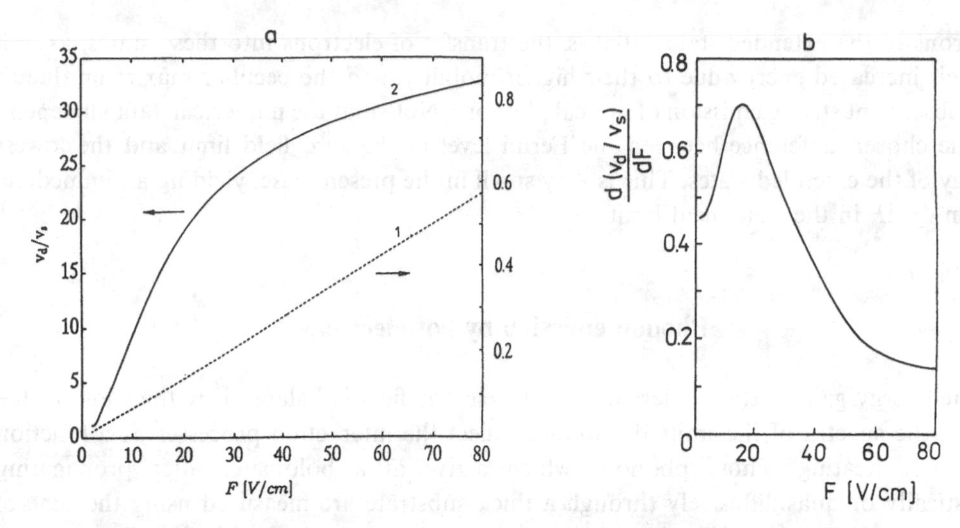

Fig. 18.7. (a) Field dependence of the 2D and 3D carrier drift velocities. (b) Differential drift velocity of the 3DEG as a function of electric field.

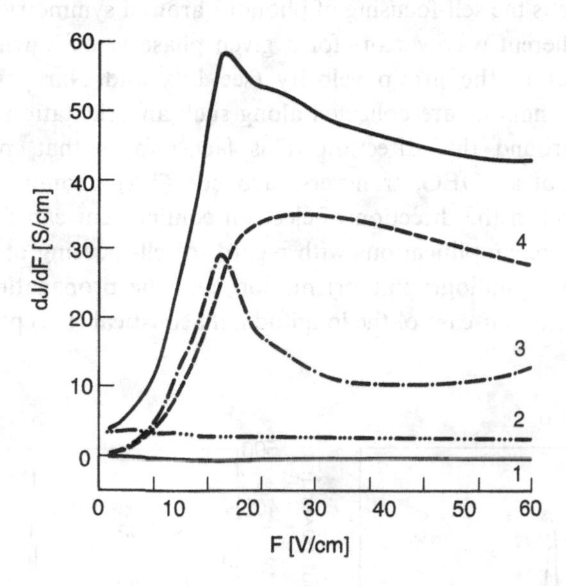

Fig. 18.8. Differential conductivity dj/dF versus electric field strength for the parameters of Fig. 18.4. Curves 1–4 show the partial contributions: 1: $v^{2D} \times dn^{2D}/dF$, 2: $n^{2D} \times dv^{2D}/dF$, 3: $v^{3D} \times dn^{3D}/dF$, 4: $n^{3D} \times dv^{3D}/dF$.

resulting differential conductivity depicted in Fig. 18.8, together with its four constituents, demonstrates the specific peculiarities as observed for sample A (compare Fig. 18.2). A comparison of the slope of the drift velocities of the electrons in the extended and confined states, respectively, demonstrates that the contributions from the 3DEG dominate: they exhibit a strong initial rise (according to regions I and II of Fig. 18.2) and a weak fall for further increase in field strength. The experimental curve thus reflects the behavior of the electrons in the extended states, that is, the transfer of electrons into these states, as well as their increased energy due to their higher mobility, and the peculiar maximum due to the subsequent strong emission of optical phonons. Note that the numerical data all depend on the chosen difference between the Fermi level in the zero field limit and the lowest energy of the extended states. This is very small in the present case, yielding an immediate rise in dj/dF in the weak field limit.

Phonon emission by hot electrons

As the energy gain of the carriers due to the electric field is balanced by their loss to the lattice, the spectra of the emitted phonons reflect the interaction processes as a function of carrier heating. Those phonons which arrive at a bolometer after propagating ballistically or quasidiffusively through a thick substrate are measured using the time of flight method (Fig. 18.9) in order to resolve the modes on account of their different group velocities (Danilchenko *et al.*, 1993a, b).

An essential feature is the self-focusing of phonons around symmetry axes because there are phonons with different wavevectors for a given phase velocity which have the same direction with respect to the group velocity (see Lax and Narayanamurti, 1980, for instance). Therefore, phonons are collected along such an orientation which are emitted in a certain cone around this direction. This fact implies that, perpendicular to a (001)-orientied layer of a 2DEG, transverse acoustic (TA) phonons with a significant wavevector component in the direction of electron confinement can be registered by the bolometer. Furthermore, amplifications with regard to self-focusing of the various modes differ with respect to crystallographic orientation, and the propagation of a mode may even be suppressed, as in our case of the longitudinal acoustical (LA) phonon in the $\langle 001 \rangle$ direction.

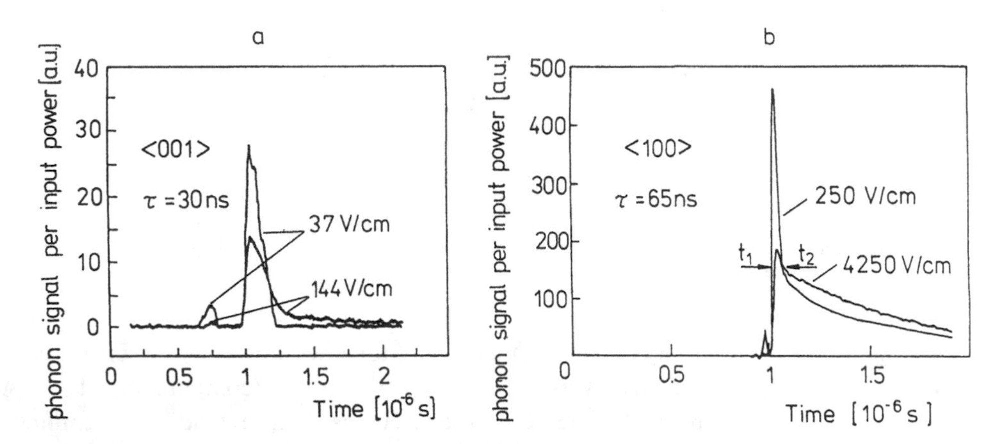

Fig. 18.9. Time of flight spectra of nonequilibrium phonons at different electric fields. The bolometer signals are divided by the input electrical power to the δ-structure. (a) and (b) correspond to bolometers of different sensitivity.

Because of the finite lifetime of the phonons, proportional to ν^{-5}, where ν is the phonon frequency, high energy phonons do not arrive at the bolometer when the path is too long. In particular, the optical phonons emitted by the hot electrons (see Fig. 18.5) have a lifetime of some ps only, leading to a cascade of repeated decay processes and conversion of the resulting transverse acoustical phonons into the longitudinal phonons that are necessary to fulfil conservation of energy and wavevector in the next decay process. In the cascade from an 8 THz phonon down to about 1 THz, the last step is performed within 23 ns, as reported by Tamura and Maris (1985) for instance, whereas the further decay time is given by 0.7 μs, which is comparable to the propagation time for ballistic motion of the phonons with the velocity of sound.

A further important fact is the isotope scattering which is proportional to ν^4 and thus influences the propagation of the high energy phonons significantly. Tamura (1984) obtained a scattering time of 0.14 μs for a 1 THz phonon on isotopes, which is still shorter than the time taken to ballistically traverse the distance from the origin to the bolometer.

All these characteristic features lead to the time of flight spectra shown in Fig. 18.9 which exhibit a sharp front of the phonon signal at about 1 μs following the application of short electric pulses for carrier heating with a distance of 3.4 mm between the conducting layer and the bolometer and a group velocity of TA phonons along ⟨001⟩ of 3.34 × 10^5 cm s^{-1}. The spread of the phonon flux for the lowest input power might be due to variation of the path length by isotope scattering and reflection at the sample boundary, as well as to the variable path lengths of the phonons emitted by the total area of the conducting layer and gathered by the bolometer area of about 1 mm^2. With increasing carrier heating the phonon flux divided by the input power can be seen to diminish, while the signal becomes broader over time, indicating increasing emission of phonons with higher energies which suffer a cascade of decay and conversion processes as well as isotope scattering. At strong electric fields, the tail dominates (see Fig. 18.9(b)) and for discussion purposes the spectra are divided into a narrow peak up to t_2 and a signal in the broad

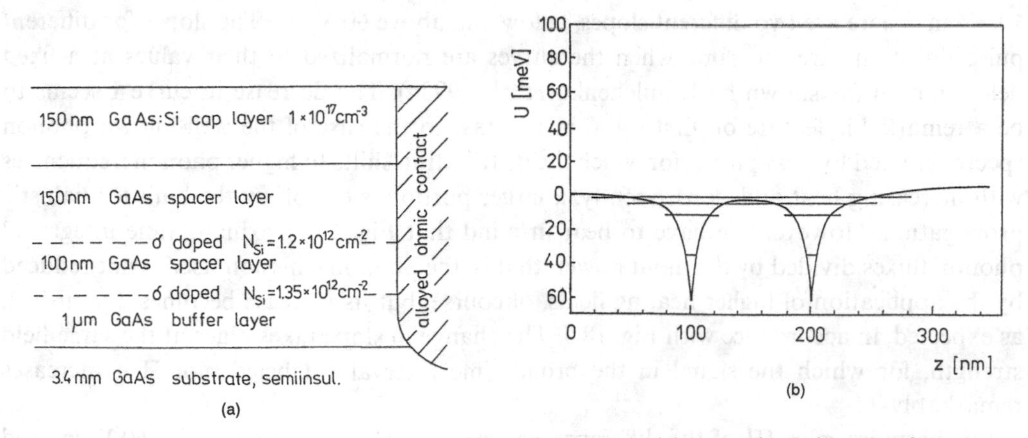

Fig. 18.10. (a) Schematic presentation of the vertical layout. (b) Sublevels of the Γ-conduction band.

time interval above t_2; both parts are integrated over time, where the upper limit of the long tail is given by the signal to noise ratio. High sensitivity could be obtained by accumulating up to 10^4 signals.

With respect to the small peak for phonons preceding the pronounced peak for TA phonons, the ratio of the arrival times is consistent with the ratio of the sound velocities of LA and TA phonons along $\langle 001 \rangle$. Bearing in mind that the great difference in the amplification factor due to self-focusing calculated by Philip and Viswanathan (1978) is 0.186 and 10.086, respectively, the conclusion can be drawn that at low heating fields an appreciable number of LA phonons are emitted along with TA phonons. The LA phonons are created by deformation potential interaction and piezoelectric interaction, while the TA phonons are not allowed for deformation potential interaction in parabolic bands. The fact that the LA phonons were not observed in other investigations (e.g. Hawker *et al.*, 1989) can be explained by the high sensitivity obtained by Danilchenko *et al.* (1993b).

The phonon emission shown in Fig. 18.9 was measured in a MBE grown GaAs sample containing two δ-layers with 1.2×10^{12} and 1.35×10^{12} Si atoms per cm^2, respectively, at a distance of 100 nm (for details see Fig. 18.10). The doping was performed at substrate temperatures of 500 °C and 520 °C, respectively, with a growth interruption of 120 s and 60 s before and after the δ-doping. The self-consistently calculated potential is depicted in Fig. 18.10(b) (lhn *et al.*, 1992) and shows four subbands populated with 72%, 21%, 6%, and 1% of the carriers respectively.

The 2-mm-wide samples were equipped with current contacts at a distance of 0.25 mm by alloying Au:Ge at a temperature of 430 °C for 4 min in a N_2/H_2 atmosphere. The smallness of the contact resistance was checked by four point measurements on analogous samples. The bolometer was fabricated by evaporating In at 300 °C on the back side. Two bolometers with different sensitivites were prepared (Fig. 18.9(a, b)). The measurements were performed by immersing the sample in superfluid helium at 2 K.

The time integrated phonon fluxes per input power are shown in Fig. 18.11 as functions of the applied field strength, as investigated by Danilchenko *et al.* (1993b). In the plot for the narrow TA peak – labeled I – a decrease with increasing field strength is seen up to 1 kV/cm. There are two different slopes, below and above 60 V/cm. The slopes for different pulse durations are the same when the curves are normalized to their values at a fixed field strength (as shown by Danilchenko *et al.*, 1993a). The decrease in curve I seems to be a remarkable feature on first sight, in contrast to the case of the well-known phonon spectra created by heat pulses, for which the distribution shifts to higher phonon frequencies with increasing heat and, consequently, a larger portion is cut off by the limit for ballistic propagation. However, we have to bear in mind that Fig. 18.11 exhibits time integrated phonon fluxes divided by the input power, that is, the phonon emission itself is not reduced by the application of higher heating fields, of course, but its increase becomes diminished, as expected, in accordance with Fig. 18.5. The change in slope takes place at the same field strength, for which the signal in the broad time interval – labeled plot II – increases remarkably.

Furthermore, plot III of the difference between signal I measured above 60 V/cm and the extrapolation of the lower field branch (straight line in Fig. 18.11) is similar to plot

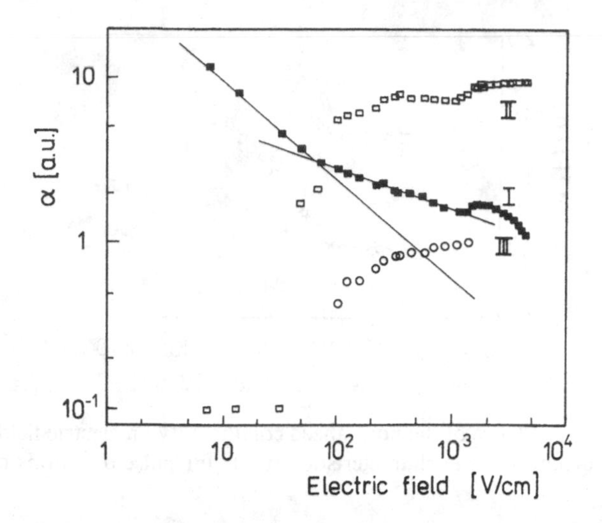

Fig. 18.11. The time integrated narrow and broad TA-phonon signals divided by the input power as functions of electric field for pulse durations of 15 ns. Plot I refers to the narrow and plot II to the broad time interval signal. Plot III is the difference between data set I and the straight line prolongation from low fields.

II. These facts suggest that for electric fields >40 V/cm, optical phonon emission becomes significant and the decay products of these optical phonons contribute in part to the TA signals measured in the narrow peak. They thus increase the flux consisting of the directly emitted acoustical phonons, but mainly contribute to the spectra in the broad interval of arrival times. While the additional ballistically propagating phonons should be created in the first steps of the cascade of decay and conversion processes, and further on escape isotope scattering, the phonons arriving with delay should be due to the lower energy decay products generated later in the cascade, as well as to those of the earlier generation, which are propagating quasidiffusively due to isotope scattering. Taking the energy dissipation due to optical phonon emission into account, calculated for a multiple δ-layer with only $5 \times 10^{11}/\text{cm}^2$ dopants per sheet (Fig. 18.5), the electric field strength value of about 40 V/cm for the current, more strongly doped, double δ-layer seems plausible. Furthermore, the conductivity as a function of field strength (Fig. 18.12) shows a pronounced increase above 10 V/cm, suggesting electron transfer from the low energy levels to the higher, more extended, states (Fig. 18.10(b)) in which, again, the carriers are more mobile due to reduced scattering on ionized impurities. This transfer promotes optical phonon emission, of course, as described in the preceding section.

With respect to electric fields above 1 kV/cm, both plots in Fig. 18.11 show a remarkable increase, followed by saturation for plot II and a decrease at higher field strengths for plot I. The increase in the phonon flux per input power results from the opening of a new channel of phonon emission by carrier heating. The inset is exhibited in the field region in which the conductivity begins to decrease (Fig. 18.12). Thus Danilchenko *et al.* (1993a) suggest that phonon assisted carrier transfer occurs from the Γ band edge to the

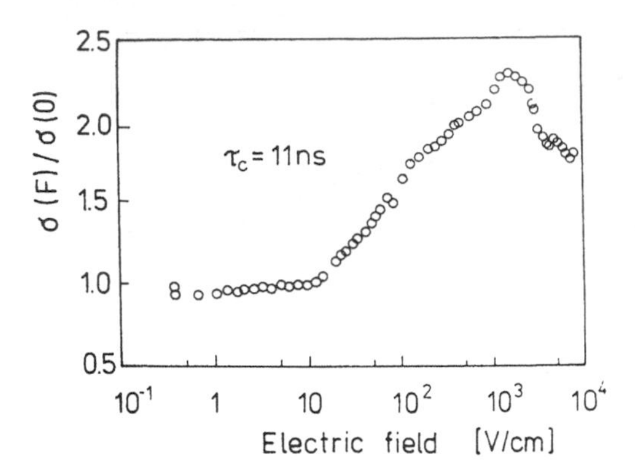

Fig. 18.12. Dependence of the normalized conductivity on electric field derived from the current–voltage characteristics at 2 K for pulse durations of 11 ns.

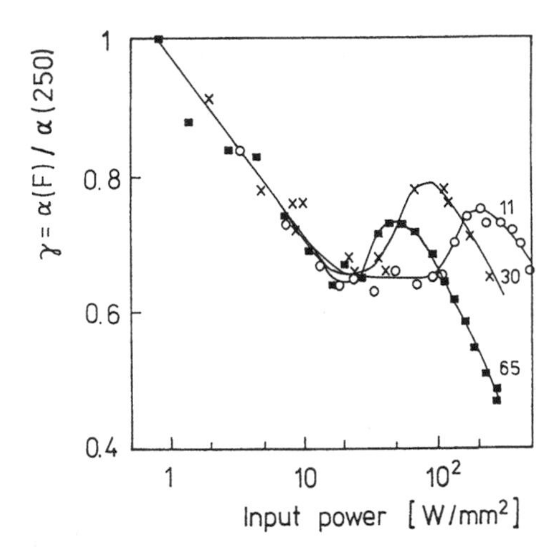

Fig. 18.13. The ratios of the time integrated narrow peak signals normalized to their values at $F = 250$ V/cm as functions of electric input power per unit sheet for different pulse durations.

L-valleys of the energetically higher conduction band in GaAs. As is known from bulk GaAs, this leads to an effective reduction in mobility due to the higher effective mass in the L-valleys, thus explaining the fall in conductivity. Such a Γ–L-transition is assisted by the deformation potential interaction of the electrons with LO-phonons in order to conserve momentum. This process thus explains the observed new burst in the time of flight spectra. Also, the maximum in plot I shifts to lower fields with growing pulse duration (Fig. 18.13). The decrease in the time integrated phonon flux per input power can be caused by the formation of a 'hot spot', as described by Kazakovtsev and Levinson (1985). This phenomenon accounts for inelastic phonon–phonon interaction when the density of nonequilibrium phonons reaches a critical value. The maxima of the narrow peak signal as functions of input power deliver 2.5, 2.7, and 2.8 μJ/mm², respectively, for

11, 30, and 65 ns pulse duration, in agreement with the value of 3 μJ/mm^2 obtained by Danilchenko *et al.* (1989) in the case of photoexcitation of GaAs and interpreted as formation of a hot spot.

For the Γ–L-transition of the carriers in quantum wells, due to the higher effective mass at the L-minimum and the resulting energy positions of the quantized levels, the energy difference between the Γ- and L-states is smaller than in bulk GaAs for the same subband number, and is still further reduced with respect to transitions from a higher Γ-subband to the zeroth L-subband. This fact and the stronger heating in the interlayer space of the relevant δ-doped structures in comparison with bulk material could explain why the critical field for Γ–L-transitions is observed at about 1 kV/cm. Possibly both effects lead to the smoother effect of scattering from Γ- to L-states as a function of electric field strength and could explain why the differential conductivity shows a decrease but no change in sign.

Summary

Experimental investigations of multiple δ-doped layers in GaAs with different separation distances between these doping planes and with various doping concentrations lead to conclusions on the carrier transfer from the quantized level of lowest energy to the higher and more extended subbands. In these higher energy levels, and especially above the potential barriers, the electrons are more mobile due to weaker scattering on ionized impurities, and, consequently gain more energy from the applied electric field. Thus, the probability of emission of optical phonons increases rapidly and is the main contribution balancing the electric input power. This effect diminishes the increase in the differential conductivity with increasing field strength and yields a maximum in the differential conductivity with successive saturation behaviors.

Numerical simulations by solving the system of coupled equations for the force and energy balances for a model with a two- and a three-dimensional electron gas at constant total carrier concentration elucidate the situation and the conclusions drawn from the experimental results.

Time of flight measurements of ballistically propagating phonons emitted by hot electrons in a double δ-layer system combined with current–voltage characteristics exhibit details with respect to the electron–phonon interaction process. This concerns the character of the emitted modes of acoustic phonons, which can clearly be separated due to their different sound velocities. Furthermore, the onset of optical phonon emission and their dominating role with increasing electric field can be established. The decay of optical phonons is discussed and a contribution of these decay products to the ballistically propagating phonons is pointed out, which could be analyzed with the help of Monte Carlo calculations (for instance Danilchenko (1994a,b)).

At heating electric fields above 1 kV/cm a new channel of phonon emission is opened. The combined investigations of the time of flight method for the detection of emitted phonons and of current–voltage characteristics for the determination of the conductivity of the electrons allow to explain the situation by transitions of the carriers from Γ- to L-band minima assited by deformation potential interaction of electrons and phonons.

References

Asche M., Friedland K. J., Kleinert P., and Kostial H. (1992) *Semicond. Sci. Technol.* **7**, 923.
Asche M., Hey R., Höricke M., Ihn T., Kleinert P., Kostial M., Danilchenko B., Klimashov A., and Roshko S. (1994) *Semicond. Sci. Technol.* **9**, 835.
Brennan K. F. and Park D. H. (1989) *Journ. Appl. Phys.* **65**, 1156.
Danilchenko B., Kazakovtsev D. V., Slutskii M. (1989) *Phys. Lett.* A **138**, 77.
Danilchenko B., Roshko S., Asche M., Hey R., Höricke M., and Kostial H. (1993a) *Journ. Phys. C, Condensed Matter* **5**, 3169.
Danilchenko B., Klimashov A., and Roshko S. (1993b) private commun.
Danilchenko B., Kazakovtsev D., and Obuchov I. (1994a) *Zhurn. Eksper. Teor. Fiz.* **106**, 1439.
Danilchenko B., Klimashov A., Rashko S., and Asche M. (1994b) *Phys. Rev.* B **50**, 5725.
Davies R. A., Uren M. Y., and Pepper M. (1981) *J. Phys. C, Condensed Matter* **14**, 531.
Gribnikov Z. S., Hess K., and Kosinovsky G. A. (1995) *J. Appl. Phys.* **77**, 1337.
Hawker P., Kent A. J., Henini M., and Hughes O. H. (1989) *Solid State Electrons* **32**, 1755.
Ihn T., Friedland K. J., and Zimmermann R. (1992) private commun.
Karl H., Dietsche W., Fischer A., and Ploog K. (1988) *Phys. Rev. Lett.* **61**, 2360.
Kazakovtsev D. V. and Levinson I. B. (1985) *Sov. Phys. JETP* **61**, 1318.
Kleinert P. and Asche M. (1994) *Phys. Rev.* B **50**, 11022.
Kostial H., Ihn T., Kleinert P., Hey R., Asche M., and Koch F. (1993) *Phys. Rev.* B **47**, 4485.
Lax M. and Narayanamurti V. (1980) *Phys. Rev.* B **22**, 4876.
Lax M. and Narayanamurti V. (1981) *Phys. Rev.* B **24**, 4692.
Lei X. L. and Ting C. S. (1985) *Phys. Rev.* B **32**, 1112.
Lei X. L., Birman J. L., and Ting C. S. (1985) *J. Appl. Phys.* **58**, 2270.
Manion S. J., Artaki M., Emanuel M. A., Coleman J. J., and Hess K. (1987) *Phys. Rev.* B **35**, 9203.
Mares J. J., Xiao Mei Feng, Ihn T., and Koch F. (1993) *The Physics of Semiconductors* (Beijing 1992, World Scientific, eds. Ping Jiang and Hou-Zhi Zheng), p. 1226.
Moško M. and Novák I. (1990) *J. Appl. Phys.* **67**, 890.
Philip J. and Viswanathan K. S. (1978) *Phys. Rev.* B **17**, 4969.
Tamura S. (1984) *Phys. Rev.* B **30**, 849.
Tamura S. and Maris H. J. (1985) *Phys. Rev.* B **31**, 2595.
Vickers A. J. (1992) *Phys. Rev.* B **46**, 133131.

19

Ordered delta-doping

R. L. HEADRICK, L. C. FELDMAN,
AND B. E. WEIR

19.1 Introduction

Tailored doping profiles are a critical component of the future semiconductor devices envisioned for the twenty-first century. Advances in thin film growth, primarily molecular-beam epitaxy, have demonstrated that such tailoring is possible almost to the monolayer level. Most examples of dopant control correspond to layers confined in the growth direction, that is, sharp two-dimensional (2D) structures as measured perpendicular to the plane of the surface. An equally important goal of growth is custom definition within the surface plane.

Before embarking on a discussion of the fabrication of such structures it is worth describing some general arguments as to the advantages one can expect from a periodic array of dopants. The first is associated with the reduction in ionized impurity scattering that one can anticipate from an ordered array of dopants. This is described more completely in Section 19.3. Basically, such an ordered array forms a Bragg lattice and carriers can be transported through the lattice with minimal ionized impurity scattering. Such mobility enhancements will require near perfection in the 2D dopant lattice, and thus present a formidable technical challenge.

A second motivation concerns the statistics of random doping. As devices approach the $(0.1\ \mu m)^3$ size they may contain as few as 10^3 dopant atoms. Variation in device properties will result from variation in dopant incorporation, which, for random doping, is (number of dopants)$^{1/2}$. Obviously, this will limit the performance of a device array, designed for uniform device performance. The tailored, 2D array will limit this randomness. A peridic 2D array will have (almost) no variability in dopant incorporation. Note that the reduction of 'random statistics' does not require the same degree of crystalline perfection as does the mobility enhancement.

A third concept is motivated by the desire to form lower-dimensional structures such as quantum wires and dots. One can envision single structures, or arrays of nanometer-scale wires or dots with unique properties. The formation of lower-dimensional

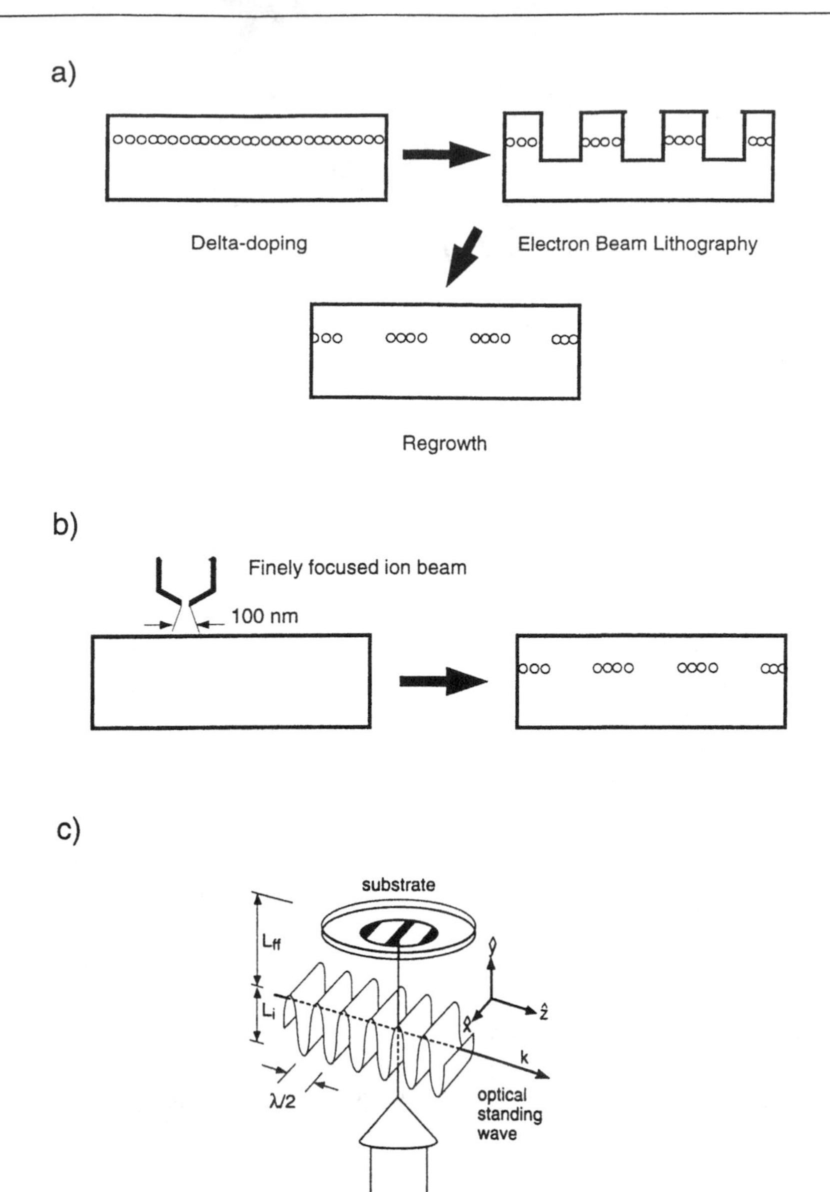

Fig. 19.1. Lithography methods for producing laterally tailored structures.

structures does not depend on atomic scale ordering, but requires control at the 100-Å-length scale.

A few methods can be envisioned to produce laterally ordered δ-doping structures. One class of fabrication procedures makes use of sophisticated lithographic and patterning techniques. The first of these (Fig. 19.1(a)) requires lithography on the 100 Å scale to produce a quasiperiodic dopant array. Such a doping profile may be created by electron beam writing of interesting submicron devices. A second approach (Fig. 19.1(b)) uses finely focused ion beams for direct implantation of dopants on the ~1000 Å scale. A third scheme, proposed by Timp *et al.* (1992), uses optical techniques and the diffraction pattern incurred by interacting laser beams (Fig. 19.1(c)). This concept uses the force exerted by light to deflect a neutral atomic beam during deposition onto a substrate. The feature length is limited by the wavelength of laser light, of the order of 1 μm. Each of these processes depends on lithographic concepts and does not produce lateral ordering of the doping on the atomic scale.

A different approach makes use of the concepts of surface science for lateral tailoring (Fig. 19.2). One suggestion (Fig. 19.2(a)) takes advantage of the step modulated growth model to have deposited dopants 'snuggle up' to a step edge. For example, Ramsteiner *et al.* (1994) studied Si δ-doping layers on 2° miscut GaAs substrates and attributed polarization asymmetries in Raman scattering measurements to wirelike incorporation of deposited Si along step edges. This technique results in a periodicity determined by the initial step spacing. However, perfect ordering may not be achievable by this method because terrace lengths are not perfectly periodic. Aperiodic step spacing also leads to a different number of atoms impinging on each terrace, so that each 'wire' may have a

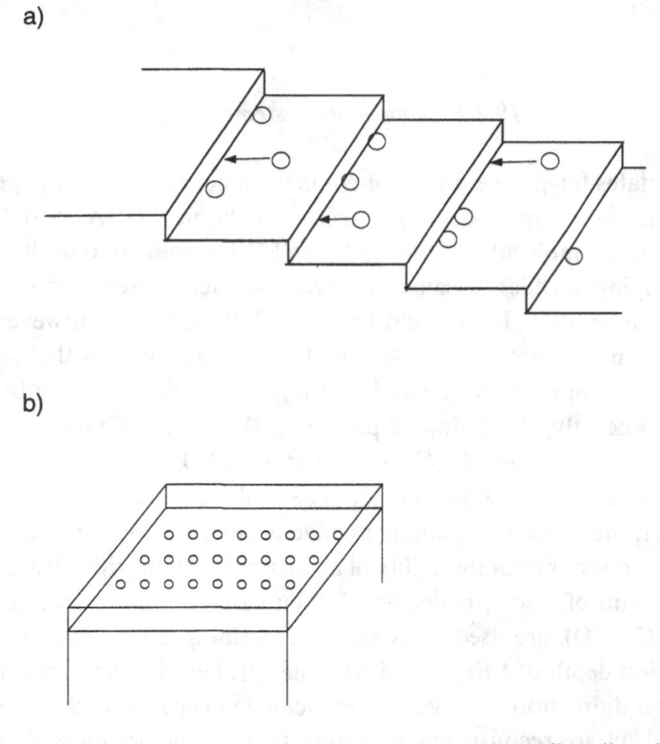

Fig. 19.2. Surface science methods for producing laterally tailored structures. In (a) deposited dopant atoms are envisioned as attaching to steps. Part (b) illustrates an ordered adsorbate (dopant) array.

slightly different doping density. Another surface science oriented scheme (Fig. 19.2(b)) makes use of the absorbate structure created when some dopants are deposited onto a semiconductor surface. Interaction of deposited atoms with the surface and with each other produces an ordered structure, with a length scale typically <10 Å. In both the 'step attachment' case and the 'ordered adsorbate' case, high quality crystal overgrowth without perturbation of the dopant structure is required for the δ-doping configuration. The last example, formation of adsorbed dopant structures (Fig. 19.2(b)), is most easily envisioned and tested, and is the central focus of this chapter.

19.2 Preserved surface reconstructions

As discussed in the introduction, surface science techniques offer the opportunity to form an ordered doping layer by overgrowth of the host semiconductor material. Preserved structures of this type are far from equilibrium, and can only be fabricated under very specific growth conditions. However, they represent the ultimate δ-doping layers both because of the possibility of confinement to one monolayer, and because of the unprecedented high doping concentrations. Most of the work on these structures has been on silicon and Si–Ge alloys. In the following sections, we will confine our discussion to group IV based structures.

19.2.1 Candidate systems

In silicon, the candidates for p-type δ-doping come from column III of the periodic table (B, Al, Ga); candidates for n-type δ-doping come from column V (P, As, Sb). Early studies of doping during silicon molecular-beam epitaxy (MBE) pointed out the difficulty of producing sharp doping profiles because of severe surface segregation of the dopant (Thomas and Francombe, 1969; Becker and Bean, 1977; Bean 1978). However, these early studies used growth temperatures of 550–900 °C. It is now recognized that segregation is a strong function of growth temperature and that high quality epitaxy can be achieved at lower temperatures. Recently, sharp doping profiles of B, Ga, and Sb have been produced at growth temperatures below 400 °C (Gossmann *et al.*, 1990).

Clean silicon surfaces and a variety of dopant-stabilized surfaces form ordered configurations. However, many reconstructions involve geometries that are not amenable to overgrowth. In order to test the compatibility of an ordered system with silicon overgrowth, a submonolayer amount of silicon is deposited, then surface probes, such as low energy electron diffraction (LEED), are used to assess the remaining degree of order. Because of the limited penetration depth of LEED (<5 Å), other probes such as X-ray diffraction or transmission electron diffraction are used to characterize deeper buried δ-layers.

Typically, ordered surface reconstructions are unstable in the presence of excess surface silicon; however, several have been found to be stable. For example, Table 19.1 shows that ordered boron structures (on Si(111), Akimoto *et al.*, 1987; on Si(100), Headrick *et al.*,

Table 19.1. *Stability with respect to deposited silicon of several dopant-induced surface reconstructions on silicon as determined by low-energy electron diffraction, X-ray diffraction, and transmission electron diffraction*

Substrate orientation	Dopant	Structure	Preserved under α-Si	Reference
(111)	None	(7 × 7)	Y	Gibson *et al.* (1986).
(111)	Boron	($\sqrt{3} \times \sqrt{3}$)	Y	Akimoto *et al.* (1987).
(111)	Gallium	($\sqrt{3} \times \sqrt{3}$)	N	Headrick *et al.* (1989a).
(111)	Antimony	($\sqrt{3} \times \sqrt{3}$)	N	Zotov *et al.* (1992a).
(111)	Antimony	(2 × 2)	N	Zotov *et al.* (1992a).
(111)	Antimony	($5\sqrt{3} \times 5\sqrt{3}$)	Y	Zotov *et al.* (1992a).
(100)	None	(2 × 1)	N	Gibson *et al.* (1986).
(100)	Boron	(2 × 1)	Y	Headrick *et al.* (1990a).
(100)	Gallium	(2 × 2)	N	Headrick (unpublished).

1990a) and an antimony ordered structure (on Si(111), Zotov *et al.*, 1992a) can be preserved under amorphous silicon. Crystalline overgrowth of silicon has also been studied extensively for the boron reconstructions, but has not yet been observed for the antimony ($5\sqrt{3} \times 5\sqrt{3}$) reconstruction on Si(111). A number of other dopant-induced surface reconstructions have been identified, but their stability with respect to silicon overgrowth has not been determined. These include Sb reconstructions on Si(110) (Zotov *et al.*, 1992b) and Al reconstructions on Si(110) (Zotov *et al.*, 1992c). Table 19.1 also demonstrates that the preservation of ordered reconstructions, that is, the absence of in-plane redistribution upon amorphous Si overgrowth, depends on the atomic arrangement of the ordered surface. For example, antimony forms three different reconstructions on Si(111), but only one retains its order after Si deposition at room temperature.

19.2.2 First observation of ordered δ-doping

The first ordered δ-doping layer to be formed was the Si(111)-B$\sqrt{3} \times \sqrt{3}$/Si(111). The B$\sqrt{3} \times \sqrt{3}$ surface reconstruction is an extremely stable surface reconstruction and has been formed by deposition of HBO_2, B_2O_3, elemental boron, decaborane, or by segregation from a heavily doped specimen. Akimoto *et al.* (1987) found that a small fraction of the boron stabilized $\sqrt{3} \times \sqrt{3}$ structure could be preserved under epitaxial silicon. Two samples were prepared, one in which the Si(111)–B$\sqrt{3} \times \sqrt{3}$ was covered by 100 Å of silicon with the substrate at room temperature producing an amorphous cap layer, and another with the substrate at 650 °C during the silicon deposition, producing an epitaxial cap layer. A reflection high energy electron diffraction pattern (RHEED) showed that the epitaxial cap layer had a 7 × 7 surface structure indicative of a clean, crystalline (111)-oriented surface layer. *In-situ* grazing angle X-ray diffraction measurements showed that the B induced $\sqrt{3} \times \sqrt{3}$ reconstruction was still present within both samples. However, the intensity of the $(-\frac{2}{3}, \frac{4}{3})$ diffraction peak was smaller on the sample with epitaxial overgrowth by a factor of ≈ 100. These results suggested that a fraction (perhaps as much as 10%) of the ordered boron originally on the surface had been preserved in a buried, ordered configuration. The remainder presumably segregated along with the growing surface as the cap layer was deposited, forming a partially ordered δ-layer with a distribution along the growth direction. In Sections 19.4 and 19.5 we describe studies designed to understand the mechanisms of ordering in these Si based layers.

19.3 Electrical properties of ordered δ-doping layers

High speed heterojunction bipolar transistors which utilize a thin, very heavily doped p-type base have been proposed (Levi, 1988). The performance of these devices is limited by elastic scattering of minority carriers injected from the emitter, so that a reduced elastic scattering rate, τ_{el}^{-1}, could improve their properties significantly. Replacing the uniformly (randomly) doped base with an ordered dopant superlattice should reduce the scattering rate. For example, Levi *et al.* (1989) predicted more than a threefold decrease in τ_{el}^{-1} in GaAs-AlGaAs bipolar transistors. This application would take advantage of the reduced scattering for low energy carriers, since the injected minority carriers have a very low energy compared to the majority carriers in heavily doped layers. Below, we briefly discuss the theory of the elastic scattering rate for ordered δ-doping structures.

Suppose that we are interested in a particle with initial wave vector $\mathbf{k}$, energy $E(\mathbf{k})$, and velocity $v_{\mathbf{k}}$. It has a scattering rate τ_{el}^{-1} from donor (or acceptor) ions and a mean free path $l_{\mathbf{k}} = v_{\mathbf{k}}\tau_{el}$. If $\mathbf{k}l_{\mathbf{k}} \gg 1$ then the Born approximation gives a good approximation of the scattering rate. Using the first term in the Born series, we obtain

$$\tau_{el}^{-1} = \frac{2\pi}{\hbar} \int \frac{d^3\mathbf{q}}{(2\pi)^3} |V(\mathbf{q})|^2 \delta(E(\mathbf{k}) - E(\mathbf{k} - \mathbf{q})). \tag{19.1}$$

Here, $\mathbf{q}$ is the momentum transfer in a scattering event, and $V(\mathbf{q})$ is the Fourier transform of $V(\mathbf{r})$, the interaction potential of the dopant ions. The term $V(\mathbf{q})$ can be rewritten as

$$|V(\mathbf{q})|^2 = s(\mathbf{q})|v(\mathbf{q})|^2, \tag{19.2}$$

where $s(\mathbf{q})$ is the structure factor

$$s(\mathbf{q}) = \left| \sum_{i=1}^{n} \exp(i\mathbf{q} \cdot \mathbf{R}_i) \right|^2 \tag{19.3}$$

and

$$v(\mathbf{q}) = \frac{4\pi e^2}{q^2 \varepsilon(\mathbf{q})}. \tag{19.4}$$

Here, $\mathbf{R}_i$ define the usual lattice vectors and $\varepsilon(\mathbf{q})$ is the dielectric screening function.

Levi *et al.* (1989) originally envisioned a δ-doping layer confined to a *single monolayer*, but randomly arranged on lattice sites in that layer. They pointed out that if the fraction, f, of impurity-occupied sites is large, then

$$s(\mathbf{q}) = n(1 - f). \tag{19.5}$$

In the dilute limit $f \to 0$, Eqn. (19.5) gives the classical result

$$s(\mathbf{q}) = n, \tag{19.6}$$

where n is the total number of impurity atoms. The factor $(1 - f)$ appears because impurity atoms sit on discrete lattice sites and double occupancy of a site is not allowed.

The implications of this analysis can be made apparent by a simple example. Suppose that we have a δ-doping layer with $\frac{1}{2}$ monolayer coverage confined to a single monolayer. If the impurity atoms are confined to lattice sites, but are distributed randomly among these lattice sites, then $f = 0.5$ and the scattering rate is reduced by a factor of two compared to completely random doping (for small k, as discussed below).

Ordered configurations can give even more dramatically-enhanced electronic effects. If we assume that the $\frac{1}{2}$ monolayer of atoms occupies every other site, then we have an ordered (2×1) arrangement. Figure 19.3(a) shows a schematic of this configuration. Here, the dopant atoms form a perfectly ordered lattice, and the structure factor $s(\mathbf{q})$ is the corresponding reciprocal lattice (of δ-functions) with lattice vectors $a^* = 2\pi/a$, $b^* = 2\pi/b$, where a, b are the dimensions of the unit cell. The maximum momentum transfer of a scattering event is $|\mathbf{q}_{max}| = 2|\mathbf{k}|$. For low energy charge carriers with $q_{max} < a^*$, b^* the ionized impurity scattering vanishes ($\tau_{el}^{-1} \to 0$). Thus a large enhancement in low temperature mobility is expected for a perfectly ordered arrangement.

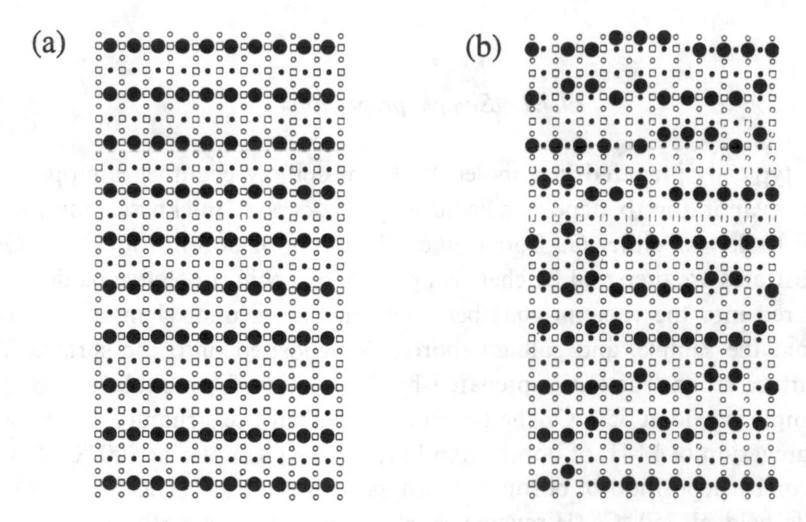

Fig. 19.3. (a) Top view of an ordered (2×1) arrangement with the dopants confined to one monolayer. (b) Partially ordered (2×1) arrangement with atoms distributed over three layers as described in the text.

A more realistic situation is shown in Fig. 19.3(b). Starting from the perfectly ordered arrangement of Fig. 19.3(a), we have allowed atoms to segregate (move out of the plane of the paper) over a total of three layers, resulting in an arrangement with one-third of the atoms in the original sites, and the others displaced by either one or two 'hops'. Since three layers are accessible, each layer has $\frac{1}{6}$ of a monolayer, and the appropriate value for f in Eqn. (19.5) is $\frac{1}{6}$, leading to a modest (17%) reduction in the elastic scattering rate. This arrangement is considered to be partially ordered since the (2×1) periodicity is not completely lost. Therefore, the dopant atoms do not occupy random sites and, by the arguments outlined above, an additional reduction in the scattering rate is expected. Specifically, the scattering is strongly suppressed at small $\mathbf{q}$, but not entirely eliminated as in the ideal case. Further results for a partially ordered δ-doping layer can be found in Section 19.5.4.

19.4 Boron vs gallium in Si(111)

The first existence of ordered δ-doping layers was observed in boron doped Si(111). In this section we describe the surface science which results in successful boron doped layers and the unsuccessful activity in the case of Ga δ-doping. The chemically similar elements boron and gallium form a $\sqrt{3} \times \sqrt{3}$ reconstruction at $\frac{1}{3}$ monolayer coverage on Si(111). However, these reconstructions have very different stability with respect to deposited silicon. We now discuss the atomic scale mechanism that causes lateral disorder in the Ga system, but not in the B system. These results have been reviewed in a previous publication (Headrick *et al.*, 1991).

19.4.1 Sample preparation

Samples are typically prepared in a molecular-beam epitaxy chamber equipped with an electron gun evaporator to deposit silicon, a quartz-crystal thickness monitor, and a Knudsen cell to deposit either Ga from elemental gallium or boron from B_2O_3. Oriented $\langle 111 \rangle$ Si substrates are prepared by chemical growth of a thin protective oxide layer, and then transferred into the vacuum chamber. Once in the vacuum chamber, the oxide is desorbed from the sample, and gallium (boron) is deposited onto the surface. The Ga $\sqrt{3} \times \sqrt{3}$ surface reconstruction is prepared by deposition of $\frac{1}{3}$ monolayer (ML) of Ga while the sample is held at 550 °C. The boron $\sqrt{3} \times \sqrt{3}$ surface structure can be prepared by either segregation of $\frac{1}{3}$ ML of boron from boron doped samples at 900 °C (Korobtsov *et al.* 1988), or by deposition of boron onto n-type samples from HBO_2 (or B_2O_3) while the sample is held at 750 °C. (Hirayama *et al.*, 1988). More recently, elemental boron sources have been used so that deposition can be accomplished at lower temperatures (Weir, 1993).

19.4.2 *Surface structure*

Glancing angle X-ray diffraction, first-principles theoretical calculations, and scanning tunneling microscopy investigations have established the structure of the Si(111)–B$\sqrt{3} \times \sqrt{3}$ reconstruction as subsurface substitutional boron, arranged in a $\sqrt{3} \times \sqrt{3}$ configuration (Headrick *et al.*, 1989b; Bedrossian *et al.*, 1989; Lyo *et al.*, 1989). Figure 19.4 shows this site and compares it to the more common $\sqrt{3} \times \sqrt{3}$ configuration (T_4 site) of Ga and other adsorbates (Nicholls *et al.*, 1987). The nominal B(Ga) coverage is $\frac{1}{3}$ monolayer, that is, one of every three sites in a single atomic layer is occupied by B(Ga). The stability of boron in the subsurface site relative to the T_4 adatom site is related to relief of subsurface strain by the mechanism of substituting a smaller boron atom for silicon.

19.4.3 *LEED and X-ray diffraction results*

Figure 19.5 shows LEED patterns for the Ga($\sqrt{3} \times \sqrt{3}$) surface structure and the B($\sqrt{3} \times \sqrt{3}$) surface structure. After $\approx$1 Å of Si deposition on these surfaces at room temperature, LEED patterns (c) and (d) were observed. The Ga superstructure disappeared, while the B($\sqrt{3} \times \sqrt{3}$) was still clearly visible. Further LEED experiments showed that the Ga($\sqrt{3} \times \sqrt{3}$) disappeared between 0.3 and 0.4 ML silicon coverage, that is,

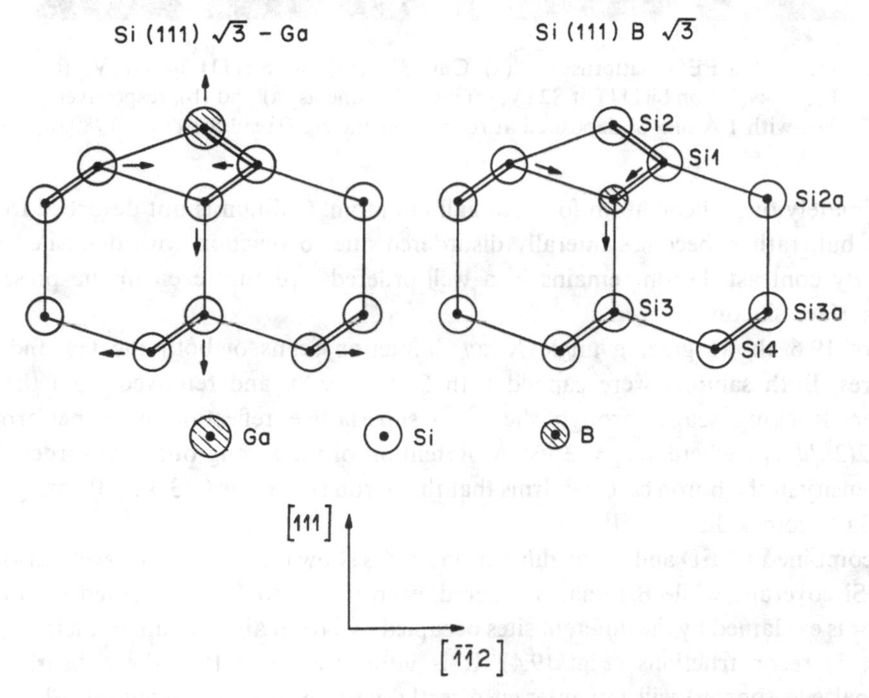

Fig. 19.4. Side view of the Si(111)–($\sqrt{3} \times \sqrt{3}$)Ga structure and the Si(111)–B($\sqrt{3} \times \sqrt{3}$). Arrows indicate the direction of displacements from the ideal tetrahedrally bonded configuration (Headrick *et al.*, 1989b).

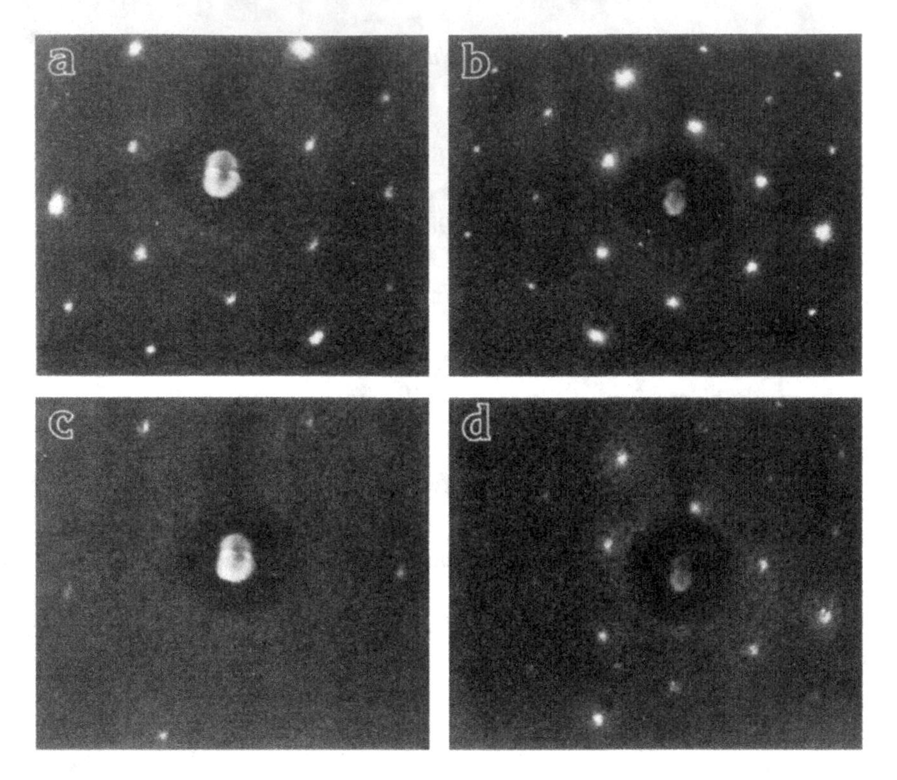

Fig. 19.5. LEED patterns for (a) Ga($\sqrt{3} \times \sqrt{3}$) on Si(111) at 66 eV, (b) B($\sqrt{3} \times \sqrt{3}$) on Si(111) at 82 eV, (c) and (d) same as (a) and (b), respectively, but with 1 Å of α-Si deposited at room temperature (Headrick *et al.*, 1989a).

approximately one silicon atom for each gallium atom. Gallium is not desorbed from the surface, but, rather, becomes laterally disordered due to reaction with deposited silicon atoms. By contrast, boron remains in a well-ordered structure even in the presence of excess surface silicon.

Figure 19.6 shows grazing angle X-ray diffraction scans of both the Ga and the B structures. Both samples were capped with 50 Å of α-Si, and removed from the MBE chamber. Rocking scans through the $(\frac{2}{3}, \frac{2}{3})$ superlattice reflection were performed at $q_\perp = 0.2(2\pi/d_{111})$, where $d_{111} = 3.135$ Å. Retention of the strong one-third-order diffraction intensity in the boron case confirms that the boron retains its ($\sqrt{3} \times \sqrt{3}$) configuration while Ga becomes disordered.

The combined LEED and X-ray diffraction results show that the Ga surface is disordered at low Si coverage while B remains ordered, even under 50 Å of deposited silicon. This behavior is explained by the different sites occupied by boron and gallium in their respective ($\sqrt{3} \times \sqrt{3}$) reconstructions (Fig. 19.4). It is intuitively clear that the subsurface substitutional site (boron) will not interact directly with deposited Si, while the T_4 adatom site (gallium) will be readily displaced by deposited Si. The subsurface substitutional site of boron becomes a normal bulk substitutional site upon growth of a silicon overlayer.

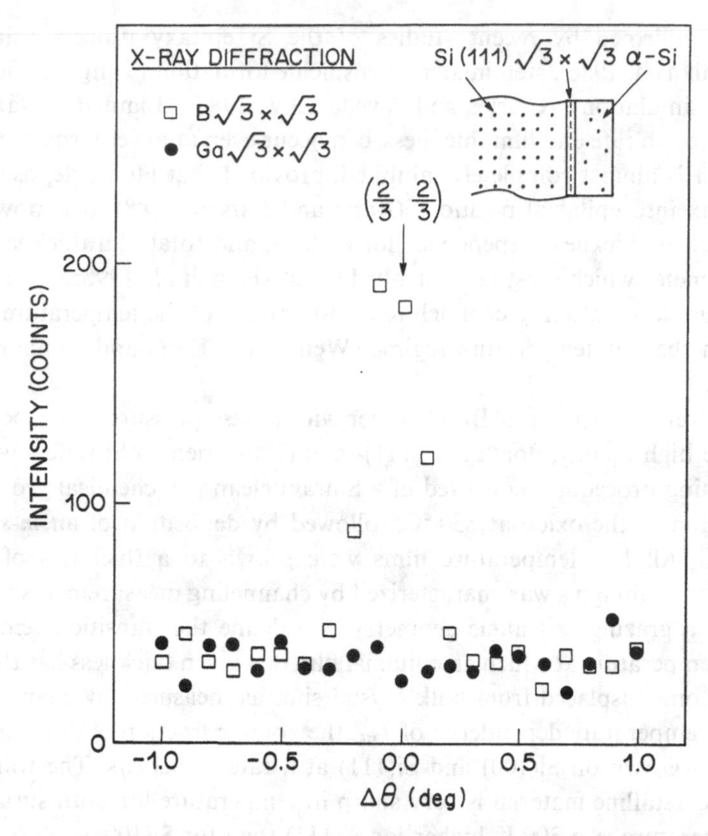

Fig. 19.6. Rocking scans through the $(\frac{2}{3}, \frac{2}{3})$ surface X-ray diffraction rod for buried boron and gallium reconstructions on Si(111) (Headrick *et al.*, 1989a).

19.5 Boron in Si(111) and Si(100)

In this section, we will describe recent attempts to fabricate ordered δ-doping layers of boron in silicon. Boron forms ordered structures on both the (111) and the (100) surface of silicon; however high quality epitaxial overgrowth, while preserving the ordering, has only been achieved for the Si(100) case. This preservation is associated with the capability of low temperature epitaxial growth for Si(100), thus minimizing boron diffusion and surface segregation.

19.5.1 MBE growth of Si on Si(111) vs Si(100)

We can think of epitaxial growth as taking place in two temperature regimes: (i) high temperature, where the growth mode is a step flow mechanism giving rise to high quality material; (ii) intermediate temperatures, where growth is pictured as arising from island formation on terraces and the eventual coalescence of these islands to form uniform layers.

This picture is affirmed by recent studies of the Si epitaxy process using RHEED oscillations (Ichikawa, 1986), step imaging via silicide formation (Tung and Schrey, 1989), and computer simulations (Clarke and Vvedemsky, 1987). Limited epitaxial growth, depending on growth rate and film thickness, can occur even at lower termperatures, where surface diffusion is almost completely inhibited, provided that atoms deposited onto the surface can relax into epitaxial positions (Aarts and Larsen, 1988). It is now recognized that there is a clear thickness dependence for epitaxy, and total film thickness is thus an important parameter which must be controlled (Eaglesham *et al.*, 1990; Jorke *et al.*, 1989). Below, we discuss ion scattering/channeling investigations of the temperature dependence of Si epitaxy in the low temperature regime (Weir *et al.*, 1991) and compare Si(111) to Si(100) growth.

Silicon films were grown in a MBE chamber with a base pressure of $<3 \times 10^{-11}$ Torr. Substrates were high quality, float zone (111)- and (100)-oriented Si wafers with a miscut of $<0.5°$. Cleaning procedures consisted of a Shiraki clean (i.e. chemical growth of a thin oxide), desorption of the oxide at 850 °C, followed by deposition of an *in-stiu* Si buffer layer at 700 °C. All low temperature films were grown to a thickness of 300–350 Å. Crystallinity of the thin films was characterized by channeling measurements with 1.8 MeV ^{4}He ions using a grazing exit angle geometry. We define the transition temperature for epitaxy as the temperature at which an epitaxial film of a given thickness has the equivalent of 50% of its atoms displaced from bulk crystal sites as measured by channeling. Figure 19.7 shows the temperature dependence of f_{dis}, the volume fraction of displaced atoms for 300–350 Å films grown on Si(100) and Si(111) at a rate of 0.3 Å/s. The transition from amorphous to crystalline material is very sharp in temperature for both surfaces, but the transition temperature is $\approx$300 K higher for Si(111) than for Si(100).

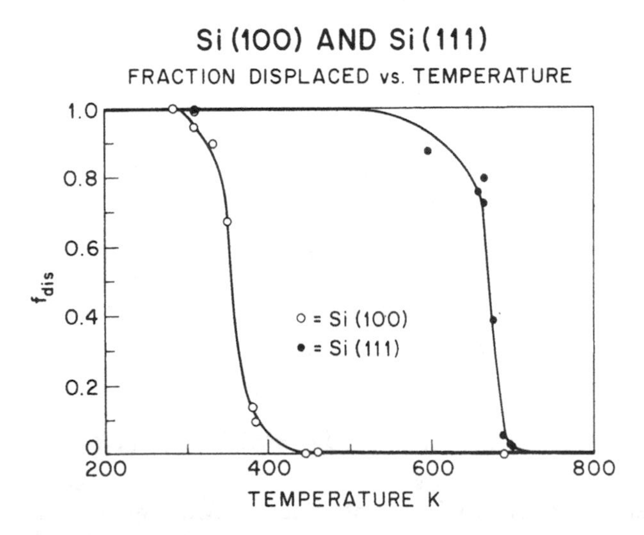

Fig. 19.7. Temperature dependence of f_{dis}, the fraction of displaced atoms, as measured by ion channeling for Si(100) and Si(111). The values of f_{dis} represent an average over the deposited film thickness (300 Å) (from Weir *et al.*, 1991).

We can understand the relatively high temperature for Si(111) epitaxy qualitatively in the light of other observations. The observed transition temperature for Si(111) epitaxy reported here is close to that observed for the $2 \times 1 \rightarrow 7 \times 7$ transition on the cleaved surface (Bäuerle *et al.*, 1972). Evidently, Si(111) epitaxial growth and the surface phase transition require a sufficiently high temperature to induce a transformation requiring the rearrangement of many atoms. Also, the large difference in epitaxy temperatures between Si(111) and Si(100) is reminiscent of the differences in the solid phase epitaxial regrowth rates for these two surfaces (Csepregi *et al.*, 1978). The difference in the regrowth temperature for these two surfaces has been explained in terms of their bonding geometries, which may also be related to the difference seen in MBE transition temperatures. For example, growth on Si(111) in the intermediate temperature range, that is, island growth and coalescence, may allow the formation of islands in a ⟨111⟩ stacking faulted configuration, while high temperature growth uses existing step edges, thereby insuring a proper stacking sequence.

The low temperature regime for Si(100) growth is between ≈50 °C and ≈400 °C. In this regime, the quality of growth depends on the thickness of the Si(100) layer. This has been shown most explicitly by Eaglesham *et al.* (1990), who characterize Si(100) epitaxial growth in terms of a 'critical thickness'. Using transmission electron microscopy as a measure of crystal perfection they show that epitaxy of a few monolayers of Si(100) is even possible at ≈50 °C. At such low temperatures, Si(100) films exhibit a transition from epitaxial to amorphous structure as growth proceeds.

Si(111) shows an analogous breakdown in epitaxy with increasing film thickness, but rather than a direct transition from high quality crystalline to amorphous material, the defect density, consisting mainly of stacking faults around ⟨111⟩ axes, gradually increases, and eventually epitaxy breaks down to amorphous growth (Weir *et al.*, 1991). Prior to the onset of amorphous Si growth, the defect density becomes so high that the channeling yield from the film is indistinguishable from that of an amorphous layer. Epitaxial growth on Si(111) to significant thickness only occurs at temperatures above 400 °C. This is above the 'low temperature' growth regime found on Si(100).

Molecular dynamics simulations of the MBE process can provide insight into the crystallization process, although computer time limitations require the use of atomic beam intensities orders of magnitude greater than those used in practice (Gilmer *et al.*, 1990). Simulations confirm that the temperature required for epitaxy on the Si(111) face is several hundred degrees higher than that on the Si(100) face (Fig. 19.8). The molecular dynamics calculations show that a propensity for growth of a stacking fault region in the Si(111) case appears to be the determining factor in the transition temperature difference for Si(111) and Si(100).

19.5.2 Surface segregation

A perfectly sharp δ-doping layer requires that surface segregation and diffusion during growth be eliminated on the scale of a single atomic layer along the growth direction.

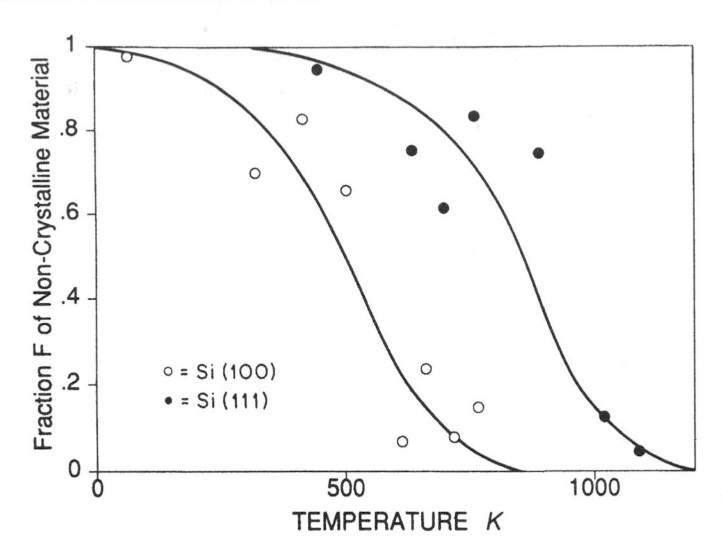

Fig. 19.8. Molecular dynamics simulation results for the temperature dependence of Si MBE growth on the (111) and (100) surfaces. The vertical axis is a parameter that measures the displacement of atoms from bulk-like positions, analogous to f_{dis} (Fig. 19.5). The simulations are for films ≈ 17 Å thick. The areas in the simulations are: 15.4 × 15.4 Å for Si(100), and 26.7 × 30.8 Å for Si(111) (from Weir *et al.*, 1991).

Auger electron spectroscopy is a convenient method of measuring segregation of high concentration δ-doping layers because of the surface sensitivity of the method. The experiment is performed by forming a surface layer with $\frac{1}{3}$ ML of boron on Si(111) or $\frac{1}{2}$ ML of boron on Si(100), in a MBE chamber equipped with a silicon source, a boron source, and an electron spectrometer. Data is collected by measuring the intensity of the boron Auger line repeatedly and depositing a small amount of silicon.

In the absence of any segregation, the data is well approximated by an exponentially decreasing function with a decay constant determined by the characteristic escape length of the Auger electrons. The envelope of the decay curve is described by:

$$I(t) = I_0 \exp(-t/\Lambda), \tag{19.7}$$

Here, t is the thickness of deposited silicon, and Λ is the effective escape depth of boron Auger electrons in silicon.

Figure 19.9 shows the boron Auger intensity as a function of total deposited silicon. For growth on Si(111) at room temperature, the data follows Eqn. (19.7) with a decay constant of 7 Å, consistent with no segregation. For growth at 673 K, that is, high enough for growth of a 300–350 Å epitaxial Si layer, limited segregation over $\sim$20 Å is detected. At 813 K, segregation is strong, resulting in a broad doping distribution.

For growth on Si(100), we choose somewhat lower growth temperatures. This reflects

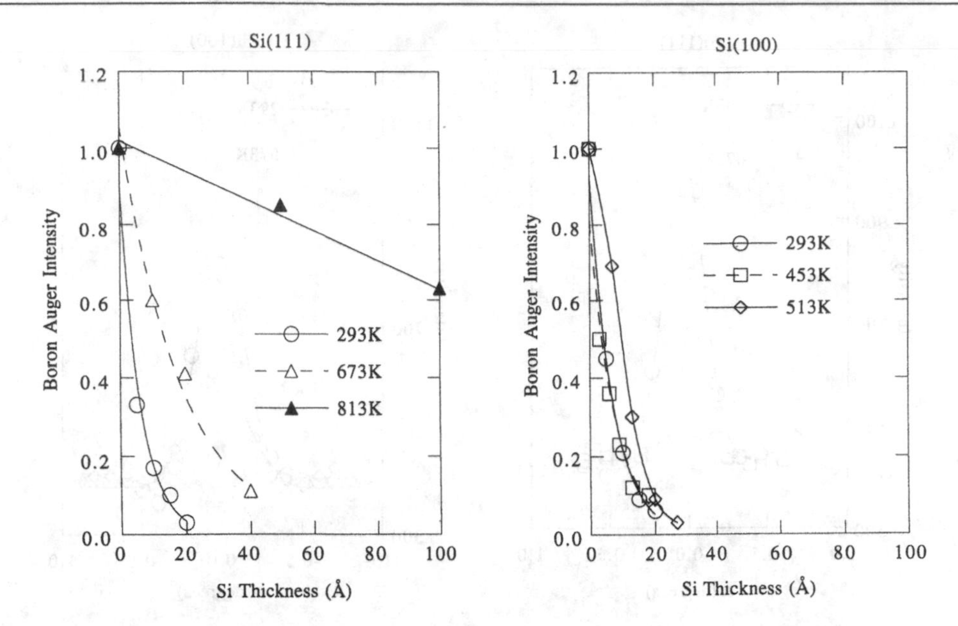

Fig. 19.9. Intensity of the boron KLL Auger line at 178 eV for Si film growth on Si(111):B and Si(100):B. The initial boron coverage is $\frac{1}{3}$ monolayer on Si(111) and $\frac{1}{2}$ monolayer on Si(100). Three different growth temperatures are shown for each surface orientation: (i) room temperature, (ii) near the minimum temperature required to grow a 300–350 Å film (see Fig. 19.7), and (iii) a higher temperature. The effective escape depth of 178 eV electrons at the analyzer angle is ≈ 7 Å. The data are from Headrick *et al.* (1990b) and Weir *et al.* (1992).

the fact that high quality epitaxy can be achieved below 673 K on Si(100), but not on Si(111). Figure 19.9 shows that no segregation occurs during growth at room temperature and 453 K. At 513 K, limited segregation over 5–10 Å is detected.

These results suggest that growth on the (100)-oriented surface is more favorable in terms of achieving high quality epitaxial growth, while simultaneously minimizing boron segregation. It should be pointed out that there is comparable boron segregation at the same temperature for the two surfaces. However, low temperature epitaxial growth is possible on Si(100), and this allows the fabrication of essentially monolayer profiles. In the next section, we discuss the in-plane ordering of these nearly ideal δ-doping layers.

19.5.3 *X-ray diffraction results*

We first discuss early grazing angle X-ray diffraction results for ordered δ-doping layers on (111)-oriented silicon. Rocking scans through the $(\frac{2}{3}, \frac{2}{3})$ reflection for Si(111) ordered boron layer with a $\sqrt{3} \times \sqrt{3}$ unit cell show the preservation of the buried ordered phase

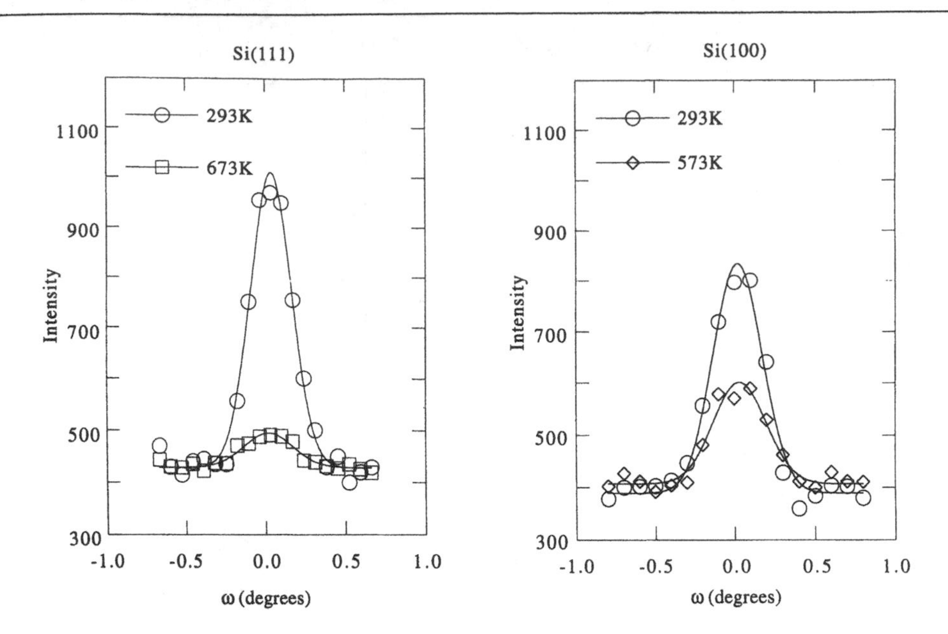

Fig. 19.10. Comparison of rocking scans through the $(\frac{2}{3}, \frac{2}{3}, 0.2)$ reflection for ordered δ-doped layers in Si(111), and rocking scans through the $(\frac{3}{2}, 0, 0.2)$ reflection for ordered δ-doped layer in Si(100). Reflections are indexed to the appropriate surface unit cells for the two surfaces, that is, a hexagonal cell on the (111) orientation and a square cell on the (100) orientation. A rotating anode source and conventional four-circle spectrometer were used for the measurements (from Headrick *et al.*, 1990b; and Headrick *et al.*, 1990a).

and that the intensity is diminished for Si overgrowth at 673 K relative to room temperature growth (Fig. 19.10). These results are similar to the results of Akimoto *et al.* (1987), except that the ratio of intensities is approximately 1:10 instead of 1:100. This is attributed mainly to the lower growth temperature for the epitaxial layer, that is, 673 K above, as opposed to 873 K by Akimoto *et al.* (1987). From the segregation curve in Fig. 19.9, we can estimate that approximately 10% of the boron initially deposited onto the surface remains at the plane of the original surface after film growth at 673 K. It is possible that the segregated boron contributes to the diffracted intensity, since the segregating species may order on the growing surface, forming an ordered structure over several atomic layers along the growth direction.

Ordered boron δ-doping layers formed on (111)-oriented SiGe alloy surfaces have also been reported. These structures are similar to those on pure Si, except that Ge and B atoms both occupy preferred sites in the ordered structure (Tatsumi *et al.*, 1990; Tweet *et al.*, 1992). Tweet *et al.* (1992) have also found that the Si–Ge–B structures are more stable than Si–B structures under annealing.

Figure 19.10 also shows results for ordered boron δ-doping on the (100) orientation. The diffracted intensity for Si overlayer growth at 573 K is $\approx 50\%$ of the room temperature intensity, confirming that the (2×1) ordered configuration can also be preserved under a

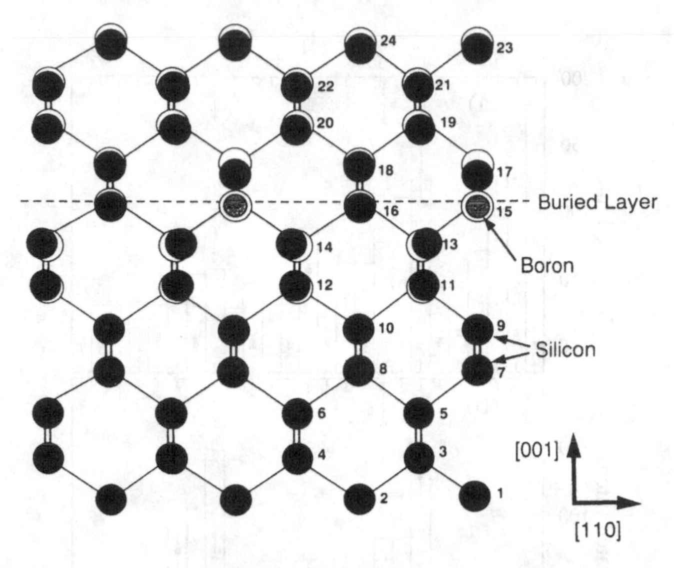

Fig. 19.11. Side view of the (2 × 1) ordered δ-doping layer in (100) oriented Si with displacements obtained from total-energy minimization. As a reference, the undistorted Si structure is represented by open circles and bonds; displaced Si atoms are shaded. Atoms in one unit cell are numbered for reference (from Weir *et al.*, 1992).

relatively thick epitaxial layer ($\approx$50 Å). Recently, Weir *et al.* (1992) have also produced (2 × 1) δ-doping layers with Si growth at 453 K, with 100% diffracted intensity relative to room temperature deposition. This is consistent with the segregation curves shown in Fig. 19.9, since no segregation is detected at 453 K.

A complete X-ray analysis of the optimized Si(100)/B/Si(100) structure has also been carried out (Weir *et al.*, 1992). The refinement allows for displacement of Si atoms due to strain induced by the small Si–B bond length and the possibility of limited boron segregation in the growth direction. The ideal configuration is shown in Fig. 19.11, while the results of the refinement are shown in Fig. 19.12. Figure 19.12(c) gives the closest approximation to the data, and is considered to be the most accurate representation of the true structure. This model consists of boron distributed over a total of three layers with realistic strain-induced bond distortions around each atom. This analysis reveals the sharpest dopant profile ever established, that is, boron confined to $\approx$3 monolayers or $\approx$4 Å.

19.5.4 Electrical activity and carrier mobility

The high density boron doped Si layers have been characterized electrically using standard Hall effect and resistivity measurements. Table 19.2 gives carrier densities and mobilities from low temperature (T = 4.2 K) Hall effect measurements for silicon overlayers grown at 293 °C (labeled amorph.) and 573 °C (labeled cryst.). Units of monolayers are used for

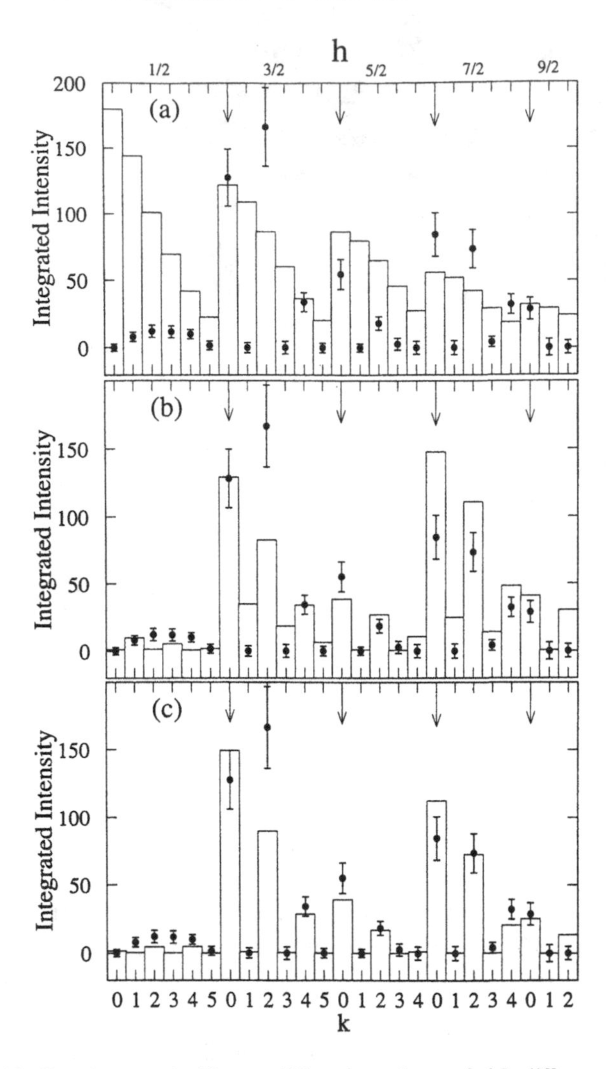

Fig. 19.12. Grazing angle X-ray diffraction data of 25 different half-order reflections are shown with three sets of calculated intensities: (a) substituted in a 2×1 arrangement with no Si displacements; (b) 2×1 structure, shown in Fig. 19.11, using displacements predicted from first-principles theory and boron confined to one monolayer; (c) partially disordered 2×1 structure with $\frac{1}{2}$ ML boron distributed over three monolayers. The indices *h*, *k* refer to the two-dimensional diffraction coordinates (from Weir *et al.*, 1992).

clarity, where one monolayer is defined as $6.8 \times 10^{14}\,\mathrm{cm}^{-2}$ for Si(100), and $7.8 \times 10^{14}\,\mathrm{cm}^{-2}$ for Si(111). Table 19.2 also shows data for several structures, including the 2×1 reconstruction of Si(100):B, the $\sqrt{3} \times \sqrt{3}$ reconstruction of Si(111) capped with amorphous silicon, and a control experiment in which the (undoped) Si(111)–(7×7) reconstruction was capped with amorphous Si. The values for carrier density are representative of a larger data set. As discussed later in this section, the high carrier density

Table 19.2. *Carrier densities and mobilities from low temperature Hall effect measurements (Headrick et al., 1991)*

Surface orientation	Reconstruction	Overlayer	Boron coverage (ML)	Carrier density (ML)	Mobility ($cm^2 V^{-1} s^{-1}$)
(100)	(2 × 1)	Cryst.	0.44	0.45	21
(100)	(2 × 1)	Amorph.	0.50	0.24	1
(111)	($\sqrt{3} \times \sqrt{3}$)	Amorph.	0.34	0.22	30
(111)	(7 × 7)	Amorph.	0.01	0	—

represents an exceedingly high electrically active concentration; the mobility is in the range expected for very high p-type doping for 3D layers.

As discussed earlier in this chapter an ideal ordered monolayer should yield a substantial increase in mobility over that expected for random doping. We now discuss a quantitative prediction of the mobility for the (2 × 1) boron structure. As demonstrated from the X-ray structure factor analysis in Fig. 19.12(c), the boron layer is only partially ordered. In order to estimate the predicted effect of this partially ordered doping on the carrier scattering rate we have applied the charge carrier structure factor in Eqn. (19.3) to the arrangement shown in Fig. 19.3(b). Equation (19.3) is not identical to the X-ray structure factor since charge carriers only scatter from charged impurities, while the X-ray structure factor represents the total electron density. Figure 19.13 shows $s(\mathbf{q})$ divided by the total number of scatterers, n,

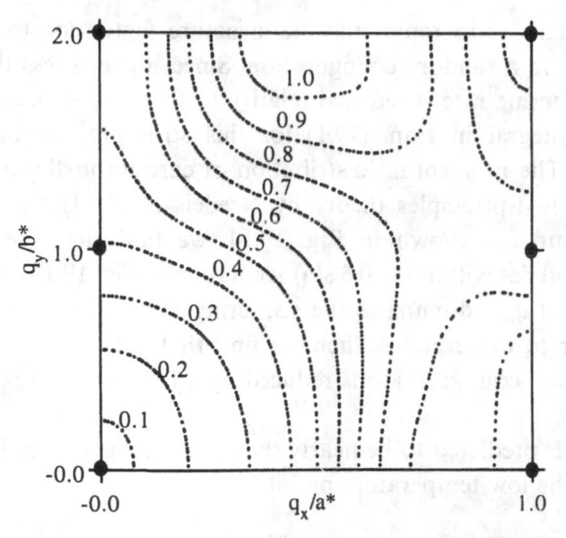

Fig. 19.13. Structure factor $s(\mathbf{q})$ as a function of $(q_x|a^*, q_y|b^*)$ for the partially ordered (2 × 1) structure shown in Fig. 19.3(b). Contours are for $s(\mathbf{q})$ calculated from Eqn. (19.3) relative to random doping (Eqn. (19.6)). The structure factor is related to the charged carrier elastic scattering rate, τ_{el}^{-1}, by Eqn. (19.2). Large circles mark the reciprocal lattice points.

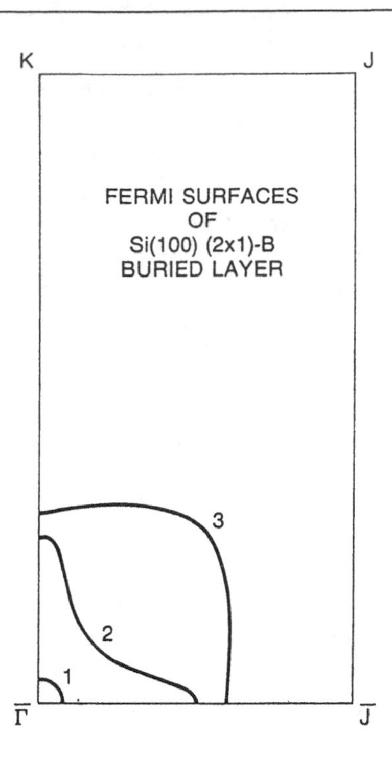

Fig. 19.14. Fermi surfaces of ordered (2 × 1) boron δ-doping layer in Si(100). The surfaces labeled 1, 2, 3 are associated with two-dimensional bands with p_x, p_y, and p_z symmetry respectively (from Needels *et al.*, 1992).

as a function of **q**. This ratio represents the structure factor for the partially ordered configuration relative to a random configuration. Since $s(\mathbf{q})/n$ is less than one at small **q** in Fig. 19.13, the scattering rate is reduced relative to the random case.

To evaluate the integral in Eqn. (19.1) for this structure, we must integrate over allowed values of **q**. The momentum distribution of carriers in the δ-doping layer have been predicted from first-principles theory by Needels *et al.* (1992). By estimating $q_{\max}$ from $2|k_f|$ for the surfaces shown in Fig. 19.14, we find that the $q_{\max}$ is $\approx 75\%$ of $b^*(b^* = 0.82\ \text{Å}^{-1})$, and lies within the 0.3 $s(\mathbf{q})/n$ contour of Fig. 19.13. Monroe *et al.* (1993) show that scattering at $q_{\max}$ dominates the scattering rate. If we assume that only holes with $q = q_{\max}$ scatter to other states, then we find that the elastic scattering rate, τ_{el}^{-1}, for the partially ordered configuration is reduced by a factor of ≈ 3 relative to a random configuration.

The effective mass is predicted to be nearly the same for the ordered structure as in the dilute doping case. The low temperature mobility

$$\mu = \frac{e\tau_{\text{el}}}{m} \tag{19.8}$$

is therefore predicted to be enhanced by a factor of three. Below, we discuss measured values of the mobility for the partially ordered (2 × 1) doping layers.

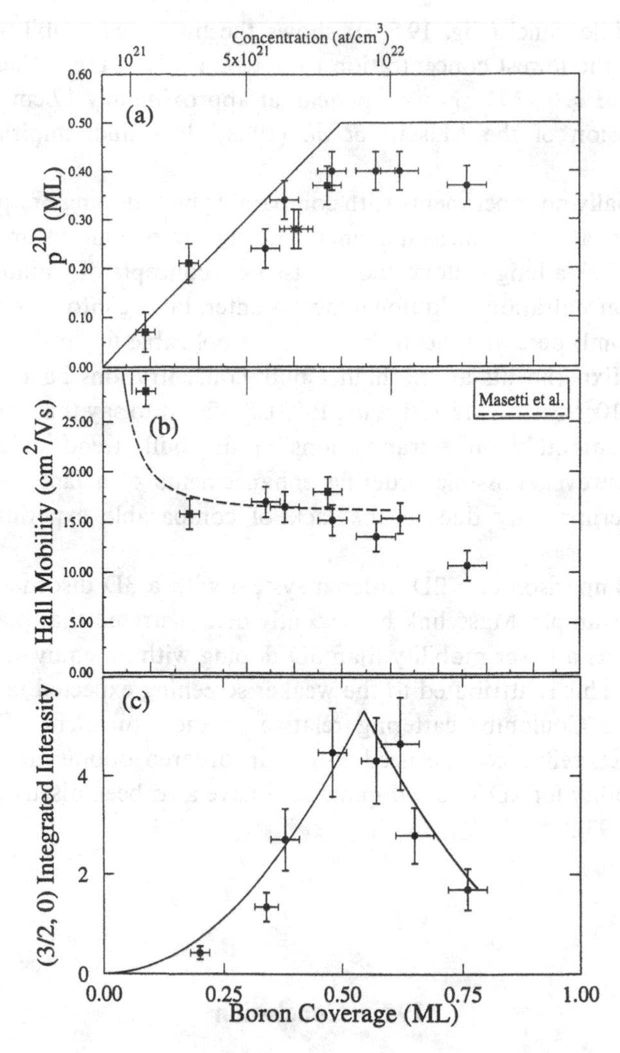

Fig. 19.15. Electrical carrier density (a), mobility (b) and X-ray diffraction intensity (c) as a function of doping concentration for δ-doping of boron in silicon (from Weir *et al.*, 1992, and Weir *et al.*, 1994).

The full set of electrical measurements correlated to the structural information is shown in Fig. 19.15 (Weir *et al.*, 1994). The X-ray intensity (Fig. 19.15(c)) shows that the in-plane ordering peaks at approximately $\frac{1}{2}$ monolayer coverage corresponding to the (2×1) configuration. The upper panel (Fig. 19.15(a)) illustrates the scaling of the electrical activity. Nearly full activation is observed for converages up to $\approx \frac{1}{2}$ monolayer. Beyond $\frac{1}{2}$ monolayer there is a loss of activation, perhaps due to boron clustering. The concentration scale is calculated assuming that the boron is spread over three monolayers, so that the (2×1) configuration corresponds to a doping concentration of 8×10^{21} cm^{-3}. Note that this configuration demonstrates the highest electrically active concentration of boron in Si ever observed. The middle panel (Fig. 19.5(b)) shows the measured mobility values varying from 30 cm^2/Vs at the lowest concentration to ≈ 12 cm^2/Vs at the highest coverage. The critical value around ≈ 0.5 ML shows a plateau at approximately 17 cm^2/Vs. The dashed curve is an extension of the Masetti *et al.* (1983) data and empirical fit at lower concentrations.

There are essentially no experiments with comparable bulk doping or appreciable theory with which to compare these measurements. The theory of ionized impurity scattering in semiconductors has a long history and is based on concepts of Coulomb scattering of carriers. At high concentrations additional factors enter, taking into account the screening effect on the Coulomb potential. Such theories are applicable up to the 10^{19} cm^{-3} range (Lin *et al.*, 1981). Experiments at still higher bulk concentrations have been reported to reach the range of 10^{21} cm^{-3} (Masetti *et al.*, 1983). Suffice it to say that the mobility values reported here fit smoothly on extrapolations of the bulk trend to exceedingly high concentrations. However, possible ordering enhancements of a factor of three are not demonstrable experimentally due to the lack of comparable experiments or reliable theory.

Furthermore, comparison of a 2D ordered system with a 3D disordered system is not totally valid. For example, Masselink has recently demonstrated that planar doping (not ordered) actually has a lower mobility than 3D doping with an equivalent concentration (Masselink, 1991). This is attributed to the weaker screening expected in two dimensions, thus enhancing the Coulomb scattering relative to the equivalent 3D case. Such an argument, if correct, reinforces the need to obtain ordered doping in these thin layers. Differences in mobility for 2D vs 3D doping in Si have also been discussed by Gossmann and Unterwald (1993).

19.6 Conclusion

In this chapter, we have reviewed recent developments in ordered δ-doping. We have discussed the model systems, boron in Si(100) and Si(111). Two-dimensional ordered layers of boron, originally formed as surface reconstructions, can be preserved under silicon cap layers. In the Si(100) case high quality Si can be grown atop the ordered structure due to the low temperature required for (100) epitaxy. These systems show a high p-type electrical activity, between 50% and 100% electrically active. Majority carrier mobilities at 4 K are close to that predicted by extrapolation of random doping mobilities to high concentrations. Definite assignment of mobility enhancements of factors of 2–3 due to ordering, or reductions due to screening modifications, await more complete theory in this high concentration regime. It is fascinating that the electrical properties of this semiconductor system follow the general behavior of doping to levels in excess of 50% atomic concentration of the doping species.

Acknowledgments

We wish to acknowledge useful discussions and collaborations with J. Bevk, D. J. Eaglesham, B. Freer, G. H. Gilmer, M. Hybertsen, A. Levi, M. Needels, and I. Robinson. Michael Schlüter (deceased) was an early collaborator and we wish to note in particular his important contributions to this project.

References

J. Aarts and P. K. Larsen, in *RHEED and Reflection Electron Imaging of Surfaces*, eds. P. K. Larsen and P. J. Dobson, Plenum, NY (1988).

K. Akimoto, J. Mizuki, I. Hirosawa, T. Tatsumi, H. Hirayama, N. Aizaki and J. Matsui, *Extended Abstracts of the 19th Conference on Solid State Devices and Materials* (Business Center for Academic Societies, Tokyo, 1987), p. 463.

F. Bäuerle, W. Mönch and M. Henzler, *J. Appl. Phys.* **43**, 3917 (1972).

J. C. Bean, *Appl. Phys. Lett.* **33**, 654 (1978).

G. E. Becker and J. C. Bean, *J. Appl. Phys.* **48**, 3395 (1977).

P. Bedrossian, R. D. Meade, K. Mortensen, D. M. Chen and J. A. Golovchenko, *Phys. Rev. Lett.* **63**, 1257 (1989).

S. S. Clarke and D. D. Vvedemsky, *Phys. Rev. Lett.* **58**, 2235 (1987).

L. Csepregi, E. F. Kennedy, J. W. Wagner and T. W. Sigmon, *J. Appl. Phys.* **49**, 3906 (1978).

D. J. Eaglesham, H.-J. Gossmann and M. Cerullo, *Phys. Rev. Lett.* **65**, 1227 (1990).

J. M. Gibson, H.-J. Gossmann, J. C. Bean, R. T. Tung and L. C. Feldman, *Phys. Rev. Lett.* **56**, 355 (1986).

G. H. Gilmer, M. H. Grabow and A. F. Bakker, *Materials Sci. & Engr.* B **6**, 101 (1990).

H.-J. Gossmann, E. F. Schubert, D. J. Eaglesham and M. Cerullo, *Appl. Phys. Lett.* **57**, 2440 (1990), and references therein.

H.-J. Gossmann and F. C. Unterwald, *Phys. Rev.* B **47**, 12618 (1993).

R. L. Headrick, L. C. Feldman and I. K. Robinson, *Appl. Phys. Lett.* **55**, 442 (1989a).

R. L. Headrick, I. K. Robinson, E. Vlieg and L. C. Feldman, *Phys. Rev. Lett.* **63**, 1253 (1989b).

R. L. Headrick, B. E. Weir, A. F. J. Levi, D. J. Eaglesham and L. C. Feldman, *Appl. Phys. Lett.* **57**, 2779 (1990a).

R. L. Headrick, B. E. Weir, J. Bevk, B. S. Freer, D. J. Eaglesham and L. C. Feldman, *Phys. Rev. Lett.* **65**, 1128 (1990b).

R. L. Headrick, B. E. Weir, A. F. J. Levi, B. Freer, J. Bevk and L. C. Feldman, *J. Vac. Sci. Technol.* A **9**, 2269 (1991).

H. Hirayama, T. Tatsumi and N. Aizaki, *Surf. Sci.* **193**, L47 (1988).

H. Jorke, H.-J. Herzog and H. Kibbel, *Phys. Rev.* **40**, 2005 (1989).

M. Ichikawa, *Mat. Sci. Rep.* **4**, 147 (1986).

V. V. Korobtsov, V. G. Lifshits and A. V. Zotov, *Surf. Sci.* **195**, 467 (1988).

A. F. J. Levi, *Electron. Lett.* **24**, 1273 (1988).

A. F. J. Levi, S. L. McCall and P. M. Platzman, *Appl. Phys. Lett.* **54**, 940 (1989).

J. F. Lin, S. S. Li, L. C. Lineares and K. W. Teng, *Sol. State Electron.* **24**, 827 (1981).

I.-W. Lyo, E. Kaxiras and Ph. Avouris, *Phys. Rev. Lett.* **63**, 1261 (1989).

G. Masetti, M. Severi and S. Solmi, *IEEE Trans. Electron Dev.* **30**, 764 (1983).

W. T. Masselink, *Phys. Rev. Lett.* **66**, 1513 (1991).

D. Monroe, Y. H. Xie, E. A. Fitzgerald, P. J. Silverman and G. P. Watson, *J. Vac. Sci. Technol.* B **11**, 1731 (1993).
M. Needels, M. S. Hybertsen and M. Schlüter, *Mater. Sci. Forum* **83–87**, 1391 (1992).
J. M. Nicholls, B. Reihl and J. E. Northrop, *Phys. Rev.* B **35**, 4137 (1987).
M. Ramsteiner, J. Wagner, D. Behr, G. Jung, L. Däweritz and R. Hey, *Appl. Phys. Lett.* **64**, 490 (1994).
T. Tatsumi, H. Hirayama, J. Mizuki and J. Matsui, *Appl. Phys. Lett.* **56**, 1225 (1990).
R. W. Thomas and M. H. Francombe, *Solid State Electron.* **12**, 799 (1969).
G. Timp, R. E. Behringer, D. M. Tennant, J. E. Cunningham, M. Prentiss and K. K. Berggren, *Phys. Rev. Lett.* **69**, 1636 (1992).
R. T. Tung and F. Schrey, *Phys. Rev. Lett.* **63**, 1277 (1989).
D. J. Tweet, K. Akimoto, T. Tatsumi, I. Hirosawa, J. Mizuki and J. Matsui, *Phys. Rev. Lett.* **69**, 2236 (1992).
B. E. Weir, B. S. Freer, R. L. Headrick, D. J. Eaglesham, G. H. Gilmer, J. Bevk and L. C. Feldman, *Appl. Phys. Lett.* **59**, 204 (1991).
B. E. Weir, R. L. Headrick, Q. Shen, L. C. Feldman, M. S. Hybertsen, M. Needels, M. Schlüter and T. R. Hart, *Phys. Rev.* B **46**, 12861 (1992).
B. E. Weir, PhD Thesis, Stevens Institute of Technology, Hoboken NJ (1993).
B. E. Weir, L. C. Feldman, D. Monroe, H.-J. Gossmann, R. L. Headrick and T. R. Hart, *Appl. Phys. Lett.* **65**, 737 (1994).
A. V. Zotov, V. G. Lifshits, Z. Z. Ditina and P. A. Kalinin, *Surf. Sci.* **273**, L453 (1992a).
A. V. Zotov, V. G. Lifshits and A. N. Demidchik, *Surf. Sci.* **274**, L583, (1992b).
A. V. Zotov, E. A. Khramtsova and V. G. Lifshits, *Surf. Sci.* **277**, L77 (1992c).

PART SIX

Electronic and optoelectronic devices

20

Delta-doped channel III–V field-effect transistors (FETs)

BRIAN W. P. HONG

20.1 Introduction

Much attention has been paid recently to δ-doping for high-performance device applications because this doping technique can provide two-dimensional electron gas (2DEG) systems with high carrier concentration in compound semiconductors. It has been shown (Drummond *et al.*, 1981; Schubert *et al.*, 1984) that, although the low-field mobility of the 2DEG in modulation-doped heterostructures is very high, it decreases significantly at fields higher than $E = 100$ V/cm. In addition, it is generally agreed that the 2DEG properties such as the carrier confinement and the resulting high channel conductivity at high fields are more critical than the low-field mobility in determining the performance of the devices. Therefore, it attracts considerable interest in investigation of devices fabricated from materials having a δ-doped channel, especially heterostructures for which an additional carrier confinement can be provided by the heterostructure barriers. It is expected that, compared with conventional metal–semiconductor field-effect transistors (MESFETs), field-effect transistors (FETs) fabricated from δ-doped materials will have a (i) high drain current capability, (ii) large breakdown voltage, (iii) easy control of the threshold voltage, and (iv) high intrinsic transconductance because of the small distance between the channel and the gate. These advantages will influence the performance of short-channel FETs considerably. The δ-doped materials have been developed primarily by using molecular beam epitaxy (MBE) and organometallic chemical vapor deposition (OMCVD), which are emerging as growth techniques for materials for advanced device applications. In this chapter, recent investigations of the δ-doped channel III–V FETs, including material growth, device design, and device characteristics, are discussed.

20.2 Homostructure δ-doped GaAs FETs

Device parameters

The GaAs FETs having a δ-doped channel have been studied by many groups. Schubert *et al.* (May, 1986) performed a simple analysis of the transconductance of the δ-doped FET using a model originally developed by Hower and Bechtel (1983). The model assumes a two-region approximation for the velocity–field ($v(\mathbf{E})$) characteristics and a velocity saturation at the drain end of the gate. The maximum intrinsic transconductance of a δ-doped FET can be expressed by (see Fig. 20.1)

$$g_m = \frac{1}{L_g}(q\mu W_g n_{2DEG})\left[1 + \left(\frac{q\mu n_{2DEG} d}{\varepsilon v_s L_g}\right)^2\right]^{-1/2} \tag{20.1}$$

$$n_{2DEG} = N_D^{2D} - \frac{\varepsilon}{qd}\phi_B \tag{20.2}$$

where L_g and W_g are the gate length and width, respectively, μ is the mobility, v_s is the electron saturated drift velocity, n_{2DEG} is the free carrier concentration of the 2DEG, N_D^{2D} is the total doping density of the δ-doped layer, ϕ_B is the Schottky barrier height, and d is the distance of the center of the 2DEG from the gate. Equation (20.1) does not, however, include transient transport phenomena such as velocity overshoot and ballistic transport. According to Eqn. (20.1) an optimized FET should have (i) a high 2DEG concentration,

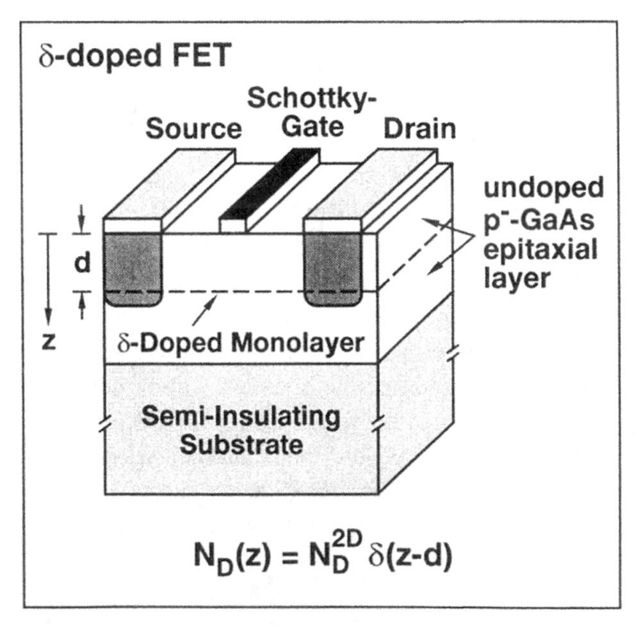

Fig. 20.1. Schematic illustration of a δ-doped GaAs FET grown on a semi-insulating GaAs substrate.

(ii) the 2DEG close to the metal Schottky contact, (iii) a high low-field mobility, as well as (iv) a high saturation velocity. For short-gate length (where L_g goes to zero), Eqn. (20.1) reduces to the well-known saturated velocity model according to

$$g_m = \frac{\varepsilon v_s W_g}{d}. \tag{20.3}$$

This simple equation shows that, in addition to the saturation velocity, the distance d of the 2DEG from the Schottky gate determines the transconductance of a short-gate FET. In a δ-doped FET, a distance, d, of less than 300 Å can easily be achieved.

The threshold voltage of a δ-doped FET can be given by

$$V_p = \frac{qd}{\varepsilon} n_{2DEG} \tag{20.4}$$

where the quantum size effect if neglected. The validity of Eqn. (20.4) is limited to long gate lengths. In a conventional, homogeneously doped MESFET, the threshold voltage depends quadratically on the channel thickness, t, according to

$$V_p = \frac{t^2}{2\varepsilon} qN_D. \tag{20.5}$$

Clearly, this quadratic dependence makes the threshold voltage control more sensitive to the gate recess depth as compared with the linear dependence of the δ-doped FET, as indicated by Eqn. (20.4). This represents another advantage of δ-doped FETs over conventional MESFETs.

MBE-grown devices

Initial research efforts to realize FET structures having a δ-doped channel were focused on MBE. When MBE is used, the material quality is mainly affected by impurities and defects. The growth of δ-doped channel GaAs FET structures using MBE is achieved by the evaporation or sublimation of heated elemental sources, such as Al, Ga, and As, thereby forming molecular beams that impinge on heated GaAs substrate in an ultra-high vacuum environment. An elemental Si source is most widely used for n-type δ-doping.

Figure 20.1 shows the schematic device structure of typical δ-doped GaAs FETs grown by MBE, in which the channel is formed by a single δ-doped layer. As discussed earlier, the threshold voltage of the δ-doped FETs can easily be controlled by either changing the doping level of the δ-doped layer or changing the distance between the gate and the channel. For the enhancement mode, the devices having a gate length of 1.3 μm showed a transconductance of 150 mS/mm (Schubert *et al.*, Dec., 1986). On the other hand, a transconductance as high as 290 mS/mm has been achieved for the depletion mode devices having the same gate length (Hong *et al.*, July, 1989). In addition, as illustrated in

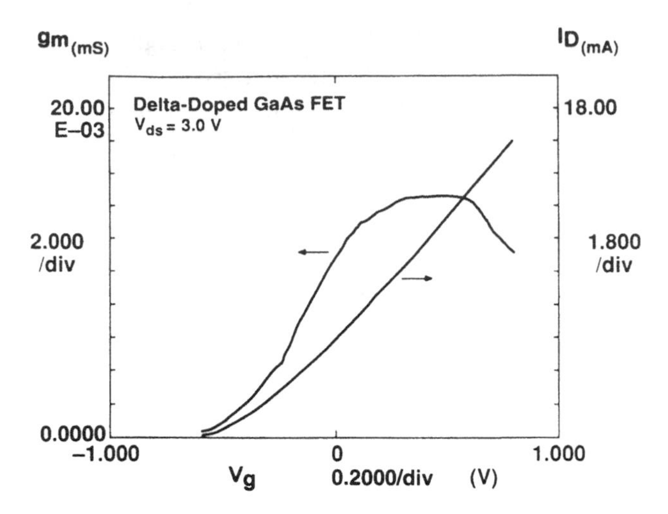

Fig. 20.2. Extrinsic transconductance and drain current versus gate voltage of a MBE-grown δ-doped GaAs FET at a drain voltage of 3 V.

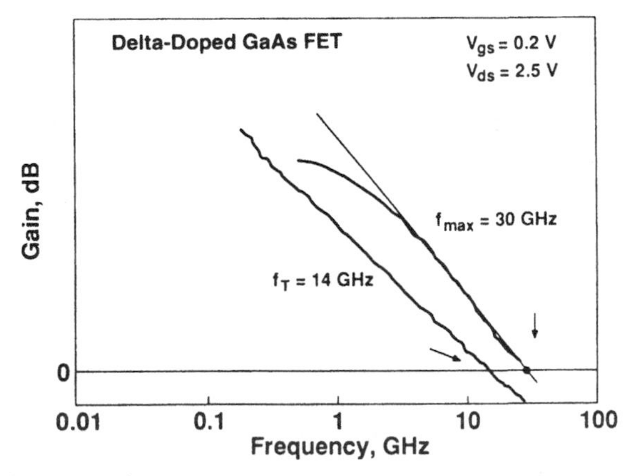

Fig. 20.3. Current gain and power gain versus frequency at a gate voltage of 0.2 V and a drain voltage of 2.5 V.

Fig. 20.2, the transconductance as a function of gate voltage shows a broad plateau around its peak, in contrast to FETs using a 2DEG channel formed by a modulation doping technique, called modulation-doped FETs (MODFETs). This is attributed to strong confinement of the carriers in a V-shaped potential well that creates a size quantization, while large forward gate bias can be applied because the gate is on the undoped GaAs. Furthermore, the δ-doped FETs have relatively high gate breakdown voltages of typically -10 to -15 V. The origins of high breakdown voltages are the linear potential drop between the gate and the channel and the absence of doped material between the gate and the channel. Figure 20.3 shows the current gain and power gain of the same depletion-mode

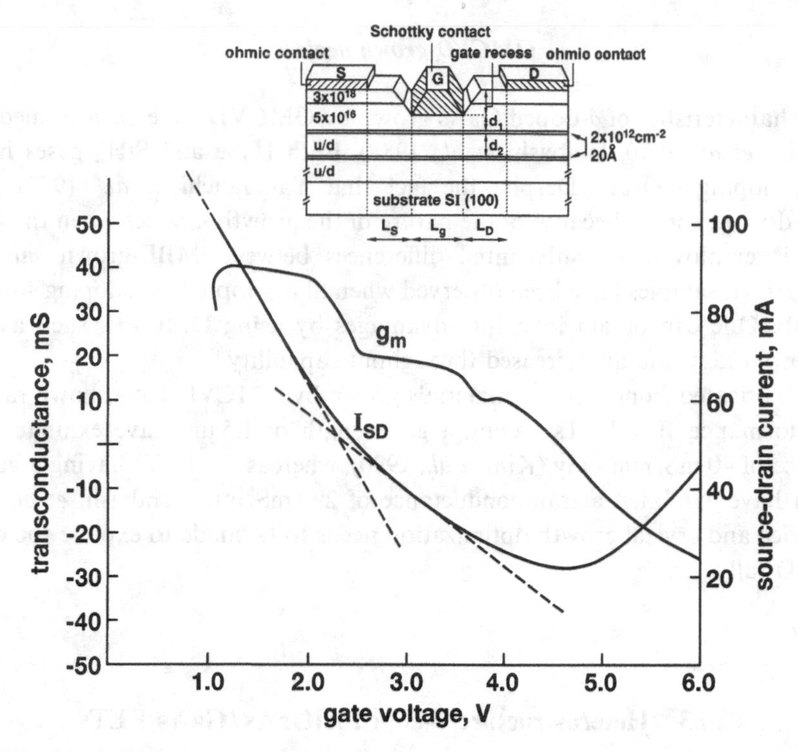

Fig. 20.4. Extrinsic transconductance and drain current versus gate voltage of a MBE-grown double δ-doped GaAs FET.

δ-doped GaAs FETs vs frequency obtained from microwave S-parameter measurements (Hong *et al.*, July, 1989). An f_T of 14.5 GHz and an f_{max} of 30 GHz have been obtained. These results are among the best reported for GaAs FETs having a similar gate geometry (gate length = 1.3 μm).

The basic concept of the GaAs FETs employing a single δ-doped channel has been extended to a FET structure containing two such δ-doped channels, where the two δ-doped channels are stacked vertically and separated sufficiently so that conduction through time occurs in two distinct regions of the transfer characteristics (Board and Nutt, 1992). Figure 20.4 shows the device structure grown by MBE, and measured drain current and transconductance against gate voltage of such a double δ-doped GaAs FET. It is expected that the uppermost δ-doped channel will be pinched off as the gate voltage is increased, with only a very small change in the number of carriers contained in the lower channel. Only as the gate voltage is increased beyond a certain point would the lower channel then be depleted. As a result, in this device the transconductance can be engineered in a more precise way as compared with conventional FETs. The two regions and the stepped form of the transconductance characteristics can clearly be seen in Fig. 20.4. An f_T of 20 GHz has been obtained with devices having a gate length of 0.5 μm.

OMCVD-grown devices

The basic characteristics of δ-doped GaAs grown by OMCVD have been studied by a few groups (Kim *et al.*, 1990; Ishibashi *et al.*, 1988). Both H_2Se and SiH_4 gases have been utilized as doping sources. Despite the fact that Yamaguchi *et al.* (1989) predicted significant dopant spread because of the nature of the growth kinetics when the OMCVD technique is employed, no substantial differences between MBE-grown samples and OMCVD-grown samples have been observed when Si is adopted as a doping source (Kim *et al.*, 1990). One can obtain inherent advantages by using OMCVD, such as superior surface morphology and an increased throughput capability.

Devices fabricated from δ-doped materials grown by OMCVD have shown rather poor device performance. The FETs having a gate length of 1.5 μm have exhibited a transconductance of 40 mS/mm only (Kim *et al.*, 1990), whereas the FETs having a gate length of 0.15 μm have exhibited a transconductance of 290 mS/mm (Ishibashi *et al.*, 1988). A further device and crystal growth optimization needs to be made to explore the capability of OMCVD fully.

20.3 Heterostructure δ-doped AlGaAs/GaAs FETs

A major part of the theoretical and experimental investigations on δ-doped materials and FETs have focused on GaAs homostructures. However, as discussed in Part 5 of this book, 2DEG systems in the quantum wells of δ-doped heterostructures have very interesting transport properties and, moreover, the quantum-well device structure has some desirable features. Figure 20.5 shows the schematic cross-section of a FET structure using a δ-doped

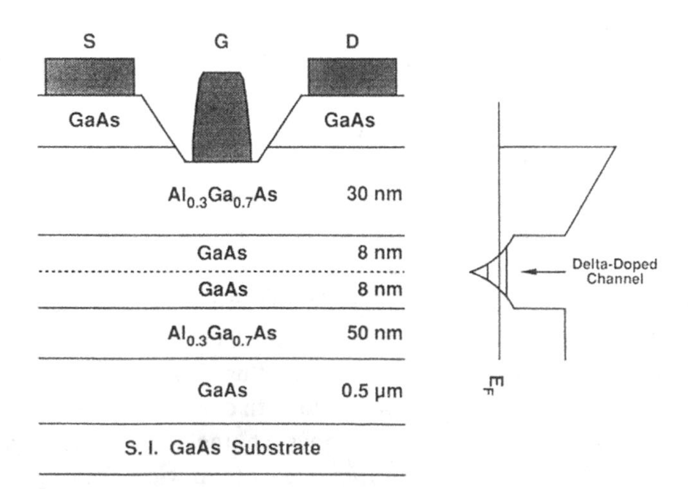

Fig. 20.5. Schematic of a δ-doped AlGaAs/GaAs FET structure.

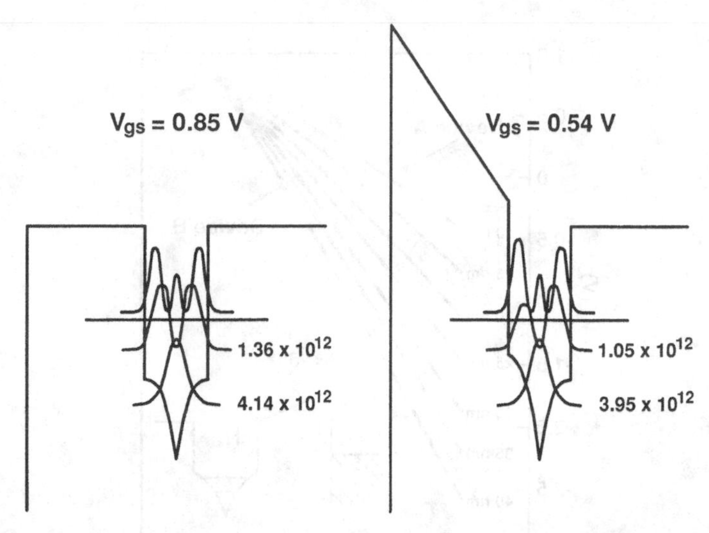

Fig. 20.6. Real-space energy-band diagram of quantum-well δ-doped heterostructures. Energy subbands and carrier occupation of the 2DEG were determined from self-consistent calculation for $N_{\text{tot}} = 5.5 \times 10^{12}\ \text{cm}^{-2}$.

AlGaAs/GaAs heterostructure where the δ-doped channel is located at the center of the GaAs well. The barriers surrounding the well provide the carriers with additional confinement, which will contribute not only to reducing the output conductance but also to extending the flatness of the transconductance over a wide range of gate voltages. In addition, an ideal Schottky diode can be formed by evaporating metals on top of the undoped barrier material, leading to high gate breakdown voltages.

The subband structure of the device shown in Fig. 20.5 has been calculated using a self-consistent method that takes account of the nonparabolicity of the GaAs band structure on the assumption that the doping is a true δ-doping (Hong *et al.*, 1990). The electron energies are quantized for motion perpendicular to the growth direction. Figure 20.6 shows the real-space energy-band diagram of a material structure schematically, where the total number of carriers is assumed to be $5.5 \times 10^{12}\text{cm}^{-2}$. From the calculation, the energy difference between the ground state and the top of the AlGaAs barrier is about 300 meV. This large difference indicates a strong carrier confinement in this device structure. Figure 20.7 shows the relationship between gate voltage and carrier concentration, with the variation of the top AlGaAs layer thickness ranging from 100 Å to 400 Å. In this case, the total carrier concentration from impurities is assumed to be $5 \times 10^{12}\ \text{cm}^{-2}$. The threshold voltage of such a device structure, as shown in the inset of Fig. 20.7, can be expressed by

$$V_{\text{p}} = \frac{qN_{\text{2DEG}}d}{\varepsilon} + \frac{qN_{\text{2DEG}}ad}{2\varepsilon} + \Delta E_{\text{c}} - \phi_{\text{B}} \tag{20.6}$$

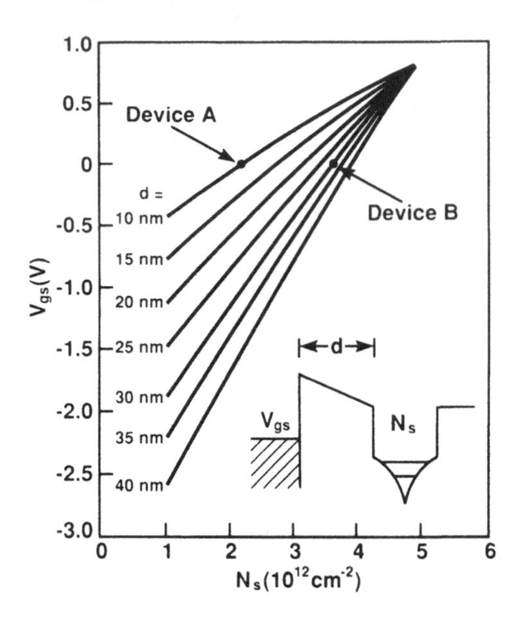

Fig. 20.7. Calculated relation between the gate voltage and total carrier concentration in the quantum-well channel with the variation of top layer thickness (*d*).

where N_{2DEG} is the total intrinsic 2DEG density, d is the top layer thickness, ΔE_C is the conduction-band discontinuity between the AlGaAs barrier and the GaAs well, and ϕ_B is the Schottky barrier height. As can clearly be seen from the plot, the drain current is saturated, independent of the top layer thickness, when the gate voltage is larger than 0.8 V.

Both MBE and OMCVD have been used to grow AlGaAs/GaAs heterostructures with a δ-doped quantum well. The first demonstration of such devices was realized using MBE (Hong *et al.*, 1990). A transconductance of 340 mS/mm and a drain saturation current of 550 mA/mm have been obtained from a MBE-grown device having a gate length of 1.3 μm. The average electron saturation velocity estimated from the drain saturation current is 7.1×10^6 cm/s. This velocity is smaller than those in typical MODFETs. This is believed to be due to enhanced impurity scattering because of the high carrier concentration in the δ-doped channel. The transconductance of the same device also shows a broad plateau around its peak, which is due to the strong carrier confinement in the V-shaped potential well created by δ-doping and additional AlGaAs barriers. An f_T of 15 GHz and an f_{max} of 30 GHz have been obtained from the same device. Very recently, OMCVD has been used to grow a similar device structure (Jeong *et al.*, 1992). The FETs fabricated from OMCVD-grown materials showed performances comparable with MBE-grown devices, again demonstrating the capability of OMCVD for growing advanced device materials.

20.4 δ-doped FETs using In-based materials

Although research in the areas of δ-doped materials and FETs has concentrated on GaAs and AlGaAs/GaAs, a few reports have dealt with the transport properties of δ-doped indium-containing materials, either strained InGaAs on GaAs or materials that are lattice-matched to InP. Furthermore, the preliminary device characteristics of FETs fabricated from these materials have also been studied. To date, only OMCVD has been used to grow δ-doped materials that are lattice-matched to InP, whereas the growth of δ-doped strained InGaAs on GaAs has been realized using MBE.

Transport studies of δ-doped In-based materials

The transport properties of δ-doped $In_{0.53}Ga_{0.47}As$ (on InP) grown by OMCVD have been studied using Hall and Schubnikov–de Haas measurements (Hong *et al.*, Jan., 1989). Hall mobilities of 9300 and 14 600 cm^2/V-s have been measured for carrier concentrations of 3.7×10^{12} and 3.0×10^{12} cm^{-2} at 300 and 77 K respectively. By comparison with homogeneously doped $In_{0.53}Ga_{0.47}As$, the mobility enhancement is estimated to be 2.5 at this doping level at 300 K. In addition, it has been shown that one can obtain carrier concentrations as high as 1.4×10^{13} cm^{-2} from the same materials. The transport properties of δ-doped InP grown by OMCVD have also been investigated (di Forte-Poisson *et al.*, 1989). δ-doped InP exhibited Hall mobilities of 1350 cm^2/V-s with a carrier concentration of 1.78×10^{12} cm^{-2} at 4.2 K.

Recently, strained $In_xGa_{1-x}As$ has been actively studied theoretically and experimentally because the addition of indium to GaAs will give rise to (i) a higher band offset in the heterostructures, leading to better carrier confinement, and (ii) enhanced device performance due to an increase in electron velocity. The transport properties of 2DEG systems in δ-doped strained $In_xGa_{1-x}As$ quantum wells have been investigated (Hong *et al.*, 1990). The study included comparison of experimental measurements of the effective mass with theoretical data obtained from self-consistent calculations.

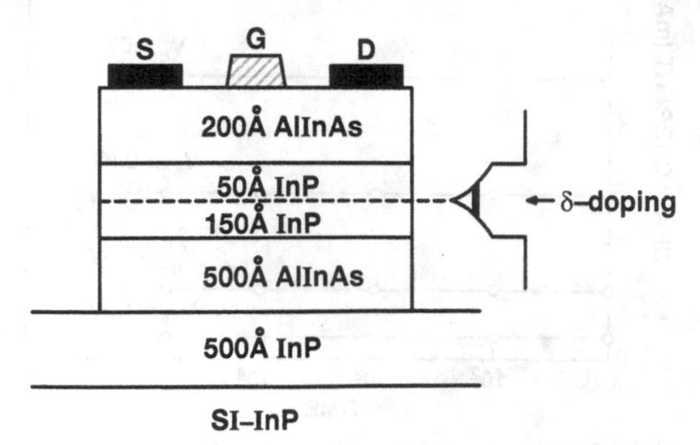

Fig. 20.8. Schematic diagram of a δ-doped InAlAs/InP FET structure.

Heterostructure δ-doped InAlAs/InP FETs

A device concept similar to that of δ-doped AlGaAs/GaAs FETs has been applied to InAlAs/InP heterostructures, where the δ-doped channel is located in the InP quantum well (Jeong and Hong *et al.*, 1992). Figure 20.8 shows the schematic diagram of the FET structure. The InAlAs/InP heterostructure is a type II heterostructure and, moreover, the conduction-band discontinuity of 0.4 eV is so large that carriers will be strongly confined

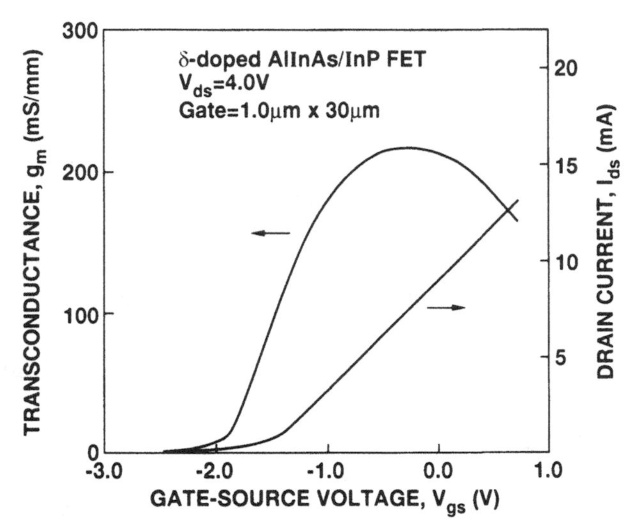

Fig. 20.9. Extrinsic transconductance and drain current versus gate voltage of an OMCVD-grown δ-doped InAlAs/InP FET.

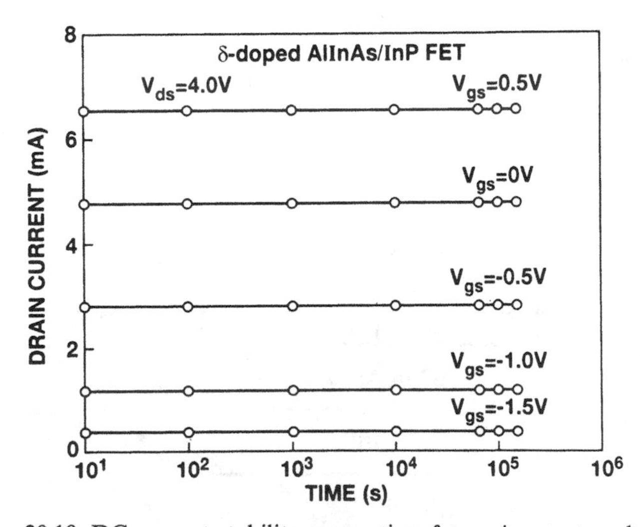

Fig. 20.10. DC current stability versus time for various gate voltages.

(Aina *et al.*, 1990). In addition, InP has a large energy bandgap of 1.35 eV and a high electron saturation velocity, leading to large breakdown voltages and high-speed performance respectively. As shown in Fig. 20.9, the extrinsic transconductance remains above 210 mS/mm at a current density of 190 mA/mm over a wide range of gate voltages. From a preliminary stability test, as shown in Fig. 20.10, no drift in drain current has been observed, demonstrating the stability of the FETs in contrast with conventional InP MISFET's.

20.5 Conclusion

In this section, the recent advances of FETs based on δ-doped homo- and heterostructures have been briefly reviewed. Continuing improvements in device processing and material growth technologies will make newly proposed high current-density and high speed performance δ-doped FETs feasible. Such devices will be based on δ-doped GaAs and InGaAs homostructures, AlGaAs/GaAs heterostructures, and InAlAs/InP heterostructures grown by MBE and OMCVD. Nevertheless, further studies on device characteristics such as parasitic effects limiting device performance, noise behavoir, and device reliability need to be carried out in order to establish the practical application of this new class of device.

References

L. Aina et al., *Appl. Phys. Lett.*, **53**, 1620, 1990.
K. Board and H. C. Nutt, *Electron. Lett.*, **28**, 469, 1992.
M. A. di Forte-Poisson *et al.*, *J. Appl. Phys.*, **66**(2), 867, 1989.
T. J. Drummond, M. Keever, W. Kopp, H. Morkoc, K. Hess, B. G. Streetman, and A. Y. Cho, *Electronics Lett.*, **17**, 545, 1981.
W-P. Hong *et al.*, *Appl. Phys. Lett.*, **54**(5), 457, 1989a.
W-P. Hong *et al.*, *IEEE Electron Device Lett.*, **10**, 310, 1989b.
W-P. Hong *et al.*, *IEEE Trans. Electron Devices*, **37**, 1924, 1990a.
W-P. Hong *et al.*, *Appl. Phys. Lett.*, **57**(11), 1117, 1990b.
P. L. Hower and G. Bechtel, *IEEE Trans. Electron Devices*, **20**, 243, 1983.
A. Ishibashi *et al.*, *Electron. Lett.*, **24**, 1034, 1988.
D. Jeong *et al.*, *IEEE Electron Device Lett.*, **13**, 270, 1992a.
Y. Jeong and W-P. Hong *et al.*, *Jpn J. Apply. Phys.*, **31**, L66, 1992b.
Kim *et al.*, *J. Appl. Phys.*, **68**(6), 2747, 1990.
E. F. Schubert, K. Ploog, H. Dambkes, and K. Heime, *Appl., Phys.*, A-**33**, 63, 1984.
E. F. Schubert *et al.*, *IEEE Trans. Electron Devices*, **33**, 625, 1986a.
E. F. Schubert *et al.*, *Appl. Phys. Lett.*, **49** (25), 1729, 1986b.
Y. Yamaguchi *et al.*, *Jpn. J. Appl. Phys.*, **28**, L1689, 1989.

21

Selectively doped heterostructure devices

E. F. SCHUBERT

Field-effect transistors (FETs) can be fabricated from either semiconductor *homostructures* or from *heterostructures*. In this section, FETs based on *selectively doped heterostructures* will be discussed. Such selectively doped heterostructures consist of a doped wide-gap semiconductor and an adjacent, undoped, narrow-gap semiconductor. We assume that barriers are formed in the valence band and in the conduction band at the interface between the two semiconductors. Free carriers originating from dopants in the wide-gap semiconductor can diffuse to the narrow-gap semiconductor, where they occupy states at lower energies. Free carriers do not diffuse back into the wide-gap semiconductors due to the heterojunction barrier. Thus, free carriers are *spatially separated* from their parent ionized impurities. This unique situation, which cannot be achieved in bulk semiconductors, leads to a reduction in ionized impurity scattering for carriers in the narrow-gap semiconductor. The resulting increase in free-carrier mobility is approximately a factor of two in the $Al_xGa_{1-x}As/GaAs$ material system at room temperature. At low temperatures, where ionized impurity scattering is the dominant scattering mechanism in bulk semiconductors, the mobility in selectively doped heterostructures can be many orders of magnitude higher as compared with bulk semiconductors.

Selectively doped heterostructures are used in field-effect transistors, which are frequently called *selectively doped heterostructure transistors* (SDHTs) or *high electron mobility transistors* (HEMTs). An excellent review of the physics and technology of SDHTs was recently given by Ali and Gupta (1991). Selectively doped heterostructure transistors take advantage of the higher free-carrier mobilities in such structures. The high free-carrier mobility makes possible an enhanced transconductance as well as reduced parasitic resistances in the regions between source and gate, and gate and drain. In addition, SDHTs exhibit reduced short-channel effects due to the spatial confinement of the free-carrier channel at the semiconductor interface, and due to the proximity of the electron channel to the gate metal. Today, selectively doped heterostructure transistors are mainly used in low-noise microwave applications.

Employment of the δ-doping technique in a selectively doped heterostructure results in the optimization of the structure in terms of its electronic properties. We will show that the use of δ-function-like doping profiles yields (i) the highest achievable free-carrier densities, (ii) the highest free-carrier mobilities, and (iii) well-defined electron channels with reduced short-channel effects. The epitaxial layer sequence of a selectively δ-doped heterostructure and of a selectively δ-doped quantum well structure are shown in Fig. 21.1(a) and (b) respectively. The selectively δ-doped heterostructure (see Fig. 21.1(a)) consists of a substrate, an undoped narrow-gap material, and a wide-gap material that is δ-doped close to the interface of the two semiconductors. The materials most widely used for the wide-gap and the narrow-gap materials are $Al_xGa_{1-x}As$ and GaAs respectively. The selectively δ-doped quantum well structure (see Fig. 21.1(b)) consists of a substrate, a wide-gap semiconductor layer, a thin narrow-gap semiconductor quantum well layer, and another wide-gap semiconductor layer. Both wide-gap semiconductor layers are δ-doped in close proximity to the two quantum well interfaces.

The electron density in δ-doped heterostructures is, as will be shown below, higher as compared with homogeneously doped heterostructures. The density is even higher in selectively δ-doped quantum well structures, where free carriers are provided from the doped layers on both sides of the quantum well. The free-carrier concentration can be up to a factor of two higher in the quantum well structure as compared with the single-interface heterostructure.

The properties of selectively δ-doped heterostructures and quantum well structures can be analyzed in terms of the band diagrams shown in Fig. 21.2(a) and (b) respectively. All the doped layers are assumed to be depleted of free carriers because they have transferred to either localized surface states or delocalized states in the well. The doped regions therefore appear as asymmetric or symmetric V-shaped regions of the band diagram.

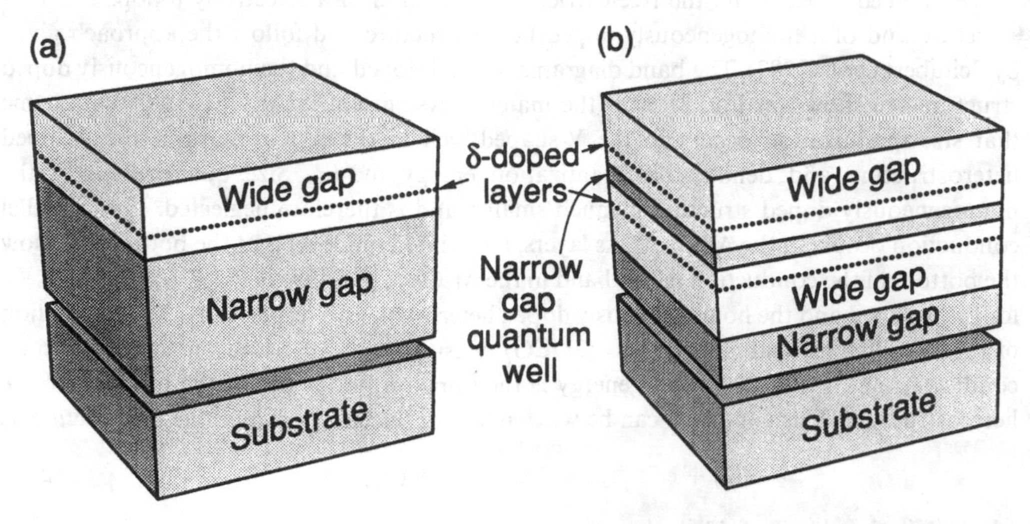

Fig. 21.1. Layer sequence of (a) a selectively δ-doped single-interface heterostructure and (b) a selectively δ-doped quantum well structure.

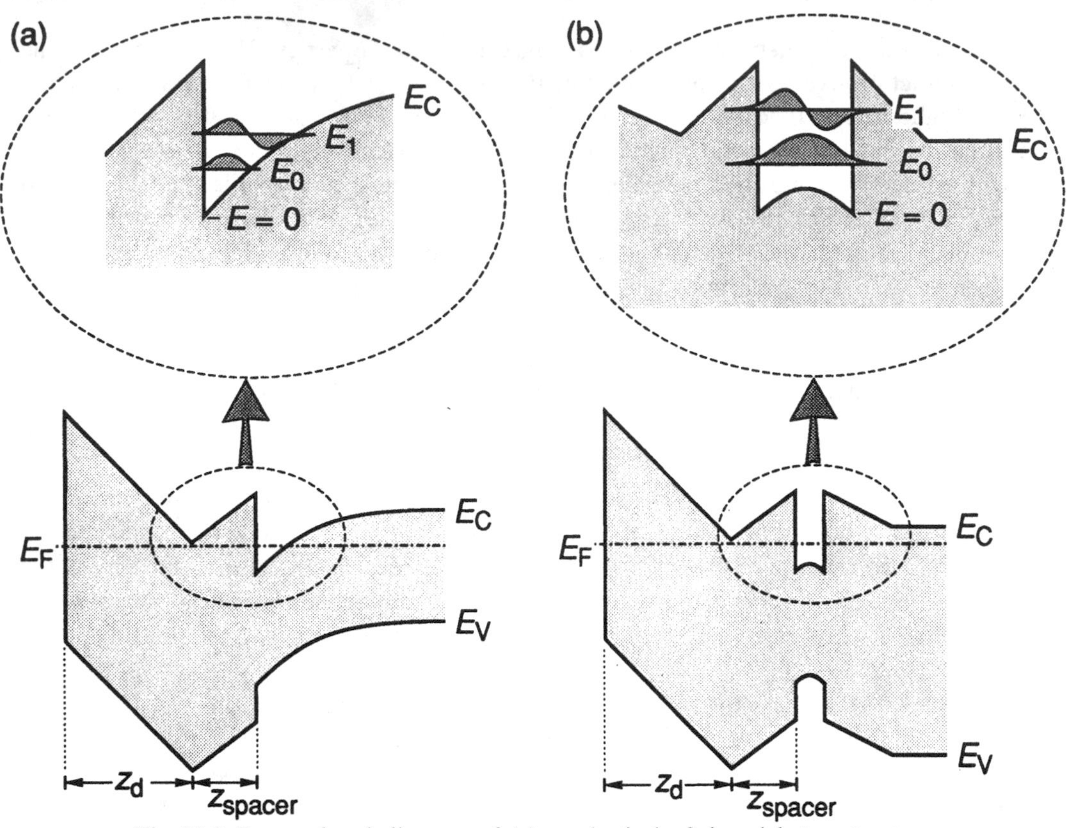

Fig. 21.2. Energy band diagram of (a) a selectively δ-doped heterostructure and (b) a selectively δ-doped quantum well structure.

We proceed to calculate the free-carrier concentration of a selectively δ-doped heterostructure and of a homogeneously doped heterostructure and follow the approach given by Schubert *et al.* (1987). The band diagrams of the δ-doped and the homogeneously doped structure are shown in Fig. 21.3 for the material system GaAs/$Al_xGa_{1-x}As$. We assume that size quantization occurs in the V-shaped well in the $Al_xGa_{1-x}As$ of the δ-doped heterostructure and denote the quantization energy by E_0^δ. Size quantization in the homogeneously doped structure is much smaller and is therefore neglected. If no parallel conduction occurs in the $Al_xGa_{1-x}As$ layers, then the Fermi level is at the bottom or below the bottom of the conduction (sub-) band in the $Al_xGa_{1-x}As$, that is, $E_F \leqslant E_0^\delta$ and $E_F \leqslant E_C$ for the δ-doped and the homogeneously doped heterostructure respectively. The calculation of the two-dimensional electron gas (2DEG) density, n_{2DEG}, is based on the equilibrium condition, that is, that the Fermi energy is the same on both sides of the interface of the heterostructure. This condition can be written for the selectivity *δ-doped heterostructure* as

$$-E_0^\delta + e\mathscr{E}z_{spacer} - \Delta E_C + E_0 + (E_F - E_0) = 0 \tag{21.1}$$

where $\mathscr{E}$ is the electric field in the spacer, z_{spacer} is the thickness of the spacer layer, ΔE_C is the magnitude of the conduction band discontinuity, E_0 is the quantized energy of the

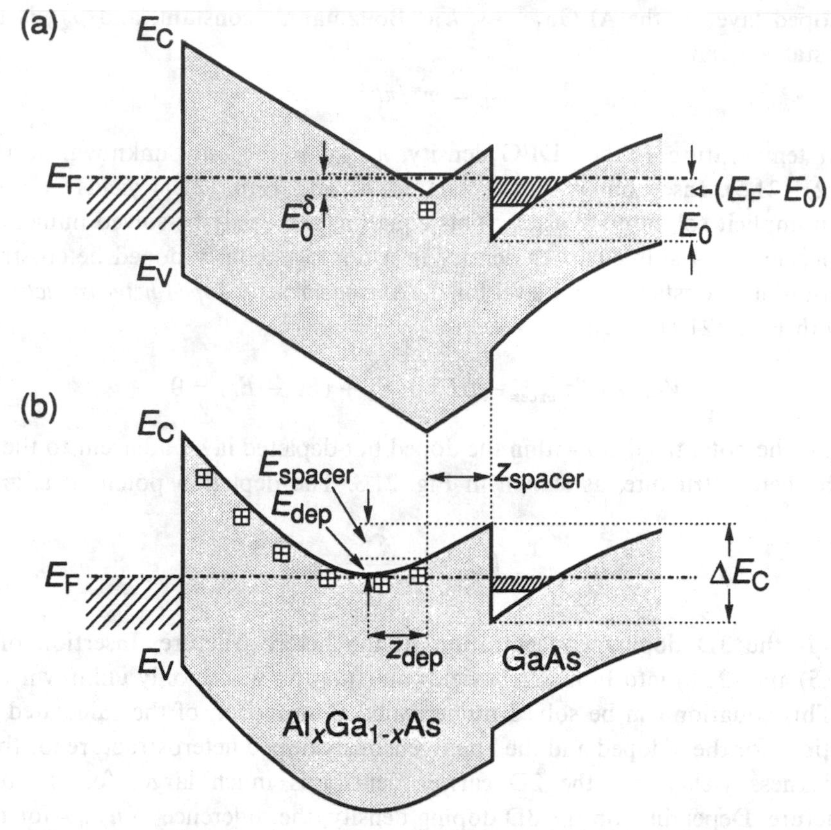

Fig. 21.3. Energy band diagram of (a) a selectively δ-doped $Al_xGa_{1-x}As/GaAs$ heterostructure and (b) a heterostructure homogeneously doped in the $Al_xGa_{1-x}As$.

lowest state of the 2DEG, and $E_F - E_0$ is the degeneracy energy of the 2DEG. All energies in Eqn. (21.1) are either constants or can be expressed as a function of the free electron density using the following simple equations based on electrostatics and quantum mechanics:

$$E_0^\delta = \frac{3}{10}\left(\frac{9^2}{2^3}\right)^{1/3}\left(\frac{e^2\hbar N^{2D}}{\varepsilon_{AlGaAs}\sqrt{(2m^*)}}\right)^{2/3} \tag{21.2}$$

$$\mathscr{E} = (e/\varepsilon_{AlGaAs})n_{2DEG} \tag{21.3}$$

$$E_0 = \frac{3}{2}\left(\frac{3}{2}\frac{e^2}{\varepsilon_{GaAs}}n_{2DEG}\frac{\hbar}{\sqrt{(m^*)}}\right)^{2/3} \tag{21.4}$$

$$E_F - E_0 = kT\ln(e^{n_{2DEG}/kT\rho_{2D}} - 1) \tag{21.5}$$

where ε is the permittivity of the semiconductor, $\hbar$ is Planck's constant divided by 2π, m^* is the effective mass of the free carriers, N^{2D} is the two-dimensional (2D) doping density of the δ-doped layer in the $Al_xGa_{1-x}As$, k is Boltzmann's constant, and ρ_{2D} is the 2D density of states given by

$$\rho_{2D} = m^*/\pi\hbar^2. \tag{21.6}$$

At a given temperature T, the 2DEG density, n_{2DEG}, is the only unknown variable in Eqns. (21.2)–(21.5). Insertion of Eqns (21.2)–(21.5) into Eqn. (21.1) yields an equation which is an implicit solution for n_{2DEG}. This equation can easily be solved numerically.

We next calculate the free-carrier density in a homogeneously doped heterostructure. The condition of a constant Fermi level for the *homogeneously doped heterostructure* is, by analogy with Eqn. (21.1), given by

$$eV_{dep} + e\mathscr{E}z_{spacer} - \Delta E_C + E_0 + (E_F - E_0) = 0 \tag{21.7}$$

where V_{dep} is the potential drop within the doped but depleted layer adjacent to the spacer layer of the heterostructure, as shown in Fig. 21.3. The depletion potential is given by

$$V_{dep} = \frac{\varepsilon}{2e}\frac{\mathscr{E}^2}{N_D} \tag{21.8}$$

where N_D is the 3D doping concentration in the heterostructure. Insertion of Eqns. (21.3)–(21.5) and (21.8) into Eqn. (21.7) yields an equation whose only unknown variable is n_{2DEG}. This equation can be solved numerically. Comparison of the calculated 2DEG concentrations for the δ-doped and the homogeneously doped heterostructure for the same spacer thickness yields that the 2D carrier density is much larger for the δ-doped heterostructure. Depending on the 3D doping density, the difference in n_{2DEG} for the two structures can be a factor of two or even larger (Schubert *et al.*, 1987). A numerical comparison between the δ-doped and the homogeneously doped heterostructure was performed by Kim *et al.* (1991) and Tian *et al.* (1991) using ensemble Monte Carlo simulations. The simulations revealed that the δ-doped structures are clearly superior, especially at increased gate biases. A comparison between different heterostructures was also reported by Liu *et al.* (1991), who found that parallel conduction is greatly reduced in selectively δ-doped heterostructures.

We next discuss the physical reasons for the enhanced 2D electron densities attainable in selectively δ-doped heterostructures. Inspection of Eqns. (21.1) and (21.7) reveals the following two differences between the two structures. First, the term E_0^δ, that is, the quantization energy in the V-shaped well of the δ-layer, enters Eqn. (21.1) but not Eqn. (21.7). It does not enter Eqn. (21.7) since size quantization in the wide parabolic wells of the homogeneously doped heterostructure is very small and can therefore be neglected. The term E_0^δ increases the lowest continuum state in $Al_xGa_{1-x}As$ from the bottom of the conduction band to a higher-lying quantized state. It is thus energetically more favorable for electrons to transfer to the GaAs layer, since the difference in energy between the lowest states in the two semiconductors is enhanced by E_0^δ. Note that the term E_0^δ in Eqn. (21.1) adds to ΔE_C. Therefore, the sum $E_0^\delta + \Delta E_C$ can be understood as an enhanced *effective conduction-band discontinuity* which in turn increases n_{2DEG}. Second, the potential drop in the depletion region (see Eqn. (21.7)), V_{dep}, does *not* enter Eqn. (21.1). This is because the thickness of the depletion region is zero in the δ-doped structure, due to the localization of donors in a sheet. The absence of a depletion layer potential drop in Eqn. (21.1) is advantageous and leads to a further increase of n_{2DEG} in the selectively δ-doped heterostructure. To summarize the comparison, the selectively δ-doped heterostructure has two advantages, namely an effective band discontinuity enhancement due to size quantization in the $Al_xGa_{1-x}As$, and the absence of a depletion layer potential drop due to the localization of donor impurities in the δ-doped layer. Both characteristics result in the desired increase in the 2DEG density.

The enhanced electron density in selectively δ-doped heterostructures, including the first observation of the population of two subbands in the GaAs/$Al_xGa_{1-x}As$ material system, has been demonstrated by Shubnikov–de Haas measurements (Schubert *et al.*, 1987). The magnetoresistance of such a structure with a spacer thickness of 25 Å is shown in Fig. 21.4(a) for a measurement temperature of $T = 4.2$ K. The onset of resistance oscillations of ρ_{xx} start at a magnetic field of $B = 0.5$ T. A second superimposed oscillation starts at 1.5 T and dominates the oscillatory behavior for $B > 3$ T. The two distinct

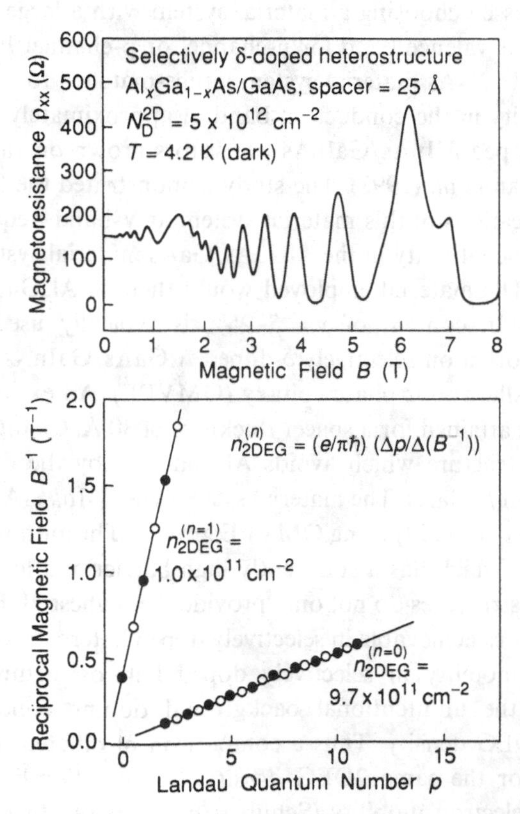

Fig. 21.4. Shubnikov–de Haas oscillations of an $Al_xGa_{1-x}As$/GaAs selectively doped heterostructure measured at $T = 4.2$ K. The total free-carrier concentration deduced from the period of the oscillation is $n_{2DEG} = 10.7 \times 10^{11}$ cm^{-2}.

oscillations are attributed to the occupation of the ground state and the first excited state of the 2DEG. Note that the population of two subbands in the $GaAs/Al_xGa_{1-x}As$ material system has been demonstrated only for δ-doped structures but not for homogeneously doped structures. The free-carrier density in each individual subband can be evaluated by plotting the Landau quantum numbers of the minima (solid circles) and maxima (open circles) versus the reciprocal magnetic field, as shown in Fig. 21.4(b). The slope of this plot yields concentrations of 9.7×10^{11} cm^{-2} and 1×10^{11} cm^{-2} for the lowest and first excited subband respectively. The total concentration is then $n_{2DEG} = 1.07 \times 10^{12}$ cm^{-2} at $T = 4.2$ K. The corresponding mobility is $\mu = 37\,000$ cm^2/Vs. At room temperature, a concentration of $n_{2DEG} = 1.7 \times 10^{12}$ cm^{-2} and a mobility of 8900 cm^2/Vs have been measured by the Hall effect on the same sample.

According to Shockley's *gradual channel approximation* for FETs, the transconductance of the device increases with the density of the electron gas. Even though the gradual channel approximation is strictly valid only for transistors with long gate lengths, the predicted trend holds even for FETs with short gate lengths. It is thus desirable to increase the 2DEG density. As shown above, the δ-doping technique allows one to achieve the highest 2DEG density among all possible doping configurations. Furthermore, the 2DEG concentration increases by choosing a material system with a large band discontinuity in the conduction band or valence band for n-channel or p-channel FETs respectively. The $Al_{0.48}In_{0.52}As/Ga_{0.47}In_{0.53}As$ material system lattice-matched to InP offers an especially large band discontinuity in the conduction band of approximately 500 meV. The dc and rf performance of δ-doped AlInAs/GaInAs transistors grown on InP substrates has been reported by Matloubian *et al.* (1993). The study demonstrated the feasibility of low-noise and high-power applications of this material system at V-band frequencies. A method for increasing the band discontinuity in the AlGaAs/GaAs material system is to add In to the GaAs channel layer. The material employed would then be $Al_xGa_{1-x}As/Ga_{1-y}In_yAs$ on GaAs substrates. An In content of $y = 5$–25% is typically used in these structures. Yang *et al.* (1992) reported on selectively δ-doped AlGaAs/GaInAs heterostructure FETs grown by organometallic vapor-phase epitaxy (OMVPE). An excellent 2DEG density of 2.25×10^{12} cm^{-2} was attained for a spacer thickness of 30 Å. Chang *et al.* (1991) reported on a GaInAs/GaAs structure which avoids Al and thereby the doping problems (DX-centers) associated with AlGaAs. The material system $Ga_{0.47}In_{0.53}As/InP$ on InP has been employed by Kusters *et al.* (1993) using OMVPE growth. The authors used p-type InP as a quasi-Schottky barrier which has a sufficiently high barrier and low leakage current.

Delta-doped heterostructures do not only provide the highest 2DEG concentrations but also the *highest mobilities* achievable in selectively doped heterostructures. It is well known that the free-carrier mobility in selectively doped heterostructures depends on many parameters such as the unintentional background doping concentration, the spacer thickness, and the 2DEG density. Thus a comparison of different doping configurations must be performed for the same 2DEG *density*. A δ-function-like doping distribution provides the highest electron mobility (Schubert *et al.*, 1989). In this study, the electron mobility in selectively doped $Al_xGa_{1-x}As/GaAs$ heterostructures was investigated for different doping profiles, all of them resulting in the *same* free-carrier density. The doping profiles were pulse (top-hat) shaped with a width of Δz_d and 3D donor concentration of N_D. Whereas the width Δz_d was varied, the product $\Delta z_d N_D$ was kept constant. The centroid of the doping distribution was kept at a constant distance from the interface. The authors showed that such a variation in Δz_d does not change the free-carrier density, n_{2DEG}. Furthermore, they calculated the potential fluctuations introduced by random dopant distribution within the doped layer and showed that the potential fluctuations are minimized for δ-function-like dopant distributions. Because the minimization of potential fluctuations is equivalent to maximizing the electron mobility (minimized coulombic scattering), the highest free-carrier mobilities are obtained for selectively δ-doped heterostructures. Experimental study of different doping distributions on the electron mobility has confirmed that the highest electron mobilities are indeed obtained for δ-doped heterostructures.

The free-carrier mobility influences the characteristics of FETs in two respects. First, the transconductance of FETs is, according to Shockley's gradual channel approximation, directly proportional to the free-carrier mobility. Even though this model is known to be inapplicable to FETs with very short gate lengths, the transconductance still increases sub-linearly with mobility for such FETs. Second, the parasitic resistances of FETs are inversely proportional to the free-carrier mobility. The low electric field regions of the channel between the source and gate contacts and between the gate and drain contacts are regions with high parasitic resistances. Lowering these resistances improves the noise figure of FETs. In addition, the reduction of the source resistance increases the transconductance. Thus, maximizing the mobility for a given concentration, n_{2DEG}, is beneficial for the operation of FETs.

The schematic structure of a selectively doped heterostructure FET is shown in Fig. 21.5. The structure comprises the source and drain ohmic contacts and the rectifying

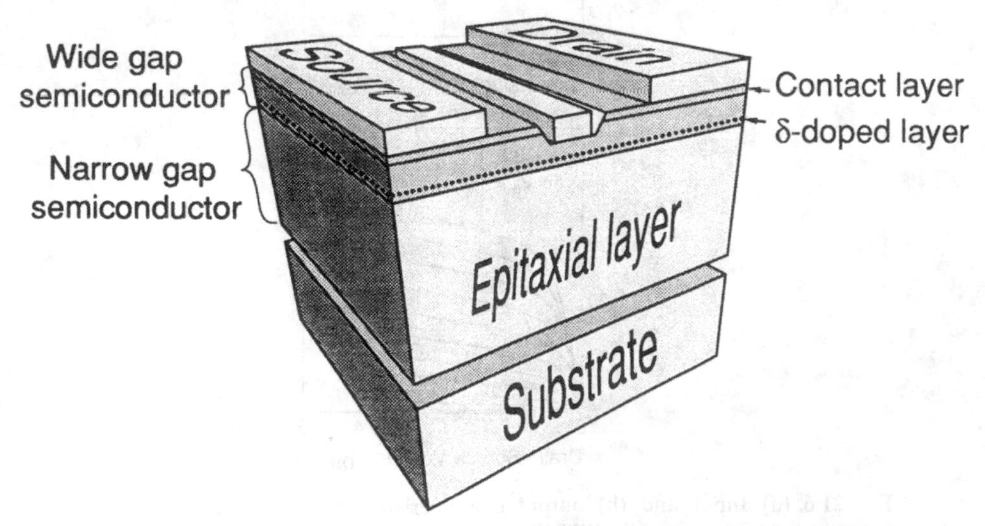

Fig. 21.5. Schematic structure of a selectively doped heterostructure FET.

Schottky-gate electrode. In addition to the layers already shown in Fig. 21.1, a highly doped contact layer is included in typical FET structures, as illustrated in Fig. 21.5. The purpose of this contact layer is (i) to reduce the contact resistance of the ohmic source and drain contacts and (ii) to reduce the parasitic resistance in the low electric field regions between the source and gate contacts and between the gate and drain contacts. The gate electrode is recessed, which makes the 2DEG the only conductive channel below the gate. The conductive channel can be either the 2DEG at the interface (single interface structure; Schubert *et al.*, 1987) or the 2DEG in a quantum well with a doping layer on both sides of the quantum well (Kuo *et al.*, 1988). In addition to doping during epitaxy, ion implantation has been employed in SDHTs. However, the post-implantation, high-temperature anneal, which reduces the damage introduced during implantation, also broadens the δ-function doping profiles by diffusion processes. This usually results in a degradation of the properties of the 2DEG.

The input and output current–voltage characteristics of a selectively δ-doped heterostructure transistor (SΔDHT) is shown in Fig. 21.6 (Schubert *et al.*, 1987). The epitaxial layers for the transistors were grown by molecular-beam epitaxy (MBE) and consist of a 1-μm-thick undoped GaAs buffer layer, an undoped 25-Å-thick $Al_{0.30}Ga_{0.70}As$ spacer layer, an n-type δ-doped sheet of density $N_{Si}^{2D} = 5 \times 10^{12}$ cm^{-2}, a 375-Å-thick n-type

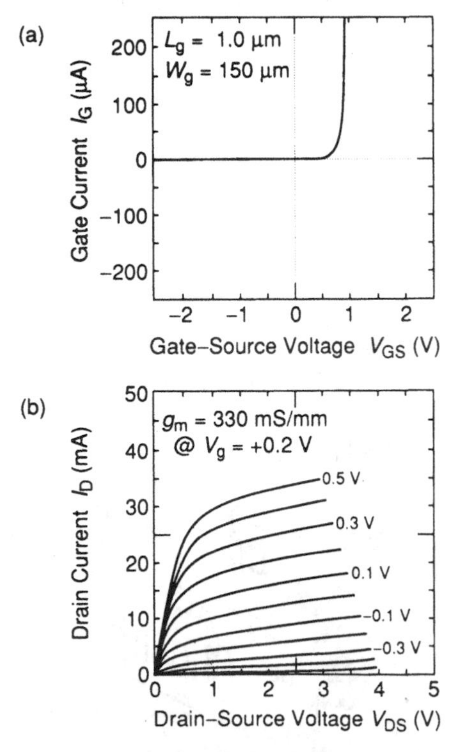

Fig. 21.6. (a) Input and (b) output characteristic of a selectively δ-doped heterostructure FET measured at room temperature.

$Al_{0.30}Ga_{0.70}As$ layer, and a 750-Å-thick n^+-type GaAs top layer with a Si concentration of $N_{Si} = 1 \times 10^{19}$ cm^{-3}. The SΔDHTs have a nominal gate length of 1.0 μm, a gate width of 150 μm, and a source-to-drain electrode spacing of 4.0 μm. AuGe/Ni/Au (1500 Å/500 Å/1500 Å) and Ti/Au (500 Å/1500 Å) were used for the ohmic source and drain contacts and for the Schottky gate metallization respectively. The ohmic contacts were alloyed at 415 °C for 25 s.

The input current–voltage characteristic of the SΔDHT shown in Fig. 21.6 displays a forward threshold voltage of 0.75 V, a low leakage current, and a large reverse breakdown voltage of $V = -6$ V. Figure 21.6 also shows the current–voltage output characteristic of an SΔDHT at room temperature. The SΔDHT exhibits a low ON resistance ($R_{ON} = 1.8$ Ωmm), excellent saturation properties, a low differential output conductance in the saturation regime, and very good pinch-off properties. The particular device shown has a maximum transconductance of 327 mS/mm at a gate voltage of $V_g = +0.3$ V. Transconductances as high as 360 mS/mm where measured on several transistors on the same wafer. Cooling the wafers to a temperature of $T = 77$ K yielded a maximum transconductance of 420 mS/mm. The increased transconductance at $T = 77$ K can be attributed to the higher electron mobility of the heterostructures at low temperatures. The room-temperature contact resistance of the devices was determined to be $R_{contact} = 0.07$ Ωmm. The results demonstrate that selectively δ-doped heterostructure transistors have superior properties when compared to their homogeneously doped counterparts.

Selectively doped heterostructure transistors have also been realized in the Si-based material system SiGe/Si (People *et al.*, 1985). Certainly, the δ-doping technique would also improve Si-based devices. However, materials problems in the SiGe/Si material system have prevented the exploitation of such transistors. First, the SiGe material system is mismatched to the Si substrates, which results in the formation of misfit dislocations and traps at the substrate–epilayer interface and in the epilayer itself. Second, the doping process in this material system is complicated and requires growth at relatively low temperatures which, again, introduces defects and deep levels.

We finally point out that selectively δ-doped heterostructure transistors exhibit reduced *short-channel effects* as compared with their homogeneously doped counterparts. At very short gate lengths ($L_g < 0.5$ μm), a lack of pinch-off and an increase in output conductance in the saturation regime are frequently found. Short-channel effects arise when the gate length becomes comparable to the gate-to-channel distance or to the spatial extent of the electron channel in the epitaxial growth direction. It is desirable, in order to reduce short-channel effects, to minimize the gate-to-channel distance as well as the spatial extent of the channel. The minimum gate-to-channel distance in Schottky gate FETs is determined by the surface depletion region and, in the case of the devices discussed here, by the interface depletion region and the spacer layer. In order to keep these layers thin, very high doping concentrations must be employed. Thus, the natural choices for reduced short-channel effects are δ-function-like doping profiles. This intuitively reached conclusion has been confirmed by theoretical Monte Carlo simulations (Tian *et al.*, 1991).

References

Ali F. and Gupta A., eds. *HEMTs and HBTs* (Artech House, Massachusetts, 1991).
Chang C. Y., Lin W., Hsu W. C., Wu T. S., Chang S. Z., and Wang C. (1991) *Jpn J. Appl. Phys.* **30**, 1158.
Kim K. W., Tian H., and Littlejohn M. A. (1991) *IEEE Trans. Electron Devices* **38**, 1737.
Kuo T. Y., Cunningham J. E., Schubert E. F., Tsang W. T., Chiu T. H., Ren F., and Fonstad C. G. (1988) *J. Appl. Phys.* **64**, 3324.
Kusters A. M., Kohl A., Muller R., Sommer V., and Heime K. (1993) *IEEE Electron Dev. Lett.* **14**, 36.
Liu D. G., Chin T. C., Lee C. P., and Hwang H. L. (1991) *Solid State Electronics* **34**, 253.
Matloubian M., Brown A. S., Nguyen L. D., Melendes M. A., Larson L. E., Delancey M. J., Pence J. E., Rhodes R. A., Thompson M. A., and Henige J. A. (1993) *IEEE Electron Device Letters* **4**, 188.
People R., Bean J. C., and Lang D. V. (1985) *J. Vac. Sci. Technol.* A **3**, 846.
Schubert E. F., Cunningham J. E., Tsang W. T., and Timp G. L. (1987) *Appl. Phys. Lett.* **51**, 1170.
Schubert E. F., Pfeiffer L., West K. W., and Izabelle A. (1989) *Appl. Phys. Lett.* **54**, 1350.
Tian H., Kim K. W., and Littlejohn M. A. (1991) *J. Appl. Phys.* **70**, 4593.
Yang G. M., Park S. G., Seo K. S., and Choe B. D. (1992) *Appl. Phys. Lett.* **60**, 2380.

22

Silicon atomic layer doping field-effect transistors (FETs)

KIYOKAZU NAKAGAWA AND KEN YAMAGUCHI

22.1 Introduction

The great progress in very large scale (VLSI) technology has been supported by the shrinking or scaling-down theory[1]. The miniaturization guided by this theory, however, is nearing its physical limit. As is well known, the major obstacles inhibiting further miniaturization are (i) punchthrough; (ii) theshold voltage V_{TH} lowering, (iii) the hot carrier injection effect, and (iv) the statistical fluctuation of doped impurity distributions. Further, the operation voltage is determined from the application systems, and is independent of the scaling-down guide-line. Item (iv) is a serious problem because it is physically unavoidable even if the fabrication process is perfectly controlled.

In this chapter, a new device structure used to overcome the above obstacles is presented. The new device is fabricated using atomic-layer-doping (ALD) technology, or so-called delta-doping technology. Device characterization is carried out using numerical simulation and a design guide-line is clarified. The advantage of the device presented is demonstrated by a reduction in punchthrough, V_{TH}-lowering, and statistical fluctuation.

22.2 Proposal of ALD-device and theoretical study

22.2.1 Miniaturization limit due to impurity fluctuation

Statistical fluctuation of the doped atom distribution is physically unavoidable, irrespective of the progress of the process technology. It is a serious problem in ultra-small devices because the number of impurity atoms between the anode and the cathode is ten or less.

Assuming a metal-oxide–semiconductor field-effect transistor (MOSFET) structure and also assuming several sets of point charges corresponding to the doped atoms, current–

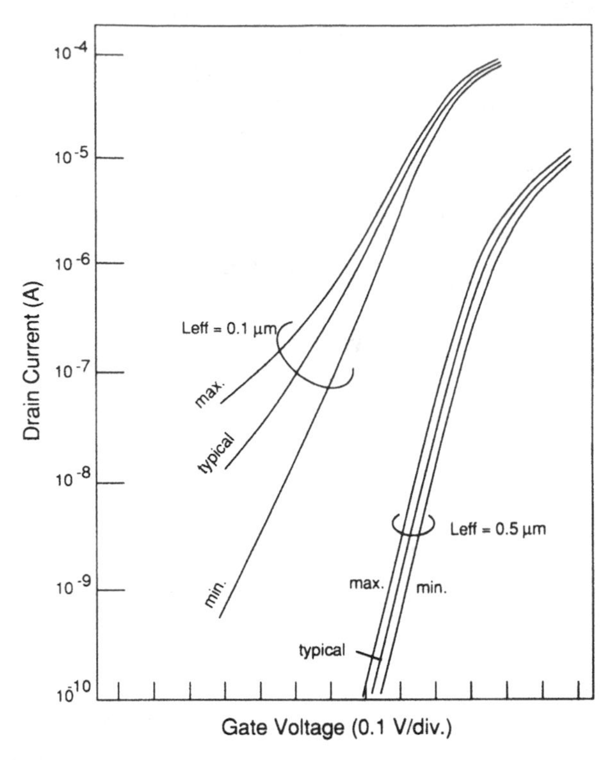

Fig. 22.1. Statistical fluctuation effect on doped impurity distribution in conventional structure MOSFETs. Assuming several cases of statistical fluctuation, current–voltage characteristics are theoretically analyzed by two-dimensional numerical simulation. See Refs. (3) and (5).

voltage characteristics are analyzed theoretically with the aid of two-dimensional numerical simulation. Poisson's and the current continuity equations are numerically solved in two dimensions[2]:

$$\varepsilon \nabla^2 \Psi = -q(N_D - N_A - n + p), \quad (22.1)$$

$$\nabla \cdot \mathbf{J} = 0, \quad (22.2)$$

where N_D and N_A stand for the donor and acceptor concentration distributions, Ψ and $\mathbf{J}$ are the electrostatic potential and current density vector, and n and p are the electron and hole densities.

Calculated current–voltage characteristics are represented in Fig. 22.1, where several cases of statistical fluctuation are studied using random numbers[3]. When the effective channel length is 0.5 μm, the I_D–V_D curves are still insensitive to fluctuation in the impurity distribution. However, at 0.1 μm, they change drastically, depending on the local distribution. Deviation in the threshold voltage of such a device due to statistical fluctuation amounts to 0.5 V. This is a big problem for developing VLSIs.

22.2.2 *Structure of ALD-MOSFET*

According to the scaling-down theory, the doping concentration in the substrate of MOSETs has to increase with a decrease in the channel length. The heavily doped substrate, however, results in large capacitance and, consequently, high-speed performance deteriorates. From the microscopic viewpoint, statistical fluctuation is unavoidable. Thus changes in the design concept are necessary.

To solve the above problem, localized distribution of doped impurity in a narrow region is desirable, as is illustrated schematically in Fig. 22.2. This is because the impurity concentration in the substrate is very low and the fluctuation of doped atoms in the substrate plays a minor role in the device operation. Of course, the capacitance becomes small. Further, statistical fluctuation in the heavily doped layer is not sensitive to the device operation because the mean value of the distance between impurities is very small compared with the channel length.

A heavily doped and very thin layer is expected to be easily fabricated using molecular beam epitaxy (MBE) technology. This fabrication technology is termed an atomic-layer-doping (ALD) or delta-doping technique and is described in Section 22.3. It can be said that one of the best application fields of MBE is the ALD device. The ALD-MOSFET structure is illustrated in Fig. 22.3. This structure has ALD regions, namely doped thin layers controlled within a range of 1 nm, embedded in a low-impurity-concentration substrate.

The major features of the ALD structure are summarized as follows:

(a) heavily doped, but very thin layers to suppress punchthrough;
(b) a small minimum channel length in the submicron range;
(c) insensitivity to statistical fluctuation; and
(d) small junction capacitance.

These features are evaluated using numerical simulation.

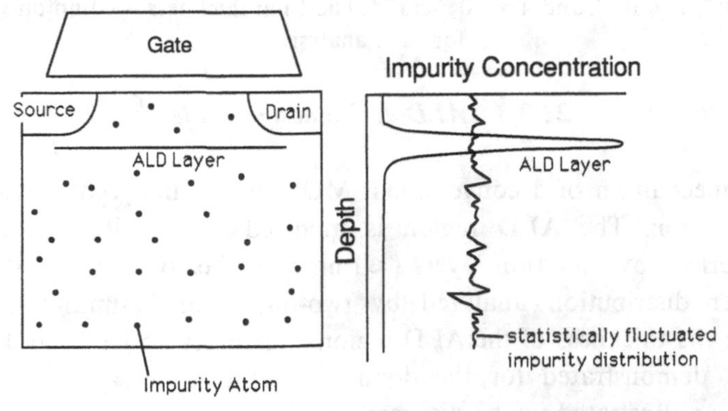

Fig. 22.2. Schematic expression of the atomic-layer-doping (ALD) technology concept from the viewpoint of the device application.

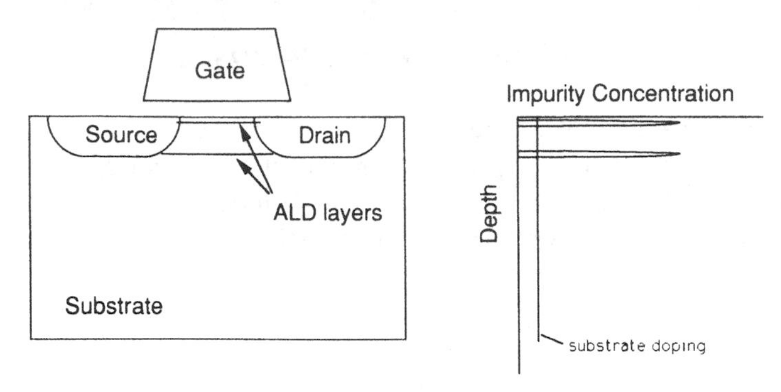

Fig. 22.3. Cross-sectional view and doping profile of an ALD-MOSFET. See Ref. (5).

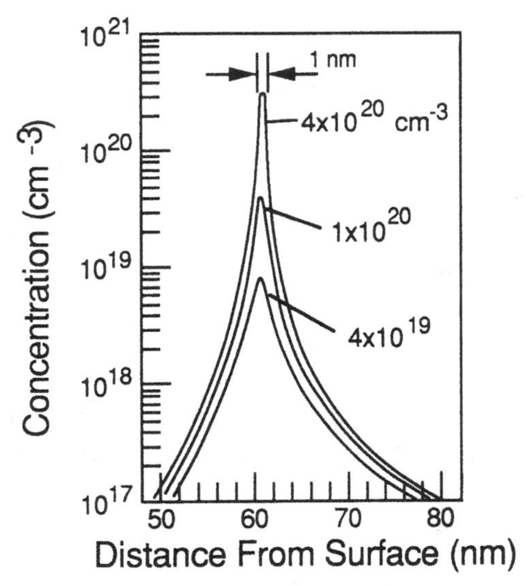

Fig. 22.4. Carrier profile analyzed by device simulation. The doping profiles of the ALD regions are shown by solid lines. Doping levels assumed here are 4×10^{19}, 1×10^{20}, and 4×10^{20} cm^{-3}. The 1 nm thickness is commonly used for each analysis.

22.2.3 ALD-device design guide

The operation mechanism of a conventional MOSFET is analyzed by two-dimensional numerical simulation. The ALD region is modeled as a delta-function-like doping profile, characterized by very thin layers ($\approx$1 nm) and heavy doping (10^{18}–10^{20} cm^{-3}). The free carrier distribution analyzed by two-dimensional simulation is shown in Fig. 22.4, where the thickness of the ALD region is assumed to be 1 nm. In this analysis, three cases are demonstrated for the doping concentration: 4×10^{19}, 1×10^{20}, and 4×10^{20} cm^{-3}. As illustrated in the diagram, carriers are confined within a very narrow region ($\approx$1 nm). However, the peak values do not coincide with the doping level, and a

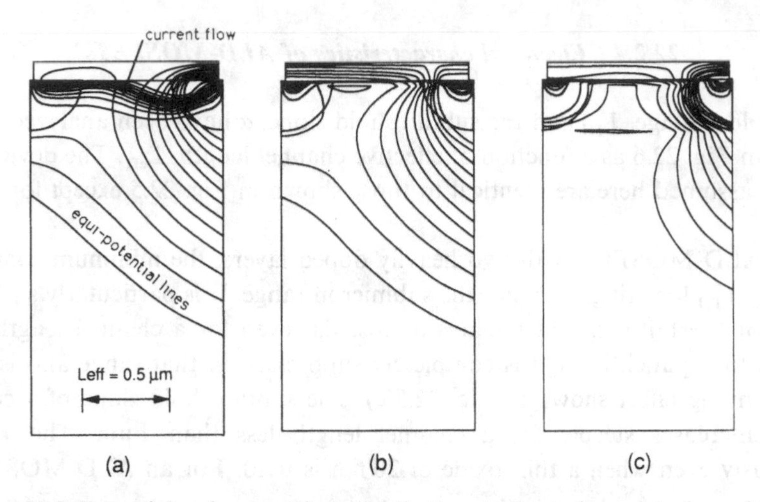

Fig. 22.5. Potential and current flow distributions in short-channel devices with $L_{eff} = 0.5$ µm and $T_{OX} = 20$ nm for: (a) a conventional structure MOSFET, (b) a delta-MOSFET with a single layer, and (c) an ALD MOSFET with two heavily doped layers. Equipotential lines are drawn at 0.5 V step. $V_{DS} = 5$ V. See Ref. (5).

long tail exists in the carrier distribution profile. These profiles were evaluated by C–V analysis[4].

As shown in Fig. 22.5(a), for a conventional MOSFET the actual current path deviates from the device surface and spreads two-dimensionally. Further, the contour lines of the electrostatic potential spread widely from the drain to the source region. This is the so-called punchthrough mechanism.

In order to suppress the punchthrough and also to suppress any increase in the capacitance value, heavily doped ALD regions are introduced into the lightly doped substrate. First, an ALD region is used to suppress the two-dimensional current flux expansion from the device surface to the substrate. That is, a heavily doped ALD region is designed at a location close to the device surface. Actually, the current flux penetration is effectively suppressed, as is shown in Fig. 22.5(b). However, the potential expansion from the drain to the source is still present. Thus, there remains the possibility for punchthrough.

In the second step, a structure with two ALD regions is considered. That is, the heavily doped ALD region is introduced to suppress the potential expansion from the drain to the source, as well as to suppress the current flux penetration into the substrate.

An ALD-MOSFET, shown in Fig. 22.5(c), is constructed using two ALD regions. One is situated at a deep region where the potential expansion becomes maximum. The other is situated close to the device surface. Potential contours, as illustrated in Fig. 22.5(c), are pinned near the drain and the current flux is confined within a very thin region near the surface. The potential and current flow distributions are effectively modulated by adopting the ALD structure and, as a result, punchthrough is suppressed[5].

22.2.4 *Electrical characteristics of ALD-MOSFETs*[5]

The threshold voltage, V_{TH} and the subthreshold slope, α have been analyzed. The results are shown in Fig. 22.6 as a function of effective channel length, L_{eff}. The device structural parameters assumed here are identical to those shown in Fig. 22.5 except for the channel length.

For the ALD-MOSFET with two heavily doped layers, the minimum channel length, predicted by V_{TH} lowering, falls into the submicron range. It is particularly significant that the curve for the tail constant remains almost flat even for a channel length of 0.2 μm. This shows that punchthrough is completely suppressed in that range and is due to the potential pinning effect shown in Fig. 22.5(c). The subthreshold slope of a conventional MOSFET increases steeply for a channel length less than 1 μm. The V_{TH} changes synchronously even when a thin oxide of 20 nm is used. For an ALD-MOSFET with a

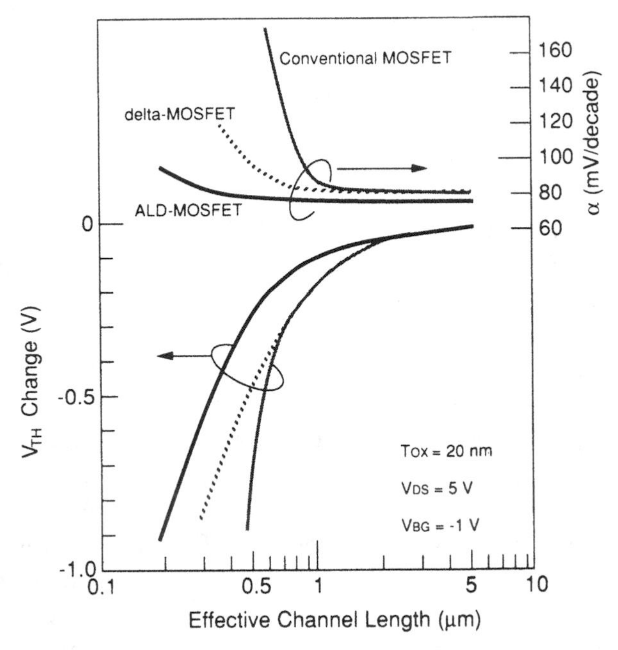

Fig. 22.6. Threshold voltage, V_{TH}, and tail constant, α, as functions of effective channel length, in three devices: an ALD-MOSFET with two heavily doped layers or a delta-MOSFET with one heavily doped layer, and a conventional structure MOSFET corresponding to the structures shown in Fig. 22.5. V_{DS} and V_{BG} respesent drain-to-source and back-gate biases. Device performance of the ALD structure is evaluated under the 5 V operation conditions. Of course, in recent LSIs, the operation voltage tends to be lower than 5 V. Low operation voltage reduces punchthrough effect and the margin against avalanche breakdown becomes wide. Therefore, there is no problem in extending the design concept of 5 V operation into a lower volt operation range. See Ref. (5).

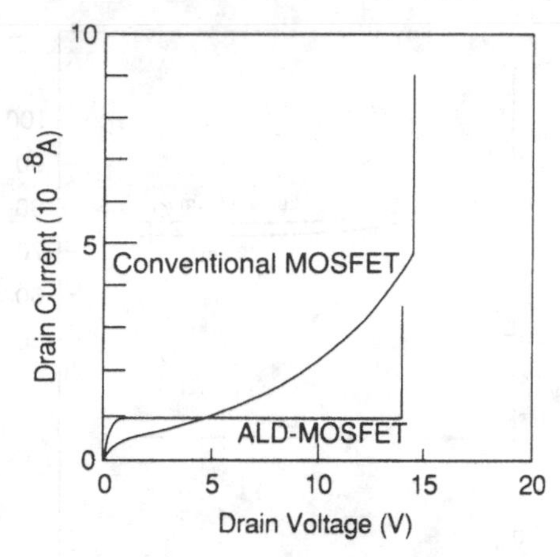

Fig. 22.7. Breakdown characteristics of conventional structure MOSFETs and ALD-MOSFETs. See Ref. (5).

single, heavily doped layer, that is, a delta-MOSFET, the flat region for α also extends into the submicron range. However, in this case, the minimum channel length restricted by V_{TH}-lowering is the same as that of conventional structure MOSFETs. When the second ALD region is designed to suppress the expansion of potential widening, that is, the second one is situated at a deep position, the V_{TH}–L_{eff} curve shifts to the left side, as shown in Fig. 22.6. This is the great advantage of the present ALD devices.

Breakdown characteristics are just as important as threshold voltage shift in evaluating device performance. Drain breakdown characteristics are shown in Fig. 22.7. Drain current in conventional MOSFETs increases gradually as a function of drain voltage; that is, a soft breakdown occurs due to punchthrough. On the other hand, in ALD-MOSFETs, drain current is kept at a constant value in the saturation region and punchthrough does not occur. The breakdown voltage, determined by avalanche multiplication, is estimated to be as high as 14 V. The thickness of the ALD region is of the order of 1 nm. It is thought that MBE technology is a unique, suitable means for fabricating such a very thin layer, since the epitaxy is processed at a low temperature and the thickness of the ALD region can be finely controlled down to the range of a few atomic monolayers. Thus, an arbitrary impurity profile is easy realized using this technology.

However, there is still unavoidable fluctuation in the doping profile in actual fabrication processes, even though the controllability is high. Assuming the thickness of the ALD region to fluctuate in a 1 nm layer, the induced fluctuation in the threshold voltage and tail constant are analyzed by simulation. The results are plotted in Fig. 22.8, where the threshold voltage change, ΔV_{TH}, is defined as the difference between V_{TH} for $L_{eff} = 2$ μm and that for 0.5 μm. As shown in the diagram, ΔV_{TH} and α are almost constant. Thus, it can be said that device performances in the ALD structure are insensitive to statistical fluctuation. In this sense, the propose ALD-MOSFET structure is realistic and would be practicable from the viewpoint of the process margin.

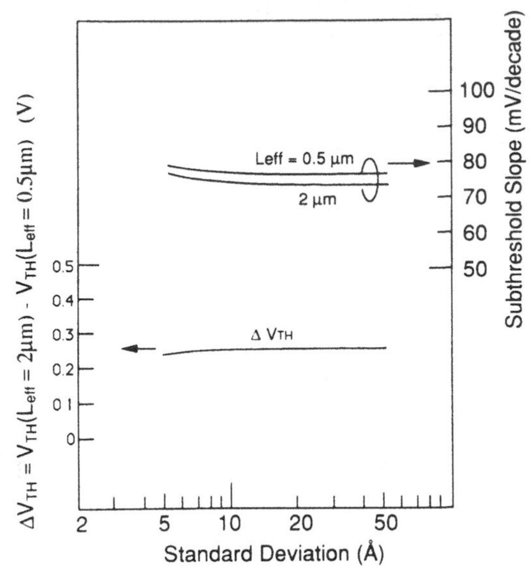

Fig. 22.8. Process sensitivity analysis of ALD-MOSFETs. Threshold voltage and tail constant are analyzed as a function of the standard deviation, σ, when the ALD profile is assumed to be Gaussian. See Ref. (5).

22.3 ALD-MOSFET fabrication and electrical characteristics

22.3.1 Control of dopant concentration and distribution

In order to obtain a sharp doping profile with controlled concentration, we have utilized Si molecular beam epitaxy (MBE) as mentioned above. Molecular beam epitaxy enables low-temperature crystal growth and was expected to be a powerful tool with which to realize arbitrary doping profiles because thermal diffusion hardly occurs at low temperatures. However, recently, studies have shown that in Si-MBE such dopant atoms as antimony and gallium segregate on surfaces during epitaxial growth to form a dopant reservoir which inhibits abrupt changes in dopant profiles, even if dopant fluxes are suddenly changed[6–8]. In order to overcome the segregation problem, we have studied the phenomenon in detail, and have developed a method for fabricating a delta-doping structure using a combination of MBE and solid phase epitaxy (SPE). In this section, we describe how to control dopant concentration and segregation.

In MBE, doping is carried out using an effusion cell and the doping level is controlled by changing vapor pressure. In order to control the doping concentration, the beam intensity of the dopant flux must be precisely controlled, which is difficult especially for dopant atoms such as Sb. A new method of controlling the number of deposited dopant atoms has been developed.[6,9] This method is based on the fact that the stable one-monolayer Ga (p-type dopant) and Sb (n-type dopant) atoms adsorb on the Si surface when the substrate temperature is kept at around 600 °C and 650 °C respectively. In

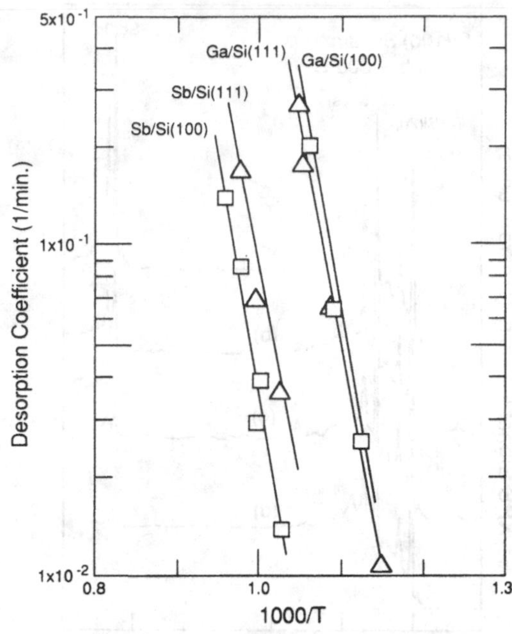

Fig. 22.9. Thermal desorption rates of antimony and gallium atoms on Si(100) and Si(111) substrates as a function of substrate temperatures.

this process, the excess dopant atoms simply sublimate at these temperatures. This is because the Ga–Si and Sb–Si bonds are stronger than Ga–Ga and Sb–Sb bonds, respectively, and the latter bonds are unstable in these temperature ranges. The concentration of the saturated layers of Sb and Ga can be reduced by raising the substrate temperature. The number of these atoms on the substrate surface were measured as a function of time at various substrate temperatures. The thermal desorption rates of Sb and Ga on Si(100) and (111) substrates are shown in Fig. 22.9. Their activation energies correspond to those of covalent bonds between Si and dopant, and the values are about 2.6 eV irrespective of the dopant species. The thermal desorption enables accurate control of the dopant atoms since it depends solely on the substrate temperature.

After controlling the dopant concentration, the dopant atoms must be buried in Si. Figure 22.10 shows an example of the change in XPS (X-ray photoelectron spectroscopy) spectra after Si overgrowth on one-monolayer Sb adsorbed surfaces. The signal at around 530 eV comes mainly from Sb atoms segregated on the epitaxial surfaces, since the escape depth of photelectrons is very small (≈ 1 nm). The temperature range of the experiment is so low that thermal desorption of the dopant atoms hardly occurs. The decrease in dopant surface concentration is due to the dopant incorporation in the growing silicon film. The dopant surface concentration C_s (atoms/cm^2) decreases exponentially with silicon deposition thickness, as shown in Fig. 22.11, and the relationship can be written as

$$C_s = C_0 \exp\left(-\frac{x}{\lambda}\right) \tag{22.3}$$

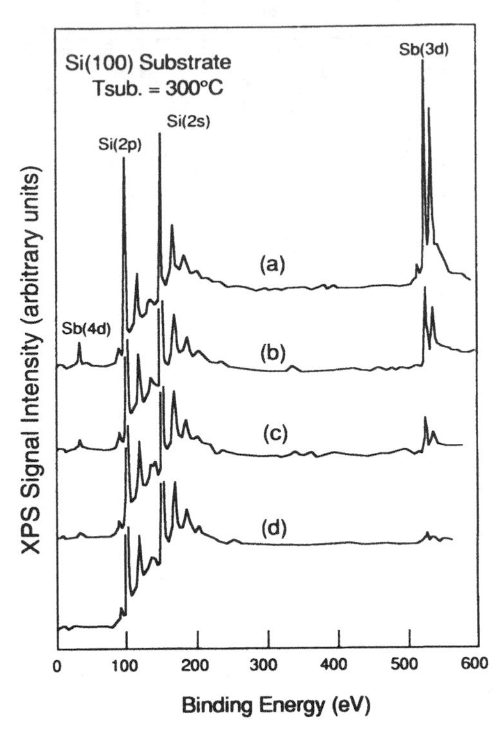

Fig. 22.10. XPS spectra for (a) 1 ML antimony deposited on Si(100), (b) 10 nm silicon film grown at 300 °C on 1 ML antimony, (c) 20 nm silicon film grown at 300 °C on 1 ML antimony and (d) 30 nm silicon film grown at 300 °C on 1 ML antimony. See Ref. (6).

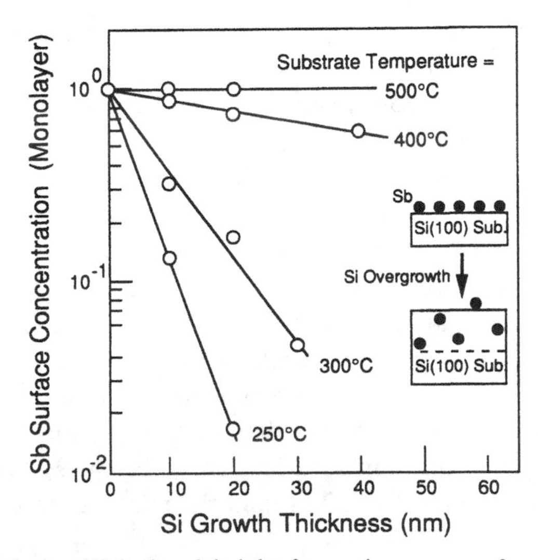

Fig. 22.11. Relative XPS signal height for antimony as a function of silicon deposition thickness. See Ref. (6).

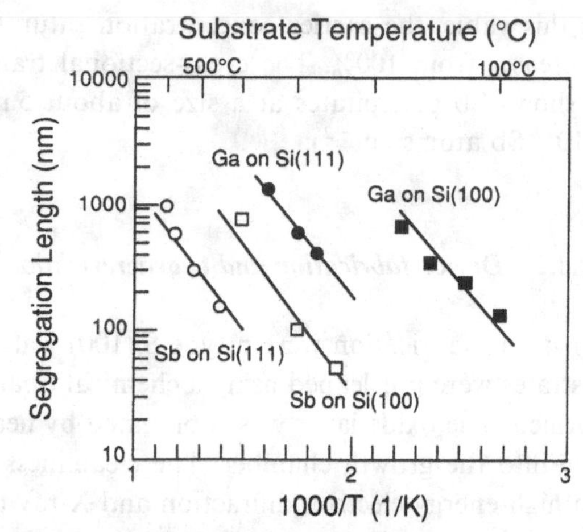

Fig. 22.12. Temperature dependence of segregation lengths for antimony and gallium atoms on Si(100) and Si(111). See Ref. (6).

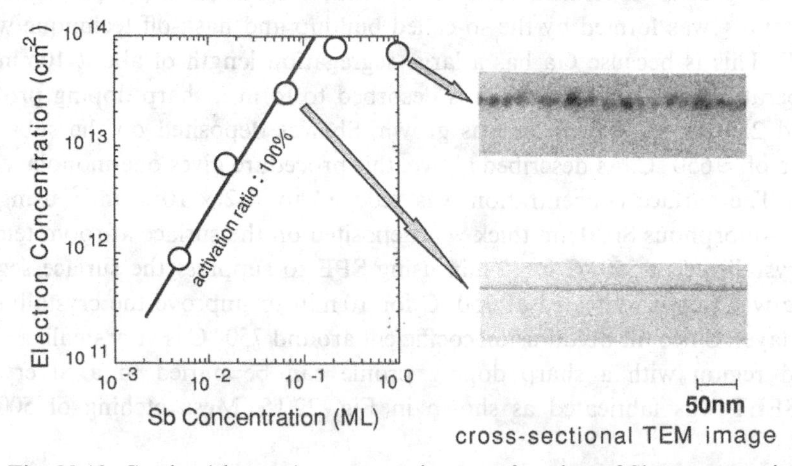

Fig. 22.13. Carrier (electron) concentration as a function of Sb concentration at room temperature as derived from Hall data. Cross-sectional TEM pictures of samples (a) and (b) are also shown. See Ref. (9).

where C_0 is the initial value of C_s, λ (nm) is the segregation length and x (nm) is the thickness of the Si growth layer. Figure 22.12 shows the temperature dependence of the antimony and gallium segregation lengths on Si(100) and (111) substrates, obtained from the results given in Fig. 22.11. Figure 22.12 indicates that the growth temperature should be lowered to obtain a sharper doping profile with smaller surface segregation. We carried out atomic layer doping (delta-doping) by depositing Si on the dopant atom adsorbed surfaces at room temperature. A silicon layer grown at room temperature, however, is amorphous and is crystallized by heating to a temperature of around 750 °C (solid phase epitaxy). As can be seen in Fig. 22.13, the activation ratio of delta-doped Sb atoms is 100% when the adsorbed Sb concentration is 1×10^{14} cm^{-2}, which corresponds to about

1×10^{20} cm^{-3}. Above this value, the carrier concentration saturates, which means that the activation ratio decreases from 100%. The cross-sectional transmission microscopic image in the diagram shows Sb precipitates at a size of about 5 nm in the case of one monolayer Sb ($\approx 6 \times 10^{14}$ Sb atoms cm^{-2}).

22.3.2 Device fabrication and characterization[10]

The substrates used for device fabrication were p-type Si(100) wafers with a resistivity of 5–10 Ω cm. The Si substrates were precleaned using a chemical treatment and a protective thin oxide layer was formed. The oxide layer was sublimated by heat treatment at 850 °C for 20 min after loading into the growth chamber. The cleanliness of the substrates was confirmed by reflection high-energy electron diffraction and X-ray photoelectron spectroscopy. The preparation procedure of substrates for device fabrication is shown in Fig. 22.14, with the exception of the Ga doping process. An undoped 300-nm-thick Si buffer layer was first grown on cleaned substrates. Then, a 5-nm-thick p-type doped layer of $\approx 2 \times 10^{11}$ cm^{-2} was formed by the so-called buildup and flash-off technique with a Ga effusion cell. This is because Ga has a large segregation length of about 10 nm, even at room temperature, and segregated Ga is desorbed to form a sharp doping profile. After an undoped 250-nm-thick Si layer was grown, Sb was deposited on the substrate at a temperature of ≈ 650 °C. As described above, this procedure gives one-monolayer adsorption of Sb. The surface concentration was reduced to $\approx 2 \times 10^{12}$ cm^{-2} using thermal desorption. Amorphous Si 30 nm thick was deposited on the surface at room temperature and was crystallized at 550 °C for 5 min using SPE to suppress the surface segregation. The sample was finally annealed at 750 °C for 10 min to improve the crystalline quality of the SPE layer. Since the Sb diffusion coefficient around 750 °C is very small, a controlled delta-doped region with a sharp doping profile can be buried in a Si crystal. The ALD-MOSFET was fabricated as shown in Fig. 22.15. Mesa etching of 500 nm was

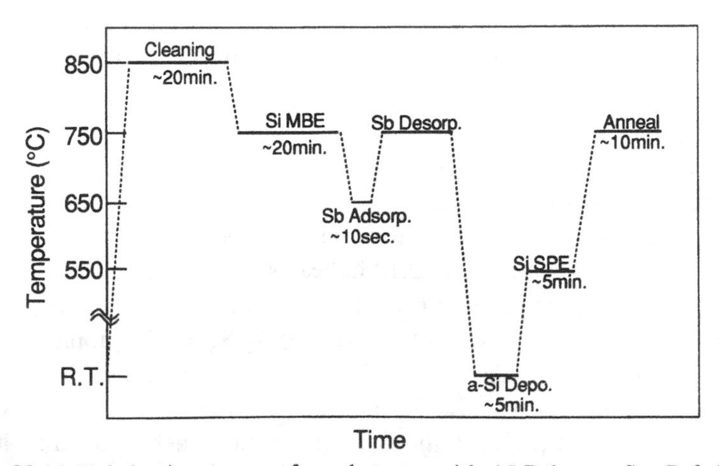

Fig. 22.14. Fabrication process for substrates with ALD layers. See Ref. (10).

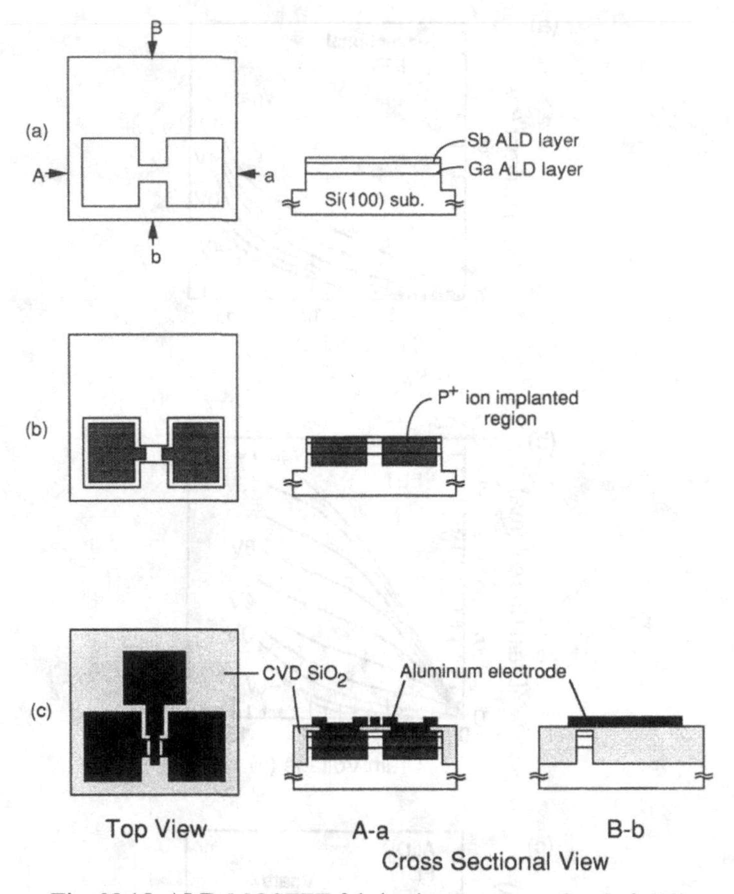

Fig. 22.15. ALD-MOSFET fabrication sequence. See Ref. (10).

carried out for device isolation (Fig. 22.15(a)). P^+ ions were implanted at 30 keV with 1×10^{15} cm^{-2} to form source and drain regions (Fig. 22.15(b)). Low-temperature annealing at 600 °C for the activation of implanted atoms was employed for 1 hr to prevent the delta-doped layer from smearing due to diffusion. The activation ratio of implanted P atoms was about 80% at the annealing temperature. The 70-nm-thick gate oxide was deposited at 400 °C by the chemical vapor deposition (CVD) method (Fig. 22.15(c)). Steps formed by the mesa etching were planarized by CVD SiO_2 deposition to avoid the formation of cracks in the aluminum electrodes.

Measured current–voltage characteristics of conventional MOSFETs, and ALD-MOSFETs with channel lengths of 2 μm are shown in Fig. 22.16. It can be seen that the delta-FET possesses a rather higher transconductance (≈20 mS/mm) as compared with the conventional MOSFET, and the punchthrough current is reduced, which is due to the fact that the delta-FET has an Sb delta-doped layer for the conductive channel which forms a potential well that confines the electrons. The threshold voltage, however, usually shifts to a lower value of around −4 V due to the high channel doping. Moreover, it was

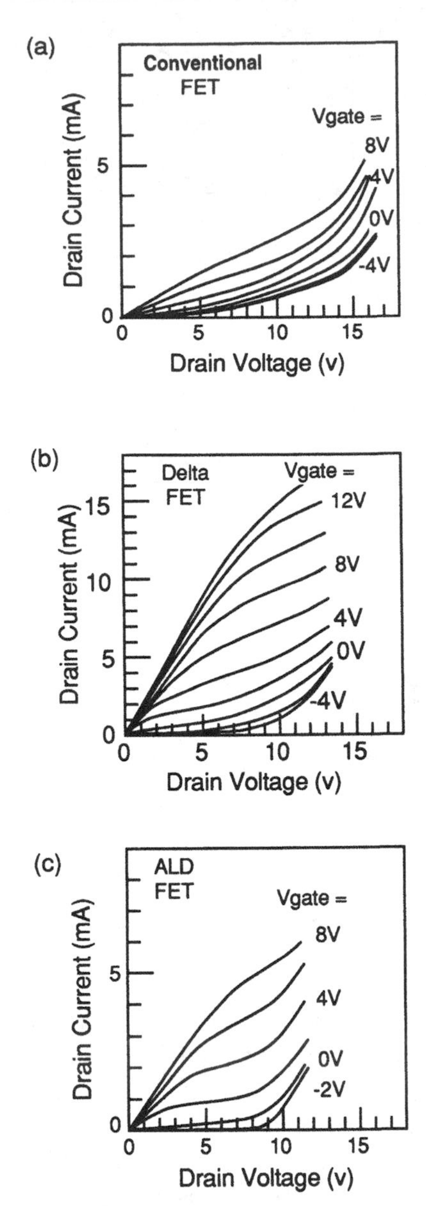

Fig. 22.16. Current–voltage characteristics (a) a conventional MOSFET, (b) a delta-MOSFET and (c) an ALD-MOSFET. See Ref. (10).

pointed out in the preceding section that punchthrough cannot be avoided completely, even by the delta-FET structure, since the delta-doped layer cannot confine electrons completely in a submicron range. On the other hand, in the case of the ALD-MOSFET structure, the triangular potential is formed as a result of the existence of the depleted Ga-doped layer. Thus electrons are expected to be sufficiently confined in the conduction channel even in the submicron range of channel length. The critical length, at which the punchthrough effect takes place, depends on the concentration of the p-type dopant in the substrate. In the case of fabricated MOSFETs, the punchthrough is expected to occur even at a gate-length of several micrometers. As can be seen in Fig. 22.16, punchthrough is efficiently suppressed in the ALD-MOSFET. Although the transconductance of ALD-MOSFETs is smaller than that of delta-FETs because of the reduction in the electron density by compensation, the total device performances of ALD-MOSFETs are attractive because of the suppression of punchthrough, the low subthreshold slope, and the controllability of the threshold voltage. Thus, the potential of the proposed ALD-MOSFET is demonstrated.

For the next step towards improving the device performance of MOSFETs, the technology to form a high-quality, thin oxide layer with a thickness of several nanometers below about 750 °C should be developed. Since complementary structures of MOS devices must be formed on the same substrate for the fabrication of VLSI, the method, which enabled us to grow Si layers with different types of dopant on different areas of the substrate, must also be developed.

22.4 Summary

A new device with an atomic-layer-doped (ALD) structure has been proposed and fabricated utilizing MBE. The device performance has been evaluated from $I-V$ characteristics and also by computer simulation. It has been shown that punchthrough is suppressed and that the output characteristics of ALD-MOSFETs are insensitive to fluctuations in doped impurity distribution. The ALD structure is attractive and promising for developing devices with very short channels and for achieving very-high-speed operation.

References

[1] R. H. Dennard, F. H. Gaensslen, H. N. Yu, V. L. Rideout, E. Bassous and A. R. LeBlanc, Design of ion-implanted MOSFET's with very small physical dimensions, *IEEE J. Solid-State Circuits*, **SC-9**, 256–68, 1974.

[2] K. Yamaguchi, An extended stream-function method for computer analysis of nonplanar structures *Solid-State Electron.*, **29**, (11), 1129–36, 1986.

[3] T. Hagiwara, K. Yamaguchi and S. Asai, Threshold voltage deviation in very small MOS transistors due to local impurity fluctuations, *1982 Symposium on VLSI Technology, 1–3 Sep. 1982.*

[4] A. A. van Gorkum and K. Yamaguchi, Classical calculations of CV profiles for atomic layer doping structures in silicon, *IEEE Trans. Electron Devices*, **36**, (2) 410–15, 1989.

[5] K. Yamaguchi, Y. Shiraki, Y. Katayama and Y. Murayama, A new short channel MOSFET with an atomic-layer-doped impurity-profile (ALD-MOSFET), *Proc. 14th Conf. (1982 International) Solid State Devices, Tokyo, 1982; Japanese J. Appl. Physics*, vol. 22, Sup. 22–1, pp. 267–70, 1983.

[6] K. Nakagawa, M. Miyao and Y. Shiraki, Influence of substrate orientation on surface segregation process in silicon-MBE, *Thin Solid Films*, **183**, 315–22, 1989.

[7] J. C. Bean, Arbitrary doping profiles produced by Sb-doped Si MBE, *Appl. Phys. Lett.* **33**, 654–6, 1978.
[8] T. Sakamoto and H. Kawanami, RHEED studies of Si(100) surface structures induced by Ga evaporation, *Surf. Sci.* **111**, 177–88, 1981.
[9] A. A. van Gorkum, K. Nakagawa and Y. Shiraki, Growth and characterization of atomic layer doping structures in Si, *J. Appl. Phys.* **65**, 2485–92, 1989.
[10] K. Nakagawa, A. A. van Gorkum and Y. Shiraki, Atomic layer doped field-effect transistor fabricated using Si molecular beam epitaxy, *Appl. Phys. Lett.* **54**, 1869–71, 1989.

23

Planar-doped barrier devices

ROGER J. MALIK

23.1 Introduction

The advent of modern epitaxial growth methods such as molecular beam epitaxy (MBE) (Cho and Arthur, 1975) and metal–organic chemical vapor deposition (MOCVD) (Manasevit, 1981) has revolutionized the fabrication of the high speed electronic and photonic devices which are now being manufactured. This is especially true in the area of III–V compound semiconductor devices where AlGaAs/GaAs modulation-doped high electron mobility transistors (HEMTs) and InP/InGaAsP quantum well (QW) lasers are routinely used in high frequency microwave and optical communication systems today. The ability to synthesize epitaxial semiconductor layers on an atomic scale, which led to new device physics due to quantum size effects, has been termed 'band-gap engineering' (Capasso and Cho, 1994). The planar-doped barrier (PDB) (Malik *et al.*, 1980) concept is another device structure which relies upon the precise control of layer thicknesses and doping levels that can be grown using advanced epitaxial techniques such as MBE.

The PDB structure is formed in a n^+–i–p^+–i–n^+ doping profile in which a highly doped, ultra-thin, acceptor layer (typically $\leqslant 50$ Å) is placed within nominally undoped (intrinsic) regions bounded by two donor contact layers. The acceptor layer is so thin that it is fully ionized and forms a negative space charge sheet within the semiconductor. The resultant triangular potential barrier impedes electron flow from the contact regions. Electron transport over the barrier is controlled by thermionic emission which is similar to the transport mechanism in metal–semiconductor contacts, which are also known as Schottky barriers. The key features of the PDB are that the zero-bias barrier height and degree of asymmetry in the rectifying *I–V* characteristics can be continuously and independently controlled. The zero-bias barrier height can be effectively varied from zero to slightly less than the band-gap of the semiconductor by adjusting the doping-thickness product of the acceptor charge sheet and the undoped region widths. The asymmetry in the *I–V* characteristics is controlled by the positioning of the acceptor charge sheet within the undoped region. The capacitance of the PDB is virtually constant and is determined by the geometrical capacitance determined from the undoped region widths. Although most of the PDB work has been demonstrated in GaAs, it is applicable to all semiconductors in general.

This designability of the PDB structure should be compared to the essentially fixed barrier height found in Schottky barriers which is related to the difference in metal work function and the semiconductor electron affinity. In addition, the metal–semiconductor interface is always non-ideal and interface states and defects also play a significant role in determining the barrier height. These factors are quite problematic in the fabrication of Schottky diodes with reproducible I–V characteristics and capacitances. This causes considerable difficulties in the manufacture of Schottky diodes, resulting in low yields and poor matching of diodes in different wafer lots. Also, whereas the high electron mobility in GaAs is desirable for making mm-wave frequency detectors and mixers, the high barrier height of GaAs Schottky diodes necessitates a DC bias for detector diodes and high local oscillator power drive for mixer diodes. The reliability of Schottky diodes is also a major concern and is determined by the stability of the metallurgical contact. It has been shown experimentally that PDB diodes have superior reliability characteristics with a much higher electrostatic discharge (ESD) damage threshold, as compared with Schottky diodes, by virtue of their rugged, all-semiconductor structure (Kearney *et al.*, 1990; Anand and Malik, 1992). For these reasons, PDB diodes are rapidly emerging to replace Schottky diodes as the preferred device for high frequency detector and mixer applications.

This chapter is organized into the following sections. A brief review of Schottky barriers is described, along with methods to modify the effective barrier height. Some alternative unipolar barrier structures are discussed and the PDB concept is formulated. The physical model for the PDB is derived, along with analytic expressions for the key device parameters. The electron transport mechanisms are also described. A brief description of the growth of PDB structures by MBE and MOCVD is given. The use of carbon doping to form a stable acceptor spike region is highlighted. The principal use of PDB diodes as high performance mixer and detector diodes is discussed. The use of the PDB structure in a number of other novel device applications and in the study of the physics of electron transport is also discussed. Finally, the chapter contains a summary and details some future trends.

23.2 Unipolar barrier devices

A brief review of unipolar barrier devices which provide current rectification is presented. Unipolar barrier devices can be operated at very high switching speeds due to the absence of minority carrier charge storage effects. In bipolar barrier devices which include p–n junctions, the current switch-off time is determined by the minority carrier recombination lifetimes, which can be relatively long (ns or greater). For unipolar barriers, the current switch-off time is fundamentally limited by the dielectric relaxation time constant, which can be very short (of the order of ps). The most common unipolar barrier device is the metal–semiconductor contact, also known as a Schottky barrier.

23.2.1 *Schottky barriers*

The observation of current rectification in metal–semiconductor contacts was first discovered by Braun (1874). The majority carrier transport in metal–semiconductor junctions enables very high frequency operation. Schottky barriers have been used as the gate electrode in metal–semiconductor field effect transistors (MESFETs) (Hooper and Lehrer, 1967), as mixer and detector diodes (Schneider *et al.*, 1977, 1982), in impact avalanche transit time (IMPATT) diodes (Sze *et al.*, 1968), varactors (Harris and Woodcock 1980), and photodetectors (Schneider, 1966). Schottky barriers have also been proposed for use in metal base transistors (MBT) (Mead, 1960) and permeable base transistors (PBT) (Bozler *et al.*, 1979), which have theoretical operating frequencies well above 100 GHz. A number of review articles have been written on the theory of Schottky barriers (see, for example, Rideout, 1978).

Schottky (1923) first postulated that rectification in metal–semiconductor contacts is caused by an electrostatic potential barrier formed by the constituent materials. He theorized that the potential barrier arises from a space charge region in the semiconductor adjacent to the metal which is necessarily compensated by an equal charge of opposite polarity of the metal surface. Schottky proposed a diffusion model for rectification in which mobile carriers diffuse from the semiconductor across the space charge region into the metal. Bethe (1942) later proposed a thermionic emission model wherein mobile carriers in the semiconductor with sufficient kinetic energy and momentum to overcome the potential barrier are emitted over the top of the barrier into the metal. The thermionic emission model is now accepted as the primary mode of rectification in typical Schottky barriers.

The space charge, electric field and conduction band energy profiles for a Schottky barrier are shown in Fig. 23.1(a), (b), and (c) respectively. The electric field and potential in the semiconductor are found from the solution of Poisson's equation in one-dimension subject to the appropriate boundary conditions:

$$\frac{d^2\phi(z)}{dz^2} = -\frac{\rho(z)}{\varepsilon} \tag{23.1}$$

where $\phi(z)$ is the electrostatic potential, $\rho(z)$ is the space charge density, ε is the dielectric permittivity, and z is the distance variable. Over the range $0 \leqslant z \leqslant W$, $\rho(z) = qN_D =$ constant, the electric field, $\mathbf{F}(z)$, is given by

$$\mathbf{F}(z) = \frac{-d\phi(z)}{dz} = \frac{-qN_D(W - z)}{\varepsilon} \tag{23.2}$$

and the potential, $\phi(z)$, is given by

$$\phi(z) = \frac{qN_D\left(Wz - \dfrac{z^2}{2}\right) - \phi_B}{\varepsilon}. \tag{23.3}$$

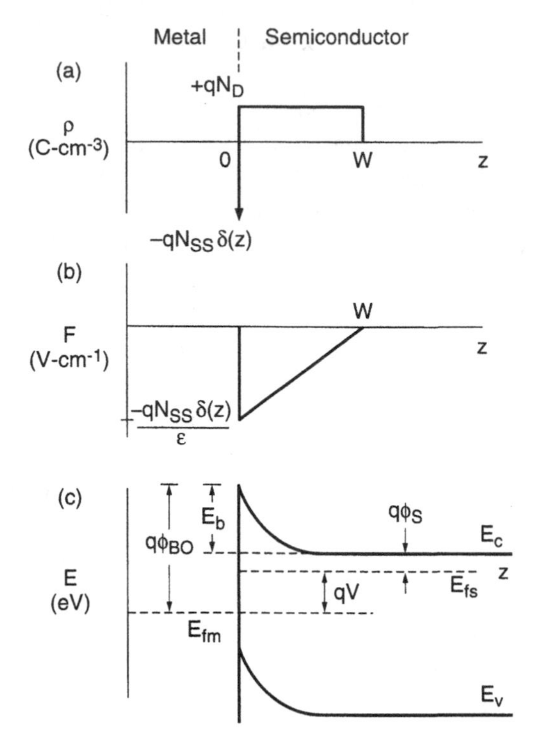

Fig. 23.1. (a) Space charge, (b) electric field, and (c) band potential profiles in a Schottky barrier.

Thus a linear electric field and parabolic potential barrier profile are formed in a Schottky barrier on a uniformly doped semiconductor. The band bending in the semiconductor can be found from the boundary conditions and is given by

$$E_B = \frac{q^2 N_D W^2}{2\varepsilon} \tag{23.4}$$

where N_D is the donor density, W is the depletion width, ε is the permittivity of the semiconductor, and q is the electronic charge. The band bending in the semiconductor can also be expressed by

$$E_B = q\phi_B - q\phi_S - qV \tag{23.5}$$

where ϕ_B is the potential barrier from the metal to the semiconductor, ϕ_S is the difference between the conduction band edge and the Fermi level in the semiconductor, and V is the applied bias potential. Since the high electron concentration in the metal precludes any potential charge within the metal, then it is clear from Eqn. (23.5) that a change in the applied bias voltage V must manifest itself as a change in the band bending in the semiconductor or, more simply, as a change in the barrier height for electrons in the semiconductor with respect to the metal.

The depletion layer width is dependent on the doping density N_D and the applied bias V and can be expressed by

$$W = \left[\frac{2\varepsilon(\phi_B - V)}{qN_D}\right]^{1/2}. \tag{23.6}$$

The capacitance per unit area, C, is given by

$$C = \frac{\varepsilon}{W}. \tag{23.7}$$

Bethe's thermionic emission model was derived using Maxwell–Boltzman statistics to arrive at the distribution of electrons which have sufficient energy and momentum to be emitted over the barrier. The following assumptions are made in the analysis:

1. The barrier height is much greater than the thermal energy of the electrons, $q\phi_B \gg kT$.
2. Electron collisions in the depleted space charge region are neglected.
3. The image force lowering of the barrier peak is neglected.

The forward current density from the semiconductor to the metal is given by

$$J_f = J_r \exp\left(\frac{qV}{kT}\right) \tag{23.8}$$

where

$$J_r = A^* T^2 \exp\left(\frac{-q\phi_B}{kT}\right). \tag{23.9}$$

J_r is the reverse current density from the metal to the semiconductor which is essentially fixed since the barrier height ϕ_B is constant. In Eqn. (23.9), A^* is the effective Richardson constant, T is the absolute temperature, and ϕ_B is the barrier height from the metal to the semiconductor.

The total current density, which is the difference between the forward and reverse components, is thus given by the familiar expression

$$J = A^* T^2 \exp\left(\frac{-q\phi_B}{kT}\right)\left[\exp\left(\frac{qV}{kT}\right) - 1\right]. \tag{23.10}$$

It has been found experimentally that the current–voltage characteristics of actual Schottky barriers do not follow the theoretical $\exp(qV/kT)$ dependence exactly. This is primarily due to image force lowering, but it is also related to interfacial layers, defects, and deep level traps in the barrier (Sze, 1981). This is sometimes taken into account empirically by introducing an ideality factor n with $n \geqslant 1$ for which

$$J \propto \exp\left(\frac{qV}{nkT}\right). \tag{23.11}$$

The ideality factor serves as a figure of merit for Schottky barriers and good diodes have $n \approx 1.0$.

In spite of the wide use and applications of Schottky barriers, they suffer from several inherent limitations. The principal disadvantage is the inability to control the Schottky barrier height in many metal–semiconductor systems (Morgan and Frey, 1977). This is a particular problem since the barrier heights are often not optimized for certain devices. For instance, GaAs is desirable for high frequency mixer diodes because of its high electron mobility and velocity, but its high barrier height of 0.7–0.9 eV necessitates relatively high local oscillator power levels in mixers. By contrast, InGaAs and InP are attractive for high frequency MESFETs but the low barrier heights on these materials leads to 'leaky' gates and low breakdown voltages which degrade device performance.

The relative insensitivity of barrier heights to the types of metal used in forming Schottky barriers was first explained by Bardeen (1947) in his hypothesis of the existence of surface states. He theorized that in highly covalent group IV and III–V semiconductors, high densities of surface states effectively 'pin' the Fermi level at the semiconductor surface. Thus when contact is made between a metal and semiconductor, charge transfer takes place only between the surface states and the metal. It has been determined empirically by Mead and Spitzer (1964) that the Fermi level pinning results in barrier heights roughly equal to two-thirds of the energy gap. This effect is shown in Fig. 23.2 for Au Schottky barriers on a number of different semiconductors.

In the manufacture of real Schottky barriers, thin interfacial oxide layers and surface contamination can increase Schottky barrier heights and ideality factors. It has been demonstrated that *in-situ* deposition of Al on an atomically clean GaAs surface in an MBE growth chamber results in slightly lower barrier heights and ideality factors than are obtained by conventional processing (Cho and Dernier, 1978).

Another significant problem with Schottky barriers is the metallurgical stability, especially at elevated temperatures. Degradation of the rectifying properties of AlGaAs Schottky barriers has been observed at temperatures as low as 275 °C (Sleger and Christou,

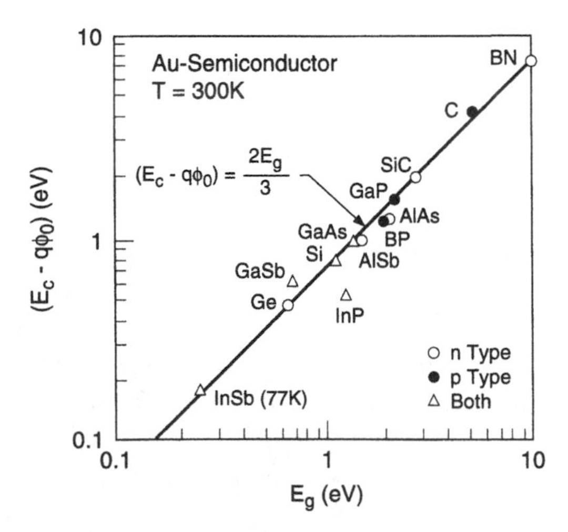

Fig. 23.2. Schottky barrier height versus band-gap energy for various Au-semiconductor contacts (after Mead and Spitzer, 1964).

1978). Refractory metals must be employed for high temperature operation or where high temperature processing is required such as for ion implant activation in self-aligned gate field effect transistors (SAGFETs) (Kohn, 1979).

Due to the limitations of Schottky barriers, some alternative barrier structures have been proposed and these are discussed in the following section.

23.2.2 Alternative barrier structures

Shannon (1976) first demonstrated the control of the effective Schottky barrier height through the use of highly doped surface layers. This was realized experimentally by shallow, low energy ion implantation into n-type Si, which is illustrated in Fig. 23.3. The thin ($\leqslant$150 Å), fully depleted surface layers are used to alter the electric field profile adjacent to the metal contact. For the case of a highly doped acceptor layer, a negative space charge region is formed, which leads to an increased depletion width in the n-type semiconductor resulting in an increased barrier height. Conversely, a highly doped donor layer increases the surface electric field leading to a reduction in the depletion width. This will enhance quantum mechanical tunneling or thermionic field emission through the barrier effectively lowering the barrier height. This approach does allow some modification of the effective Schottky barrier height. However, the technology is very difficult to reproduce in order to control the I–V characteristics of the diodes. This technique is critically dependent upon the dose and range of the implant species as well as on the activation of the dopant atoms through thermal annealing after ion implantation. There also remains some residual damage after annealing, which affects the stability of the doping profile and the metallurgical junction adversely.

Shannon (1979) proposed another unipolar barrier structure, which is termed the 'camel' barrier due to the shape of the band potentials which are shown in Fig. 23.4. The camel barrier has an n^{++}–p^{+}–n doping profile and can be considered as an all-semiconductor analog of a Schottky barrier. The p^{+} layer remains depleted under all bias conditions and

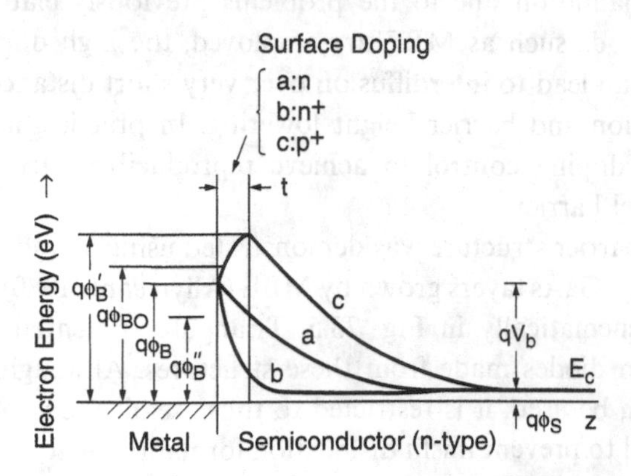

Fig. 23.3. Modification of effective Schottky barrier heights using highly doped surface layers (after Shannon, 1976).

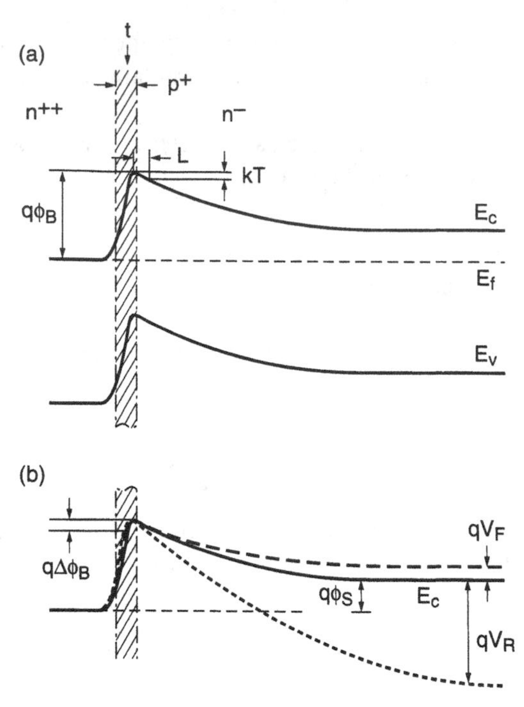

Fig. 23.4. Band diagrams for a 'camel' barrier under (a) equilibrium and (b) conduction-band edge under forward and reverse bias (after Shannon, 1979).

the asymmetry of the donor densities, and hence the depletion widths, modifies the effective barrier height for electrons from the respective donor regions. A large range of barrier heights in Si (Shannon, 1979) and GaAs (Woodcock and Harris, 1983) have been demonstrated using this technique. GaAs camel barrier mixer diodes have also been fabricated (Battersby and Harris, 1987). Unfortunately, it is quite difficult to fabricate camel diodes using ion implantation due to the problems previously elaborated. Even where epitaxial growth methods such as MBE are employed, the high doping densities in the n^{++} and p^{+} regions can lead to interdiffusion over very short distances (a few tens of Å), leading to compensation and barrier height lowering. In practice, it is quite difficult to achieve the requisite doping control to achieve reproducible barrier heights and I–V characteristics in camel barriers.

Another rectifying barrier structure was demonstrated using a graded $Al_xGa_{1-x}As$ layer sandwiched between n^{+} GaAs layers grown by MBE (Allyn *et al.*, 1980). The band potential profiles are shown schematically in Fig. 23.5. Triangular potential barriers resulted in current rectification in diodes made from these structures. Although the graded heterojunction approach can be used, it is restricted to those semiconductor systems which are closely lattice-matched to prevent misfit dislocation formation. Also, undoped $Al_xGa_{1-x}As$ layers tend to have high densities of deep electron traps which increases the measured barrier height.

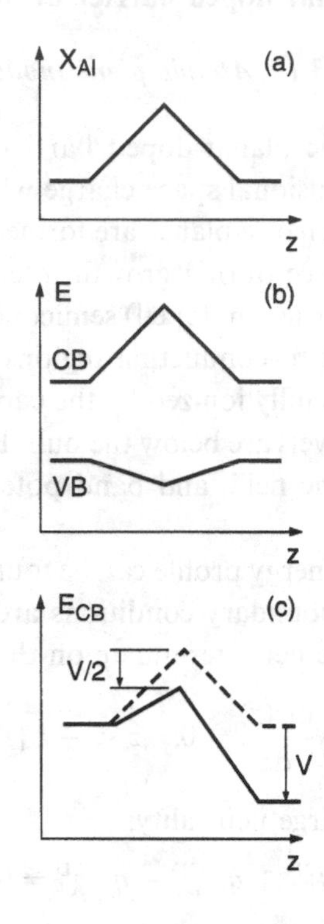

Fig. 23.5. (a) Al mole fraction, (b) band potentials, and (c) conduction band edge under forward bias for a graded $Al_xGa_{1-x}As$ barrier (after Allyn *et al.*, 1980).

Some methods of obtaining low resistance, tunneling ohmic contacts have also been investigated. Popovic (1978) studied the use of shallow diffusion, highly doped donor layers to obtain tunneling ohmic contacts. Non-alloyed ohmic contacts to GaAs have been formed using MBE by degenerate Sn doping to $n = 5 \times 10^{19}$ cm^{-3}, with a contact resistivity of $\rho < 2 \times 10^{-6}$ Ω cm (Barnes and Cho, 1979). Molecular beam epitaxy was also used to grow n^+–Ge/n^+–GaAs heterojunctions with extremely low contact resistance of $\rho = 1 \times 10^{-8}$ Ω cm (Stall, *et al.*, 1979). Another method uses a very high density, delta-doped donor layer placed a few lattice constants away from a metal–semiconductor junction grown by MBE to form a tunneling contact (Schubert *et al.*, 1986).

A new type of barrier was invented to overcome the limitations of these previously demonstrated unipolar barrier structures. The planar-doped barrier (PDB) concept was first demonstrated in GaAs by placing planar-doped charge sheets of opposite polarity in close proximity to one another within an undoped semiconductor using MBE growth. The following section describes the PDB device model and the electron transport physics in these structures.

23.3 Planar-doped barrier device model

23.3.1 Atomic plane model

A simplified doping model for the planar-doped barrier is illustrated in Fig. 23.6. The barrier is formed by the two-dimensional space charge which resides on the planar-doped (delta-doped) atomic planes. The charge planes are formed by depositing dopant impurity atoms on the semiconductor surface during growth interruption by MBE. The acceptor plane N_A^{2D} is placed within an intrinsic (undoped) semiconductor region and the two donor planes N_{D1}^{2D} and N_{D2}^{2D} are adjacent to conducting regions $n(1)$ and $n(2)$ respectively. It is assumed that the acceptor plane is fully ionized by the capture of electrons from the donor planes since the acceptor energy levels are below the bulk Fermi level in the semiconductor. The resulting space charge, electric field, and band potential profiles at equilibrium are shown in Fig. 23.7.

The electric field and potential energy profile can be found from the solution of Poisson's equation in one dimension. The boundary conditions are that the electric field is zero in both n-regions, that is, the electric fields terminate on the donor dopant planes:

$$\mathbf{F}(z) = -\frac{\mathrm{d}\phi(z)}{\mathrm{d}z} = 0; \quad z < -L_1,\ z > L_2. \tag{23.12}$$

This implies the condition for charge neutrality:

$$qN_{D1}^{2D} + qN_{D2}^{2D} - qN_A^{2D} = 0. \tag{23.13}$$

The electric field in region 1 is constant and is given by

$$\mathbf{F}_1(z) = -\frac{\mathrm{d}\phi_{B1}}{\mathrm{d}z} = \frac{qN_{D1}^{2D}}{\varepsilon}; \quad -L_1 < z < 0. \tag{23.14}$$

The electric field in region 2 is constant and is given by

$$\mathbf{F}_2(z) = -\frac{\mathrm{d}\phi_{B2}}{\mathrm{d}z} = -\frac{qN_{D2}^{2D}}{\varepsilon}; \quad 0 < z < L_2. \tag{23.15}$$

The potential reaches a maximum at the acceptor plane where there is a discontinuity in the electric field:

$$\phi(0) = \phi_{BO} \tag{23.16}$$

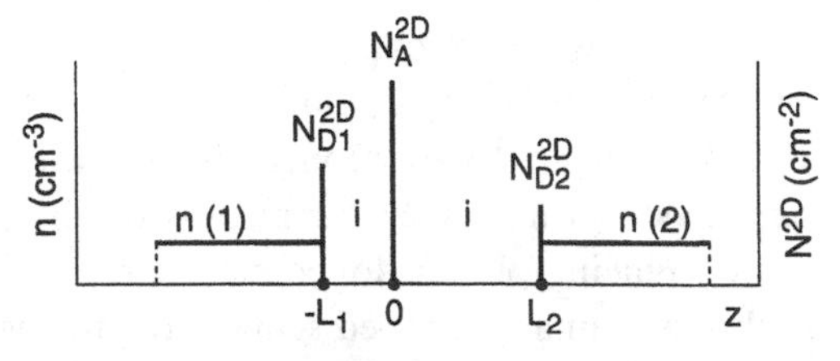

Fig. 23.6. Simplified atomic plane doping model for a PDB (after Malik *et al.*, 1981).

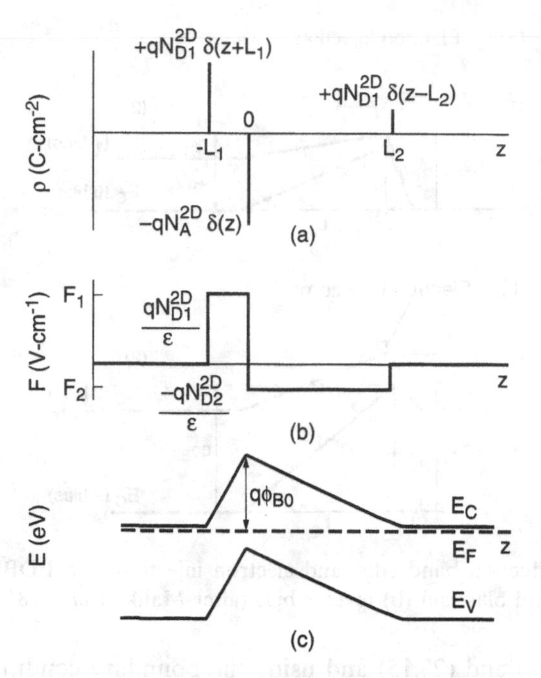

Fig. 23.7. (a) Space charge, (b) electric field, and (c) equilibrium band potential profiles in the atomic plane doping model of a PDB (after Malik *et al.*, 1981).

Integrating the electric fields and setting them equal at $\phi(0)$ yields

$$\phi_{BO} = \frac{qN_{D1}^{2D}}{\varepsilon} L_1 = \frac{qN_{D2}^{2D}}{\varepsilon} L_2. \tag{23.17}$$

This expression can be simplified using Eqn. (23.13), the condition for charge neutrality to yield

$$\phi_{BO} = \frac{qN_A^{2D}}{\varepsilon} \frac{(L_1 L_2)}{(L_1 + L_2)}. \tag{23.18}$$

Thus the zero-bias barrier height is given by a simple expression which is directly related to the two-dimensional acceptor charge density and the undoped region widths. This is an extremely important result which demonstrates that the barrier height can be effectively varied from zero to approximately the band-gap of the semiconductor, $0 < \phi_{BO} < E_G/q$, through the appropriate choice of these parameters. This expression is used routinely in the design criteria of planar-doped barriers.

The response of the barrier under an applied bias which causes current rectification can be explained with the aid of Fig. 23.8. The applied bias is dropped linearly across the barrier region, that is, the changes in the two-dimensional donor densities are equal in magnitude and opposite in sign. The difference in the barrier heights for electrons in the two donor regions is equal to the applied bias

$$\phi_{B1} - \phi_{B2} = V. \tag{23.19}$$

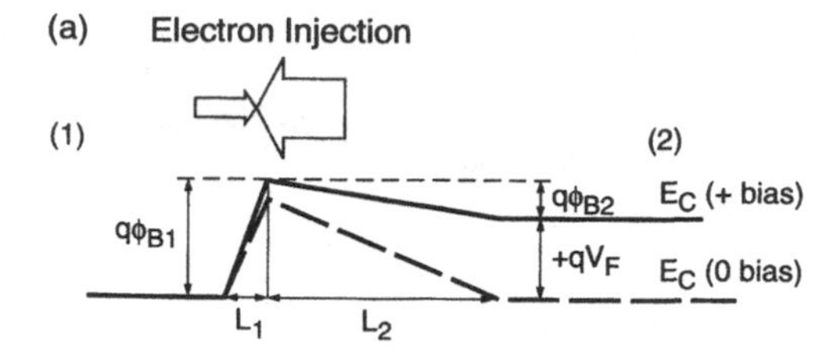

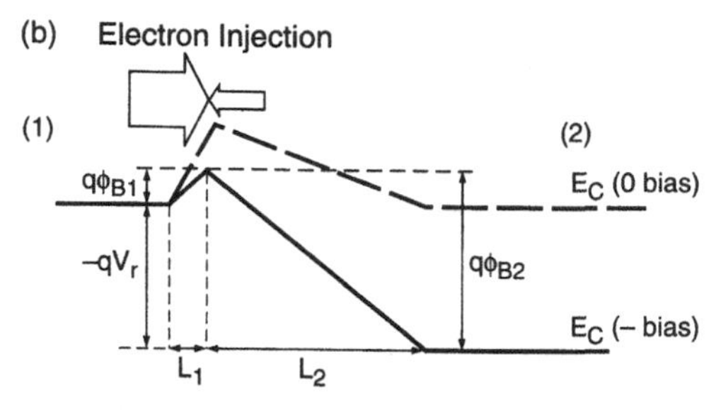

Fig. 23.8. Conduction band edge and electron injection in a PDB under (a) forward bias and (b) reverse bias (after Malik *et al.*, 1981).

Integrating Eqns. (23.14) and (23.15) and using the boundary condition of Eqn. (23.19), the barrier heights can be expressed as

$$\phi_{B1} = \phi_{BO} + \left(\frac{L_1}{L_1 + L_2}\right)V \tag{23.20}$$

$$\phi_{B2} = \phi_{BO} - \left(\frac{L_2}{L_1 + L_2}\right)V. \tag{23.21}$$

The current transport is assumed to be by thermionic emission over the barrier, as is normally seen in Schottky barriers, and is schematically depicted in Fig. 23.8. The current injection from region 1 to region 2 is given by

$$J_{1-2} = A^* T^2 \exp\left(\frac{-q\phi_{B1}}{kT}\right). \tag{23.22}$$

And the current injection from region 2 to region 1 is given by

$$J_{2-1} = A^* T^2 \exp\left(\frac{-q\phi_{B2}}{kT}\right). \tag{23.23}$$

The total current density over the barrier is thus equal to

$$J = J_{2-1} - J_{1-2} = A^* T^2 \exp\left(\frac{-q\phi_{BO}}{kT}\right)\left[\exp\left(\frac{q\alpha_2 V}{kT}\right) - \exp\left(\frac{-q\alpha_1 V}{kT}\right)\right] \tag{23.24}$$

where

$$\alpha_1 = \frac{L_1}{L_1 + L_2} \tag{23.25}$$

and

$$\alpha_2 = \frac{L_2}{L_1 + L_2}. \tag{23.26}$$

The geometric ratios α_1 and α_2 are sometimes called the 'barrier leverage' ratios and they determine the rate of change of the respective barriers with applied bias voltage. They can be considered similar to the inverse of the ideality factor found in Schottky barriers although their physical origin is quite different. Since $\alpha_1 + \alpha_2 = 1$, it is obvious that the greater the difference in these two ratios, the more highly asymmetric the $I-V$ characteristics will be from Eqn. (23.24). For the special case of $\alpha_1 = \alpha_2 = 0.5$, the $I-V$ characteristics will be antisymmetric about the origin (i.e. equal in magnitude and opposite in sign). Thus the choice of L_1 and L_2 affects the asymmetry of the $I-V$ characteristics.

The capacitance of the PDB can be calculated from the total depletion width. Since the change in applied bias voltage occurs through modulation of the two-dimensional charge density on the donor planes, the depletion width is simply $L_1 + L_2$. Thus the capacitance per unit area is simply given by

$$C = \frac{\varepsilon}{L_1 + L_2}. \tag{23.27}$$

The capacitance is essentially constant and independent of the bias voltage.

A comparison of the simplified atomic plane models and experimentally measured $I-V$ and $C-V$ measurements of some representative PDB diodes are shown in Figs. 23.9 and 23.10, respectively. Although the trends in the current rectification and capacitance can be seen, the differences are mainly due to the band bending in the distributed space charge regions of the PDB, as discussed further in the following section.

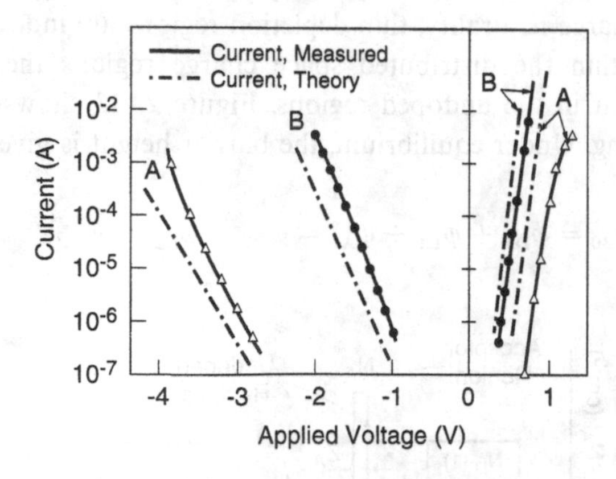

Fig. 23.9. Comparison of experimental and theoretical $I-V$ characteristics of two PDB diodes using simplified atomic plane doping model. For both diodes $a = 7.8 \times 10^{-5}$ cm^2, $N_D = 1.0 \times 10^{18}$ cm^{-3}, $N_A = 2.0 \times 10^{18}$ cm^{-3}; diode A, $L_1 = 250$ Å, $L_2 = 2000$ Å, $Z_A = 100$ Å; diode B, $L_1 = 500$ Å, $L_2 = 2000$ Å, $Z_A = 50$ Å (after Malik *et al.*, 1980).

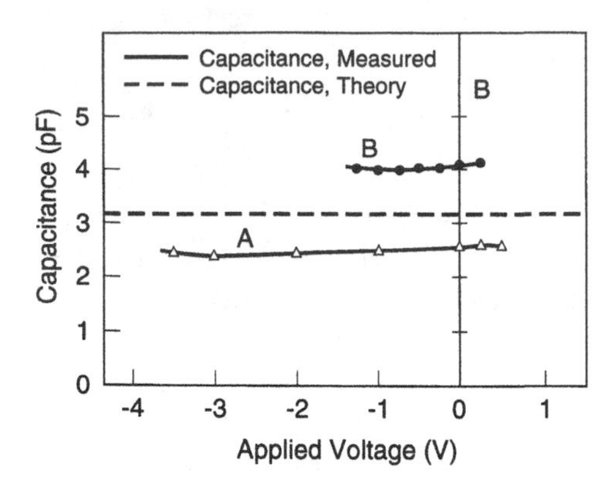

Fig. 23.10. Comparison of experimental and theoretical $C-V$ characteristics of two PDB diodes using simplified atomic plane model. Same diode parameters as given in Fig. 23.9 (after Malik *et al.*, 1980).

23.3.2 Distributed space charge model

Although the first PDB structures were grown by MBE using atomic plane doping (Wood *et al.*, 1980; Metze, 1981), it was found that many of the diodes showed poor, 'leaky' $I-V$ characteristics. It is believed that impurities or defects may have been incorporated during MBE growth interruption which resulted in degraded performance. By using relatively high donor and acceptor concentrations ($>10^{18}$ cm^{-3}) it is possible to form a PDB using distributed space charge. The PDB uses an n^+–i–p^+–i–n doping configuration, as illustrated in Fig. 23.11. The space charge, electric field, and band potential profiles for the PDB are shown in Fig. 23.12. The acceptor layer thickness is typically $Z_A \leqslant 50$ Å. In order to satisfy space charge neutrality, thin depletion regions are induced at the edges of the donor regions. Within the distributed space charge regions the band bending is parabolic, and it is linear in the undoped regions. Figure 23.13 shows the details of the conduction band bending. Under equilibrium, the barrier height is given by

$$\phi_{B0} = \phi_{D1} + \phi_{L1} + \phi_{A1} = \phi_{D2} + \phi_{L2} + \phi_{A2}. \tag{23.28}$$

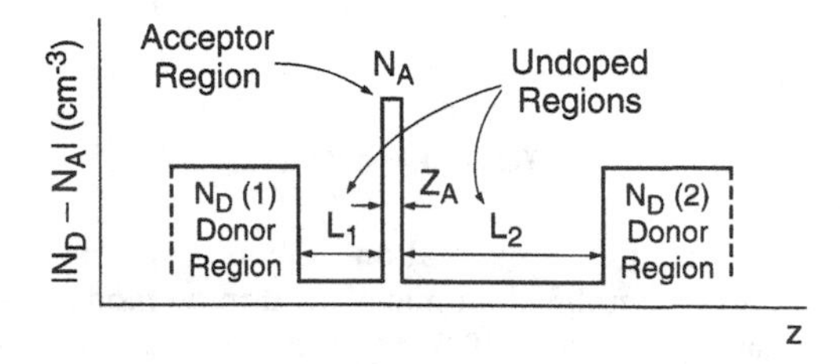

Fig. 23.11. Distributed space charge model for a PDB (after Malik *et al.*, 1980).

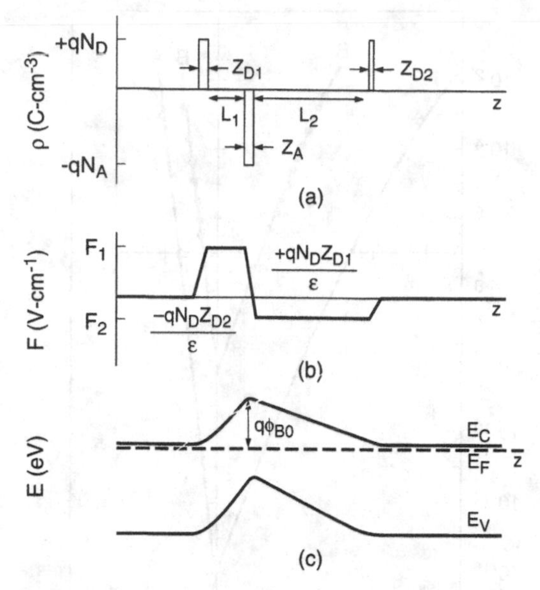

Fig. 23.12. (a) Space charge, (b) electric field, and (c) band potential profiles for a PDB with distributed space charge (after Malik *et al.*, 1980).

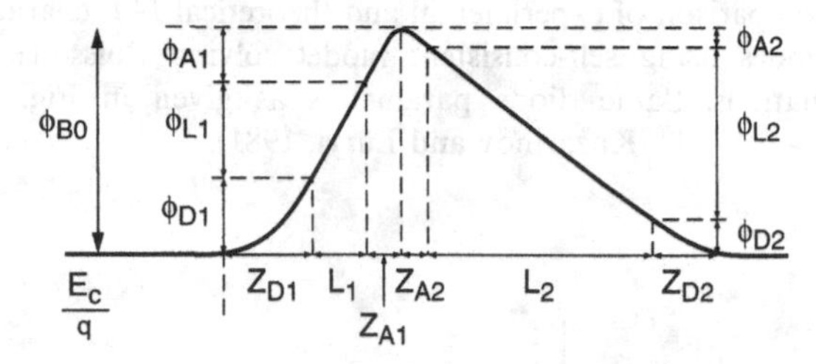

Fig. 23.13. Detailed band bending of conduction band edge for a PDB with distributed space charge (after Malik, 1982).

Introducing the constant K given by the acceptor to donor density ratio

$$K = \frac{N_A}{N_D}. \tag{23.29}$$

The zero-bias barrier height then becomes

$$\phi_{B0} = \frac{qN_A Z_A}{\varepsilon}\left[\frac{L_1 L_2 + Z_A(K+1)\dfrac{L_1}{2}}{L_1 + L_2 + Z_A(K+1)} + Z_A\frac{(K+1)}{2}\left[\frac{L_2 + Z_A\left(\dfrac{K+1}{2}\right)}{L_1 + L_2 + Z_A(K+1)}\right]^2\right]. \tag{23.30}$$

In addition to the band bending due to the space charge, there are contributions from the diffusion potential of electrons diffusing in from the n-regions (Shur, 1985) and the

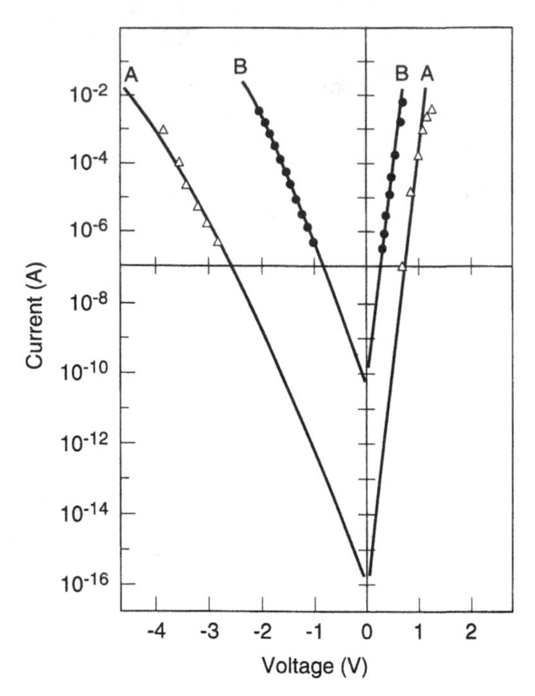

Fig. 23.14. Comparison of experimental and theoretical $I-V$ characteristics of two PDB diodes using self-consistent model solving Poisson's and drift-diffusion equations. Same diode parameters as given in Fig. 23.9 (after Kazarinov and Luryi, 1981).

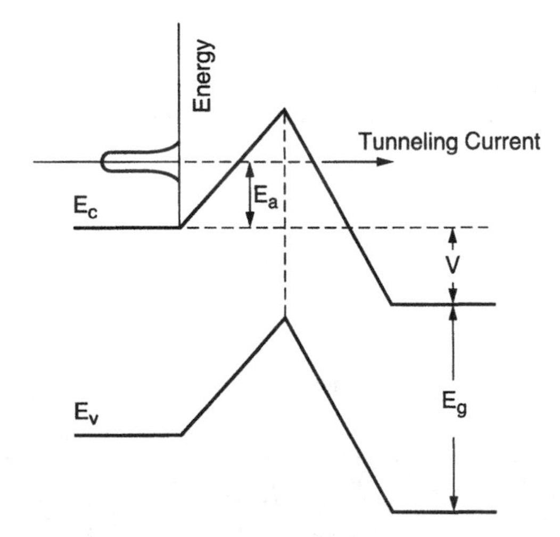

Fig. 23.15. Schematic diagram of thermionic field emission electron tunneling current in thin PDB structures (after Gossard *et al.*, 1982b).

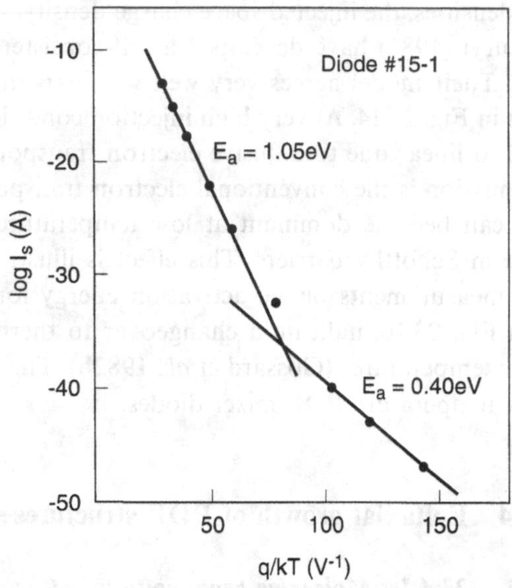

Fig. 23.16. Saturation current versus inverse temperature to determine activation energy of electron transport as a function of temperature. PDB diode #15-1 parameters (after Gossard *et al.*, 1982b).

difference between the Fermi level and the conduction band edge. The effective barrier height is then

$$\phi_{\rm B0}^{\rm eff} = \phi_{\rm B0} + \phi_{\rm dif} + \phi_{\rm S} \tag{23.31}$$

where

$$\phi_{\rm dif} = \frac{kT}{q} \tag{23.32}$$

and

$$\phi_{\rm S} = \frac{E_{\rm c} - E_{\rm f}}{q}. \tag{23.33}$$

The effective barrier leverage ratios become

$$\alpha_1^{\rm eff}(V) = \frac{L_1 + Z_{\rm D1} + Z_{\rm A1}}{L_1 + L_2 + Z_{\rm D1} + Z_{\rm D2} + Z_{\rm A}} \tag{23.34}$$

$$\alpha_2^{\rm eff}(V) = \frac{L_2 + Z_{\rm D2} + Z_{\rm A2}}{L_1 + L_2 + Z_{\rm D1} + Z_{\rm D2} + Z_{\rm A}}. \tag{23.35}$$

The capacitance of the structure becomes

$$C = \frac{\varepsilon}{L_1 + L_2 + Z_{\rm D1} + Z_{\rm D2} + Z_{\rm A}}. \tag{23.36}$$

At high injection current densities, the injected space charge density will alter the potential profile. Kazarinov and Luryi (1981) have developed a self-consistent analytic model for transport in PDB diodes. Their model agrees very well with experimental data reported in the literature, as shown in Fig. 23.14. At very high injection conditions, the $I-V$ relation changes from exponential to linear due to diffusive electron transport (Buot *et al.*, 1982).

Although thermionic emission is the conventional electron transport mechanism in the PDB, electron tunneling can become dominant at low temperatures with short barrier widths as is also the case in Schottky barriers. This effect is illustrated schematically in Fig. 23.15. Experimental measurements of the activation energy for transport in a thin barrier PDB, depicted in Fig. 23.16, indicate a changeover to thermionic field emission through the barrier at low temperatures (Gossard *et al.*, 1982b). This can be an important effect in the design of low temperature PDB mixer diodes.

23.4 Epitaxial growth of PDB structures

23.4.1 Molecular beam epitaxy

The ability to synthesize planar-doped barrier structures was made possible by the development of molecular beam epitaxy (MBE) which was pioneered by Cho and Arthur (1975) at AT&T Bell Laboratories in the 1970s. The majority of research has focused on the MBE growth of GaAs and the first demonstration of the PDB concept was made in this material system (Malik *et al.*, 1980). While metal–organic chemical vapor deposition (MOCVD) has also been used to grow GaAs PDB diodes (Kearney *et al.*, 1990), it does not provide the precise control of layer thickness and doping level afforded by MBE. The reader is referred to several chapters in this book which review MBE and other epitaxial technologies which have been used in the growth of PDB structures.

Molecular beam epitaxy can be defined as the growth of stoichiometric, single crystal films through condensation of evaporated molecular beams on a heated substrate in ultra-high vacuum. A schematic diagram of an MBE system is shown in Fig. 23.17 (Cho, 1983). A variety of vacuum pumps and liquid nitrogen cryo-shrouding are employed to reduce unintentional impurities in the grown films. The molecular beam fluxes are derived from heated furnaces or effusion cells which have individual shutters to control the arrival of the beam fluxes on the heated substrate.

The molecular beam flux densities at the growth position can be described by the Knudsen (1909) equation

$$J_x(T) = \frac{aP_x(T)}{\pi d^2(2\pi MRT)^{1/2}} \tag{23.37}$$

where $J_x(T)$ is the flux density of element x, $P_x(T)$ is the temperature dependent vapor pressure of the source, a is the aperture of the effusion cell, d is distance between the cell and substrate, M is the molecular weight, R is the universal gas constant, and T is the absolute temperature. The temperature dependence for the beam flux density can be

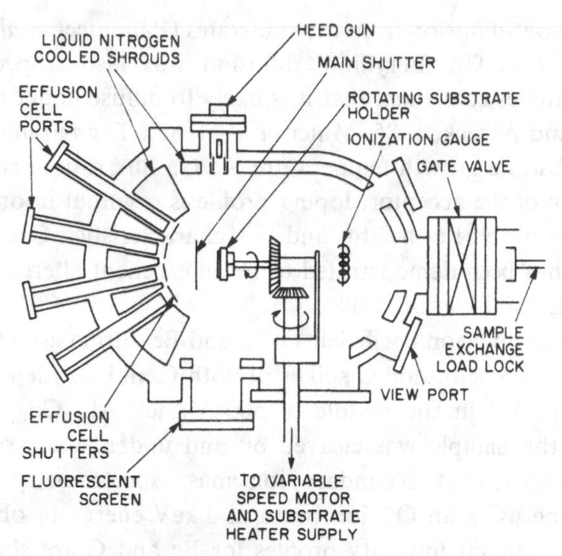

Fig. 23.17. Schematic diagram of molecular beam epitaxy (MBE) system (after Cho, 1983).

expressed by

$$J_x(T) = J_0 \exp\left(\frac{-E_{Ax}}{kT}\right) \tag{23.38}$$

where J_0 is a constant, E_{Ax} is the activation energy for vaporization of element x, k is Boltzmann's constant, and T is the absolute temperature. The typical activation energy for the vaporization of Ga, Si, Be, and C used in the MBE growth of GaAs PDB diodes approximately results in the following temperature dependence of the beam fluxes:

$$\frac{\partial J_x(T)}{\partial T} = 0.02 J_x(T). \tag{23.39}$$

Thus by controlling the cell temperature to within 1 K, which is easily achieved, the beam flux is stable to within a few percent. In addition, the GaAs growth rate can be very accurately measured using reflection high energy electron diffraction (RHEED) oscillations (Foxon and Joyce, 1975). Typical GaAs growth rates used are about one monolayer/s = 2.83 Å/s. The typical donor and acceptor doping densities are in the range 10^{18} cm^{-3}. With this precise control of molecular beam fluxes, MBE is ideally suited for the growth of PDB structures.

23.4.2 *Stability of acceptor doping planes*

The original work on PDB diodes utilized Be, which is the conventional acceptor dopant in solid source MBE of GaAs. However, it was found experimentally to be problematic to grow PDB diodes with reproducible barrier heights. Some of these difficulties were ascribed to impurity out-diffusion from the substrates (Palmateer *et al.*, 1983; Brown *et al.*, 1984; Schubert *et al.*, 1990). Beryllium diffusion was also suspected as a potential problem and later studies confirmed that it is likely to diffuse under certain MBE growth conditions (Miller and Asbeck, 1985; Miller *et al.*, 1985). The parameters which affect Be diffusion include substrate growth temperature, As/Ga flux ratio, and Be doping concentration. The stability of the acceptor doping profile is essential in order to achieve PDB diodes with reproducible barrier heights and I–V characteristics. Carbon evaporated from a graphite filament has been demonstrated to be an excellent alternative acceptor dopant in solid source MBE (Malik *et al.*, 1988).

A comparison of the diffusion coefficients of C and Be during growth has been analyzed (Malik *et al.*, 1993). Two delta-doped spikes of both C and Be (deposited during growth interruption) were placed in the middle of 500-Å-thick $Al_{0.3}Ga_{0.7}As$ layers grown at 580 °C. A piece of this sample was cleaved off and underwent a rapid thermal anneal (RTA) at 900 °C for 15 s. A secondary ion mass spectrometry (SIMS) analysis of the samples was done using an O^+ ion beam at 3 keV energy to obtain high resolution depth profiles. The resultant impurity profiles for Be and C are shown in Fig. 23.18(a) and (b) respectively. It can be seen that although the nominal sheet acceptor concentrations are the same, the peak Be concentration is approximately five times lower than for C. Also the leading edges of the C profiles are parallel with the Al signal. Together, this implies that the C profile is atomically abrupt (as is known for the Al profile) and that significant diffusion of Be has occurred during MBE growth. After the RTA, there is no indication of C diffusion. By contrast, after the RTA there is further additional diffusion of Be leading to a broader profile. This data suggests that C is preferred over Be as the acceptor dopant in GaAs PDB diodes since hyperabrupt and stable doping spikes can be obtained.

A SIMS depth profile of a typical PDB diode is shown in Fig. 23.19 (Anand and Malik, 1992). The C-doping spike is 35 Å wide and the undoped region widths are 100 Å and 2000 Å, resulting in a highly asymmetric rectifying diode. The finite width of the C-doping profile is due to SIMS instrumental broadening associated with the sputtering rate and 'knock-on'. Also to be noted is the large unintentional C contamination at the epilayer–substrate interface. This can be minimized by ozone cleaning of the substrate prior to epitaxial growth.

The control of the C-doping concentration and barrier height has also been examined (Anand and Malik, 1992). A series of PDB diodes was as grown using a fixed C-doping level and varying the growth time of the acceptor spike region. The measured log I–V characteristics are shown in Fig. 23.20. The barrier height, ϕ_B, can be determined from the extrapolated saturation current at zero-bias, I_0. From the inset in Fig. 23.20, it can be seen that the calculated barrier heights are directly proportional to the growth time of the acceptor spike regions. This is expected since the barrier height is proportional to the doping-thickness product of the acceptor spike region. From this data, it is estimated that the C-doping level is controlled to within +/−2%, which corresponds to +/−1 °C control in the graphite filament temperature. This level of doping control is very important in order to obtain PDB diodes with reproducible I–V characteristics.

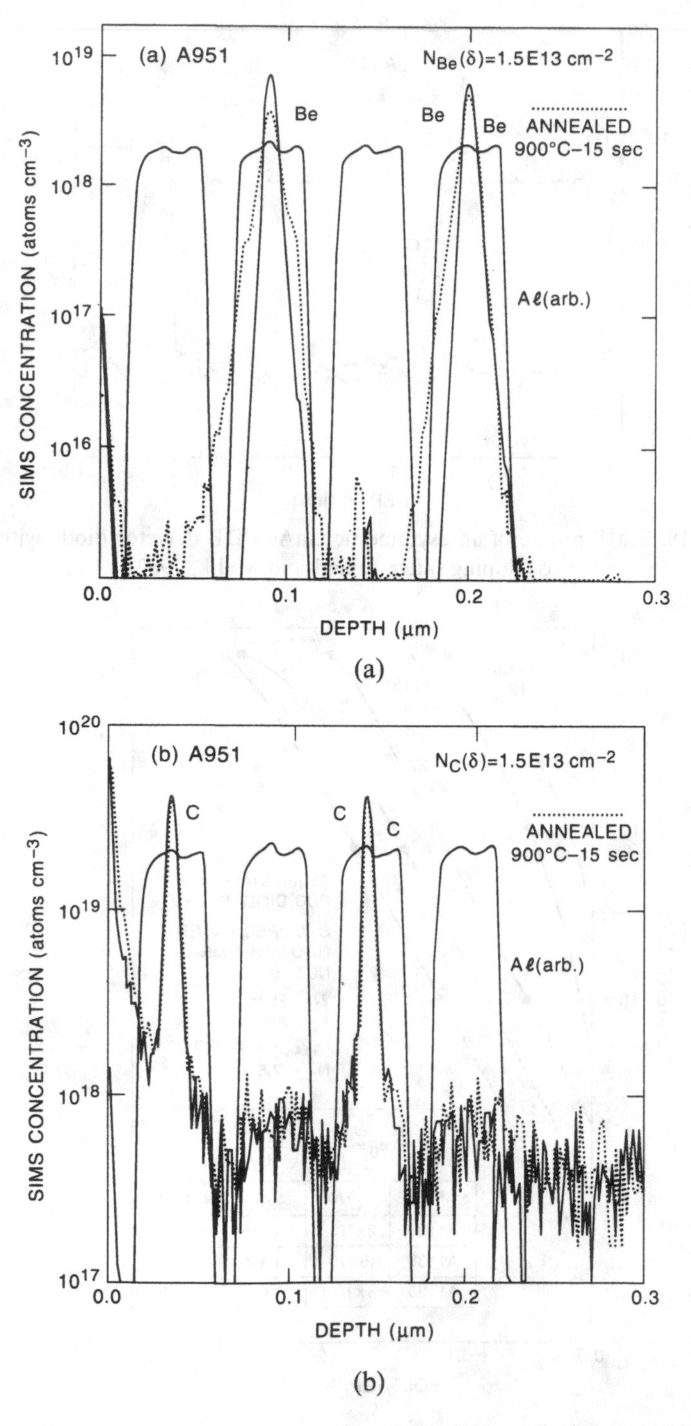

Fig. 23.18. SIMS depth profiles for delta-doped (a) Be and (b) C in $Al_{0.3}Ga_{0.7}As$ layers to examine acceptor diffusion during MBE growth and after postgrowth thermal annealing (after Malik *et al.*, 1993).

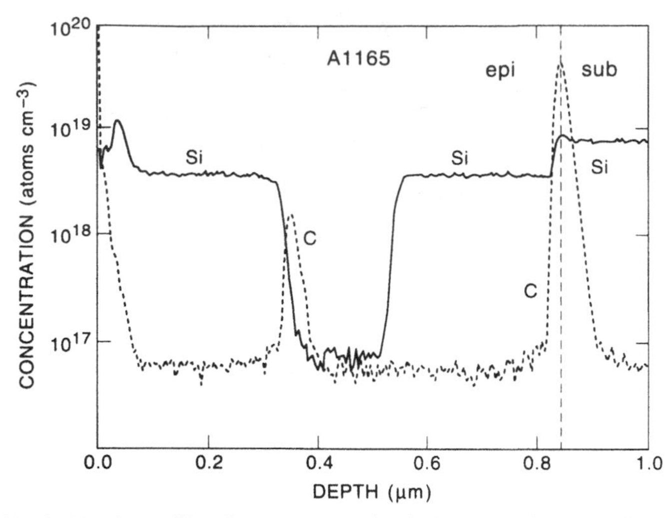

Fig. 23.19. SIMS profile of an asymmetric GaAs PDB detector diode with C acceptor doping (after Anand and Malik, 1992).

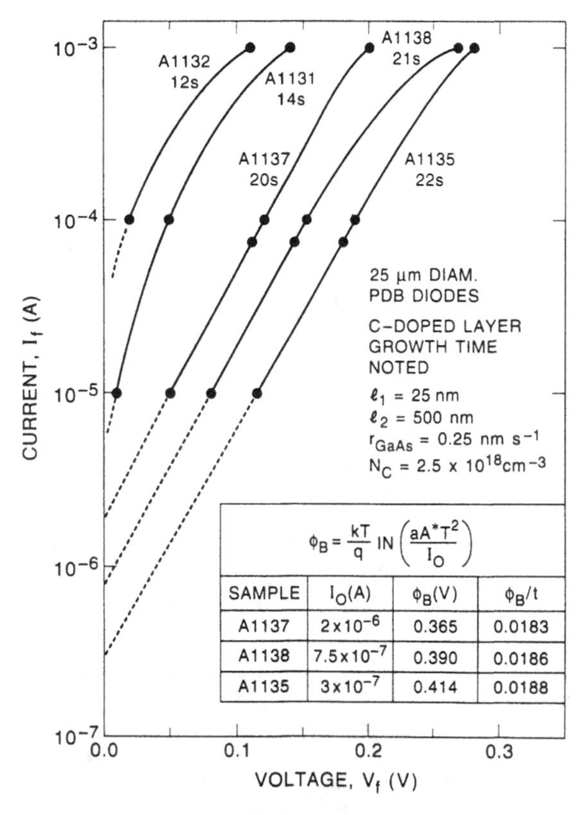

SAMPLE	I_O(A)	ϕ_B(V)	ϕ_B/t
A1137	2×10^{-6}	0.365	0.0183
A1138	7.5×10^{-7}	0.390	0.0186
A1135	3×10^{-7}	0.414	0.0188

Fig. 23.20. Log $I-V$ characteristics for a series of C-doped GaAs PDB diodes in which the growth time of the C-doping spike was varied (after Anand and Malik, 1992).

23.5 Planar-doped barrier devices

The planar-doped barrier concept has been utilized in a variety of novel device applications. The PDB diode was first conceived as a replacement for the Schottky diode for use as microwave/millimeter wave detector and mixer diodes (Malik, 1983). The principal advantage of the PDB diode over the Schottky diode is the ability to design the barrier height, the asymmetry of the $I-V$ characteristics, and the capacitance of the PDB diode to optimize device performance independently. The PDB diode has also been found to have other superior properties compared with Schottky diodes, including lower $1/f$ noise, reduced temperature sensitivity of the $I-V$ characteristics, and much higher resistance to burn-out from nanosecond RF pulses and accidental electrostatic discharge (ESD). Planar-doped barrier detector and mixer diodes are now being manufactured commercially due to their desirable properties. When high volume production of PDB diodes is achieved, they may surpass the use of Schottky diodes in microwave detector and mixer circuits. This section will describe the characteristics of PDB detector and mixer diodes, along with some other interesting devices which incorporate the PDB concept.

23.5.1 Detector diodes

PDB diodes have been investigated, produced, and marketed by a number of microwave semiconductor companies including GEC/Marconi Electronic Devices, Hewlett-Packard, M/A-COM, Alpha Industries, and Hughes for use as high performance microwave and millimeter wave detector and mixer diodes. The designability of the PDB device parameters has resulted in superior detector characteristics at frequencies up to 94 GHz (Kearney, *et al.*, 1991).

A schematic cross-section of a typical PDB detector diode is shown in Fig. 23.21. A

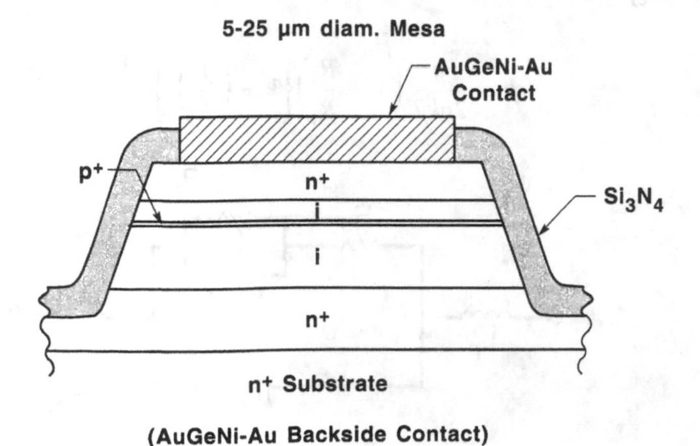

Fig. 23.21. Schematic cross-section of small-diameter PDB diode which uses an etched mesa configuration.

shallow circular mesa is formed by etching with a diameter of 2–25 μm depending on the desired diode capacitance which is determined by the required cut-off frequency of the diode. Silicon nitride is used to passivate the sidewall of the mesa; Au/Ge/Ni ohmic metallization is used for the top anode contact and for the backside contact. After dicing the wafer, the diode is soldered into a microwave package and the anode is contacted using either a bonded Au wire or a W whisker probe. An equivalent circuit model for the packaged PDB diode is shown in Fig. 23.22. The intrinsic diode junction resistance and capacitance are denoted by R_j and C_j respectively; R_s is the diode series resistance which is due to the ohmic contact resistance as well as to the injected space charge and hot electron scattering resistances (Couch and Kearney, 1989). The parasitic package inductance and capacitance are denoted by L_p and C_p respectively. A conventional figure of merit for the diode is the cut-off frequency, which is given by

$$f_c = (2\pi R_s C_j)^{-1}. \tag{23.40}$$

Cut-off frequencies above 1500 GHz have been obtained in small diameter (2 μm) PDB diodes (Lee *et al.*, 1993).

A detector diode is used to convert a modulated high frequency signal voltage into a time-varying current which is followed by further stages of amplification. Another figure of merit for the detector diode is the tangential sensitivity, S_t, which is equivalent to the minimum detectable signal level. The tangential sensitivity can be expressed by

$$S_t = \frac{[4YkT\Delta f(F_v + \gamma - 1)]^{1/2}}{\beta R_v} \tag{23.41}$$

where β is the current sensitivity, R_v is the video impedance, F_v is the amplifier noise figure, γ is the noise temperature ratio, and Y is an empirical fitting parameter. The current sensitivity is given by

$$\beta = \frac{q}{2nkT}\left(1 + \frac{R_s}{R_j}\right)\left\{1 + \left(\frac{f}{f_c}\right)^2\right\}^{-1} \tag{23.42}$$

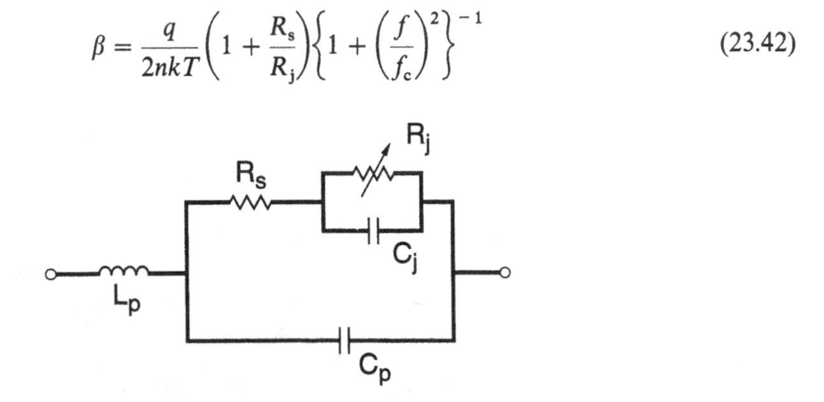

Fig. 23.22. Equivalent circuit diagram of a PDB diode (after Kearney *et al.*, 1990).

and the video impedance is given by

$$R_v = R_s + R_j$$
$$= R_s + \left\{\frac{dI(V)}{dV}\right\}^{-1}. \tag{23.43}$$

In most cases, it is desirable to use a zero-bias detector diode since this avoids the complication of providing a DC bias for the diode in the detector circuit. A zero-bias detector diode typically has a barrier height in the range 0.05–0.15 eV. Fortunately, the designability of PDB barrier height can easily accomplish this, unlike Si Schottky diodes, where very low barrier heights are difficult to obtain and reproduce.

Table 23.1 shows a comparison of the DC characteristics of some X-band PDB and Schottky diodes from GEC-Marconi Electronic Devices (Kearney *et al.*, 1990). The zero-bias PDB GB322 is seen to have a lower junction capacitance and series resistance than the zero-bias silicon Schottky DC7059. This results in an improved detector sensitivity of −57 dBm versus −55 dBm, respectively, at 9.375 GHz. Also, the high turn-on voltage of the GaAs Schottky DC1301 results in no detectable output voltage at signal levels below approximately −10 dBm without a DC bias voltage. The linear I–V characteristics of two different barrier height GaAs PDB diodes are shown, along with a GaAs Schottky diode, in Fig. 23.23. It can be seen that I–V curves are highly asymmetric for the PDB diodes which also have significantly lower turn-on voltages in the forward bias direction. The sharp reverse breakdown voltage of the Schottky diode is due to avalanche multiplication, whereas the PDB diodes exhibit much softer reverse I–V characteristics due to thermionic emission over the barrier.

Table 23.1. *DC characteristics of X-band PDB and Schottky diodes*

Device	Reverse voltage (V_R) $I_R = 10\ \mu A$ (V)	Forward voltage (V_F) $I_F = 100\ \mu A$ (V)	Junction capacitance (C_j) (fF)	Series resistance (R_S) $I_F = 10$ mA (Ω)	Ideality factor (n) $I_F = 1 - 10\ \mu A$
PDB GB322	0.45	0.13	53	10	–
PDB CB46	4.10	0.38	60	12	1.17
'Zero bias' silicon Schottky DC7509	0.35	0.06	70	45	–
Low-barrier silicon Schottky DC1630	2.00	0.24	60	25	1.18
GaAs Schottky DC1301	7.00	0.59	70	4	1.17
NIP/Schottky DC1596	0.70	0.24	110	25	1.18

Table 23.2. *DC characteristics of zero-bias PDB Diodes (10 GHz–40 GHz pill package)*

V_f @ 10 μA	0.06 volts
V_f @ 100 μA	0.14 volts
V_b @ 10 μA	1.0 volts
C_j (calculated)	60 fF
R_S	15 Ω

Table 23.3. *RF performance of zero-bias PDB diodes (10 GHz–40 GHz pill package)*

Frequency	10 GHz	35 GHz
Video impedance R_v	2–50 kΩ	2–50 kΩ
Tangential sensitivity S_t	−50 to −58 dBm	−50 to −58 dBm
Voltage output @ −10 dBm	80 mV	65 mV

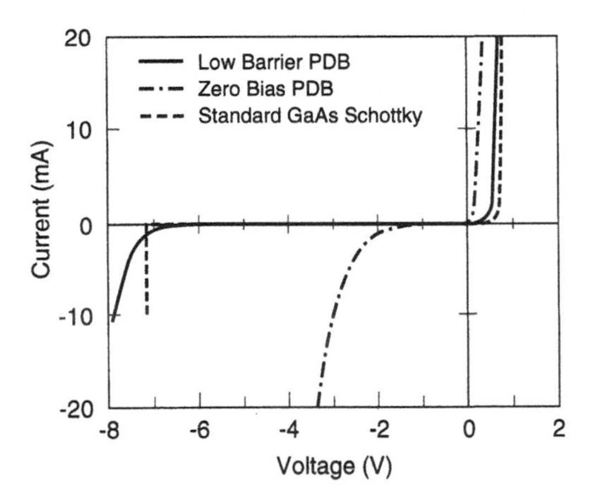

Fig. 23.23. Linear I–V characteristics of zero-bias and low-barrier GaAs PDB diodes and a standard GaAs Schottky diode (after Kearney *et al.*, 1990).

The DC and RF characteristics for some X–Ku band zero-bias PDB detector diodes from M/A-COM are given in Tables 23.2 and 23.3 respectively (Anand and Malik, 1992). The diodes exhibit very high reverse/forward bias ratio asymmetry (>10:1) and have very low forward turn-on voltages. The video impedance can be varied over a wide range depending upon the barrier height. These zero-bias detectors exhibit excellent tangential sensitivity up to −58 dBm. The output voltage as a function of input power for these

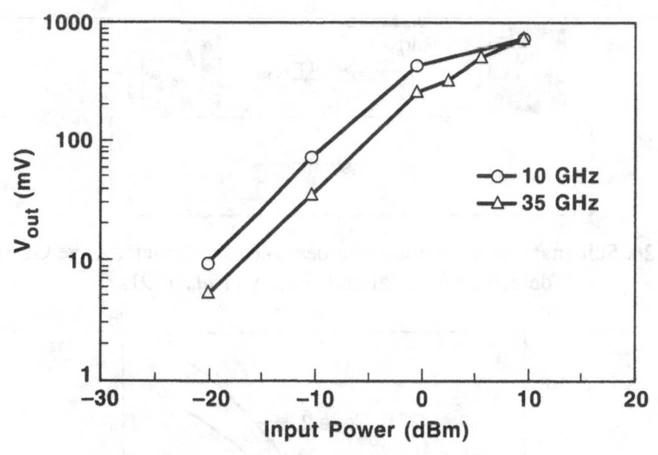

Fig. 23.24. Output voltage versus RF input power for zero-bias GaAs PDB detector diodes at 10 and 35 GHz (after Anand *et al.*, 1991).

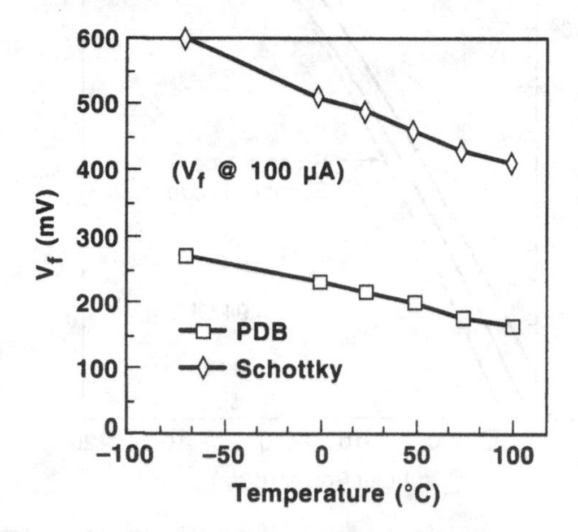

Fig. 23.25. Temperature dependence of detector output voltage for zero-bias GaAs PDB and Si Schottky diodes (after Anand *et al.*, 1991).

detectors at 10 and 35 GHz is shown in Fig. 23.24. The temperature dependence of the output voltage of these detectors compared to a Si Schottky diode is shown in Fig. 23.25. Planar-doped barrier diodes have been found to have a weaker temperature dependence in their $I-V$ characteristics than do Schottky diodes. The reasons for this are not yet fully understood. It has been proposed that carrier spill-over from the n^+ contact regions in the PDB results in a positive temperature coefficient for the effective barrier height leading to a reduced temperature dependence of the output voltage (Kearney *et al.*, 1990). This effect in the PDB can simplify or eliminate the need for temperature compensation in detector circuits.

The performance advantages of PDB diodes over Schottky diodes have also been demonstrated at millimeter wave frequencies. Figure 23.26 illustrates a cross-section of a

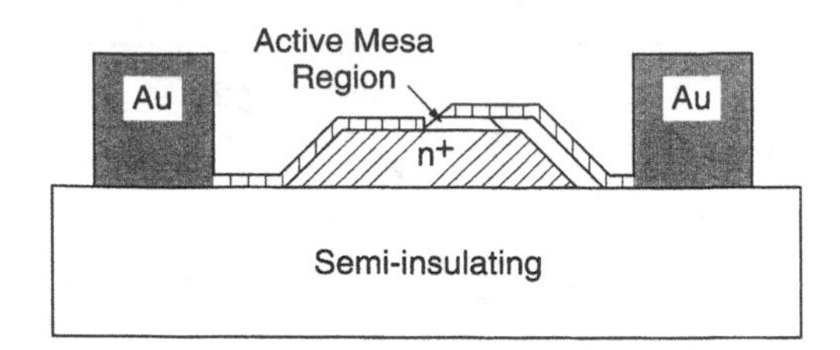

Fig. 23.26. Schematic cross-section of a beam-lead millimeter wave GaAs PDB detector diode (after Kearney *et al.*, 1991).

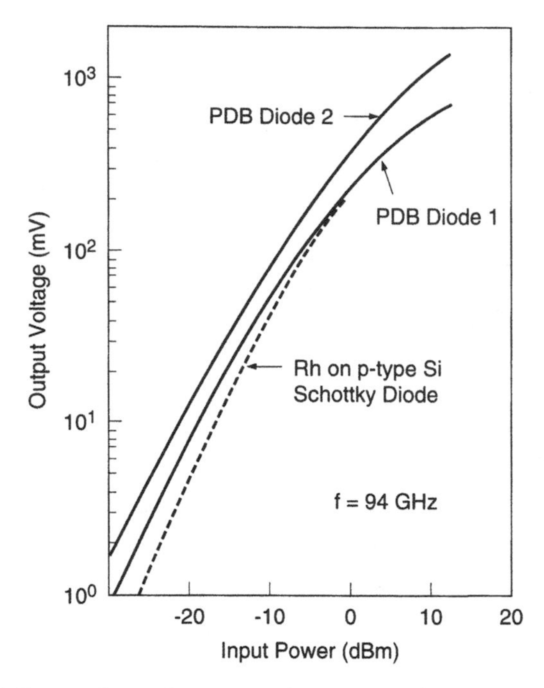

Fig. 23.27. Comparison of output voltage versus input power for millimeter wave GaAs PDB and Si Schottky detector diodes (after Kearney *et al.*, 1990).

beam-lead diode structure which uses an air bridge contact to the small area anode in order to reduce the parasitic capacitances. The measured output voltages of two different PDB diodes compared with state-of-the-art Rh on p-type Si Schottky diodes are shown in Fig. 23.27. Overall, the PDB diodes exhibit better tangential sensitivity, dynamic range, and voltage sensitivity than does the Si Schottky diode.

The attractive RF characteristics of PDB diodes have been used in coaxial detector assemblies produced by Hewlett-Packard (Fraser, 1987). Peak power sensors have also been fabricated using two PDB diodes monolithically integrated in GaAs with RC matching networks and transmission lines in a dual balanced detector circuit design (Scherer, 1990; Fischer *et al.*, 1992). The peak power sensor is calibrated by measuring the output voltage as a function of both frequency and temperature and storing the resultant three-dimensional matrix in an EEPROM.

Table 23.4. *Electrostatic and CW burnout at 10 GHz*

	Low-barrier Si-Schottky	Med.-barrier Si-Schottky	High-barrier Si-Schottky	GaAs Schottky	GaAs PDB (zero bias)
V_f @ 1 mA (volts)	0.4	0.55	0.7	0.72	0.2
V_b @ 10 μA (volts)	3	5	7	7	1
C_t (pF)	0.1	0.1	0.1	0.1	<0.1
Electrostatic pulses (3)					
Forward (volts)	800	1100	1500	900	3500
Reverse (volts)	300	300	300	200	3500
CW (watts)	0.2–0.4	0.3–0.5	0.4–0.6	0.4–0.6	0.4–0.6

Another key advantage of PDB diodes over Schottky diodes is their much higher resistance to damage or failure due to accidental electrostatic discharge (ESD) or high peak power RF pulses. Table 23.4 shows the results of a study comparing ESD and CW burnout of X-band detector diodes (Anand and Malik, 1993). It can be seen that the zero-bias GaAs PDB diode has a much higher ESD threshold voltage as compared with all the Schottky diodes, especially in the reverse bias direction. It is theorized that the single crystal nature of the PDB diode render it much more rugged against ESD damage. By contrast, in the Schottky diodes, the I–V characteristics are intimately related to the stability of the metallurgical interface, which can easily be damaged by high voltage stressing. Planar-doped barrier diodes also exhibit a much higher tolerance to microsecond high peak power RF pulses as compared with Si Schottky diodes, as shown in Fig. 23.28 (Kearney *et al.*, 1990). Also, breakdown in a PDB diode is by thermionic emission in contrast to avalanche breakdown in Schottky diodes which thus limits the peak electric field in the undoped region of the PDB (Anand, 1993). The much greater reliability of PDB diodes under high voltage stressing is another important factor for their use as replacements for Schottky diodes.

23.5.2 Mixer diodes

Heterodyne receivers, also known as mixers, are used to increase the sensitivity considerably for detection of low level RF signals. This technique relies on the non-linearity of the mixer diode to multiply the low level signal with a local oscillator (LO) signal. This produces a beat or difference frequency at some intermediate frequency (IF) which is amplified by additional circuitry in the receiver. The IF is chosen to be above the l/f noise corner frequency of the mixer diode in order to increase the receiver sensitivity.

A figure of merit for the receiver is the conversion loss (Schneider, 1982), which is given by

$$L_c = L_{RF} L_{IF} \tag{23.44}$$

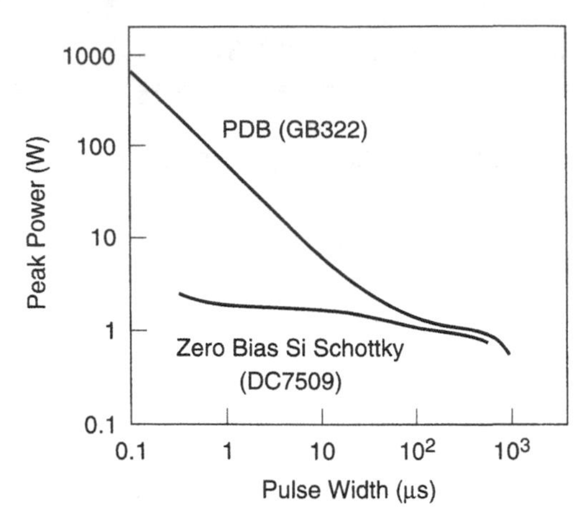

Fig. 23.28. Peak RF pulse power versus pulse width failure threshold comparison for GaAs PDB and Si Schottky detector diodes (after Kearney *et al.*, 1990).

where L_{RF} is the conversion loss at the signal frequency and L_{IF} is the conversion loss at the intermediate frequency, which are determined by the parasitic diode series resistance.

The RF conversion loss L_{RF} is given by

$$L_{RF} = 1 + \frac{R_s}{R_d} + \frac{R_d}{R_s}\frac{f_s^2}{f_c^2} \tag{23.45}$$

where R_s is the diode series resistance, R_d is the average junction resistance (approximately 50 Ω), f_s is the signal frequency, and f_c is the cut-off frequency given in Eqn. (23.40).

The IF conversion loss L_{IF} is given by

$$L_{IF} = 1 + \frac{R_s}{R_o} \tag{23.46}$$

where R_o is the IF output impedance.

The overall noise factor F_o for the receiver is given by

$$F_o = L_{IF}(F_{IF} + \gamma - 1) \tag{23.47}$$

where F_{IF} is the noise factor of the IF amplifier and γ is the diode noise temperature ratio. In order to achieve good performance of the mixer with low conversion loss and low overall noise figure, the mixer diode should have a low series resistance, high cut-off frequency, and low noise temperature.

Another important consideration in the mixer is the required LO power needed to obtain the minimum conversion loss and noise figure. Low barrier PDB mixer diodes require considerably lower LO powers for optimum mixer performance than do equivalent Schottky diodes. A comparison of the overall noise figure as a function of LO power for a variety of mixer diodes is shown in Fig. 23.29 (Kearney *et al.*, 1990). The PDB mixer diode has a minimum noise figure of 6 dB with only 0.28 mW of LO power. By contrast,

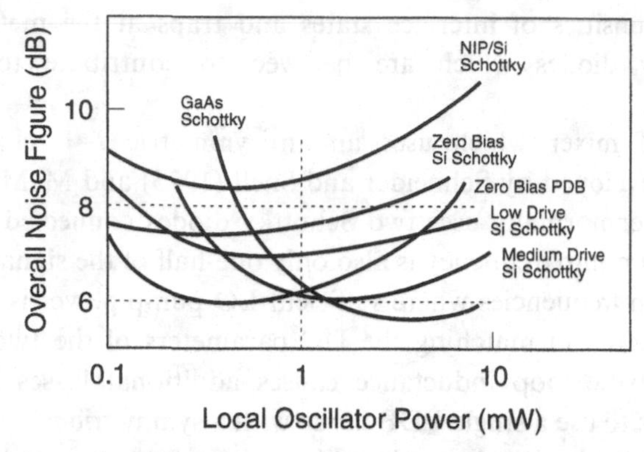

Fig. 23.29. Overall noise figure versus local oscillator power comparison for PDB and Schottky detector diodes: $f = 9.375$ GHz, IF amplifier noise figure = 1.5 dB (after Kearney *et al.*, 1990).

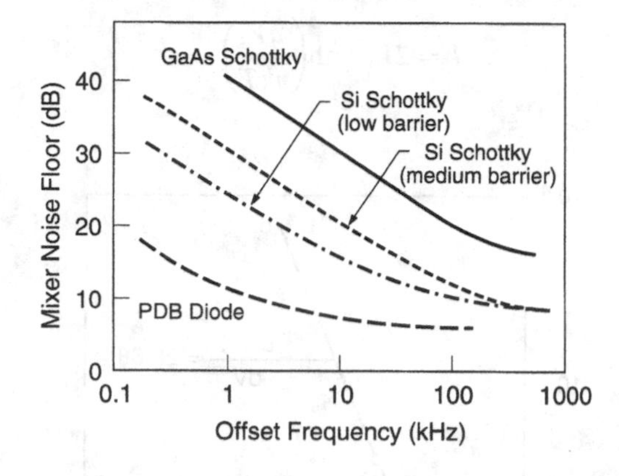

Fig. 23.30. Mixer noise floor versus offset frequency for PDB and Schottky mixer diodes; $f = 9.2$ GHz (after Kearney *et al.*, 1990).

the GaAs Schottky diode requires over 3.0 mW of LO power for a minimum noise figure. The reduced LO power requirement for PDB mixer diodes is a significant advantage over Schottky diodes, especially at higher frequency where it is difficult to obtain high LO power. This simplifies the design of the LO circuit and reduces external LO radiation.

Planar-doped barrier mixer diodes have also been found to have lower l/f noise levels and corner frequencies than do Schottky diodes. Figure 23.30 shows the mixer noise floor as a function of offset frequency for a PDB and several Schottky mixer diodes (Kearney *et al.*, 1990). The corner frequency for the PDB diode is only 10 kHz as compared with 100 kHz for the Si Schottky and 500 kHz for the GaAs Schottky. The measured noise floor for the PDB diode is only 6.6 dB above the -174 dBm/Hz single sideband noise floor at 50 kHz offset frequency. The origin of the lower l/f noise properties of the PDB diode are not completely understood but are thought to be related to the single crystal structure of the diode, which is devoid of defects and traps. By comparison, it is known that there are high densities of interface states and traps at the metal–semiconductor interface in Schottky diodes which are believed to contribute to their $1/f$ noise spectrum.

A different type of mixer which uses an antisymmetric I–V characteristic is the subharmonic mixer developed by Schneider and Snell (1975) and McMaster *et al.* (1976). The subharmonic mixer normally uses two Schottky diodes connected in an antiparallel configuration. The LO pump frequency is also only one-half of the signal frequency, which is an advantage at high frequencies where sufficient LO pump power is difficult to obtain. Problems are encountered in matching the DC parameters of the two Schottky diodes exactly, and the parasitic loop inductance causes additional losses in the mixer. An alternative approach is to use a single PDB diode with a symmetric n^+–i–p^+–i–n^+ doping profile where the p^+ spike is placed exactly in the center of the undoped region. This leads to an antisymmetric I–V characteristic, as shown in Fig. 23.31. The I–V relation is given by

$$I = 2I_0 \sinh\left(\frac{qV}{nkT}\right) \tag{23.48}$$

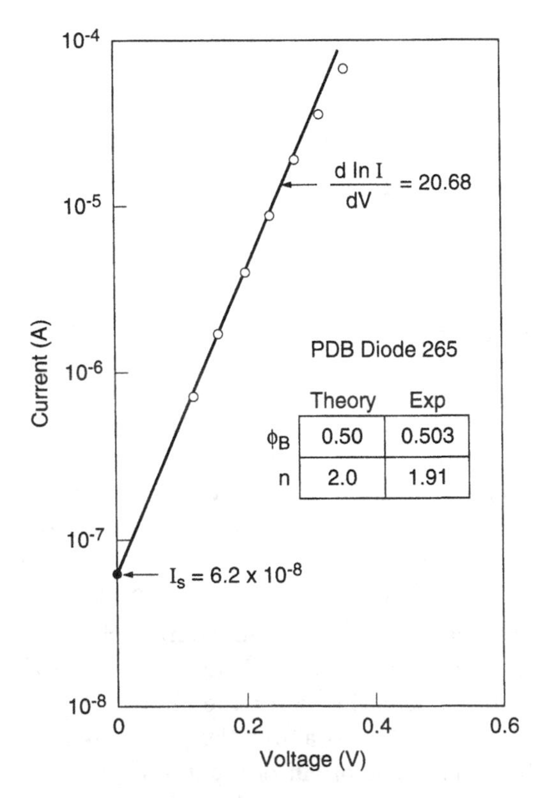

Fig. 23.31. Curve tracer I–V characteristics of an antisymmetric PDB diode used in a subharmonic mixer (after Malik and Dixon, 1982).

where I_0 is the extrapolated saturation current at zero-bias which is given by

$$I_0 = aA^*T^2 \exp\left(\frac{-q\phi_{B0}}{kT}\right) \tag{23.49}$$

The log $I-V$ characteristics of a symmetric PDB diode are shown in Fig. 23.32, where good agreement is obtained for the measured and calculated barrier height and n-factor (Malik and Dixon, 1982).

For a symmetric PDB diode, $n = 2$, which is a consequence of the voltage dividing equally across the barrier region. Normally, $n = 1$ is desired for a mixer to enhance the diode non-linearity. Thus the PDB diode would be expected to give poorer performance in a subharmonic mixer. However, it has been demonstrated experimentally that excellent subharmonic mixers can be made using a single symmetric PDB diode. A conversion loss of 10.8 dB with a LO power of 9 dBm was obtained at 126 GHz using a Si PDB diode with a barrier height of 0.38 eV (Güttich *et al.*, 1991). GaAs PDB subharmonic mixer diodes with 2-μm-diameter junction areas have also been evaluated at 182 GHz (Lee *et al.*, 1993). Figure 23.33 shows the conversion loss as a function of LO power for two different barrier height PDB diodes. A minimum conversion loss of 10 dB is obtained using a LO power of 3 mW. The measured noise temperature for these diodes as a function of LO power is shown in Fig. 23.34. A minimum noise temperature of 2450 K is obtained at a LO power of about 3 mW. The excellent conversion loss performance of these PDB subharmonic mixers is probably related to the excellent matching of the forward and reverse $I-V$ characteristics, simpler circuit implementation, and reduced LO power requirement. Dixon and Malik (1983) have shown that exact matching of the diode forward and reverse $I-V$ characteristics is critical to minimization of the conversion loss. In subharmonic mixers employing two Schottky diodes, current and phase mismatches result in degraded mixer performance.

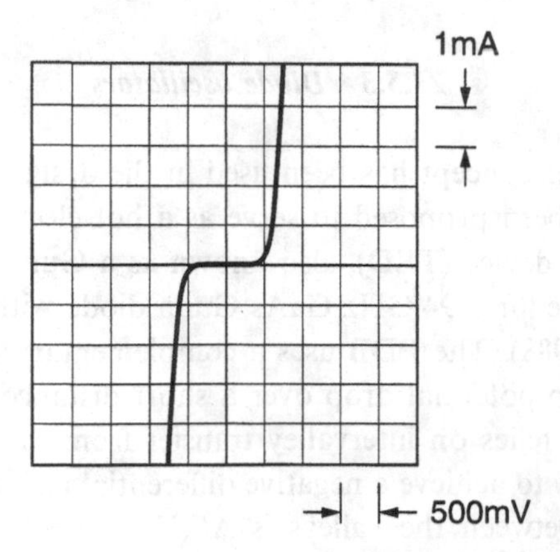

Fig. 23.32. Log $I-V$ characteristics for an antisymmetric PDB diode used in a subharmonic mixer (after Malik and Dixon, 1982).

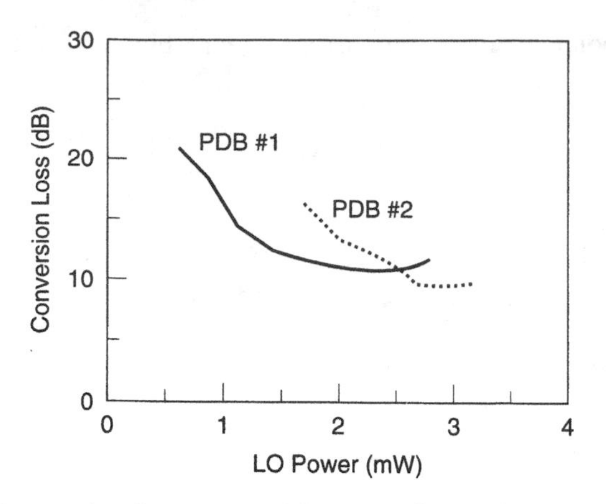

Fig. 23.33. Conversion loss versus LO power for GaAs PDB subharmonic mixers measured at $f = 182$ GHz (after Lee *et al.*, 1993).

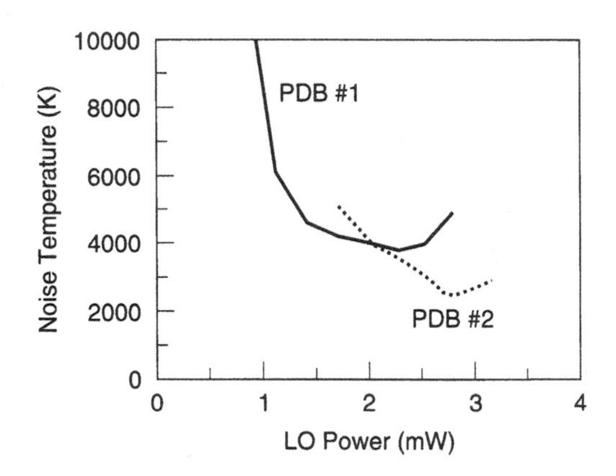

Fig. 23.34. Noise temperature versus LO power for GaAs PDB subharmonic mixers measured at $f = 182$ GHz (after Lee *et al.*, 1993).

23.5.3 Diode oscillators

The planar-doped barrier concept has been used in the design of some microwave diode oscillators. A PDB has been proposed to serve as a hot electron injector in the cathode in a transferred electron device (TED), also known as a Gunn diode (Malik and Iafrate, 1985). The doping profile for a 94 GHz GaAs Gunn diode with a PDB injector is shown in Fig. 23.35 (Ondria, 1988). The PDB uses a complementary pair of p^+ and n^+ doping spikes to provide a steep potential drop over a short distance to heat electrons injected by the cathode. A TED relies on intervalley transfer from the high mobility Γ valley to the low mobility L valley to achieve a negative differential resistance (Sze, 1992). In GaAs, the energy separation between the valleys is $\Delta E(\Gamma - L) = 0.31$ eV. In a normal Gunn diode, with an n^+–n–n^+ doping profile, there is a long acceleration region up to 1 μm which is required to heat the electron distribution sufficiently for electron transfer to the

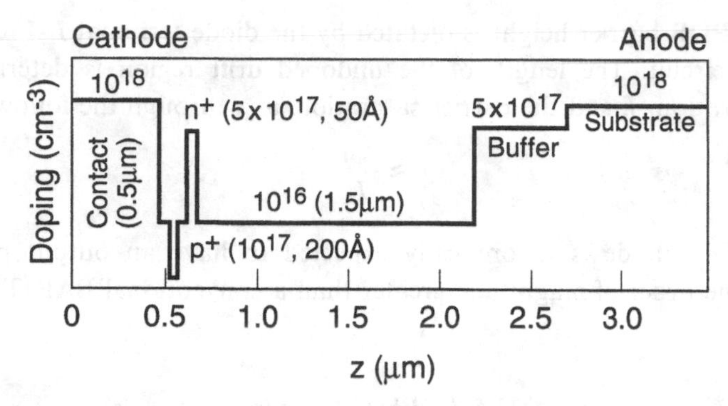

Fig. 23.35. Doping profile for a 94 GHz GaAs Gunn diode with a PDB hot electron injecting cathode (after Ondria, 1988).

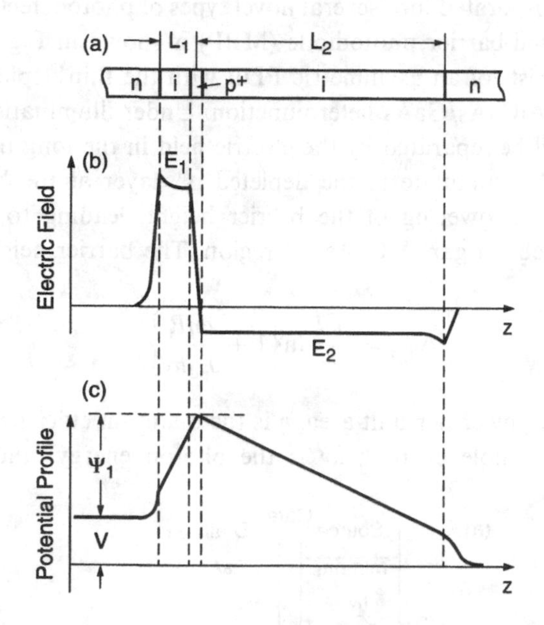

Fig. 23.36. (a) Doping, (b) electric field, and (c) conduction band potential profiles for a PDB BARITT diode (after Luryi and Kazarinov, 1982).

L valley. This parasitic region or 'dead-zone' results in a higher series resistance which degrades the RF/DC conversion efficiency. This parasitic region also contributes to excess noise due to carrier scattering and statistical variation in the intervalley transfer process. The use of a PDB injector in a GaAs Gunn diode has resulted in low noise 94 GHz oscillators with record power levels (Ondria, 1988).

A PDB has also been proposed in the design of a barrier injection transit time (BARITT) diode (Luryi and Kazarinov, 1982). The doping profile, electric field, and potential profile of the PDB used in the BARITT diode are shown in Fig. 23.36. The BARITT has lower power and also lower noise than does an impact avalanche transit time (IMPATT) diode since only electrons contribute to the current in a BARITT by thermionic emission over the barrier. The PDB barrier height is dictated by the diode terminal $I-V$ characteristics in the resonant circuit. The length of the undoped drift region is determined by the frequency of operation, f, and the carrier saturation, v_{sat}, through the following relation:

$$f \approx \frac{v_{sat}}{L} \tag{23.50}$$

The PDB BARITT diode is theoretically expected to have an output power density approximately one order of magnitude greater than a conventional BARITT diode.

23.5.4 Photodetectors

The PDB has been incorporated into several novel types of photodetector diode. The band diagram for a modulated barrier photodiode (MBP) is shown in Fig. 23.37 (Chen *et al.*, 1981a). The MBP consists of an asymmetric PDB with the thin depleted p^+-GaAs layer placed adjacent to an AlGaAs/GaAs heterojunction. Under illumination, photogenerated electron–hole pairs will be separated by the electric field in the long undoped region and some of the holes will accumulate in the depleted p^+ layer at the heterojunction. The accumulated holes cause lowering of the barrier height, leading to enhanced electron injection from the wideband gap AlGaAs-n^+ region. The barrier height lowering can be expressed by

$$\Delta\phi_b = \frac{nkT}{q}\ln\left\{1 + \frac{q\eta P_i}{J_{pd}h\nu}\right\} \tag{23.51}$$

where P_i is the incident power per unit area, η is the quantum efficiency of the absorption process, J_{pd} is the dark hole current, $h\nu$ is the photon energy, and n is the effective

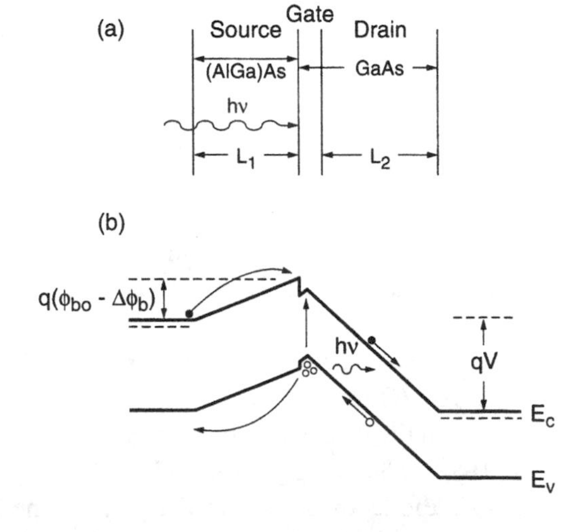

Fig. 23.37. (a) Device structure and (b) band potential profiles for a modulated barrier photodiode (MBP) (after Chen *et al.*, 1981a).

ideality factor ($n \sim 1$ in an asymmetric diode). The sensitivity of the MBP is then given by

$$S = \frac{J_{\mathrm{nd}}}{P_{\mathrm{i}}}\left[\exp\left(\frac{q\Delta\phi_{\mathrm{b}}}{kT}\right) - 1\right] \tag{23.52}$$

where J_{nd} is the dark electron current which depends upon the zero-bias barrier height and the applied reverse bias, V. The effective lifetime of the accumulated holes which affects the optical gain of the MBP is determined by the emission and recombination rates of the holes (Barnard *et al.*, 1982). For an n-type GaAs MBP, an optical gain of 1000 was obtained with 1.5 nW incident power, with associated rise and fall times of 50 ps and 600 ps respectively (Chen *et al.*, 1981a). For a p-type GaAs MBP, an optical gain of 90 was obtained with an incident power level of 10 nW, with associated rise and fall times below 30 ps (Chen, 1981b). The faster response of the p-type MBP is not fully understood but is thought to be related to the faster transit time and re-emission time from the barrier of the minority carrier electrons. The MBP has also been used to make an opto-electronic picosecond sampling system (Bethea *et al.*, 1982).

Another related device which can also be triggered optically is the regenerative feedback switch or triangular barrier switch (TBS), which is illustrated in Fig. 23.38 (Board *et al.*, 1982). The TBS consists of an n-type PDB integrated with a p–n junction placed within a minority carrier diffusion length. The device, which essentially behaves like a thyristor, relies upon barrier height lowering of the PDB by accumulated holes diffusing in from the p–n junction, which in turn causes increased electron emission over the PDB (Habib and Board, 1983). The device will switch from a high impedance 'off' state to a low impedance 'on' state when the closed loop regenerative feedback gain exceeds unity. The electron barrier height in the PDB can be expressed by

$$\phi = \frac{q(N_{\mathrm{A}}Z_{\mathrm{A}} - p\delta)}{\varepsilon}\frac{L_1L_2}{L_1 + L_2} + V\frac{L_1}{L_1 + L_2} \tag{23.53}$$

where p and δ are, respectively, the average density and width of holes which accumulate at the barrier peak. The switching voltage can be adjusted by choosing the barrier height and L_1/L_2 ratio (Najjar *et al.*, 1982) or by using a three-terminal device with a gate contact to the n-region between the PDB and the p–n junction (Rees and Barnard, 1985). Typical switching current densities are of the order of 1 $\mathrm{A/cm^2}$.

Alternatively, the device can be triggered optically through incident illumination of the PDB using the optical gain in the MBP (Chen *et al.*, 1981a). Another similar device structure which can be switched optically is the double heterostructure opto-electronic switch (DOES) device (Simmons and Taylor, 1988). The DOES can also be constructed to give light output from a single quantum well laser when the device is switched to the on state (Claisse *et al.*, 1992).

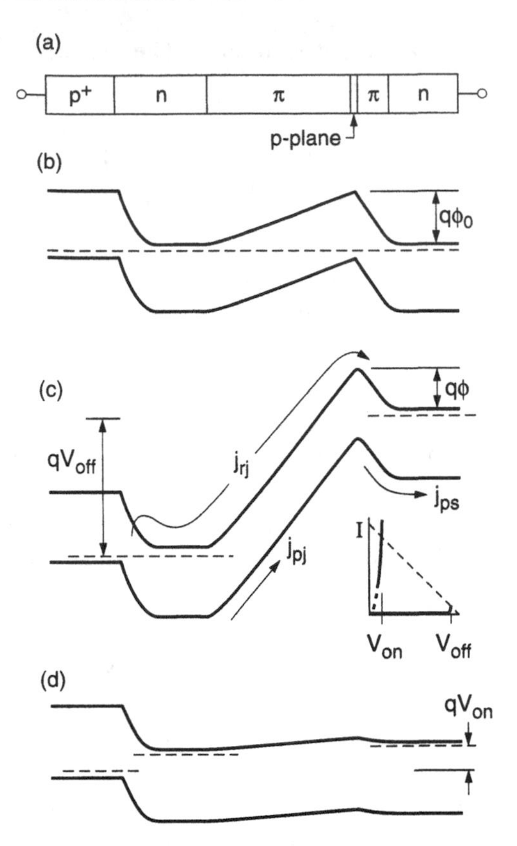

Fig. 23.38. (a) Doping structure, (b) equilibrium band potentials, (c) band potentials under reverse bias 'off' state, and (d) band potentials after switched 'on' state for a regenerative feedback switch (after Board *et al.*, 1982).

23.5.5 Hot electron transistors

After the invention of the PDB, it was soon recognized that it could be used to accelerate electrons to high kinetic energies over very short distances (100–200 Å) to serve as a hot electron injector for ballistic transistors (Eastman, 1981). This concept was incorporated into the design of a planar-doped barrier transistor (PDBT) whose conduction band diagram is illustrated schematically in Fig. 23.39 (Malik *et al.*, 1981). The PDBT consists of two PDBs joined together by a thin n^+ region which serves as the base layer to bias the emitter and collector barriers, as shown in Fig. 23.39(b). Hot electrons injected by the emitter barrier traverse the thin base layer (< 1000 Å) with either no collisions (ballistic) or with just a few collisions (quasi-ballistic). Only electrons with sufficient energy and momentum to overcome the collector barrier after crossing the base can contribute to the collector current. The rest of the injected hot electron distribution will be scattered and will thermalize in the base and contribute to the base current. There have been a number

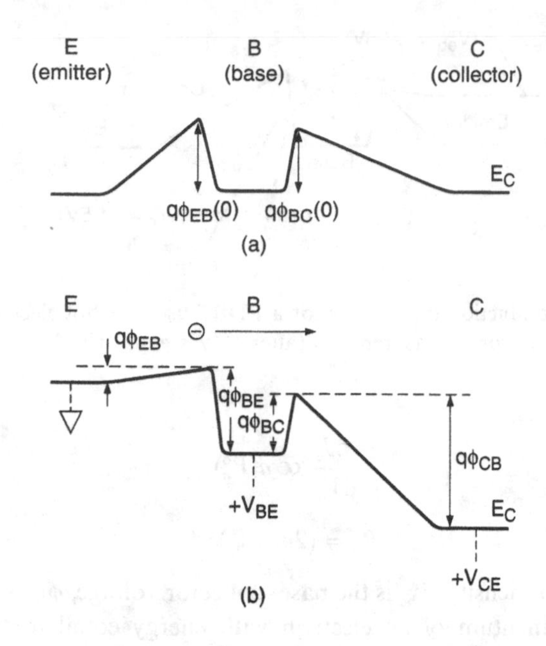

Fig. 23.39. Conduction band edge at (a) equilibrium and (b) under forward bias in a planar-doped barrier transistor (PDBT) (after Malik *et al.*, 1981).

of theoretical studies on electron transport in the PDBT, including Monte Carlo simulations (Wang *et al.*, 1985; Lugli and Ferry, 1985; Littlejohn *et al.*, 1983). The injection energies of the hot electrons from the emitter must be below the Γ–L intervalley separation of 0.31 eV in GaAs to prevent intervalley scattering. Thus the low emitter and collector barrier heights of $\phi < 0.3$ eV require low temperature $T \leqslant 77$ K operation of the PDBT. This is necessary due to the very high thermal emission rates of electrons from the contact regions over the barriers at room temperature. Measurements at 4.2 K on the PDBT demonstrated a base transport factor of $\alpha = 0.1$–0.5 for a base width of 1000 Å and n-type doping level of 1×10^{18} cm^{-3}. The poor transport of hot electrons through the base region even at very low temperatures was explained by coupled phonon–plasmon scattering modes in which the hot carriers lose energy to the Fermi sea of electrons in the base (Hollis *et al.*, 1983).

Although the preceding performance limitations of the PDBT preclude any practical use, it has been a useful device with which to study electron transport in semiconductors. The conduction band diagram of a specially designed PDBT, as shown in Fig. 23.40, was used to perform hot electron spectroscopy (Hayes *et al.*, 1984, 1985). A hot electron distribution is injected by the emitter with an energy approximately equal to the emitter barrier height. The injected hot electrons will interact with electrons at the Fermi level in the base, resulting in a scattered distribution of hot electrons incident upon the collector barrier. The height of the collector barrier can be varied by the collector bias voltage to act as an energy analyzer for the hot electron distribution, which can be described by the

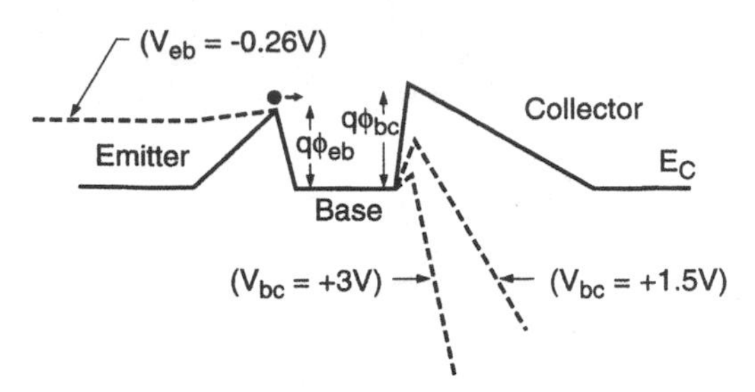

Fig. 23.40. Conduction band edges for a PDBT used for hot electron spectroscopy measurements (after Hayes *et al.*, 1984).

following relations:

$$\frac{dJ_c}{dV_{bc}} \propto n(P_n^o) \tag{23.54}$$

$$P_n^o = (2m_e\phi_{bc})^{1/2} \tag{23.55}$$

where J_c is the collector density, V_{bc} is the base–collector voltage, ϕ_{bc} is the collector barrier height, P_n^o is the momentum of an electron with energy equal to the collector barrier height, and $n(P_n^o)$ is the density of an electron with momentum P_n^o. Figure 23.41(a) and (b) show the electron spectra for a PDBT measured at 4.2 K with respective base widths of 1200 Å and 1700 Å. It can be seen that the 1200 Å base exhibits a pronounced peak which is further removed from the Fermi level energy as compared with the 1700 Å base. This indicates that there is less scattering of hot electrons in the PDBT with the smaller base width.

23.5.6 Hot electron launchers

The linear potential change over short distances made possible with the PDB has been used in some other applications for hot electron launchers. Complementary pairs of donor and acceptor charge sheets have been used to effectively alter the barrier height for carriers across a heterojunction interface. The doping interface dipole (DID) concept is depicted in Fig. 23.42 (Capasso *et al.*, 1985). The band line up for an intrinsic heterojunction is shown in Fig. 23.42(a). The addition of a complementary dipole pair across the heterojunction (Fig. 23.42(b)), results in a linear potential drop between the charge sheets. Assuming that the thickness of the charge sheets is much less than the distance d separating them, the potential drop $\Delta\phi$ is given approximately by

$$\Delta\phi = \frac{\sigma}{\varepsilon} d \tag{23.56}$$

where σ is the two-dimensional sheet charge density. Figure 23.42(c) shows how the DID can effectively lower the heterojunction discontinuity where the potential drop occurs within a tunneling distance. This effect has been used to enhance carrier collection efficiency

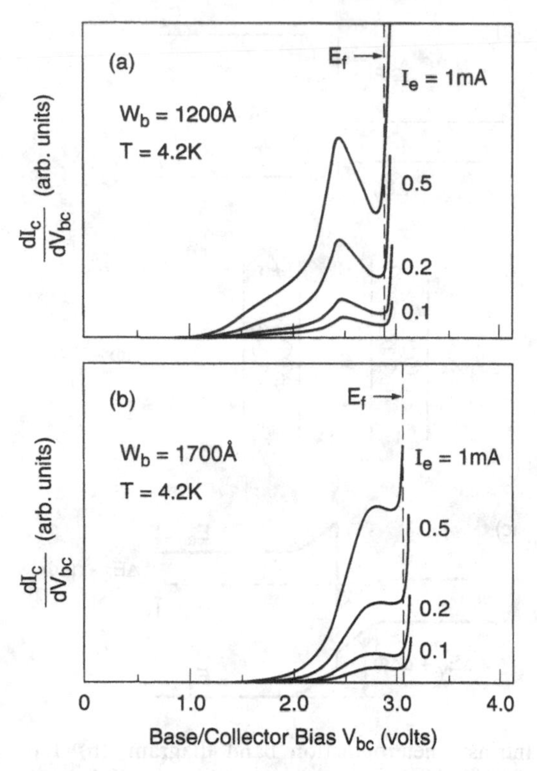

Fig. 23.41. Hot electron spectra for a PDBT measured at 4.2 K with (a) 1200 Å and (b) 1700 Å base-width (after Hayes *et al.*, 1985).

in pin photodiodes. Conversely, the polarity of the charge sheets may be reversed to raise the heterojunction barrier height effectively.

Another novel application for PDB hot electron launchers is in GaAs PDB vacuum microelectronic electron emitters (Jiang *et al.*, 1993). The band diagram and physical structure for this device are illustrated in Figs. 23.43 and 23.44 respectively. The PDB is used to accelerate electrons to high kinetic energies (2–6 eV) in order to overcome the surface work function of the semiconductor for emission into vacuum. Cesium is applied to the semiconductor surface to lower the work function. Current densities of approximately 100 A cm^{-2} and injection efficiencies of several percent have been obtained in preliminary devices. These cold cathode emitters may eventually find applications in displays and in microwave amplifier tubes.

23.5.7 Field effect transistors

A PDB hot electron launcher has also been employed in a vertical field effect transistor (VFET). A schematic cross-section of the VFET is shown in Fig. 23.45 (Won *et al.*, 1990). The VFET consists of a source region on top of which a PDB is used to launch electrons ballistically into the low doped channel region. The channel conductance is modulated

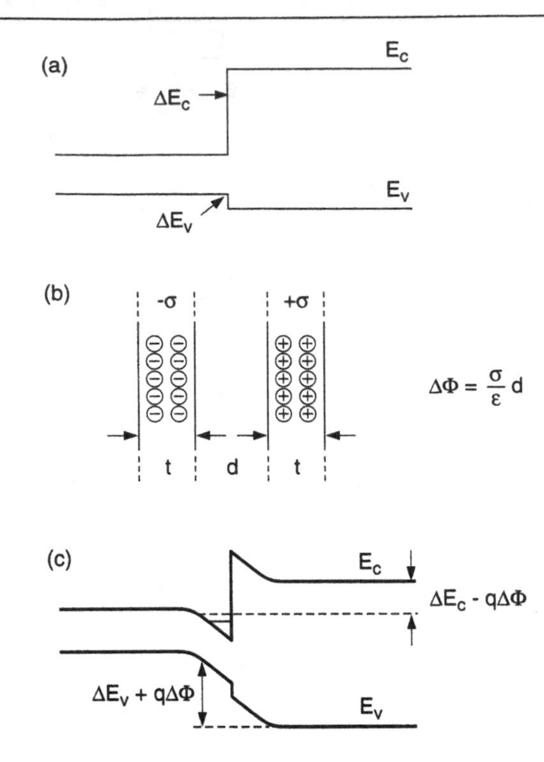

Fig. 23.42. (a) Intrinsic heterojunction band diagram, (b) doping interface dipole structure and (c) heterojunction band diagram including a doping interface dipole.

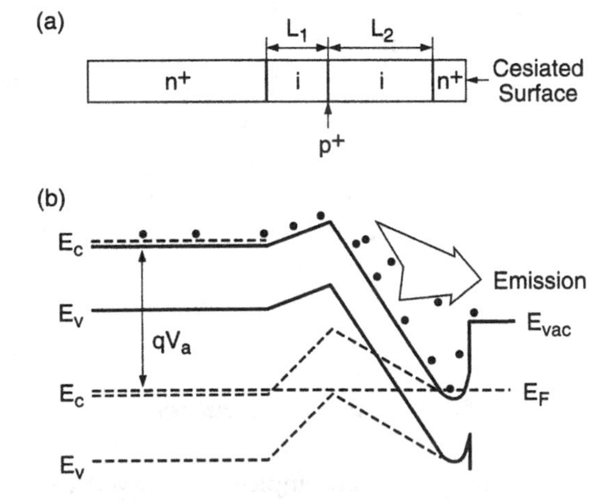

Fig. 23.43. (a) Doping profile and (b) band potential profiles for a GaAs PDB vacuum microelectronic electron emitter (after Jiang *et al.*, 1993).

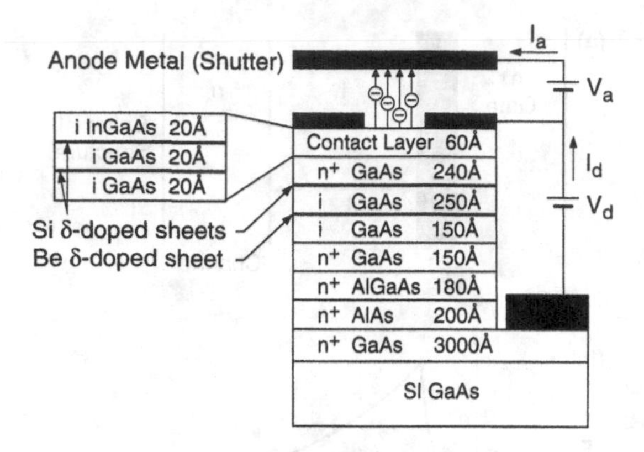

Fig. 23.44. Epitaxial structure and bias conditions for a GaAs PDB vacuum microelectronic electron emitter (after Jiang *et al.*, 1993).

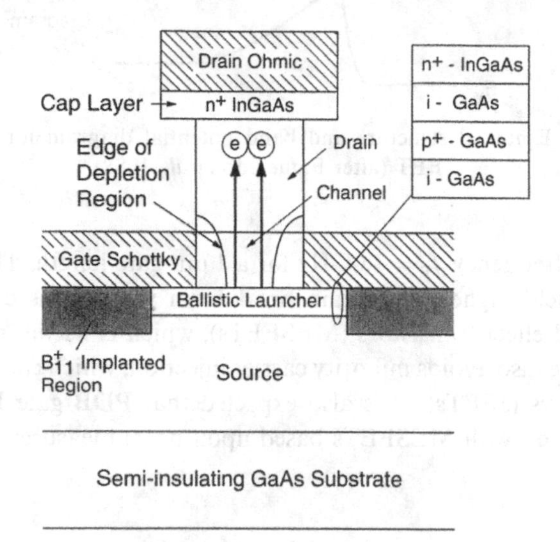

Fig. 23.45. Schematic cross-section of a vertical field effect transistor (VFET) with an integral PDB ballistic electron launcher (after Won *et al.*, 1990).

by Schottky gates deposited on the sidewalls of the channel and a top-side drain contact is used to collect the electrons. Channel conductances of 400 and 177 mS/mm were measured at 77 and 300 K, respectively, for a 0.1 μm channel length. Hot electron spectroscopy performed at 77 K in specially constructed VFETs with energy analyzer barriers on the drain side confirmed that high energy injection by the PDB results in non-equilibrium transport with a hot electron distribution in the channel which reduces the electron transit time (Daniels-Race *et al.*, 1992).

Another proposed application for the PDB was to replace a conventional metal Schottky gate in a field effect transistor (Malik and AuCoin, 1984). The structure and band diagram for a PDB gate FET are given in Figure 23.46 (Figueredo *et al.*, 1988). An asymmetric PDB with a barrier height of $\phi_B \approx 1.2$ V was used with an ohmic contact to the gate. The

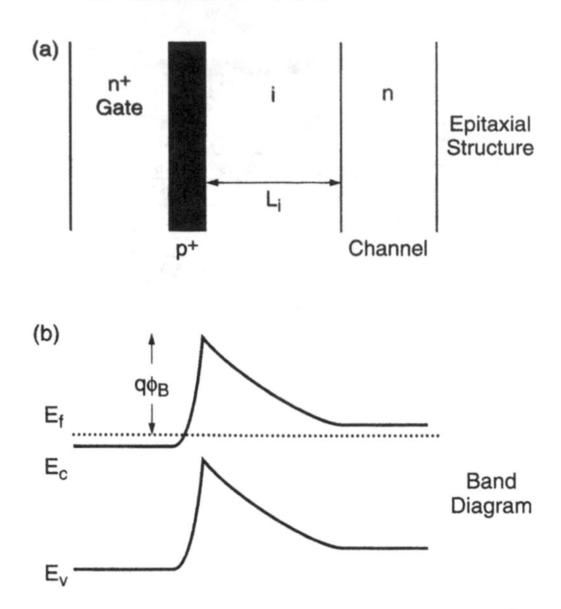

Fig. 23.46. Epitaxial structure and band potential diagram for a PDB gate FET (after Figueredo *et al.*, 1988).

FET had a cut-off frequency $f_T = 18$ GHz for a 1 μm gate length. The high barrier height of the PDB will yield higher gate–drain breakdown voltages as compared with metal–semiconductor field effect transistors (MESFETs), which is useful in high power applications. The PDB gate also avoids minority carrier injection, which can occur in p–n junction field effect transistors (JFETs). It is also expected that PDB gate FETs will have lower l/f noise as compared with MESFETs based upon noise measurements of PDB detector diodes.

23.5.8 Potential barrier transistors

The PDB has also been incorporated in the design of three-terminal transistors in which a thermionic emission current is controlled by modulating the barrier height of the PDB. The doping profile and band diagram for a PDB GaAs V-groove barrier transistor are shown in Fig. 23.47 (Chang *et al.*, 1985). The device relies on hole injection from a metal contact deposited on the mesa sidewall of the PDB (hence the V-groove designation) to accumulate in the $\delta(p^+)$ layer, thus lowering the barrier for electron injection. The V-groove transistor operates in an analogous manner to the modulated barrier photodiode (Section 23.5.4), where in the latter case the hole injection originates from the photon absorption process. The V-groove transistor is also very similar to a transistor structure formed in Si by an ion implantation process which is known as the bulk barrier transistor (Mader *et al.*, 1983).

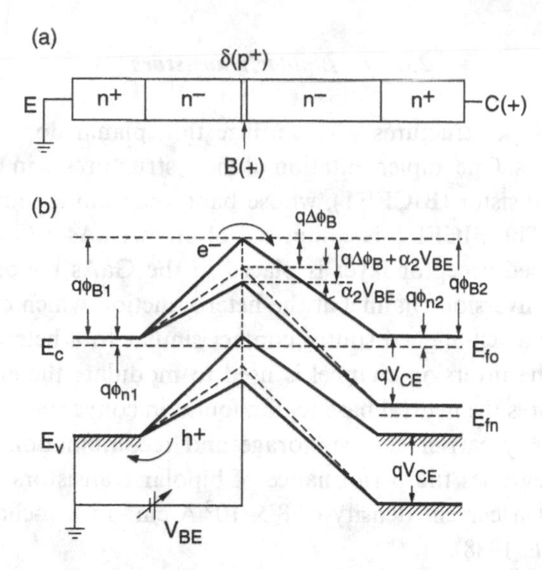

Fig. 23.47. (a) Doping structure and (b) band potential diagrams for a GaAs PDB V-groove transistor (after Chang *et al.*, 1985).

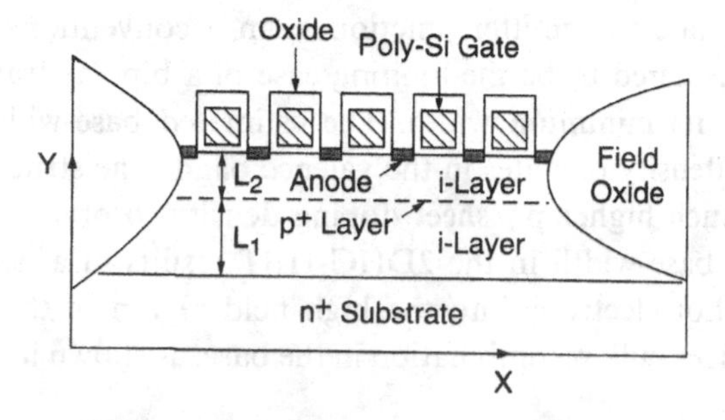

Fig. 23.48. Schematic cross-section for a Si PDB thermionic emission transistor (after Kazarinov and Luryi, 1982).

A different approach to modulation of the PDB barrier height in a thermionic transistor structure is through a gate control potential. A schematic cross-section of a proposed voltage-controlled thermionic emission transistor (TET) is shown in Fig. 23.48 (Kazarinov and Luryi, 1982). The transistor uses a MOS poly-Si gate formed on top of a Si PDB. Modulation of the gate voltage results in a change in the barrier height of the PDB. The potential of the PDB has a saddle-shaped potential minimum directly below the gate when the transistor is turned on. This results in vertical injection of current from the cathode towards the gate which is collected by the anode contacts adjacent to the gates. Some modified TET structures have been predicted to operate at high switching speeds (5 ps) using low supply voltages (0.5 V) (Luryi and Kazarinov, 1982).

23.5.9 *Bipolar transistors*

There are also PDB-type structures which utilize thin planar-doped p^+ layers to form novel bipolar transistors. One implementation of these structures is in the bipolar inversion channel field effect transistor (BICFET), whose band diagram is illustrated in Fig. 23.49 (Taylor *et al.*, 1986). The BICFET is comprised of an AlGaAs-n/GaAs-n heterojunction in which a planar-doped acceptor layer is placed in the GaAs region near the interface. This results in a hole inversion channel at the heterojunction which can be modulated by high field injection by a self-aligned source contact similar to a heterojunction field effect transistor (HFET). The inversion channel is used to modulate the emitter barrier height and effectively eliminates the neutral base region found in conventional bipolar transistors. This eliminates minority carrier charge storage and recombination associated with the neutral base which degrades the performance of bipolar transistors. A current gain of 8 has been measured at a current density of 8×10^3 A/cm^2 in a p-channel AlGaAs/GaAs BICFET (Taylor *et al.*, 1988).

Another type of bipolar transistor can be formed by using a very thin, heavily doped p^+ layer placed at an n-type heterojunction interface. The band diagram of a two-dimensional hole gas heterojunction bipolar transistor (2DHG-HBT) is shown in Fig. 23.50 (Malik *et al.*, 1988; Kuo *et al.*, 1990). The base consists of a high conductivity hole gas which is used to modulate the emitter junction as in a conventional bipolar transistor. This device can be considered to be the limiting case of a bipolar transistor in which the base is compressed to its minimum width. The estimated base-width is approximately 100 Å due to the high density of states in the valence band. The structure is similar to the BICFET but uses a much higher p^+ sheet doping density to obtain a high conductivity 2DHG base. The thin base-width in the 2DHG-HBT results in a negligible base transit time and injection of hot electrons into the high field region of the collector. The thin base-width also minimizes bulk recombination in the base, as shown in the common emitter

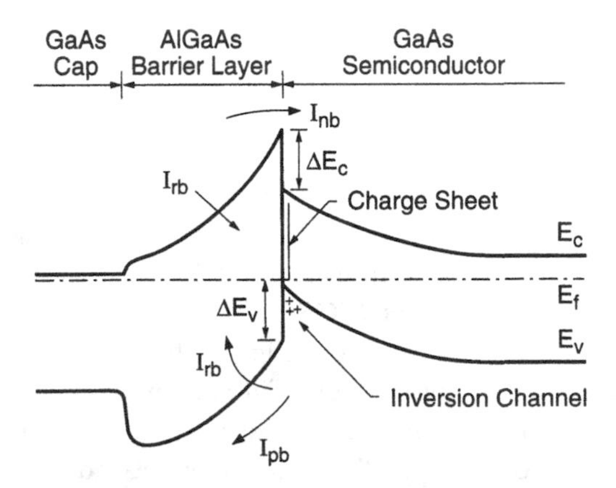

Fig. 23.49. Band potential diagram for a bipolar inversion channel field effect transistor (BICFET) (after Taylor *et al.*, 1986).

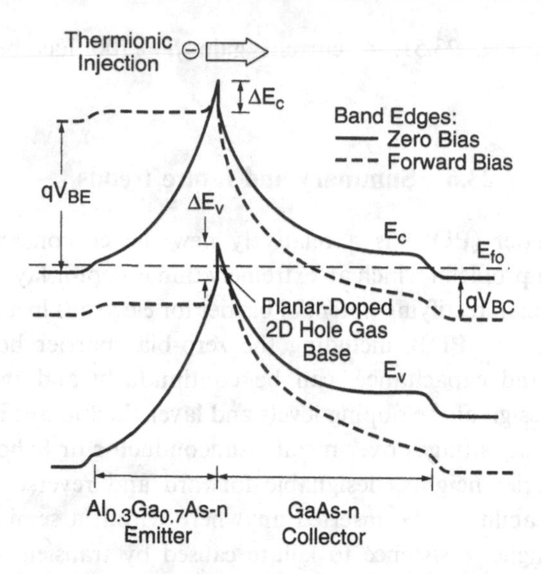

Fig. 23.50. Band potential diagram for a planar-doped two-dimensional hole gas heterojunction bipolar transistor (2DHG-HBT) (after Malik *et al.*, 1988).

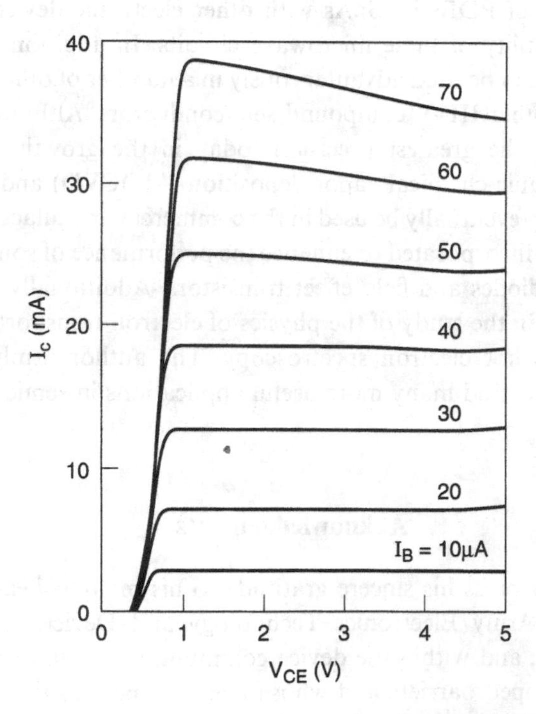

Fig. 23.51. Common emitter I–V characteristics of a 2DHG-HBT (after Malik *et al.*, 1988).

I–V characteristics in Fig. 23.51. A current gain $\beta = 700$ has been measured in a 2DHG-HBT.

23.6 Summary and future trends

The planar-doped barrier (PDB) is a relatively new device concept comprised of an n^+–i–p^+–i–n^+ doping profile in which an extremely thin acceptor layer (in the limit $\delta(p^+)$) is used to form a triangular, rectifying potential barrier for electrons in a bulk semiconductor. The key parameters of the PDB, including the zero-bias barrier height, asymmetry in current rectification, and capacitance, can be continuously and independently varied through appropriate design of the doping levels and layer thicknesses in the structure. The PDB has a number of advantages over metal–semiconductor or Schottky barriers which include adjustable barrier heights, designable forward and reverse I–V characteristics, fixed capacitance, the ability to be inserted anywhere within a semiconductor, very low l/f noise, and much higher resistance to failure caused by transient high voltage spikes.

The PDB concept has been demonstrated in a number of useful and novel device applications. The principal commercial use of the PDB today and in the future is in high performance PDB mixer and detector diodes as replacements for conventional Schottky diodes. The utilization of PDB diodes is expected to grow dramatically as the markets for new wireless communications products in the microwave frequency range expand rapidly. Monolithic integration of PDBs in GaAs with other electronic devices will enhance the functionality and versatility of these microwave circuits. In addition to GaAs, it is also expected that the PDB can be used advantageously in a number of other materials systems, including Si, Ge, and other III–V compound semiconductors. Although molecular beam epitaxy (MBE) affords the greatest precision today in the growth of PDB structures, advances in metal–organic chemical vapor deposition (MOCVD) and the related atomic layer epitaxy (ALE) may eventually be used in the commercial manufacture of PDB devices. The PDB has also been incorporated to enhance the performance of some other microwave devices such as Gunn diodes and field effect transistors. Additionally, the PDB has been used as a research tool in the study of the physics of electron transport in semiconductors via techniques such as hot electron spectroscopy. The author firmly believes that the planar-doped barrier will find many more useful applications in semiconductor devices in the future.

Acknowledgements

The author wishes to express his sincere gratitude to his many colleagues at AT&T Bell Laboratories, the US Army Electronics Technology and Devices Laboratory, Cornell University, M/A-COM, and within the device community, who have made contributions in the field of planar-doped barriers and whose names appear in the reference list of this chapter. The expert technical assistance of J. Preckwinkle and D. Sowinska-Khan in the preparation of this chapter is also gratefully acknowledged.

References

Allyn, C. L., A. C. Gossard, and W. Wiegmann (1980) *Appl. Phys. Lett.*, **36**, 373.
Anand, Y. and R. J. Malik (1992) *Proc. GaAs Reliability Workshop*, paper IV–4, IEEE, New York.
Anand, Y. and R. J. Malik (1993) *Proc. 15th Annual EOS/ESD Symp.*, p. 103, IEEE, New York.
Anand, Y. (1993) private communication.
Anand, Y., J. Hillson, A. Torabi, and J. East (1991) *Proc. 2nd Intl. Conf. on Space Terahertz Technology*, p. 467, IEEE, New York.
Arnold, D. and K. Hess (1987) *J. Appl. Phys.*, **61**, 5178.
Bardeen, J. (1947) *Phys. Rev.*, **71**, 717.
Barnard, J. A., F. E. Najjar, and L. F. Eastman (1982) *IEEE Trans. Electron Dev.*, **ED-29**, 1396.
Barnes, P. A. and A. Y. Cho (1979) *J. Appl. Phys.*, **49**, 91.
Battersby, S. J. and J. J. Harris (1987) *IEEE Trans. Electron Dev.*, **ED-34**, 1046.
Bethe, H. A. (1942) *MIT Rad. Lab. Rep.*, **43-12**, 1.
Bethea, C. G., C. Y. Chen, A. Y. Cho, and P. A. Garbinski (1982) *Appl. Phys. Lett.*, **40**, 591.
Board, K., K. Singer, R. Malik, C. E. C. Wood, and L. F. Eastman (1982) *Electron. Lett.*, **18**, 676.
Board, K. and M. Darwish (1982) *Solid State Electron.*, **25**, 571.
Bozler, C. O., C. D. Alley, R. A. Murphy, D. C. Flander, and W. T. Lindley (1979) *Proc. 7th Biennial Cornell Conf. on Active Microwave Semiconductor Devices*, IEEE, New York.
Braun, F. (1874) *Ann. Physik Chem.*, **153**, 556.
Brown, A. S., S. C., Palmateer, G. W. Wicks, L. F. Eastman, and A. R. Calawa (1984) *Proc. 5th Molecular Beam Epitaxy Workshop, J. Vac. Sci. Tech.* B **2**, 194.
Buot, F. A., J. A. Krumhansl, and J. B. Socha (1982) *Appl. Phys. Lett.*, **40**, 814.
Capasso, F. and A. Y. Cho (1994), *Surf. Science*, **300**, 878.
Capasso, F., A. Y. Cho, K. Mohammed, and P. W. Foy (1985) *Appl. Phys. Lett.*, **46**, 664.
Carlson, E. R., M. V. Schneider, and T. F. McMaster (1978) *IEEE Trans. Microwave Theory Tech.*, **MTT-26**, 706.
Chandra, A. and L. F. Eastman (1982) *J. Appl. Phys.*, **53**, 9165.
Chang, C. Y., Y. H. Wang, W. C. Liu, and S. A. Liao (1985) *IEEE Electron Dev. Lett.*, **EDL-6**, 123.
Chen, C. Y., A. Y. Cho, P. A. Garbinski, C. G. Bethea, and B. F. Levine (1981a) *Appl. Phys. Lett.*, **39**, 340.
Chen, C. Y. (1981b) *Appl. Phys. Lett.*, **39**, 979.
Chen, C. Y., A. Y. Cho, P. A. Garbinski, and C. G. Bethea (1981c) *IEEE Electron Dev. Lett.*, **EDL-2**, 290.
Cho, A. Y. (1983) *Thin Solid Films*, **100**, 291.
Cho, A. Y. and J. R. Arthur (1975) *Progress in Solid State Chemistry*, **10**, 163.
Cho, A. Y. and P. D. Dernier (1978) *J. Appl. Phys.*, **49**, 3328.
Claisse, P. R., G. W. Taylor, D. P. Docter, and P. W. Cooke (1992) *IEEE Trans. Electron Dev.*, **39**, 2523.
Cook, R. K. (1983) *Appl. Phys. Lett.*, **42**, 439.
Couch, N. R. and M. J. Kearney (1989) *J. Appl. Phys.*, **66**, 5083.
Dale, I., S. Neylon, A. Condie, and M. J. Kearney (1989) *Proc. 19th European Microwave Conf*, **1**, 237, Microwave Exhibition and Publishers, Tunbridge Wells, UK.
Dale, I., S. Neylon, A. Condie, M. Hobden, and M. J. Kearney (1990) *Proc. 20th European Microwave Conf.*, **1**, 605, Microwave Exhibitions and Publishers, Tunbridge Wells, UK.
Daniels-Race, T., K. Yamasaki, W. J. Schaff, P. J. Tasker, and L. F. Eastman (1992) *J. Appl. Phys.*, **72**, 5735.
Dixon, S. Jr and R. J. Malik (1982) *IEEE Trans. Microwave Theory Tech.*, **MTT-31**, 155.
Dixon, S. Jr and R. J. Malik (1983) *IEEE Trans. Microwave Theory Tech.*, **MAT-31**, 155.

Dixon, S., T. R. AuCoin, R. L. Ross, and L. T. Yuan (1984) *Proc. 8th Intl. Conf. on Infrared and Millimeter Waves Conf. Dig.*, p. T6.9/1-2, ed. R. J. Temkin, IEEE, New York.
Eastman, L. F. (1981) *Proc. 3rd Intl. Conf. on Hot Carriers in Semiconductors, J. de Phys. Colloque*, **42**, 263.
Eastman, L. F. (1982) *Proc. 9th Intl. Symp. on GaAs and Related Compounds*, p. 245, ed. T. Sugano, Inst. of Phys., Bristol, UK.
Escobosa, A., H. Kräuthe, and H. Beneking (1982) *Electron. Lett.*, **18**, 888.
Figueredo, D. A., M. P. Zurakowski, S. S. Elliott, W. J. Anklam, and S. R. Sloan (1988) *Appl. Phys. Lett.*, **52**, 1395.
Fischer, M. C., M. J. Schoessow, and P. Teng (1992) *Hewlett-Packard J.*, **4**, 90.
Foxon, C. T. and B. A. Joyce (1975) *Surface Sci.*, **50**, 434.
Fraser, A. A. (1987) *Microwave J.*, **5**, 309.
Gossard, A. C., W. Brown, C. L. Allyn, and W. Wiegmann (1982a) *J. Vac. Sci. Tech.* B **20**, 694.
Gossard, A. C., R. F. Kazarinov, S. Luryi, and W. Wiegmann (1982b) *Appl. Phys. Lett.*, **40**, 832.
Gunn, J. B. (1963) *Solid State Comm.*, **1**, 88.
Güttich, U., K. M. Strohm, and F. Schäffer (1991) *IEEE Trans. Microwave Theory Tech.*, **39**, 366.
Habib, S. E.-D. and K. Board (1983) *IEEE Trans. Electron Dev.*, **ED-30**, 90.
Harris, J. S. and J. M. Woodcock (1980) *Electron. Lett.*, **16**, 319.
Hayes, J. R., A. F. J. Levi, and W. Wiegmann (1984) *Electron. Lett.*, **20**, 851.
Hayes, J. R., A. F. J. Levi, and W. Wiegmann (1985) *Phys. Rev. Lett.*, **54**, 1570.
Hollis, M. A., S. C. Palmateer, L. F. Eastman, N. V. Dandekar, and P. M. Smith (1983) *IEEE Electron Dev. Lett.*, **EDL-4**, 440.
Hooper, W. W. and W. I. Lehrer (1967) *Proc. IEEE*, **55**, 1237.
Jiang, W. N., D. J. Holcombe, M. M. Hashemi, and U. K. Mishra (1993) *IEEE Electron Dev. Lett.*, **14**, 143.
Kaushik, S. B., R. K. Purobit, and B. L. Sharma (1983) *Infrared Physics*, **23**, 15.
Kazarinov, R. F. and S. Luryi (1981) *Appl. Phys. Lett.*, **38**, 810.
Kazarinov, R. F. and S. Luryi (1982) *Appl. Phys.* A **28**, 151.
Kearney, M. J. and I. Dale (1990) *GEC J. Res.*, **8**, 1.
Kearney, M. J., M. J. Kelly, A. Condie, and I. Dale (1990) *Electron. Lett.*, **26**, 671.
Kearney, M. J., M. J. Kelly, R. A. Davies, T. M. Kerr, P. K. Rees, A. Condie, and I. Dale (1989) *Electron. Lett.*, **25**, 1454.
Kearney, M. J., A. Condie, and I. Dale (1991) *Electron Lett.*, **27**, 721.
Knudsen, M. (1909) *Ann. der Physik*, **28**, 999.
Kohn, E. (1979) *Proc. IEDM*, 469, IEEE, New York.
Kuo, T. Y., K. W. Goossen, J. E. Cunningham, A. Ourmazd, C. G. Fonstad, R. Len and W. Jan (1990) *Electron. Lett.*, **26**, 1187.
Lee, T. H., J. R. East, and G. I. Haddad (1991) *Microwave and Optical Tech. Lett.*, **4**, 53.
Lee, T. H., J. R. East, C-Y Chi, R. Dengles, I. Medhi, P. Siegel, and G. I. Haddad (1993) *Proc. 4th Intl. Conf. on Space Terahertz Technology*, IEEE, New York.
Littlejohn, M. A., R. J. Trew, J. R. Hauser, and J. M. Golio (1983a) *Proc. Intl. Conf. on Metastable and Modulated Semiconductor Structures, J. Vac. Sci. Tech.* B **1**, 449.
Littlejohn, M. P., R. J. Treug, J. R. Hauser and J. M. Golio (1983b) *J. Vac. Sci. Tech.*, **1**, 449.
Lugli, P. and D. K. Ferry (1985) *IEEE Electron Device Lett.*, **EDL-26**, 25.
Luryi, S. and R. F. Kazarinov (1982) *Solid-State Electron.*, **25**, 933.
Mader, H. (1982) *IEEE Trans. Electron. Dev.*, **ED-29**, 1766.
Mader, H., R. Muller, and W. Beinvogl (1983) *IEEE Trans. Electron Dev.*, **ED-30**, 1380.
Malik, R. J. (1982) *Mat. Lett.*, **1**, 22.
Malik, R. J. (1983) Us Patent no. 4, 410,902.
Malik, R. J. and T. R. AuCoin (1985) US Patent no. 4, 442,445.
Malik, R. J. and S. Dixon (1982) *IEEE Electron Dev. Lett.*, **EDL-3**, 205.

Malik, R. J., K. Board, L. F. Eastman, C. E. C. Wood, T. R. AuCoin, and R. L. Ross (1980) *Electron. Lett.*, **16**, 836.
Malik, R. J., K. Board, L. F. Eastman, C. E. C. Wood, T. R. AuCoin, and R. L. Ross (1981) *Proc. 8th Intl. Symp. on GaAs and Related Compounds*, ed. H. W. Thim, Inst. of Phys., p. 697, Bristol UK.
Malik, R. J. and G. J. Iafrate (1985) US Patent no. 4, 539,581.
Malik, R. J., R. N. Nottenberg, E. F. Schubert, J. F. Walker, and R. W. Ryan (1988a) *Appl. Phys. Lett.*, **53**, 2661.
Malik, R. J., L. M. Lunardi, J. F. Walker, and R. W. Ryan (1988b) *IEEE Electron Dev. Lett.*, **9**, 7.
Malik, R. J., Y. Anand, M. Micovic, M. Geva, and R. W. Ryan (1993a) *Proc. 14th Biennial Conf. Advanced Concepts in High Speed Semiconductor Devices and Circuits*, Cornell University.
Malik, R. J., J. Nagle, M. Micovic, R. W. Ryan, T. Harris, M. Geva, L. C. Hopkins, J. Vandenberg, R. Hull, R. F. Kopf, Y. Anand, and W. D. Braddock (1993b) *J. Crystal Growth*, **127**, 686.
Manasevit, H. M. (1981) *J. Crystal Growth*, **55**, 1.
McMaster, T. F., M. V. Schneider and W. W. Snell Jnr (1976) *IEEE Trans. Microwave Theory Tech.*, **MMT-24**, 948.
Mead, C. A. (1960) *Proc. IRE*, **48**, 359.
Mead, C. A. and W. G. Spitzer (1964) *Phys. Rev.*, **134**, A717.
Metze, G. M. (1981) Ph.D dissertation, Cornell University.
Miller, D. L. and P. M. Asbeck (1985) *J. Appl. Phys.*, **57**, 1816.
Miller, J. N., D. M. Collins and N. J. Mall (1985) *Appl. Phys. Lett.*, **46**, 960.
Morgan, D. V. and J. Frey (1977) *Solid State Electron.*, **22**, 865.
Najjar, F. E., J. A. Barnard, S. C. Palmateer, and L. F. Eastman (1982) *Proc. IEEE Intl. Electron Dev. Meet.*, p. 177, IEEE, New York.
Ondria, J. (1988) *IEEE Colloquium on Solid State Components for Radar*, p. 1, IEEE, New York.
Palmateer, S. C., P. A. Maki, M. A. Hollis, L. F. Eastman, and I. Ward (1983) *Proc. 10th Intl. Symp. on GaAs and Related Compounds*, Inst. of Phys., p. 149.
Popovic, R. S. (1978) *Solid State Electron.*, **21**, 1133.
Rees, P. K. and J. A. Barnard (1985) *IEEE Trans. Electron Dev.*, **ED-32**, 1741.
Rideout, V. L. (1978) *Thin Solid Films*, **48**, 261.
Scherer, D. (1990) *Microwaves & RF*, **2**, 113.
Schneider, M. V. (1966) *Bell Sys. Tech. J.*, **45**, 1611.
Schneider, M. V. (1982) *Infrared and Millimeter Waves*, Vol. 6, Academic Press, p. 208.
Schneider, M. V., R. A. Linke, and A. Y. Cho (1977) *Appl. Phys. Lett.*, **31**, 219.
Scheider, M. V. and W. W. Snell Jnr (1975) *IEEE Trans. Microwave Theory Tech.*, **MMT-23**, 271.
Shottky, W. (1923) *Z. Phys.*, **74**, 63.
Schubert, E. F., J. F. Cunningham, W. T. Tsang and T. H. Chiu (1986) *Appl. Phys. Lett.*, **49**, 292.
Schottky, W. (1938) *Natorwissenschaften*, **26**, 843.
Schubert, E. F., J. E. Cunningham, W. T. Tsang, and T. H. Chiu (1987) *Appl. Phys. Lett.*, **49**, 292.
Schubert, E. F., J. M. Kuo, R. F. Kopf, H. S. Luftman, L. C. Hopkins, and N. J. Saver (1990) *J. Appl. Phys.*, **67**, 1969.
Shannon, J. M. (1976) *Solid State Electron.*, **19**, 537.
Shannon, J. M. (1979) *Appl. Phys. Lett.*, **35**, 63.
Shur, M. (1985) *Appl. Phys. Lett.*, **47**, 869.
Sleger, K. and A. Christou (1978) *Solid State Electron.*, **21**, 677.
Stall, R. A. (1980) Ph.D. dissertation, Cornell University.
Stall, R., C. E. C. Wood, K. Board, and L. F. Eastman (1979) *Electron Lett.*, **15**, 800.
Sze, S. M., M. P. Lepselter, and R. W. MacDonald (1968) *Solid State Electron.*, **11**.
Sze, S. M. (1981) *Physics of Semiconductor Devices*, 2nd edn, p. 279, John Wiley and Sons, New York.
Sze, S. M. (eds.) (1982) *High Speed Semiconductor Devices*, John Wiley and Sons, New York, p. 576.

Szubert, S. M. and K. E. Singer (1985) *Proc. 3rd Intl. Conf. on Molecular Beam Epitaxy, J. Vac. Sci. Tech.* B **3**, 794.
Taylor, G. W., M. S. Lebby, A. Izabelle, B. Tell, K. Brown-Goebeler, T.-Y. Chang, and J. G. Simmons (1988) *IEEE Electron Dev. Lett.*, **9**, 84.
Taylor, G. W., J. G. Simmons, A. Y. Cho, and R. S. Mand (1986) *J. Appl. Phys.*, **59**, 596.
Trew, R. J., R. Sultan, J. R. Hauser, and M. A. Littlejohn (1984) Proc. of a Workshop: Physics of Submicron Structures, p. 177, eds. H. L. Grubin, K. Hess, G. J. Iafrate and D. K. Ferry, Plenum, New York.
Tuyen, V. V. and B. Szentpali (1990) *J. Appl. Phys.*, **68**, 2824.
Wang, T., J. P. Leburton, and K. Hess (1984) Proc. of a Workshop: Physics of Submicron Structures, p. 239, eds. H. L. Grubin, K. Hess, G. J. Iafrate and D. K. Ferry, Plenum, New York.
Wang, T., K. Hess, and G. J. Iafrate (1985) *J. Appl. Phys.*, **58**, 857.
Wang, T., K. Hess, and G. J. Iafrate (1986) *J. Appl. Phys.*, **59**, 2125.
Wood, C. E., C., G. Metze, J. Berry, and L. F. Eastman (1980) *J. Appl. Phys.*, **51**, 383.
Woodcock, J. M. and J. J. Harris (1983) *Electron. Lett.*, **19**, 181.
Won, Y. H., K. Yamasaki, T. Daniels-Race, P. J. Tasker, W. J. Schaff, and L. F. Eastman (1990) *IEEE Electron Dev. Lett.*, **11**, 376.

24

Silicon interband and intersubband photodetectors

I. EISELE

Introduction

Delta- (δ-) doping structures in silicon open up new possibilities for electronic and optoelectronic device structures. Within this scope, the application of interband and intersubband transitions for photodetectors will be discussed in this chapter. Two phenomena are inherent in δ-doping layers and cannot be obtained with conventional doping:

1. For an alternating series of donor and acceptor delta profiles (n–i–p–i structures) with low doping concentrations a depleted structure can be obtained. In this case the total donor impurity concentration compensates the total acceptor impurity concentration, and the free charge carriers recombine. The resulting potential is sawtooth-like. Photogeneration of free charge carriers due to interband transitions between the valence and the conduction bands leads to high efficiency photodetectors due to the fact that the spatial separation of the carriers leads to a long lifetime.
2. For a highly doped single δ-layer the free charge carrier concentration is equal to the electrically active impurity concentration, and the free carriers are bound within the V-shaped potential of the dopants. If the spatial extent of the potential wells is comparable to the free carrier de Broglie wavelength, quantum mechanics rather than classical physics must be considered. The confinement of the free charge carriers leads to quantization effects [1], and the corresponding intersubband transitions can be utilized for photodetection. These transitions are within the same band (conduction or valence band) in the two-dimensional systems at the δ-doped layers.

Interband photodetectors

The properties of modulation doped semiconductors (n–i–p–i structures) have been described extensively [2]. One of the interesting properties is the prolonged recombination lifetime τ for photogenerated carriers according to [3]

$$\tau_{\text{nipi}} = \tau_{\text{bulk}} \cdot \exp(V_{\text{bi}}/kT)$$

where V_{bi} is the built-in superlattice potential barrier. This means that, even at room temperature, n–i–p–i structures should exhibit nearly infinite lifetimes. In practice, the limiting factors are caused by junction defects and leakage currents through the bulk. The idealized doping profiles and the resulting modulation of the valence and conduction bands is presented in Fig. 24.1 for completely depleted δ-layers. In order to study the electrical, as well as optoelectronic behavior of such a structure, Zeindl *et al.* [4] prepared Sb-δ/i/Ga-δ/i modulation doped samples with selective contacts to the n- and p-type layers respectively. It was demonstrated that by using mechanical masks inside the molecular beam epitaxy (MBE) system, locally doped areas, several mm in size, can be grown. Two different masks (one for n- and one for p-type doping) allowed $5 \times 1.5\,\text{mm}^2$ δ-doping bars perpendicular to each other (see inset in Fig. 24.2). A sequence of alternating δ-layers results in a $1.5 \times 1.5\,\text{mm}^2$ n–i–p–i structure (shaded area) with individual connections to the contacts. On a 50 Ωcm (100) silicon substrate five Sb δ-layers with $N_{\text{Sb}}^{2\text{D}} = 1.8 \times 10^{13}\,\text{cm}^{-2}$, and five Ga δ-layers with $N_{\text{Ga}}^{2\text{D}} = 1.8 \times 10^{13}\,\text{cm}^{-2}$ separated by 100-nm-thick intrinsic regions have been grown. The doping concentrations have been chosen to be quite high in order to guarantee the ohmic behavior of the δ-layers for electrical transport investigations at low temperatures [5]. It should be noted that

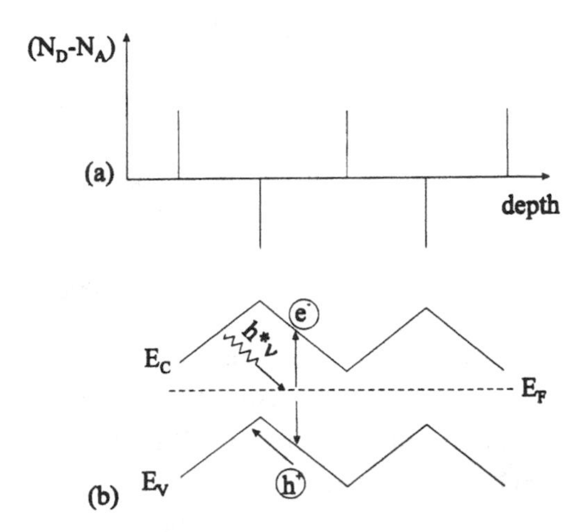

Fig. 24.1. Idealized δ-doping n–i–p–i structure as a function of depth: (a) doping profile, and (b) band diagram corresponding to the completely depleted case.

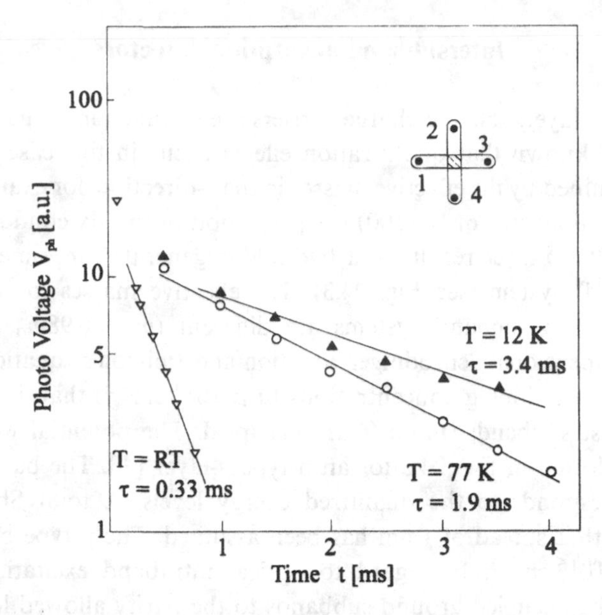

Fig. 24.2. Time dependent photovoltage decay for different temperatures. The corresponding carrier lifetime τ is indicated [4]. The inset shows the masked patterns of MBE grown structures: the n-type deltas can be measured between contacts 1 and 3, the p-type deltas between 2 and 4. The n–i–p–i structure occurs in the shaded area.

this degeneracy restricts the lifetime of the photoexcited charge carriers because the individual layers cannot be completely depleted.

Photovoltage measurements have been carried out by generating electron–hole pairs with a chopped He–Ne laser ($\lambda = 633$ nm). Due to the sawtooth potential of the n–i–p–i structure, the generated electrons are collected in the minimum of the conduction band, and the holes in the maximum of the valence band. The lifetime of the excess carriers has been determined from the exponential decay of the photovoltage. The time dependence of the photovoltage decay is plotted in Fig. 24.2 for different temperatures. For $T = 300$ K a lifetime of 330 μs has been determined, which is a factor of 5–20 higher than that of bulk MBE silicon [6]. In a n–i–p–i structure with thicker n- and p-type regions (35 nm) Landheer *et al.* [7] measured lifetimes of 1.5 ms at room temperature. The lower value for the δ-superlattice as compared with [7] probably arises from the nondepleted layer doping. At low temperatures ($T = 12$ K) the recombination lifetime amounts to 3.4 ms, which is probably limited by crystal defects originating from Ga-rich clusters in the p-type δ-layers. A modified p-type δ-doping layer with boron as the dopant material should be suited to overcoming this problem. Optimizing the growth process and decreasing the δ-layer concentrations should allow lifetimes of several seconds, as has been published for GaAs n–i–p–i superlattices [8].

Intersubband absorption detectors

For highly doped δ-layers the free charge carriers are confined in a quasi-two-dimensional manner. It is well known that quantization effects occur in this case [9]. The subband energies are determined by the effective masses in the z-direction, for example perpendicular to the plane of the δ-layer. For Si (100) the projection of the six conduction band valleys into the plane of the δ-layer results in a two-fold degenerate (unprimed) and a four-fold degenerate (primed) system (see Fig. 24.3). The effective masses for the binding in the z-direction of the two subband systems are different ($m_z = 0.98m_0$ and $m'_z = 0.19m_0$). Solving the one-dimensional Schrödinger equation and Poisson's equation self-consistently, it turns out for high δ-doping concentrations ($n \geqslant 10^{13}$ cm^{-2}) that, in the main, the two energetically lowest subbands 0 and 0′ are occupied. The potential well and the charge distribution are plotted in Fig. 24.3 for an n-type δ-layer [1]. The baseline of the charge distributions corresponds to the quantized energy levels. A total Sb concentration of 2×10^{13} cm^{-2} with a spread of 1 nm has been assumed. The p-type background doping amounts to 4×10^{16} cm^{-3}. In regard to optical intraband excitation we expect two transitions from the occupied ground subbands to the parity allowed levels 1 and 1′. The next allowed transitions are to subbands 3 and 3′, but the absorption strengths of these are negligibly small.

Tempel *et al.* [10, 11] have measured the infrared resonance excitation of single n-type δ-layers for different electron densities between 2.0×10^{13} cm^{-2} and 1.4×10^{14} cm^{-2}. The antimony doped δ-layers in silicon (100) substrates have been prepared according to the solid phase epitaxy (SPE) growth mode described in Chapter 6 by Eisele and first developed by Zeindl *et al.* [12]. Measurements of the IR absorption were carried out over a temperature range between 10 K and 290 K using a Fourier transform infrared (FTIR) spectrometer. Because the intersubband resonance can be excited only by polarization of the electromagnetic wave perpendicular to the δ-layer [13], the samples were provided with

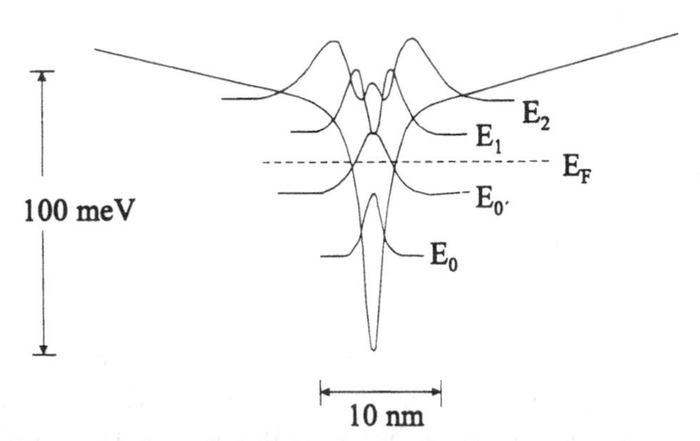

Fig. 24.3. Energy levels and electron charge distributions for an n-type delta layer on (100) silicon. The total Sb doping amounts to 2×10^{13} cm^{-2} with a spread of 1 nm. The background doping is p-type with 4×10^{16} cm^{-3}. E_0 and $E_{0'}$ are occupied.

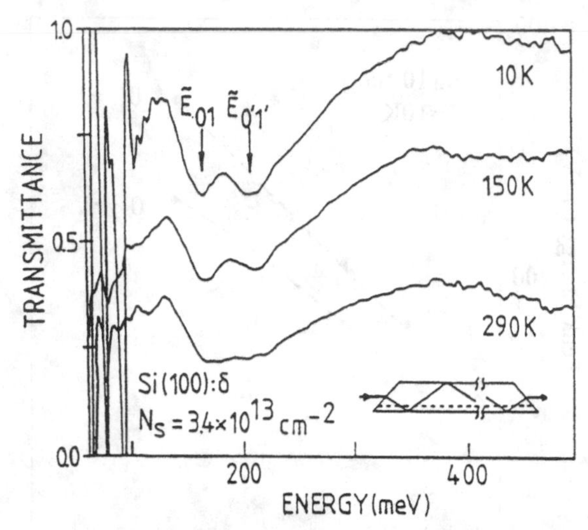

Fig. 24.4. Transmission spectra of an n-type δ-doped Si (100) sample as a function of temperature. The antimony concentration amounts to 3.4×10^{13} cm^{-2}. The inset depicts the light passing through the sample [10].

bevelled 57° edges in order to obtain several double passes of the light through the sample. The experimental set-up is shown in the inset of Fig. 24.4. Because, on Si (100), excitation by parallel polarization is not allowed and not observed, this polarization has been used for normalization of the FTIR spectrum. Typical transmission spectra for an n-type δ-layer with a Sb concentration of 3.4×10^{13} cm^{-3} are presented in Fig. 24.4 [14]. At a temperature of 10 K a distinct double structure is present in the transmission spectrum. The two absorption peaks correspond to transitions $0 \rightarrow 1$ and $0' \rightarrow 1'$, as given on the diagram. The dominant structures below 100 meV can be attributed to multi-phonon absorption lines. The individual peaks smear out slighty with increasing temperature, but they can clearly be resolved. It should be noted that the absorption amounts to about 40%, which is comparable to the values for doped single quantum wells in GaAs [15].

The resonance energies of differently doped δ-layers as a function of electron density, as determined by Hall measurements at 4.2 K, are depicted in Fig. 24.5. The absorption measurements at a temperature of 10 K prove that the resonance absorption energy is tuneable from 120 to 350 meV by varying the doping concentration. The lower limit is given by the dominant absorption of multi-phonon modes. The upper limit amounts to about 9×10^{13} cm^{-2}. Beyond these values a broad absorption without any prominent structures is always seen. The reason for this behavior might be the beginning of cluster formation, which has been described in Chapter 6 by Eisele on solid phase epitaxy. It is believed that precipitates or density fluctuations cause a fluctuating potential and broaden the well-defined quantization.

In order to obtain a better understanding of the quantization behavior of δ-layers the experimental results have been compared with self-consistent subband calculations [14]. But even when including depolarization and excitonic terms [16], the calculated values

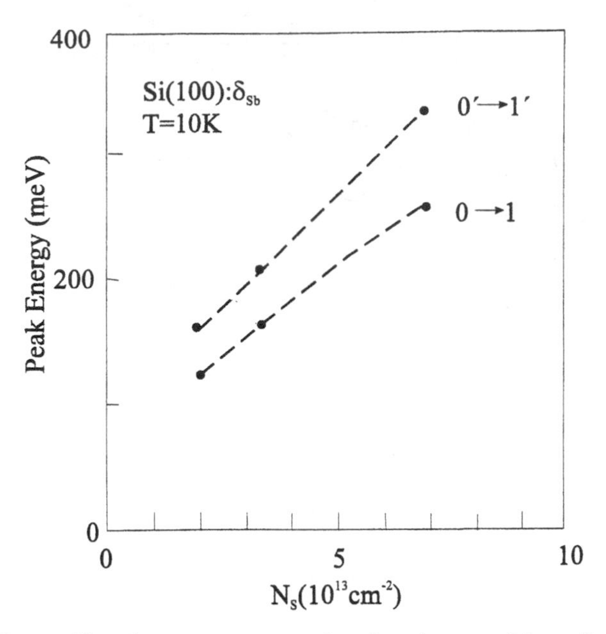

Fig. 24.5. Intersubband resonance energies for the transitions $0 \rightarrow 1$ and $0' \rightarrow 1'$ as a function of electron density in n-type δ-layers for absorption measurements at 10 K [11].

were always less than the measured data by as much as a factor of two. This is not the case for recent investigations into Sb δ-doping on (110) substrates and boron deltas, which will be described below. The explanation for this discrepancy could be the presence of a negatively charged background near the δ-layer. Assuming that 30% of the electrons are trapped in deep levels with a homogeneous spread of 3 nm leads to good agreement with the experimental data. The assumption of the negatively charged background could be based on the following reasons:

1. The existence of vacancies in combination with antimony has been demonstrated by diffusion experiments [17, 18] and theoretical calculations [19]. Furthermore, it has been shown that vacancies are responsible for the generation of electrically inactive antimony [20].
2. The vacancies are caused by the SPE growth mode. If the recrystallization step is not carried out carefully, segregation effects as well as point defects could lead to a disturbed lattice [21].

Except for the (100) surface orientation there have also been interesting results with respect to intersubband absorption in Sb δ-doped $Si/Si_{1-x}Ge_x$ quantum well structures grown on (110) silicon [22]. In this case the absorption is allowed for optical field components that are perpendicular as well as parallel to the quantum wells due to the tilted ellipsoidal of the constant energy surfaces. For the experiments, structures consisting of ten periods of 5-nm-wide Si-layers doped with Sb have been grown. The spacer between the δ-layers was undoped $Si_{1-x}Ge_x$ with a thickness of 30 nm. Due to the combined effects of conduction band offset and δ-doping, an increased energy level separation between subbands occurs. Absorption peaks ranging from 4.9 to 5.8 μm have been observed for various samples. The transition energy has also been calculated using the Hartree–Fock potential including depolarization and exciton-like effects, and close agreement with experimental data has been achieved.

Hole-intersubband absorption in boron δ-doped multiple quantum wells has been obtained by Park *et al.* [23]. They used an FTIR set up to detect the transition between heavy hole subbands. The structure was grown on high resistivity 100 Ωcm Si(100) substrates in order to avoid absorption within the substrate. After the growth of an undoped buffer, ten periods of 3.5-nm-wide, heavily boron doped δ-layers spaced by 30 nm undoped silicon were grown. The doping concentration in the δ-layers was about 2.7×10^{13} cm^{-2} and 5.4×10^{13} cm^{-2} respectively. The calculated potential well, the subband energies, and the wavefunctions for the heavy hole system are shown in Fig. 24.6 for the lower doping concentration [23]. For measurements 5 mm long and 0.5 mm thick, waveguide structures with bevelled 45° edges were prepared according to the inset in Fig. 24.7. The absorption spectra were taken at room temperature as a function of photon energy. The polarization effect of the excitation is demonstrated in Fig. 24.8 for sample B. For an angle of 0° the polarization is perpendicular to the (100) δ-planes, and thus subband transitions are allowed because the quantization is also perpendicular to the plane. Tilting the polarization leads to a decreasing signal that vanishes if the polarization is in the δ-plane.

The peak positions in Fig. 24.7 are found near 205 meV (6 μm) and 360 meV (3.4 μm).

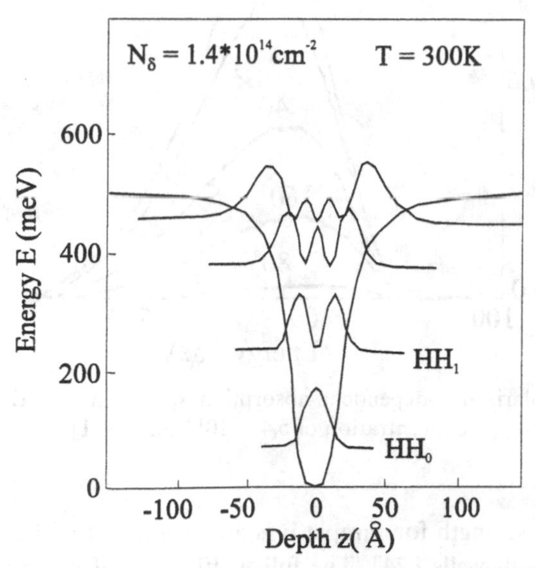

Fig. 24.6. Calculated potential well, subband energies, and wavefunctions for the heavy hole system. The boron δ-concentration amounts to 5.4×10^{13} cm^{-2}. The hole energy is taken as positive for convenience [21].

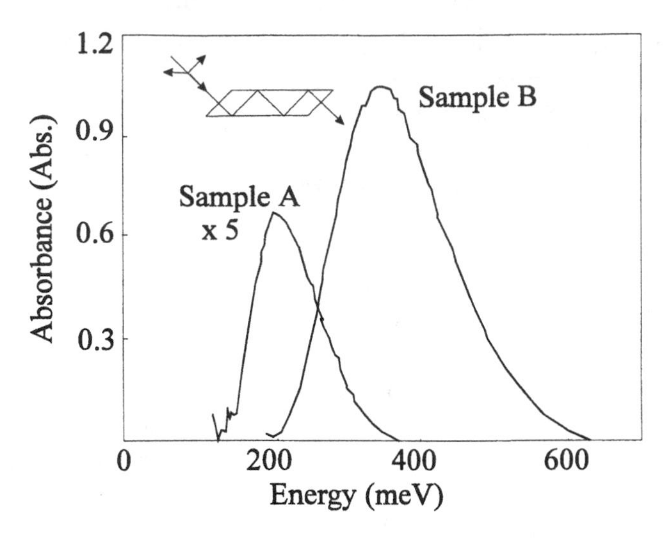

Fig. 24.7. Absorption spectra at 300 K for boron δ-layers with concentrations of $2.7 \times 10^{13}\ \mathrm{cm}^{-2}$ (A) and $5.4 \times 10^{13}\ \mathrm{cm}^{-2}$ (B). The inset depicts the light passing through the sample [21].

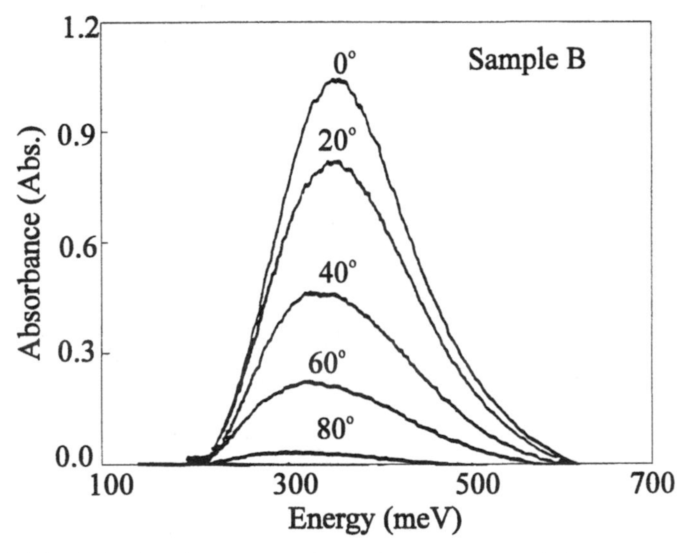

Fig. 24.8. Polarization-dependent absorption spectra at 300 K for a boron concentration of $5.4 \times 10^{13}\ \mathrm{cm}^{-2}$ [21].

The peak absorption strength for sample B is about a factor of 4 above that reported for GaAs/AlGaAs quantum wells [24]. The full widths at half maximum of the absorption peaks are 97 and 174 meV respectively, which are considerable larger than the FWHM observed for intersubband resonance peaks in GaAs/AlGaAs multiple quantum wells

($\approx$15 meV) but comparable to the Sb doped δ-layers (see Fig. 24.4) and δ-doped layers in GaAs [25]. This broadening could be partially due to nonparabolic effects in the valence band. At high doping concentrations the Fermi level is a few hundred meV above the valence band minima, causing holes to occupy states away from the Γ point, where the nonparabolicity is increased [26]. These effects have been considered using the $k \cdot p$ approximation for the calculation of the subband dispersion [27]. The results show that the subband separation increases up to 100 meV or more with increasing transverse momentum, and produces a broad absorption spectrum, just as in the case of interband transitions. Since the holes are only filled up to the Fermi level, the absorption eventually decreases at high photon energies. In this case the onset of the absorption peak should determine the energy separation rather than the peak maximum. If this is assumed, recent results show that for boron doped δ-profiles, measured and calculated transition energies agree quite well [27]. Structures consisting of ten periods of 3.5-nm-thick boron-doped Si layers and 30 nm undoped Si spacers have been grown. Subband separation as well as peak width is depicted in Fig. 24.9 as a function of doping density. For comparison, calculations using the Hartree, as well as the Hartree–Fock, approximation, are depicted. The doping concentration is assumed uniformly over a 3.5 nm layer of silicon and the occupation of the light hole band is taken into account by combining the light hole density of states with the heavy hole density of states. It can be seen that the calculated values are considerably less than the experimental data for the Hartree approximation. However, including many-body effects, such as the exchange term in the Hartree–Fock equation, yields good agreement with experimentally observed values.

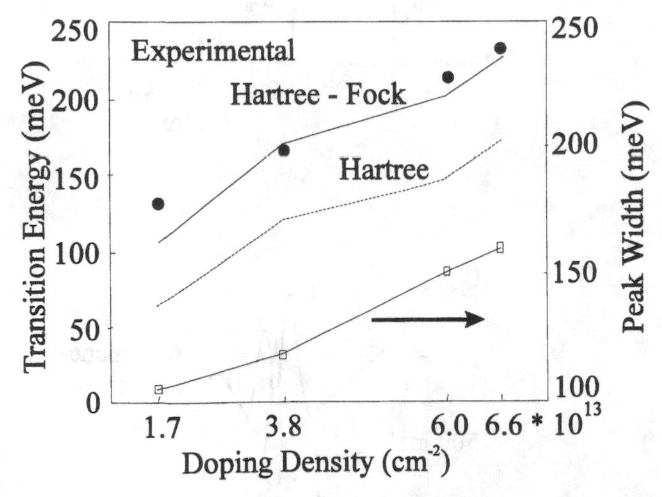

Fig. 24.9. Subband separation and peak width of the absorption as a function of doping density. Experimental data (●, □), as well as calculations using Hartree approximation (---), and Hartree–Fock approximation including hole–hole exchange correlation effects (——), are presented [27].

Detector concepts

To produce an infrared detector it is necessary to measure the current proportional to the photoexcited subband electron or hole density. This has been demonstrated by Levine *et al.* [28], who applied an electric field to a quantum well, thus causing tunneling of photoexcited electrons out of the well. For δ-layers the simplest realization of such a device is the combination of an infrared resonance device with tunneling devices, as depicted in Fig. 24.10 [14]. In both cases the photoexcited charge carriers 'see' a reduced potential barrier and tunnel to either the Schottky gate or the avalanche region. If, in the latter case, it is possible to accelerate tunneled electrons in an electric field of sufficient strength, avalanche multiplication occurs, and an intergrated electron multiplier can be designed [29]. However, the simplest technological realization is the combination of an IR resonance device with a Schottky tunneling diode used for tunneling spectroscopy [30]. First results for a Sb δ-layer with a concentration of 7×10^{13} cm^{-2} are shown in Fig. 24.11 [14]. The maximum of the tunnel current corresponds to the transition of the primed subband system. This is expected because for n-type δ-layers the primed subband system has a lower effective mass perpendicular to the δ-plane ($m_z = 0.19m_0$ as compared to $m_z = 0.98m_0$ for the

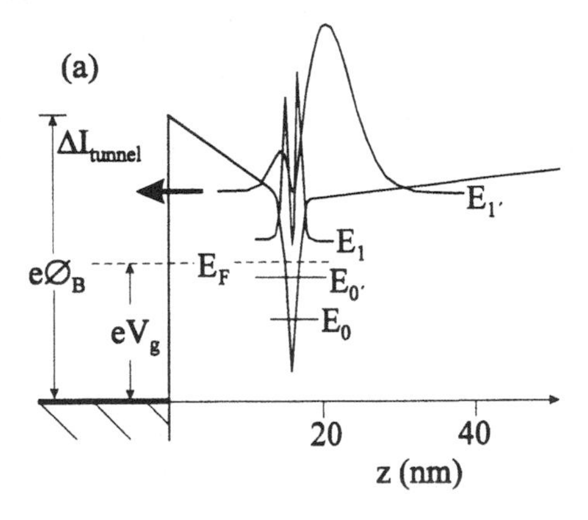

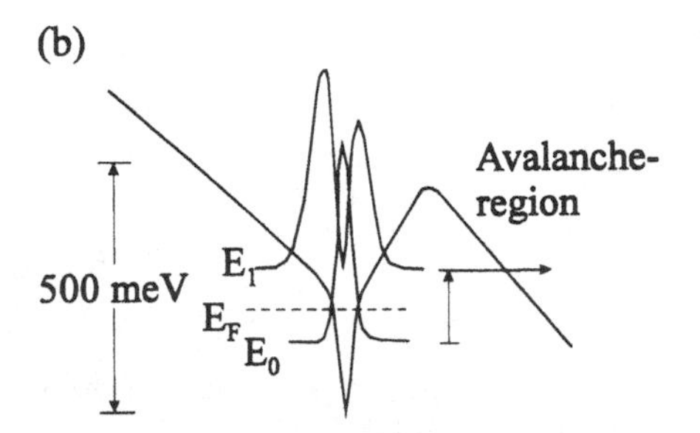

Fig. 24.10. Principle of a FIR detector using (a) a Schottky tunneling diode, and (b) tunneling in combination with avalanche multiplication.

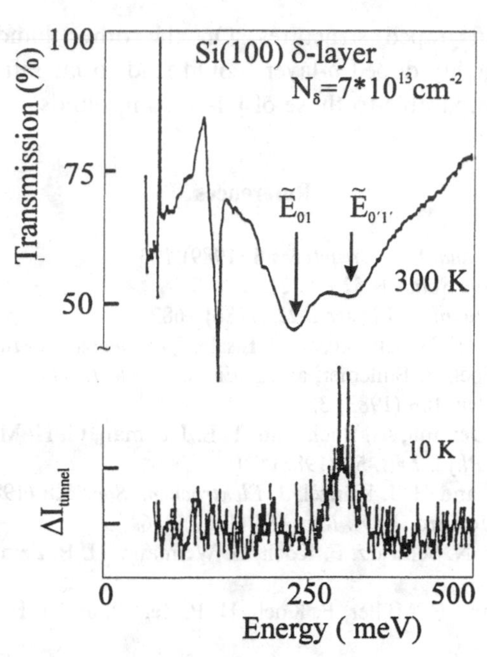

Fig. 24.11. Signal of a FIR detector based on a Sb δ-layer in (100) silicon and a Schottky tunnel device according to Fig. 24.10(a). The doping concentration amounts to 7×10^{13} cm^{-2}. For comparison the corresponding absorption signal is shown [14].

unprimed system). Thus the tunneling probability will be much higher for the primed system.

The signal is quite low, which is due to the small Schottky gate area of 0.2 mm^2. The absorption strength in this case is weak because only a small portion of the photoexcited electrons will contribute to the tunneling current.

It should be noted that the electromagnetic wave polarization perpendicular to the δ-layer requires complicated methods of light propagation. By contrast, the intersubband responance on Si (110) and Si (111) can be excited simply by the normal incidence of light, for example parallel polarization. This is caused by the tilted Fermi spheres at these orientations [31]. For (110) surface orientations this has been shown [22], whereas no results have been reported for (111) due to the reduced crystal quality for δ-layers grown by (SPE), which has been described in Chapter 6 by Eisele earlier in the book.

Summary

Delta-doping in silicon opens up new possibilities for optoelectronic device concepts. Interband transitions between valence and conduction bands can lead to photodetectors with a long lifetime of the photoexcited charge carriers, and to a high absorption efficiency in modulation doped structures. On the other hand, the exploitation of quantization effects in highly doped δ-layers could lead to far-infrared detectors with an absorption efficiency comparable to those of III–V compounds.

References

[1] I. Eisele, *Superlattices and Microstructures* **6** (1989) 123.
[2] G. Döhler, *Phys. Status Solidi* B **52** (1972) 79.
[3] G. Döhler, *IEEE J. Quantum Electron.* **22** (1986) 1682.
[4] H. P. Zeindl, E. Hammerl, W. Kiunke, and I. Eisele, *J. of Electronic Materials* **19** (1990) 1119.
[5] H. P. Zeindl, G. Tempel, B. Bullemer, and I. Eisele, *J. Electrochem. Soc.* **88–8** (1987) 515.
[6] Y. Ota, *Thin Solid Films* **106** (1983) 3.
[7] D. Landheer, M. W. Denhoff, M. Buchanan, T. E. Jackman, G. H. McKinnon, K. H. Teo, and J. A. Jackman, *Appl. Phys. Lett.* **52** (1988) 910.
[8] K. Ploog, A. Fischer, and H. J. Künzel, *J. Elektrochem. Soc.* **128** (1981) 400.
[9] F. Stern and W. E. Howard, *Phys. Rev.* **163** (1967) 816.
[10] G. Tempel, F. Müller, N. Schwarz, F. Koch, G. Weimann, H. P. Zeindl, and I. Eisele, *Surf. Sci.* **228** (1990) 247.
[11] G. Tempel, N. Schwarz, F. Müller, F. Koch, H. P. Zeindl, and I. Eisele, *Thin Solid Films* **184** (1990) 171.
[12] H. P. Zeindl, T. Wegehaupt, I. Eisele, H. Oppolzer, H. Reisinger, G. Tempel, and F. Koch, *Appl. Phys. Lett.*, **50** (1987) 1164.
[13] A. Kamgar, P. Kneschaurek, G. Dorda, and F. Koch, *Phys. Rev. Lett.*, **32** (1974) 1251.
[14] G. Tempel, Ph.D. Thesis, Technische Universität München (1990).
[15] M. J. Kane, M. T. Emeny, N. Apsley, C. R. Whitehouse, and D. Lee, *Semicond. Sci. Technol.* **3** (1988) 722.
[16] T. Ando, *Z. Phys.* B **26** (1977) 263.
[17] P. Fahey, S. S. Iyer, and G. J. Scilla, *Appl. Phys. Lett.*, **54** (1989) 843.
[18] F. Shimura, W. Dyson, J. W. Moody, and R. S. Hockett in W. M. Bullis and S. Broydo (eds.), *VLSI Science and Technology 1985*, Electrochemical Soc., Pennington, N.J. (1985), p. 507.
[19] C. S. Nichols, C. G. van de Walle, and T. Pantelides, *Phys. Rev. Lett.* **62** (1989) 1049.
[20] A. Nylandsted Larsen, F. T. Pedersen, G. Weyer, R. Galloni, R. Rissoli, and A. Armigliato, *J. Appl. Phys.* **59** (1986) 1908.
[21] W. X. Ni, G. V. Hansson, J. E. Sundgren, L. Hultman, L. R. Wallenberg, J. Y. Yao, L. C. Markert, and J. J. E. Greene, *Phys. Rev.* B **46** (1992) 7551.
[22] C. Lee and K. L. Wang, *Appl. Phys. Lett.* **60** (1992) 2264.
[23] J. S. Park, R. P. G. Karunasiri, Y. J. Mii, and K. L. Wang, *Appl. Phys. Lett.* **58** (1991) 1083.
[24] B. F. Levine, R. J. Malik, J. Walker, K. K. Choi, C. G. Bethea, D. A. Kleinman, and J. M. Vandenberg, *Appl. Phys. Lett.* **50** (1987) 273.
[25] N. Schwarz, F. Müller, G. Tempel, F. Koch, and G. Weimann, *Semicond. Sci. Technol.* **4** (1989) 571.
[26] J. Ohkawa and Y. Uemura, *Suppl. Progress of Theoretical Physics* **57** (1975) 164.
[27] K. L. Wang, R. P. G. Karunasiri, and J. S. Park, *Surf. Sci.* **267** (1992) 74.
[28] B. F. Levine, K. K. Choi, C. G. Bethea, J. Walker, and R. J. Mailk, *J. Phys.* C **5** (Suppl. 11) (1987) 611.
[29] M. D. Petroff, M. G. Stapelbroek, and W. A. Kleinhans, *Appl. Phys. Lett.* **51** (1987) 406.
[30] I. Eisele, *Appl. Surf. Sci.* **36** (1989) 39.
[31] S. M. Nee, U. Claessen, and F. Koch, *Phys. Rev.* B **29** (1984) 3449.

25

Doping superlattice devices

E. F. SCHUBERT

Doping superlattices consist of semiconductors containing alternating regions of n- and p-type conductivity. If the n- and p-type regions are closely spaced and if the total donor density equals the total acceptor density, then electrons and holes of extrinsic origin recombine and the entire doping superlattice is depleted of free carriers. The band-edge potential variations along the superlattice growth direction are then determined solely by the charges of the ionized donor and acceptor impurities. Among the intriguing properties of doping superlattices, which were first proposed by Esaki and Tsu (1970), is reduced bandgap energy. That is, the bandgap of any semiconductor can be shifted to lower energies by using the doping superlattice concept. Furthermore, the energy gap depends on the level of excitation, that is, the energy gap can be *tuned* by means of the excitation level. The basic properties of doping superlattices are discussed in detail in Chapter 14 by Yao and Schubert and will not be repeated here.

If delta- (δ-) doping rather than homogeneous doping is employed for doping superlattices, several advantages result. First, the period of the superlattice can be made very short. If dopants are spatially well confined and do not spread during growth, then the period of the superlattice can be reduced to very small values. Second, the modulation of the band edges is larger for δ-doping, as compared with homogeneous doping, if the same doping density is used in both cases. Third, potential fluctuations arising from random dopant distribution are minimized by using δ-function-like doping profiles. These advantages have led to the first observation of quantum-confined interband transitions in doping superlattices (see Chapter 14 by Yao and Schubert).

Optoelectronic doping superlattice devices include light-emitting diodes (LEDs), lasers, and modulators. I will discuss the properties of these devices in this chapter. *Electronic* devices based on doping superlattices will not be discussed here and I refer the reader to the literature (Schubert *et al.*, 1987; Baillargeon *et al.*, 1989; Sun and Liu 1991; Wang *et al.*, 1987, 1991). All the electronic devices exhibit an S-shaped negative differential conductivity effect which is observed in doping superlattices for a current flow direction normal to the superlattice planes. Such devices may be suitable for high-speed switching applications and high-speed oscillators. A structure consisting of two triangular barriers and a V-shaped well has been proposed by Houng *et al.* (1992), who reported a negative differential resistance effect in this structure.

25.1 Light-emitting diodes

Doping superlattice light-emitting diodes (LEDs) have two unique characteristics which distinguish them from conventional LEDs. First, doping superlattice LEDs emit light at an energy below the gap of the host semiconductor. Second, the emission energy is tunable by the excitation energy (see Chapter 14 by Yao and Schubert). The magnitude of the energy gap reduction and of the tunability depend on the superlattice period. Specifically, the gap reduction and the tunability decrease considerably as the period of the doping superlattice is reduced.

A schematic illustration of an edge-emitting LED with a sawtooth-shaped band diagram in the active region (Schubert *et al.*, 1985) is shown in Fig. 25.1, along with a δ-function-like doping profile. The dopant density in the superlattice is $N^{2D} = N_A^{2D} = N_D^{2D} = 5 \times 10^{12}\ \text{cm}^{-2}$. The period of the superlattice is 200 Å. The edge-emitting diode has $Al_xGa_{1-x}As$ confinement layers. The GaAs doping superlattice region is located between the n-type and the p-type confinement region. Thus free carriers will recombine in the doping superlattice region.

Electroluminescent spectra of two doping superlattice LEDs operating at 300 K are presented in Fig. 25.2. The peak wavelengths of the two samples are at $\lambda = 925$ nm and

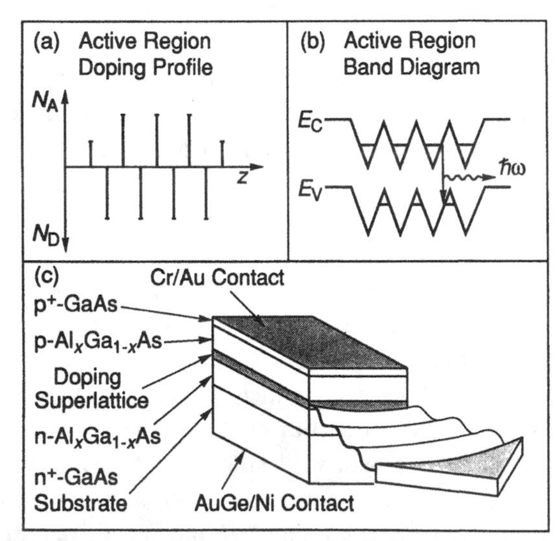

Fig. 25.1. (a) Active region doping profile, (b) active region band diagram, and (c) schematic structure and layer sequence of a doping superlattice light-emitting diode.

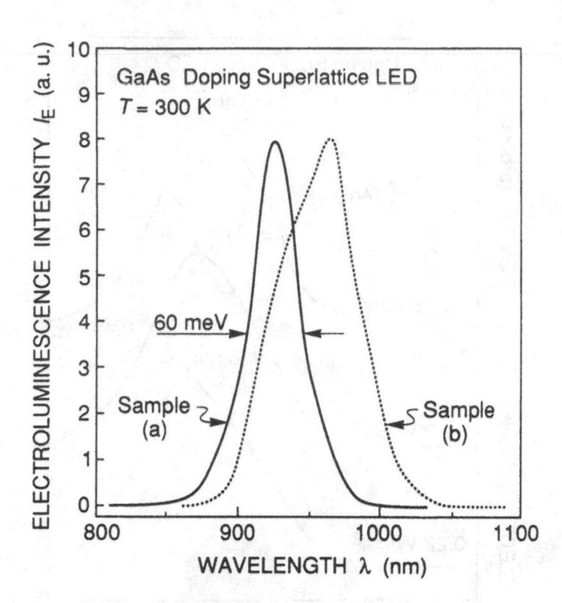

Fig. 25.2. Room-temperature spectra of two doping superlattice light-emitting diodes with an emission energy below the bandgap of GaAs.

965 nm, which is well below the bandgap of GaAs. These wavelengths are inaccessible to conventional GaAs LEDs. At the wavelength corresponding to the bandgap energy of the GaAs host material, that is, $\lambda = 870$ nm, no luminescence signal is detected. The longer emission wavelength is attributed to the superlattice character of the active region of the LED.

The full-width at half-maximum (FWHM) of the electroluminescence spectrum shown in Fig. 25.2 is 60 meV for sample (a). The linewidth is approximately 2.2 kT for $T = 300$ K, which is slightly broader than the theoretical linewidth of 1.8 kT, which is expected for band-to-band transitions. The broader experimental linewidth is probably due to random doping density fluctuations, which cause broadening in addition to the thermal broadening of the luminescence line. The integrated emission intensity was investigated as a function of the injection current. Linear dependence was found over a wide current range, as expected for spontaneous electroluminescence.

The shift of luminescence energy with excitation intensity is relatively small. Figure 25.3 shows electroluminescence spectra for three different injection currents. The peak wavelength shifts from 966 nm at the lowest excitation intensity to 959 nm at the highest excitation intensity. The total shift in energy is relatively small when compared to the total linewidth of the luminescence line ($\Delta\lambda > 50$ nm). The shift of the luminescence line can be explained as follows. At low injection currents, carriers recombine at the superlattice bandgap energy. As the injection current is increased, the free carrier concentration increases in the active superlattice region of the LED. The free carriers screen the charges of ionized dopants, which reduces the band modulation of the superlattice. As a result, the energy gap of the superlattice increases along with the photon emission energy. The

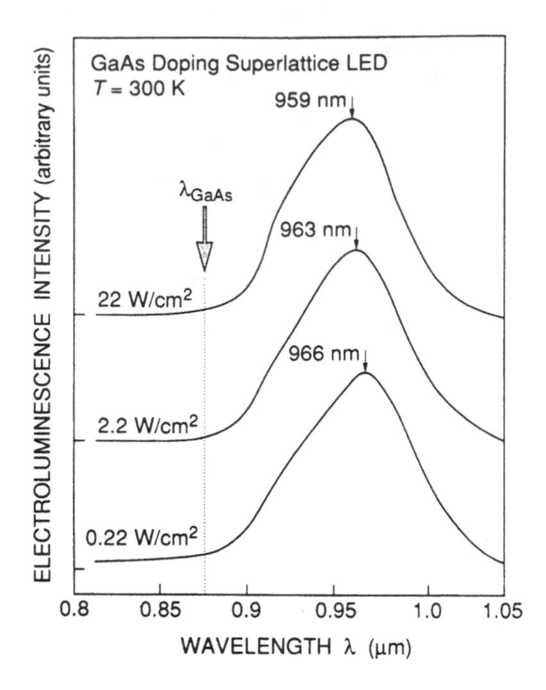

Fig. 25.3. Luminescence spectra of a doping superlattice light-emitting diode at three different injection currents corresponding to excitation densities of 0.22, 2.2 and 22 W/cm^2.

relatively small magnitude of the shift displayed in Fig. 25.3 is due to the small period of the superlattice ($z_p = 200$ Å). The magnitude of the the tunability can be increased by a larger period of the doping superlattice (see Chapter 14 by Yao and Schubert). At the present time it is not clear if the physically interesting phenomenon of tunability has any practical relevance.

25.2 Lasers

Semiconductor lasers are the most interesting device applications of doping superlattices. Tunable lasers, which can be produced from doping superlattices, are desirable components for wavelength-division multiplex systems. A drawback of doping superlattices is their inherently weaker optical oscillator strength, which is due to the spatial separation of electrons and holes. The separation is especially large in superlattices with a long period. Non-radiative transitions are therefore more prevalent in doping superlattices as compared with bulk material. However, the significance of the weak oscillator strength can be reduced by very-high-quality material with a low-defect concentration, in which non-radiative recombination processes via traps are less likely.

The first doping superlattice current injection laser (Schubert *et al.*, 1985) was achieved in GaAs using the δ-doping technique. The spontaneous and stimulated emission spectra are shown in Fig. 25.4 for different injection currents. Below threshold (Fig. 25.4(a)), a

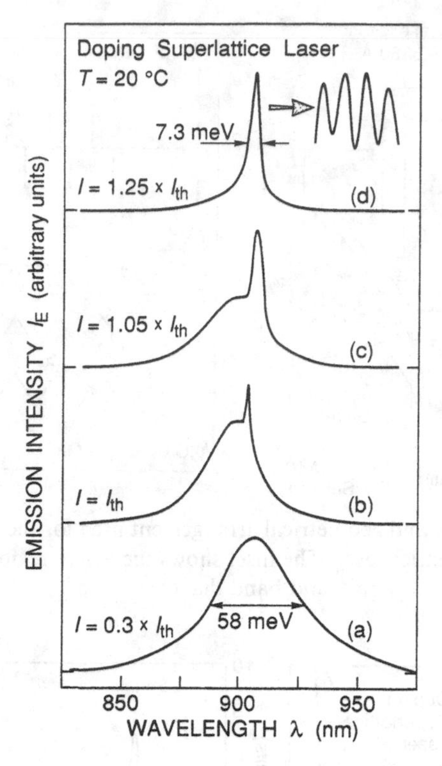

Fig. 25.4. Emission spectrum of a doping superlattice laser below, at, and above threshold. Also shown is a high-resolution scan which resolves the Fabry–Perot modes in the lasing regime.

wide, spontaneous spectrum is observed. As the current density is increased above threshold (Fig. 25.4(b–d)), a narrow line arises and completely dominates the spectrum. Longitudinal Fabry–Perot modes are found under high-resolution detection, as shown in the inset of Fig. 25.4. The peak energy of the laser emission occurs at $\lambda = 905$ nm, well below the bandgap of GaAs. Although laser emission was achieved, the emission wavelength was found to be constant, that is, independent of the injection current density. The lack of tunability in the stimulated emission regime is due to the very short carrier lifetimes in this regime. Short lifetimes prevent the accumulation of carriers in the V-shaped wells of the superlattice and thus do not result in a reduction in the band-edge modulation via the screening of dopant charges.

Laser emission from a doping superlattice structure was also reported by Vojak *et al.* (1986). The photopumped structures lased at an energy below the bandgap of the GaAs host semiconductor. The authors found no evidence of tunability.

The first tunable doping superlattice laser was realized by inhomogeneous optical excitation of the Fabry–Perot cavity (Schubert *et al.*, 1989). A schematic sketch of the layer sequence, the active region doping profile, and the corresponding band diagram are shown in Fig. 25.5. After epitaxial growth, the layers were cleaved into bars of nominal width 250 μm and length 1 cm. A frequency-doubled, neodymium-doped, Q-switched YAG

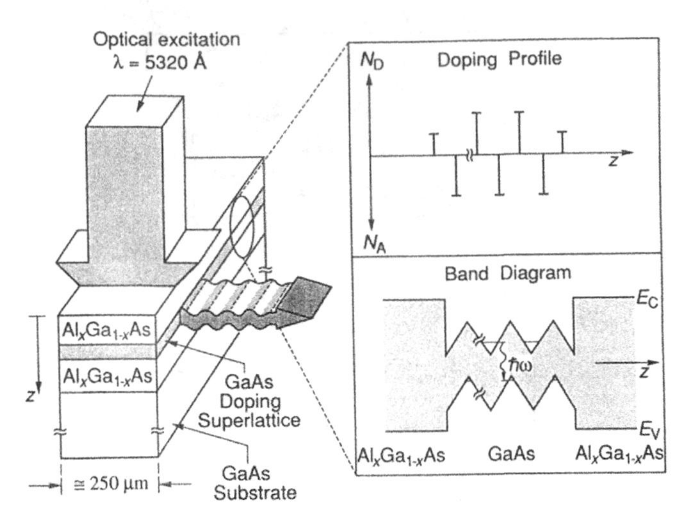

Fig. 25.5. Structure and geometrical arrangement used for the optical pumping of a doping superlattice laser. The inset shows the active region doping profile and band diagram.

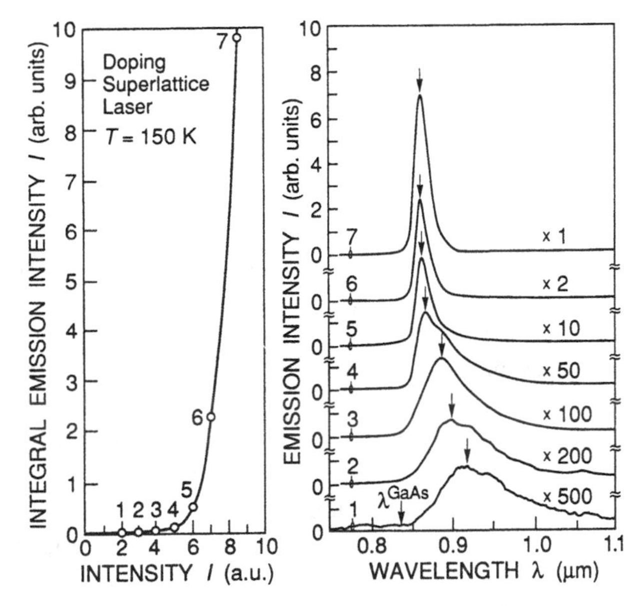

Fig. 25.6. Light intensity (L) versus current (I) and optical spectra at several points on the L–I curve for an optically pumped doping superlattice laser.

laser ($\lambda = 532$ nm) was used for optical excitation. Light emission from the sample was detected by a silicon detector using gated detection. The laser samples were cooled in a variable-temperature helium cryostat.

The spontaneous and stimulated emission spectra, along with the light output vs excitation intensity curve, are shown in Fig. 25.6 for a temperature of $T = 150$ K. In the spontaneous emission regime (low excitation intensity), the peak wavelength shifts to shorter wavelengths with increasing excitation intensity. At higher excitation intensities, stimulated emission occurs which is accompanied by the characteristic kink in the light output curve of Fig. 25.6 and a narrowing of the emission spectrum to values below the thermal energy kT. However, the peak wavelength does not change in the stimulated emission regime, as can clearly be seen in Fig. 25.6(b). Such a constant emission energy is not unexpected, since, upon reaching the laser threshold, the Fermi level remains constant and additional carriers undergo stimulated recombination with a correspondingly short lifetime. Thus, the emission energy remains constant with excitation energy in the stimulated emission regime.

The dependence of the emission wavelength on the excitation intensity is shown in greater detail in Fig. 25.7 for three different temperatures. In the spontaneous regime, the emission shifts to shorter wavelengths. However, beyond the laser threshold intensity, the emission remains constant. The stimulated emission wavelength of the devices is typically 30 nm below the bandgap of GaAs.

We next show that the emission can be *tuned* continuously by inhomogeneous excitation of the Fabry–Perot cavity. Such inhomogeneous excitation is achieved by displacing the

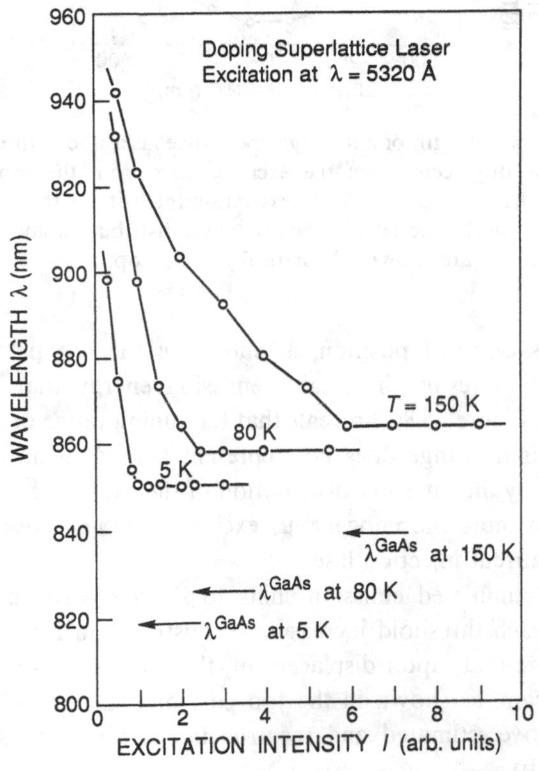

Fig. 25.7. Peak wavelength of a doping superlattice laser below and above threshold for three different temperatures. The peak wavelength does not depend on the excitation intensity in the lasing regime.

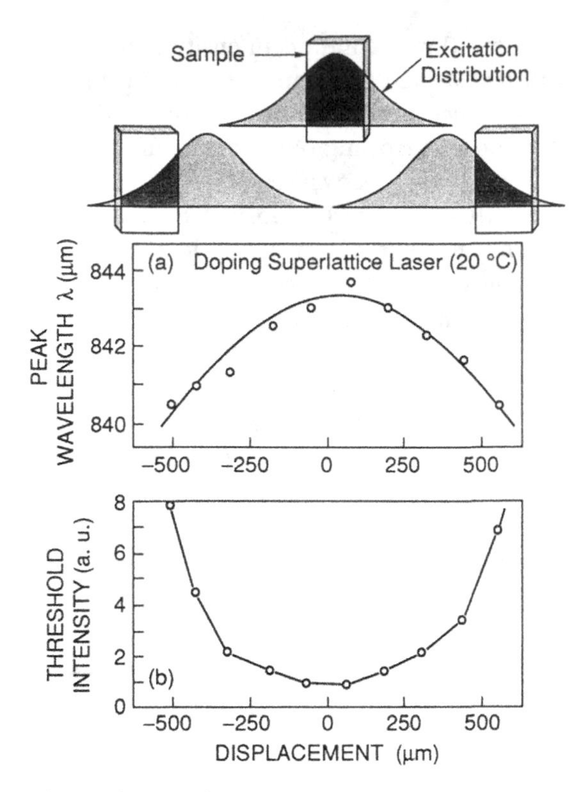

Fig. 25.8. Peak wavelength of a doping superlattice laser above threshold as a function of the displacement of the exciting laser from the center of the semiconductor bar. Also shown is the excitation intensity at threshold versus the displacement of the excitation. The excitation distribution and the sample are shown schematically at the top.

exciting beam from its centered position, as shown in the top part of Fig. 25.8. The inhomogeneous excitation results in a laser emission energy that is higher than the symmetrical excitation. Figure 25.8(a) reveals that the tuning range of the laser is approximately 3.5 nm. This tuning range does not represent a fundamental limit. The current tuning range is limited by the intensity distribution of the exciting source. A wider tuning range is expected for a more inhomogeneous excitation, which could be achieved in a two- or three-section current-injection laser.

As the peak of the stimulated emission shifts to shorter wavelengths, the excitation intensity required to reach threshold increases, as illustrated in Fig. 25.8(b). However, it is important to visualize that, upon displacement, the sample is excited by only a small part of the exciting beam, as shown in the top part of Fig. 25.8. Thus, the increase in threshold intensity is overestimated and the true increase is not as pronounced as is suggested by Fig. 25.8(b).

A high-resolution spectrum of the doping superlattice laser is shown in Fig. 25.9 for the transverse electric (TE) and the tranverse magnetic (TM) polarization modes. In analogy

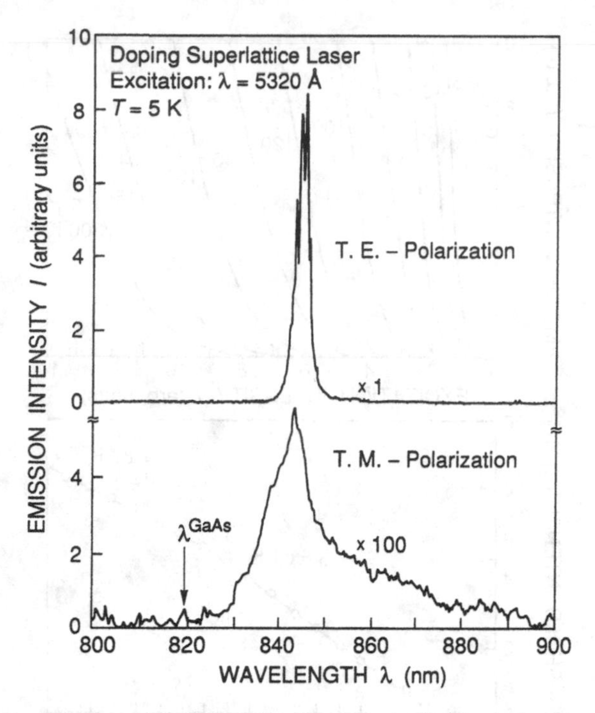

Fig. 25.9. Emission spectrum of a doping superlattice laser for the transverse electric (TE) and the transverse magnetic (TM) polarization.

with conventional $Al_xGa_{1-x}As$/GaAs double-heterostructure lasers with the cleavage planes being $\{110\}$ planes, the TE mode dominates the laser emission, whereas the TM mode is comparatively weak.

Under homogeneous excitation conditions, the stimulated emission energy is below the bandgap of GaAs. However, the emission energy is much higher than the spontaneous emission energy at low excitation intensities ($E \approx 1.35$ eV). The superlattice modulation is reduced by photoexcited electrons and holes, which screen the ionized dopant charges of donors and acceptors respectively. Even though the modulation is reduced, a residual band modulation is maintained, as suggested by the low emission energy. Thus, stimulated emission is achieved before the bands are completely flat.

The physical mechanism leading to tunability of the semiconductor laser can be understood on the basis of increased loss induced by inhomogeneous excitation, that is, reduced excitation intensity in one part of the Fabry–Perot cavity. In order to reach laser emission, the other section must be subjected to even higher excitation. As a result, the band modulation decreases and the superlattice energy gap increases in this section, due to an enhanced density of carriers. Once the intentionally induced loss is overcome, stimulated emission occurs. The corresponding emission energy is increased as compared with the homogeneously excited cavity. The fundamental limit of the tuning range is reached when the band edges are completely flat in one part of the laser. The

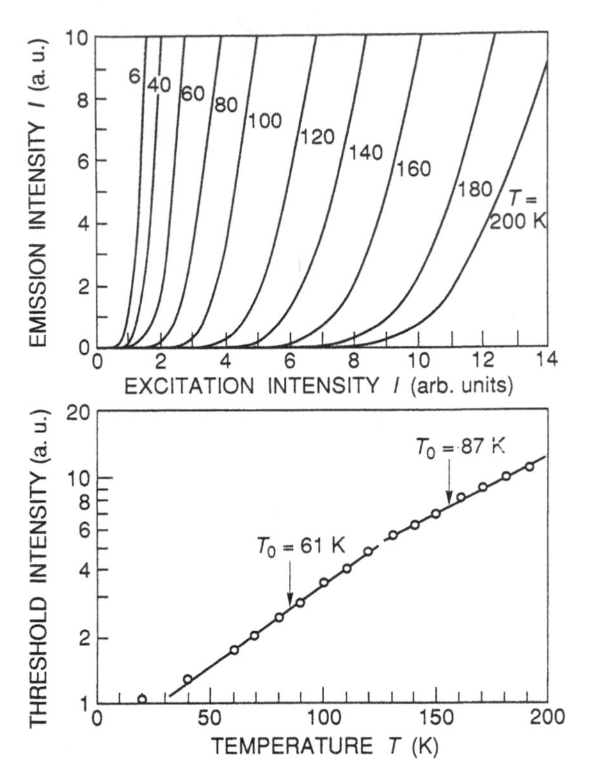

Fig. 25.10. *L–I* characteristics and threshold intensity of a doping superlattice laser at different temperatures.

corresponding tuning range is approximately 25.0 nm at low temperatures for the laser samples discussed here.

The temperature-dependence of the threshold intensity is shown in Fig. 25.10. The threshold intensity increases at higher temperatures and can be expressed by

$$I_{th} \propto \exp(T/T_0) \tag{25.1}$$

where T is the absolute temperature. The parameter T_0 describes the increase in threshold current with temperature. Evaluation of T_0 from the slope of I_{th} against T yields a T_0 of 61 and 87 K for low and intermediate temperatures respectively. The relatively low T_0 is probably due to the low substrate temperatures employed during crystal growth. The low growth temperatures were chosen in order to reduce dopant diffusion and segregation during growth.

25.3 Modulators

Modulation and switching of near-infrared radiation by means of an electrical control will be an important characteristic for devices in future photonic switching systems. It is desirable that such photonic switching devices (i) have a broad wavelength range in which the light intensity can be modulated, (ii) have a large contrast ratio between the transparent and the opaque states, (iii) can operate at voltages compatible with electronic integrated circuits, and (iv) have high-speed capability.

Modulators with such characteristics can be fabricated from doping superlattices with long periods. Contacting the n- and p-type regions of the superlattice allows the band modulation to be changed, that is, the *internal* electric field of the superlattice can be changed by means of an external bias. A schematic of the band diagram is shown in Fig. 25.11 for different voltages. Forward bias of the p–i–n structure results in a decrease in the band modulation and in the strength of the internal electric field. Reverse bias results in an increase in the band modulation, that is, an increase in the strength of the internal electric field. Tunneling-assisted absorption (Franz–Keldysh absorption), which occurs at energies below the bandgap of the semiconductor, depends exponentially on the electric field. Thus, doping superlattices can be used to modulate the intensity of transmitted light by the Franz–Keldysh effect.

Experimental results of intensity-modulation experiments are shown in Fig. 25.12 (Schubert and Cunningham, 1988) for different voltages applied to the modulator. Incident light is absorbed by the substrate at energies higher than the fundamental gap, that is, $\lambda < 870$ nm. The long-wave decay of the transmission signal ($\lambda \geqslant 1000$ nm) is due to the decreasing sensitivity of the silicon photodetector used for the measurement. A striking feature of the modulator is the wide wavelength range ($\Delta\lambda > 100$ nm) over which a

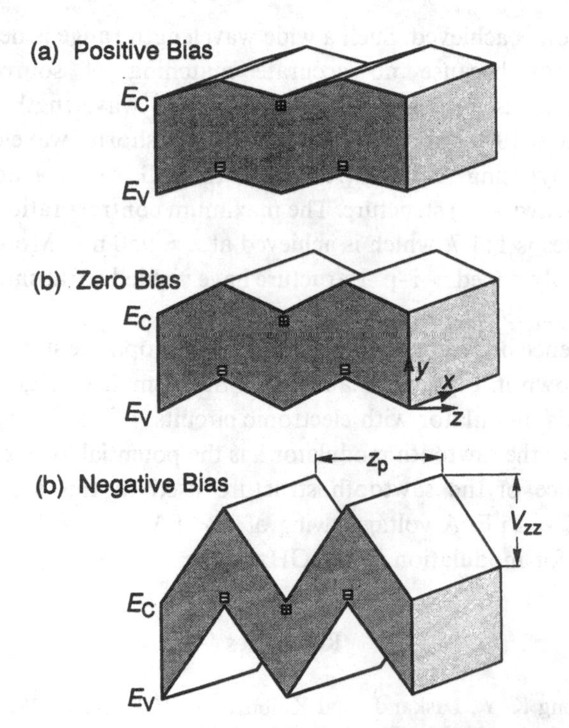

Fig. 25.11. Schematic band diagram of a long-period doping superlattice for different voltages applied to the n-layers and p-layers of the structure.

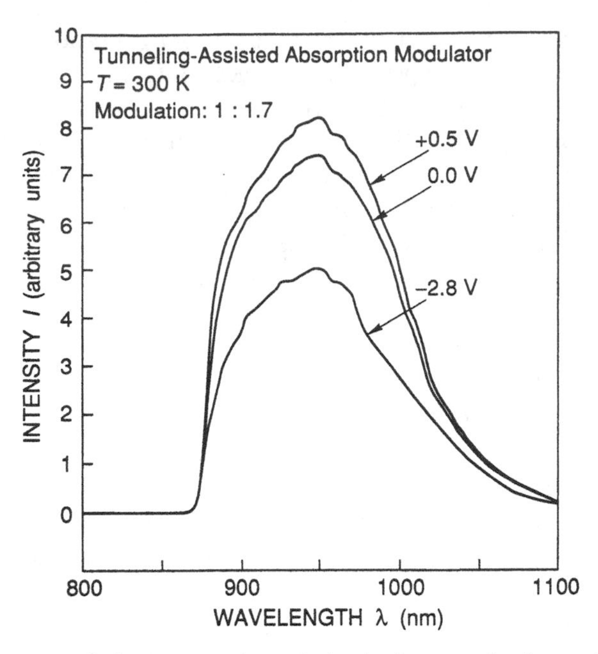

Fig. 25.12. Transmission spectra through the doping superlattice modulator at different voltages. The transmitted light was measured with a Si detector.

significant modulation is achieved. Such a wide wavelength range is desirable for photonic switching applications because no accurate matching of source wavelength and modulator wavelength is required. Furthermore, the wavelength range suitable for modulation ($900 \leqslant \lambda \leqslant 1000$ nm) can be shifted towards shorter wavelengths (for example, $800 \leqslant \lambda \leqslant 900$ nm) by using $Al_xGa_{1-x}As$ rather than GaAs as a host material for the doping superlattice (sawtooth) structure. The maximum contrast ratio between the opaque and transparent states is 1 : 1.7, which is achieved at $\lambda \approx 950$ nm. Modulation experiments with the conventionally doped n–i–p–i structure have yielded a transmission change of 22% (Chang-Hasnain *et al.*, 1987).

The voltage difference between the transparent and the opaque states is 3.3 V for the sawtooth modulator shown in Figs. 25.11 and 25.12. Such small voltage swings are desirable for integration of this modulator with electronic circuits.

We finally note that the sawtooth modulator has the potential for high-speed modulation. The total capacitance of the sawtooth structure used in this study with an area of 10 μm × 10 μm is $C < 2$ pF. A voltage swing of $V < 5$ V makes the sawtooth modulator potentially suitable for modulation in the GHz range.

References

Baillargeon J. N., Cheng K. Y., Laskar J., and Kolodzey J. (1989) *Appl. Phys. Lett.* **55**, 663.

Chang-Hasnain G., Johnson N. M., Doehler G. H., Miller J. N., Whinnery J. R., and Dienes A. (1987), *Appl. Phys. Lett.* **50**, 915.

Esaki L. and Tsu R. (1970) *IBM J. Res. Dev.* **14**, 61.
Houng M. P., Wang Y. H., Chen H. H., and Pan C. C. (1992) *Solid-State Electronics* **35**, 67.
Schubert E. F. and Cunningham J. E. (1988) *Electron. Lett.* **24**, 980.
Schubert E. F., Cunningham J. E., and Tsang W. T. (1987) *Appl. Phys. Lett.* **51**, 817.
Schubert E. F., Fischer A., Horikoshi Y., and Ploog K. (1985), *Appl. Phys. Lett.* **49**, 1357.
Schubert E. F., Fischer A., and Ploog K. (1985) *Electron. Lett.* **21**, 411.
Schubert E. F., van der Ziel J. P., Cunningham J. E., and Harris T. D. (1989) *Appl. Phys. Lett.* **55**, 757.
Sun C.-Y. and Liu W.-C. (1991) *Appl. Phys. Lett.* **59**, 2823.
Vojak B. A., Zajac G. W., Chambers F. A., Meese J. M., Chumbley P. E., Kaliski R. W., Holonyak N., and Nam D. W. (1986) *Appl. Phys. Lett.* **48**, 251.
Wang Y. H., Houng M. P., Chen H. H., and Wei H. C. (1991) *Electron. Lett.* **27**, 1668.
Wang Y. H., Yarn K. F., Chang C. Y., and Jame M. S. (1987) *Electron. Lett.* **23**, 874.

Index